DEVELOPMENTS IN SEDIMENTOLOGY 35

INTERNATIONAL CLAY CONFERENCE 1981

FURTHER TITLES IN THIS SERIES
VOLUMES 1, 2, 3, 5, 8 and 9 are out of print

4 *F.G. TICKELL*
THE TECHNIQUES OF SEDIMENTARY MINERALOGY
6 *L. VAN DER PLAS*
THE IDENTIFICATION OF DETRITAL FELDSPARS
7 *S. DZULYNSKI* and *E.K. WALTON*
SEDIMENTARY FEATURES OF FLYSCH AND GREYWACKES
10 *P.McL.D. DUFF, A. HALLAM* and *E.K. WALTON*
CYCLIC SEDIMENTATION
11 *C.C. REEVES Jr.*
INTRODUCTION TO PALEOLIMNOLOGY
12 *R.G.C. BATHURST*
CARBONATE SEDIMENTS AND THEIR DIAGENESIS
13 *A.A. MANTEN*
SILURIAN REEFS OF GOTLAND
14 *K.W. GLENNIE*
DESERT SEDIMENTARY ENVIRONMENTS
15 *C.E. WEAVER* and *L.D. POLLARD*
THE CHEMISTRY OF CLAY MINERALS
16 *H.H. RIEKE III* and *G.V. CHILINGARIAN*
COMPACTION OF ARGILLACEOUS SEDIMENTS
17 *M.D. PICARD* and *L.R. HIGH Jr.*
SEDIMENTARY STRUCTURES OF EPHEMERAL STREAMS
18 *G.V. CHILINGARIAN* and *K.H. WOLF, Editors*
COMPACTION OF COARSE-GRAINED SEDIMENTS
19 *W. SCHWARZACHER*
SEDIMENTATION MODELS AND QUANTITATIVE STRATIGRAPHY
20 *M.R. WALTER, Editor*
STROMATOLITES
21 *B. VELDE*
CLAYS AND CLAY MINERALS IN NATURAL AND SYNTHETIC SYSTEMS
22 *C.E. WEAVER* and *K.C. BECK*
MIOCENE OF THE SOUTHEASTERN UNITED STATES
23 *B.C. HEEZEN, Editor*
INFLUENCE OF ABYSSAL CIRCULATION ON SEDIMENTARY ACCUMULATIONS IN SPACE AND TIME
24 *R.E. GRIM* and *GUVEN*
BENTONITES
25A *G. LARSEN* and *G.V. CHILINGAR, Editors*
DIAGENESIS IN SEDIMENTS AND SEDIMENTARY ROCKS, I
26 *T. SUDO* and *S. SHIMODA, Editors*
CLAYS AND CLAY MINERALS OF JAPAN
27 *M.M. MORTLAND* and *V.C. FARMER, Editors*
INTERNATIONAL CLAY CONFERENCE 1978
28 *A. NISSENBAUM, Editor*
HYPERSALINE BRINES AND EVAPORITIC ENVIRONMENTS
29 *P. TURNER*
CONTINENTAL RED BEDS
30 *J.R.L. ALLEN*
SEDIMENTARY STRUCTURES
31 *T. SUDO, S. SHIMODA, H. YOTSUMOTO* and *S. AITA*
ELECTRON MICROGRAPHS OF CLAY MINERALS
32 *C.A. NITTROUER, Editor*
SEDIMENTARY DYNAMICS OF CONTINENTAL SHELVES
33 *G.N. BATURIN*
PHOSPHORITES ON THE SEA FLOOR
34 *J.J. FRIPIAT, Editor*
ADVANCED TECHNIQUES FOR CLAY MINERAL ANALYSIS

DEVELOPMENTS IN SEDIMENTOLOGY 35

INTERNATIONAL CLAY CONFERENCE 1981

Proceedings of the VII International Clay Conference 1981 held in Bologna and Pavia, Italy, September 6—12, 1981, organized by the Italian Group of AIPEA, under the auspices of Association Internationale Pour l'Etude des Argiles

Edited by

H. VAN OLPHEN

Doorwerth, The Netherlands

and

F. VENIALE

Instituto Mineralogia Petrografia, University of Pavia, Italy

ELSEVIER SCIENTIFIC PUBLISHING COMPANY
Amsterdam — Oxford — New York 1982

ELSEVIER SCIENTIFIC PUBLISHING COMPANY
Molenwerf 1
P.O. Box 211, 1000 AE Amsterdam, The Netherlands

Distributors for the United States and Canada:

ELSEVIER SCIENCE PUBLISHING COMPANY INC.
52, Vanderbilt Avenue
New York, NY 10017

ISBN 0-444-42096-7 (Vol. 35)
ISBN 0-444-41238-7 (Series)

Printed in The Netherlands

CONFERENCE ORGANIZING COMMITTEE

Chairman	F.Veniale	University of Pavia
Secretary	F.Soggetti	
Administrative Secretary	Miss C.Caucia	
Social functions Program (Pavia)	Mrs. L.Rampulla-Miradoli	
Local organization	N.Morandi	University of Bologna
Ceramic District of Sassuolo	C.Palmonari	
Press and Proceedings Bureau	Mrs. S.Wentworth-Rossi	
Commercial Exhibits	L.Tomadin	
Conference Center	G.Vianelli	Regione Emilia-Romagna
Social functions Program (Bologna)	Mrs.C.Cremonini	
Pedological Field Trip	G.Vianelli	
Research Center of ENI (San Donato Milanese)	L.Mattavelli G.L.Morelli	
Editor-in-Chief	H.van Olphen	
Co-Editor	F.Veniale	
Scientific Committee	A.Alietti (University of Modena) F.Palmieri (University of Naples) G.Ristori (University of Florence)	
Field Excursions Sardinia/Central Italy	G.Lombardi P.Mattias	University of Rome
Southern Italy	A.Pozzuoli P.Violante	University of Naples
Sicily	R.Alaimo P.Ferla	University of Palermo

PREFACE

The 7th Conference of the Association Internationale Pour l'Etude des Argiles (AIPEA) was held in Bologna and Pavia, Italy, Sept.6-12, 1981. This volume of Proceedings contains in addition to selected submitted papers, the opening addresses by the Chairman of the Italian Organizing Committee, Professor F.Veniale, and the President of the AIPEA, Dr.R.C. Mackenzie, as well as the texts of several plenary and introductory lectures presented at various sessions of the Conference. During the Conference, Dean Georges Millot of the Université Louis Pasteur of Strasbourg, France, was awarded an honorary doctorate by the University of Pavia. The address he delivered on this occasion is also included. More than 260 papers were delivered at the Conference. One session, on "Equilibria and kinetics of clay minerals in aqueous solution" was organized jointly with the Group on Water-Rock Interaction of the International Association of Geochemistry and Cosmochemistry (IAGC). The selection of papers for these Proceedings was based on the recommendations of reviewers and on the consideration of achieving a proper balance between the many topics dealt with at the Conference. Many of the papers that could not be published in this volume will undoubtedly appear in the general literature. Two additional publications from the Conference are: "Occurrences and Genesis of Sepiolite/Palygorskite Sediments", edited by A.Singer and E.Galan, and "Advanced Techniques for Clay Mineral Analysis", edited by J.J.Fripiat. Both Monographs are also published by Elsevier. The reviewed and edited manuscripts for the Proceedings were finally typed by the authors in camera-ready form and this volume was produced by photo-offset.

Thanks are due to authors, session chairmen and referees. It is a special pleasure to acknowledge the able assistance of Mrs. Sally Wentworth-Rossi at the time of the Conference.

H.van Olphen
Editor-in-Chief

F.Veniale
Co-editor

OPENING SPEECH

Prof.F.Veniale, Chairman of the Organizing Committee

Ladies and gentlemen, dear friends.

On behalf of the Organizing Committee and of the Italian Group of AIPEA, I welcome you to the Seventh International Clay Conference.

We have been working for three years to develop a scientific program, excursions, and social activities which we hope will provide an interesting, enjoyable, stimulating and memorable experience for all of you.

We especially tried to provide extensive opportunities for general review and discussion. The scientific program includes 33 invited Plenary and Introductory lectures, 140 oral presentations, and more than 90 poster presentations. The booklet of Abstracts was mailed to you later than planned. How could this have happened? - Lazy authors!: By the 15th of April, the deadline for submitting abstracts, we had received only 40% of the abstracts of announced papers. By the 15th of May this number had risen to 70%, and by June 15th to 90%. This, then, is the reason why listings of both "Late Abstracts" and "Last (but not least) Abstracts" were necessary.

At this time, registered participants number more than 350, plus nearly 80 accompanying persons, representing 47 different countries. We are pleased that for the first time an official delegation of our Chinese colleagues is attending the Conference. We are sorry to have to tell you that we have received notice from three of our "youngest" colleagues that they are not able to attend this meeting as they originally planned. I am, of course, referring to Professor Brindley, Professor Grim, and Professor Keller.

We are honored by the presence of the Italian Minister of Scientific Research, On. Tesini. We have also received telegrams from the Minister of Public Works, from the Presidents of the Italian Mineralogical-Petrological and Geological Societies, expressing their regrets for not being able to attend because of prior commitments.

The Italian Group of AIPEA is pleased to have this opportunity to bring together clay scientists from many other countries for discussion and exchange of information and ideas. We hope participants also will use the full range of social activities to renew old friendships, make new ones and establish personal and scientific contacts that will continue long after this Conference.

Without doubt, you will find some flaws in our organizing ability (perhaps also in the weather of "Sunny Italy"); however, I am sure that our hospitality and friendship will be warm.

And now, as we show the new flag of AIPEA, I declare the Seventh International Clay Conference officially open.

CONTENTS

SECTION V. APPLICATIONS

PRESIDENT'S ADDRESS

Signor Sindaco di Bologna, Signor Presidente della Regione Emilia-Romagna, Onorevole Ministro della Ricerca Scientifica, Magnifico Rettore dell'Università di Bologna, Professore Veniale, Signori e Signore:

È un mio piacere vivissimo poter dare a tutti i partecipanti alla Settima Conferenza Internazionale delle Argille il più cordiale benvenuto oltre che a titolo personale anche a nome del Consiglio della AIPEA. Anche se conosco bene molte parti dell'Italia e moltissimi Italiani, la mia conoscenza dell'Italiano è veramente scarsa e continuo quindi questo mio saluto in Inglese.

And now, having taken the President's privilege of breaking all the rules and welcoming you in what I hope was understandable Italian, I shall obey the regulations and revert to English which is the official language of this Conference.

First of all, I should like on behalf of Council and participants to thank the previous speakers for their good wishes and to ask them and Professor Veniale to express our sincere gratitude to the bodies they represent and indeed to all sponsors for everything they have done towards the success of this conference. We really do appreciate it. I should like, too, to say how pleased Council are to have with us a group from a sister Affiliated Association of the International Union of Geological Sciences - the International Association of Geochemistry and Cosmochemistry. We look forward to further collaboration in the future.

For me, it is a great pleasure to be at another International Clay Conference - for several reasons. First, there is the interdisciplinal nature of Clay Conferences, so well commented on by Prof. Bailey at Oxford and which is most stimulating. Then, there is the delight of being again in Italy, a country of which I have so many pleasant memories and which has contributed much to clay mineralogy. In addition, there is the pleasure of meeting old friends, the privilege of making new ones, the prospect of learning of new developments in one's own field - and so on - but above all there is the friendly atmosphere always found at clay conferences and which has already been particularly noticeable here in Italy. I can only hope that you will experience all these benefits for yourselves and treasure them in the years to come.

Officially, this is the Seventh International Clay Conference - but, if my arithmetic is correct, it is in fact the 13th international conference on clays since CIPEA was founded at the international meeting on clays organized by Drs Brindley and MacEwan in association with the 1948 International Geological Congress in London. We very much regret the absence of Professor Grim and Professor Brindley, who played so large a part in the 1948 meeting, but the fact that they are still active attests not only to the addictive nature of clay mineralogy but also to the longevity of clay mineralogists.

Finally, I know Council would wish me to pay tribute to the

enormous effort put into the organization of this Conference by our present hosts, the Italian Section of AIPEA, and particularly by Professor Veniale and his Organizing and Scientific Committees. I feel too that those organizing the field excursions deserve special mention for working under, at times, very adverse and indeed dangerous conditions - one group, for instance, was caught near the epicentre of an earthquake and were lucky to escape. The result of their combined efforts is well seen in the comprehensive programme which has, literally, something for everyone connected with clays.

And so, I wish all here a very fruitful conference.

R.C. Mackenzie
AIPEA President 1978-81

AIPEA COUNCIL 1978-1981

President	Dr.R.C.Mackenzie (Scotland)
Vice President	Prof.A.Weiss (German Federal Rep.)
Secretary General	Dr.A.Breeuwsma (The Netherlands)
Treasurer	Dr.H.Kodama (Canada)
Past-President	Prof.S.W.Bailey (U.S.A.)
Editor-in-Chief	Dr.H.van Olphen (The Netherlands)
Councillors	Prof.J.Konta (Czechoslovakia)
Ordinary	Mr.D.Mitchell (England)
	Prof.H.Minato (Japan)
	Dr.L.de Pablo Galan (Mexico)
	Dr.G.Pedro (France)
Ex-Officio	Prof.F.Veniale (Italy) Int.Conf.
	Prof.C.Palmonari (Italy) Int.Conf.
	Prof.S.W.Bailey (Nomenclature Comm.)
	Dr.J.Thorez (Comm.Standardization of Preparation Techniques)

Remerciement du Doyen Georges MILLOT
prononcé le I2 Septembre I98I,
à l'Université de PAVIE.

Monsieur le Recteur Magnifique,
Messieurs les Membres du Sénat Académique,
Mesdames et Messieurs les Autorités de la Région et de la Cité,
Mesdames, Messieurs,

Vous me faites beaucoup d'honneur.

"Comment peut-on être persan"?, se demandait MONTESQUIEU au XVIIIe siècle. Voici une question que nous ne nous posons plus aujourd'hui, alors que nous avons exploré le monde.

Comment peut-on avoir passé sa vie à étudier les argiles ? Voici la question que vous vous posez aujourd'hui, devant un étrange personnage.

Je vais essayer de vous répondre de trois façons.

La première raison réside dans l'usage extraordinairement varié de cette matière innocente. Dans votre vieille ville universitaire de Pavie, vos maisons, vos églises et vos tours sont en argile. De même, les tuiles de vos toits, les céramiques de vos cuisines, les carreaux de vos chambres, et même les moules qui ont servi à fondre vos cloches et vos statues équestres.

Dans la vie quotidienne, c'est en argile qu'ont été tournés vos assiettes de porcelaine, vos plats de faïence, et vos cruches où le vin se tient frais. Vos boissons ont été filtrées sur des gâteaux d'argile qui a aussi des centaines d'usages industriels, comme la fabrication du ciment, la charge des papiers et des pneumatiques automobiles, la confection de nombre de produits d'entretien et d'engrais, mais aussi de certaines variétés de médicaments et de catalyseurs.

Ce n'est pas tout.

D'un autre côté, celui de la nature, c'est à l'argile que vos jardins, vos champs et vos vignes doivent la mise en réserve de leur nourriture. Ces minuscules particules, qui se mesurent en millièmes de millimètre, sont capables de saisir, en période d'abondance, l'eau, les éléments minéraux et les molécules organiques, de vos irrigations, de vos engrais et de vos fumures. Après plusieurs semaines, si la sécheresse et la disette surviennent, les argiles restituent à la plante ce qu'elles ont mis en réserve.

De cette façon, les argiles sont les "banques" des réserves naturelles des sols. Ce sont les plus anciens des banquiers, bien avant que les prêteurs lombards, vénitiens ou florentins, ne soient autorisés, à certaines heures, à faire le prêt sur le banc de l'Eglise ou le banc public. Et c'est de cette première pratique que vient le nom de "banque".

Revenons à notre propos. Vous devinez que tous les secrets de nos argiles aux multiples usages, doivent être compris, expliqués, si possible améliorés et perfectionnés, pour le progrès des industries et de l'agriculture. C'est notre travail et la raison des journées que nous passons dans les murs de votre cité.

Une deuxième raison pour avoir étudié les conditions de genèse et les conditions de la mort des argiles dans le grand cycle de la nature est leur statut particulier.

Les argiles ne sont pas des cailloux rigides et blessants, mais une matière souple et capable de s'adapter. Nous y rencontrons à la fois, dans leur extrême petitesse, l'ordre et le désordre.

Ordre et désordre, c'est l'image de la nature, de la vie, et même de nos sociétés, qui doivent être assez ordonnées, mais assez inventives.

Oui, les argiles sont des cristaux, dans lesquels les éléments sont régulièrement disposés et ces silicates ordonnés peuvent résister longtemps à la percolation des eaux, et aux atteintes du temps qui se mesure en millions d'années.

Mais, en même temps, l'ordonnance de ces minuscules cristallites est souvent troublée, avec des fautes dans l'empilement des atomes, avec des désordres de couche à couche... Et ceci donne des propriétés électriques capables d'attirer l'eau, les ions, les molécules...

Voici un état intermédiaire entre la rigueur géométrique et le jeu d'échecs aux millions de solutions. C'est ce statut intermédiaire, que j'ai décrit ailleurs comme à mi-chemin "entre le cristal et la fumée", entre le cristal robuste et raide, et la fumée aux volutes pleines de fantaisie.

A ce statut intermédiaire, sont dues les propriétés curieuses des argiles naturelles, que le monde méditerranéen a ennoblies dans son habitat et ses arts, tandis que les peuples barbares du Nord, auxquels j'appartiens, habitant la grande forêt, utilisaient le bois.

La troisième raison de mon attention aux argiles est celles de nos origines.

On a dit et écrit que les argiles ont joué un rôle dans la synthèse de certaines premières molécules organiques, caractéristiques des êtres vivants. Ce n'est pas démontré et divers arguments inclinent nombre de spécialistes à songer que ceci est difficile à concevoir.

De toutes façons, nous n'en savons rien actuellement et peut-être ne le saurons-nous jamais. En effet, pour les biologistes qui sont ici et pour vous tous, je relis aujourd'hui un texte que j'aime, en vous laissant deviner provisoirement son auteur. Il s'agit d'un texte sur la fragilité des origines dans l'Evolution.

"Rien n'est délicat et fugitif, par nature, comme un commencement. Aussi longtemps qu'un groupe zoologique est jeune, ses caractères demeurent indécis. Son édifice est tendre. Ses dimensions sont faibles. Peu d'individus le composent, et qui changent rapidement...

Comment va donc agir le Temps sur cette zone faible?

Inévitablement en la détruisant dans ses vestiges.

Irritante, mais essentielle fragilité des origines, dont le sentiment devrait pénétrer tous ceux qui s'occupent d'histoire !

Lorsque, en tous domaines, une chose vraiment neuve commence à poindre autour de nous, nous ne la distinguons pas, - pour la bonne raison qu'il nous faudrait voir dans l'avenir son épanouissement, pour la remarquer à ses débuts. Et quand, cette même chose ayant grandi, nous nous retournons pour en retrouver le germe et les primes ébauches, ce sont ces premiers stades à leur tour qui se cachent, détruits ou oubliés. Où sont, si proches de nous pourtant, les premiers Grecs et les premiers Latins? ... Où sont les premières navettes, les premiers chars et les premiers foyers? En Biologie, en Civilisation, en Linguistique, partout : telle la gomme entre les mains d'un artiste, le Temps efface chaque ligne faible dans les dessins de la Vie".

Vous aurez reconnu le style d'un des plus grands paléontologistes de ce siècle : il s'appelait TEILHARD DE CHARDIN.

Si la Science ne nous indique pas si l'argile est aux aurores de la Vie, nous ne sommes pas démunis pour réfléchir par un autre chemin,celui de nos traditions nourricières.

Tout le monde sait que l'homme "Adam" fut fait d'une motte d'argile. Ceci paraîtrait cocasse, si les exégètes ne nous enseignaient que le mot hébreu qui veut dire argile, n'est pas aussi pauvre que le nôtre. Il ne s'agit pas de matière terrestre, mais de l'homme lui-même qui est argile, c'est-à-dire poussière, c'est-à-dire innombrable et fécond.

C'est parce que l'homme avait oublié, par orgueil et par égoïsme, sa mission essentielle, que Dieu lui dit : "Tu es poussière et tu retourneras à la poussière". Il le renvoie après la chute, à la fécondité innombrable de sa nature.

Je conclus. Mesdammes, Messieurs, étudier les argiles, c'est étudier la fécondité de la Terre, mais aussi contribuer à la fécondité des hommes.

Je l'ai fait pendant 42 ans et je suis au terme de ma route. C'est le moment où votre illustre Université, fondée en I36I, l'une des plus anciennes d'Europe, me range parmi ses docteurs, pour une nouvelle et aléatoire fécondité...

Vous me faites beaucoup d'honneur...Malgré la peine que j'ai prise, je ne pense pas apporter à votre Université un renom comparable à celui que vous donna l'Abbé SPALLANZANI voici 200 ans, par d'autres travaux sur la fécondité.

Mais, je vous remercie vivement et, par ma voix, les Spécialistes des argiles du Monde entier vous saluent avec respect, au son de vos cloches, fondues dans des moules d'argile.

Je vous présente, Monsieur le Recteur Magnifique, Messieurs du Sénat Académique, Mesdames, Messieurs, au nom de tous mes amis de l'Association Internationale pour l'Etude des Argiles", et en mon nom, tous mes voeux pour votre Université, afin qu'elle soit multiple et féconde. J'ai dit.

Georges MILLOT.

ACKNOWLEDGMENT ADDRESS OF DEAN GEORGES MILLOT

University of Pavia
September 12th, 1981

Your Honour the Rector of the University,
Gentlemen Members of the Academic Senate,
Town Councillors and Local Authorities,
Ladies and Gentlemen,

It is a great honour for me to be among you today.

In the 18th century, Montesquieu wondered how one could be a Persian. That is a question we no longer ask ourselves now that we have explored the world. Today, it may seem strange to you how a man can devote his life to studying clays. I shall try to give you three different answers.

The first reason lies in the wide range of uses of this simple material. In your old university town of Pavia, your houses, churches and towers are made of clay, as well as the tiles of your roofs, the ceramics in your kitchens, the tiled floors of your bedrooms and even the moulds used to cast your bells and your equestrian statues.

In everyday life potters use clay to shape china plates, your "faience" dishes and the jugs to keep your wine cool. Your drinks have been filtered through beds of clay, which also has hundreds of uses in industry, such as a filler for paper and car tires, in manufacture of cement, household products, fertilizers, and even medicines and catalysts.

Let us consider nature now. Your gardens, your fields and your vineyards can keep nutrients in store thanks to clay. Its tiny particles,that can only be measured in thousandths of a millimetre, are able, in times of plenty, to retain the water, the mineral elements and the organic molecules from irrigation, fertilizers and manure. Several weeks later, should drought or dearth occur, clays will supply plants with what has been stored for them, so that clays can be said to be the "banks" of the natural reserves of soils. They are the oldest bankers, long before Lombardian, Florentine and Venetian money-lenders were authorized to do their job, at certain hours of the day, in a church's pew or on a public bench. The word "bank" stems from this early practice (French : "banque" from Italian "banca", bench, moneychanger's table).

But let us come back to our subject. You may guess that all the secrets of our multipurpose clays must be understood and explained and that clay should possibly be improved and perfected for foster progress in industry and agriculture. This is our task and also the reason for our stay within the walls of Pavia.

Another reason why I studied the conditions of genesis and "death" of clays in the vast cycle of nature lies in their peculiar structure.

Clays are not solid, sharp stones but they are changeable and adaptable materials. In their tiny particles, we find simultaneously order and disorder. Order and disorder are intrinsically the very image of nature, of life and even of our societies, which must be fairly organized, but at the same time allow creativeness.

Clays are indeed crystals in which the elements are regularly distributed and these well-ordered silicates can for a long time withstand water percolation and weathering actions, measurable only in millions of years. At the same time, however, the organization of these tiny crystallites is often disturbed, combining errors in the stacking of atoms and disturbances from layer to layer, which create electric properties enabling the adsorption of water, ions, molecules...

This state could be described as halfway between geometric rigour and a chess game with its thousands of solutions. It is this intermediate state I described earlier as being halfway "between crystal and smoke", between the robust solid crystal and the smoke plumes of fantasy.

It is from this intermediate state that the remarkable properties of natural clays arise. Clays that the people of the Mediterranean world ennobled in their houses and in their arts, at the time when the Barbarians of the North I stem from, living in the forest, were using wood.

The third reason for my interest in clays is linked with our origins.

It has been said and written that the clays have played a part in the synthesis of some primary organic molecules characteristic for living matter. This has not yet been proved and various arguments led many experts not to accept this theory. Anyway, we actually do not know and perhaps we never will. For the biologists present and for all of you, I shall read a text I am fond of, letting you guess, for the time being, who the author is. It deals with the fragility of origins in Evolution.

" Nothing is so delicate and elusive, by its very nature, as a beginning. As long as a zoological group is young, its characters remain indeterminate, its structure precarious and its dimensions scant. It is composed of relatively few individual units and these change rapidly...

What, then, will be the effect of time of this elusive adolescence ?

Inevitably to destroy all vestiges of it.

Beginnings have an irritating, but essential, fragility and one that should be taken to heart by all who concern themselves with history.

It is the same in every domain : when anything really new begins to germinate around us we cannot distinguish it for the very reason that it could only be recognized in the light of what it is going to be. Yet , when it has reached full growth, we look back to find its starting point and we only find that the starting point itself is now hidden from view, destroyed or forgotten. Close as they are to us, where are the first Greeks and Romans ? Where are the first shuttles, chariots and hearth-stones ?... In biology, in civilization, in linguistics , as in all things, time, like a draftsman with an eraser, rubs out every weak line in the design of life".

You will have recognized this quotation from one of the greatest palaeontologists of our century. His name : Pierre Teilhard de Chardin.

If science does not clearly tell us whether clay was at the dawn of life or not, the traditions that nurtured our minds can yet help us look at it in a different way. Everybody knows that Man "Adam" was made out of a lump of clay. This would seem preposterous should not exegetes teach us that the meaning of the word "clay" in Hebrew is not as restricted as ours. It does not refer to the matter of earth but to Man himself, who is clay, that is dust, that is to say innumerable and fecund.

Because of pride and selfishness Man has forgotten his essential mission and God reminded him : " Dust thou art and unto dust thou shalt return". After the Fall, He ordered him back to the innumerable fecundity of his nature.

In conclusion, Ladies and Gentlemen, to study clays implies to study the fecundity of the earth and also to contribute to the fecundity of Man.

That is what I have been doing for 42 years and I am now reaching the end of the span of my career. This is the time when your renowned university, founded in 1361 and one of the oldest in Europe, did choose to appoint me as one of its doctors for a new, though uncertain, fecundity...

It is a great honour for me. In spite of all my efforts, I do not think that my contribution to the fame of your university can in anyway be compared with that of Father Spallanzani, two hundred years ago, thanks to his renowned works dealing with fecundity.

I gratefully thank you and, with me all the world experts on clays respectfully greet you, enjoying with you the distant chime of your bells which were cast in clay moulds.

Your Honour, Gentlemen Members of the Academic Senate, Ladies and Gentlemen, in my name and on behalf of all my friends belonging to the "International Association for the Study of Clays", I should like to express all my good wishes for your University. Be it fecund and multiple !

OPENING LECTURE

THE PRESENT STATE AND DEVELOPMENT TRENDS OF CLAY SCIENCE

JIRI KONTA
Department of Petrology, Charles University, Prague 2, Albertov 6, Czechoslovakia

ABSTRACT

Clay science or "argillology" has developed into a huge interdisciplinary science. Clay science today represents a vast range of areas of research and production. From the mid-20th century, when the first specific clay journals and monographs appeared, and when the first national clay groups as well as AIPEA were established, one can say that a new science, "argillology" emerged. To provide an objective answer to several fundamental questions this review introduces a new approach based on a statistical evaluation of 500 contemporaneous papers. In the near future interest will be strongly influenced by pressures from industrial and agricultural technology. Argillology will focus more on surface characteristics of various clay substances, primarily swelling phyllosilicates and kaolinite, including organoclay complexes and the properties of interlayer materials. New methods based on the effects of various types of radiation on clay substance will be more intensively used to refine and supplement classical methods of crystallography, surface chemistry and analytical chemistry in argillology.

INTRODUCTION

Each of the international conferences of the AIPEA /Association Internationale pour l'Etude des Argiles/ usually starts with a general lecture in which the author sketches a profile of our science, simultaneously presenting his own thoughts and style of work. At the Sixth International Clay Conference in Oxford in 1978 we listened with great interest to Dr. R.C. Mackenzie's introductory plenary lecture "Clay mineralogy - whence and whither?" /1979/ and tho the introductory lecture of section 5, delivered by Dr. P. De Souza-Santos "Recent developments in applied clay mineralogy" /1979/. Also we all appreciate the well conceived paper by Professor G.W. Brindley "Current and future trends in clay mineralogy - a review", prepared on the occasion of the centenary of the Mineralogical Society of Great Britain and Ireland and the thirtieth anniversary of its Clay

Minerals Group /1976/.

It is an honour for me to be entrusted with the introductory plenary lecture of the 7th International Clay Conference held in northern Italy, one of the most beautiful parts of Europe. For argillologists from all over the world, all roads these days do not lead to Rome, but to Bologna and Pavia.

In my contribution I present a statistical survey of the contents of 500 papers published between the years 1976 and 1980, primarily in the journals "Clays and Clay Minerals" and "Clay Minerals" and partially in the proceedings "Kora vyvetrivaniya" and the journal "Clay Science". The data obtained provide the mindful reader with a picture of the contemporaneous state of argillologists' activities, and an indication of where the pace of research has accelerated and where there is a need to increase our efforts.

THE PLACE OF ARGILLOLOGY AMONG OTHER SCIENCES

Clay substance represents nearly half of the bulk of the sedimentary lithosphere, including the weathering crust of the Earth /Fig. 1/. This fact alone would suffice for this substance to become the object of interest of an independent science among the geological sciences. Instead, clay science or argillology actually developed into a huge interdisciplinary science. Clay science today represents a vast field of research with industrial and agricultural application including the solving of environmental problems. Various areas of research and production contribute to or use argillological research. They include, for example, mineralogy, crystallography, petrology, geochemistry, geology, soil science, inorganic and organic chemistry, the food, paper and chemical industries, in particular manufacture of plastics, rubber, cables etc., the ceramic and glass industry, the foundry and oil industries, the pencil industry and numerous others. Clay minerals are all around us, their domain being the uppermost skin of the Earth's crust. If we accept the term argillology as a one-word designation for the science concerning the research of any material containing clay minerals, then we may define as "argillosphere" the vertical zone of the Earth involving the residual and sedimentary accumulations rich in clay minerals, i.e. the reaction products of the interaction between the lithosphere, the hydrosphere and the atmosphere.

Modern clay research began in the period just after the application of x-ray diffraction in identifying and interpreting crystal structures between 1925 and 1930. Further development continued mainly within the geological sciences and in the field of colloid

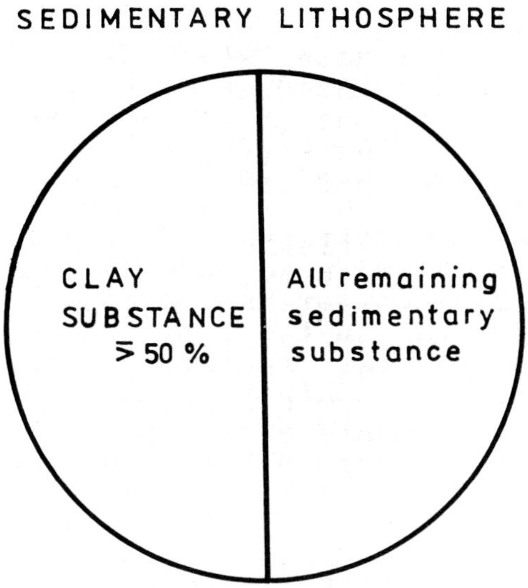

Fig. 1. The bulk contents of clay substance in the sedimentary lithosphere.

chemistry. From the middle of this century, when the first specific clay journals or special conference proceedings pertaining to clay substance /Table 1/ and the first specialized monographs were published, one can say that a new science - argillology emerged. The volumes of the international clay conferences organized by the AIPEA appeared a little later, namely from 1963.

A STATISTICAL EVALUATION OF THE PRESENT STATE OF ARGILLOLOGY

An assessment of the present state of argillology and of ideas concerning its future developments entails the examination of many different views. A statistical evaluation of a large number of contemporaneous papers represents a first attempt to answer several questions through the tabulation of the frequencies with which a variety of topics are dealt with.

/1/ Methods applied in contemporaneous clay research and clay properties studied

Table 2a shows the number of papers out of every five hundred which included the use of such basic methods as size fractionation, especially separation of the size fractions below 2 µm or 1 µm, the basic techniques of x-ray identification and methods used for refinement of the crystal structure, where infrared spectroscopy always plays a major role. Electron diffraction has had only limited use. The most striking fact is that x-ray determination of the contents of clay minerals, including associated non-clay crystalline minerals is still used infrequently.

Table 2b shows the frequency of use of other types of radiation that especially complement the x-ray diffraction method with the aim of further refinement of the crystal structure of clay minerals.

Column c shows the frequency with which optical methods are used, among which scanning electron microscopy has recorded a substantial development in the past ten years. Transmission electron microscopy still remains a useful method, applied in almost every eighth paper. The study of thin sections of various clay aggregates under the polarizing microscope represents a classical method used in earth sciences from the mid-19th century. Its advantage and strength lies in the rapidity of the determination of structural and textural features, the identification of non-clay minerals and the recognition of their shape and distribution. Sometimes an experienced petrologist can recognize geological history according to traces of processes which occurred in the aggregate during initial formation up to late diagenesis and which are preserved as changes occurring in the rock. Though the study of thin sections under the polarizing microscope has been overshadowed in the investigation of clay aggregates by the effective methods of electron microscopy, its use still remains invaluable.

Table 2d shows a survey of separation techniques, among which size separation from suspensions in a centrifugal field dominates; next is chemical destruction, i.e. the selective chemical dissolution of some phases either in acid or alkaline environments, then thermal destruction and complete dispersion by means of ultrasonic treatment. Other separation techniques are applied sporadically. We do need new methods for separation of the purest possible mineral phases from clay aggregates.

Chemical analysis /Table 3/ still remains a widely used method permitting a chemical characterization of a clay substance: nearly every fourth paper contains the results obtained by various analytical techniques. Of these the most frequently applied method is atomic absorption spectroscopy. Next is the electron microprobe method or energy dispersion analysis of x-rays /EDAX/, then colorimetry, x-ray fluorescence, flame emission spectrophotometry and some less frequently used techniques. The crystallochemical formula of any clay mineral now represents an important characteristic reported in more than one fifth of all papers.

Table 3 gives information about the methods by which the surface region and interlayer space of clay minerals may be characterized and the relations between the solid mass and water or other polar substances in clay minerals elucidated. The study of exchangeable cations is almost seven times more frequent than that of exchangeable anions. Better experimental work on anion exchange is desirable and perhaps some interesting practical applications will be developed.

Preparation and the study of homoionic clays is an integral part of the research on the surface and the interlayer space of clay minerals /Table 3c/. Authors most frequently prepare sodium-, calcium-, potassium-, magnesium-, and in descending order lithium-, barium- and caesium-clays. For a further twenty different cations the order of frequency is also given in table 3c. The saturation of a sample with a single suitable cation is also common practice for x-ray identification of some clay minerals, in the determination of the specific surface area and the swelling capacity, and in the study of the orientation of water molecules or other polar liquids in the interlayer space, and other clay properties.

Differential thermal analysis remains the most frequently used thermal method /Table 3d/, followed by thermal gravimetric analysis, both being applied as complementary identification methods or for the study of reaction temperatures of newly formed phases and for checking the content of molecular and hydroxyl water or other sorbed volatile substances. The remaining thermal methods are used in a small number of papers.

/2/ <u>Investigation of individual clay mineral groups</u>

Table 4 shows that the most frequently studied clay mineral group is the smectite group, dealt with in more than half of the papers. Next are the minerals of the kaolinite group, then the group of micas, artificially prepared organoclay derivatives, minerals of the vermiculite group, minerals with irregular and regular interstratifications, and chlorites. Substantially less frequently studied are sepiolite with palygorskite, serpentine with chrysotile, pyrophyllite and talc, in descending order. Interest in minerals of the allophane group has grown considerably during the last five years.

/3/ <u>Non-clay constituents connected with clay substance, whose identification or research frequency above all depends on the frequency of their occurrence</u>

Quartz is the most frequently studied non-clay mineral occurring in clay accumulations, followed by oxides and oxyhydroxides of trivalent iron, and by calcite /Table 5/. In descending order were studied organic combustible substance, gibbsite, potassium feldspar, amorphous silica, biotite, hematite, volcanic glass and further constituents listed in Table 5.

/4/ <u>The natural accumulations in which the studied clay minerals occur</u>

Table 6 indicates that the most frequently studied clay accumulations are residual rocks, predominantly as old weathering crusts and soils. Then follow clays and claystones, then chemically altered volcanic rocks and volcanoclastites, hydrothermal associations, shales and clastic sediments, in this order. Only slight attention is devoted to clay minerals in bauxites, though we don't know yet whether there exist relations between gibbsite, boehmite or diaspore and the degree of crystallinity of kaolinite or to various forms of iron oxides and oxyhydroxides.

/5/ <u>Natural genetic processes studied</u>

During past five years considerable attention was paid to the processes occurring in various stages of diagenesis in the sedimentary rocks containing clay materials, especially to the progressive transformations that always lead in a final phase of late diagenesis to two final species, namely illite and/or chlorite /Table 7/. In 27 papers the authors deduce on the basis of the properties of clay materials or of the whole mineral association the geological environment of the formation of a sediment or a residual rock. In every twentieth paper the authors attempt to establish or also directly observe the natural conditions of the clay deposition in diverse environments. Less attention is paid to the solubilities of clay minerals in natural environments or under simulated conditions in laboratories, and further to the surface chemical modifications or reactivity of clay minerals. Some authors at the same time apply thermodynamics and calculate Gibbs energies of formation, study kinetics of clay mineral formation or decomposition, or present stability phase diagrams or solubility diagrams of clay minerals. A relatively small number of papers are devoted to swelling, compaction, the geochemistry of interstitial water in connection with the last evolutional stage of the sediment, and to CO_2 partial pressure. Interest in direct studies or laboratory simulation of natural processes in clay accumulations and especially in explanations of the transformations of clay minerals during diagenetic processes has increased parallel with the rising interest of the oil industry in these problems.

/6/ <u>Fields of practical human activity in argillological papers</u>

The greatest attention is paid to the study of clay minerals in soils and application to the needs of agriculture, and less for engineering geology /Table 8/. Second place in the last five years has been occupied by papers dealing with the problems of the recent environment. Study of clays and shales used as raw materials in the ceramic industry share third and fourth place with investigations on the catalytic activity of clays. The relatively small number of papers concerning the application of argillology in the ceramic industry can be explained by the appreciable number of specialized ceramic journals which attract papers on argillaceous raw materials and their properties. These journals were not included in my survey, for example the American Ceramic Society Bulletin, the Journal of the American Ceramic Society, the Ceramic Age, the Transactions of the British Ceramic Society, Steklo i keramika, the Berichte der Deutschen Keramischen Gesellschaft, the Keramische Zeitschrift, Interceram, the Sprechsaal, Ceramurgia, Silicates Industriels, Silikáty and others. Similarly, papers on other practical applications of argillology listed in Table 8 appear in corresponding specialized journals. The readers of argillological journals would certainly welcome more frequent information about new discoveries in applied argillology of importance to various industries, agriculture and environmental research, appearing in these other journals.

/7/ General contents of the statistically evaluated papers

The ratio of the numbers of papers in three categories A, B, and C is A : B : C = 388 : 154 : 20, where category A contains purely experimental work or methodology, B contains papers dealing predominantly with natural clay accumulations, mostly with geological or pedological conclusions, and C represents papers with formal information, such as nomenclature, awards, etc. A number of papers in groups A and B contain substantial information of both kinds and were placed in both group A and group B.

/8/ Kinds of institutions in which authors are employed and study clay materials

The great majority of individual authors or groups of authors of the analyzed 500 argillological papers are employed at universities or colleges /Table 9/. Persons employed at national or state institutions take up second place and those in industrial laboratories occupy third place in a relatively small number. There are only 18 private authors or without indication of address of employer, 6 authors from museums and a single author from an international research institute.

FUTURE DEVELOPMENTS

The destiny of life on this planet and the modern expansion of mankind is linked with clay materials. Argillological research, therefore, will be strongly influenced in the nearest future by population growth and by the growing pressure of industrial and agricultural technologies. Argillology will orient itself more to the surface characteristics of various clay substances, among them mainly to differently prepared swelling phyllosilicates and kaolinite, including organoclay complexes. Research will lead to a deeper knowledge of the interlayer space in clay minerals under different experimental conditions. In recent research in the natural sciences interest in the development of methods based on the interaction of solids and various types of radiation is growing. The application of these methods will also be further developed in argillology, and improve our knowledge of structural crystallography, crystal chemistry, surface chemistry and analytical chemistry of clay materials.

Original results in the field of applied argillology are dispersed in a fair number of journals and conference proceedings. Considering the large interest for this kind of information in industry one might expect the future foundation of a new journal, e.g. a Journal of Applied Argillology.

The terms "clay science" and "argillology"

The literal translation of the English term "clay science" into several other languages is not possible because the term "clay" has a distinct limited meaning, involving only plastic clay material, and not a material in which clay minerals predominate, such as some shales or hard hydrothermal clay accumulations etc. The two-word term "clay science" is not used after all even in English if it should stand as an attribute where it is shortened to a mere "clay", e.g. "clay groups", "clay research", "clay conference". The one-word term for our science, argillology, created from the Latin word argilla = clay, and Greek logos = science can be used in desirable modifications whether in the form of noun or of adjective. Its first part can be connected with another term derived from classical Greek or Latin, e.g. the argillosphere. We can therefore expect that a further development of the science of clay accumulations will bring along a common and natural use of terms such as argillology, argillologist, argillological research, argillological group, argillological conference, International Conference of Argillologists, Journal of Argillology, applied argillology, argillosphere etc.

CONCLUSION

The procedure of the statistical evaluation applied to the recent argillological literature is explained in detail in the paper by Konta, 1978. This survey may serve not only to establish the contemporaneous state of the art and to indicate development trends of argillology, but they may also be helpful for institutions planning new laboratories and research programs, as well as for persons in industry using clay raw materials and who need reference sources for consultation.

The importance of the research of clay accumulations within the framework of the whole sedimentary lithosphere can also be seen from the fact that the number of specialized journals in argillology is now larger than in sedimentary petrology. In the Journal of Sedimentary Petrology in the years 1972 to 1977 the greatest attention was devoted to arenites and other clastic sediments, and clay accumulations with lutites only come in second place.

I hope that the compilation of these statistics will be continued in the following years and will cover not five but more volumes of our most advanced clay journals. I have not yet considered the excellent proceedings of the International Clay Conferences because they contain papers directed by the pre-set program. Nevertheless I intend to evaluate those statistically too, and will correlate their content with that obtained from regular clay journals.

REFERENCES

Brindley, G.W., 1976. Current and future trends in clay mineralogy - a review. Clay Min., 11: 257-268.

Konta, J., 1978. Present state of sedimentary petrology and trends of its further developments. /In Czech with English and Russian summaries./ Acta Univ. Carol., Geologica, Kratochvíl Vol., No. 3-4, 225-256, Praha.

Mackenzie, R.C., 1979. Clay mineralogy - whence and whither? Internat. Clay Conf., Oxford, Ed. M.M. Mortland and V.C. Farmer, p. 1-14, Elsevier.

De Souza Santos, P., 1979. Recent developments in applied clay mineralogy. Internat. Clay Conf., Oxford, Ed. M.M. Mortland and V.C. Farmer, p. 427-436, Elsevier.

TABLE 1.

A list of specialized argillological journals and national and international conference proceedings dealing with clay materials.

Title	Year	Notes
Clay Minerals Bulletin /UK/	- 1947	Succeeded by the journal Clay Minerals - 1965
Bulletin du Groupe Francais des Argiles /F/	- 1949	Fusion with Clay Minerals - 1971 /Journal of the European Clay Groups/
Kora vyvetrivaniya /USSR/	- 1952	/fifteen volumes/
Reports from the Swedish Society for Clay Research	- 1954	
Clays and Clay Minerals /USA/	- 1955	/1952 1st conference; fifteen volumes and 6 numbers each year from 1961, the largest argillological journal today/
Clay Science /Japan/	- 1962	
Proceedings of the Conference on Clay Mineralogy and Petrology /Czechoslovakia/	- 1958	/eight volumes/
Genèse et Synthèse des Argiles /F/	- 1962	/Symposium held in 1961/
International Clay Conference Proceedings /AIPEA/	- 1963	/six conferences, nine volumes: 1963 Stockholm, 2 vol. 1966 Jerusalem, 2 vol. 1969 Tokyo, 2 volumes 1972, Madrid, 1 vol. 1975 Mexico, 1 volume 1978 Oxford, 1 volume
Tonmineraltagung proceedings /Greifswald, DDR/	- 1965	/four conferences and three volumes/
First National Clay Conference Proc. /Bucharest, Rumania/	- 1973	/1st volume/
First Conference on Clay Minerals and Raw Materials /Boleslawiec, Poland/	- 1978	/1st volume/
Kaolin Correlation Program of the IUGS and UNESCO, symposia proceedings	- 1972	/several symposia volumes edited in different countries/

Table 1 continued:

The Kaolin Correlation Program was based on the preceeding symposia of the 23 Session of the International Geological Congress, Prague, 1968 /ed. 1969/, where the following volumes were published: "Genesis of Kaolin Deposits", "Kaolin Deposits of the World" and "Kaolin and Laterite Weathering Crust in Europe".

TABLE 2.

Survey of applied methods, properties studied and their frequency, based on a statistical evaluation of 500 argillological papers from 1976 to 1980.

a.

Method	Frequency
Size fractionation	188
XRD of oriented preparations	160
XR powder diffraction	149
IR spectroscopy	145
Crystal structure of CM	52
Crystallinity state	45
Fourier an., synth., projection	32
XR single crystal method	22
Electron diffraction	20
XR quant. determination	18

b.

Other radiations:	
Mössbauer spectra	34
Electron spin resonance	27
Neutron powder diffract.	7
Ultraviolet spectroscopy	7
Proton nuclear magnetic resonance	2
Electronic parameter reflectance	2
Luminiscence spectroscopy	2

c.

Optical methods:	
Megascopic description	79
Scanning electron microscopy	77
Transmission electron microscopy	61
Microscopic study of thin sections	32

d.

Separation techniques:	
Centrifugation	97
Chemical destruction /removal of var. substances/	88
Thermal destruction	57
Ultrasonic dispersion	57
Electromagnetis separation	11
Isodynamic separation	5
In heavy liquids	5
Hand picking	2

TABLE 3.

Frequency of chemical data, applied chemical methods, investigation of the surface and interlayer space, preparation and application of homoionic clays, and thermal methods.

a.

Chemical analyses data obtained by:	118
Atomic absorption spectrophotometry	77
Electron microprobe /or EDAX/	45
Colorimetry	43
XR fluorescence	32
Flame emission spectrophotometry	18
Spectral	14
Gas and liquid chromatography	14
XR photoelectron spectroscopy	7
Titration	7
Proton bombardment and XR detection	5
Electron spectroscopy for chemical analysis	5
Quantitative mineral analysis by chemical allocation	5
Gravimetric	4
Volumetry	2
Electron paramagnetic resonance	2
Crystallochemical formula calculation	108

b.

Surface and interlayer space:	
Cation exchange capacity	106
Exchangeable cations	95
Adsorption or absorption	86
Specific surface area	68
Water in clay minerals	52
Relative humidity controlled	32
Gas adsorption isotherms	27
Exchangeable anions	16
Clay-water interface	11

c.

Homoionic clays, saturated by:			
Na	131	NH_4	9
Ca	84	Rb	7
K	70	Cd, Au; Co, Be; Cu, Zn	à 4 or 6
Mg	63		
Li	41		
Ba	29		
Cs	25	Pb, Ag, Fe^{2+}, Cr, UO_2^{2+}, Mn^{2+}, VO_2^{2+}	à 2 or 3
H	18		
Sr	14		
Al	13		
Ni	11		
La	10		

d.

Thermal methods:			
DTA	63	Thermal stability	5
TGA	34	Heat of immersion	4
DTGA	14	Thermomagnetism	2
Differential scanning calorimetric cell	9	Thermal reactions /mullite etc./	2

TABLE 4.

Frequency of investigation of the individual clay mineral groups.

Smectites	255	Chlorites	50
Kaolinite	206	Sepiolite, palygorskite	23
Micas /+ glauconite/	142	Serpentine, chrysotile	16
Organoclay derivatives	97	Pyrophyllite	9
Vermiculites	70	Talc	7
Interstratifications:		Allophane-like minerals	32
Irregular:	52		
Regular:	43		

TABLE 5.

Identification or research frequency of non-clay constituents occurring together with clay materials.

Quartz	84	Pyrite	14
Iron oxides and oxy-hydroxides	41	Boehmite	11
Calcite	34	Anatase, Na-feldspar, Gypsum, Bayerite, Hydrotalcite... each	9
Organic combustible substance	29	Amphibole, Rutile, Ilmenite, Tridymite, Alunite, Siderite, Apatite, Brucite, Amorphous Al/OH/3, Mn-oxides, Barite, Magnetite, Andalusite, Tourmaline, Diaspore, Norstrandite, Various sulphides, Remnants of organic skeletons... each	5 to 2
Gibbsite	27		
K-feldspar	25		
Amorphous silica	23		
Biotite	23		
Hematite	18		
Volcanic glass	18		
Feldspar only	18		
Dolomite	16		
Zeolites	16		
Cristobalite	16		

TABLE 6.

Frequency of natural accumulations in which the clay minerals studied occur.

Residual rocks and soils	120	Shales	29
Clays and claystones	59	Clastic sediments	29
Volcanic rocks and volcanoclastites	52	Bauxites	2
Hydrothermal associations	38	Other sediments	38

TABLE 7.

Frequency of studies of the natural genetic processes.

Diagenesis	50	Swelling	11
Geological environments	27	Compaction	7
Natural deposition	25	Chemistry of interstitial water	7
Solubilities	14	CO_2 partial pressures	2
Surface chemical modification /reactivity/	14		

TABLE 8.

Fields of practical human activity and their frequencies in the recent argillological papers.

Soils	72	Medical area, biology	9
Recent environment	29	Pesticides	7
Ceramic clays and shales	20	Economic geology	5
Catalytic activity	20	Magnetic filtration	5
Petroleum reservoirs	14		
Filler, paper, print, plastics	14		

TABLE 9. Institutions in which authors are employed and study clay materials.

Universities and colleges	700
National and state research institutions	258
Industrial laboratories	41
Private persons /or not given/	18
Museums	6
International Research Institute	1

THE TEACHING OF CLAY MINERALOGY

G.W. BRINDLEY, Professor Emeritus of Mineral Sciences, The Pennsylvania State University, University Park, PA 16802

The invitation of Dr. F. Veniale and the Organizing Committee of the 1981 International Clay Conference to present a plenary lecture on this subject is an honor which I have accepted with some misgivings. I wonder what I can contribute which may be of interest to this large and distinguished audience. My views will reflect my early training as a physicist in the laboratory of Sir Lawrence Bragg and R.W. James in the University of Manchester and with this background I lean naturally towards basic X-ray diffraction studies and crystal structure analysis. My interest in clays arose almost by accident after 15 or more years of research in other directions, but the broad interest of clay mineral research has kept me busy for 35 years and it still continues. Scientific interests grow from interactions and collaborations with students and fellow scientists; the list of those to whom I should like to tender thanks is very long but if one name is to be selected, then it is that of Jacques Mering, to whose memory I dedicate this presentation.

The problems to be considered here are not those of clay mineralogy but of education, and are more difficult than those of identifying a clay or studying its occurrence or its properties. Teaching involves interactions of minds and stimulation of new thinking and the methods to be used are much less clearly defined than those used for example in analyzing a crystal structure.

INTRODUCING CLAY MINERALOGY TO STUDENTS

The subject is one which is likely to be taught to senior undergraduates and to graduate students. It is a subject which often does not have an immediate appeal; no one would describe it as being glamorous. However, those who work in this area recognize that it has very wide interests. Therefore it seems desirable in the introduction to the study of clays to indicate that the subject involves a wide variety of basic and applied sciences. Many areas of chemistry (inorganic, colloid, organic, and physical chemistry) crystallography, physics, and geology form the core of the subject. In the natural world, sedimentar and diagenetic processes are basically important. Clays are intimately related to soil science and agronomy, and are of great importance to the petroleum geologist and technologist. The interactions of clays with organic materials is another broad area of study and the possible relation of clay-related processes to the development of living forms can nc longer be dismissed as science fiction. Clays have a multitude of applications in the chemical industry, ceramic applications being only one of many. An initial general lecture indicating the depth and breadth of clay science and technology will indicate to students the wide interest and importance of the subject and the many possibilities for future employment. This preliminary lecture also must indicate the available textbooks and monographs, and the various journals where clay mineral studies are published. Many sources of information are now available in many languages.

THE SECOND STAGE - A BROAD SURVEY

A distinction must be made between teaching a subject to a group of students with various interests, backgrounds, and capabilities, and writing a text-book or monograph in a logical manner. A broad survey o the subject is desirable before details of stuctural and chemical variations are discussed. I find it useful to survey the principal mineral groups and their characteristic properties, notably the exchangeability of cations and reversible swelling behavior, prior to any detailed crystallographic considerations. The structures of tetrahedral and octahedral sheets and their combination in 1:1 and 2:1 type layers are described. The structure of muscovite and projections seen in different directions are studied. It is then easy to pass to other layer structures using simplified structure diagrams. A general lecture on clays as seen in the electron microscope with SEM and TEM techniques and the possibilities of chemical analysis by electron microprobe methods can usefully be given. The variability of crystal form, including the rolled forms of halloysite and chrysotile and the undulating structure of antigorite, will be mentioned. Likewise the intermediate forms between smectites, vermiculites and chlorites may be pointed out as well as the interstratifications of layers of more than one kind. The content of this part of the course can be varied to suit the main interests of the students; the chief objective will be to show the diversity of clay minerals and their properties and to develop an interest in the subject.

METHODS USED IN STUDYING CLAY MINERALS: X-RAY DIFFRACTION ANALYSIS

The introductory lectures will have said little about the methods of studying clay minerals. X-ray diffraction analysis and chemical analysis by one or more instrumental methods are of major importance. Many fine points of detail are involved in the diffraction study of clays, but most students studying clay mineralogy are unlikely to be x-ray crystallographers or physicists with a good understanding of diffraction phenomena. Therefore a compromise between a useful working knowledge and a full theoretical treatment is necessary.

The elementary principles of x-ray diffraction by powdered substances must be clearly understood and often will be taught in other courses, but when students do not have such a background, then the elementary principles must be explained. It is questionable how far Fourier analysis and reciprocal lattice concepts should be introduced. It is evident that one-dimensional Fourier synthesis has been a very useful method in studying layer silicates, and without reciprocal lattice methods of treating x-ray diffraction it is impossible to present satisfactory explanations of diffraction by disordered systems and especially of two-dimensional diffraction bands. Therefore I lean towards introducing these methods and concepts if a useful proportion of the students are able to follow.

A clear understanding of the Bragg law and of integral orders of reflection, and of the so-called "d-spacings" is essential; it must be emphasized that "d-spacings" are not identical with lattice spacings. Relations between polytypic varieties (previously not emphasized) lattice parameters, hkl indices, and powder diffraction patterns must be clearly presented. Computer programs for determining unit cell parameters can be usefully included.

The difference between x-ray reflection and the more general concept of diffraction is important in many respects in the study of clay minerals in that diffraction maxima are observed which cannot be interpreted by simple application of the Bragg law. The displacement of broad low-angle diffractions giving anomalous "d-spacings" must be discussed. Caution in applying the Scherrer equation must be emphasized. Diffraction by mixed-layer systems and by systems inhomogeneously hydrated, an the influence of variable degrees of order and disorder require careful discussion. The qualitative treatment described by Mering supplemented by reference to more detailed treatments by MacEwan and his collaborators, and by Reynolds will probably suffice.

In teaching these topics I often devote two or three lectures to the relevant aspects of x-ray powder diffraction and then pass to a detaile discussion of the principal mineral groups, usually in the sequence kaolinite group minerals, smectites, chlorites, intergrade and interstrati fied minerals, illites, and finally sepiolite and palygorskite. Under the heading of smectites, I like to introduce one or two lectures on organic complexes of clays with discussion of both cationic and netural complexes. An attempt must be made to break down the inhibitions of many students towards any consideration of organic materials. Throughou these discussions, the chemistry as well as the structure of clay minerals is considered. Particular attention is given to the problems involved in passing from a chemical analysis to a structural formula. I emphasize the great variety of ordered and disordered forms, especially for students leaning towards geological work. The interstratified minerals in particular cannot be considered without reference to diagenesis. Usually I find it necessary to restrain my personal enthusiasms for these aspects of clay mineralogy in order that other aspects are not neglected.

At some stage of the course, discussion of the preparation of clay samples for diffraction study is necessary. This involves a consideration of dispersion and flocculation of clays and the separation of size fractions by sedimentation. Although logically these topics should come at the beginning of the course, I often pass over these topics rather quickly in the early lectures and then return to them after the students are familiar with the swelling clays. Again I emphasize that the learning process depends more on developing an interest in the subject than on a strictly logical procedure.

OTHER METHODS OF STUDYING CLAY MINERALS

Personal enthusiasms for the crystal-structural aspects of clay minerals must be restrained so that other aspects are not neglected. Thermal methods, notably differential thermal analysis and thermogravimetric analysis, deserve at least one lecture because of their historical importance and their relevance to the high temperature behavior of clays. The usefulness of thermogravimetric data in the chemical analysis of clays must be mentioned. The kinetics of dehydroxylation processes can be discussed and the many difficulties inherent in the interpretation of kinetic data mentioned.

The application of infrared spectroscopic analysis must be considered. The amount of detail which can be introduced will depend on the special interests of the students and perhaps also of the professor teaching the course. It may be possible also to mention briefly the potential applications of other spectroscopic techniques including Mössbauer spectroscopy, x-ray photoelectron spectroscopy, and electron spin resonance spectroscopy. These and other modern techniques are usefully surveyed in the papers presented at the NATO Advanced Study Institute, 1979; "Advanced Chemical Methods for Soil and Clay Minerals Research," J.W. Stucki and W.L. Banwart, editors, Reidel Publishing Co., 1980, as well as in the monograph "Advanced Techniques for Clay Mineral Analysis" J.J. Fripiat, editor, Elsevier Scientific Publishing Co., 1981 which contains papers presented at the present conference.

GEOLOGY OF CLAYS

Mineralogy should not, and indeed cannot, be divorced from geology. The association of the finer points of the mineralogy of clays with particular geologic conditions opens up broad vistas of research in both subject areas. The study of the diagenetic changes of clay mineral structures and compositions has developed into a major segment of clay mineralogy and is likely to become of increasing practical importance as shown, for example, by a recent publication, "Clay Mineralogy and the Resource Geologist," F.J. Longstaffe, editor, Mineralogical Association of Canada, 1981.

LEARNING BY DOING

It is a good question how far any branch of science can be taught at the university level, or indeed at any level, without parallel practical studies. A laboratory assistant is almost essential for running a practical experimental course. In the absence of such assistance, a partial solution is achieved by providing experimental data for evaluation by students. Diffractometer patterns of clay mineral mixtures can be made available for mineral analysis. Chemical analysis data for some typical minerals can be supplied for evaluation of structural formulae. An exercise in the use of silicon and quartz standard patterns for calibration of an x-ray diffractometer can be combined with data for a chlorite and for a long-spacing mineral such as rectorite to illustrate the determination of basal spacings. By providing data for a range of samples to different members of a class, a feeling for research in clay mineralogy may be developed. For example, by using diffraction patterns for a range of analyzed chlorites, the class could 'explore' the variations of d(001) and of the b-parameter with composition. However, it is not necessary here before the present audience to discuss at length such exercises. It might be useful for an AIPEA committee to compile a program of experiments and exercises relating to the practical teaching of clay mineralogy.

It will be agreed that a true insight into clay mineralogy can be achieved only by participation in a research program as well as by attendance at lectures. Such a course of study would be very suitable for a one-year M.Sc. course in a British university, or a one and one-half-year M.S. study in a United States university (a longer period because often the undergraduate curriculum is less intensive). The advanced study involved in achieving a Ph.D. or equivalent degree completes the production of a clay mineralogist.

CONCLUDING REMARKS

There is no single way of teaching a subject, but there is one essential requirement: the course must be interesting to the students who attend. Without such interest, no fresh thinking will be generated.

I regret very much being unable to attend this international conference. I thank Dr. F. Veniale and the Council of AIPEA for the cordial invitation to present this plenary lecture, and to submit the text for publication, and Dr. J.J. Fripiat for presenting it in my absence.

CHAIRMAN'S INTRODUCTION

SIGNIFICANT FEATURES OF CONTEMPORARY STRUCTURE STUDIES OF PHYLLOSILICATES

B.B.ZVYAGIN
Academy of Sciences, Moscow, USSR

Ladies and gentlemen,

It is a great honour for me and Prof. C.Tchoubar to be chairmen of this section. Crystal chemistry and crystal structure have had the greatest consequences for the understanding and practical use of solid matter in general and of clays in particular. Looking back one may see how step by step the knowledge of the structure of phyllosilicates, these main constituents of clays, has developed and deepened. The consecutive stages of this process are defined by the following ideas and investigations each of which was dominating for a period of time: a) The basic deductions and generalizations made by Linus Pauling to represent the phyllosilicate structures as combinations of ideal tetrahedral and octahedral sheets. b) The quantitative characterization of the lattices in terms of unit cells (the kaolinite triclinic cell determination by G.W.Brindley and K.Robinson was an astonishing and important achievement for that time). c) The recognition of basal reflexions as precious identification criteria. d) The discovery of mixed-layered structures and deduction of their diffractional features connected with the names of S.B.Hendricks and E.Teller, J.Mering, G.Brown and D.M.C.Mc. Ewan, C.E.Weaver, Yu.S. Dyakonov. e) The summary of accumulated experience and knowledge of the early fifties in the excellent book written by prominent specialists and edited by G.W.Brindley (later by G.Brown). Its significance for disseminating methods of investigation and for the true understanding of classification relationships of the clay minerals may hardly be overrated. f) The structural refinements and the modification of ideal models on the basis of more precise data on atomic positions and interatomic distances obtained through the analysis of reflexion intensities. g) A more perfect operation with ideal models for deduction and identification of polytypes. Finally, h) the generalization of structural data obtained and the deduction of law-governed relations permitting the prediction of structural details according to the composition and unit cell data as presented by E.W.Radoslovich and V.A.Drits.

These earlier developments established a beneficial background for the further development of structure studies bringing them to their contemporary state which has the following significant features.

1. The level of the X-ray structure analysis has highly increased, as demonstrated for example by the work of S.W.Bailey and his co-authors who succeeded to refine crystal structures in sub-groups of ideal symmetry and to reveal the ordered distribution of isomorphic replacements as a widespread phenomenon and, in favorable cases, to localize the hydrogen positions.

2. Much attention (especially in France) is focused on the structures of vermiculites, and this makes good sence because of their peculiar structural variability in result of temperature and cation-exchange treatments and because of the possibility to reveal the configuration of interlayer cation-water complexes.

3. a great contribution to the structural mineralogy of phyllosilicates is made with the use of oblique-

texture electron diffraction. Most important results are obtained by means of high voltages (B.B.Zvyagin and his co-authors) and with direct intensity measurements (V.A. Drits and S.I.Tsipursky) for a wide range of minerals varying in type and stacking of layers, and octahedral and interlayer cations. These studies have revealed the structural consequences of changes in composition and layer stacking and the significance of Si-O bond-length variations for the stability of these minerals. New examples of ordered cationic distributions were found. In some cases electron diffraction analysis made it possible to locate hydrogen positions. New structures like the unexpected Fe^{3+}-kaolin analogues,- chapmanite and bismuthoferrite, were discovered and a complete determination of their structure was carried out.

4. The successful applications of X-ray and electron diffraction are accompanied by an intensification of neutron diffraction studies (R.Rothbauer, W.Joswig a.o.) realizating the advantages of this method, particulary in the determination of the hydrogen positions necessary for the knowledge of the nature of interlayer bonds and for the calculation of the bonding energies.

5. In its applications the polytypism theory proved to be a powerful tool. It permits not only a simple and rapid identification of such essential features as the type and stacking of layers, and the discovery of new polytype structures, but also the understanding of phyllosilicate structures as members of an integral set of interrelated structures of a united crystallochemical system. Using beforehand known diffraction properties of a polytype family one may recognize structures even when they are imperfect and inaccessible for direct determination. Especially effective is the polytypism approach in combination with oblique-texture electron

diffraction which for example permitted F.V.Chukhrov and his co-authors to establish original polytypes like halloysite 2M, ferripyrophyllite 2M, Zn-serpentine (zinalsite) A. The elaboration of analytical fully descriptive polytype symbols made it possible to realize easily different symmetric operations with complicated structural arrangements relative to different settings and coordinate systems, distinguishing equivalent and independent polytypes as shown in the recent book by B.B.Zvyagin and his co-authors on high-voltage electron diffraction analysis of layered minerals. Very promising is the combination of the polytypism approach with the use of detailed structural data. It may help to estimate the relative stability of different polytypes. This was demonstrated by R.Newnham who could explain IR- and DTA-properties and the relative abundance of kaolin polytypes on the basis of the refinement of the structure of dickite 20 years ago.

6. Owing to the matrix-approach of J.Kakinoki and Y.Komura the study of mixed-layered structures is now in a new stage. Its contemporary state which is applicable to structures with more than two kinds of layers with different scattering powers and with low-range interaction factors $s > 1$ is presented in the excellent book of V.A.Drits and B.A.Sakharov.

7. Significant results are obtained in the study of crystal structure defects. The successors of the outstanding French scientist J.Mering, C.Tchoubar and his coworkers, have developed a mathematical formalism for comparison of experimental and calculated intensities in the whole range of continious scattering, both in the vicinity of the central beam and between the Bragg reflexions. In this way they have established the nature and the quantitative distribution of the defects in the

structures of kaolinites and smectites. They also found some regularities in the arrangement of microparticles, and interrelations between solid and liquid gels which (for smectites) have common sub-units consisting of 2:1 layers with two molecular planes of water in the interlayers.

8. The diffraction studies of the real structure of the minerals are greatly reinforced by fantastic achievements in high-resolution electron microscopy, best presented by the work of Japanese authors. It became possible to obtain direct images of structural projections in which real atomic arrangements (e.g. in the octahedral and tetrahedral sheets) are visible, concretizing the averaged pattern of distortions deduced from diffractional data. The combination of diffraction with high-resolution electron microscopy promises an important tool for further progress in structure analysis of minerals in general and of phyllosilicates in particular.

9. The calculations of structural energies like those published by R.F.Giese a.o. became a peculiar kind of structure study providing an independent reliability control of structural data and permitting the comparison of different variants of isomorphic substitutions and layer stackings. These studies allow one to predict probable hydrogen positions,- so important for the evaluation of interactions between both adjacent layers and layers lying further apart. Such calculations may in principle serve for the interpretation of spectroscopic data.

10. Spectroscopic methods (Mössbauer, IR, EPR, NMR) as sources of structural information concerning local structural variations independent of their periodic or aperiodic distribution are becoming more and more important. It should be noted however that trustworthy results are obtained only when diffractional data are also taken

into account and when structural modelling is applied. Without diffractional information one may not be sure about the structural nature of the object, and without modelling one might not avoid conclusions which are more accessible and only apparently obvious, but in fact are not corresponding to the truth. A persuasive example of such a situation was presented in the case of celadonite and nontronite. Calculating the atomic coordinates for different surroundings of Fe^{3+} and the corresponding electric gradients L.G.Daynyak and V.A.Drits have disproved the existing interpretation of Mössbauer data which favored occupation of trans-octahedra. They have established an ordered distribution of tri- and divalent cations among the cis-octahedra of celadonite and of different tetrahedral cations in nontronite. In turn Mössbauer data have served to refine the diffractional data showing that the ordering inside separate layers is higher than in the whole structure. Thus the combination of both kinds of methods is most effective and desirable.

It follows from the above-statements that in spite of increasing completeness of structural knowledge, or rather because of that, the structure studies of phyllosilicates become even more important and neccessary. Let us hope that this conference and our sessions will promote their further development.

X-RAY AND ELECTRON DIFFRACTION STUDY OF THE STRUCTURE OF THE GARFIELD NONTRONITE

G. BESSON, C. DE LA CALLE*, M. RAUTUREAU, C. TCHOUBAR[1], S.I. TSIPURSKI and V.A. DRITZ[2]

[1] Université d'Orléans, Laboratoire de Cristallographie, E.R.A. 841, 45046 Orléans, France.

[2] Geological Institute of Academy of Sciences, Pijevski 7, Moscow, S.S.S.R.

* Permanent address : C.S.I.C., Serrano 113, Madrid 6, Espagne.

ABSTRACT.

A study of the Garfield nontronite by electron and X-ray diffraction in order to determine the repartition of the cations inside the octahedral sheet and the stacking arrangement of the layers is presented.

From the electron microdiffraction patterns of single layers, we observed a diagram consistent with a tendancy of the structure to be centrosymmetric (trans position vacant).

The study of the 02ℓ and 11ℓ reflexions of the X-ray diffraction pattern for K-nontronite partially organized by several wetting and drying cycles can give evidence of the centrosymmetry of the structure.

Moreover it is possible to determine the peculiarity of the stacking faults and their abundances.

INTRODUCTION.

It is well known that diffraction methods applied to a single crystal without defects permit the establishment of the atomic coordinates of different atoms with good precision. But, when the mineral is in microcrystalline form and in addition contains effects to a great extent, it is often very difficult to determine precisely all the structural characteristics by X-ray diffraction. This problem arises in the case of smectites whose powder patterns contain (hk) diffraction bands having scarcely any hkℓ modulations (biperiodic turbostratic character). It is particularly difficult for these minerals to determine, for example : (i) what is the precise distribution of certain cations in the different sheets (tetrahedral or octahedral) ; (ii) which sites are occupied by certain cations (in the case of dioctahedral smectites whether

positions "cis" or "trans" are vacant). One difficulty is that the profiles and intensities of the unmodulated diffraction bands are not very sensitive to the variations of the positions of the cations in the sheets, at least when the patterns are recorded with an arbitrary scale. However in crystallography the problem can be solved in two ways :
- either it is possible to handle single domains by means of the selectes area electron diffraction technique. Moreover one ignores the exact number of layers as well as their precise orientations in the stacking. But these factors may cause large variations in the intensities of reflections from one particle to the other (contrary to the average effect of X-ray powder diffraction) ;
- or to reorganize the smectite so as to obtain, in the powder pattern, distinct $hk\ell$ reflections which will permit determination of all the structural characteristics. For instance, a potassium exchange of interlayer cations in montmorillonites and wetting and drying cycles (Joffre *et al.*, 1938 ; Chaussidon, 1963 ; Mamy and Gaultier, 1976) make $hk\ell$ modulations reappear in patterns. Recently, this technique has been used in studies of other smectites (Tsipurski *et al.*, 1978) and with other interlayer cations (Besson, 1980).

Similar difficulties appear when physical methods other than diffraction are employed for the study of short distance order-disorder parameters : the powder effect, the great number of the structural defects and the very small size of domains cause difficulties in the interpretation of certain observed spectra of smectites (EPR, Mössbauer...).

One needs to explain the apparent disagreements observed in recent years, particularly on the localization of iron in nontronites, between the results based on the data of diffraction and Mössbauer spectroscopy. For instance, on the basis of the study of isolated layers of Pfaffenreuth nontronite (Germany) by electron diffraction, Mering and Oberlin (1967) have concluded that the layers of nontronite should have a centro-symmetric structure. Such a symmetry is due to a distribution of iron cations in the octahedral sheet in which the M_1 sites are always vacant (with the OH groups in "trans" positions). Very recently Tsipurski *et al.* (1978) have reached a similar conclusion by using the oblique texture electron diffraction data obtained from a potassium nontronite taken from Lake Baîkal (U.S.S.R.). However the Mössbauer spectra of nontronites are generally interpreted* by considering occupancy in large proportion of the "trans" positions (Goodman *et al.*, 1976 ; Rozenson and Heller-Kallai, 1976,1977).

The aim of the present work is the study by different methods (X-ray and electron diffraction) of Garfield nontronite in order to attempt to settle the various propositions over the localization of iron and to determine the degrees of stacking order acquired by the mineral after potassium exchange and wetting and drying cycles.

EXPERIMENTAL CONDITIONS AND PREPARATION OF SAMPLES.

The study of structural characteristics of nontronite was carried out simultaneously by three methods : by selected area diffraction (S.A.D.), by oblique texture electron diffraction, and by X-ray diffraction.

Selected Area Diffraction.

The aim is to obtain the patterns of isolated layers of nontronite, as has been done by Mering and Oberlin. The advantage of the isolated layers is to have biperiodic diffraction patterns. Observed intensities then are directly related to the distribution of atoms in the planar cell of the layer. The intensities would not be affected by interference among various layers of the stacking sequence.

The recording of patterns was done with JEOL 100CX equipment using two techniques : (i) classical S.A.D. technique (using the selective apertures) ; (ii) diffraction by a very fine beam (≃ 250 Å ; technique called "micromicrodiffraction").

Oblique texture electron diffraction.

It consists of obtaining an overall pattern of an oriented deposit of particles ; this deposit is obliquely exposed to the electron beam in a microscope or an electron diffractograph (Pinsker, 1953 ; Zvyagin,

*In 1978 Goodman proposed another interpretation which is in agreement with the conclusion of Mering and Oberlin. Recent works (Daîniak,1980 ; Bookin *et al.*, 1978) on the interpretation of Mössbauer spectra have also shown that the occupation of M_1 "trans" is not necessary.

1967). This technique allows separate observation of the hkℓ reflections on different ellipses which would superimpose in a classical powder pattern.

X-ray diffraction.

The recordings are carried out with powders in which the degree of particle orientation has been determined experimentally from the intensity of the 001 reflection (Plançon and Tchoubar, 1977 ; de Courville *et al.*, 1980). The radiation used is MoKα_1 (curved quartz monochromator). The apparatus equipped for energy discrimination enables suppressing about 90% of the fluorescence of iron. The recordings are carried out with the aid of a linear detector whose resolution is 3.10^{-4}Å^{-1}. These recordings are then corrected for absorption, polarization, and background effects due to the experimental conditions.

Preparation of samples.

The nontronite studied comes from Garfield, Washington (U.S.A.). Taking into account the chemical analysis carried out at the Laboratory of Applied Geology of the University of Paris VI, and the Mössbauer spectroscopy results of Dritz, the derived structural formula is :

$$M.(Si_{6.91}A\ell_{0.77}Fe_{0.32})(Fe_{3.67}A\ell_{0.29}Mg_{0.04})(OH)_4O_{20}\ .$$

In the natural state M, the interlayer cations, are essentially calcium.

According to the diffraction technique used, the samples studied are prepared from the starting mineral in different ways :

- For the S.A.D., natural nontronite is first exchanged with Na^+ ions. From an extremely dilute suspension of the Na-mineral in water, sufficient isolated layers can be obtained to be studied by S.A.D..
- For the X-ray and oblique texture electron diffractions experiments, the natural nontronite is directly exchanged with K^+. Then it is purified by decantations. Finally the K-mineral is subjected to a series of 100 cycles of wetting and drying at a temperature not exceeding 50°C (under an infrared lamp).

QUALITATIVE STUDY OF DIFFRACTION CHARACTERISTICS.

On the basis of the structural formula of nontronite, we have taken three models for the distribution of cations in the octahedral sheet.

a) All the octahedral cations are localized in M_2 sites (OH groups are in "cis" positions) leaving vacant all the M_1 sites, conforming to the proposition of Mering and Oberlin (1967).

b) All the M_1 sites are occupied by a cation. This corresponds to another extreme case proposed for the montmorillonite (Mering and Oberlin, 1967).

c) The octahedral cations are distributed equally between the M_1 and the two M_2 sites. This corresponds to an average occupation of 2/3 of a cation per site. This model has been included because the percentage of octahedral cations occupying the M sites is of the order of magnitude of the one proposed by different authors (Goodman *et al.*, 1976 ; Rozenson and Heller-Kallai, 1976) in order to explain the characteristic features of Mössbauer spectra.

The atomic coordinates of the cell have been calculated by applying the regression formulae proposed by Dritz (1971,1975) which correspond to a modification of the formulae of Donnay *et al.* (1964), Brown *et al.* (1963) and Radoslovich *et al.* (1962). The use of these formulae requires the b parameter of the unit cell (experimentally determined) and the structural formula to be known. Furthermore these are valid only in the case of the potassium mineral in the collapsed state.

We have regrouped in table I the coordinates so obtained of centrosymmetric (M_1 sites vacant : column 1) and non centrosymmetric cells (M_1 sites occupied : column 2). In the case of an average occupation of all the sites, the formalism used for the calculation of coordinates is no longer applicable ; we have then employed the averaged coordinates (column 3) of the first two models.

In employing these coordinates we have calculated : (i) the intensities of hk reflections which would be obtained by electron microdiffraction at the level of $Z = 0$ Å^{-1} of each hk line for an isolated layer (fig.1a) ; (ii) the intensity distribution which would be obtained in oblique electron diffraction (fig.1b) and in a powder pattern for reflections $02(\ell)$,$11(\ell)$ (fig.2).

We find that a distribution of iron leaving the M_1 sites vacant leads to the following main characteristics :

- In microdiffraction from an isolated layer, the intensities of

Atoms	Trans vacant (M_1 vacant) x/a	y/b	Cis vacant (M_2 vacant) x/a	y/b	Average occupation (M_1, M_2, M_2' = 2/3) x/a	y/b	z (Å)
OH	0.3754	0	0.3123	0.9790	0.3333	0	1.11
O_1	0.3123	0.6877	0.3754	0.6667	0.3333	0.6666	1.08
O_2	0.3123	0.3123	0.3123	0.3544	0.3333	0.3333	1.08
O_3	0.3286	0.5	0.3610	0.5105	0.3496	0.5	3.16
O_4	0.0542	0.7419	0.0857	0.7524	0.0752	0.7419	3.37
O_5	0.0542	0.2581	0.0857	0.2686	0.0752	0.2581	3.37
T_1	0.3139	0.6719	0.3509	0.6738	0.3333	0.6666	2.72
T_2	0.3139	0.3281	0.3353	0.3457	0.3333	0.3333	2.72
M_2'	0.5	0.8333	0.5	0.8333	0.5	0.8333	0
M_2	0.5	0.1666	0.5	0.1666	0.5	0.1666	0
M_1	0	0	0	0	0	0	0
K	0.3168	0	0.3416	0.0083	0.3333	0	5.05

Table I The atomic coordinates as a function of the occupation of sites

the reflections 02 are stronger than those of reflections 06, 33 and $\bar{3}3$ (the latter are used as standard in the scheme), whereas the reflections 11 and $\bar{1}1$ are practically eliminated.

- In oblique diffraction, on the first ellipse, the reflections $\bar{1}12$ and 112 are clearly more intense than the reflections 111, 022, $\bar{1}13$ (similar conclusions have been made by Zvyagin, 1979).
- In the X-ray powder pattern, the same effect as in the texture pattern is observed (in the model, 25% random stacking faults was considered).

On the contrary, when the M_1 sites are occupied, one obtains :

- In microdiffraction of a single layer, the intensities of the reflections 02, 11 and $\bar{1}1$ are practically equal and are of about 2/3 of those of reflections 06, 33 and $\bar{3}3$.
- In oblique diffraction, the sequence of the strong and weak reflections is opposite to that of the previous model.
- In the X-ray pattern, the same contrary tendancy as for oblique diffraction is observed.

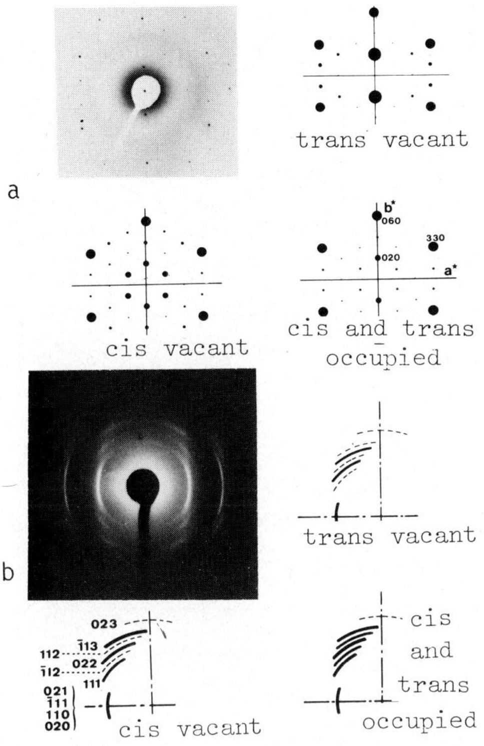

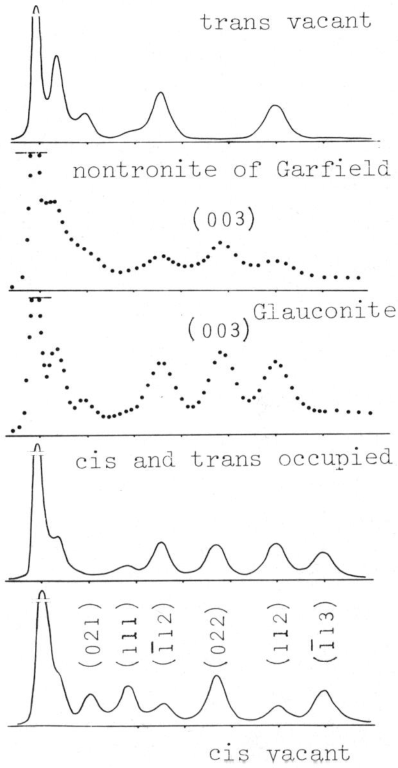

Figure 1.a-Selected area diffraction, b-Oblique texture diffraction.

Figure 2 - X-ray diffraction bands of (02ℓ,11ℓ) calculated (full lines) and observed (dotted lines).

Finally, for each octahedral site with 2/3 cation, one observes in microdiffraction a very clear weakening of 02 reflections in comparison with the 06, 33 and $\bar{3}3$ reflections while the reflections 11 and $\bar{1}1$ have appreciable intensities. Regarding the characteristics of patterns of X-ray and oblique diffraction, we find that the reflections 111, 022, 112 and $\bar{1}13$ have nearly the same intensity.

If the experimental patterns of Garfield nontronite are compared with the calculated ones on a qualitative basis, we find that the data better correspond to the model with trans positions vacant.

In figure 2 is also given the pattern of glauconite of Villers (France) which presents more intense $\bar{1}12$ and 112 reflections than in nontronite, owing to the smaller proportion of stacking faults.

THE QUANTITATIVE STUDY OF THE STRUCTURAL CHARACTERISTICS OF K-NONTRONITE, 100 HD.

The recordings made correspond to the 02(ℓ),11(ℓ) and 20(ℓ),13(ℓ) bands.

The principle of analysis of experimental patterns consists in specifying both the atomic coordinates and determining the nature and amount of the stacking faults in the powder sample. The analysis involves fitting the theoretical profiles to the experimental ones according to a model which *must have a physical meaning and be in agreement with previously confirmed results established by the methods of physics, physical-chemistry, or chemistry*. The search for the best agreement is made by modifying in the model the nature and the amount of stacking faults as well as the coordinates of the atoms (if necessary). The mathematical approach to the problem has been described in the case of kaolinites by Plançon and Tchoubar (1975,1976,1977), and recently adapted for the 2:1 silicates (Plançon *et al.*, 1978 ; Besson, 1980).

For the quantitative study of the characteristics of the 100 W-D K-nontronite we started from the coordinates given in the table I for a centrosymmetric layer structure ("trans" sites vacant : column 1). The parameter $b = 9.14 \pm 0.01$ Å has been simultaneously determined both from the S.A.D. and oblique texture patterns as well as from the comparison of the positions of calculated and experimental (02,11) bands. The iron and other octahedral cations were taken to be randomly distributed only among the M_2 sites. In the same way the substitutions in the tetrahedral sheets were supposed to be at random. Finally, the amount of orientation of particles in the powder was determined experimentally.

The final agreements obtained (in relative scale) between experimental and calculated profiles are given in figures 3a,b where the points represent the observed intensities. The coordinates, characteristic structural parameters, and defect parameters obtained for this nontronite are given in table II.

- $\vec{t}_{KK}$ represents the translation between two first neighbour layers in their plane ;
- P_T and P_R are the probabilities of the existence of $\pm\frac{b}{3}$ translative or $n\frac{2\pi}{6}$ rotative faults ;

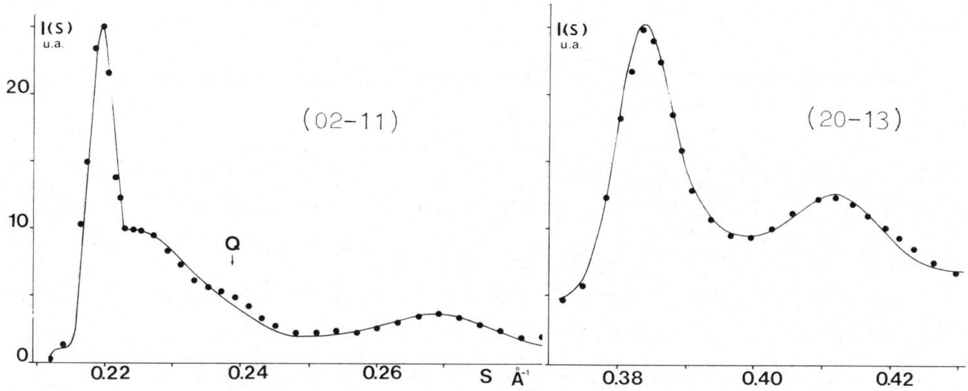

Figure 3 - Agreement between observed and calculated profiles.

Atoms	x	y	z (Å)
OH	0.3754	0	1.11
O_1	0.3123	0.6877	1.08
O_2	0.3123	0.3123	1.08
O_3	0.3387	0.5	3.16
O_4	0.0491	0.7368	3.37
O_5	0.0491	0.2632	3.37
T_1	0.3148	0.6719	2.72
T_2	0.3148	0.3281	2.72
M'_2	0.5	0.8333	0
M_2	0.5	0.1666	0
M_1	0	0	0
K	0.3168	0	5.05

b	9.14 ± 0.01 Å
α	6 ± 1 °
P_A	65 ± 5 %
P_T	0
P_R	0
$\vec{t}_{KK}$	- 0.367 $\vec{a}$
d_{001}	10.1 ± 0.05 Å
M	10 ± 1 layers
R	130 ± 10 Å 80 ± 10 Å

Table II - Coordinates and structural parameters obtained for Garfield nontronite.

- P_Λ is the probability of random stacking faults (translation or/and rotation) between successive layers.
- The coherent domain has an average thickness of 10 layers and a radius in the layer plane of 130 ± 10 Å and 80 ± 10 Å determined respectively from the 02(ℓ),11(ℓ) and 20(ℓ),13(ℓ) bands.
- If one compares the coordinates obtained with those initially considered (table I, column 1), one finds that they differ for the basal oxygens. This is due to the fact that the rotation of tetrahedra is taken as $\alpha = 6° \pm 1°$ whereas the regression formulae give $\alpha \sim 4°$. In

order to interpret this difference, three propositions may be made : (1) the effect of the presence of Fe^{3+} ions in the tetrahedral sheet is not taken into account in the regression formalism used ; (2) these formulae are not valid for this type of mineral ; (3) chemical analysis is incorrect.

In fact, the introduction of a corrective term taking into account the tetrahedral iron in the regression formulae together with $\alpha = 6°$ (determined from profile fitting) and $b_{exp.} = 9.14$ Å would give an amount of tetrahedral iron of 0.34 Fe^{3+} per cell. This is very close to the value accepted in the structural formula.

CONCLUSIONS.

The use of various diffraction methods has shown that the Garfield nontronite must be considered as a mineral in which nearly all of the octahedral cations occupy only the M_2 "cis" sites.

The analysis of the data proves the impossibility of the average occupation of nearly 2/3 of each type of octahedral site as has been proposed for interpretation of the Mössbauer spectra. However, although the fundamental characteristics of the Garfield nontronite are those of a mineral with vacant "trans" positions, it is not possible to exclude the presence of a small fraction of occupied "trans" sites relying only on the investigation of diffraction data. Taking into account the precision of recording and the amount of defects in the mineral (even after reorganization), it can be shown that the occupation of no more than 10% of all the M_1 sites would not be detectable by profile fitting. A similar conclusion can be obtained on the basis of the intensity measurements of hk reflections in S.A.D. patterns of single layers.

Finally we remark that the Garfield nontronite shows a very small volume of coherence. A domain of coherence cannot systematically be defined by certain boundaries (for example, slight waviness of layers). The smallness of domains in the nontronite enables us to suppose that the mineral contains distortions in a relatively large proportion which can eventually lead to a perturbation of certain characteristics of local order.

REFERENCES

Besson, G., 1980. Structures des smectites dioctaédriques. Thèse. Orléans. France.

Bookin, A.S., Dainyak, L.G. and Dritz, V.A., 1978. Interpretation of the Mössbauer spectra of layer silicates on the basis of structural modeling. IMA Symp. Physics of Minerals. Phys. Chem. Minerals, 3: 58-59.

Brown, B.E. and Bailey, S.W., 1963. Chlorite polytypism. Am. Miner., 48: N1,2.

Chaussidon, J., 1963. Evolution des caracteristiques chimiques et cristallographiques des montmorillonites bioniques K-Ca au cours d'alternances répétées d'humectation et de dessiccation. Proc. Int. Clay Conf. Stockholm, 1: 195-201.

Courville, J. de, Tchoubar, D. and Tchoubar, C., 1979. Détermination expérimentale de la fonction d'orientation. J.A.C., 12: 332-338.

Daĩnyak, L.G., 1980. Interpretation of Mössbauer spectra of some Fe^{3+} layer silicates using modeling (in Russian). Doctoral thesis. Moscow.

Donnay, G., Donnay, J.D.H., Takeda, H., 1964. Trioctahedral one layer micas. Acta Cryst., 17: N11,1369.

Dritz, V.A., 1971. Crystallochemical rules in the trioctahedral silicates (in Russian). Epigenesis and its mineral indicators. Publ. Office Nauka, Moscow, series 221: 96-110.

Dritz, V.A., 1975. Structural and crystallochemical characteristics of layer silicates (in Russian). Crystallochemistry of minerals and geological problems. Nauka, Moscow, 35-52.

Goodman, B.A., Russel, J.D., Fraser, A.R. and Woodhams, F.W.D., 1976. A Mössbauer and I.R. spectroscopic study of the structure of nontronite. Clays Clay Miner., 24: 53-59.

Goodman, B.A., 1978. The Mössbauer spectra of nontronites : consideration of an alternative assignment. Clays Clay Miner., 26: 176-177.

Joffe, J.S., Kolodny, L., 1938. Effect of alternate drying and wetting on the base exchange complex with special reference to the behaviour of the potassium soil. Soil Sci., Soc. Amer. Proc., 3: 107-111.

Mamy, J. and Gaultier, J.P., 1976. Les phénomènes de diffraction des rayonnements X et électroniques par les réseaux atomiques. Application à l'étude de l'ordre cristallin dans les minéraux argileux. Ann. Agron., 27: 1-16.

Mering, J. and Oberlin, A., 1967. Electron-optical study of smectites. Clays Clay Miner., 15: 3-25.

Pinsker, Z.G., 1953. Electron diffraction. Butterworths Scient. Publ., London, 443 pp.

Plançon, A. and Tchoubar, C., 1975. Etude des fautes d'empilement dans les kaolinites partiellement désordonnés. I. J.A.C., 8: 582-588.

Plançon, A. and Tchoubar, C., 1976. Etude des fautes d'empilement dans les kaolinites partiellement désordonnées. II. J.A.C., 9: 279-285.

Plançon, A. and Tchoubar, C., 1977. Determination of structural defects in phyllosilicates by X-ray powder diffraction. Parts I and II. Clays Clay Miner., 25: 430-435 and 435-450.

Plançon, A., 1980. Calculation of intensities diffracted by partially oriented powder of layer structures. J.A.C., 13: 524-528.

Radoslovitch, E.W. and Norrish, K., 1962. The cell dimensions and symmetry of layer-lattice silicates. Amer. Miner., 47: 599-604.
Rozenson, I. and Heller-Kallai, L., 1976. Reduction and oxidation of Fe^{3+} in dioctahedral smectites. Clays Clay Miner., 24: 271-282.
Rozenson, I. and Heller-Kallai, L., 1977. Mössbauer spectra of dioctahedral smectites. Clays Clay Miner., 25: 94-101.
Tsipurski, S.I., Dritz, V.A. and Chekin, S.S., 1978. Study of the structural ordering in nontronites by oblique texture method (in Russian). Isvestia Akad. Nauk. SSSR, Geol. series, 10: 105-113.
Zvyagin, B.B., 1979. High voltage electron diffraction in the study of lamellar minerals (in Russian). Ed. Nauka, Moscow, SSSR.
Zvyagin, B.B., 1967. Electron diffraction analysis of clay mineral structures. Plenum Press, New-York, 364 pp.

THE CRYSTAL STRUCTURE OF 2:1 LAYER SILICATES

H.M. KÖSTER

Lehrstuhl für Mineralogie, Technische Universität München, Lichtenbergstrasse 4, D-8046 Garching, Federal Republic of Germany.

ABSTRACT

Structural formulas of 2:1 layer silicates are calculated from chemical analyses of purified mineral samples on the basis of 10 oxygens plus 2 hydroxyls or fluorines and a total charge of 22. The distribution of the 22 cationic charges in the structure of 2:1 layer silicates on tetrahedral, octahedral, and interlayer positions is presented in a triangular diagram where the corners represent the end members muscovite-celadonite-pyrophyllite or phlogopite-taeniolite-talc. The limits of variation of 2:1 layer silicate species are determined by statistical methods and plotted in the triangular diagram. The differentiation and classification of 2:1 layer silicates is based on the sharing of the cationic charges among the three positions of cations in the crystal structure.

INTRODUCTION

Most clay mineral samples contain accessory minerals and amorphous materials. Calculation of reliable structural formulas and crystal chemistry studies require processing the raw clay mineral samples to extract and purify the mineral of interest.

A processing operation for 2:1 layer silicates developed in our laboratory includes the following steps (Köster et al., 1973; Kohler and Köster, 1976; Vogt and Köster, 1978):

1. Treatment of the sample with o.1 M EDTE-solution of pH 4.5 for the decomposition of carbonates.

2a. Treatment of the sample with o.1 M EDTE-solution of pH 8.0 or 1 M NaCl-solution (pH 7) to exchange the naturally adsorbed cations (Mg included) by sodium for destruction of clay aggregates and to get a clay easily dispersible in water.

2b. Iron-rich clay minerals may show a sprinkling of iron oxides. Theses samples must be treated with Na-dithionite (Mehra and Jackson, 1960) for decomposition of the oxidic iron minerals.

3. Preparation of size fractions from the clay by settling and centrifugation of the clay suspensions. Generally the finest clay size fractions smaller than 0.2 µm Ø or smaller o.6 µm Ø are free of quartz, feldspar, sericite, and other accessory minerals.

4. The finest clay size fractions contain the 2:1 layer minerals and small amounts of amorphous compounds of silica, alumina, and iron, which will be determined by the methods of Hashimoto and Jackson (1960)and Roth et al. (1969).

For the present study purified samples of glauconite (Kohler and Köster, 1976), montmorillonite (Vogt and Köster, 1978; Saghrawanian, 1977), nontronite (Kerschreiter, 1975), and hectorite (Saghrawanian, 1977) were used. Selected samples of pure or almost pure 2:1 layer minerals reported from other authors are also discussed.

The structural formulas of these 2:1 layer silicates are calculated from chemical analyses on the basis of 10 oxygens plus 2 hydroxyls or fluorines and a total charge of 22. (It would be more accurate to describe the structural formulas on the basis of one unit cell. But this would make comparison with data sets from the literature difficult because almost all cited authors used the smallest unit.) The distribution of the cations (22 cationic charges) in tetrahedral, octahedral, and interlayer positions is calculated taking into account the directly measured interlayer charge (Lagaly and Weiss, 1971) or alternatively the measured cation exchange capacity as proposed by Foster (1951). A mathematically correct and simple manner of converting chemical analyses to structural formulas of layer silicates is illustrated by Köster (1977). Various formulas reported by other authors are recalculated with the procedures proposed by Köster (1977).

The relationship between 2:1 layer silicates can be explained on the basis of solid solutions between the end members muscovite, celadonite, and pyrophyllite for the dioctahedral minerals (Hower and Mowatt, 1966) and the end members phlogopite(biotite), taeniolite, and talc for the trioctahedral minerals. The total cationic charge of the formula unit is 22. It is shared among cations in tetrahedral positions (15-16), octahedral positions (5-6), and interlayer positions (0-1). The best manner of presenting the distribution of cationic charges in the structure of 2:1 layer silicates is in a triangular diagram (fig.1) where the corners represent the end members muscovite-celadonite-pyrophyllite or phlogopite-taeniolite-talc. Each point in the triangular diagram presents the total cationic charge (22) divided among the three structural positions of the cations. From the pyrophyllite corner to the celadonite/muscovite connection the charge of the interlayer cations increases from 0.0 to 1.0, from the celadonite corner to the muscovite/pyrophyllite connection the charge of the octahedral cations increases from 5.0 to 6.0, from the muscovite corner to the pyro-

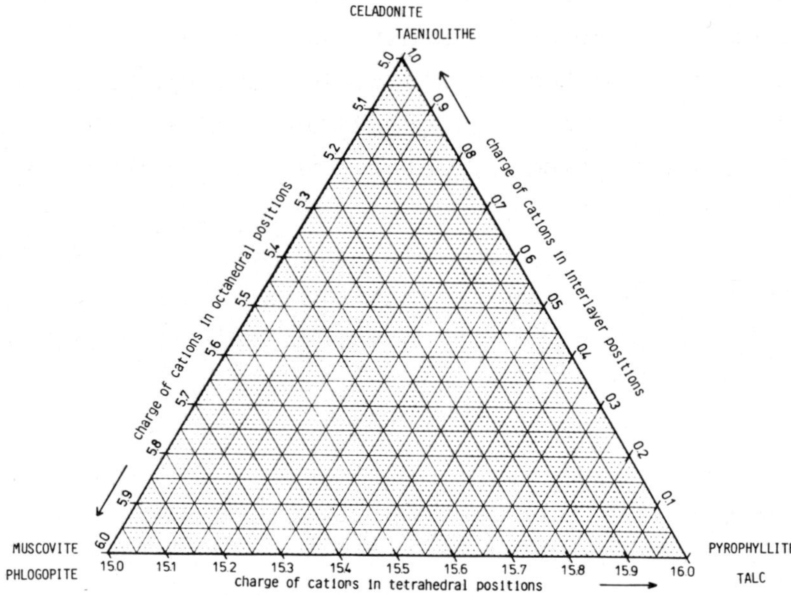

Fig. 1 Triangular diagram for the charge distribution among the three positions of cations in the structure of 2:1 layer silicates.

phyllite /celadonite connection the charge of the tetrahedral cations increases from 15.0 to 16.o. This triangular diagram is a mathematically correct modification of similar diagrams used by Hower and Mowatt (1966) and Yoder and Eugster (1955). It is suitable for comparing the complex substitutions in 2:1 layer silicates (Köster, 1977).

The limits of variation of 2:1 layer silicates can be determined by statistical methods (Youden, 1957; Doerffel, 1959, 1962). From a group of chemical analyses of a mineral species, the average $\bar{x}$, the standard deviation s, and the interval t(n,S).s for the charges of interlayer, octahedral, and tetrahedral cations can be calculated for the individual cations in the three positions of the structure. The result is the mean structural formula of a mineral species and the limits of variation of this species. Samples exceeding the limits of variation generally show contamination with other minerals or amorphous materials, or were later found to be based on incorrect chemical analyses and were therefore eliminated from the data set.

average: $\bar{x} = (x_1 + x_2 + x_3 + ... + x_n)/n$ (1)

deviation: $d = x - \bar{x}$ (2)

standard deviation: $s = [d^2/(n-1)]^{0.5}$ (3)

interval: $\Delta x = t(n,S) . s$ (4)

confidence limits for the average: $\bar{x} \pm \Delta x$ (5)

DIOCTAHEDRAL MINERALS

Table 1 shows the statistical analyses from sets of celadonites, glauconites, illites, montmorillonites, and pyrophyllites. The same relations are shown graphically in the triangular diagrams of figures 4 and 5. The sides of the hexagonal areas of the different mineral species (fig. 2-5) are the upper and lower 95 (or 99) per cent confidence limits for the average of the charges of cations in the tetrahedral, octahedral, and interlayer positions of the structure.

In figure 2 data for glauconites and montmorillonites are presented which were analysed in our institute. Three plots of glauconites outside the glauconite area are for samples contaminated with about 5 to 10 per cent montmorillonite or nontronite. These samples were eliminated from the original set of 19 glauconites investigated by Kohler and Köster (1976). The montmorillonite plot outside the montmorillonite area is for a sample containing about 5 per cent of sericite (Vogt and Köster, 1978). These examples show the influence of impurities on the calculated structural formula of clay minerals and show that it is impossible to get reliable information about the chemical structure of the crystal from unpurified samples of raw clay.

The largely coincident hexagons of glauconite sets from different authors (Buckley et al., 1978; Kohler and Köster, 1976; Cimbalnikova, 1971) in figure 3 show the validity of the statistical calculations if pure clay minerals are used for comparison of their crystal chemistry. Only three from originally 22 glauconite analyses reported by Buckley et al. (1978), three from the 19 analyses reported by Kohler and Köster (1976), but none of the analyses reported by Cimbalnikova (1971) were eliminated. The small divergences of the glauconites of the three authors can be easily explained: Buckley et al. (1978) used microprobe analyses from K-rich sections of their samples, resulting in a relatively high interlayer charge of their glauconites. Cimbalnikova (1971) selected and analysed glauconite pellets only. Kohler and Köster (1976) analysed the glauconites from both the pellets and the clay size fractions of the rocks. The glauconites in the pellets and the clay of the same rock differ a little in their crystal chemistry, which is reflected in the enlarged variation area in the triangular diagram (fig. 3). For the following investigations the glauconite sets of the three authors are combined in one representative set of 55 samples.

In figure 4 the 95 per cent confidence intervals of the dioctahedral minerals celadonite, glauconite, illite, montmorillonite, and pyrophyllite are compared. The results are: Celadonite (Buckley et al., 1978) and glauconites differ largely in the distribution of the cationic charges over the three structural positions. Statistically

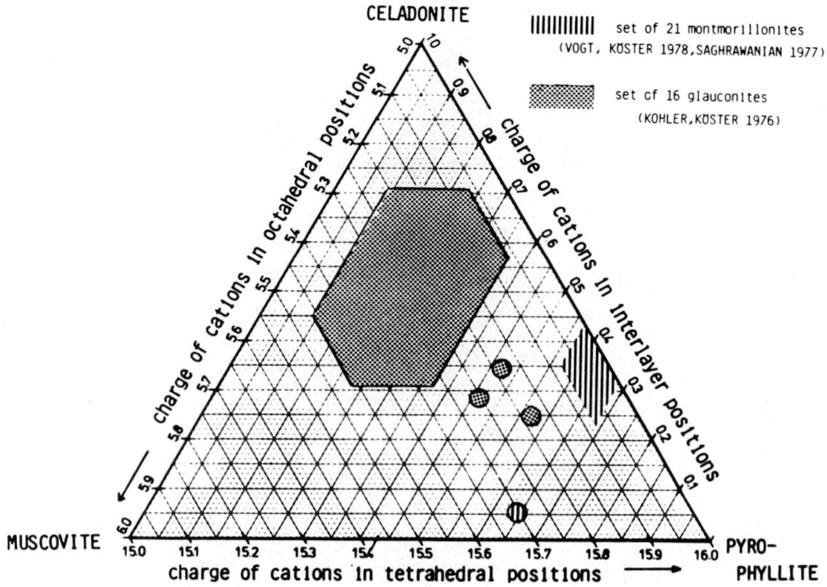

Fig. 2 Sets of 16 glauconites and 21 montmorillonites with 95 per cent confidence limits for the average.

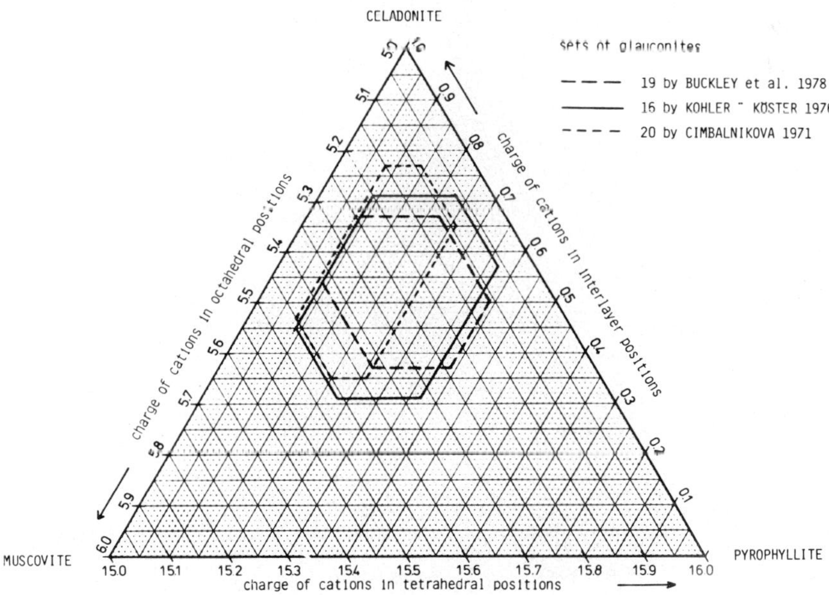

Fig. 3 Three sets of glauconites from different authors with 95 per cent confidence limits for the average.

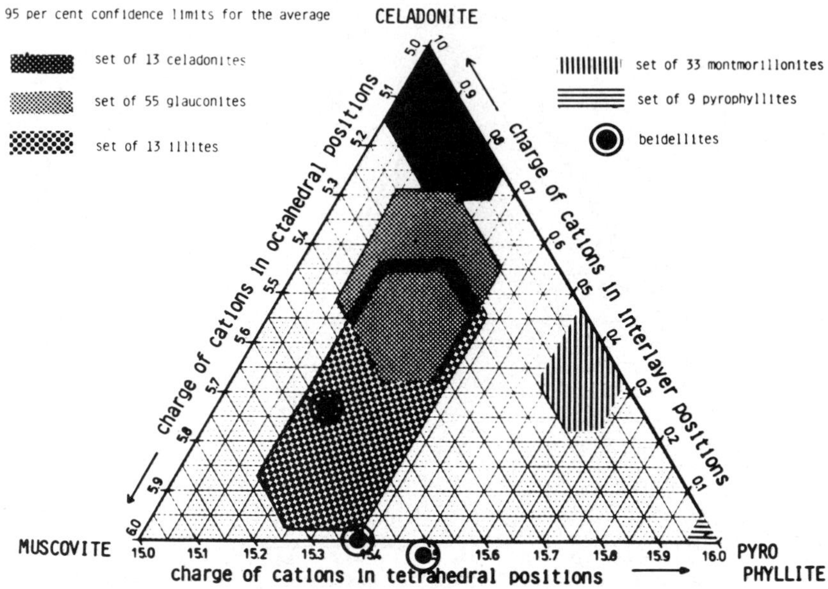

Fig. 4 Sets of dioctahedral minerals (95 per cent confidence limits for the average)

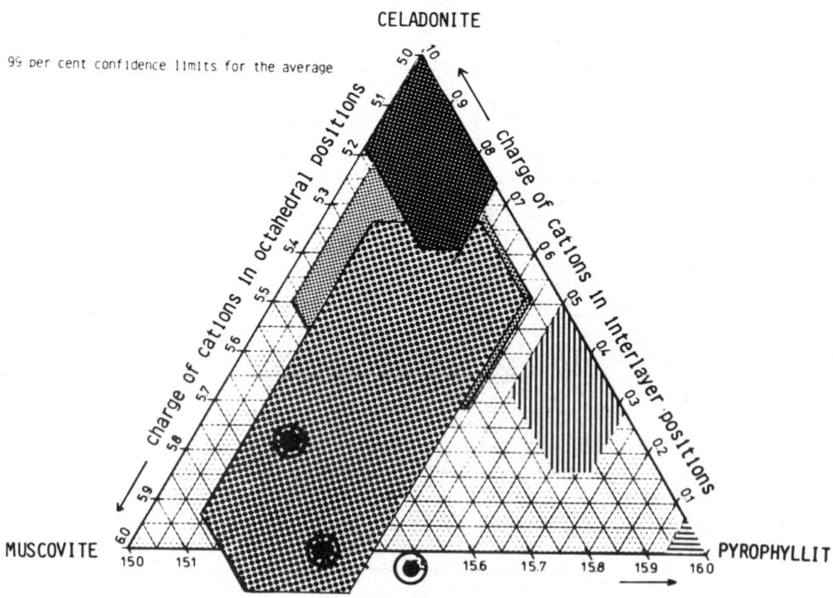

Fig. 5 Sets of dioctahedral minerals (99 per cent confidence limits for the average)

TABLE 1 Dioctahedral 2:1 layer silicates (clay minerals).

	tetrahedral sites			octahedral sites						interlayer sites				
	Si	Al	charge	Al	Fe^{3+}	Fe^{2+}	Mg	Σ	charge	Ca	Na	K	Σ	charge
celadonite BUCKLEY et al.1978														
average $\bar{x}$	3.96	0.04	15.96	0.29	0.81	0.22	0.69	2.01	5.12	0.02	0.02	0.86	0.90	0.92
stand.deviation s	0.05	0.05	0.05	0.18	0.21	0.06	0.17	0.07	0.09	0.02	0.03	0.09	0.07	0.06
interval t(n,S).s for n=13, S=95%	0.11	0.11	0.11	0.39	0.45	0.12	0.37	0.16	0.19	0.05	0.07	0.19	0.16	0.13
, S=99%			0.15						0.27					0.18
glauconite BUCKLEY et al.1978														
average $\bar{x}$	3.73	0.27	15.73	0.30	1.06	0.24	0.44	2.04	5.44	0.03	0.01	0.76	0.80	0.83
stand.deviation s	0.09	0.09	0.09	0.15	0.17	0.14	0.11	0.04	0.10	0.02	0.03	0.06	0.04	0.05
interval t(n,S).s for n=19, S=95%	0.18	0.18	0.18	0.32	0.35	0.30	0.23	0.08	0.21	0.04	0.06	0.12	0.09	0.09
, S=99%			0.25						0.28					0.13
glauconite CIMBALNIKOVA 1971														
average $\bar{x}$	3.76	0.24	15.76	0.52	0.97	0.11	0.39	1.99	5.48	0.04	-	0.68	0.72	0.76
stand.deviation s	0.06	0.06	0.06	0.09	0.11	0.04	0.04		0.07	0.03	-	0.06		0.07
interval t(n,S).s for n=20, S=95%	0.13	0.13	0.13	0.19	0.22	0.09	0.09		0.15	0.05	-	0.13		0.15
, S=99%			0.17						0.21					0.21
glauconite KOHLER, KÖSTER'76														
average $\bar{x}$	3.74	0.26	15.74	0.63	0.84	0.13	0.41	2.01	5.49	-	0.06	0.71	0.77	0.77
stand.deviation s	0.09	0.09	0.09	0.25	0.27	0.04	0.06		0.10	-	0.02	0.08		0.06
interval t(n.S).s for n=16, S=95%	0.20	0.20	0.20	0.55	0.56	0.08	0.12		0.20	-	0.05	0.17		0.14
, S=99%			0.27						0.28					0.19

TABLE 1 (continued)

	tetrahedral sites			octahedral sites						interlayer sites				
	Si	Al	charge	Al	Fe^{3+}	Fe^{2+}	Mg	Σ	charge	Ca	Na	K	Σ	charge
illite														
HOWER, MOWATT 1966														
GAUDETTE et al.65														
average $\bar{x}$	3.55	0.45	15.55	1.46	0.21	0.08	0.28	2.03	5.71	0.04	0.02	0.64	0.70	0.74
stand.deviation s	0.13	0.13	0.13	0.20	0.20	0.06	0.07	0.04	0.13	0.02	0.01	0.06	0.06	0.06
interval t(n,S).s														
for n=13, S=95%	0.28	0.28	0.28	0.44	0.44	0.13	0.16	0.08	0.27	0.03	0.02	0.13	0.12	0.12
, S=99%			0.39						0.38					0.17
montmorillonite														
VOGT, KÖSTER 1978														
SAGHRAWANIAN 1977														
average $\bar{x}$	3.98	0.02	15.98	1.49	0.10	0.02	0.41	2.02	5.64					0.37
stand.deviation s	0.03	0.03	0.03	0.06	0.03	0.01	0.10		0.07					0.03
interval t(n,S).s														
for n=21, S=95%	0.06	0.06	0.06	0.13	0.05	0.03	0.21		0.15					0.06
, S=99%			0.09						0.20					0.08
montmorillonite														
FOSTER 1953														
average $\bar{x}$	3.93	0.07	15.93	1.53	0.13	-	0.35	2.00	5.66					0.41
stand.deviation s	0.06	0.06	0.06	0.09	0.10	-	0.09		0.09					0.05
interval t(n,S).s														
for n=12, S=95%	0.13	0.13	0.13	0.20	0.22	-	0.19		0.20					0.11
, S=99%			0.19						0.28					0.15
pyrophyllite														
average $\bar{x}$	3.88	0.12	15.88	2.01	-	0.01	0.02	2.05	6.11					0.01
stand.deviation s	0.07	0.07	0.07	0.02	-	0.01	0.02		0.07					0.02
interval t(n,S).s														
for n=9, S=95%	0.17	0.17	0.17	0.06	-	0.02	0.03		0.16					0.04
, S=99%			0.25						0.23					0.06

TABLE 1a Glauconite and montmorillonite.

	tetrahedral sites			octahedral sites						interlayer sites				
	Si	Al (Fe)	charge	Al	Fe^{3+}	Fe^{2+}	Mg	Σ	charge	Ca	Na	K	Σ	charge
glauconite														
average $\bar{x}$	3.73	0.27	15.73	0.50	0.95	0.16	0.41	2.02	5.49	0.04	0.01	0.70	0.75	0.78
stand.deviation s	0.09	0.09	0.09	0.28	0.25	0.12	0.08	0.06	0.13	0.02	0.04	0.10	0.10	0.10
interval t(n,S).s														
for n=61, S=95%	0.18	0.18	0.18	0.55	0.50	0.24	0.16	0.12	0.26	0.05	0.08	0.19	0.19	0.21
, S=99%	0.25	0.25	0.25	0.74	0.66	0.31	0.21	0.15	0.35	0.06	0.11	0.26	0.26	0.27
glauconite														
average $\bar{x}$	3.74	0.26	15.74	0.48	0.96	0.17	0.41	2.02	5.47	0.03	0.01	0.72	0.76	0.79
stand.deviation s	0.08	0.08	0.08	0.22	0.20	0.12	0.08	0.04	0.09	0.02	0.02	0.07	0.06	0.07
interval t(n,S).s														
for n=55, S=95%	0.16	0.16	0.16	0.44	0.41	0.24	0.16	0.09	0.18	0.04	0.03	0.15	0.13	0.14
, S=99%	0.21	0.21	0.21	0.58	0.54	0.31	0.21	0.11	0.24	0.05	0.04	0.20	0.17	0.18
montmorillonite														
average $\bar{x}$	3.95	0.05	15.95	1.50	0.12	0.01	0.38	2.02	5.66					0.39
stand.deviation s	0.05	0.05	0.05	0.07	0.06	0.01	0.10	0.02	0.06					0.04
interval t(n,S).s														
for n=33, S=95%	0.10	0.10	0.10	0.15	0.13	0.03	0.19	0.05	0.12					0.08
, S=99%	0.13	0.13	0.13	0.20	0.17	0.04	0.26	0.07	0.17					0.11

TABLE 2 Dioctahedral smectites.

	tetrahedral sites			octahedral sites								interlayer
	Si	Al	charge	Al	Cr	Fe^{3+}	Fe^{2+}	Ni	Mg	Σ	charge	charge
montmorillonite (21) VOGT, KÖSTER 1978 SAGHRAWANIAN 1977	3.98	0.03	15.98	1.49	-	0.10	0.02	-	0.41	2.02	5.64	0.37
montmorillonite (12) FOSTER 1953	3.93	0.07	15.93	1.53	-	0.13	-	-	0.35	2.00	5.66	0.41
beidellite (1) WEIR, GREENE-KELLY'62	3.49	0.51	15.49	1.99	-	0.02	-	-	0.01	2.02	6.05	0.46
beidellite MUCHIE, NURUYU 1975	3.38	0.62	15.38	1.87	-	0.07	0.04	-	0.05	2.03	6.00	0.52
beidellite, iron-rich TOMITA, ONISHI 1976	3.46	0.54	15.46	0.96	-	0.68	-	-	0.40	2.04	5.72	0.82
nontronite (3) Al-rich	3.46	0.54	15.46	1.12	-	0.73	-	-	0.29	2.14	6.13	0.41
nontronite (5) Fe-rich	3.64	0.36	15.64	0.19	0.04	1.56	0.01	-	0.30	2.09	5.99	0.37
nontronite from HUNDSANGEN KÖSTER 1960	3.68	0.32	15.68	0.33	0.06	1.15	0.01	n.d.	0.57	2.12	5.78	0.54
KERSCHREITER 1975	4.00	-	16.00	0.20	0.01	1.22	0.03	0.02	0.61	2.09	5.61	0.39
volchonskoite (8)	3.51	0.49	15.51	0.97	0.23	0.47	0.09	0.01	0.47	2.23	6.15	0.34

significant are the different charges of the cations in octahedral positions with values smaller than 5.3 for celadonites and greater than 5.3 for glauconites. Less characteristic are the charges of the cations in tetrahedral positions which overlap with less than 15.9 for glauconites and greater than 15.8 for celadonites (table 1). Other crystal chemical values reported by the AIPEA nomenclature committee (Bailey et al., 1979) are statistically not significant and are insufficient for the classification of a single mineral sample. Nineteen samples of celadonites reported by Wise and Eugster (1964) from the older literature were found to be extraordinary heterogenous materials. From six samples it was impossible to calculate a mica formula from their chemical analyses. The plots of the other samples are distributed over the celadonite, glauconite, and illite areas in the triangular diagram. As a result these celadonite samples were excluded.

The areas of glauconites and illites in figure 4 overlap to a large extend. The 13 useful analyses of illite samples are too small for extensive interpretation, but the samples of 1M-illite polytype are concentrated in the celadonite corner and of $2M_1$-polytype in the muscovite corner of the triangular diagram. Therefore, the glauconites represent the iron-rich equivalents of the 1M-illites.

The montmorillonite samples from Vogt and Köster (1976) and from Foster (1953) are combined to make a set of 33 samples which vary over only a small area at the celadonite/pyrophyllite connection in the triangular diagram. The area of montmorillonite is separated by broad gaps from the areas of pyrophyllite and the mica minerals. The gaps indicate the lack of a homogenous solid solution between these species. The plots of the three pure beidellites (Weir and Greene-Kelly, 1962; Muchie and Nuruyu, 1975;

Tomita and Onishi, 1976) are far from the montmorillonite area but nearer the illite area. A solid solution series between illites and beidellites is more likely than between montmorillonite and beidellite. This is confirmed by the triangular diagram with 99 per cent confidence limits for the cationic charges in the three structural positions of the different mineral species (figure 5). The overlapping of the celadonite area with the glauconite and illite areas is due to analytical errors. Data reported for silica and alumina are the primary sources of this error (Köster, 1977, 1979).

From the celadonite corner to the central part of the triangular diagram the potential for charge variation between the three structural positions increases and the structural disorder must increase consequently. But the structural order-disorder from celadonite to glauconite must increase discontinously because celadonite and glauconite can be distinguished by their infra-red spectra according to Buckley et al. (1978).

For many smectites the charge of the cations in the octahedral positions exceeds 6. The triangular diagram must be completed by a scond triangle to make a rhombic diagram. Figure 6 shows a rhombic diagram with the plots for dioctahedral smectites. Lacking a sufficient number of analyses of nontronites and volchonskoites for statistical analysis, the average values for 5 iron-rich nontronites (Besnus et al., 1975; Henning and Helmchen, 1967; Kerr, 1951; Kerschreiter, 1975; Schneiderhöhn, 1965), 3 Al-rich nontronites (Alietti, 1960; Stringham and Taylor, 1950; Wieden, 1960), and 8 volchonskoites (Besnus et al., 1975; Sartori, 1967; Serdiuchenko, 1933; Weiss et al., 1954) are plotted. Most of these mineral samples are reported in the older literature. The purity of the samples is doubtful and references to cation exchange

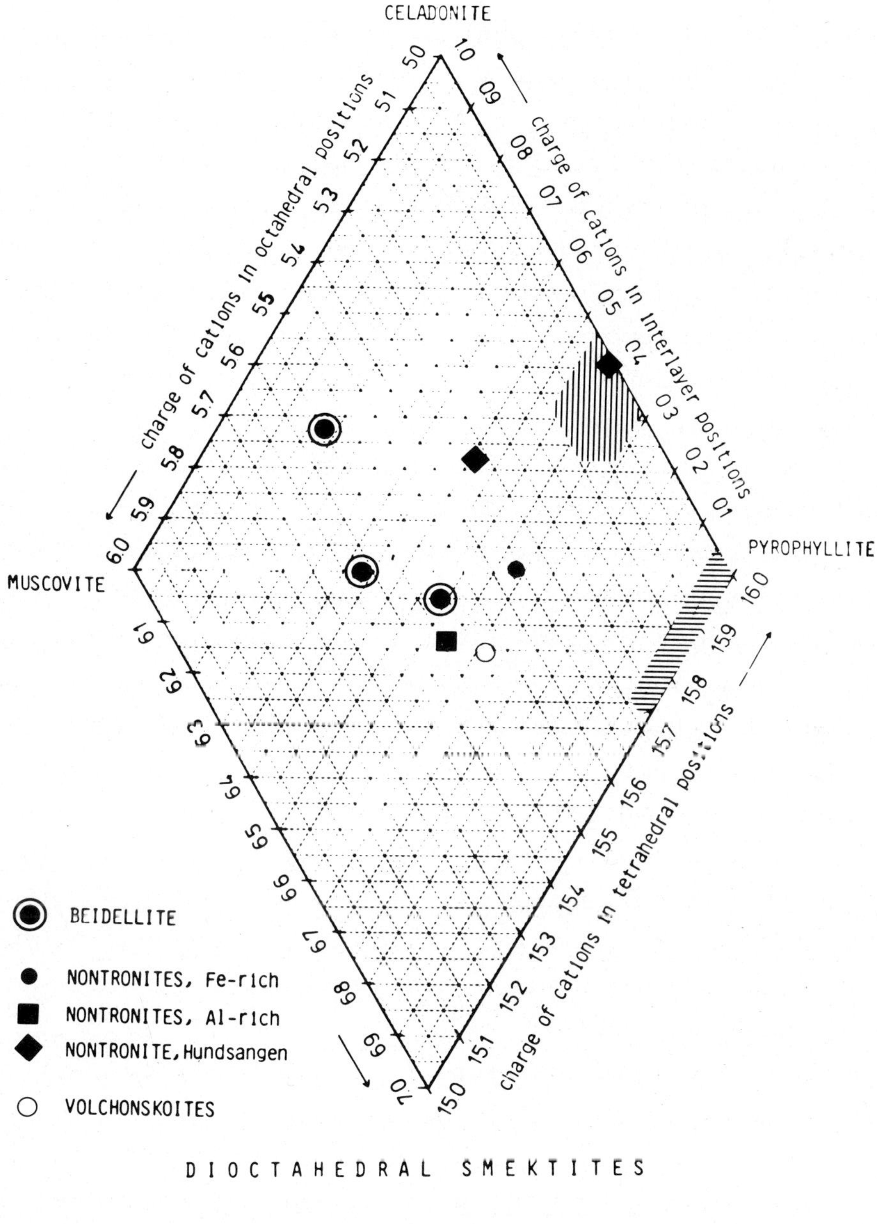

Fig. 6 Rhombic diagram for the charge distribution among the three positions of cations in the structure of dioctahedral smectites and pyrophyllite.

capacity are lacking. The nontronite of Hundsangen was first analysed by Köster (1960). A second investigation by Kerschreiter (1975), who took into consideration admixed iron oxides, shows a montmorillonite-type mineral. The plot for the Hundsangen nontronite in the triangular diagram is thereby shifted from the central part of the triangle to the middle of the montmorillonite area at the pyrophyllite/celadonite connection. Minerals of the montmorillonitic and beidellitic type may exist among the nontronites. Furthermore, there are two different types of nontronites with Al or Fe(III) dominating in the octahedral positions. The volchonskoites closely resemble the Al-rich nontronites. If analysed for chromium, the nontronites (Besnus et al., 1975; Kerschreiter, 1975; Köster, 1960) give Cr-values like those of the volchonskoites. Differentiation between nontronites and volchonskoites, therefore cannot be based exclusively on the Cr-content. Thus Al-rich nontronites and volchonskoites may be identical and representative of a solid solution series with beidellite. Previously Fleischer (1971) has defined volchonskoite as a chromian nontronite. For many nontronites and volchonskoites the calculated high charge of the cations in octahedral positions largely exceeding 6 may be due to admixed iron oxides in the clay samples. Further detailed investigations on nontronites and volchonskoites are necessary.

Most of the pyrophyllites approach the ideal structural formula (Black, 1975; Kodama, 1958; Papezik and Keats, 1976; Sykes and Moody, 1978). The elongation of the pyrophyllite area in the diagram (fig. 6) may be caused by admixed kaolinite or aluminum oxide minerals which accompany pyrophyllite in its deposits.

TRIOCTAHEDRAL MINERALS

Only a few analyses of trioctahedral smectites, but many analyses of talc, are reported in the literature. It is best to give only average values of the trioctahedral smectites. Hectorite and stevensite (table 3, fig. 7) are montmorillonite-type minerals with regard to the distribution of the cationic charges in the structure. Saponite and sauconite are beidellite-type minerals (table 3, fig. 7).

Hectorite (Ames and Goldich, 1958; Faust et al., 1959; Saghrawanian, 1977) is correctly called a trioctahedral smectite. About 10 % of the octahedral sites are occupied by lithium substituting for magnesium. With regard to the distribution of the cationic charges in the structure, hectorite corresponds to dioctahedral montmorillonite.

The average values of 5 stevensite samples (Bradley and Fahey, 1962; Faust et al., 1959; Millot, 1954; Randall, 1959; Shimoda, 1971) were calculated. The interlayer charge of stevensite results from a small deficit of cations in octahedral positions. The relatively low interlayer charge of the stevensites may be due to interstratification with talc-layers (Brindley, 1955).

The lithium-mineral, swinefordite(Tien et al., 1975), seems actually to be a trioctahedral smectite (table 3) related to stevensite and hectorite. The formula was recalculated from samples 1 and 3 reported by Tien et al. (1975) taking into account the reported cation exchange capacity.

About half of the saponites (Cahoon, 1954; Kitahara, 1957; Mackenzie, 1957; Midgley and Gross, 1956; Schmidt and Heystek, 1953; Weiss et al., 1955; Whelan and Lepp, 1961) show only Mg in octahedral positions and a cationic charge for the octahedral positions close to 6 (table 3). The average values of these 7 saponites are very close to

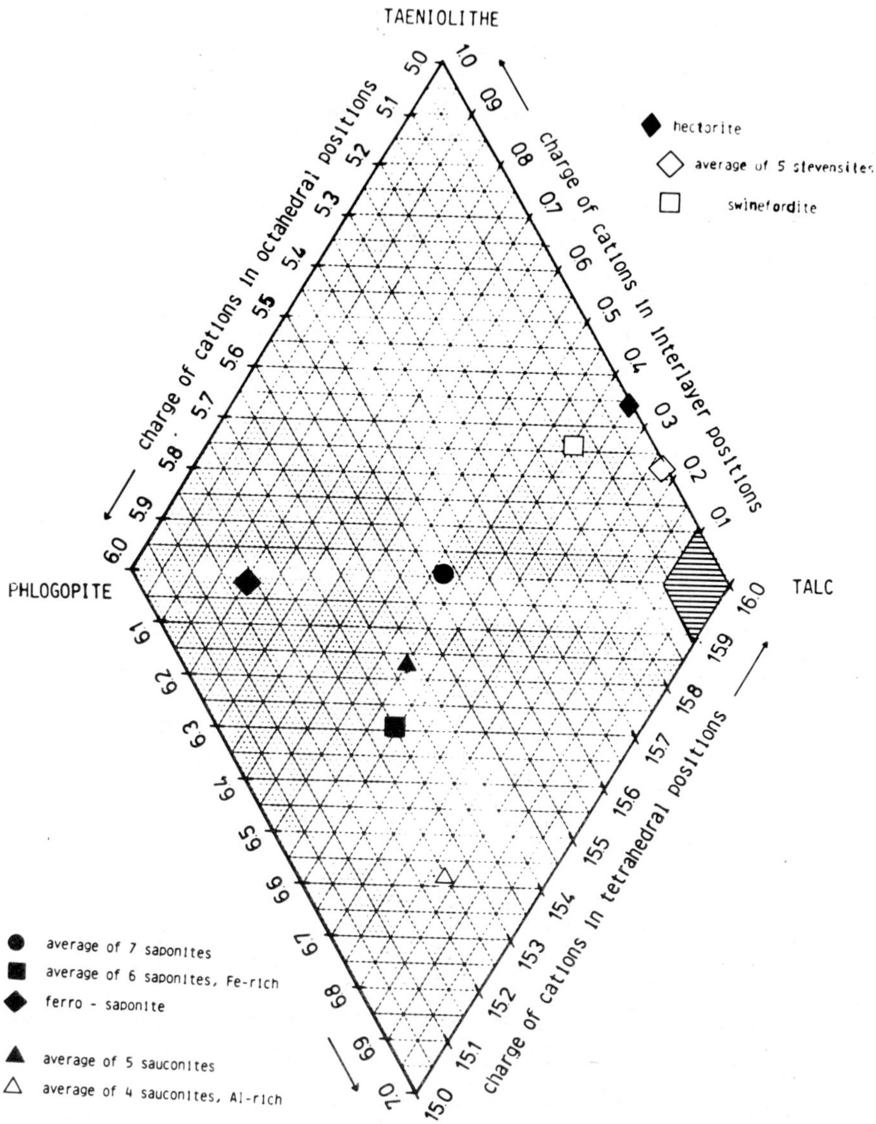

Fig. 7 Rhombic diagram for the charge distribution among the three positions of cations in the structure of trioctahedral smectites and talc.

TABLE 3 trioctahedral smectites.

	tetrahedral sites			octahedral sites									interlayer
	Si	Al	charge	Al	Fe^{3+}	Fe^{2+}	Ni	Zn	Mg	Li	Σ	charge	charge
talc													
average $\bar{x}$	3.97	0.03	15.97	0.02	0.01^{+}	0.07	0.03^{++}	-	2.89	-	2.99	6.01	0.02
stand.deviation s	0.06	0.06	0.06						0.12			0.05	0.04
interval t(n,S).s													
for n=11, S=95%			0.13						0.27			0.12	0.09
, S=99%			0.18						0.38			0.17	0.13
stevensite (5)	3.99	0.01	15.99	0.03	0.02	-	0.03	-	2.77	0.03	2.89	5.78	0.23
hectorite (3)	3.99	0.01	15.99	-	-	-	-	-	2.67	0.32	2.99	5.66	0.35
swinefordite (2)	3.87	0.13	15.87	0.39	0.10	-	-	-	0.70	1.06	2.85	5.73	0.40
saponite (7)	3.53	0.47	15.53	0.03	0.02	-	-	-	2.92	-	2.97	5.99	0.48
saponite, Fe-rich (6)	3.30	0.70	15.30	0.04	0.54	0.27	0.01	-	2.00	-	2.86	6.30	0.40
ferro-saponite (1)	3.18	0.82	15.18	0.01	-	1.99	-	-	1.00	-	3.00	6.02	0.80
sauconite (5)	3.38	0.62	15.38	0.15	0.10	-	-	2.55	0.16	-	2.96	6.17	0.45
sauconite, Al-rich (4)	3.25	0.75	15.25	0.81	0.23	-	-	1.57	0.16	-	2.77	6.58	0.17

+) Fe^{3+} + Cr^{3+} ++) Ni^{2+} + Cu^{2+} + Mn^{2+}

the beidellite from the Black Jack Mine (Weir and Greene-Kelly, 1962). Six other saponite samples (Faust, 1951; Konta and Sindelar, 1955; Mel'nik, 1965; Miyamoto, 1957; Mongiorgi and Morendi, 1970; Sudo, 1954; Wilson et al., 1968) are rich in di- and trivalent iron. The charge of their cations in octahedral positions considerably exceed 6. These saponite samples may partly be mixtures of saponite with iron oxides. Small amounts of iron minerals hardly detectable by x-rays will effectively falsify the structural formulas of 2:1 layer silicates. The ferrosaponite reported by Kohyama et al. (1973) originated from biotite and shows a mica-like formula.

The samples of sauconites (Faust, 1951; Ginzburg and Vitovskaya, 1956; Semenov, 1956) may be mixtures of clay with ore minerals. But two different types of sauconites can be differentiated by their Al-values. The first type with Zn predominating in octahedral positions resembles the saponites with only Mg in octahedral positions. The second type of sauconite has approximately one-third of the octahedral sites occupied by Al and trivalent iron. Extremely high cationic charges are calculated for the octahedral positions (table 3) and very low charges are calculated for the interlayer positions. This result may be attributed to impurities in the sauconite samples. An investigation of sauconites with modern methods seems to be necessary.

The average values of 9 talc samples (Abraham and Schreyer, 1976; Basta and Kamel, 1970; Besnus et al., 1975; Evans and Misch, 1976; Juan and Lo, 1975; Rayner and Brown, 1973) approach the ideal structural formula (table 3). Therefore, talc shows only a small variation area in figure 7.

Chemical analyses of 25 vermiculites are reported by Foster (1963) and the analyssis of an iron-rich vermicu-

TABLE 4 Vermiculite.

	tetrahedral sites			octahedral sites						interlayer
	Si	Al	charge	Al	Fe^{3+}	Fe^{2+}	Mg	Σ	charge	charge
vermiculite FOSTER 1953										
average $\bar{x}$	2.87	1.13	14.87	0.14	0.34	0.09	2.40	2.97	6.42	0.71
stand.deviation s	0.14	0.14	0.14	0.14	0.26	0.10	0.43		0.22	0.18
interval t(n,S).s for n=25, S=95%	0.28	0.28	0.28	0.28	0.52	0.20	0.89		0.46	0.38
vermiculite, iron-rich (1) BLACK 1975	3.09	0.91	15.09	0.59	-	1.28	1.13	3.00	6.59	0.32

lite by Black (1975). Vermiculites are reported to have high values of interlayer charge (Brindley and Brown, 1980). The average value is 0.7 indeed (table 4), but the interlayer charges of the vermiculites vary from low values of about o.3 to high values of 1.1 as calculated from the chemical analyses of the reported vermiculite samples. Unfortunately interlayer charges or cation exchange capacities of these vermiculite samples were not measured and are not reported by Foster (1963). The calculated low interlayer charges of some samples may be due to errors of chemical analyses or the samples are smectites. The charges of cations in octahedral positions vary from about 6 to 7. The average value for the charge of the cations in tetrahedral positions is 14.9 and lower than the values of other 2:1 layer silicates. The average formula of the vermiculites (table 4) illustrates the formation of the vermiculites from biotites by oxidation of iron in octahedral positions connected with an equivalent lowering of the interlayer charge.

A rhombic diagram presenting all species of 2:1 layer silicates, brittle-micas and vermiculites included, is shown in figure 8. The diagram presents interlayer charges from o to 2, charges of octahedral cations from 4 to 8, and charges of tetrahedral cations from 14 to 16. The area of brittle-micas is situated in the leftcorner of the rhomb and the area of vermiculites in the central part of the lower triangle.

CONCLUSIONS

The different species of 2:1 layer silicates and their limits of variation are definable by means of statistical methods, if analyses of a sufficient number of pure mineral samples are available. The classification of 2:1 layer silicate species based on their crystal chemistry is con-

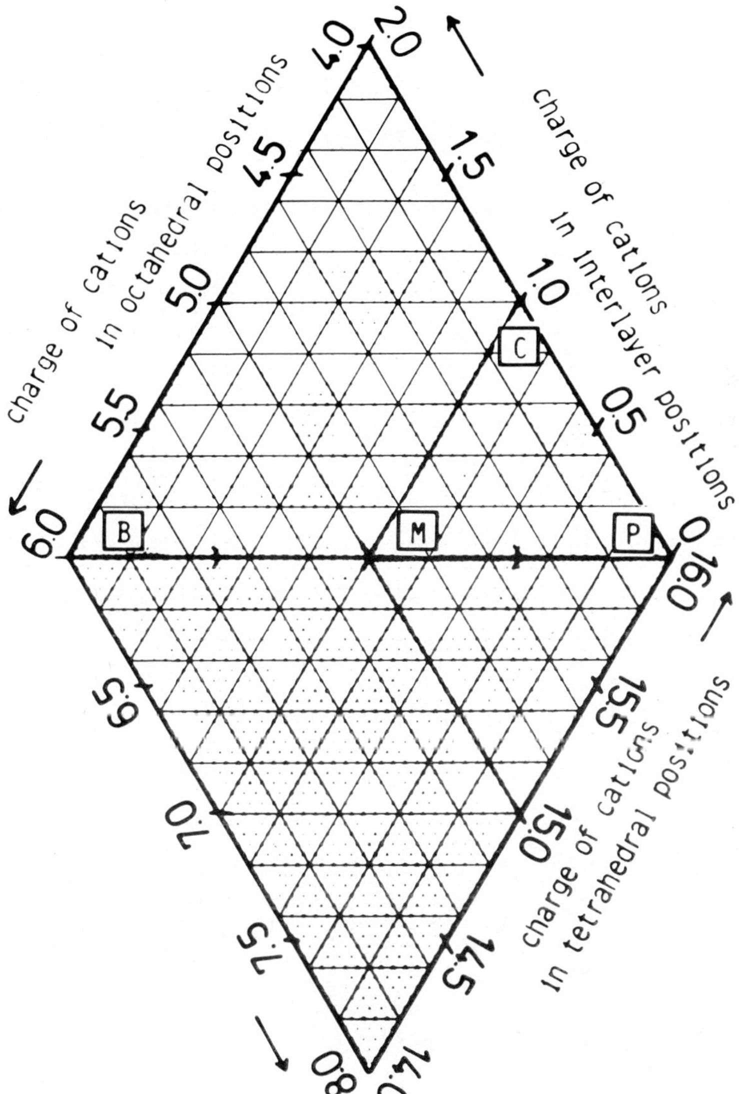

Fig. 8 Rhombic diagram for the charge distribution among the three positions of cations in the structure of 2:1 layer silicates presenting all the different species of these minerals.

TABLE 5 Calculated average formulas.

Mineral	Interlayer	Octahedral	Tetrahedral
DIOCTAHEDRAL MICAS			
Celadonite	$K^+_{0.9}$	$(Al_{0.3}Fe^{3+}_{0.8}Fe^{2+}_{0.2}Mg_{0.7})$	$(Si_{4.0}O_{10}(OH)_2)$
Glauconite	$K^+_{0.8}$	$(Al_{0.5}Fe^{3+}_{1.0}Fe^{2+}_{0.1}Mg_{0.4})$	$(Al_{0.3}Si_{3.7}O_{10}(OH)_2)$
Illite	$M^+_{0.1}\ K^+_{0.6}$	$(Al_{1.5}Fe^{3+}_{0.2}Mg_{0.3})$	$(Al_{0.4}Si_{3.6}O_{10}(OH)_2)$
DIOCTAHEDRAL SMECTITES			
Montmorillonite	$M^+_{0.4}$	$(Al_{1.5}Fe^{3+}_{0.1}Mg_{0.4})$	$(Si_{4.0}O_{10}(OH)_2)$
Nontronite, Hundsangen.	$M^+_{0.4}$	$(Al_{0.2}Fe^{3+}_{1.2}Mg_{0.7})$	$(Si_{4.0}O_{10}(OH)_2)$
Beidellite	$M^+_{0.5}$	$(Al_{2.0})$	$(Al_{0.5}Si_{3.5}O_{10}(OH)_2)$
Nontronite	$M^+_{0.4}$	$(Al_{0.2}Fe^{3+}_{1.6}Mg_{0.3})$	$(Al_{0.4}Si_{3.6}O_{10}(OH)_2)$
Nontronite, Al-rich.	$M^+_{0.4}$	$(Al_{1.1}Fe^{3+}_{0.7}Mg_{0.3})$	$(Al_{0.4}Si_{3.6}O_{10}(OH)_2)$
Volchonskoite	$M^+_{0.4}$	$(Al_{1.0}Cr^{3+}_{0.2}Fe^{3+}_{0.5}Mg_{0.5})$	$(Al_{0.5}Si_{3.5}O_{10}(OH)_2)$
TRIOCTAHEDRAL SMECTITES			
Stevensite	$M^+_{0.2}$	$(Mg_{2.9})$	$(Si_{4.0}O_{10}(OH)_2)$
Hectorite	$M^+_{0.3}$	$(Mg_{2.7}Li_{0.3})$	$(Si_{4.0}O_{10}(OH)_2)$
Swinefordite	$M^+_{0.2}$	$(Al_{1.0}Fe^{3+}_{0.1}Mg_{0.7}Li_{1.0})$	$(Al_{0.1}Si_{3.9}O_{10}(OH)_2)$
Saponite	$M^+_{0.5}$	$(Mg_{3.0})$	$(Al_{0.5}Si_{3.5}O_{10}(OH)_2)$
Saponite, Fe-rich.	$M^+_{0.4}$	$(Fe^{3+}_{0.5}Fe^{2+}_{0.4}Mg_{2.0})$	$(Al_{0.7}Si_{3.3}O_{10}(OH)_2)$
Ferro-Saponite	$M^+_{0.8}$	$Fe^{2+}_{2.0}Mg_{1.0})$	$(Al_{0.8}Si_{3.2}O_{10}(OH)_2)$
Sauconite	$M^+_{0.4}$	$(Al_{0.1}Fe^{3+}_{0.1}Zn_{2.6}Mg_{0.2})$	$(Al_{0.6}Si_{3.4}O_{10}(OH)_2)$
Sauconite, Al-rich.	$M^+_{0.2}$	$(Al_{0.8}Fe^{3+}_{0.2}Zn_{1.6}Mg_{0.2})$	$(Al_{0.8}Si_{3.2}O_{10}(OH)_2)$

trolled

1. by the sharing of the cationic charges among the three positions of cations in the crystal structure.

In nomenclature rules one considers also

2. the di- or trioctahedral occupation of the octahedral positions, and
3. the dominant cations in the octahedral positions (Al or Fe(III), Mg or Fe(II) or Zn) or in the interlayer positions of the mica minerals particularly (K, Na, Ca or Mg).

If insufficient analyses are available for statistical determination of the limits of variation, as was the case for trioctahedral smectites, the differentiation of mineral species can be discussed only in a qualitative fashion based on the average values of the structural formulas. In this case, classification of a given mineral sample may be difficult or impossible.

ACKNOWLEDGEMENTS

I am indebted to Prof. S.W. Bailey and to Mr. D.G. Schulze for critically reading the manuscript.

REFERENCES

Abraham, K. and Schreyer, W. 1976. A talc-phengite assemblage in piemontite schist from Brezovica, Serbia. J.Petrol., 17: 421-439.

Alietti, A., 1960. Sudi una nontronite poco ferrifera di Chiampo (Lessini). Soc.Tipogr.Editr.Modense, 1960: 33-44.

Ames, L.L. and Goldich, S.S., 1958. A contribution on the Hector, California, bentonite deposit. Econ.Geol., 53: 22-37.

Bailey, S.W., Brindley, G.W., Kodama, H. and Martin, R.T.,

1979. Report of the Clay Minerals Society Nomenclature Committee for 1977 and 1978. Clays Clay Min., 27: 238-239.

Basta, E.Z. and Kamel, M., 1970. Mineralogy and ceramic properties of some Egyptian talcs. Bull.Fac.Sci.Univ. Cairo, 43: 285-299.

Besnus, Y., Fusil, G., Janot, C., Pinta, M. and Sieffermann, G., 1975. Characteristics of some weathering products of chromitic ultrabasic rocks in Bahia State, Brazil: Nontronites, chlorites and chromiferous talc. Proc.Intern. Clay Conf., Mexico, 27-34.

Black, P.M., 1975. Mineralogy of New Caledonian metamorphic rocks: IV. Sheet silicates from the Quégoa district Contr.Min.Petr., 49: 269-284.

Bradley, W.H. and Fahey, J.J., 1962. Occurance of stevensite in the Green River formation of Wyoming. Am.Miner., 47: 996-998.

Brindley, G.W., 1955. Stevensite, a montmorillonite-type mineral showing mixed-layer characteristics. Am.Miner., 40: 239-247.

Brindley, G.W. and Brown, G., 1980. Crystal structures of clay minerals and their x-ray identification. Mineralogical Society, London.

Buckley, H.A., Bevan, J.C., Brown, K.M., Johnson, L.R. and Farmer, V.C., 1978. Glauconite and celadonite: two separate mineral species. Min.Mag., 42: 373-382.

Cahoon, H.P., 1954. Saponite near Milford, Utah. Am.Miner., 39: 222-230.

Cimbálniková, A., 1971. Chemical variability and structural heterogeneity of glauconites. Am.Miner., 56: 1385-1392.

Doerffel, K., 1962. Beurteilung von Analysenverfahren und Ergebnissen. Springer-Verlag, Berlin-Göttingen-Heidelberg.

Doerffel, K., 1959. Anwendung der Statistik in der analytischen Chemie. Chem.Techn., 11: 579-582.

Evans, B.W. and Misch, P., 1976. A quartz-aragonite-talc schist from the lower Skagit Valley, Washington. Am. Miner., 61: 1005-1oo8.

Faust, G.T., 1951. Thermal analysis and x-ray studies of sauconite and of some zinc minerals of the same paragenetic association. Am.Miner., 36: 795-822.

Faust, G.T., Hathaway, J.C. and Millot, G., 1959. A restudy of stevensite and allied minerals. Am.Miner., 44: 342-370.

Fleischer, M., 1971. Glossary of mineral species 1971. Bowie, Maryland (Mineralogical Record Inc.).

Foster, M.D., 1963. Interpretation of the composition of vermiculites and hydrobiotites. Clays Clay Min., 10: 70-89.

Foster, M.D., 1953. Geochemical studies of clay minerals: II. Relation between ionic substitution and swelling in montmorillonites. Am.Miner., 38: 994 1006.

Foster, M.D., 1951. The importance of exchangeable magnesium and cation exchange capacity in the study of montmorillonite clays. Am.Miner., 36: 717-730.

Gaudette, H.E., Eades, J.L. and Grim, R.E., 1965. The nature of illite. Clays Clay Min., 13: 33-48.

Ginzburg, I.I. and Vitovskaya, I.V., 1956. Sauconite in the crust of weathering of the lead-zinc deposits of central Kazakhstan. In "Crust of Weathering", Acad. Sci.USSR, no. 2, 184-187.

Hashimoto, I. and Jackson, M.L., 1960. Rapid dissolution of allophane and kaolinite-halloysite after dehydration. Clays Clay Min., 7: 102-113.

Henning, K.H. and Helmchen, H., 1967. Nontronit von Bautzen-Seidau. Ber.deutsch.Ges.geol.Wiss., B, Miner.Lagerstättenf., 12: 195-199.

Hower, J. and Mowatt, T.C., 1966. The mineralogy of illites and mixed illite-montmorillonite. Am.Miner., 51: 825-854.

Juan, V.C. and Lo, H.J., 1975. The stability field of talc. Proc.Geol.Soc. China (Formosa), 18: 29-38.

Kerschreiter, H., 1975. Mineralogische und kristallchemische Untersuchungen am Nontronit von Hundsangen (Westerwald) und den Montmorilloniten von Göpfersgrün und Groschlattengrün (Nordostbayern). Diplomarbeit TU München (unpublished).

Kitahara, Jun-Ichi, 1957. Saponite from the Wakamatsu Mine. Tari district, Tottori Prefecture. Journ. Japanese Assoc.Min.Petr. and Econ.Geol., 41: 148-151.

Kodama, H., 1958. Mineralogical study on some pyrophyllites. Mineral. J. (Tokyo), 2: 236-244.

Köster, H.M., 1979. Die chemische Silikatanalyse. Spektralphotometrische, komplexometrische und flammenspektrometrische Analysenmethoden. Springer-Verlag.

Köster, H.M., 1977. Die Berechnung kristallchemischer Strukturformeln von 2:1-Schichtsilikaten unter Berücksichtigung der gemessenen Zwischenschichtladungen und Kationenumtauschkapazitäten, sowie die Darstellung der Ladungsverteilung in der Struktur mittels Dreieckskoordinaten. Clay Miner., 12: 45-54.

Köster, H.M., 1960. Nontronit und Picotit aus dem Basalt des Ölberges bei Hundsangen, Westerwald. Beitr.Mineral. Petrogr., 7: 71-75.

Köster, H.M., Kohler, E.E., Krahl, J., Kröger, J. and Vogt, K., 1973. Veränderungen am Montmorillonit durch Einwirkung von 0,1 m AeDTE-Lösungen, 1 n NaCl-Lösung und 0,1 n Salzsäure. N.Jb.Miner., Abh., 119: 83-1oo.

Kohler, E.E. and Köster, H.M., 1976. Zur Mineralogie, Kristallchemie und Geochemie kretazischer Glaukonite. Clay Miner., 11: 273-302.

Kohyama, N., Shimoda, S. and Sudo, T., 1973. Iron-rich saponite (ferrous and ferric forms). Clays Clay Min., 21: 229-237.

Konta, J. and Sindelar, J., 1955. Saponite from the fissure of the amphibolites of Caslav. Acta Universitatis Carolinae, Geologica, 1: 177-186.

Lagaly, G. and Weiss, A., 1971. Neue Methoden zur Charakterisierung und Identifizierung quellungsfähiger Dreischichttonminerale. Z.Pflanzenern.u.Bodenk., 130: 9-24.

Mackenzie, R.C., 1957. Saponite from Allt Ribhein, Fiskavaig Bay, Skye. Min.Mag., 31: 672-680.

Mehra, O.F. and Jackson, M.L., 1960. Iron oxide removal from soils and clays by a dithionite-citrate system buffered with sodium bicarbonate. Clays Clay Min., 7: 317-327.

Mel´nik, Yu.M., 1965. Iron saponite from the weathered crust of basic rocks from the Korosten complex, Ukrainian shield. Min.Sborn.L´vovsk.Univ., 19: 101-105.

Midgley, H.G. and Gross, K.A., 1956. Thermal reactions of smectites. Clay Min. Bull., 3: 79-90.

Millot, G., 1954. La ghassoulite, pôle magnésien de la seric des montmorillonites. Compt.rend.Acad.Sci., Paris, 238: 257-259.

Miyamoto, N., 1957. Iron-rich saponite from Maźe, Niigata Prefecture, Japan. Min.Journ. (Japan), 2: 193-195.

Mongiorgi, R. and Morendi, N., 1970. Al saponite e strati misti clorite-Al saponite nelle idrotermaliti di una breccia di contatto coi diabasi si Rossena nell´Appenino reggiano. Min.Petr. Acta, 16: 139-151.

Muchi, M. and Nuruyu, K., 1975. Beidellite-type mineral from Hazami-cho, Nagasaki Prefecture, Japan. Bull. Fukuoka Uni. Educ., 25: Pt. III, Nat.Sci., 101-108.

Papezik, V.S. and Keats, H.F., 1976. Diaspore in a pyrophyllite deposit on the Avalon Peninsula, Newfoundland.

Can.Min., 14: 442-449.

Randall, B.A.O., 1959. Stevensite from the Whin Sill in the region of the North Tyne. Min.Mag., 32: 218-225.

Rayner, J.H. and Brown, G., 1973. The crystal structure of talc. Clays Clay Min., 21: 103-114.

Roth, C.B., Jackson, M.L. and Syers, J.K., 1969. Deferration effect on structural ferrous-ferric iron ratio and cec of vermiculites and soils. Clays Clay Min., 17: 253-264.

Saghrawanian, B., 1977. Mineralogische und kristallchemische Untersuchungen an drei amerikanischen Bentonitproben aus Chambers/Arizona, Otay/Californien, Santa Rita/ New Mexico und Hectorit aus Hector/Californien. Diplomarbeit TU München (unpublished).

Sartori, F., 1967. La wolchonskoit di Castiglioncello (Livorno). Periodico Min., Roma, 36: 103-124.

Schmidt, E.R. and Heystek, H., 1953. A saponite from Krugersdorp district, Transvaal. Min.Mag., 30: 201-210.

Schneiderhöhn, P., 1965. Nontronit vom Hohen Hagen und Chloropal vom Meenser Steinberg bei Göttingen. Tschermaks Min.Petr.Mitt., 10: 385-399.

Semenov, E.I., 1956. Sauconite in alkali pegmatites. In "Crust of Weathering", Acad.Sci.USSR, no. 2, 179-183.

Serdiuchenko, D.P., 1933. Chrom-nontronites and their genetical relations with the serpentines of the northern Caucasus. Soc. russe minéralogie Mem., 2d sér., 62: 376-390.

Shimoda, S., 1971. Mineralogical studies of a species of stevensite from the Obori Mine, Yamagata Prefecture, Japan. Clay Miner., 9: 185-192.

Stringham, B.F. and Taylor, A., 1950. Nontronite at Bingham, Utah. Am.Miner., 35: 1060-1066.

Sudo, T., 1954. Iron-rich saponite found from Tertiary iron sand beds of Japan (re-examination on "Lembergite"). Journ.Geol.Soc. Japan, 60: 18-27.

Sykes, M.L. and Moody, J.B., 1978. Pyrophyllite and metamorphism in the Carolina slate belt. Am.Miner., 63: 96-108.

Tien, Pei-Lin, Leavens, P.B. and Nelen, J.A., 1975. Swinefordite, a dioctahedral-trioctahedral Li-rich member of the smectite group from Kings Mountain, North Carolina. Am.Miner., 60: 540-547.

Tomita, K. and Onishi, K., 1976. An iron-rich beidellite-like mineral found in an altered andesite at Yamada, Hishikari-cho, Kagoshima Prefecture, Rept.Fac.Sci., Kagoshima Univ. (Earth Sci. Bio. l.) no.9: 1-7.

Vogt, K. and Köster, H.M., 1978. Zur Mineralogie, Kristallchemie und Geochemie einiger Montmorillonite aus Bentoniten. Clay Miner., 13: 25-43.

Weir, A.H. and Greene-Kelly, R., 1962. Beidellite. Am. Miner., 47: 137-146.

Weiss, A., Koch, G. and Hofmann, U., 1955. Zur Kenntnis von Saponit. Ber.Dtsch.Keram.Ges., 32: 12-17.

Whelan, J.A. and Lepp, H., 1961. An occurance of saponite near Silver Bay, Minnesota. Am.Miner., 46: 430-433.

Wieden, P., 1960. Ein eisenarmer Nontronit. Tschermaks Min.Petr.Mitt., 7: 186-199.

Wilson, M.J., Bain, D.C. and Mitchell, W.A., 1968. Saponite from the Dalradian meta-limestone of Northeast Scotland. Clay Miner., 7: 343-349.

Wise, W.S. and Eugster, H.P., 1964. Celadonite; synthesis, thermal stability and occurance. Am.Miner., 49: 1031-1083.

Yoder, H.S. and Eugster, H.P., 1955. Synthetic and natural muscovites. Geochim.Cosmochim. Acta, 8: 225-280.

Youden, W.J., 1957. Statistical methods for chemists. J.Wiley and Sons, Inc., New York (3.ed.).

PROTONIC STRUCTURE OF KAOLINITE DURING DEHYDROXYLATION STUDIED BY PROTON NUCLEAR MAGNETIC RESONANCE

C.OTERO-AREAN, M.LETELLIER, B.C.GERSTEIN and J.J.FRIPIAT
C.R.S.O.C.I.- C.N.R.S.,1B,rue de la Férollerie,45045 Orléans,France

ABSTRACT

The second moment of the nuclear magnetic resonance line of protons in kaolinites at various degrees of dehydroxylation (α) has been measured from the Fourier transform of the free induction decay after a $\frac{\pi}{2}$ pulse (M_2). In addition, the contribution of the homonuclear contribution to the second moment, $M_2(^1H - {}^1H)$ has been obtained from the decay of the echo amplitudes under the Carr-Purcell sequence.

For 0% < α < 55%, the resonance line is Gaussian, and M_2 remains constant. Above 70% dehydroxylation, the proton signal is composed of two Gaussian lines with second moments, M_2' and M_2'', decreasing as α increases.

The difference between M_2 and $M_2(^1H - {}^1H)$ for 0 < α < 70% or the difference between M_2' and $M_2(^1H - {}^1H)$ for α > 70% is constant. This difference represents the contribution of the $^1H - {}^{27}Al$ interaction to the second moment. If this contribution is removed from M_2 or M_2' for 0 < α < 70%, one obtains a proton-proton second moment equal to 6 gauss^2, the calculated value for kaolinite being 5.5 gauss^2. The implication of these results is that dehydroxylation occurs layer by layer from 0 to 70% dehydroxylation.

For α > 70% the protons left in the solid are distributed among two ensembles P' and P", P" increasing at the expense of P' as α increases. P' is identified with patches of protons still linked to octahedrally co-ordinated aluminum. These patches have about the area of one unit cell. Between these patches remain isolated protons which constitute ensemble P". These protons are perhaps linked to tetrahedrally co-ordinated aluminum.

INTRODUCTION

Since potteries have been found in the oldest prehistoric graves, the dehydroxylation of kaolinite probably has been among the very first solid state reactions carried out by man. The mechanism by which kaolinite looses its constitutional water has attracted much attention in the last century.The structure of the initial material being rather well characterized, numerous studies have been devoted to the kinetic rate process. The space allowed for this paper would not suffice for a simple summary of them all.

The structural arrangement in the dehydroxylation product, i.e. metakaolinite, was studied by several experimentators, the most classical work being that of Brindley and Nakahira (1959) who concluded that the original ordering is maintained along the a-and b-axes but that it disappears along the c-axis. According to Brindley et al. (1958-1959) the coordination of the aluminum changes from six-fold to four-fold during the dehydration reaction :

$$Al_2O_3.2SiO_3.2H_2O \longrightarrow Al_2O_3.2SiO_2 + 2H_2O \text{ (vap)}$$

Kaolinite Metakaolinite

and the reaction proceeds by successive and complete dehydroxylation of the octahedral layers. This idea was primarily substantiated by the observation that the intensity of the [001] reflection decreases linearly with the dehydroxylation degree α.

Gastuche et al. (1963) have tried to complement these conclusions by measuring the coordination number of Al for various values of α, using the shift of the Al K_α line. They observed a linear increase of the relative content in Al^{IV} with respect to α. Later on Wardle et al. (1974) pointed out the ambiguous character of this type of measurement since the shift of the fluorescence line is actually a function of the average Al-O distance.

More conclusively Gastuche et al. measured the second moment of the nuclear magnetic resonance line of protons in kaolinite at increasing α and they observed that up to about 60% dehydroxylation it was practically constant. Since the main contribution to this second moment was postulated to be associated with proton-proton interactions they concluded that the average proton-proton distance remained constant until 60% dehydroxylation.

If dehydroxylation would proceed by the growth of randomly distributed nuclei, the average distance between hydroxyls should continuously increase with α : accordingly the second moment of the proton NMR signal should decrease very rapidly since it is inversely proportional to the sixth power of that distance. Therefore these NMR experiments indicated a reaction process of successive and complete dehydroxylation of octahdedral layers. However, the data had a limited significance for the following reasons :

1. Because of the lack of sensitivity of the instrument available in 1963, it was impossible to obtain a signal-to-noise ratio large enough to observe the proton resonance for samples with $\alpha > 60\%$.

2. The total second moment M_2 of the proton signal is the sum of two contributions, namely $M_2(^1H - {}^1H)$ (the homodipolar contribution), and M_2 $(^{24}Al - {}^1H)$ (the heterodipolar contribution). The proton-aluminum contribution in the work of Gastuche et al. was not measured but calculated assuming that all the interlayer OH bonds have the same orientation. Since the result of Giese and Datta (1973) agree with the infrared dichroïsm of the OH-stretching vibrations, we now know that this assumption was not correct.

It was, therefore, desirable to reexamine the NMR data, and

specifically to measure the proton signal at higher α in order to separate experimentally the two contributions to the proton signal. Moreover, the calculated value of $M_2(^1H - ^1H)$ should take into account the discrete orientations of the OH bond. These were the two objectives of the present work.

It should also be recognized that the degree of crystallinity must be considered as a factor in the dehydration process, since structure defects could influence the dehydroxylation mechanism. However, the present results indicate clearly that for the three chosen samples (which were low enough in iron to remove the supplementary complication due to proton-paramagnetic interactions) no influence of crystallinity on the protonic structure of partially dehydroxylated samples has been observed. Nevertheless, the dynamic mechanism of dehydroxylation could well be structure-sensitive. That aspect is beyond the scope of the present work.

EXPERIMENTAL AND RESULTS

Samples

Three different kaolinites with low iron contents (as determined by chemical analysis) and different degrees of crystallinity (characterized by the corresponding values of Hinckley index) were selected. Table 1 lists some characteristic parameters of these kaolinites.

TABLE 1

Sample (*)	wt.% Fe	Hinckley index	A_{BET} (m^2g^{-1})	% H_2O
F	0.39	0.79	18.0	13.39
US	0.05	0.88	11.7	13.62
GB	0.24	1.43	11.8	13.18

(*) The code refers to the origin of the samples (F = France; US = United States; GB = Great Britain).

Portions of the kaolinites, previously dried at 383 K, were heated at 693 K for different periods of time in order to prepare homogeneous samples with increasing degrees of dehydroxylation (Toussaint et al., 1963), as determined by the weight loss. The samples were then sealed in the 10 mm diameter glass tubes used for NMR spectroscopy.

NMR measurements

Room temperature pulse NMR spectra were taken at 90.070 MHz using a Bruker SXP-90 spectrometer. 10^2 10^3 scans, at a scan rate of 1 pulse s^{-1}, were accumulated for each sample using a dwell time of 2 μs, a filter width of 0.3 MHz and a delay time of 5 μs. The total spectral width was 0.5 MHz. To minimize the effect of acoustic ringing and other phase-coherent extraneous signals, an empty sample tube was inserted into the probe immediately following the accumulation with the sample and the signal was accumulated for the same number of scans given to the sample. The latter accumulation was substracted from the sample signal before Fourier transforming.

For Carr-Purcell measurements used to separate the contributions of homonuclear and heteronuclear interactions to the proton linewidth (vide infra), the width of the π pulses was 4 μsec. The filter was set at 1 MHz. Spectrometer hardware limitationsonly allowed a 2τ pulse spacing of 18 μsec. The dead time associated with transmitter and probe ringdown and receiver saturation was 7 μsec. There were 4 μse in the 2τ observation window when an echo amplitude could be observed. Only the first five echos yielded an observable decay, and the echo spacing was not sufficient to determine whether there were two Gaussian decays present. Tuning of the 180° pulse width was accomplished using a $(\pi,\tau)^n$ sequence, with τ = 200 μsec, and n = 100. At the decay following the 99th pulse, there was no observable deviation from the 180° condition, so it is assumed that pulse width errors correspond to an error in the condition at less than 0.5 degrees. The width of the $\pi/2$ pulse was adjusted to obtain the maximum of an on-resonance decay with appropriate phase detection. The relative phases of the initial $\pi/2$ pulse and the π_y pulse were adjusted by changing the pulse width of the y channel by roughly a factor of two (using a step attenuator) and adjusting the phase of the y channel, while observing the response of the sequence $(P_y,\tau)^n$, with $P_y \simeq \pi/2$, τ = 200 μsec, to obtain a null.

The spectra of the three series of samples are given in Fig.1. The number following the code letters represents the percent of dehydroxylation. The line profile is approximately gaussian down to 50-60% dehydroxylation. However, after 65% dehydroxylation the PMR lines can be clearly decomposed into two gaussians, as shown for one of the samples in Fig.2.

From the line profiles, second moments, M_2, were calculated according to the expression :

$$M_2 = \frac{\int_{-\infty}^{\infty}(\Delta\nu)^2 g(\nu)d\nu}{\int_{-\infty}^{\infty}(g(\nu)d\nu}$$

and also from the slope of the semi-logarithmic plots of the decays in the time domain; both sets of results showed a good agreement. Table 2 lists the values obtained. For the samples with a degree of dehydroxylation over 65%, M_2' and M_2'' represent the second moments corresponding to each of the two gaussian signals in which the original line can be decomposed.

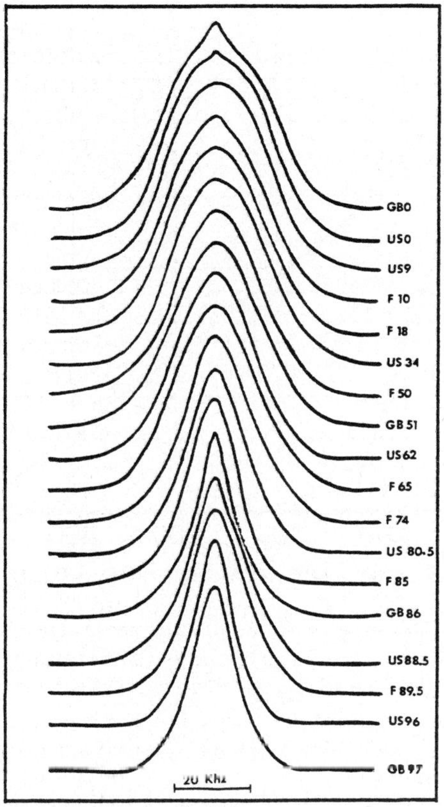

Fig. 1. Proton magnetic resonance spectra observed in the course of dehydroxylation of kaolinites (see text).

CONTRIBUTION OF THE $^{27}Al-^{1}H$ HETERONUCLEAR DIPOLAR INTERACTIONS TO THE ^{1}H LINE WIDTH ; THEORY AND RESULTS

Utilizing pulse decoupling, it is possible to separate the contributions of heteronuclear dipolar (plus shielding anisotropy) broadening from homonuclear dipolar broadening. In this section we develop the theory of these experiments in the density matrix formulation.

Theory

The expectation value of a transverse component of nuclear spin, i.e. $< I_y >$, is given by (Slichter, 1963)

$$< I_y(t) > = Tr\rho(t)I_y \qquad (1)$$

with

$$\psi = \Sigma\ C_i\ \varphi_i \qquad (2)$$

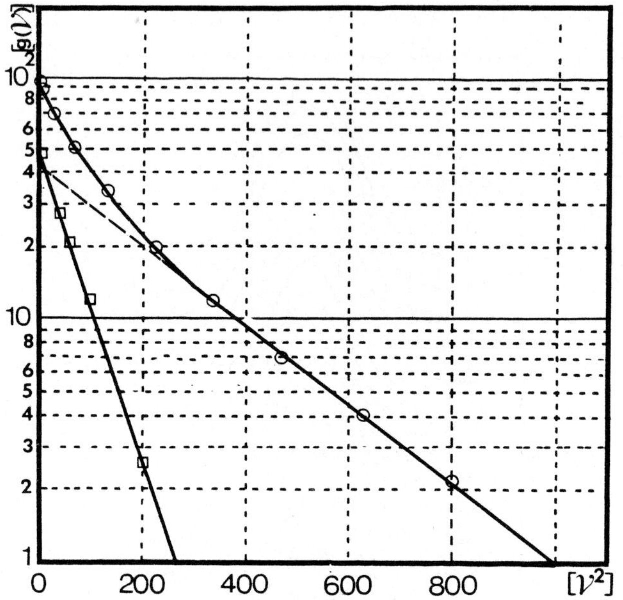

Fig. 2. Splitting of the line profile into two gaussians for F.74 (see Table 2 for the corresponding second moments M_2' and M_2''). $g(\nu) = \frac{1}{\sigma\sqrt{2\pi}} \exp(-\nu^2/2\sigma^2)$ where ν is the frequency (KHz), $g(\nu)$ is the amplitude (arbitrary units) and $2\sigma^2 = (\text{slope})^{-1}$ and $M_2 = \sigma^2$.

TABLE 2.

Sample		Degree of dehydroxylation (%)	M_2(gauss^2)	M_2'(gauss^2)	M_2'' (gauss^2)
GB	0	0	7.9	-	-
US	0	0	7.7	-	-
US	9	9	7.9	-	-
F	10	10	7.4	-	-
F	18	18	7.7	-	-
US	34	34	7.8	-	-
F	50	50	7.6	-	-
GB	51	51	7.2	-	-
US	62	62	6.1	-	-
F	65	65	6.4	-	-
F	74	74	5.5	7.3	1.9
US	80.5	80.5	4.4	5.5	1.7
F	85	85	3.9	5.0	1.5
GB	86	86	3.8	5.2	1.2
US	88.5	88.5	3.8	4.7	1.3
F	89.5	89.5	3.7	4.4	1.2
US	96	96	2.4	3.1	0.7
GB	97	97	2.6	3.3	0.7

and

$$\rho = \overline{C_i \; C_i^*}$$

The bar denotes an ensemble average. The time dependence of ρ in the frame of observation is given by the time dependent Schrödinger equation (with $\hbar = 1$).

$$i \frac{\partial\rho}{\partial t} = \left[(H_{rf} + H_{Int}), \rho \right] \tag{3}$$

where H_{rf} is the Hamiltonian associated with the radiofrequency pulses, and H_{int} is the Hamiltonian associated with all internal (and in particular the heteronuclear dipolar) interactions. If the radiofrequency Hamiltonian is that associated with a sequence of π pulses on the I spin, e.g. along y in the rotating frame, separated by time 2τ, and if H_{int} is the heteronuclear dipolar interaction between spins I and S, then (Mehring, 1976) :

$$H_{int} = -\omega_D I_z S_z$$

and at times $t = 4n\tau$ between pairs of π pulses the density matrix develops in time according to Gerstein (1980),

$$\rho(t) = U_{rf} U_{int} \rho(o) U_{int}^{-1} U_{rf}^{-1} \tag{4}$$

where

$$U_{rf} = \exp - i \int_o^{4\tau} H_{rf}(t)dt \tag{5}$$

For ideal delta function π pulses along y in the frame of observation, Eq.5 becomes

$$U_{rf} = \exp - i \frac{I_y}{4\tau} \pi \int_o^{4\tau} (2+2)dt = \exp -i \; 2\pi \; I_y \tag{6}$$

This is just the operator corresponding to a 2π rotation about y, which is an identity operation. Phsically, this means $U_{rf} = 1$.

Now consider U_{int}. This operator is determined by the Magnus expansion (Wilcox, 1967),

$$U_{int} = \exp - i \left[\bar{H}_{int}^{(o)} + \bar{H}_{int}^{(1)} + \ldots \right] \tag{7a}$$

where

$$\bar{H}_{int}^{(o)} = \frac{1}{t_c} \int_o^{t_c} \tilde{H}_{int}(t)dt \tag{7b}$$

$$\bar{H}^{(1)}_{int} = \frac{i}{2t_c} \int_o^{ti} dt \int_o^{t} dt' \left[\tilde{H}_{int}(t), \tilde{H}_{int}(t') \right] \tag{7)c$$

and higher order terms are commutators of cummutators : $\tilde{H}_{int}(t)$ is given by

$$\tilde{H}_{int}(t) = U^{-1}_{rf}(t,o)\, H_{int}(t)\, U_{rf}(t,o) \tag{8}$$

where $U_{rf}(t,o)$ is the operator related to the radiofrequency Hamiltonian between time zero and t by an equation of the form of (5). More specifically, under a sequence of ideal π_y pulses the behaviour of $H_{rf}(t)$ and $\tilde{H}_{int}(t)$ is shown in Fig.3.

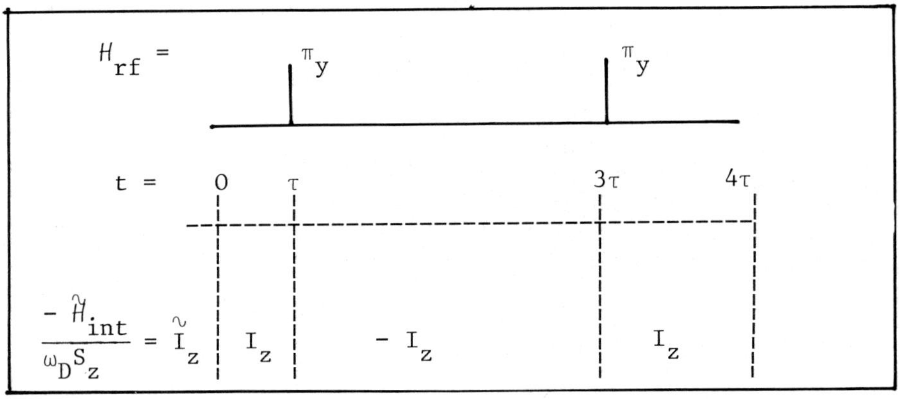

Fig. 3. Schematic representation of the Carr-Purcell sequence (see text).

Here π_y represents a 180° pulse on spin I along y in the rotating frame, i.e. an r.f. field H_{1y} along y such that

$$\int_o^{t_p} H_{r.f.}(t)dt = -\gamma_1 \int_o^{t_p} H_{1y}(t)\, I_y\, dt = -\pi\, I_y \tag{9}$$

with t_p being the pulse width, and $\gamma_1 H_1(t) = \omega_1$ such that $\omega_1 t_p = \pi$. By (7)b, therefore.

$$\bar{H}^{(o)}_{int} = \frac{\omega_D S_z I_z}{4\tau} \int_o^{4\tau} [1-2+1]dt = 0 \tag{10}$$

since $\tilde{H}^{(o)}_{int}(t)$ involves only the products $I_z S_z$, all commutators are zero, and all higher order terms vanish.Therefore :

$$U_{int}(4\tau,0) = U_{int}(4n\tau,0) = 1 \tag{11}$$

Also, since one is free to choose the zero of time between any two π_y pulses, it must be true that

$$U_{int}(2n\tau,0) = 1 \tag{12}$$

Therefore at each "echo time" $t = 2n\tau$, $n = 0, 1,2, \ldots$

$$\rho(2n\tau) = \rho(0) \tag{13}$$

and the system behaves as if it did not feel the effect of the heteronuclear dipolar interaction. In addition, since the shielding anisotropy is also proportional to I_z, at echo times $2n\tau$, the line width determined from the echo envelope will only be associated with homonuclear dipolar and lifetime broadening.

Results

Semi-log plots of echo intensity versus $(\text{time})^2$ were linear, indicating that the echo envelope was that of a Gaussian decay. This decay represents only $^1H - {^1H}$ homonuclear dipolar interactions, since both heteronuclear and shielding anisotropy broadening are removed at the echo times under the Carr-Purcell sequence. From the slope of the obtained straight lines the "C.P. second moment" ($^1H - {^1H}$ interaction) was calculated, for some of the US series samples. The results are shown in Table 3. Once the magnitude of the $^1H - {^1H}$ interaction is known (second column of Table 3) the magnitude of the $^1H - {^{27}Al}$ interaction can be obtained from the total line width by difference. The values in the third column of Table 3 represent this difference, which is constant to within experimental uncertainty. Of course, in obtaining this value for clays with greater than 50% dehydration, the highest value of the total line width was used. As already emphasized, instrumental limitations did not allow measurement of the C.P. $^1H - {^1H}$ second moment that would contribute to M_2'' (Table 2).

TABLE 3

Sample	C.P. ($^1H - {^1H}$) second moment (gauss^2)	Difference between M_2(or M_2') and ($^1H - {^1H}$) C.P. second moment (gauss^2)
US 0	6.4	1.3
US 34	6.3	1.5
US 62	4.6	1.5
US 80.5	4.0	1.5
US 96	2.0	1.1

According to Mestdagh et al. (1973), $M_2(^1H - ^{27}Al)$ = 1.2 $gauss^2$ for an Al-H distance of 2.18 Å and an Al-O-H angle of 108°. Keeping the Al-O and O-H distances constant, $M_2(^1H - ^{27}Al)$ obtained in Table 3 is very close to that obtained for the tetrahedral angle based on the sp^3 hybridation of oxygen.

Figure 4 summarizes all the experimental results.

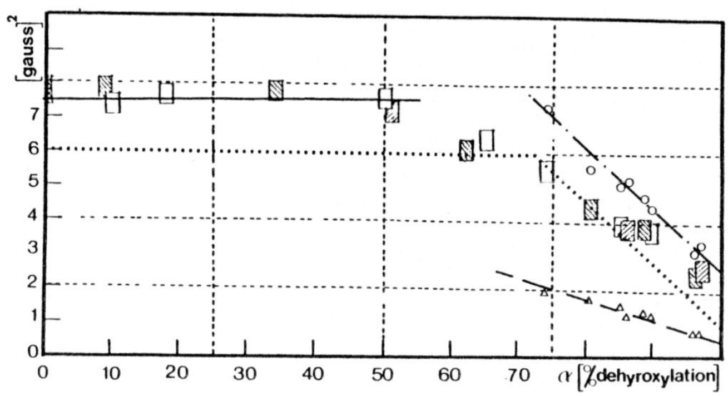

Fig. 4. ^1H NMR second moments. — average value of M_2 for $\alpha < 55\%$; —.— M_2' ; — — M_2'' ; .... M_2-$M_2(^1H - ^{27}Al)$ for $\alpha < 70\%$, or $M_2' - M_2(^1H - ^{27}Al)$ for $\alpha > 70\%$.

▨ , G.B. ; ▧ , US ; □ , F samples.

Discussion

Since the main interest in this work is concerned with the proton-proton contribution, the evolution of $M_2(^1H - ^1H)$ with α is the most important parameter to be considered. From $0 < \alpha < 70\%$, $M_2(^1H - ^1H)$ is ≃ 6 $gauss^2$, and there is a single ensemble P of protons.

From α ≃ 70% up to α < 100%, two ensembles P' and P" with high (M_2') and small (M_2'') second moments respectively are observed. The break appearing in the overall experimental second moment for α ≃ 55% is due to the superimposition of the P' and P" contributions.

If we substract $M_2(^1H - ^{27}Al)$ from the experimental M_2 for ensemble P or from the experimental M_2' for ensemble P', the break in $M_2(^1H - ^1H)$ occurs for α = 70%. According to Ackar et al. (1964) the variation of α with respect to the time of dehydroxylation at constant temperature no longer fits the two-dimensional diffusion equation for α > 60%. In this range $M_2'(^1H - ^1H)$ decreases from 6 to 1 $gauss^2$ for α increasing towards 100%. Since the Al-H contribution to M_2'' cannot be measured, the variation of $M_2''(^1H - ^1H)$ with α is unknown.

A model which is in general agreement with experimental observation is the following : between 0% and 50% dehydroxylation, protons are removed layer by layer leaving all remaining $^1H - ^1H$ and $^1H - ^{27}Al$ distances constant, and maintaining constancy of both homonuclear and heteronuclear dipolar interactions (and thus of second moments) characteristic for a single ensemble P. Beyond 70% dehydration, protons begin to be removed from within layers, leaving isolated patches and isolated hydroxyl groups behind. The result is that on average, there are two types of $^1H - ^1H$ interactions: that due to hydroxyl groups within patches (ensemble P'), and the other associated with relatively isolated hydroxyl groups (ensemble P"). In ensemble P', the $^1H - ^{27}Al$ interaction remains constant, down to a very high degree of dehydroxylation, indicating that the H-O-Al angle remains constant.

The above interpretations suggest that the hydroxylic structure in ensemble P' remains the same as in the original kaolinite but the patches are "shrinking" as α increases from 70% to $\leq$ 100%. The type of theoretical predictions that can be proposed for the evolution of $M_2'(^1H - ^1H)$ in ensemble P' will be demonstrated later.

As far as ensemble P" is concerned one can only speculate on the meaning of the decrease of M_2'' with increasing dehydroxylation. If it is assumed that ensemble P" is made of protons, sufficiently isolated for the $^1H - ^1H$ interactions to be negligible, then the variation of M_2'' with α should represent mainly the decreasing contribution of $M_2''(^1H - ^{27}Al)$. We note that for α = 70% M_2'' = 2.3 gauss^2 and that for α = 100%, its extrapolated value is 0.5 gauss^2. These two limits are indeed quite close to the calculated $^1H - ^{27}Al$ second moments, for Al-O-H angles of 90° and 180° respectively. It could then be suggested that ensemble P" is made from isolated OH which change orientation with respect to the Al-O bond. If these OH are linked to Aluminum in four-fold coordination (Al^{IV}), there are two distinct possibilities. Al^{IV} tetrahedra share either corners or edges. There are not enough oxygen atoms left upon complete dehydroxylation to have Al^{IV} tetrahedra sharing corners only. Thus, as α increases an increasing number of these tetrahedra must share edges, resulting in a change of the Al-O-H angle of the remaining isolated hydroxyls. In tetrahedra sharing edges, the OH orientation should be such that the proton is at the largest distance from Al^{3+}. Thus, for $\alpha \rightarrow$ 100%, the Al-O-H angle should be close to 180° to minimize the accumulation of positive charge. This interpretation is purely speculative but it might explain the variation of M_2'' with α.

The deduction concerning the size of the patches constituting ensembles P' is founded on a safer basis since the constancy of the proton-aluminum interaction suggests that the original hydroxyl arrangement is conserved.

In order to explain the variation of $M_2'(H_1-H_1)$ from 6 to 1 gauss^2 the most sensible way is to evaluate the contribution to the second moment of some structural units taking into account the various proton-proton distances and OH bond orientations given by Giese and Datta (1973). The nearest neighbor interproton distances are 2.7, 2.9, 3.33 and 3.92 Å. The first value is the distance between protons tilted by 77° with respect to the kaolinite layer, the second is the distance between one proton tilted by 77° and the nearest proton tilted by 14° with respect to the kaolinite layer. The third value corresponds to the interatomic distance between one 77° proton and the second nearest 14° proton and the fourth is the distance between one 77° proton and most distant 14° neighbour. Using these data, second moments were calculated according to the classical equation :

$$M_2(^1H - ^1H) = 3.56\ 10^{-6}\ \frac{1}{N_h}\ \sum_i \sum_j r_{ij}^{-6} (\text{gauss}^2) \qquad (14)$$

where N_h is the number of protons in the structural unit and r_{ij} is the interprotonic distance, in nm. Table 4 shows the results of some of these calculations. The value of $M_2(^1H - ^1H)$ for the "initial kaolinite" was calculated on the basis of an octahedron and six neighboring octahedra contributions. Contributions of more distant protons are negligible.

TABLE 4 - Theoretical proton-proton second moment

Structural unit	$M_2(^1H - ^1H)$ (gauss^2)	
Initial kaolinite	5.5	
Unoccupied isolated octahedron	4.1	
Occupied isolated octahedron	0.8	
	max.	min.
3 interlayer protons + 1 inner proton	4.1	0.7
2 interlayer protons + 1 inner proton	2.8	0.3

The inner proton contributes by about 0.2 gauss^2 to the total value. "Unoccupied" or "occupied" means the absence or the presence of octahedral Al^{3+}. The large difference between $M_2(^1H - ^1H)$ calculated for unoccupied and occupied isolated octahedra is due to the proton locations in these structural units. The source of these differences is the spread of the H^1-H^1 (interproton) distances indicated above.

As one can see the theoretical value computed for kaolinite is 5.5 gauss^2 instead of the observed 6 gauss^2. The difference may be due to both experimental errors and uncertainty of the interproton distances.

As dehydroxylation proceeds above 70%, nothing is known about the type of protons which will first nucleate in the formation of water ; but, as shown in Table 4, very small patches such as those made from unoccupied octahedrons may already account for a value of $M_2(^1H - ^1H)$ of the order of 4 gauss^2. When $M_2(^1H - ^1H)$ is lower than 4 gauss^2, the calculated second moment drops rapidly, but because of the probable randomness of the dehydroxylation process in this domain it is impossible to suggest even an approximate size for the patches. For instance,

3 interlayer protons and one inner proton may yield a second moment between 4 (e.g. that of an unoccupied octahedron) and 0.7 $gauss^2$. Patches with a second moment between these two values may well be more extended, but since the $Al-^1H$ interaction remains practically constant within a patch, it must be admitted that the minimum number of OH necessary for an octahedral coordination is retained. It should be noted that the minimum value for an ensemble of 3 interlayer and 1 inner protons is near the minimum experimental value obtained for $M_2'(^1H-^1H)$.

Unfortunately it is impossible to calculate the ratio of the proton populations in ensembles P' (patches) and P" (isolated protons) because the line width is continuously varying with α, and in order to get absolute measurements, calibration curves should be carried out. This is indeed difficult to perform but qualitatively population P" increases at the expense of P' as α increases. This is of course expected since as the patches are "shrinking", more isolated protons are produced.

AKNOWLEDGMENTS

The authors like to thank Dr.J.M.Cases for the gift of the samples and for the characterization shown in Table 1.

REFERENCES

Ackar, B.N.M., Brindley, G.W. and Sharp, J.H., 1966. Kinetics and mechanism of dehydroxylation processes, III. Applications and limitations of dynamic methods. Proceeding of the Int. Clay Conf. Jerusalem, Israël Program for Scientific Translations, Jerusalem, 1:67-73.

Brindley, G.W. and Nakahira, M., 1958. Further considerations of the crystal structure of kaolinite.Miner. Mag. 31:781-786.

Brindley, G.W. and Nakahira, M., 1959. J. Amer. Ceram. Soc. 42:311.

Gastuche, M.C. Toussaint, F., Fripiat, J.J., Touilleaux R. and Van Meersche, M., 1963. Study of intermediate stages in the kaolin-metakaolin transformation. Clay Minerals Bull. 5:227-236.

Gerstein, B.C. High Resolution NMR in Solids with strong Homonuclear Dipolar Broadening : Combined Multiple Pulse NMR and Magic Angle Spinning ,1981. Phil. Trans. R. Soc. Lond. A.299:521-546.

Giese, R.F. and Datta, P. 1973. Hydroxyl orientations in Kaolinite, Dickite and Nacrite. Am. Min., 58:471-479.

Mehring, M. High Resolution NMR spectroscopy in Solids, in NMR-II,Basic Principles and Progress, Ed.Diehl, Fluck and Kosfeld, Springer-Verlag.,1976.p.2.2,2.23.

Mestdagh, M., Stone, W.E.E. and Fripiat, J.J., 1973. Proton mobility in solids. IV study of proton motions in the decationated Y zeolites by nuclear magnetic resonance. J. Phys. Chem., 76:1220-1226.

Slichter, C. Principles of Magnetic Resonance ,Harper and Row.1963. Chap.5, Section 5.4.

Wardle, R. and Brindley, G.W.,1974. The dependence of the wavelength of Al K_α radiation from alumino-silicates on the Al-O distance. Am.Min.56: 2123-2128.

Willcox, R.M.,1967.Exponential Operators and Parameter Differentiation in Quantum Physics,J. Math. Phys. 8:962-982.

A NEW APPROACH TO ELECTROSTATIC CALCULATIONS FOR COMPLEX SILICATE STRUCTURES AND THEIR APPLICATION TO VERMICULITES CONTAINING A SINGLE LAYER OF WATER MOLECULES

H DONALD B JENKINS[(1)] and P HARTMAN[(2)]

(1) Department of Chemistry and Molecular Sciences, University of Warwick, Coventry CV4 7AL, West Midlands, UK

(2) Instituut voor Aardwetenschappen, Department of Crystallography, Rijksuniversiteit Utrecht, Budapestlaan 4, 3508TA Utrecht, The Netherlands.

ABSTRACT

Applying a novel approach described earlier to electrostatic energies of expansion of micas, it is concluded that:

(i) it takes more energy to expand a phyllosilicate when the layer charge is in octahedral sites than when it is in the tetrahedral sites;

(ii) the expansion energy of fluormicas is slightly larger than that of hydroxymicas;

(iii) when the substitution is tetrahedral, the dioctahedral micas have a slightly larger expansion energy than the trioctahedral micas.

Application to the energetics of incorporation of water molecules into an expanded phlogopite structure leading to a vermiculite intercalate, intercalate, $K_{2x}Mg_6(Si_{4-x}Al_x)_2O_{20}(OH)_4.(H_2O)_4$ leads us to conclude that:

(i) at large values of x, the expansion energy required to separate the interlayer of the phlogopite is larger than the stability provided by the hydration of the interlayer K^+ ions;

(ii) at low values of x, the expansion energy is unimportant and very small and the predominant term is the K^+-H_2O interaction;

(iii) the interaction of the water molecules with the silicate layer only becomes important when x>0.5;

(iv) for intermediate values of x intercalation occurs due to the large K^+-H_2O interaction energy.

INTRODUCTION

There are two features of our recent work (Jenkins and Hartman, 1979, 1980, 1982; Hartman and Jenkins, 1980) concerning the energetics of complex phyllosilicate structures which constitute a novel approach. The first concerns the adaption of the standard Madelung procedure to give an account of the interaction energies in a particular class of silicate structures utilising the basic similarity of the crystal structures of the members of the class. The technique not only reduces the computational labour, which can be considerable in these types of structures, but it also enables one to gain valuable insight into the energetics of substitutional changes when converting one structure to another. The second feature relates to the evolution of the driving force behind the intercalation process in vermiculites containing a single layer of water molecules by using plausible model calculations.

The advent of modern digital computers coupled with the techniques we have developed has enabled electrostatic contributions to the energy to be calculated with relative ease.

Quite recently we have embarked on a programme of study of the silicates with the following objectives in mind:

(i) to consider computational aspects of the theory involving the treatment of large structures at the electrostatic level;

(ii) to examine the energetics of silicate layer expansion, bearing in mind the growing interest in intercalction phenomena whereby small molecules and ions are incorporated within the layers of expanded silicates;

(iii) to set up models for vermiculites in which water molecules are incorporated within the layers, as a specific case study.

COMPUTATIONAL ASPECTS

We consider a general phyllosilicate $YX_2X'T_4O_{10}(OH)_2$ where Y is a univalent ion, X' corresponds to a trioctahedrally occupied (extra site occupied by an atom in a trioctahedral mica but vacant in the dioctahedral case), X corresponds to a dioctahedrally occupied site and T refers to the tetrahedral site (T_4 is either Si_4 or Si_3Al etc).

A conventional computation of the electrostatic energy of this mineral would input cell parameters and atomic coordinates into a suitable program (eg: MADPROG or LATEN in Jenkins and Pratt, 1978, resp 1980) with assigned charges (usually formal oxidation state charges) $q_Y = +1$, $q_X = q_{Mg} = +2$, $q_{X'} = q_{Mg} = +2$, $q_T = +3.75$ and $q_O = q_{O'} = -2$, $q_H = +1$ (where q_O and $q_{O'}$ are the charges on the oxygen atom of the silicate layer and of the hydroxyl oxygen respectively). Using the crystal structure data for this mineral (McCauley et al, 1973) an electrostatic energy of 74915 kJ mol^{-1} is computed. More recently however, Jenkins and Hartman (1979), recognising that the electrostatic energy for $YX_2X'T_4O_{10}(OH)_2$ can be written:

$$U_{elec} = \sum_{i=0}^{2} \sum_{j=0}^{(2-i)} \sum_{k=0}^{(2-j)} \sum_{l=0}^{(2-k)} \sum_{m=0}^{(2-l)} \sum_{n=0}^{(2-m)} \sum_{p=0}^{(2-n)} A_{ijklmnp} q_Y^i q_X^j q_{X'}^k q_T^l q_O^m q_{O'}^n q_H^p \quad (1)$$
$$i+j+k+l+m+n+p = 2$$

considered the 1M-Al mica structure reported by Siderenko et al (1975, 1975 a), (having C2 symmetry) modified by the introduction of a mirror plane and the addition of two atoms X' in positions (0,½,½) and (½,0,½) (the trioctahedral site positions). They then

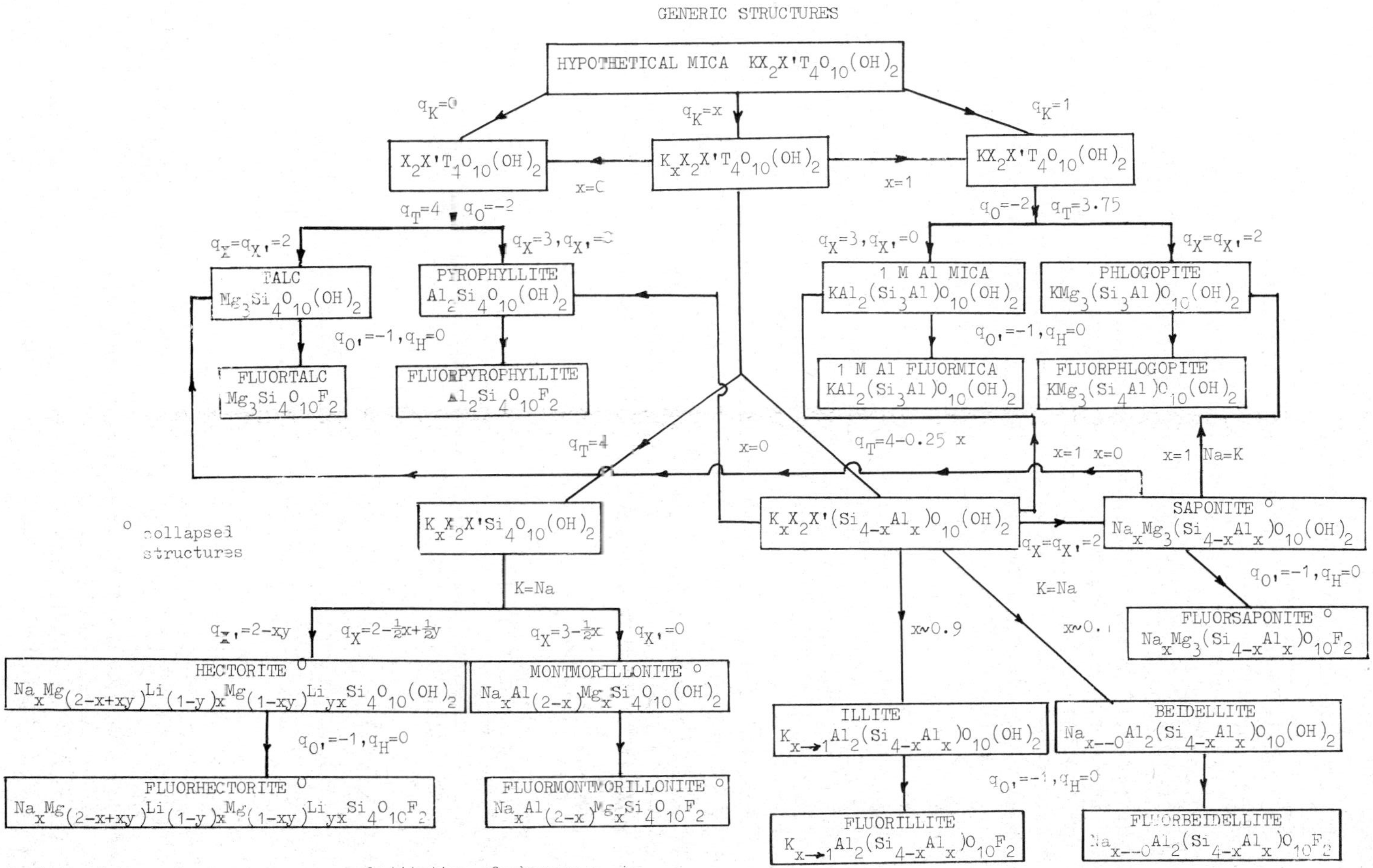

Table 1. Scheme showing necessary substitution of charges, q, in equation (1) in order to give energies of related silicates.

obtained the converged values of the coefficients $A_{ijklmnp}$ for this 'generic' hypothetical mica structure.

Using the values obtained for these coefficients, we obtained from this single calculation, the structural electrostatic energies of all the minerals shown in the Scheme given in Table I by adopting in equation (1) the various values of the charges indicated in that Table for each specific case.

The separation of the structurally dependent coefficients, $A_{ijklmnp}$ from the product of the magnitudes of the interacting charges enables us to examine the electrostatic energy change on substitution of one atom for another without corresponding relaxation of the positions of the surrounding atoms. Comparison of the energies of the various phyllosilicates, calculated using precise crystal structure data in each case, with the energies of the same phyllosilicates using the Scheme of Table 1 have enabled Jenkins and Hartman (1979) to evaluate energies of relaxation of the structures, change in the energies brought about by changes in angle between hydroxyl groups and the mica layer, energies of transition of trioctahedral to dioctahedral phyllosilicates, site potentials and surface energies. The results obtained in these studies indicate:

(i) that insertion of an OH group in place of an F atom in a trioctahedral mica is easier than in a dioctahedral mica;

(ii) that the energy of replacement of Al in the tetrahedral layer by Si to convert Si_3Al into Si_4 and consequent removal of the interlayer Y^+ ion is similar in both di- and trioctahedral micas;

(iii) dioctahedral to trioctahedral conversion is easier in a fluor-mica than in a hydroxymica;

(iv) in the presence or absence of Y^+ ions in the phyllosilicate has only a small influence on the OH orientation with respect to the mica layer;

(v) the surface energies of hydroxy- and fluormicas are similar (for 001 faces);

(vi) the surface energies of Y^+ containing micas are almost the same irrespective of whether they are di- or trioctahedral (for 001 faces).

EXPANSION OF PHYLLOSILICATES

The first calculation of the electrostatic contribution to interlayer bonding was made by Giese (1974) for the dioctahedral mica muscovite $2M_1$. His method is based on the comparison of the energies of a normal mica and of an expanded mica. The expansion is carried ou in such a manner as to move the layers apart from each other until no further energy change is obtained on still further expansion of the layers. Any interlayer Y^+ ions are divided between the two separating layers in an ordered fashion. The difference between the fully expanded and normal mica is called the surface energy.

Using our hypothetical structure, $KX_2X'T_4O_{10}(OH)_2$ at the head of Table 1 we have now obtained the coordinates of the expanded structure by separating the layers along the direction of the c-axis. We can operate the basic scheme of Table 1 to investigate octahedral and tetrahedral substitutional effects on expansion energies. For an expansion in which interlayer K^+ ions are divided between the separating layers, we considered interlayer separations of 2.5Å, 4.5Å and 10.0Å. Of course an alternative expansion process can be considered in which expansion takes place leaving the interlayer cations in the centre of the expanded interlayer gap.

The influence of the interlayer charge on the expandability of micas has been the subject of several studies. Appelo (1978, 1978a) considered the expansion of the di- and trioctahedral micas. Giese (1974) also considered the cases in which the potassium ions remain all on one side of the separating layers and in which half of the ions remained on one side and half on the other side of the layers in an ordered arrangement. The inherent importance of this topic in relation to intercalation properties of micas has led us to examine this question as a first step for an extended study on intercalation of micas.

In this work we have considered the cases where the layer charge is either in the tetrahedral sites, $K_xX_2X'(Si_{4-x}Al_x)O_{10}(OH)_2$, or in the octahedral sites, $K_xAl_{2-x}Mg_xSi_4O_{10}(OH)_2$, both formulae being derived from the 'generic' $K_xX_2X'T_4O_{10}(OH)_2$, where $T_4 = (Si_{4-x}Al_x)$ and X is the octahedral cation present in both di- and trioctahedral micas, and X' is the octahedral cation present only in trioctahedral micas. Calculations have been presented (Jenkins and Hartman, 1979, 1980) for the unexpanded micas and the micas expanded by 1.0Å, 2.5Å and 4.5Å. The conclusions listed below are generated from the results of the calculations, adopting a mode of expansion in which the silicates move apart in the direction of c^*, while the K^+ ions remain exactly midway between the layers.

For an expansion by ΔÅ, we define the expansion energy, ΔU^{Δ}_{elec}, as:

$$\Delta U^{\Delta}_{elec} = U_{elec}(\text{unexpanded}) - U_{elec}(\text{expanded}) \qquad (2)$$

The conclusions are:

(i) It takes more energy to expand a mica when the layer charge is in the octahedral sites than when it is in the tetrahedral sites, the difference being about 27 kJ mol^{-1} at most.

(ii) The expansion energy of fluor micas is larger than that of O-H micas by a few kJ mol^{-1}.

(iii) When the substitution is tetrahedral, the dioctahedral micas have a slightly larger expansion energy than the trioctahedral micas, the difference being about 1-2 kJ mol^{-1}.

CALCULATIONS ON A MODEL POTASSIUM VERMICULITE

We can model (Jenkins and Hartman, 1982) a potassium vermiculite, $K_{2x}Mg_6(Si_{4-x}Al_x)_2O_{20}(OH)_4 \cdot (H_2O)_4$, by taking a phlogopite based on the structure proposed by McCauley, Newnham and Gibbs (1973) and expanding this structure in a direction perpendicular to the (001) plane such that a = 5.308Å, b = 9.183Å, c = 12.608Å, β = 98.08°, and incorporating the four molecules of H_2O per unit cell in the interlayer. For vermiculites in which $x<1$ specific consideration must be given to the arrangement of the interlayer cations and the water molecules and these interactions are then calculated separately. When x is low isolated K^+ ions are surrounded by large numbers of water molecules. In each case we must pay careful attention to the arrangement of the water dipoles.

As the interlayer charges, $q_K = x$, are decreased from unity the vermiculite is correspondingly modified:

(i) in the interlayer by geometrical changes caused by creation of vacant cation sites and the corresponding adjustments caused by the orientations of the water molecule dipoles; and

(ii) in the silicate framework by corresponding occupational changes in the tetrahedral layer sites ($q_T = 4 - 0.25x$) required to maintain electroneutrality.

The latter modifications are readily incorporated using our 'generic' approach, and the former changes affect the K-H_2O, H_2O-H_2O and K-K interaction energies within the interlayer. For the purposes of calculations when $x<1$ all interlayer cations are assumed to be located in the central position within the interlayer.

INTERCALATION ENERGIES

The intercalation energy of the process:

$$2K_xMg_3(Si_{4-x}Al_x)O_{10}(OH)_4(c) + 4H_2O(g) \xrightarrow{U_{inter}(g)} K_{2x}Mg_6(Si_{4-x}Al_x)_2O_{20}(OH)_4\cdot(H_2O)_4(c)$$

PHLOGOPITE → VERMICULITE INTERCALATE

is found to be

$$\Delta U_{inter}(g) = U_W + \Delta U^{\Delta}_{exp} \qquad (3)$$

where ΔU^{Δ}_{exp} is the expansion energy of phlogopite as given by equation (2) and U_W is given by:

$$U_W = -E(H_2O\text{-}H_2O) - E(K\text{-}H_2O) - E(H_2O\text{-silicate layer}) \qquad (4)$$

where $E(H_2O\text{-}H_2O)$, $E(K\text{-}H_2O)$ and $E(H_2O\text{-silicate layer})$ are the H_2O-H_2O and K-H_2O interlayer interactions and $E(H_2O\text{-silicate layer})$ is the interaction of the water molecules with the silicate layer.

For the process:

$$2K_xMg_3(Si_{4-x}Al_x)O_{10}(OH)_4(c) + 4H_2O(l) \xrightarrow{\Delta U_{inter}(l)} K_{2x}Mg_6(Si_{4-x}Al_x)O_{20}(OH)_4\cdot(H_2O)_4(c)$$

PHLOGOPITE → VERMICULITE INTERCALATE

where liquid water is taken up by the expanded phlogopite crystal:

$$\Delta U_{inter}(l) = \Delta U_{inter}(g) + \Delta H_{vap}(H_2O)(l) \qquad (5)$$

where $\Delta H_{vap}(H_2O)(1)$ is the enthalpy of vapourisation of water, which has a value of 44 kJ mol^{-1}.

Figure 1 shows the variation of $\Delta U_{inter}(g)$ with x for various choices of the charge $q_{H''}$ on the hydrogen atoms of the intercalated water molecules from which we see that for $q_{H''} \geq 1/3$ no uptake of water is predicted as being possible while for $q_{H''} = \frac{1}{2}$ (a value commonly chosen to model water in crystal structure work on hydrates) no intercalation is possible when $x \leq 0.1$ or $x \geq 0.8$.

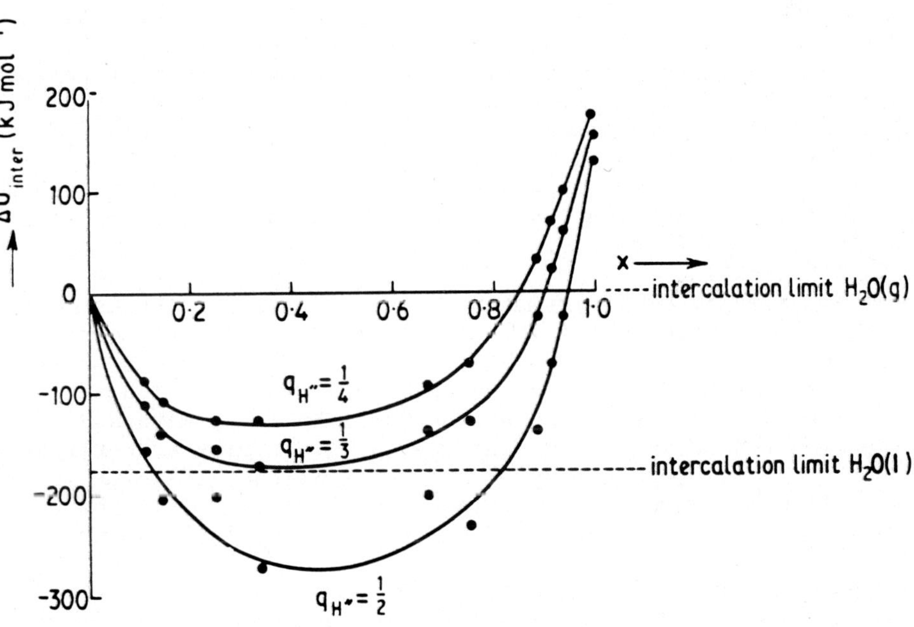

Fig 1 Variation of $\Delta U_{inter}(g)$ and $\Delta U_{inter}(1)$ as functions of x for hydrogen atom charges $q_{H''}$ of $\frac{1}{4}$, 1/3 and $\frac{1}{2}$.

The following conclusions emerge:

(i) At large values of x, the large expansion energy required, ΔU_{exp}, to separate the interlayer of the phlogopite is greater than the stability provided by U_W and $\Delta U_{inter}(g)$ (and hence $\Delta U_{inter}(1)$) rises above the intercalation limit and no uptake of water takes place.

(ii) At low values of x, the expansion energy is unimportant and very small and $\Delta U_{inter}(g) \simeq U_W$. The predominant term in U_W is the K-H_2O interaction energy, while the intercalation energy of the water molecules with the silicate layer is vanishingly small.

(iii) The interaction of water molecules with the silicate layer becomes a relatively important factor only when $x>0.5$.

(iv) For intermediate values of x intercalation occurs due to the large K-H_2O interaction energy.

Much work on these systems remains to be done and we are currently examining further questions: the effect of inclusion of replusion energy and dispersion energy on these calculations and a closer estimation of $q_{H''}$ (the hydrogen charge on the water molecules) necessary to interpret the results more fully. The fact that the layer charge is important to determine whether intercalation takes place has already been recognised experimentally (Schultz, 1969). On the basis of the present work it seems clear that this is indeed the case and that factors regarding the choice between tetrahedral and octahedral substitution is very much of secondary importance.

ACKNOWLEDGEMENTS

The provision of a six-month visiting fellowship by the ZWO (Netherlands Organisation for Scientific Research) to carry out this work at the University of Leiden, The Netherlands, and the assistance of the Science Research Council is acknowledged. The provision of a travel grant by the Royal Society, with assistance also from the University of Warwick, is gratefully acknowledged by HDBJ to enable presentation of this work at the AIPEA 7th International Clay Conference in Bologna and Pavia, Italy. The kind invitation of the organising Committee is gratefully acknowledged.

REFERENCES

Appelo, C A J, 1978. Layer deformation and crystal energy of micas and related minerals. I Structural models for 1M and $2M_1$ polytypes. Am Mineral, 63: 782-792.

Appelo, C A J, 1978a. Aspects of mica-related clay minerals in hydrogeochemistry. Thesis, Vrije Universitat, Amsterdam.

Giese, Jr, R F, 1974. Surface energy calculations for muscovite, Nature (London) Phys Sci, 248: 580-581.

Hartman, P and Jenkins, H D B, 1980. On the swelling of phyllosilicates: Lattice Energy Considerations. Abstr Int Mineral Assocn, 12th General Meeting, Orleans, France: p 152.

Jenkins, H D B and Hartman, P, 1979. A new approach to the calculation of electrostatic energy relations in minerals: the dioctahedral and trioctahedral phyllosilicates. Phil Trans Roy Soc (London), A293: 169-208.

Jenkins, H D B and Hartman, P, 1980. Application of a new approach to the calculation of electrostatic energies of expanded di- and trioctahedral micas. Phys Chem Minerals, 6: 313-235.

Jenkins, H D B and Hartman, P, 1982. Calculations on a model intercalate containing a single layer of water molecules: A study of potassium vermiculite, $K_{2x}Mg_6(Si_{4-x}Al_x)_2O_{20}(OH)_4.(H_2O)_4$, for $1 \leq x \leq 0$. Phil Trans Roy Soc (London), A304: in the press.

Jenkins, H D B and Pratt, K F, 1978. Calculation of first derivatives of Madelung constants with respect to cell lengths. Comp Phys Commun, 13: 341-348.

Jenkins, H D B and Pratt, K F, 1980. LATEN - a complete lattice energy program. Comp Phys Commun, 21: 257-269.

McCauley, J W, Newnham, R E and Gibbs, G V, 1973. Crystal structure analysis of synthetic fluor-phlogopite. Am Mineral 58: 249-254.

Schultz, L G, 1969. Lithium and potassium absorption, dehydroxylation temperature and structural water content of aluminous smectites. Clays and Clay Minerals. 17: 115-125.

Siderenko, O V, Zvyagin, B B and Soboleva, S V, 1975 Crystal structure refinement for 1M dioctahedral mica. Kristallografiya, 20: 543-549.

Siderenko, O V, Zvyagin, B B and Soboleva, S V, 1975a. Crystal structure refinement for 1M dioctahedral mica. Sov Phys Crystallogr, 20: 332-335.

HISINGERITE: A REVIEW OF ITS CRYSTAL CHEMISTRY

Maria Franca BRIGATTI
Università di Modena, Istituto di Mineralogia e Petrologia, Largo S. Eufemia 19, 41100 - Modena - Italy.

ABSTRACT

The chemical properties of hisingerite are studied in connection with those of Fe-rich smectites. All data were selected from the literature. The discrimination was attempted by means of the Q-MODE multivariate analysis and possible correlations between chemical composition and b-dimension were sought.

Hisingerite appears to be a mineral with a well defined structure and chemical composition. The trioctahedral origin of the mineral is confirmed. The question of stuctural continuity between trioctahedral and dioctahedral states is discussed.

INTRODUCTION

Hisingerite, at first described by Berzelius (Strunz, 1978), has been called by Brown (1955) an "amorphous iron silicate with the ratio Fe_2O_3:SiO_2=1:2 molecularly". It has been ascertained that hisingerite is not completely X-ray amorphous but there is no agreement on its own kinds. It has been considered as a nontronite (Gruner, 1934; Sudo & Nakamura, 1952; Bowie, 1955), as an iron-rich saponite (Whelan & Goldich, 1961; Kohyama & Sudo, 1975) and even as a mica (Lindqvist & Jansson, 1962). Kohyama & Sudo (1975) suggested that hisingerite has a chemical composition variable between a trioctahedral iron-rich saponite and a dioctahedral nontronite. In other respects, the existence of intermediate phases between nontronite and Fe-saponite is confirmed by some authors. Van Breemen (1980), when studying the chemical and mineralogical compositions of clay soils with pyritic substrata from Thailand, tested the possible evolution nontronite→ saponite (and/or vice-versa) in natural systems. Bischoff (1972) emphasized that "ferrous nontronite" from the Red Sea was intermediate, in composition, between dioctahedral nontronite and trioctahedral ferrous iron end members of the smectite group.

Hisingerite occurs:ì)- with uranium, tin, copper, lead, zinc,

TABLE 1

Source and chemical analyses of the selected samples.

Sample	SiO_2	TiO_2	Al_2O_3	Fe_2O_3	FeO	MnO	MgO	CaO	Na_2O	K_2O	H_2O^+	H_2O^-	References
H. 1	37.54		0.56	37.02	4.66	0.75	2.81	1.52			9.20	6.00	Osborne & Archambault (1950)
H. 2	36.64			36.88	5.51		1.26	0.83			18.57		Lebedev (1936)
H. 3	27.99			34.25	0.54			2.33			7.11	27.89	Sudo & Nakamura (1952)
H. 4	35.57	0.12	0.38	39.20	4.80		1.60	0.85			11.60	6.00	do
H. 5	38.14			36.66	0.84		2.45				8.53	13.20	do
H. 6	35.24		2.00	35.51	2.40	0.34	3.77	1.99		0.04	10.42	9.03	Lindqvist & Jansson (1962)
H. 7	32.34		1.35	36.88	1.46		4.48	0.96			7.54	14.34	Bowie (1955)
H. 8	39.11	0.16	3.89	22.22	5.90	0.92	5.21	2.52	0.27	0.07	5.82	13.54	Whelan & Goldich (1961)
H. 9	42.35	0.13	3.65	23.28	5.54	0.39	8.22	2.06			6.85	6.95	do
H.10	38.19	0.01		19.91	24.64	0.66	2.36	0.61			8.40	5.53	do
H.11	32.85	0.26	2.64	40.70		0.74	2.05	1.50			7.90	11.40	do (canbyite)
H.12	40.35	0.20	5.11	24.99	3.55	0.22	5.48	1.32	0.18	1.44	8.51	8.51	do
H.13	38.63		1.78	32.49			6.64	2.79			17.70		do
H.14	33.60		0.84	38.92	0.89	0.27	0.44	1.18			7.24	16.58	Mozgova (1957)
H.15	36.11			31.13	4.35	0.17	5.14	0.57			7.02	14.84	Nikolsky (1953)
H.16	40.68		4.44	28.32	0.20		1.85	1.42	1.50	0.33	6.43	13.70	Kohyama & Sudo (1975)
S. 1	39.68	0.37	3.93	19.82	1.12	0.19	11.21	2.37			6.16	15.11	Sudo (1954)
S. 2	40.46		10.15	3.56	4.98	0.24	20.71	1.94	0.25	0.32	4.24	13.33	Konta & Sindelar (1955)
S. 3	35.10	0.10	7.10	13.50			21.00	3.80	0.90	0.30	9.10	10.10	Kimbara & Shimoda (1972)
S. 4	41.47		7.42	7.36	5.80		19.53	1.26	2.47	0.11	4.47	10.63	do
S. 5	37.52	0.03	12.06	14.34	5.08	0.04	9.12	3.87	1.36	0.68	5.02	11.27	Kimbara (1973)
S. 6	42.68	0.16	8.83	11.62			10.21	1.84	0.28	0.28	9.13	15.41	Kimbara & Honda (1975)
S. 7	40.53	0.20	12.20	8.79	4.20	0.20	14.42	3.24	0.48	1.04	7.44	8.30	Kimbara (1975)
S. 8	43.98	0.16	6.30	7.85	5.32	0.32	18.02	2.78			9.24	6.24	Miyamoto (1957)
S. 9	46.77	0.02	12.47	8.27	3.07	0.20	3.20	1.55	1.50	0.40	5.23	17.92	Kohyama et al. (1973)
S.10	46.05		13.57	9.15	1.56	0.14	4.32	1.99	1.71	0.36	5.17	16.00	do
S.11	50.03		13.94	9.34	2.12	0.28	3.50	1.49	1.50		5.17	12.25	do
S.12	32.83		8.41	0.93	6.75	0.20	2.58	1.38	1.42	0.21	3.46	42.04	do
S.13	47.11		4.16	3.25		0.01	25.95	1.96	0.04	0.06	17.20		Haranczyk & Prochazka (1974)
S.14	48.29	0.10	7.50	1.46	6.60		19.88	1.31			9.40	5.58	Muchi & Higashiyama (1972)
S.15	43.60		8.20	3.70	4.80		20.00	2.70			17.00		Caillère & Hénin (1957)
S.16	40.44		14.52	1.85	2.37	0.15	21.11	2.32	0.42		6.31	10.65	Mongiorgi & Morandi (1970)
N. 1	44.30	0.18	5.37	29.47	0.57		1.74	1.70			8.52	7.99	Kimbara & Honda (1975)
N. 2	41.41	0.01	5.48	29.89		0.01	0.66	2.07	0.04	0.01	20.40*		Eggleton (1977)
N. 3	42.90	0.01	4.70	29.51	0.05	0.03	1.14	1.98	0.04	0.01	19.41*		do
N. 4	41.22		5.58	29.01	0.05	0.03	0.77	2.02	0.04	0.01	20.08*		do
N. 5	46.42	0.05	5.73	26.51	0.05	0.04	1.62	0.97	0.07	0.02	18.40*		do
N. 6	49.75		10.06	22.06			0.25	2.68			6.44	9.43	Ross & Hendricks (1945)
N. 7	43.51		2.94	28.61	0.99		0.05	2.22			6.62	14.05	do
H. 8	49.50		16.18	13.08			0.61	2.57			7.66	9.35	do
N. 9	46.06	0.84	12.22	18.54	0.28		1.62	1.66			17.26		do
N.10	40.72	0.02	4.96	29.57	0.71		0.74	1.98			6.66	15.46	do
N.11	40.48		6.00	29.33	0.26		0.31	2.90			5.56	15.39	do
N.12	46.65		17.65	18.39				1.21	0.55		5.51	10.81	do
N.13	39.54		6.04	30.32				0.04	0.24	0.32	22.27		do
N.14	42.74		9.76	26.71			1.06	1.94			6.74	10.42	do
N.15	40.42		10.42	25.31	0.45		0.13	0.38			21.80		do
N.16	41.38		9.84	27.47							9.25	12.25	do
N.17	41.88		11.90	26.20			0.10	0.67	0.52		7.77	10.78	do
N.18	40.02		10.55	26.16	1.37						21.83		do
N.19	40.54		5.19	31.24	0.39		0.06	1.92	0.14	0.24	6.00	14.75	Muchi (1977)
N.20	41.56	0.17	11.35	22.48	0.62	0.01	1.02	2.23	0.35	0.33	20.45		do
N.21	40.25	0.03	5.50	29.44			0.53	2.29			7.25	15.09	do
N.22	44.00		3.60	29.00			2.10				18.70		do
N.23	49.90		0.27	29.50			1.75	0.77	0.12	0.01	16.12		Isphording (1975)
N.24	48.82		4.30	35.88			0.35				9.66		do
N.25	42.10	0.26	4.70	29.00	0.50	0.15	0.85	1.15	0.45	0.58	19.80		Muller & Forstner (1976)
N.26	52.30	0.05	4.43	22.73		0.01	2.20	1.21	1.26	0.06	15.75		Brigatti (in preparation)
N.27	51.46	0.05	2.20	24.70		0.03	3.27	1.45	1.06	0.24	15.34		do
N.28	52.08	0.06	6.92	18.40		0.06	3.39	1.31	0.31	0.07	17.40		do
M. 1	42.61		4.17	19.87	1.39	0.02	5.94	1.67	0.24	0.13	23.60		Muchi (1977)
M. 2	44.52		10.91	13.77	1.34	0.34	3.81	2.67	0.65	0.07	6.15	16.50	do
M. 3	46.51	0.01	14.80	12.48		0.04	4.78	2.46	0.29	0.66	17.97		Brigatti (in preparation)
M. 4	44.64	0.06	22.04	8.02		0.08	3.44	0.94	0.19	0.29	20.30		do
M. 5	41.94	0.08	18.59	11.98		0.03	5.08	3.30	0.13	0.27	18.60		do
M. 6	42.72	0.03	15.03	12.06		0.06	6.90	3.21	0.13	0.60	19.26		do
M. 7	45.34	0.04	14.25	9.02		0.08	7.00	3.13	0.17	0.55	20.42		do
M. 8	49.29	0.10	13.47	13.45		0.06	2.10	2.07	0.13	0.69	18.64		do
M. 9	56.91	0.65	22.65	6.29	0.11		3.62	1.47	0.18	0.74	7.34		Weaver & Pollard (1973)
M.10	50.30	0.13	14.86	8.86	0.35		3.06	2.07		0.04	6.20	15.06	Schultz (1969)
M.11	52.90	0.98	17.90	4.80	0.20		2.90	0.12	2.90	0.27	6.30	10.20	do
M.12	50.44	0.42	16.26	5.38			3.92	0.72			6.30	16.00	do
M.13	39.52	1.88	18.48	12.60			3.27	2.72	0.10	0.50	12.10	9.04	Alietti (1960)

(*) N.2: included CO_2=.51; N.3: included CO_2=.30; N.4: included CO_2=.11; N.5: included CO_2=.08.

TABLE 2
Recalculated chemical formulae of the samples used.

	IV			VI							charge unbalance		
Sample	Si	Al	Fe^{3+}	Al	Fe^{3+}	Fe^{2+}	Mg	Mn	Ti	Σ	IV	VI	tot.
H. 1	3.27	0.04	0.69		1.67	0.33	0.36	0.04		2.40	-0.78	+0.47	-0.26
H. 2	3.30		0.70		1.80	0.41	0.17			2.38	-0.70	+0.56	-0.14
H. 3	3.15		0.85		2.06	0.05				2.11	-0.85	+0.28	-0.57
H. 4	3.17	0.04	0.79		1.77	0.39	0.25		0.01	2.42	-0.83	+0.63	-0.20
H. 5	3.44		0.56		1.93	0.06	0.33			2.32	-0.56	+0.57	
H. 6	3.12	0.21	0.67		1.73	0.15	0.50	0.03		2.41	-0.88	+0.55	-0.33
H. 7	3.01	0.16	0.83		1.77	0.11	0.62			2.50	-0.99	+0.77	-0.22
H. 8	3.39	0.40	0.21		1.24	0.43	0.67	0.07	0.01	2.42	-0.61	+0.10	-0.51
H. 9	3.38	0.35	0.27		1.16	0.37	0.98	0.03	0.01	2.55	-0.62	+0.28	-0.34
H.10	3.38		0.62		0.71	1.82	0.31	0.05		2.89	-0.62	+0.49	-0.13
H.11	2.96	0.28	0.76		2.00		0.28	0.06	0.02	2.36	-1.04	+0.72	-0.32
H.12	3.36	0.50	0.14		1.43	0.25	0.68	0.02	0.01	2.39	-0.64	+0.23	-0.41
H.13	3.27	0.18	0.55		1.52		0.84			2.36	-0.73	+0.24	-0.49
H.14	3.20	0.09	0.71		2.08	0.07	0.06	0.02		2.23	-0.80	+0.54	-0.26
H.15	3.33		0.67		1.49	0.34	0.71	0.01		2.55	-0.67	+0.59	-0.08
H.16	3.52	0.40	0.03		1.81	0.01	0.24			2.06	-0.43	-0.07	-0.50
S. 1	3.37	0.39	0.24		1.03	0.08	1.42	0.01	0.02	2.56	-0.63	+0.19	-0.44
S. 2	3.16	0.84		0.10	0.21	0.33	2.41	0.02		3.07	-0.84	+0.45	-0.39
S. 3	2.87	0.68	0.45		0.38		2.56		0.01	2.95	-1.13	+0.30	-0.83
S. 4	3.21	0.68	0.11		0.32	0.38	2.25			2.95	-0.79	+0.22	-0.57
S. 5	3.03	0.97		0.18	0.87	0.34	1.10			2.49	-0.97	+0.03	-0.94
S. 6	3.56	0.44		0.43	0.73		1.27		0.01	2.44	-0.44	+0.06	-0.38
S. 7	3.13	0.87		0.24	0.51	0.27	1.66	0.01	0.01	2.70	-0.87	+0.17	-0.70
S. 8	3.39	0.57	0.04		0.42	0.34	2.07	0.02	0.01	2.86	-0.61	+0.16	-0.45
S. 9	3.79	0.21		0.98	0.50	0.21	0.39	0.01		2.09	-0.21	-0.34	-0.55
S.10	3.66	0.34		0.93	0.55	0.10	0.51	0.01		2.10	-0.34	-0.32	-0.64
S.11	3.78	0.22		1.02	0.53	0.13	0.39	0.02		2.09	-0.22	-0.27	-0.49
S.12	3.83	0.17		0.99	0.08	0.66	0.45	0.02		2.20	-0.17	-0.53	-0.70
S.13	3.55	0.37	0.08		0.10		2.91			3.01	-0.45	+0.12	-0.33
S.14	3.58	0.42		0.34	0.08	0.41	2.20		0.01	2.94	-0.42	+0.22	-0.20
S.15	3.36	0.64		0.11	0.21	0.31	2.30			2.93	-0.64	+0.18	-0.46
S.16	3.07	0.93		0.37	0.11	0.15	2.39	0.01		3.03	-0.93	+0.54	-0.39
S.17*	3.30	0.58	0.12		0.49	0.07	2.27		0.01	2.84	-0.70	+0.19	-0.51
S.18*	3.50	0.40	0.10			0.20	2.80		0.01	3.01	-0.50	+0.04	-0.46
N. 1	3.57	0.43		0.08	1.79	0.04	0.21		0.01	2.13	-0.43	+0.15	-0.28
N. 2	3.52	0.48		0.07	1.91		0.08			2.06	-0.48	+0.10	-0.38
N. 3	3.64	0.36		0.05	1.86		0.08			1.99	-0.36	-0.11	-0.49
N. 4	3.53	0.47		0.09	1.87		0.10			2.06	-0.47	+0.08	-0.39
N. 5	3.74	0.26		0.28	1.61		0.19			2.08	-0.26	+0.05	-0.21
N. 6	3.76	0.24		0.66	1.26		0.03			1.95	-0.24	+0.18	-0.42
N. 7	3.75	0.25		0.05	1.85	0.07	0.01			1.98	-0.25	-0.14	-0.39
N. 8	3.73	0.27		1.17	0.74		0.07			1.98	-0.27	-0.13	-0.40
N. 9	3.61	0.39		0.74	1.00	0.02	0.10		0.06	2.09	-0.39	+0.11	-0.28
N.10	3.51	0.49		0.01	1.92	0.05	0.10			2.08	-0.49	+0.09	-0.40
N.11	3.47	0.53		0.08	1.89	0.02	0.04			2.03	-0.53	+0.03	-0.50
N.12	3.49	0.51		1.05	1.03					2.08	-0.51	+0.24	-0.27
N.13	3.49	0.51		0.12	2.02					2.14	-0.51	+0.42	-0.09
N.14	3.44	0.56		0.37	1.62		0.13			2.12	-0.56	+0.23	-0.33
N.15	3.45	0.55		0.50	1.63	0.03	0.02			2.18	-0.55	+0.51	-0.04
N.16	3.45	0.55		0.42	1.73		0.02			2.17	-0.55	+0.49	-0.06
N.17	3.39	0.61		0.53	1.60		0.01			2.14	-0.61	+0.41	-0.20
N.18	3.40	0.60		0.46	1.67	0.10				2.23	-0.60	+0.59	-0.01
N.19	3.48	0.52			2.02	0.03	0.01			2.06	-0.52	+0.14	-0.38
N.20	3.42	0.58		0.52	1.39	0.04	0.13		0.01	2.09	-0.58	+0.11	-0.47
N.21	3.49	0.51		0.05	1.92		0.07			2.04	-0.51	+0.05	-0.46
N.22	3.72	0.28		0.08	1.84		0.26			2.18	-0.28	+0.28	
N.23	4.00			0.03	1.79		0.21			2.03		-0.15	-0.15
N.24	3.67	0.33		0.05	2.03		0.04			2.12	-0.33	+0.32	-0.01
N.25	3.58	0.42		0.05	1.85	0.04	0.11	0.01	0.02	2.08	-0.42	+0.10	-0.32
N.26	4.00			0.40	1.31		0.25			1.96		-0.37	-0.37
N.27	3.98	0.02		0.15	1.45		0.39			1.99	-0.02	-0.40	-0.42
N.28	3.98	0.02		0.60	1.06		0.39			2.05	-0.02	-0.24	-0.26
N.29*	3.66	0.06	0.28		1.95		0.06			2.01	-0.34	-0.03	-0.37
N.30*	3.41	0.47	0.12		2.00		0.03			2.03	-0.59	+0.06	-0.53
M. 1	3.69	0.31		0.12	1.29	0.10	0.77			2.28	-0.31	-0.03	-0.34
M. 2	3.64	0.36		0.69	0.85	0.09	0.46	0.02		2.11	-0.36	-0.24	-0.60
M. 3	3.56	0.44		0.90	0.72		0.55			2.17	-0.44	-0.04	-0.48
M. 4	3.41	0.59		1.40	0.46		0.39			2.25	-0.59	+0.36	-0.23
M. 5	3.25	0.75		0.95	0.70		0.59			2.24	-0.75	+0.13	-0.62
M. 6	3.36	0.64		0.75	0.71		0.81			2.27	-0.64		-0.64
M. 7	3.55	0.45		0.87	0.53		0.82			2.22	-0.45	-0.16	-0.61
M. 8	3.77	0.23		0.99	0.78		0.24			2.01	-0.23	-0.21	-0.44
M. 9	3.69	0.31		1.12	0.21	0.01	0.35		0.03	2.12	-0.31	+0.03	-0.28
M.10	3.84	0.16		1.18	0.51	0.02	0.35		0.01	2.07	-0.16	-0.15	-0.31
M.11	3.83	0.17		1.36	0.26	0.01	0.31		0.05	1.99	-0.17	-0.30	-0.47
M.12	3.88	0.12		1.35	0.31		0.45		0.02	2.13	-0.12	-0.04	-0.16
M.13	3.17	0.83		0.92	0.76		0.39		0.11	2.18	-0.83	+0.26	-0.57

(*)For the following samples only the chemical formulae were available: S.17:Quakermat,1970; S.18:Poppi,1971; N.29,N.30:Russell & Clark,1978.

iron and manganese ores mostly in the form of sulphides (Schwartz 1924; Hewett & Shaller, 1924; Osborne & Archambaut, 1950; Sudo & Nakamura, 1952; Mozgova, 1957). In these occurrences, hisingerite is an alteration product derived from hyperstene, pyrite and chalcopyrite (Schwartz, 1924); from siderite (Hewett & Shaller, 1924); from pyrrhotite and from other iron minerals (Sudo & Nakamura, 1952); ìì)- in Fe-gabbroic rocks as an alteration product of either olivine (Lebedev, 1936) or Fe-, Mn-wollastonite (Nel'nitskaya, 1967); ììì)- in ryolithic tuffs as a weathering product of Fe-rich saponite in the following successive stages: Fe^{2+}-rich saponite $\rightarrow Fe^{3+}$-rich saponite $\longrightarrow$ hisingerite (Kohyama et al. 1973, Kohyama & Sudo 1975).

The features of hisingerite, can be summarized as follows: ì)- the X-ray spectra show few reflections and the basal peaks are weak or not observable; ìì)- the b-dimension is variable between 9.00 Å and 9.30 Å; ììì)- the IR spectra are similar to those of nontronite (Kohyama & Sudo, 1975); ìììì)- the DTA curve is characterized by an endothermic effect at about 200°C and an exothermic reaction at about 800°C.

The aim of this work is to establish the range in chemical composition of hisingerite and to compare this range with those of dioctahedral and trioctahedral Fe-rich members of the smectite group (Fe-rich saponite, Fe-rich montmorillonite and nontronite); to emphasize the relationship between chemical composition and physical properties if any; and to relate the crystallochemical evidence to genetic environments.

MATERIALS AND METHODS

Chemical analyses and b-dimensions of hisingerites (H), Fe-rich saponites (S), nontronites (N) and Fe-rich montmorillonites (M), were selected from the literature. The selected data (with references) are given in table 1 and 3. In order to standardize the chemical formulae of samples analyzed by different investigators, they were recalculated by: i)- balancing the charges on the basis of 22 negative charges; ii)- assigning all of the silicon to tetrahedral positions, together with enough Al (and/or Fe^{3+} if there was not enough Al) to bring the total to 4.00; iii)- assigning Ca, Na and K to interlayer positions; iiii)- assigning all residual cations to octahedral positions. The recalculated chemical formulae are listed in table 2.

The Q-MODE multivariate analysis (Harman, 1970; Klovan & Imbrie, 1971; Davis, 1973) was used as discussed previously (Alietti et al.,1979; Brigatti & Poppi 1980, 1981), in order to: ì)- characterize hisingerite in relation to Fe-dioctahedral and Fe-trioctahedral end members of the smectite group, using all the data in table 2 (case a); ìì)- evaluate the role of the iron oxidation state, using all the data obtained by chemical analyses with iron recalculated as Fe^{3+} (case b); ììì)- correlate the hisingerite chemical data with iron recalculated as Fe^{3+} (case c.1) or Fe^{2+} (case c.2), with all other data of table 2 to solve the question of the iron oxidation state in hisingerite; ìììì)- distinguish possible different compositional ranges of hisingerites, using (H) data only (case d). The chemical variables were reduced to a percentage of their range of variability and then normalized. Data not sufficiently detailed in the quantity of information (communality <0.98), were not considered. The meaning of the vertices in the following triangular diagrams can be deduced from the varimax factor score matrix relative to each one of them (Table 5).

TABLE 3
b-dimensions of the selected samples.

sample	b(Å)	sample	b(Å)	sample	b(Å)	sample	b(Å)	sample	b(Å)	sample	b(Å)	sample	b(Å)
H. 1	9.240	H. 8	9.240	S. 3	9.216	S. 8	9.198	N. 2	9.138	N.27	9.085	M. 4	8.942
H. 3	9.000	H. 9	9.180	S. 4	9.216	S.13	9.210	N. 3	9.132	N.28	9.030	M. 5	9.020
H. 4	9.240	H.10	9.300	S. 5	9.246	S.17	9.180	N. 4	9.138	N.29	9.162	M. 6	9.015
H. 6	9.180	H.16	9.060	S. 6	9.090	S.18	9.228	N. 5	9.114	N.30	9.144	M. 7	8.952
H. 7	9.240	S. 1	9.198	S. 7	9.228	N. 1	9.114	N.26	9.070	M. 3	8.978	M. 8	8.993

As already reported in the literature, the chemical composition can be correlated with the b-dimension (Russell & Clark, 1978; Kohyama & Sudo, 1975; Brigatti, in preparation). Through the use of the BMD 02R stepwise multiple regression program possible correlations between the chemical and b-dimensions of Fe-rich smectites were studied.

RESULTS

Fig. 1 shows the variability of all Fe-rich smectites here studied. The analyses are plotted in terms of molecular percentages in a triangular diagram where the vertices represent $(R^{3+})^{IV}$, $(R^{3+})^{VI}$, $(R^{2+})^{VI}$. The percentage of $(Al+Fe^{3+})$ in tetrahedral coordination was less than 30% and constant for all samples considered. On the basis of the octahedral filling, hisingerite and Fe-rich montmorillonite are very similar and a continuity is indicated between a trioctahedral (saponite) and a dioctahedral (nontronite) structure. The compositional area of nontronite is restricted and clearly separated from that of other Fe-rich smectites.

In a diagram $[Fe_{tot.}]^{VI}$ vs. Mg^{VI} (fig. 1b) several compositional

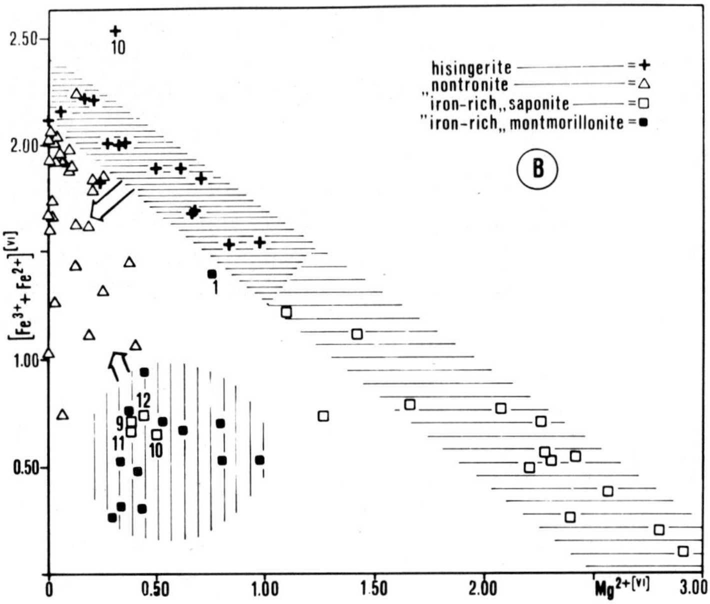

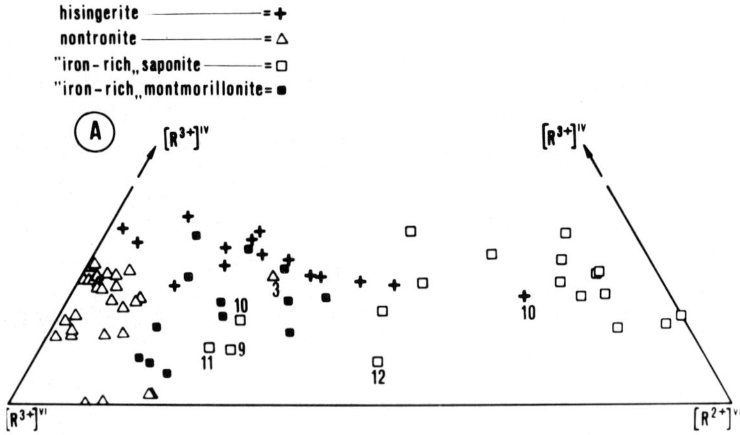

Fig. 1. a) Correlation between $(R^{3+})^{IV}-(R^{2+})^{VI}-(R^{3+})^{VI}$ in Fe-rich smectites; b) correlation between total Fe and Mg in Fe-rich smectites.

TABLE 4
Correlation matrices between b-dimension and chemical properties.

case a

variable	1	2	3	4	5	6
1	1.000	.581	.256	.222	.252	.909
2		1.000	.156	-.597	.835	.530
3			1.000	.223	-.156	.416
4				1.000	-.851	.364
5					1.000	.070
6						1.000

case b

variable	1	2	3	4	5	6	7
1	1.000	-.306	.867	.983	.993	-.742	-.387
2		1.000	-.537	-.271	-.283	.785	.509
3			1.000	.872	.887	-.941	-.426
4				1.000	.992	-.723	-.311
5					1.000	-.747	-.388
6						1.000	.527
7							1.000

Case a)Fe-rich smectites group $1=b(\mathring{A})$; $2=\Sigma$; $3=$*charge unbalance*$/\Sigma$; $4=(Fe_{tot.})^{VI}/\Sigma$; $5=Mg/\Sigma$; $6=\left[(Fe_{tot.})^{VI}/\Sigma\right]+\Sigma$.
Case b)hisingerite $1=b(\mathring{A})$; $2=(Fe_{tot.})^{VI}$; $3=\Sigma$; $4=$*charge unbalance*; $5=$*charge unbalance*$/\Sigma$; $6=(Fe^{3+})^{VI}/\Sigma$; $7=(Fe_{tot.})^{VI}/\Sigma$.

areas can be distinguished. There is a continuity between Fe-rich saponite and hisingerite, as in a hypothetical series of (Mg,Fe^{2+}) saponites in which the Fe^{2+} is subsequently oxidized to Fe^{3+}. On the basis of the percentage of total Fe in octahedral coordination, one also can see a continuity between Fe-rich montmorillonite, nontronite and hisingerite. Fe-rich saponites NO.9,10,11,12 fall in the Fe-rich montmorillonite area, but the characterization of these samples could be wrong.

The correlation matrices between the b-dimension and chemical properties of Fe-rich smectites are given in table 4. If all the data were considered, the best correlation would be b-dimension versus $\left[(Fe_{tot})^{VI} : \Sigma\right] + \Sigma$ (fig. 2a), where Σ is octahedral occupancy. Iron is a discriminant factor for nontronite and Fe-rich montmorillonite, as was shown by Russell & Clark (1978), and Brigatti (in preparation), (see figs. 2d, 2c) but not for Fe-rich saponite and hisingerite. Hisingerite shows a good correlation between the b-dimension and the ratio (total chargc umbalancc)/Σ (fig. 3, Table 4).

The results of the Q-MODE multivariate analysis are given in fig. 4, and the matrices of the three rotated eigenvalues, the variance and the cumulative variance for the cases a, b ,c.1, c.2, d are summarized in table 5. Hisingerite can be easily distinguished from other Fe-rich smectites principally on the basis of octahedral population (F.1,F.2,F.3 in case a are considerably affected by Al, Fe^{3+}, Mg respectively). Discrimination between the different

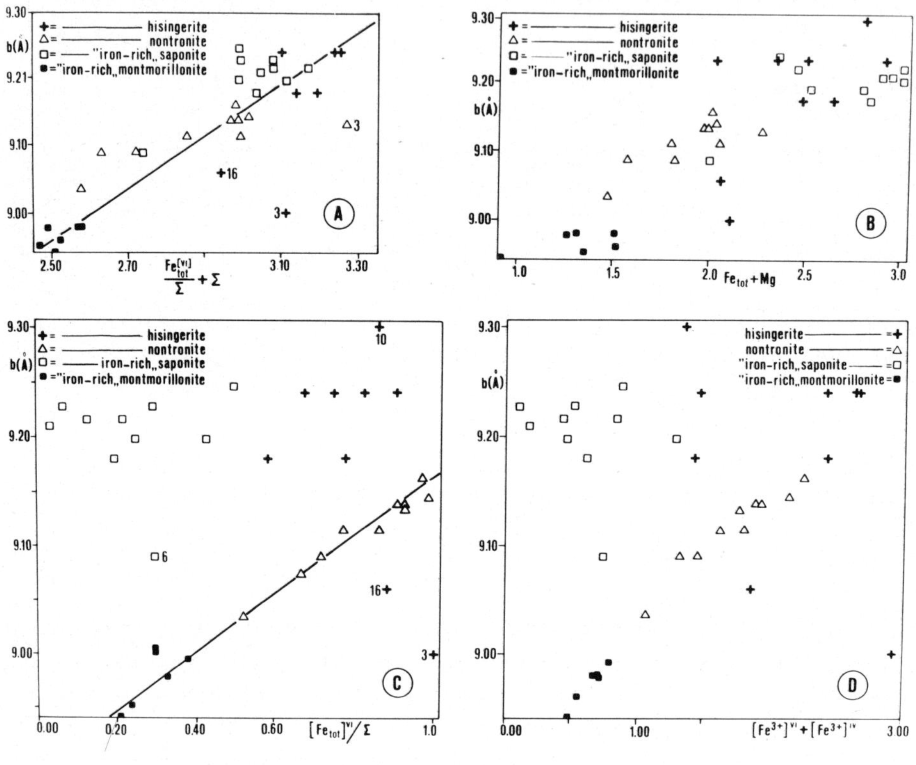

Fig. 2. Correlations between b-dimension and chemical properties of Fe-rich smectites. a) on the basis of this study; b) from Kohyama & Sudo, 1975; c) from Brigatti, in preparation; d) from Russell& Clark, 1978.

compositional ranges of hisingerite and the other Fe-rich smectites is not affected by the percentage of Fe^{2+} present in the structure. The effect of the oxidation state of iron in hisingerite is shown in fig. 4 c.1 and fig. 4 c.2; when iron is assumed to be Fe^{2+} there is a continuous compositional range between hisingerite and Fe-rich saponite; a much better correlation is found for nontronite when iron is assumed to be Fe^{3+}. Case d (fig. 4d) shows a wider spreading of hisingerite points and seems to show the existence of two populations on the basis of Al^{IV} and $(Fe^{3+})^{IV}$.

CONCLUSIONS

The chemical and crystallographic properties allow characterization of hisingerite into the Fe-rich smectites group. The chemi-

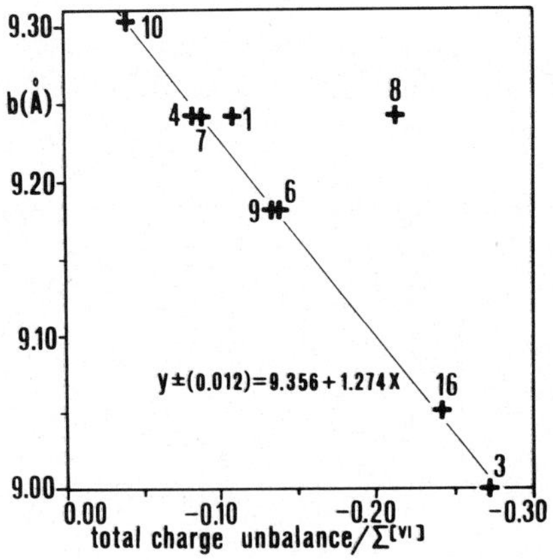

Fig. 3.Correlation between b-dimension and chemical properties of hisingerites.

cal composition is characterized by a very small Al percentage (and consequently Fe^{3+} in tetrahedral coordination) whereas the octahedral population is almost completely Fe^{3+} (table 2). The octahedral filling ( ), varies between that of a dioctahedral and of a trioctahedral state (fig.1 a, b), as in some Fe-rich montmorillonites. So there may be a continuous isomorphous series from trioctahedral smectites (Mg-saponite and Fe^{2+}-hisingerite as end members) to nontronite as a result of iron oxidation (in particular in sedimentary environments) even though with intermediate terms of low crystallinity. Similarly, Fe-rich montmorillonite is confirmed as an intermediate term between montmorillonite and nontronite (Brigatti, in preparation).

From the results of the Q-Mode multivariate analysis and the correlations between the b-dimensions and chemical properties the following conclusion may be drawn:

i)- Hisingerite is well characterized on the basis of chemical composition and there is a closer similarity between Fe^{2+}-hisingerite and Fe-rich saponite than between Fe^{3+}-hisingerite and nontronite. The good correlation between the b-dimension and a chemical variable in the hisingerites (as in many clay mineral species) warrants its individuality in the smectite group. ii)- A continuous isomorphous series exist between Fe-rich montmorillonite and nontronite, but there is no evidence of a similar continuity between hisingerite and nontronite. iii)- It is known that in a dioctahedral structure the b-dimension is affected by the octahedral population (Russell & Clark, 1978; Brigatti, in preparation) while in a trioctahedral structure the b-dimension is affected by the layer charge (Suquet et al. 1981). The correlation between b-dimension and chemical features is different in hisingerite, in Fe-rich montmorillonites and in

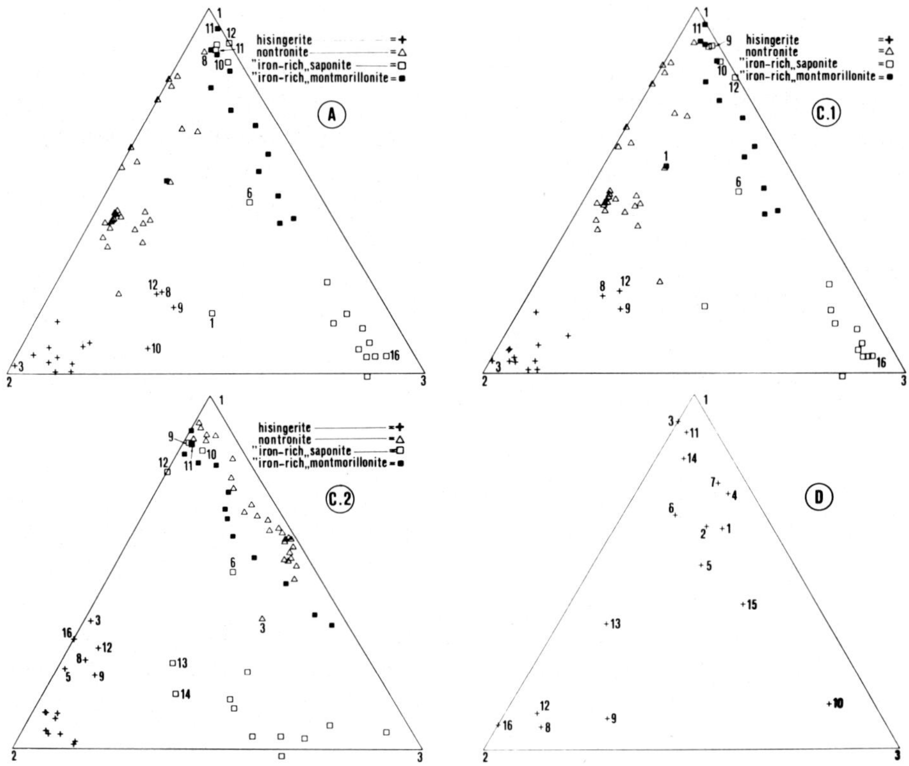

Fig. 4. Q-MODE multivariate analysis.

TABLE 5
Varimax factor score matrices.

		case a			case b			case c.1			case c.2			case d		
		F.1	F.2	F.3	F.1	F.2	F.3	F.1	F.2	F.3	F.1	F.2	F.3	F.1	F.2	F.3
IV	Si	.548	.061	-.007	.556	.054	-.016	.558	.037	-.002	.519	.283	-.142	-.012	.470	.191
IV	Al	.128	-.014	.500	.130	-.038	.529	.123	-.001	.509	.018	-.055	.632	-.006	.579	-.244
IV	Fe^{3+}	-.239	.401	.031	-.226	.437	.012	-.231	.437	.016	-.134	.303	.046	.603	-.112	.104
VI	Al	.475	-.285	.122	.446	-.288	.138	.451	-.263	.104	.366	-.056	-.067			
VI	Fe^{3+}	.132	.658	-.195	.160	.669	-.132	.150	.664	-.163	.361	-.299	.397	.635	.225	-.351
VI	Fe^{2+}	-.030	.041	.100				.007	-.078	.252	-.072	.531	-.168	-.095	-.007	.389
VI	Mg	-.044	-.087	.510	-.058	-.075	.524	-.063	-.054	.503	-.117	.176	.303	-.036	.372	.239
VI	Σ	-.122	.069	.566	-.121	.069	.560	-.117	.074	.553	-.163	.544	.172	.017	.072	.462
charge unbal.	IV	.547	.061	-.007	.556	.054	-.017	.557	.037	-.002	.549	.174	-.119	-.010	.471	.186
charge unbal.	VI	-.025	.409	.314	-.058	.373	.291	-.015	.396	.272	.020	.210	.469	.437	-.080	.257
charge unbal.	tot.	.257	.369	.084	.259	.349	.104	.257	.353	.077	.318	.223	.180	.178	-.089	.497
Var.		38.548	33.687	20.588	40.883	32.092	20.408	39.910	32.360	20.055	43.382	24.962	22.473	45.715	30.462	18.274
Cum.V.		38.548	72.236	92.824	40.883	72.975	93.383	39.910	72.270	92.325	43.382	68.344	90.818	45.715	76.177	94.451

nontronite, too: this may be only explicable by assuming a trioctahedral origin of hisingerite. The original Fe^{2+} was oxidized in sedimentary environments and partially removed, but the structure (even though reduced structural order, as shown by X-ray diffraction and thermal features) retains some features of the original trioctahedral patterns. On the contrary in nontronite, the structure is of dioctahedral origin and the b-dimension is affected by octahedral population: this is also the case for Fe-rich montmorillonite.

Even though a compositional continuity between dioctahedral and trioctahedral members is evident, a similar continuity does not seem to be possible from the structural point of view.

ACKNOWLEDGEMENT

Thanks are due to Prof. G. Gottardi and L. Poppi for critical reading of the work, and to Mr. W. Lugli for the drawing.

This work was supported by the "Consiglio Nazionale delle Ricerche" of Italy and the "Centro di Calcolo" of Modena University.

REFERENCES

Alietti, A., 1960. Su di una nontronite poco ferrifera di Chiampo (Lessini). Atti Soc. Nat. Mat. Modena, XCI: 33-44.

Alietti, A., Brigatti, M.F. and Poppi, L., 1979. An illite/montmorillonite interlayer mineral in porphyritic rock of "Alpe di Siusi" (BZ-Italy): the chemistry of I/M interlayer minerals. Clay Miner., 14: 39-48.

Bischoff, J.L., 1972. A ferroan nontronite from the Red Sea geothermal system. Clays Clay Min., 20: 217-223.

Bowie, S.H.U., 1955. Thucholite and hisingerite-pitchblende complexes from Micholson Mine Saskatchewan, Canada. Bull. Geol. Surv. Gr. Britain, 10: 45-57.

Brigatti, M.F. A chemical framing of Fe-rich montmorillonite in the smectite group. In preparation.

Brigatti, M.F. and Poppi, L., 1980. Vermiculite and its relation to parent materials as revealed by chemical features. Miner. Petrogr. Acta, 24: 123-134.

Brigatti, M.F. and Poppi, L., 1981. A mathematical model to distinguish the members of the dioctahedral smectite series. Clay Miner., 16: 81-89.

Brown, G., 1955. Report of the clay minerals group subcommittee on nomenclature of clay minerals. Clay Miner. Bull., 2: 294-302.

Caillère, S. and Hénin, S., 1957. Sur la présence a Diélette d'une saponite a texture fibreuse. Bull. Soc. Franç. Minér. Crist., 80: 343-345.

Davis, J.C., 1973. Statistics and Data Analysis in Geology. Wiley, New York.

Eggleton, R.A., 1977. Nontronite: chemistry and X-ray diffraction. Clay Miner., 12: 181-194.

Gruner, J.W., 1934. The structures of vermiculites and their collapse by dehydration. Am. Miner., 19: 557-585.

Haranczyk, C. and Prochazka, K., 1974. Uwodnione krzemiany magnezowo-niklowe z Wir na Dolnym Slasku. Prace Muz. Ziemi, 22: 3-62.

Harman, H.H., 1970. Modern factor analysis. University of Chicago Press, Chicago and London.

Hewett, D.F. and Schaller, W.T., 1924. Hisingerite from Blaine Co., Idaho. Am. Jour. Sci., Fifth Series, X, 55: 29-38.

Isphording, W.C., 1975. Primary nontronite from Venezuelan Guyana. Am. Miner. 60: 840-848.

Kimbara, K., 1973. Clay minerals in the Miocene low-grade metamorphic rocks, Tanzawa Mountains, Kanagawa Prefecture, Central Japan. Jour. Japan Assoc. Min. Pet. Econ. Geol., 68 (10): 311-328.

Kimbara, K., 1975. Chlorite, saponite and regularly interstratified chlorite-saponite in the Miocene pyroclastic sediments (Green Tuff) at Taiheizan, Akita Prefecture, Japan. Mem. Vol. Prof. T. Sudo, 42-47.

Kimbara, K. and Honda, S., 1975. An Iron-rich Saponite-like Mineral Found in the Moriyama Volcanic rocks, Gojome, Akita Prefecture, Japan. Bull. Geol. Survey Japan, 26: 37-40.

Kimbara, K. and Shimoda, S., 1972. An iron-rich saponite and randomly interstratified mineral of chlorite and saponite in a pillow lava at Nibetsu, Akita Prefecture. Jour. Clay Sci. Soc. Japan, 12: 133-141.

Klovan, J.E. and Imbrie, J., 1971. An algorithm and FORTRAN IV program for large scale Q-MODE factor analysis and calculation of factor score. J. Math. Geol., 3 (I): 61.

Kohyama, N., Shimoda, S. and Sudo, T., 1973. Iron-rich saponite (ferrous and ferric forms). Clays Clay Min., 21: 229-237.

Kohyama, N. and Sudo, T., 1975. Hisingerite occurring as a weatherig product of iron-rich saponite. Clays Clay Min., 23: 215-218.

Konta, J. and Sindelar, J., 1955. Saponite from the fissure fillings of the amphibolites of Caslov. Univ. Carolina Geologica, 1(2): 177-186. From Miner. Abstr., 13: 148.

Lebedev, P.I., 1936. Beitrag zur Petrographie and Mineralogie der basichen Paegmatite Volyniens. Vernadsky Jubilee volume Acad. Sci. USSR, 2: 199-1012. From Miner. Abstr., 7: 199.

Lindqvist, B. and Jansson, S., 1962. On the crystal chemistry of hisingerite. Am. Miner., 47: 1356-1362.

Miyamoto, N., 1957. Iron-rich saponite from Mazé, Miigata Prefecture, Japan. Miner. Journ., 2: 193-195.

Mongiorgi, R. and Morandi, N., 1970. Al saponite e strati misti Clorite - Al saponite nelle idrotermaliti di una breccia a contattto coi diabasi di Rossena nell'Appennino reggiano. Miner. Petrogr. Acta, 16: 139-154.

Mozgova, N.M., 1957. An hisingerite and stilpnomelane from the skarn polimetallic deposit of the upper mine of Tatukhe. Min. Sbornik Lvov Geol. Soc., 11: 273-287. From Miner. Abstr., 14: 272.

Muchi, M., 1977. Saponite and Related Thomsonite from Iwano, Saga Prefecture, Japan. Bull. Fukuoka Univ. Educ., 26, III: 103-115.

Muchi, M. and Higashiyama, K., 1972. Iron-rich saponite occurring in druse cavities of basalt near Yamashiro-cho, Saga Prefecture. Bull. Fukuoka Univ. Educ., 21, III: 151-163.

Muller, G. and Forstner, U., 1976. Primary nontronite from the Venezuelan Guayana additional primary occurrences. (Red Sea, Lake Malawi). Am. Miner., 61: 500-501.

Nel'nitskaya, E.F., 1967. Manganoferrous Wollastonite and its alteration. Zap. Vses Min. Obshch., 96: 927-305. From Miner. Abstr., 19: 47.

Nikolsky, A.P., 1953. On the hisingerite of the Saksagan belt. Min. Sbornik Lvov Geol. Soc., 7: 127-134. From Miner. Abstr., 13: 61.

Osborne, F.F. and Archambault, M., 1950. Hisingerite from Montauban-les-Mines. Le Naturaliste Canadien, Quebec., 77: 283-290. From Miner. Abstr.,12: 226.

Poppi, L., 1971. Su alcune saponiti appenniniche. Comportamento del reticolo alla disidratazione. Rend. SIMP., XXVII: 211-226.

Quakernat, J., 1970. A new occurrence of a macrocrystalline form of saponite. Clay Miner., 8: 491-493.

Ross, C.S. and Hendricks, S.B., 1945. Minerals of the montmorillonite group. Prof. Paper. U.S. Geol. Surv., 205-B.

Russell, J.D. and Clark, D.R., 1978. The effect of Fe- for Si-substitution on the b-dimension of nontronite. Clay Miner., 13: 133-137.

Schultz, L.G., 1969. Lithium and potassium absorption, dehydroxylation temperature and structural water content of aluminous smectites. Clays Clay Min., 17: 115-137.

Sudo, T., 1954. Iron-rich saponite found from Tertiary iron sand beds of Japan. Journ. Geol. Soc. Japan, 60: 18-27.

Sudo, T. and Nakamura, T., 1952. Hisingerite from Japan. Am. Miner., 37: 618-621.

Strunz, U., 1978. Mineralogische Tabellen. Akademische Verlagsgesellschaft, Geest & Portig K.G.

Schwartz, G.M., 1924. On the nature and origin of hisingerite from Parry Sound, Ontario. Am. Miner., 9: 141-144.

Suquet, H., Malard, C., Copin, E. and Pezerat, H., 1981. Variation du paramètre b et de la distance basale d_{001} dans une série de saponites à charge croissante:I-Etats hydratés. Clay Miner., 16: 53-68.

Van Breemen, N., 1980. Magnesium-Ferric iron replacement in smectite during aeration of pyritic sediments. Clay Miner., 15: 101-110.

Weaver, C.E. and Pollard, L.D., 1973. The chemistry of clay minerals. Elsevier, New York.

Whelan, J.A. and Goldich, S.S., 1961. New data for hisingerite and neotocite. Am. Miner., 46: 1412-1423.

MIXED-LAYERING IN Ni-CONTAINING TALC-LIKE MINERALS FROM SZKLARY, LOWER SILESIA, POLAND

A.WIEWIÓRA[x], E.DUBIŃSKA[x], I.IWASIŃSKA[xx]

[x]Institute of Geological Sciences, Polish Academy of Sciences, Warszawa

[xx]Geological Institute, Warszawa

ABSTRACT

Nickel-containing talc-like minerals from Szklary display a variety of morphological forms and crystal-chemical features. By their X-ray characteristics they may be included in a kerolite-pimelite series. Detailed X-ray investigations disclosed various degrees of susceptibility of particular samples to swelling in water and organic liquids and to a minor degree of collapse by heating. Partial swelling produced interstratified zonal structures incorporated with basically 10 A material. In some cases swelling could be increased significantly by ultra-sonic treatment.

Visual and Fourier methods applied to transform the apparent $d_{/001/}$ values of the interstratified phases into real basal spacings showed various stages of solvation of the essentially talc-like layers. They proved the interstratified structure of the glycolated Ni-kerolite /by its crystal-chemical formula/ to be formed of triplets AAB and pairs AB /where A - non solvated, B - solvated layers/ disorderly interlayered. The type of interstratification depends on the mineral origin.

INTRODUCTION

Nickel-containing talc-like layer silicates are widespread in the weathering crusts developed on ultramafic rocks. They are found in abundance in Szklary near Ząbkowice Sląskie /Lower Silesia, Poland/ the historical locality for these minerals. Talc-like minerals from Szklary were investigated by Spangenberg /1938/, Slansky /1955/, Ostrowicki /1965/, Faust /1966/ and Brindley et al. /1979/.
The present study brings new data on their expanding ability.

MATERIALS, OCCURRENCE AND MORPHOLOGY

Talc-like, clay-grade garnierites from Szklary occur mostly in veinlets, usually thinner than several centimeters. They cut altered ultramafic rocks and extend into laterite. Alteration of surrounding rocks is followed by substantial changes in the compositions of the veinlets from kerolitic-sepiolitic, frequently with carbonates in nearly fresh rocks, to talc-like silicates and quartz in laterite. Samples 6 and 42 collected from such veinlets are macroscopically similar to pimelites described by Maksimovič /1966/. Sample 11 comes from a net of irregular thin /2-3 mm/ veinlets in the nontronitic zone of the weathering profile, samples 17 and 74 come from the same type of veinlets in laterite and are composed of quartz and soft, gel-like, sometimes semi-lucent clay material, bright to emerald green in colour.

Material from veinlets being a continuation of sepiolite containing veins consists for the most part of flaky aggregates with spongy structure /Photo 1/. Some samples show a fibrous habit /Photo 2/ probably inherited from sepiolite. The homogenous mass of the flake-like individuals /Photo 3/ was found in the irregular thin veinlets. In the neighbouring rocks, coatings are also common on the primary components of the rock and most of them also show a honeycomb structure /Photo 4/. These coatings probably represent the initial stage of veinlet formation.

The clay material was carefully selected for investigations by hand picking under a binocular microscope.

X - RAY CHARACTERISTICS

X-ray studies of 5 samples coming from the described veinlets were performed using non-oriented and oriented aggregates. In some cases oriented aggregates could not be obtained because of spontaneous flocculation of the clay particles.

Diffractograms of non-oriented aggregates presented in Fig. 1 show X-ray characteristics similar to those of talc-like minerals described by Maksimovič /1966/, Brindley and Pham Thi Hang /1973/, Suzuki et al. /1976/, Brindley et al. /1977, 1979/. Unfortunately, despite careful separation, four out of the five samples are contaminated by traces of other minerals detected by X-ray observation. Samples 17 and 42 contain quartz, sample 11 - chlorite and tremolite coming from the parent rocks, and sample 6, an admixture of serpentine, and may be named deweylite /Bish and Brindley 1978/. The studied samples show very broad 001 /10 Å/ reflection with low angle "tailing" and likewise a broad third order basal

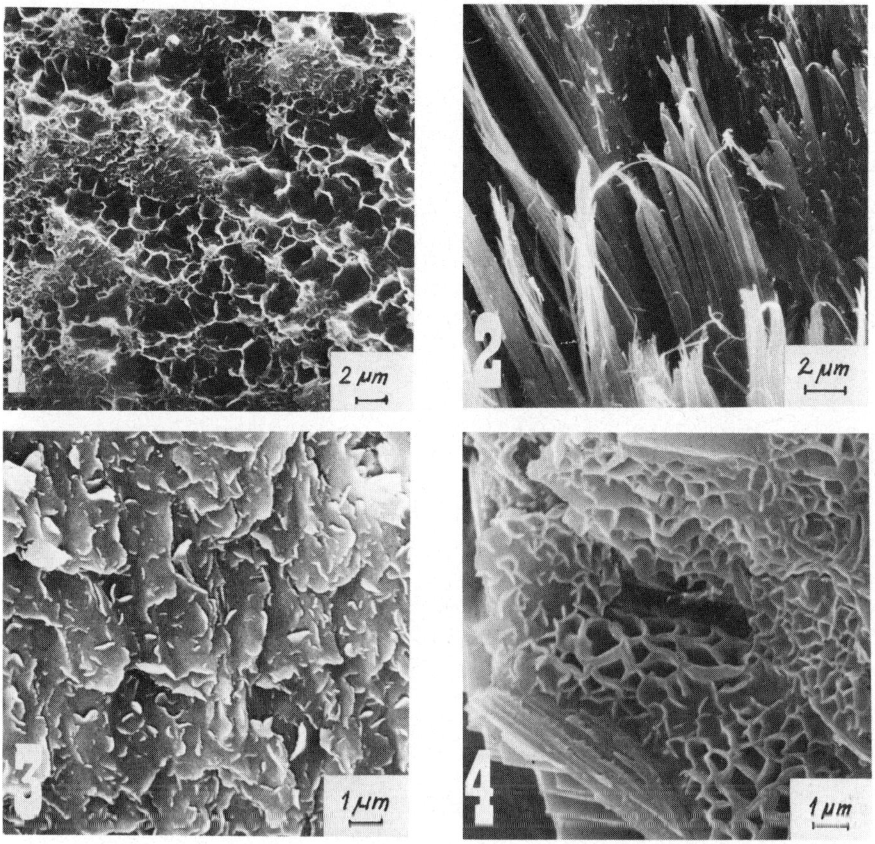

Fig. 1. Scanning electron micrographs: 1. Spongy aggregation of the flaky material in the kerolitic-sepiolitic veins. Sample 6. Magn. 2000x, 2. Fibrous habit Sample 42. Magn. 2600x, 3. Mass appearance of the flaky material. Sample 11. Magn. 4000x, 4. Honeycomb structure of the coating material. Sample 11. Magn. 4800x.

reflection /3.16 - 3.20 A/, which is usually symmetrical. Among hk reflections, 02, 11 and 06, 33 are sharp and strong, and 13, 20 is broad. All other reflections are highly degraded /Fig. 2/. The X-ray tracings indicate disordered, turbostratic structures. Variations in intensity, shape and position of the peaks /Fig. 2/

reflect some features which were observed when trying to prepare oriented aggregates. Especially, samples 6 and 42, genetically related to sepiolite-containing veins, showed flocculation, which could not be eliminated by various treatments, such as sodium saturation, changes of dielectric constant or of pH of the suspensions, as recommended by van Olphen /1977/. Flocculation made the oriented sedimentation of the samples 6 and 42 impossible. Well oriented aggregates were obtained for samples 11, 17 and 74. Most X-ray tracings presented in the literature /Maksimovič 1966, Imai et al. 1973, Suzuki et al. 1976, Yokomizo and Moriyama 1976/ are of non-oriented aggregates.

X-ray diffractograms of non-oriented aggregates /Fig. 2/ show variations in the intensities of basal reflections, which may be related directly with the nickel content as found by Brindley et al. /1979/ for the intensity of the first order of diffraction in the kerolite-pimelite series. A similar trend may be observed in our tracings /compare chemical data/. Some deviations, such as the relatively low intensity of 10 A reflection in sample 6, may be caused by greater disorientation of clay particles in the pressed aggregates due to very small size of mineral grains. A relatively high 060/001 intensity ratio supports this explanation of the observed differences.

Saturation of non-oriented aggregates with ethylene glycol /Fig. 3/ resulted in a partial swelling of the studied samples, discernible not only at small 2θ angles but also in the region of 20-30° 2 θ. Similar effects may be observed in several diffractograms presented by Maksimovič /1966/, Imai et al. /1973/, Yokomizo and Moriyama /1976/, Suzuki et al. /1976/; they seem to be characteristic of glycol solvated kerolites-pimelites and may be explained as follows. Untreated kerolites--pimelites inhibit b/3 and random translation errors. Saturation with ethylene glycol eliminates b/3 translations /appearance of the band composed of 021 and 111 peaks - fig. 3/, unchanged 20,13 band shows that random translations persist.

Low angle reflections could be studied better using diffractograms of oriented aggregates. It is interesting to compare diffractograms obtained after mild wet grinding and after ultrasonic treatment of well dispersed samples 17 and 74, which gave quasi-similar patterns as presented in Fig. 2, except for the first order reflection which has a higher intensity at a higher nickel content. In sample 17, increase in orientation, may be observed due to perfect dispersion after ultrasonic treatment /frequency 22 KHz, amplitude 2,0 m s $^{-1}$, 3 min/ as compared to grinding /Fig. 4/. Glycolation produces

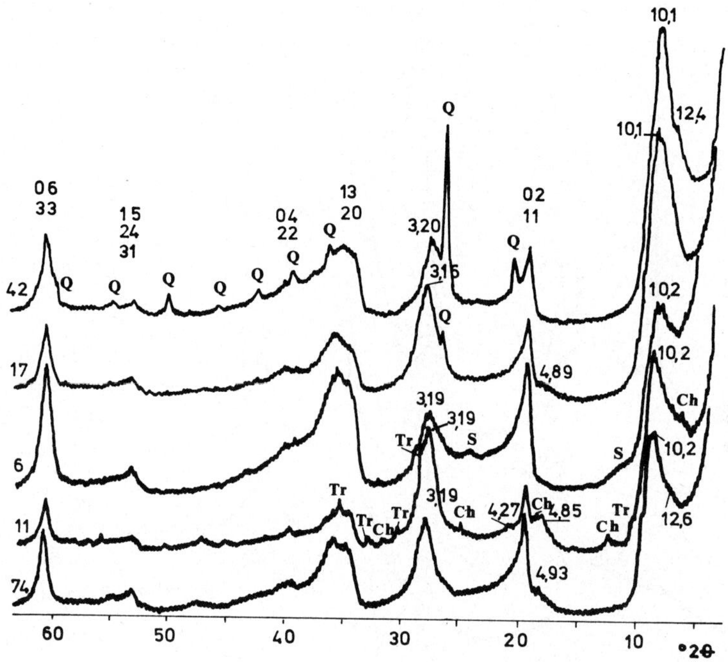

Fig. 2. XRD patterns of non-oriented aggregates of talc-like minerals; indices according to Brindley et al. /1979/, Q-quartz, Tr-tremolite, Ch - chlorite, S - serpentine

similar effects: splitting of the basal reflections /in each of three reproduced orders/ into two reflections, in both cases. In contrast with sample 17, ultrasonic treatment of sample 74 causes a very significant increase in its swelling ability in water /compare curve 1 and 3 in Fig. 5/ and in organic liquids, like ethylene glycol /compare curves 3 and 5 in Fig. 5 and curves 1 and 3 in Fig. 6/. The splitting of each of the basal reflections into two reflection, one with higher and the other with lower $d_{/001/}$ values, is much more distinct than that observed in the ground samples 17 and 74.

The disclosed swelling phenomenon of our samples of the kerolite-pimelite series could be studied in detail due to the fortunate fact, that sample 11 displays sponteneous swelling exactly similar to that of sample 74 after ultrasonic treatment. To facilitate penetration of water or organic liquid molecules between the layers,

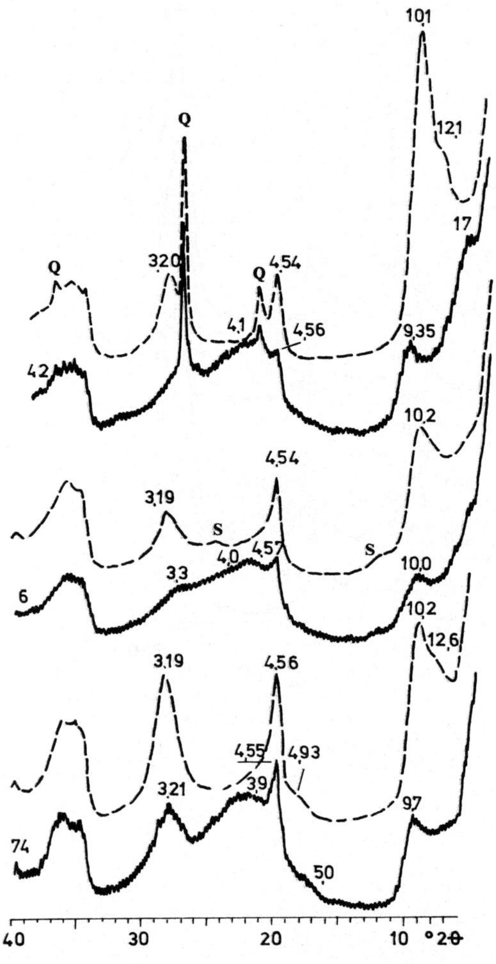

Fig. 3. XRD patterns of glycolated non-oriented aggregates of talc-like minerals, full line - glycolated samples, dashed line - untreated samples

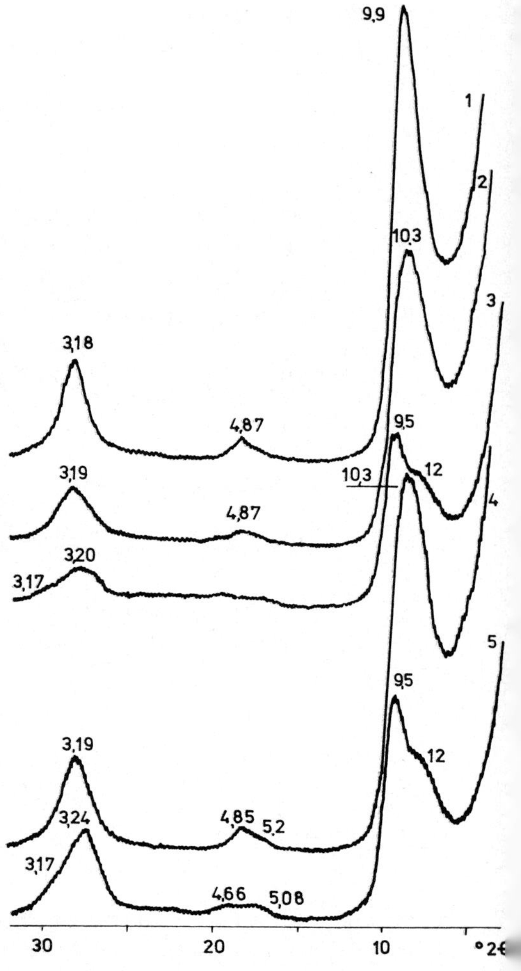

Fig. 4. XRD patterns of sample 17, oriented aggregates, 1 - heated at 300°C, 2 - at room temperature, 3 - treated with ethylene glycol, 4 - treated with ultrasonics for 4 min., 5 - glycolated after ultrasonic treatment.

the homocationic Na-sample 11 was used. The well oriented aggregates of this sample were equilibrated in various relative humidities and subsequently X-rayed. Changes

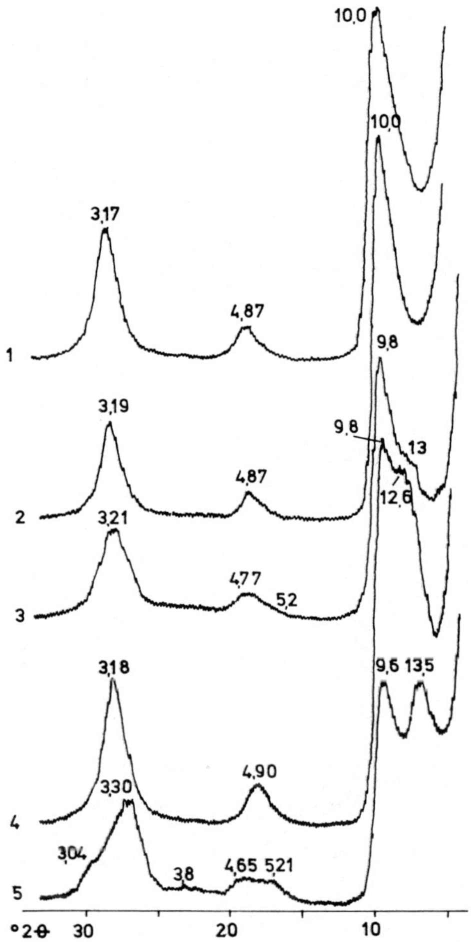

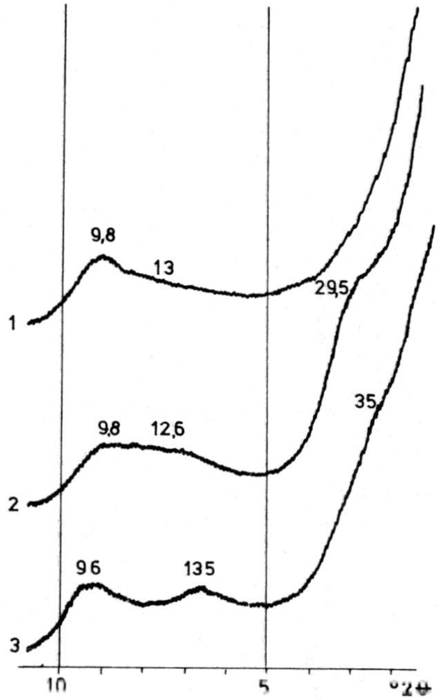

Fig. 5. XRD patterns of sample 74, oriented aggregates, 1 - heated at 300°C, 2 - at room temperature, RH = 80%, 3 - treated with ethylene glycol, 4 - treated with ultrasonics for 1 min/until thixotropy occurred/, RH = 80%, 5 - glycolated after ultrasonic treatment.

Fig. 6. XRD patterns of sample 74, low angle region, oriented aggregates, 1 - treated with ultrasonics for 1 min. /until thixotropy occurred/, RH = 80%, 3 - glycolated after ultrasonic treatment.

of the apparent interlayer distances can be seen in figure 7. Other aggregates equilibrated at 60% RH, were initially heated at 40°C, subsequently at 300°C, and first reexpanded first at 65% RH and next in ethylene glycol /Fig. 8/.

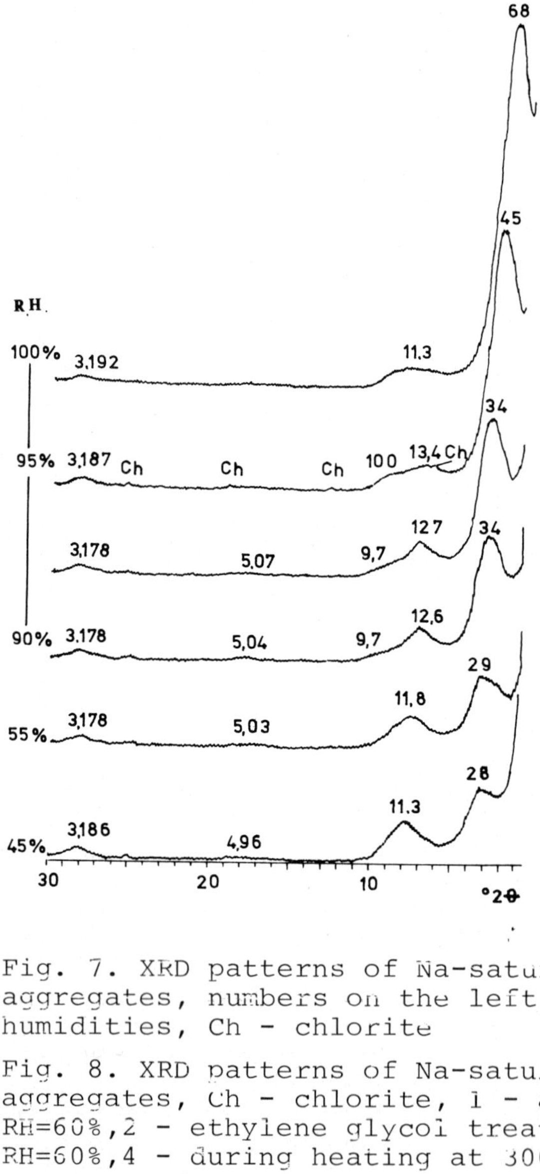

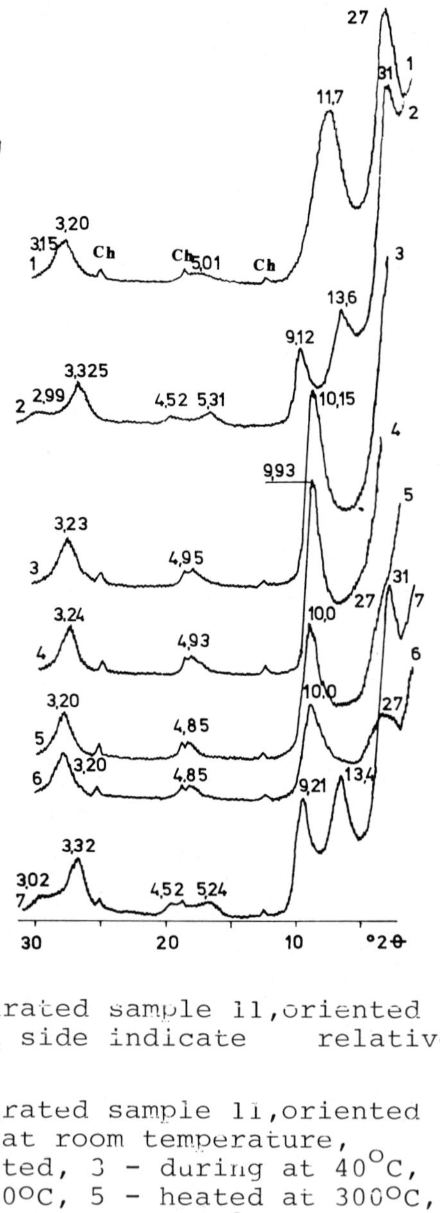

Fig. 7. XRD patterns of Na-saturated sample 11,oriented aggregates, numbers on the left side indicate relative humidities, Ch - chlorite

Fig. 8. XRD patterns of Na-saturated sample 11,oriented aggregates, Ch - chlorite, 1 - at room temperature, RH=60%,2 - ethylene glycol treated, 3 - during at 40°C, RH=60%,4 - during heating at 300°C, 5 - heated at 300°C, measured after 1h cooling, 6 - heated at 300°C, next rehydrated in desiccator with RH=100% for 24 h, and measured at rh=60%, 7 - glycolated after heating at 300°C

Fourier transform of the basal reflections of glycol - treated sample 11 is presented in Figure 9.

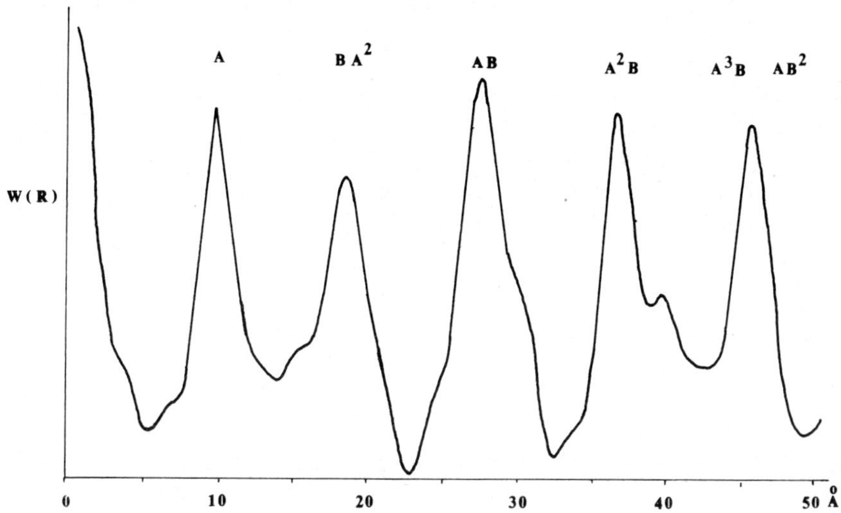

Fig. 9. Fourier transform of basal reflections according to MacEwan's /1958/ formulae, ethylene glycol treated sample 11-Na /Fig.7, curve 2/, A denotes 9,5 Å, B denotes 17 Å. Calculated probability coefficients $w_A = 0.62$, $p_{AAA} = 0$, $p_{BAB} = 1$, short distance order coefficient $g = 2$.

THERMAL ANALYSIS

DTA curves show two endothermic effects /Fig. 10/, a low temperature peak due to dehydration and a high temperature peak due to dehydroxylation of the 2:1 layer, both characteristic for the kerolite-pimelite series of minerals. Weight loss caused by dehydration in the temperature range up to 200°C varies from about 5% for sample 17 up to 10% for sample 11, and reflects the solvation ability of the clay material. Hydroxyl water loss proceeds in the temperature range 750-960°C and is nearly equal /~ 3,5%/ for all 5 samples. At temperatures between 200-750°C a continuous weight loss is observed. A hardly distinguishable DTA effect at about 600°C and a corresponding TG step in sample 6 are caused by serpentine admixture.

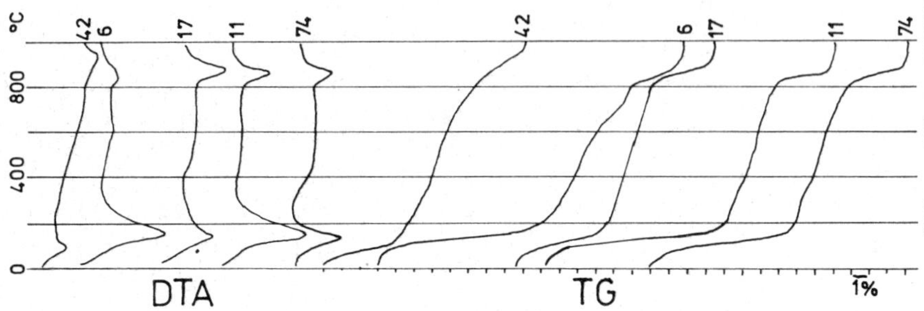

Fig. 10. DTA and TG curves of talc-like minerals

CHEMICAL ANALYSIS

Chemical analysis of talc-like minerals from Szklary are given in Table 1. Their compositions agree with those of kerolite-pimelites from the literature /Maksimovič 1966, Brindley and Pham Thi Hang 1973, Brindley at al. 1979/. Some corrections were made for the impurities present namely for quartz content in samples 17 and 42 determined by the addition method /0.48 and 8.9%, respectively/ and for serpentine contamination of sample 6 of 15,4% derived from the TG curve /Fig. 10/. In the case of sample 11 it was assumed that the total Al_2O_3 content sterns from chlorite with the composition of the sample S4 /Wiewióra 1978/; after this modification, the chemical composition given is probably still not entively correct because of admixture of tremolite.

The structural formulae /Table 1/ were calculated from corrected analysis on the basis of total charge +22. To satisfy four tetrahedral positions, Si was supplemented by Al and Fe^{+3}. Magnesium and nickel were shown to be the principal octahedral cations. Calcium ions /if present/ were placed in the interlayer space. The numbers of tetrahedral and octahedral cations slightly deviate from those for a talc-like composition. Only sample 17 shows a significant excess of octahedral cations, perhaps because of contamination by poorly crystalline serpentine and/or amorphous impurities.

Table 1
Chemical composition of talc-like minerals from Szklary

	74	11	11[x]	17	17[x]	6	6[x]	42	42[x]
SiO_2	59.57	56.56	58.14	53.41	53.25	51.42	52.97	53.71	59.57
TiO_2	0.02	0.02	0.01	-		-		-	
Al_2O_3	0.06	1.00		0.47	0.47	0.68	0.80	1.37	1.50
Fe_2O_3	0.78	1.53	1.49	tr		tr		tr	
FeO	tr	0.37	0.30	-		-		-	
MgO	27.80	27.79	27.63	19.80	19.92	20.30	16.07	6.75	7.40
NiO	4.82	5.26	5.56	17.99	18.10	18.56	21.97	29.75	32.61
MnO	tr	0.01	0.01	-		-		-	
CaO	0.85	0,55	0,58	-		-		0.25	0.27
Na_2O	0.09	0.13	0.14	0.08	0.08	-		0.01	0.01
K_2O	0.08	0.18	0.19	-		-		-	
Ign.loss	6.05	6.16	5.82	8.08	8.13	8.89	8.16	8.28	9.08
Total	100.12	99.58		99.83		99.85		100.12	
Si	3.94		3.88		3.89		3.91		3.90
Al^{IV}					0.04		0.07		0.10
Al^{VI}									0,04
Fe^{3+}	0.04		0.08						
Fe^{2+}			0.02						
Mg	2.74		2.76		2.17		1.77		0.88
Ni	0.26		0.30		1.06		1.30		2.08
Ca	0.06		0.04						0.01
Σ tetr Si,Al^{IV} Fe^{3+}	3.98		3.96		3.93		3.98		4.00
Σ oct Mg Ni,Al^{IV}, Fe^{2+}	3.00		3.08		3.23		3.07		3.00
CEC meq/ /100g	1.46	n.d.		29.7		32.4		21.4	
charge from CEC	0.086				0.131		0.144		0,102

[x] corrected analyses /explanation in text/ an.:K.Hnatyszak

IR INVESTIGATIONS

The infrared spectra of talc-like minerals from Szklary were recorded using a Beckman IR 4240 spectrometr and pressed KBr discs. Special attention was given to the hydroxyl stretching region /3400 - 3800 cm^{-1}/. Permanent presence of the three bands: 3670-3680 cm^{-1}, about 3640 cm^{-1} and 3620-3630 cm^{-1} was observed /Fig. 11/. By analogy of the IR spectra of Ni-talcs /Wilkins and Ito, 1967/ these bands were attributed respectively to the Mg_3OH, $MgNi_2OH$ and Ni_3OH arrangements in the octahedral sheet. The presence of the Mg_2NiOH arrangement was observed only in sample 42 which has the highest Ni content of all the studied samples /Table 1/. This band is absent in other samples like in the talc-like minerals described by Brindley et al. /1979/. Relative intensites of the Mg_3OH, Mg_2NiOH, $MgNi_2OH$ and of Ni_3OH bands calculated on the basis of chemical data /Table 1/ using Vedder's /1964/ expressions[x] are given in Table 2

Table 2

Intensities of OH-stretching bands calculated from Vedder's /1964/ expressions.

Cation arrangement		Mg_3OH	Mg_2NiOH	$MgNi_2OH$	Ni_3OH
sample	42	1.0	7.0	16.3	12.7
	6	1.0	2.2	1.6	0.4
	17	1.0	1.0	0.4	0.1
	11	1.0	0.3	0.04	0.001
	74	1.0	0.3	0.03	0.001

x/ $1 : \frac{3x}{1-x} : 3\left(\frac{x}{1-x}\right)^2 : \left(\frac{x}{1-x}\right)^3$ where x denotes Ni ions and 1-x denotes Mg ions

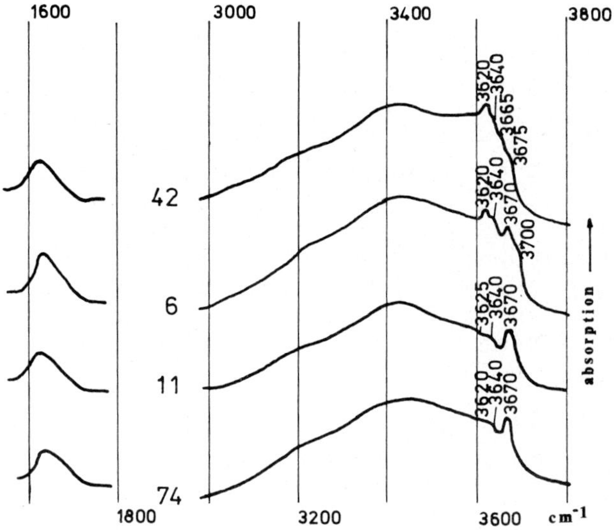

Fig. 11. IR spectra of talc-like minerals, OH stretching region, KBr discs.

These expressions correspond to random distribution of the cations in the octahedral sites. The results of calculations, as compared with the observed relative intensities of bands /Fig. 11/ confirm the inhomogenous distribution of nickel and magnesium in the octahedral sheet of talc-like minerals. Shoulders at about 3690 cm^{-1} /sample 17/ and 3700 cm^{-1} /sample 6 and 17/ were present in spectra of KBr pellets /Fig. 11/. Similar shoulders were attributed by Brindley et al. /1977/ to vibrations of surface OH groups. Chemical, thermal and X-ray data of the Szklary samples indicate however, that in samples from our collection these bands may be assigned to the admixture of serpentine and /or amorphous material.

CONCLUSIONS

Talc-like minerals from Szklary demonstrate their ability to form anhydrous layers and layers with single, double or triple water sheets. Two types of interstratification can be observed based on the response to glycol solvatation. Minerals genetically related to sepiolite containing veins show disordered mixed-layering and zonal structures whereas talc-like minerals formed indirectly from the primary components of the ultramafic rocks have more complicated structures, sometimes with g=2. They show an extreme lability of its water complexes depending on the relative humidity of the environment, similar to those observed in stevensites.

In some samples the expandibility can be increased by means of mechanical agents like ultrasonics. Therefore it is difficult to decide whether the expanding layers have a stevensite - or a talc-like nature. There are two main schools of thought regarding the expandibility of talc-like minerals . The first considers expanding layers to be of the steversite type /D'yakonov, 1963/ the second regards the expandibility as the result of a decreased interlayer bonding in the kerolite-pimelite group as compared to that in talc /Brindley et al.1977, 1979/. Our experiment with ultrasonic vibration confirms the conclusion of Brindley et al. /op.cit/, that the swelling of talc-like minerals differs from those of other layer silicates, but we cannot decide the reason for expansion in each particular case.

ACKNOWLEDGMENTS

The authors express their thanks to Dr G.W.Brindley for improving the style of the manuscript and to Mr A.Roćhewicz for scanning micrographs. This study was carried out within the framework of project M.R.I.16.

REFERENCES

Ahlrichs J.L., Serna C.,Serratosa J.M.,1975.Structural hydroxyls in sepiolites, Clays Clay Miner.23,119-124.

Brindley G.W.,Bish L.D,Wan H.M., 1979. Compositions, structures, and properties of nickel-containing minerals in the kerolite-pimelite series, Am.Miner.64, 615-625.

Brindley G.W.,Bish D.L., Wan H.M.,1977. The nature of kerolite, its relation to talc and stevensite.Miner. Mag.41,443-452.

Brindley G.W., Pham Thi Hang 1973. The nature of garnierites. I.Structures, chemical compositions and colour characteristics. Clays Clay Miner. 21, 27-40.

Dritz V.A.,Sakharow B.A.,1976. X-ray structural analysis of mixedlayer minerals. Trans.Acad.Sci.USSR,295, Nauka, /in Russian/.

D'yakonov Y.C. 1963. The results of an X-ray study of kerolites.Dokl.Akad.Nauk SSSR,148,909-911 /in Russian/.

Faust G.T.1966. The hydrous nickel-magnesium silicates - the garnierite group. Am.Miner. 51, 279-297.

Imai N.,Otsuka R.,Nakamura T.,Tsunashima A., Sakamoto T. 1973. "Hydrated talc" - an alternation product of wollastonite by reaction with magnesium-bearing hydrothermal solution.Clay Sci. 4, 173 - 189.

MacEwan D.M.C., 1958. Fourier transform methods.II. Calculation of difraction effects for different types of interstratification. Kolloid z.156, 61 - 67.

Maksimovič Z.,1966. β-kerolite-pimelite series from Goleš Mountain. Yugoslavia.Proc.Int.Clay Conf. Jerusalem, I, 97 - 105.

Olphen H. van 1977. An Introduction to Clay Colloid Chemistry, 2nd ed. J.Wiley & Sons.

Ostrowicki B., 1965. Minerały niklu strefy wietrzenia serpentynitów w Szklarach /Dolny Śląsk/. Prace Mineral. PAN 1.

Serna C., Vanscoyoc G.E., Ahlrichs J.L. 1977. Hydroxyl groups and water in palygorskite. Am.Miner.62,784-792.

Slansky F. 1955. Prispevek k poznani hydrosilikaty niklu z Kremže w Jiznich Cechach.Acta Univer.Carolinae Geologica 1. 1-28.

Suzuki S., Nishido H., Otsuka R., 1976. Hydrated talc from Chichijama, the Ogasawara Islands, Journ.Japan. Assoc.Min.Petr.Econ.Geol. 71, 238-247.

Spangenberg K. 1938. Die hydroxydischen nickel - und magnesiasilikatmineralen. Naturwiss. 26, 578-579.

Vedder W. 1964. Correlations between infrared spectrum and chemical composition of mica.Am.Miner.49, 736-768.

Wiewióra A., 1978. Ni-containing mixed-layer silicates from Szklary, Lower Silesia, Poland.Bull.BRGM, sect. 2, 247-261.

Wilkins R.W.T., Ito J., 1967. Infrared spectra of some synthetic talcs. Am Miner. 52, 1649-1661.

Yokomizo H., Moriyama Z., 1976. Stevensite from the Kiura Mine, Oita Prefecture /in Japan/, J. Japan Assoc.Min.Petr.Econ.Geol. 71, 221-228.

CLAY-SALT INTERACTIONS

L. HELLER-KALLAI, DEPARTMENT OF GEOLOGY, THE HEBREW UNIVERSITY, JERUSALEM

It is my impossible task to introduce the entire subject of clay-salt interactions, ranging from reactions in dilute solutions to those in the solid state at elevated temperatures, in fifteen minutes. In this brief period I cannot attempt to review the field, but can, at best, hope to point out a few areas of active research.

The apparently simple system "clay-salt" is highly complex, involving many variables: the clay - its type, composition, charge distribution, exchangeable cations, surface area and particle size, the salt - cation and anion, the proportion of the reagents and the experimental conditions, every detail of which may affect the course of the chemical and physico-chemical reactions and the nature of the products. The role of water in the system is very important and it is on this aspect that I shall primarily concentrate my talk. I shall not consider other solvents and shall confine myself to inorganic salts.

ION EXCHANGE REACTIONS AND ION SELECTIVITY

Many papers have been published on reactions of clays with dilute salt solutions, mostly on colloidal stability and on the subject of cation exchange, fewer on anion exchange, dealing with both theoretical aspects and practical applications. A comprehensive review of exchange reactions was recently presented by Bruggenwert and Kamphorst (1979).

Cation exchange capacity of a clay is not always a material constant; it may vary depending on the composition of the entire system, which must therefore be carefully defined. Three types of sites must be differentiated: in the interlayers, on flat external surfaces and on crystal edges. The proportion of these sites accessible to exchanging cations may vary in the course of a reaction due to changes in the size of the tactoids and to different particle interactions arising from changes in the acidity or ionic strength of the solution.

In recent years interest has been focussed on the selectivity of exchange reactions, with competing ions present either in trace amounts or as major components. Exchange reactions may be reversible or partly or completely irreversible. For reversible reactions thermodynamic constants can be established. Depending on the theoretical model adopted, different "selectivity coefficients" have been defined. At least six are in common use. The question of which model is most appropriate for any particular system and hence which selectivity constant can most reliably predict the course of an exchange reaction is still under discussion.

The electrical double layer theory predicts that the ratio of divalent to monovalent exchangeable counter ions is much higher in the double layer than in the equilibrium liquid. Hence, in the absence of chemisorption, the exchanging cation with the higher valency is preferentially adsorbed by the clay. For homovalent exchange the selectivity decreases with decreasing charge density of the clay. For heterovalent exchange higher charge density of the clay layers is expected to augment their preference for ions with higher valency. This was confirmed experimentally by Maes and Cremers (1979) using a series of reduced charge montmorillonites. The total electrolyte concentration of the solution also affects the selectivity coefficient for heterovalent exchange. Dilution results in greater preference for the ions with higher charge. This has been overlooked in some studies on clay-sea water interaction and has led to the common misconception that Mg^{2+} is concentrated in the interlayers of clays exposed to sea water, an artefact that was introduced on washing the clays, which caused preferential uptake of Mg^{2+} from the dilute solutions (Sayles and Mangelsdorf, 1977).

A very important cause of selective adsorption is the chemisorption energy of a cation on specific sites in the interlayers, on clay surfaces or on secondary sites occurring on adsorbed organic matter or oxides.

McBride (1979, 1980) contended that selectivity coefficients cannot be expected to be independent of the surface coverage of expanding clays, unless the surface mobility of the adsorbed ions is "solution-like". Site localisation of cations introduces entropy changes that are composition dependent. This leads to greater selective adsorption on homogeneous surfaces.

The concepts of the clay as a homogeneous or heterogeneous adsorber are not necessarily mutually exclusive. Depending on the reaction and on the relative number of different participating sites, either homogeneous or heterogeneous adsorption may predominate.

Although exchange reactions of clays with dilute salt solutions have received most attention, particularly from soil scientists, the effect of more concentrated solutions on selectivity is also of interest. High salt concentrations reduce the adsorbability of other ions, present in smaller amounts, on the clay. Plots of log D (the distribution coefficient for such ions) against log Na (the concentration of Na^+ ions in solution) are linear, with a negative slope for Cs^+ and alkaline earths, at least up to 4M Na^+ concentrations, but show a minimum for Eu (III) and are monotonous but nonlinear for Cd^{2+} and Co^{2+} (Shiao et al., 1979; Egozy, 1980).

It would, of course, be desirable to be able to interpret ion selectivity directly in terms of the known crystal structures of the minerals. The concept of "frayed edges", highly charged edge-interlayer sites, has been postulated to account for the selective sorption of K^+ and Cs^+ ions by weathered micas and vermiculites (for a review see Sawhney, 1972). Recently Eberl (1980) proposed a model to explain selectivity of exchange reactions of alkali ions in terms of the anionic field strength at the exchange sites and the interlayer water content. No comprehensive scheme has, however, been developed that will account for all aspects of cation exchange reactions. In view of the complexity of the system it is questionable whether this goal can be attained.

Cation exchange reactions are sensitive to variations in acidity. A small reduction in pH may convert a binary system, with two cations competing for the exchange sites, into a ternary one, with protons participating in the reactions. Such a decrease in pH may be caused by atmospheric CO_2 if the solutions are very dilute. A further decrease in pH may result in partial dissolution of the clay and sorption of various Al species in the interlayers. An increase in pH may lead to hydrolysis of interlayer cations. Charged hydroxides, in monomeric or polymeric form, may compete for the exchange positions and uncharged species may be incorporated into the clay interlayers. The effect of pH on clay-salt interactions has been very extensively investigated, both due to its importance in natural processes and to the potential industrial applications of the catalytic properties of hydroxy-clays.

Cation exchange of a clay composed of layers with different charges may lead to a change in their stacking (Lagaly, 1981). Rearrangement may also occur with uniformly charged layers, due to differences in hydration energy of the exchanging ions, e.g., in a Na/Ca montmorillonite some interlayers contained only Na^+, others only Ca^{2+} ions (Keren and Shainberg, 1979). Depending on the system, regular interstratification of alternate layers or a domain structure can be theoretically predicted (Barrer and Klinowski, 1979).

REACTIONS WITH CONCENTRATED SALT SOLUTIONS AND SALT INTERCALATION

Addition of concentrated salt solutions to clay suspensions flocculates the clay and may lead to salt entrainment. The minimum salt concentration required to cause flocculation differs greatly, depending on the particular clay, size fraction and the pH (e.g. Arora and Coleman, 1979).

When 1:1 clay minerals are brought in contact with some concentrated salt solutions or ground with the "dry" solids, intercalation (or "intersalation") may occur. Halloysites intercalate inorganic salts of K, Rb and Cs directly, forming dipole associations. Other cations can be intercalated if the anions have strong tendencies to hydrogen bonding (Wada, 1959; Garrett and Walker, 1959; Carr et al., 1978). Potassium acetate and other organic salts intercalate directly into kaolinites from concentrated solutions (e.g. Wada, 1961; Weiss et al., 1966) or from dilute solutions in alcohol-water mixtures (Deeds et al., 1966). Most inorganic salts cannot be intercalated into kaolinites directly. Grinding with Na, K and Rb chlorides or with KBr leads to delamination of the kaolinite layers, followed by hydration (Miller and Oulton, 1972; Yariv, 1975a, b), but no intercalation. All types of kaolinites (Lim et al., 1981) and dickite (Yariv, private communication, 1981) will, however, intercalate CsCl on dry grinding with the clay. It has been suggested that in halloysite this salt is retained between the oxygen and hydroxyl sheets by water bridges, the water molecules acting as both proton donors and acceptors (Yariv and Shoval, 1976). By analogy this may also apply to kaolinite and dickite.

SOLID STATE REACTIONS AT ELEVATED TEMPERATURES

Dry grinding of montmorillonite with KBr at room temperature caused cation exchange (McBride and Mortland, 1974). Other reactions were brought about by heating mixtures of montmorillonite with KBr or other potassium or sodium halides. Alkali ions were incorporated into the clay and protons derived from structural hydroxyls combined with the anions to form hydrogen halides. Small interlayer cations initially present penetrated into the octahedral sheets and were retained. The reactions commenced at 300°C or less, but were intensified at higher temperatures. The dehydroxylation temperature of the clay was lowered by the presence of the salt (Heller-Kallai, 1976; Heller-Kallai and Rozenson, 1980). Lim and Jackson (1980) noted that 1:2 phyllosilicates fused with $LiNO_3$ differed from samples saturated with Li^+ and subsequently heated. No doubt reactions analogous to those observed with alkali halides occurred. A similar mechanism may account for the transformation of sericite into interstratified minerals on fusion with $LiNO_3$, reported by Tomito and Sudo (1971). The properties of the products of such reactions, in which the clay acts as a Brønsted acid donating protons from the interior of the structure, have been insufficiently investigated.

In contrast to montmorillonites, kaolinites did not show any structural changes on heating with alkali halides or other alkali salts before dehydroxylation of the clay commenced. Only then were alkali ions incorporated into the clay and the corresponding acids given off. The rate of the reactions is related to the solubility of the salts added and depends on the availability of water, but above the dehydroxylation temperature even trace amounts suffice to sustain it. The products are generally amorphous, but with highly soluble salts crystalline products were obtained, e.g. kaliophilite on reaction with KF (Heller-Kallai, 1977; Heller-Kallai and Frenkel, 1978).

The difference between the reactions of montmorillonite and kaolinite with alkali halides is that while alkali ions are incorporated into the montmorillonite structure at temperatures far below those at which dehydroxylation takes place, this does not occur with kaolinite. The phenomenon can probably be explained in terms of the different water content of the samples at the temperature required to induce the reaction. Montmorillonite retains some interlayer water at elevated temperatures , whereas the water necessary for the reaction with kaolinite is supplied only by the dehydroxylation of the clay itself. All dehydroxylated samples appear to be very reactive and water from the surrounding atmosphere suffices for the reaction.

Preliminary experiments showed that other clay minerals, e.g. chlorites, palygorskites and biotites also react with salts on heating in the solid state, but these reactions have not yet been studied in detail.

REFERENCES

Arora, H.S. and Coleman, N.T., 1979. The influence of electrolyte concentration on flocculation of clay suspensions. Soil Science, 127: 134-139.

Barrer, R.M. and Klinowski, J., 1979. Order-disorder model of cation exchange in silicates. Geochim. Cosmochim. Acta, 43: 755-766.

Bruggenwert, M.G.M. and Kamphorst, A., 1979. Survey of experimental information on cation exchange. In: G.H. Bolt (Editor), Soil Systems in Soil Chemistry. Elsevier, Amsterdam, pp. 141-203.

Carr, R.M., Chaikum, N. and Patterson, N., 1978. Intercalation of salts in halloysite. Clays Clay Miner., 26: 144-152.

Deeds, C.T., van Olphen, H. and Bradley, W.F., 1966. Intersalation and interlayer hydration of minerals of the kaolinite group. Proc. Int. Clay Conf. Jerusalem, 1: 295-296; 2: 183-199.

Eberl, D.D. 1980. Alkali cation selectivity and fixation by clay minerals. Clays Clay Miner., 28: 161-172.

Egozy, Y., 1980. Adsorption of cadmium and cobalt on montmorillonite as a function of solution composition. Clays Clay Miner., 28: 311-318.

Garrett, W.G. and Walker, G.F. 1959. The cation-exchange capacity of hydrated halloysite and the formation of halloysite-salt complexes. Clay Miner. Bull., 4: 75-80.

Heller-Kallai, L., 1976. Interaction of montmorillonite with alkali halides. Proc. Int. Clay Conf. Mexico, 1975: 361-372.

Heller-Kallai, L., 1978. Reactions of salts with kaolinite at elevated temperatures, I. Clay Miner., 13: 221-235.

Heller-Kallai, L. and Frenkel, M., 1979. Reactions of salts with kaolinite at elevated temperatures, II. Proc. Int. Clay Conf. Oxford, 1978: 629-637.

Heller-Kallai, L. and Rozenson, I., 1980. Dehydroxylation of dioctahedral phyllosilicates. Clays Clay Miner., 28: 355-368.

Keren, R. and Shainberg, I., 1979. Water vapor isotherms and heat of immersion of Na/Ca montmorillonite systems - I: mixed systems. Clays Clay Miner., 27: 145-151.

Lagaly, G., 1981. Characterization of clays by organic compounds. Clay Miner., 16: 1-21.

Lim, C.H. and Jackson, M.L., 1980. Reaction of 2:1 phyllosilicates with lithium salts. 29th Annual Minerals Conf., Waco, Texas (Abstracts): 64.

Lim, C.H., Jackson, M.L. and Higashi, T., 1981. Intercalation of soil clays with dimethylsulfoxide. Soil Sci.Soc. Amer., 45: 433-436.

Maes, A. and Cremers, A., 1979. Cation exchange in clay minerals: some recent developments. In: G.H. Bolt (Editor), Soil Chemistry. Elsevier, Amsterdam, pp. 205-232.

McBride, M.B. and Mortland, M.M., 1974. Cu II interactions with montmorillonite: evidence from physical methods. Soil Sci. Soc. Amer. Proc., 38: 408-415.

McBride, M.B., 1979. An interpretation of cation selectivity variations in $M^{+} - M^{+}$ exchange on clays. Clays Clay Miner., 27: 417-422.
McBride, M.B. 1980. Interpretation of the variability of selectivity coefficients for exchange between ions of unequal charge on smectites. Clays Clay Miner. 28: 255-261.
Miller, J.G. and Oulton, J.D., 1972. Some effects of grinding kaolinite with potassium bromide. Clays Clay Miner., 20: 389-390.
Sawhney, B.L., 1972. Selective sorption and fixation of cations by clay minerals: a review. Clays Clay Miner., 20: 93-100.
Sayles, F.L. and Mangelsdorf, P.C. Jr., 1977. The equilibration of clay minerals with seawater: exchange reactions. Geochim. Cosmochim. Acta, 41: 951-960.
Shiao, S.Y., Rafferty, P., Meyer, R.E. and Rogers, W.J., 1978. Ion exchange equilibria between montmorillonite and solutions of moderate to high ionic strength. ACS Symposium Monograph, 100: 297-324.
Tomito, K. and Sudo, T., 1971. Transformation of sericite into an interstratified mineral. Clays Clay Miner., 19: 263-270.
Wada, K., 1959. Oriented penetration of ionic compounds between the silicate layers of halloysite. Am. Miner., 44: 153-165.
Wada, K., 1961. Lattice expansion of kaolinite minerals by treatment with potassium acetate. Am. Miner., 46: 78-91.
Weiss, A., Thielepape, W. and Orth, H., 1966. New kaolinite interlamellar complexes. Proc. Int. Clay Conf. Jerusalem, 1: 277-293.
Yariv, Sh., 1975a. Some effects of grinding kaolinite with potassium bromide. Clays Clay Miner., 23: 80-82.
Yariv, Sh, 1975b. Infrared study of interaction between CsCl and kaolinite. J. Chem. Soc. Far. Trans. I, 75: 674-684.
Yariv, S. and Shoval, S., 1976. Interaction between alkali-halides and halloysite: IR study of the interaction between alkali-halides and hydrated halloysite. Clays Clay Miner., 24: 253-261.

THERMODYNAMIC STUDY OF CATION-EXCHANGE IN A VERMICULITE WITH K^+ IONS

SAEHR, D., LE DRED, R. and WEY, R.
Laboratoire de Chimie Minérale Générale, E.N.S.C.Mu.,
3, rue Alfred Werner, 68093 MULHOUSE CEDEX.

ABSTRACT

Aqueous ionic solutions of K^+ and Ca^{2+} were percolated at constant ionic strength, at 60°C, until equilibrium, over K- or Ca-vermiculite, in the form of 1-2 mm flakes. Chemical and X-ray analysis of the vermiculites obtained show that two successive cation-exchange reactions occur.

1. INTRODUCTION

Thermodynamic study of cation-exchange reactions with K^+ions on alumino-silicates as montmorillonite, zeolite, has been usually performed excepted in the case of vermiculite. Indeed, this cation-exchange is reputed irreversible and no reproducible and only a few papers are concerned with small size particles (Goulding and Talibudeen, 1980). However a vermiculite in the form of 1-2 mm flakes swollen with water has led to reproducible results for the $Na^+ \longleftrightarrow K^+$ ion-exchange (Le Dred et al., 1978). Therefore a second exchange was studied : the $Ca^{2+} \longleftrightarrow 2K^+$ ion-exchange.

2. PREPARATION OF VERMICULITES. EXPERIMENTAL METHODS.

Vermiculite was prepared from an interstratified mica-vermiculite mineral from Palabora (Transvaal, South Africa). This mineral occurs in the form of 1-2 mm yellow-brown flakes, of micaceous appearence. Its chemical composition (Le Dred and Wey, 1978), expressed in g per 100 g is : SiO_2 : 38.59; Al_2O_3: 9.69; TiO_2: 0.98; Cr_2O_3: 0.16; Fe_2O_3: 3.08; FeO: 1.89; MgO: 26.50 ; Na_2O: 0.14; K_2O: 4.02; H_2O: 13.88; F: 0.80.

This mineral was converted to vermiculite by alternating treatment with concentrated sodium chloride solution, containing hydrogen peroxide and treatment with distilled water, until complete removal of K^+ ions (Le Dred and Wey, 1978). The Na-vermiculite thus prepared contains 93 % of water and exhibits a seemingly homogeneous macroscopical swelling perpendicular to the planes of the sheets. The periodicity in that direction, derived from the water content, should be about 300 Å. However, no periodicity higher than 15 Å is actually observed by radiocristallography. Consequently the swollen vermiculite is heterogeneous : it consists of water-filled interflakes with crystallites exhibiting a 15 Å periodicity. Its cation-exchange capacity is 190 meq per 100 g of vermiculite, calcined at 1000°C.

From this swollen Na-vermiculite swollen K- and Ca-vermiculites were prepared by complete cation-exchange, with respectively KCl and $CaCl_2$ solutions, with $[K^+] = 100$ meq/dm^3, $[Ca^{2+}] = 66,6$ meq/dm^3.

In order to prepare diionic vermiculites corresponding to various equilibrium states, mixed aqueous solutions of K^+ and Ca^{2+} ions were passed through columns packed with swollen K-vermiculite, or with swollen Ca-vermiculite. The equivalent cation fractions of the K^+ ions of these solutions (K_s) are defined as :

$$K_s = [K^+]\,([K^+] + 2[Ca^{2+}])^{-1},$$

$[K^+]$ and $[Ca^{2+}]$ are expressed in meq/dm^3 ; $0 \leqslant K_s \leqslant 1$; ionic strength: I = 100 mmol/dm^3. The columns are kept at 60°C. When the equivalent fractions K_s of the percolated solutions became identical to those of the initial ones, the latter were replaced by distilled water. The percolation was performed until absence of Cl^- in the effluent solutions. The diionic vermiculites thus prepared were dried at 40°C and analysed. Chemical analysis allows calculation of the equivalent cationic fraction of K^+ ions in the diionic vermiculites (K_v) defined as :

$$K_v = [K^+]\,([K^+] + 2[Ca^{2+}])^{-1}$$

$[K^+]$ and $[Ca^{2+}]$ being expressed in meq per 100g of vermiculite, dried at 40°C.

X-ray analysis was performed on oriented flakes in air of 52 % relative humidity.

3. EXPERIMENTAL RESULTS

The direct and reverse equilibrium isotherms

$$K_v = f\,(K_s),\ \text{(Figure 1)}$$

are reproducible and have similar shapes, the reverse curve however does not pass through point (1,1), as a small quantity of Ca^{2+}ions is not exchanged. In addition, the two curves cannot be superimposed indicating hysteresis. A similar phenomenon was observed for other ion-exchanges such as $Na^+ \longleftrightarrow K^+$ (Le Dred et al., 1978) or with other silicates such as micas (Newman, 1970) or some tectosilicates (Barrer and Falconer, 1956).

Vermiculite has three d-spacings : 10, 15 and 25 Å (Wey et al., 1974 a and b) corresponding to three minerals :

K-vermiculite, Ca-vermiculite and a regular 1/1-type-interstratified-vermiculite. Depending on the equivalent fraction K_v, one or two vermiculites are observed (Figure 2) :

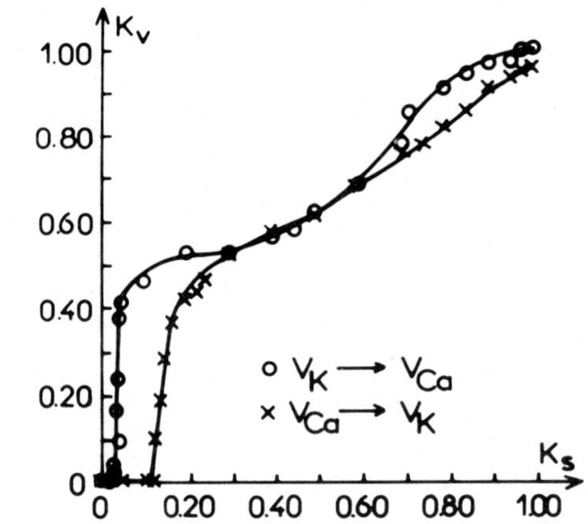

Fig. 1. $K^+ \longleftrightarrow Ca^{2+}$ exchange isotherms at 60°C: $K_v = f(K_s)$.

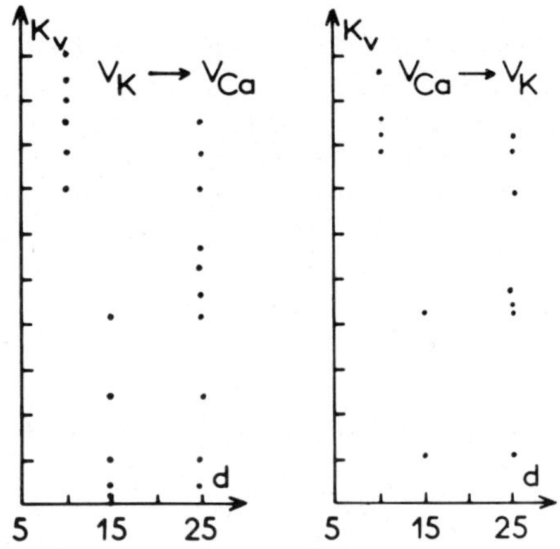

Fig. 2. $K^+ \longleftrightarrow Ca^{2+}$ exchange at 60°C: d-spacings as a function of the equivalent fraction K_v for exchange of K-vermiculite and of Ca-vermiculite.

For exchange of K-vermiculite :

$1 \geqslant K_v > 0.85$	K-vermiculite
$0.85 \geqslant K_v \geqslant 0.70$	K- and interstratified-vermiculite
$0.70 > K_v > 0.42$	interstratified-vermiculite
$0.42 \geqslant K_v > 0$	interstratified- and Ca-vermiculite
$K_v = 0$	Ca-vermiculite

For exchange of Ca-vermiculite, comparable results are obtained.

4. INTERPRETATION AND DISCUSSION

The results provide evidence for two heterogeneous systems in equilibrium, according to the domain of equivalent fractions K_v :

- system (1) for $1 \geqslant K_v > 0.42$
- system (2) for $0.42 \geqslant K_v \geqslant 0$

The almost vertical part of the direct and reverse curves and the coexistence of two well-defined vermiculites in the domain of equivalent fractions corresponding to system (2) show that this system is invariant : demixing takes place.

Although two vermiculites are present in a part of the equivalent fraction domain corresponding to system (1), the shape of direct and reverse curves are rather characteristic of a univariant system. In this case, cation-exchange is better represented by a solid solution than by demixing.

It appears that two well-defined equilibrated and reversible cation-exchange reactions take place :

$$(1)\ (K^+)_v + (0.29Ca^{2+})_s \rightleftharpoons (0.42K^+, 0.29Ca^{2+})_v + (0.58K^+)_s$$

$$(2)\ V\text{-}(K_{0.42}Ca_{0.29}) + (0.21Ca^{2+})_s \rightleftharpoons V\text{-}(Ca_{0.5}) + (0.42K^+)_s$$

v, V : vermiculite,
s : solution.

These two reactions may be considered to involve two types of cation-exchange sites in vermiculite, which would be formed by alternating sheets. Then, each isotherm $K_V = f(K_S)$ results from two curves corresponding to these two types of sites (Figures 3 and 4).

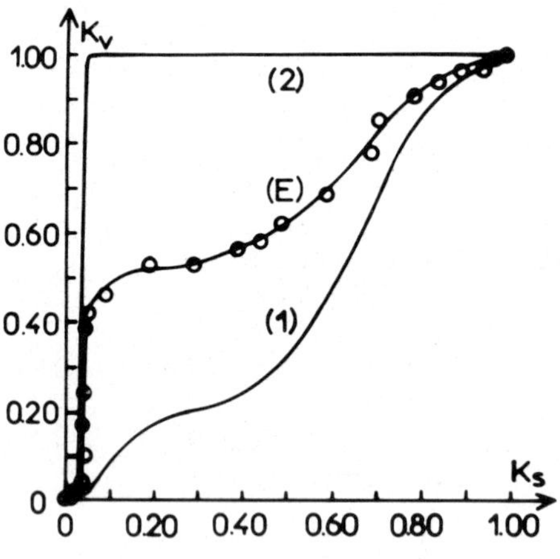

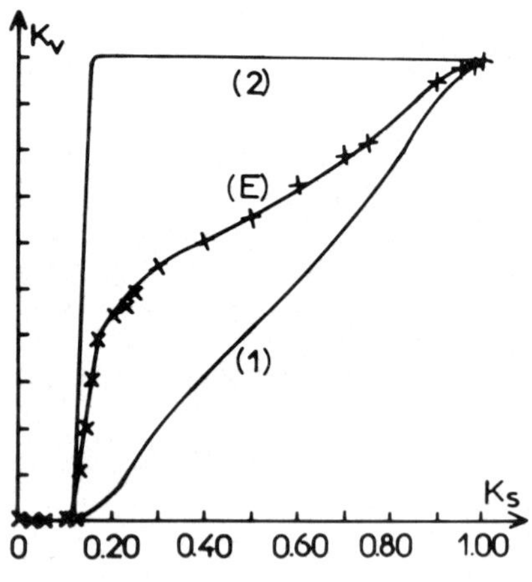

Fig. 3 Fig. 4
Analysis of the isotherm of Figure 1, in terms of two groups of sites :
for exchange of K-vermiculite for exchange of Ca-vermiculite

The resultant isotherm may be decomposed in a way similar to a procedure proposed for zeolites (Barrer et al., 1974). It is related to the component curves by

$$K_v = (1-x)\ K_{v1} + x\ K_{v2}$$

K_{v1}, K_{v2} being the equivalent fractions of K^+ ions in the sites of type (1) and (2),
and x, (1-x) the proportions of type (1) and type (2) sites : x = 0.42 for both exchanges.

Reactions (1) and (2) may be represented schematically by reaction (i) of standard Gibbs energy ΔG_i° :

$$\text{(i)}\ V_A \rightleftharpoons V_B \quad \text{with (i) = (1) or (2)}$$

The hysteresis reveals the existence of two subsidiary reactions, one for each direction, the Gibbs energies being ΔGa_i and $\Delta Ga_i'$, so that these curves correspond to the overall reactions, with ΔG_i and $\Delta G_i'$ Gibbs energies :

$$V_A \longrightarrow V_B, \quad \Delta G_i = \Delta G_i^\circ + \Delta Ga_i$$

$$V_A \longleftarrow V_B, \quad \Delta G_i' = -\Delta G_i^\circ + \Delta Ga_i'.$$

Assuming that the Gibbs energies of the subsidiary reactions are equal :

$$\Delta Ga_i = \Delta Ga_i' = \frac{1}{2}(\Delta G_i + \Delta G_i'),$$

$$\Delta G_i^\circ = \frac{1}{2}(\Delta G_i - \Delta G_i').$$

A general thermodynamic reasoning worked out by Gaines and Thomas (1953) allows calculation of the thermodynamic equilibrium constant k_1 corresponding to system (1) :

$$\log_e k_1 = -1 + \int_o^1 \log_e k_{c1}\ dK_v$$

$$k_{c1} = \frac{(1-K_{v1}) \cdot K_{s1}^2 \cdot 4I}{K_{v1}^2 \cdot (1-k_{s1})(3-K_{s1})} \cdot \frac{\gamma^2_K}{\gamma_{Ca}}$$

where γ_K and γ_{Ca} are the activity coefficients of K^+ and Ca^{2+} ions in aqueous solution (Nordstrom et al., 1979).

The value of the integral corresponds to the surface limited by the curve $\log_e k_{c1} = f(K_V)$ and the x axis. Hence :

$$\Delta G_1 = -\frac{RT}{2}\log_e k_1, \quad \text{with } R = 8.314\ \text{J K}^{-1}\text{eq}^{-1}$$
$$T = 333.2\ \text{K}$$

$$\Delta G_1 = -1{,}385 \log_e k_1\ \text{J eq}^{-1} \quad \text{for } I = 0.1\ \text{mol}^{-1}.$$

The constant k_1' may be calculated from the normalized Gaines and Thomas relation (Barrer et al., 1971) :

$$\log_e k_1' = -1 + \int_o^1 \log_e k_{c1}'\ dK_{v1}'$$

$$k'_{c1} = \frac{K'_{v1}(1-K_{s1})\cdot(3-K_{s1})}{(1-K'_{v1})\ \ K^2_{s1}\cdot 4I} \cdot \frac{\gamma_{Ca}}{\gamma^2_K}$$

$$K'_{v1} = \frac{K_{v1}}{K_{v1}max}$$

$K_{v1}max$: maximum value of the equivalent fraction K_v in the reverse exchange : 0.96.

A thermodynamic treatment of heterogeneous systems, in invariant equilibrium, like system (2) already has been proposed in the case of the exchange $Na^+ \longleftrightarrow K^+$ (Le Dred et al., 1978). Such a system exists only for a well-defined equivalent fraction K_s of the aqueous solution. The most probable equivalent fractions are the abscissae K_{s2} and K'_{s2}, corresponding to the points of ordinates :

$$K_{v2} = 0.50 \qquad K'_{v2} = 0.48$$

Then, constants k_2 and k'_2 are defined by the relations :

$$k_2 = \frac{[K^+]^2}{[Ca^{2+}]} \cdot \frac{\gamma^2_K}{\gamma_{Ca}} = \frac{K^2_{s2}\cdot 4I}{(1-K_{s2})(3-K_{s2})} \cdot \frac{\gamma^2_K}{\gamma_{Ca}}$$

$$k'_2 = \frac{(1-K'_{s2})(3-K'_{s2})}{K'^2_{s2}\cdot 4I} \cdot \frac{\gamma^2_{Ca}}{\gamma^2_K}.$$

All results are shown in Table 1.

Table 1. Thermodynamic equilibrium constants and Gibbs energies of the reactions

(i)	$\log_e k$	$\log_e k'$	ΔG_i	$\Delta G_i'$	ΔG_a	$\Delta G°$
(1)	-0.862	0.845	1,194	-1,170	12	1,182
(2)	-7.70	4.93	10,665	-6,828	1,919	8,747

Gibbs energies in J eq^{-1}

CONCLUSION

Swollen vermiculite has more affinity for K^+ ions than for Ca^2 ions. The cation-exchange goes through an intermediary state which is a regular 1/1-type-interstratified-vermiculite.

K-vermiculite has more affinity for Ca^2 ions than interstratified-vermiculite. The conversion of K-vermiculite into interstratified-vermiculite is, therefore, easier than the conversion of the latter into Ca-vermiculite. These observations might account for the natural occurence of interstratified mica-vermiculite minerals and the non-existence of vermiculite, with a few exceptions, in nature.

The subsidiary reactions are probalby initiation steps to the cation-exchange reactions.

REFERENCES

Barrer , R.M., 1974. Isomorphous replacement by ion exchange: some equilibria aspects. Bull.Soc.Fr.Mineral.Cristallogr., 97 : 89-100.

Barrer,R.M., and Falconer, J.D., 1956. Ion exchange in feldspathoids as a solid-state reaction. Proc.Roy.Soc., A 236 : 227-49.

Barrer, R.M. and Klinowski, J., 1972. Ion exchange involving several groups of homogeneous sites. J.Chem.Soc., Faraday Trans., 1 : 68 : 73-87.

Gaines, G.L. and Thomas, H.C., 1953. Adsorption studies on clay minerals. II. A formulation of the thermodynamics of exchange adsorption. J.Chem.Phys., 21 : 714-18.

Goulding, K.W.T. and Talibudeen, O., 1980. Heterogeneity of cation-exchange sites for potassium-calcium exchange in aluminosilicates. J.Colloid Interface Sci., 78 : 15-24.

Le Dred, R., Saehr, D. and Wey, R., 1978. Etude thermodynamique de l'échange d'ions $Na^+ \longleftrightarrow K^+$ dans une vermiculite. C.R.Acad.Sc.Paris, 286: 807-10.

Le Dred, R. and Wey R., 1978. Formation et applications de complexes mica-vermiculite-chlorure de sodium. Clay Min., 13: 177-86.

Newman, A.C.D., 1970. Cation exchange properties of micas. II. Hysteresis and irrversibility during potassium exchange. Clay Min., 8 : 267-72.

Nordstrom, D.K., Plummer, L.N., 17 cow., 1979, Chemical modeling in aqueous systems. A comparision of computerizd chemical models for equilibrium calculations in aqueous systems. A.C.S. symposium series, 93 : 857-91.

Wey, R., Saehr, D. and Le Dred, R., 1974 a. Mise en évidence des phases apparaissant lors de la transformation d'une vermiculite-Ca ou -Na en vermiculite-K. C. R. Acad. Sc. Paris, 278 : 2393-96.

Wey, R., Saehr, D. and Le Dred, R., 1974 b. Obtention d'un mineral interstratifié régulier à partir d'une vermiculite-K. C.R.Acad.Sc.Paris, 278 : 2493-5.

WATER ADSORPTION IN Cu AND Ca SMECTITES OF VARYING CHARGE

André MAES and Adrien CREMERS
Centrum voor Oppervlaktescheikunde en Colloïdale Scheikunde
Katholieke Universiteit Leuven, Belgium

ABSTRACT

Water desorption isotherms were measured on Cu and Ca exchanged montmorillonites with different layer charges. d001 sapcings increase with increasing charge density. The Gibbs energies of solvation or swelling, $\overline{\Delta G}_s$, increase linearly with the charge density of the clays. Extrapolation to zero charge allows correction of the $\overline{\Delta G}_s$ values for the interaction of water with the uncharged clay matrix and for the adsorption in micropores. The data are discussed in terms of the relative contributions of hydration of the exchangeable cations and of the anionic framework charges to the swelling of smectite clay minerals.

INTRODUCTION

Under conditions of free swelling in water the basal spacing of clay minerals exchanged with alkali or alkaline earth metal cations generally decreases with increasing charge density (Foster, 1953; Glaeser et al., 1971; Suquet et al., 1977 b; Maes et al., 1977). The variation with charge density is not continuous in minerals with tetrahedral isomorphic substitutions (Suquet et al., 1977 b). The changes in swelling are supposed to be mainly due to ion-solvent and/or framework-solvent interactions. In montmorillonites the apparent change in d001 is more or less continuous due to interstratification (Maes et al., 1977). Usually the same behaviour is observed for other polar liquids (Brindley et al., 1971; Suquet et al., 1977 b).

The amount of water or other polar solvents that adsorb at relative pressures below saturation increases with the charge density of the clay (Harward et al., 1966; Clementz et al., 1974; Stul et al., 1975; Annabi-Bergaya et al., 1979). The d001 spacing similarly increases with charge density in octahedrally substituted clays (Glaeser et al., 1971 & 1975; Clementz et al., 1974) but the reverse behavior is observed in synthetic beidellites with varying layer charge (Harward et al., 1966; Gillery, 1969).

We investigated the charge density dependence of the swelling in water of octahedrally charged smectites, including reduced charge montmorillonites. The relative contributions of the interactions of anionic sites and cations with water are discussed.

EXPERIMENTAL PROCEDURES

The fractions < 0.5 μm of Otay montmorillonite (OT), Camp Berteau montmorillonite (CB) and Hectorite (HEC) were obtained from their Na forms using conventional methods.

Two reduced charge montmorillonites, 0.75 RCM and 0.45 RCM were prepared from CB using a procedure based on the Hofmann-Klemen effect described previously (Maes et al., 1977). The code names refer to the fractional Na-cation exchange capacity of the original

Na-CB. The Cu and Ca exchanged clays were obtained by repeated ion exchange using 0.1 molar solutions of the corresponding nitrate salts. They were washed free of excess salts and freeze dried before storage and use.

Water desorption isotherms were determined as follows : samples of about 0.25 g were weighed in weighing bottles. They were first allowed to stand for one week over pure water and then equilibrated in dessicators containing sulfuric acid mixtures corresponding to p/p_o values 0.1, 0.2, 0.4, 0.6, 0.8 and 0.9. After 4 weeks of equilibration at 20°C, the samples were weighed and subsequently heated overnight at 110°C to determine their water content.

Capillaries filled with sample were equilibrated the same way. They were sealed before transfer to a Debye-Scherrer camera and to the R-X apparatus (Seifert-Scintag).

RESULTS AND DISCUSSION

The desorption isotherms of water on Ca and Cu loaded clay minerals of different charge density are shown in Fig. 1a,b as mg water/g ovendry (110°C) clay and in Fig. 1c,d as mol water/equivalent of exchangeable cation. The data are based on the assumption that heating at 110°C was sufficient for complete dehydration, and on the following cation exchange capacity (CEC) values in meq/g : 1.3(OT), 1.1(CB), 0.8(HEC), 0.85 (0.76 RCM), 0.55 (0.45 RCM). These CEC data were obtained by atomic absorption methods after acid breakdown of the Ca samples.

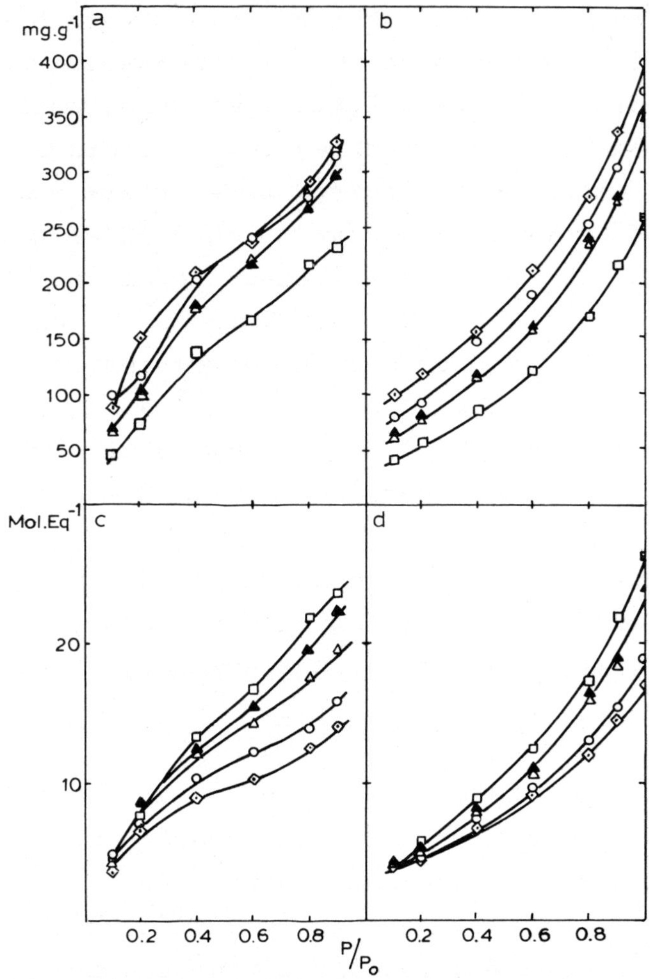

Fig. 1 Water desorption isotherms at 20°C for the Ca(a) and Cu(b) form of OT (◇), CB (o), 0.75 RCM (Δ), 0.45 RCM (□) and HEC (▲) expressed either as mg water/g clay (upper curves) or as mol water/ equivalent exchangeable cation (lower curves).

The influence of the charge density on the hydration is evident : the amount of water adsorbed per cation decreases with increasing charge density, whereas the total adsorption follows the reverse order. This demonstrates that the adsorption sinks for water are the charged-site centers which may be either the exchangeable cations alone or the cation-anionic site combination.

The foregoing results are corroborated by the d001 spacing measurements which are represented as a function of the relative humidity in Fig. 2a & b. The change in basal plane distances differs markedly with the type of ion (Cu or Ca) and with the charge density of the clay minerals.

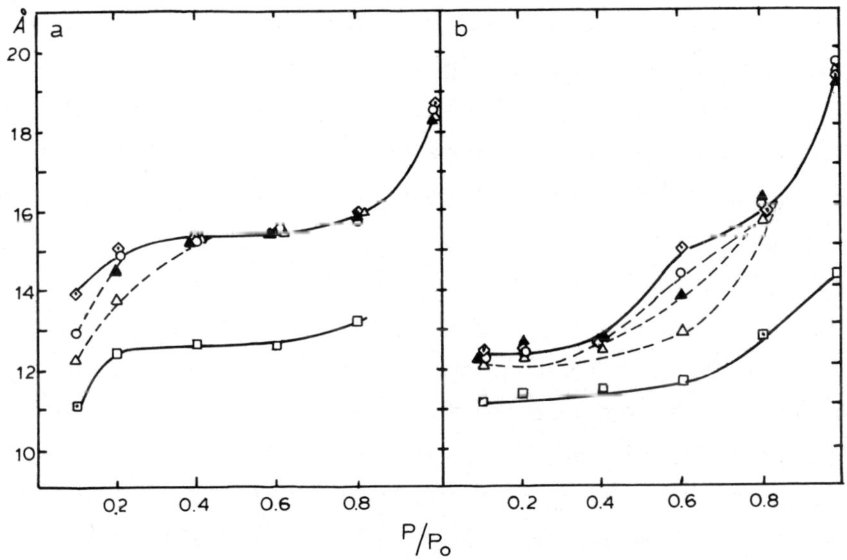

Fig.2 d001 spacings at different p/p_o values for the Ca(a) and Cu(b) form of clay minerals with various charge densities. Same symbols as in Fig.1.

Ca clays : In agreement with literature data (Mooney et al., 1952; Gillery, 1959; Keren et al., 1975) a plateau is observed at 15.4 Å for the samples with charge density > 0.8, indicating interaction with two layers of water. At lower charge density for the 0.45 RCM sample, a plateau at 12.6 Å is observed up to p/p_o = 0.6 which corresponds to interaction of only one water layer. The transition from the dry state to the two-layer hydrates occurs without the formation of a visible stability region for a one layer hydrate structure. The general pattern indicates the successive opening of the layers, which is controlled by the charge density of the minerals.

Cu clays : Two plateaus corresponding to one and two water layers are observed for the Cu clays. Below a charge corresponding to that of the 0.75 RCM sample the one layer hydrate appears between $p/p_o \simeq 0.1$ and 0.4. The two layer hydrate appears beyond $p/p_o \simeq 0.5$ and is interstratified with a three layer hydrate structure A plateau tends to develop with the highest charge densities in the region of p/p_o 0.6 to 0.8.

The Cu clays swell less readily than the Ca clays as evidenced by the stability range of the one layer hydrate which extends up to p/p_o values of ≃ 0.4 for the Cu clays, but which is absent in the Ca clays.

That Cu clays are more reluctant to swell than Ca clays is confirmed by the occurrence of a 11.5 Å spacing for the Cu 0.45 RCM sample in a range of p/p_o values of 0.2 to 0.6, in agreement with the observation of Clementz et al. (1974).

The presence of two plateau values, one at 11.6 Å for the Cu 0.45 RCM smaple and one at 12.4 Å for all other Cu clays indicates that two different configurations are possible for the one layer hydrate depending on the charge density. The reasons for the appearance of the 11.6 Å spacing are not obvious but might be due to the fact that the fraction of tetrahedral charge to the total charge becomes relatively high in the 0.45 RCM sample, leading in turn to a behaviour corresponding to tetrahedrally substituted clays. Glaeser et al. (1975) indeed observed that a 20 % degree of tetrahedral substitution is sufficient to confer the characteristic of complete tetrahedral charge origin to the mineral. Suquet et al. (1977 a) observed a 11.6 Å spacing in Ca saponite at p/p_o values smaller than 0.1.

The general increase in spacing observed with increasing charge density of the clay is especially apparent in the transition region between the one- and two-layer hydrates for both Cu and Ca clays. This trend is also shown by data for an independently prepared series given in Table 1, which shows d001 spacings and Ca CEC values at $p/p_o = 0.1$ for Ca clay, and $p/p_o = 0.6$ for Cu clay. The RCM samples in this series were however expanded for 24 hours over NH_3 vapour (Farmer et al., 1967) prior to exchange with Cu or Ca.

It is generally accepted that higher electrostatic interactions associated with higher charge densities lead to decreasing swelling with increasing charge density. This is verified when comparing montmorillonites with vermiculites e.g. (Norrish, 1954; Giese, 1978). Under conditions of free swelling in water both montmorillonites (Maes et al., 1977) and saponites (Suquet et al., 1977 b) of variable charge also behave in conformity with this rule. The present observations made in the region of crystalline swelling, however invalidate this general pattern of decreasing swelling with increasing charge density.

TABLE 1
Ca cation exchange capacity (meq/g) and d001 spacing (Å) for Ca and Cu exchanged montmorillonites at $\frac{p}{p_o}$ = 0.1 and 0.6 respectively

	CaCEC meq/g	d001 Ca($\frac{p}{p_o}$ = 0.1)	Cu($\frac{p}{p_o}$ = 0.6)
OT	1.24	13.88	15.45
CB	1.09	13.29	15.21
HEC	0.85	13.41	15.10
0.87 RCM	0.94	12.90	14.56
0.84 RCM	0.88	12.72	13.77
0.80 RCM	0.86	12.50	12.85
0.61 RCM	0.64	10.90	12.31

Gibbs energy of swelling

The Gibbs energy change upon sorption of water from the dry to the swollen state, can be obtained from the water adsorption data at different vapour pressures (Van Olphen, 1953).

$$\overline{\Delta G_s} = RT\ln\frac{p}{p_o} - RT\int_0^p \frac{n}{p}\,dp$$

which upon integration between p/p$_o$ = 0 and 1 leads to

$$\overline{\Delta G_s} = -RT\int_0^1 \frac{n}{p}\,dp$$

This value is called the Gibbs energy of swelling of the clay. It includes the total Gibbs energy change for the hydration of the exchangeable cations and the charged clay surface as well as possible energy changes involved in separating the clay layers.

The total Gibbs energies of swelling were determined graphically from plots of $\frac{np_o}{p}$ versus p/p$_o$. The data are shown in Fig. 3, expressed both in kJ/mol exchangeable cation and in kJ/cm^2, as a function of the Ca-CEC of the clay minerals. A surface area value at 750 m^2/g was assumed for all clay minerals. The uncertainty in the data is due to the scarcity of experimental points at low p/p_o values and is estimated to correspond to 14.5 kJ/mol or 1.25 10^{-9} kJ/cm^2 at most.

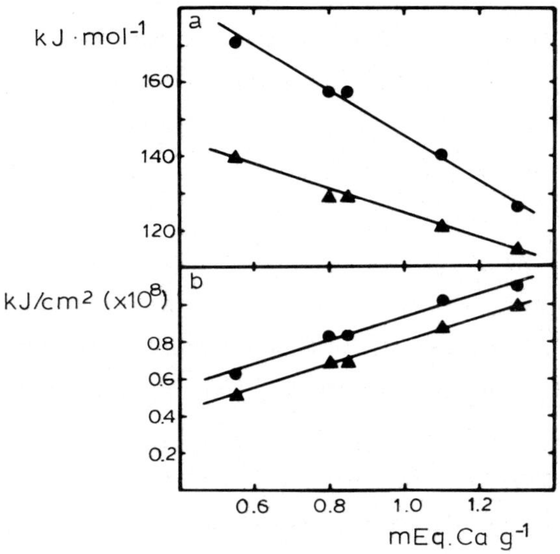

Fig. 3 Dependence of the Gibbs energy of swelling of Ca (●) and Cu (▲) clays, expressed a) as kJ/mol exchangeable cation and b) as kJ/cm^2 clay surface, on the cation exchange capacity of the clay minerals.

The figure shows that the Gibbs energy change per unit expanded surface area, increases linearly with the charge density of the mineral from about .5 10^{-8}kJ/cm^2 in the case of the Cu and Ca 0.45 RCM samples to about 1.10^{-8} kJ/cm^2 for Otay montmorillonite.

Extrapolation of the data to zero charge corresponds to the interaction of water with the (uncharged) clay mineral framework (external and internal), to adsorption in micropores and to interaction between water molecules themselves (Annabi et al., 1973), and yields energy values of 3.0 10^{-9} and 1.5 10^{-9} kJ/cm^2 for Ca and Cu clays respectively. These values correspond roughly to 30 % and 50 % of the total $\overline{\Delta G_s}$ change at high (OT) and low (0.45 RCM) charge density respectively. Annabi-Bergaya et al. (1979) using a different approach found that in the case of alcohol adsorption about half of the total adsorbed molecules are occluded in micropores.

One may correct the data for adsorption into micropores, the uncharged framework and other non-ionic adsorption sinks by simple subtraction of the extrapolated value at zero charge density. The slope of the corrected $\overline{\Delta G_s^i}$ - charge density relationship remains unchanged. Recalculation on a molar exchanger basis results in the $\overline{\Delta G_s^i}$ values given in table 2 which now remain practically independent of the charge density of the clay mineral. These $\overline{\Delta G_s^i}$ values are assigned to the interaction with water of the cation-anionic site combination in the interlamellar space only. The contribution of charges on external surfaces is negligible, due to the small external surface area, and is corroborated by the fact that identical external isotherms were obtained for methanol adsorption on a comparable series of RCM smaples (Annabi-Bergaya et al., 1979).

Role of anionic site hydration

The present results are not accurate enough in the low relative humidity region to allow a discussion on the relative importance of the hydration of the cations and the framework anionic charges in the initial process of swelling. However the overall behavior as obtained from integration up to saturation pressure may be relevant.

The Gibbs energy of interlamellar swelling ($\overline{\Delta G_S^i}$) may be viewed as resulting from two opposing forces (Norrish, 1954; Van Olphen, 1953; Mackenzie, 1964; Kittrick, 1959). Hydration of the exchangeable cation and the negatively charged clay matrix will tend to separate the layers, and is counteracted by the coulombic attraction between the cations and the anionic framework. The van der Waals attraction between the layers may be neglected with respect to these forces (Van Olphen, 1953). It is also assumed that in both one- and two layer hydrates of all the Ca and Cu clays, the cations are positioned midway between the layers. This model is comparable to the procedure used in thermodynamics of hydration of common salts (Gordon, 1975).

$$\Delta G_{S-S} = \Delta G_{I-S} - \Delta G_{el}$$

The Gibbs energy of salt-solvent interaction is the sum of the separate ion-solvent contributions of the anion and the cation (ΔG_{I-S}) and a Gibbs energy term for the crystal structure (ΔG_{el}).

In the clay water systems the term ΔG_{I-S} is the sum of the cation-hydration and the "clay-anion" hydration contribution. In order to evaluate the "clay-anion" hydration we must estimate ΔG_{S-S} and ΔG_{el} for the clay. The term ΔG_{S-S} is approximated by the experimentally obtained Gibbs energy of interlamellar swelling ($\overline{\Delta G_S^i}$). The ΔG_{el} term can be evaluated by calculating the Coulombic interaction energy of the bare cation with the negative charge centers. The cationic radii 0.82 (Cu) and 1.06 (Ca) were used together with a dielectric constant of 5 (Keren et al., 1975). Two possibilities are considered for the choice of the anionic radii in the "clay-anion".

a) A constant anionic radius for all charge densities, equal to 4.8 Å, which corresponds to a non-polarizable negative point charge residing in the center of the octahedral layer (Keren et al., 1975).
b) The more plausible alternative of considering an equivalent anionic radius (Eberl, 1980) $r_a = \frac{46.5}{8\pi C}$ in which C is the layer charge in equivalents per $O_{10}(OH)_2$ and which is calculated from the Ca CEC's and taking 750 m^2/g for the specific surface for all minerals. The anionic radii used and the electrostatic interaction terms are given in table 2.

TABLE 2

Experimental Gibbs energies of interlamellar swelling ($\overline{\Delta G}_S^i$), the Gibbs energy of electrostatic interaction ($\overline{\Delta G}_{el}$) calculated from a model, and the resulting Gibbs energy of ion-solvent interactions ($\overline{\Delta G}_{I-S}$), according to two suppositions of the equivalent anionic radius, r_a (Å), for a series of clay minerals with different octahedral charge density. All ΔG data in kJ/mol.

			supposition a			supposition b		
		$-\overline{\Delta G}_S^i$	r_a	$-\overline{\Delta G}_{el}$	$-\overline{\Delta G}_{I-S}$	r_a	$-\overline{\Delta G}_{el}$	$-\overline{\Delta G}_{I-S}$
Cu	OT	97.2	4.8	98.7	195.8	1.95	199.9	297.1
	CB	100.0	4.8	98.7	198.6	2.12	188.6	288.6
	HEC 0.75 RCM	99.8	4.8	98.7	198.4	2.49	167.5	267.3
	0.45 RCM	98.5	4.8	98.7	197.1	3.00	145.1	243.6
Ca	OT	92.3	4.8	94.5	186.8	1.95	184.0	276.3
	CB	99.6	4.8	94.5	194.0	2.12	174.4	273.9
	HEC 0.75 RCM	101.3	4.8	94.5	195.7	2.49	156.2	257.5
	0.45 RCM	90.0	4.8	94.5	184.5	3.00	136.6	226.6

Depending on the choice of r_a, the electrostatic interaction term $\overline{\Delta G}_{el}$, expressed per mole of charged site, remains constant (supposition a) or increases with the charge density (supposition b) and the resulting Gibbs energies of hydration $\overline{\Delta G}_{I-S}$ either remain practically constant or increase with charge density.

In the case of hydration of salts it is known that the contributions of the Gibbs energy of hydration of the anion and cation are independent and additive. Applying this principle to the clays we may assume that the hydration of the cation is identical at all charge densities.

a) If $\overline{\Delta G}_{I-S}$ is independent on charge density (supposition a), the hydration of the clay-polyanion will also be. The increase in swelling with charge density is then explained by the increasing number of sites per unit surface area.

b) If $\overline{\Delta G}_{I-S}$ increases with charge density (supposition b) the Gibbs energies of hydration of the polyanion will correspondingly increase. The importance of the clay anion hydration can be judged from the change with charge density and amounts to about 20 % (0.45 RMC) to 30 % (Otay) of the total $\overline{\Delta G}_{I-S}$.

Although supposition a) may not be ruled out a priori it seems logical to assume that in an isostructural series the increase in mean charge density corresponds to an overall increase of the field strength in the oxygen plane which can be described by a decreasing equivalent anionic radius (Eisenmann, 1962; Eberl, 1980). In this case the hydration of the anionic framework promotes the increase in swelling with charge density.

ACKNOWLEDGMENT

Financial support of the Fonds voor Kollektief en Fundamenteel onderzoek (FKFO) and the K.U.Leuven (Fonds Derde Cyclus) is acknowledged.

REFERENCES

Annabi-Bergaya F., Cruz M.I., Gatineau L. and Fripiat J.J., 1979. Adsorption of alcohols by smectites : I. Distinction between internal and external surfaces. Clay Miner., 14 : 249-257.

Annabi F., Glaeser R. et Mering J., 1973. Rétention des alcools par la montmorillonite. Bull. Groupe Franç. Argiles, 25 : 129-133.

Brindley G.W. and Gözen E., 1971. Preparation and solvation properties of some variable charge montmorillonites. Clays Clay Min., 19 : 399-404.

Clementz D.M., Mortland M.M. and Pinnavaia T.J., 1974. Properties of reduced charge montmorillonites : hydrated Cu(II) ions as a spectroscopic probe. Clays Clay Min., 22 : 49-57.

Eberl D., 1980. Alkali cation selectivity and fixation by clay minerals. Clays Clay Min., 28 : 161-172.

Eisenmann G., 1962. Cation selective glass electrodes and their mode of operation. Biophys. J., 2 : 259-323.

Farmer V.C. and Russell J.D., 1967. Infrared absorption spectrometry in clay studies. Clays Clay Min., 15 : 121-142.

Foster M.D., 1953. Geochemical studies of clay minerals. II. Relation between ionic substitution and swelling in montmorillonites. Am. Miner. 38 : 994-1006.

Giese R.F., 1978. The electrostatic interlayer forces of layer structure minerals. Clays Clay Min., 26 : 51-57.

Gillery F.H., 1959. Adsorption-desorption characteristics of synthetic montmorillonoids in humid atmospheres. Am. Miner., 44 : 806-818.

Glaeser R. and Mering J., 1971. Migration des cation Li dans les smectites di-octaedriques (effet Hofmann-Klemen). C.R. Acad. Sc. Paris, 273 : 2399-2402.

Glaeser R. and Mering J., 1975. Influence du taux de substitution isomorphique en couche tetraedrique sur les propriétés et l'organisation structurale des smectites dioctaedriques. Proc. Int. Clay Conf., Mexico : 173-183.

Gordon J.E., 1975. The organic chemistry of electrolyte solutions. John Wiley & Sons, New York.

Harward M.E. and Brindley G.W., 1966. Swelling properties of synthetic smectites in relation to lattice substitutions. Clays Clay Min., 13 : 209-222.

Keren R. and Shainberg I., 1975. Water vapor isotherms and heat of immersion of Na/Ca montmorillonite systems 1. Homoionic clay. Clays Clay MIn., 23 : 193-200.

Kittrick J.A., 1969. Quantitative evaluation of the Strong-Force model for expansion and contraction of vermiculite. Soil Sci. Soc. Amer. Proc., 33 : 222-225.

Mackenzie R.C., 1964. Hydratationseigenschaften von Montmorillonit. Ber. Dtsch. Keram. Ges., 41 : 696-708.

Maes A. and Cremers A., 1977. Charge density effects in ion exchange. Part 1. Heterovalent exchange equilibria. J.C.S., Faraday I, 73 : 1807-1814.

Mooney R.W., Keenan A.G. and Wood L.A., 1952. Adsorption of water vapor by montmorillonite. II. Effect of exchangeable ions and lattice swelling as measured by X-ray diffraction. J. Am. Chem. Soc., 74 : 1371-1374.

Norrish K., 1954. The swelling of montmorillonite. Disc. Faraday Soc., 19 : 120-134.

Stul M.S. and Uytterhoeven J.B., 1975. Interlamellar sorption of ethanol on montmorillonite clays with different layer charges. J.C.S., Faraday I, 71 : 1396-1401.

Suquet H., Prost R. et Pezerat H., 1977. (a) Etude par spectroscopie infrarouge de l'eau adsorbée par la saponite-calcium. Clay Miner., 12 : 113-126.

Suquet H., Iiyama J.T., Kodama H. and Pezerat H., 1977. (b) Syntheses and swelling properties of saponites with increasing layer charge. Clays Clay Min., 25 : 231-242.

Van Olphen H., 1953. Interlayer forces in bentonite. Clays Clay Min., 2 : 418-438.

CHARGE PROPERTIES OF SOME CLAYS FROM SOILS WITH VARIABLE CHARGE

L. MADRID, E. DIAZ, F. CABRERA, P. DE ARAMBARRI and J. BALBONTIN

Centro de Edafología y Biología Aplicada del Cuarto, Apartado 1052, Sevilla (Spain)

ABSTRACT

The pH-dependence of surface charge density of the clay fraction of five horizons of a variable-charge soil has been studied by adsorption of potential-determining ions (p.d.i.) and by adsorption of electrolyte ions. In samples with a high proportion of clay minerals (mainly illite) it is not possible to reach the point of zero charge, but samples with lower contents in clay minerals and a higher proportion of metal oxides show a well-defined PZC. Adsorption of p.d.i. by those samples with a low content of clay minerals can be satisfactorily described by a model which considers a three - plane distribution of the electric potential. Some deviations found at acid pH values can be explained by the presence of minerals with a permanent negative charge. Results for adsorption of electrolyte ions are also consistent with the presence of permanent charge.

INTRODUCTION

Constant surface charge, mainly due to the permanent charge of clay minerals, usually predominates in most soils from temperate regions, while in many soils from the tropics a pH-dependent charge prevails, so that the surface electric potential is largely, but not exclusively, determined by the concentration of H^+ or OH^- in solution. This electric potential is usually assumed to be determined by the Nernst relationship (Van Raij and Peech, 1972; Keng and Uehara, 1973; Laverdiere and Weaver, 1977; Parker et al., 1979; Uehara and Gillman, 1980). This assumption is not rigorous, as it implies that the activity of potential-determining ions (p.d.i.), H^+ and OH^-, on the surface is independent of the pH of the solution. A detailed description of the implications of the Nernst equation is given by Bowden et al. (1977).

The use of the Nernst equation and of erroneous versions of the Stern model make some modern studies of charge properties of soils with variable charge of questionable value (Madrid, 1980).

In this paper the model of three potential planes described by Barrow et al. (1980) is used to describe some charge properties of a soil from South Western Spain, which, unlike most mediterranean soils, shows variable charge properties. This model, which contains a number of parameters which are determined essentially by curve-fitting procedures, represents a realistic approach inasmuch as the simultaneous variation of all variables involved is considered, and its applicability has been demonstrated for various adsorbents and experimental conditions (Barrow et al., 1980; Madrid et al., 1981).

EXPERIMENTAL

Clays

The samples studied are clay fractions of five horizons of a ferritic brown earth (Eutrochrepts from Los Marines, South Western Spain). Some relevant properties of the soils are given in Table 1, and the approximate composition of the clay fractions are given in Table 2 (Sociedad Española de Ciencia del Suelo, 1966).

TABLE 1

Some properties of the soils

Sample	Horizon	Depth cm	pH KCl	SiO_2	Al_2O_3	Fe_2O_3	MgO	Clay
				per cent				
1	A_0A_1	0–30	6.0	27	29	25	2	8
2	A_1	30–60	6.3	23	31	26	2	7
3	IIAB	60–90	6.5	16	40	26	2	22
4	IIB_{11}	90–120	6.4	17	39	25	2	24
5	IIB_{12}	120–150	6.2	13	40	26	2	20

Adsorption of potential-determining ions

Volumes of 0.05N HCl or NaOH were added to polythene bottles containing suspensions of 0.1 g of clay in 20 ml of 0.01, 0.1 or 1M NaCl solutions, while flushing wet CO_2-free N_2 gas purified as described by Evans (1976) through the suspensions. Volumes of titrant were chosen so that pH readings ranged from 3.5 to 10.5 . The bottles were sealed air-tight while in N_2 atmosphere and after shaking for 1 hour pH was measured. Similar treatments were carried out with 20-ml samples of the appropriate NaCl solutions (blanks). Uptake of H^+ (or OH^-) by the clay was calculated from differences between titrant volumes added to test and blank samples at each pH. The value obtained was considered equal to the amount of positive (or negative) charges on the surface.

TABLE 2

Composition of the clay fractions

Sample	Illite	Goethite	Gibbsite	Amorphous	Other clay minerals
1	++	+++	++	+	None
2	++	+++	++	+	None
3	+	+++	+++	+	None
4	+	+++	+++	++	None
5	+	+++	+++	+++	None

+++ Abundant
++ Scarce
+ Present

The common point of intersection of curves obtained at different ionic strengths was considered the point of zero charge (PZC). The net surface charge density at each pH was calculated by subtracting from the uptake of H^+ or OH^- the value corresponding to the PZC, and dividing by the BET specific surface area determined by N_2 adsorption on the samples pretreated at 10^{-6} torr and room temperature for 24 hours. About 30 points were obtained for each curve of adsorption of p.d.i.

Adsorption of electrolyte ions

Adsorption of Na^+ and Cl^- ions was measured at three pH values (i.e. 4, 7 and 10) and two ionic strengths (i.e. 0.1M NaCl and 0.01 M NaCl) by the method of Van Raij and Peech (1972). When the electrolyte was 1M NaCl, the entrained solution caused large errors in the determination of adsorbed ions, so that no results are shown here for this ionic strength. All these experiments were done in duplicate.

DISCUSSION

Figures 1, 2 and 3 show the variation of the electric surface charge with pH for samples 3, 4 and 5. Full lines indicate measurements of adsorption of p.d.i. and crosses and circles those of electrolyte ions in 0.01 and 0.1M NaCl respectively. The results show a well-defined PZC (crossover point) at pH values which are in good agreement with PZC values of pure aluminium and iron oxides (Parks, 1965; Cabrera, 1975; Madrid et al., 1981). On the other hand, samples 1 and 2, with higher contents in SiO_2 and illite (Tables 1 and 2), gave curves of adsorption of p.d.i. which did not cross within the studied pH range (not shown). They seemed to converge towards very acid pH values, so that even though no PZC could be defined, both samples seem to have their PZC at pH below 2. This behaviour of samples 1 and 2 may be due to a higher proportion of minerals with a permanent charge, primarily illite.

When the set of equations given by Barrow et al. (1980) is solved for data of samples 3, 4 and 5, dashed lines of Figures 1, 2 and 3 are obtained. As can be seen, the basic side of the curves of adsorption of p.d.i. of these three samples can be fitted by the model, but important deviations are found on the acid side. As these samples also contain some illite, its permanent charge can have some influence on the results. This influence would be less noticeable when the variable-charge surfaces were also negative, but when these surfaces are uncharged or positively charged the heterogeneous distribution of charge would cause a definite deviation of the behaviour of the sample from that described by the model. The presence of permanent charge makes the estimated values of the parameters of the model less reliable than if only variable charges were present, as the parameters relate to an "average" surface. Therefore the information in Table 3 is only given for orientative purposes.

Results for K_H are about one order of magnitude lower than those obtained for pure iron oxides (Madrid et al., 1981). As silicate surfaces are known to show less affinity for H^+ ions than iron or aluminium oxide surfaces (Bowden, 1973), this result can be explained by the contribution of the clay minerals to the "average" surface. On the other hand, values of K_{OH} are of the same order of magnitude as those found for pure iron oxides (Madrid et al., 1981).

Results for ion adsorption (Figures 1, 2 and 3, crosses and circles) at pH 4 and 7 seem to indicate that the charge is more negative (or less positive) than that measured by adsorption of p.d.i. This could again be explained by the presence of permanent negative charge on sites of the illite not accesible to H^+ ions. Part of this permanent charge could become accesible to the ions in solution during the period of treatment and measurement, if the technique used could cause alteracion of the mineral. However, experiments at pH 10 do not show any difference between adsorption of p.d.i. and electrolyte ions. This observation could be explained by the higher stability of the mineral at alkaline pH values. It has been demonstrated that the illite contained in these samples is easily altered at acid pH (J.L. Pérez, personal communication, 1981).

TABLE 3

Estimated values of some parameters of the model of three planes of potential

Sample	R^2	$N_S \times 10^9$	$K_H \times 10^{-7}$	$K_{OH} \times 10^{-7}$	K_{Na}	K_{Cl}
3	0.98	1.7	4.00	10.08	0.40	0.30
4	0.98	2.0	0.11	1.83	0.04	0.80
5	0.97	4.3	0.13	4.35	0.01	0.30

R^2, fraction of the sum of squares accounted for by the model
N_S, maximum number of sites, eq x cm^{-2}
K_H, K_{OH}, K_{Na} and K_{Cl}, binding constant between each ionic species and the surface, l x mol^{-1}

In summary, the general trend of the variation of the electric surface charge with pH of these soil clays seems to be largely determined by their high content of minerals with variable-charge surfaces (iron and aluminium oxides), but the presence of minerals with a permanent charge should be considered when explaining the particular data, even for samples of those horizons where clay minerals are present in lower proportions.

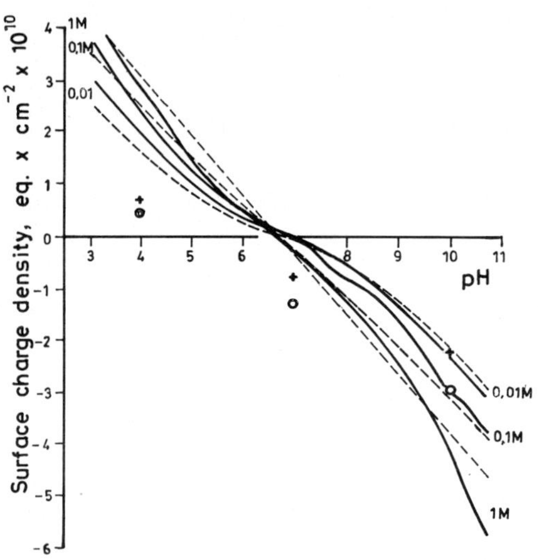

Fig. 1. Variation of surface charge of sample 3 with pH. See text for explanation of the symbols.

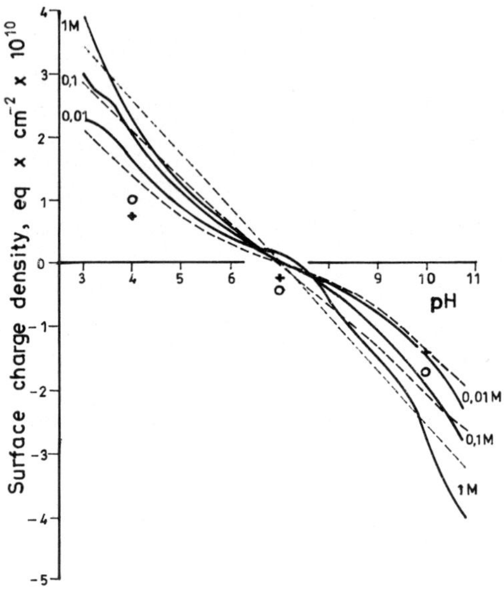

Fig. 2. Variation of surface charge of sample 4 with pH. See text for explanation of the symbols.

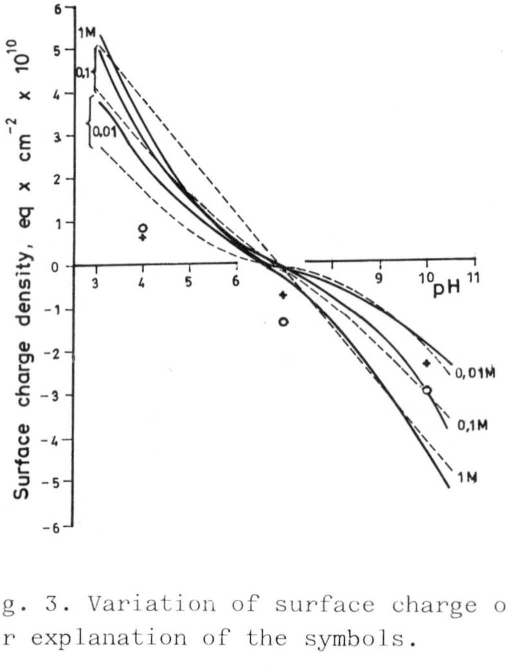

Fig. 3. Variation of surface charge of sample 5 with pH. See text for explanation of the symbols.

REFERENCES

Barrow, N.J., Bowden, J.W., Posner, A.M. and Quirk, J.P., 1980. An objective method for fitting models of ion adsorption on variable charge surfaces. Aust. J. Soil Res., 18: 37-47.

Bowden, J.W., 1973. Models for ion adsorption in mineral surfaces. Ph. D. Thesis, University of W. Australia.

Bowden, J.W., Posner, A.M. and Quirk, J.P., 1977. Ionic adsorption on variable charge mineral surfaces. Theoretical charge development and titration curves. Aust. J. Soil Res., 15:121-136.

Cabrera, F., 1975. Estudio de la adsorción isoterma de aniones ortofosfato por óxidos metálicos existentes en la Naturaleza. Tesis Doctoral, Universidad de Sevilla.

Evans, T.D., 1976. Interfacial electrochemistry of goethite (alpha -FeOOH). Ph: D. Thesis, University of Newcastle upon Tyne.

Keng, J. and Uehara, G., 1973. Chemistry, mineralogy and taxonomy of oxisols and ultisols. Soil Crop Sci. Soc. Fla. Proc., 33: 119-125.

Laverdière, M.R. and Weaver, R.M., 1977. Charge characteristics of spodic horizons. Soil Sci. Soc. Am. J., 41:505-510.

Madrid, L., 1980. Some errors found in the use of the Stern model for ion adsorption. J. Soil Sci., 31:709-711.

Madrid, L., Díaz, E., Cabrera, F. and Arambarri, P. de, 1981. Use of the modified Stern model to describe some surface charge properties of oxides. Abstracts, 7th Int. Clay Conf., Bologna, Pavia.

Parker, J.C., Zelazny, L.W., Sampath, S. and Harris, W.G., 1979. Critical evaluation of the extension of zero point of charge (ZPC) theory to soil system. Soil Sci. Soc. Am. J., 43:668-674.

Parks, G.A., 1965. The isoelectric points of solid oxides, solid hydroxides and aqueous hydroxo complex systems. Chem. Rev., 65: 177-198.

Sociedad Española de Ciencia del Suelo, 1966. Conferencia sobre suelos mediterráneos. Guía de la Excursión Española. Madrid.

Uehara, G. and Gillman, G.P., 1980. Charge characteristics of soils with variable and permanent charge minerals: I. Theory. Soil Sci. Soc. Am. J., 44:250-252.

Van Raij, B. and Peech, M., 1972. Electrochemical properties of some oxisols and alfisols of the tropics. Soil Sci. Soc. Am. Proc., 36: 587-593.

ELECTRON MICROSCOPY STUDY OF Na SMECTITE FABRIC - ROLE OF LAYER CHARGE, SALT CONCENTRATION AND SUCTION PARAMETERS

D. TESSIER and G. PEDRO
Station de Science du Sol - I.N.R.A.
78000 - Versailles (France)

ABSTRACT

Both scanning and transmission electron microscopy was applied to clay-water systems using Na-smectites with layer charges from 0.29 to 0.52 per half unit cell. The effects of salt concentration and of suction were studied. General conclusions are drawn on the smectite fabric and the location of water in clay-water systems.

INTRODUCTION

Swelling properties of clayey soils are closely related to the clay mineral type, nature of adsorbed ions and interstitial solution, and to external strains in particular those due to climatic conditions. The study of the relationship between swelling properties of clays and the organization of the solid phase of such clay water systems is essential for the understanding of soil behaviour.

The aim of this paper is to relate the Na smectite fabric with the behaviour of the clay water system.

I. DESCRIPTION OF METHODS

1. Sample preparation

The fractions smaller than 2 μm of five smectites with a layer charge from 0.29 to 0.52 were collected. These samples were saturated with Na cation and pastes were prepared by mixing the clay with 1 M NaCl and 10^{-3}M NaCl solutions.They were submitted to suction using an ultrafiltration cell as described by Tessier and Berrier (1979). Suction parameter σ is the decimal logarithm of the suction pressure in cm of water.

2. Electron microscopic observations

These were carried out with the S.E.M. Jeol JSM 35 with Cryoscan equipment and the T.E.M. Jeol 100 CX. Two methods of preparation of the samples were used :

- for the S.E.M., a method described by Tessier and Berrier (1979) was applied. A small fragment of the samples is frozen using freon 22 (which is cooled with liquid nitrogen) and in this condition, the sample is fractured. The fracture planes thus obtained are independant discontinuities in the samples. Then the sample is shadowed with gold vapor and directly observed at -100°C.

- for the T.E.M. a new method developed by Tessier and Clinard (1981), was applied. To avoid the destruction of the fabric that exists in the wet state, water is progressivly displaced by propylene oxide and subsequently by epoxy resin. After hardening microtomic sections of about 500 Å thickness can easily be made with a diamond knife, allowing observations in transmission lattice imaging.

II. RESULTS

Table 1 shows the water content of five smectites in 10^{-3} NaCl solution equilibrated at σ = 1.5. The water content decreases as the layer charge increases analogous to results of Foster (1955) who showed that the water content of montmorillonites can be related to octahedral substitution.

Plate 1 shows a S.E.M. picture of Na Greek Montmorillonite. The general organization of the system consists of a solid phase network with large pores with sizes of up to

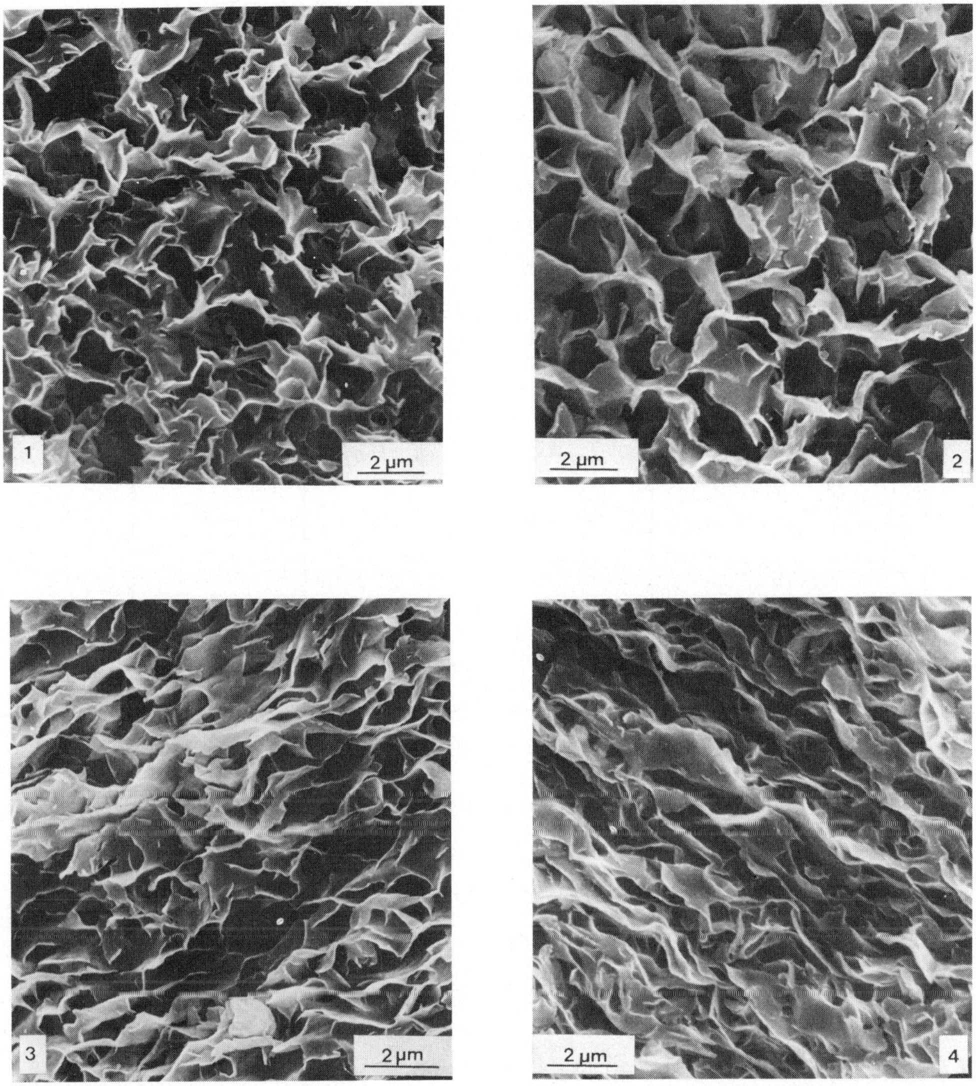

Plates 1,2,3 and 4 : Scanning Electron Microscopy micrographs for Na Greek montmorillonite. (1) 10^{-3}M NaCl, σ = 1.5 showing an isotropic network, (2) 1 M NaCl, σ = 1.5, (3) 10^{-3}M NaCl, σ = 3.0, (4) 1 M NaCl, σ = 3.0.

TABLE 1 - Water content of Na smectites prepared with 10^{-3}M NaCl and σ = 1.5 (3200 Pa - 32 mbar)

Smectite	Charge per half unit cell	Tetrahedral charge	Cation exchange capacity meq/100 g	Water content g/g
Hectorite	0.29	-	80	17
Belle Fourche *	0.35	0.10	92	12.5
Greek	0.40	0.08	108	9.3
Lorena *	0.44	0.18	127	5.1
Santa Rita*	0.52	0.01	132	5.0

* Samples from Schultz's collection (1969)

about 1.5 μm. The elements which make up the network are very thin and their orientation seems to be isotropic.

For the five clays studied, the fabric is similar and appears to depend mainly on the development of very large particles which are comparable to the "quasi crystals" described by Aylmore and Quirk (1971) for montmorillonite.

Plates 5, 6 and 7 show T.E.M. pictures of the internal organization of the quasi crystals of hectorite, Belle Fourche and Greek montmorillonites in 10^{-3}M NaCl solution, at σ = 1.5.

- In plate 5 for the hectorite sample we observe that th number of layers in the quasi crystals is about 5. This num ber may be considered as representative of the average number of layers for the hectorite sample prepared under the chosen conditions.

The contact between part of the quasi crystals is tangen tial while the orientation angle of the remaining parts of

Plate 5 - Lattice Imaging Transmission Electron Microscopy for Na Hectorite, 10^{-3}M NaCl, $\sigma = 1.5$

Plate 6 - L.I.T.E.M. for Na Belle Fourche montmorillonite 10^{-3}M NaCl, $\sigma = 1.5$ showing a thicker layer stacking the quasi crystals is about 30°. These data which agree wit S.A.S. observation by Pons (1980) indicate that there is no obvious edge-to-face bond in this system at the given conditions.

- T.E.M. pictures on the Belle Fourche sample show two kinds of quasi crystals. The major part of them (about 80 % has a layer number very comparable to that of the hectorite and the other part has a larger layer stacking, i.e. about 30 layers per quasi crystal (plate 6). Thus, this clay presents a bimodal particle size distribution.

Plate 7 shows a quasi crystal of Greek montmorillonite formed by stacking of a few layers. At the ends of the particle only a few layers are stacked and in the middle more than 20 layers. Examination of 30 microphotographs showed that on the average the quasi crystals of this smectite consist of 10 to 15 layers. This number of layers per particle agrees with S.A.S. data obtained by Pons et al (1981) who found an average number of layers per quasi crystal of 11, with a distribution from 1 to 21 layers.

Table II shows that for the five smectites in 10^{-3}M NaCl solution at a suction of σ = 1.5 the number of layers forming the quasi crystals increases as the water content decreases. From this number a theoretical external surface area was calculated assuming a total layer surface area of 780 m2/g. The table shows that there is a constant factor proportionality between the external surface area and the water content at σ = 1.5. This would suggest that the majo part of water is retained by the same mechanism for the fi clays.

2. Influence of salt concentration and suction

For the Na Greek montmorillonite prepared in 10^{-3}M and 1 M NaCl solution and a suction σ from 1 to 3, figure 1 sh that the water content of the clay decreases as the salt

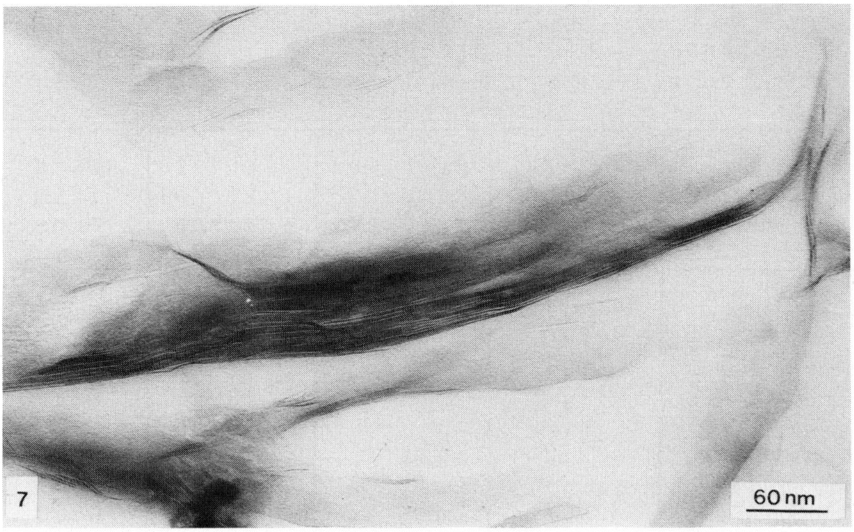

Plate 7 - L.I.T.E.M. for Na Greek montmorillonite, 10^{-3}M $\sigma = 1.5$

Plate 8 - L.I.T.E.M. for Na Greek montmorillonite, 1 M NaCl $\sigma = 1.5$

concentration and/or the suction increases. For example comparable water contents were obtained for (σ = 3 ; NaCl 10^{-3}M) and (σ = 1.5 ; NaCl 1 M).

TABLE II - Water film thickness referred to the calculated external surface for the five Na smectites ; 10^{-3}M NaCl - σ = 1.5

Smectite	Water content g/g (1)	Average number of layers	Calculated external surface area m2/g (2)	Water film tickness µm (3)
Hectorite	17.0	5	156	0.11
Belle Fourche	12.5	8	100	0.12
Greek	9.3	11	70	0.13
Lorena	5.1	20	40	0.13
Santa Rita	5.0	20	40	0.13

(3) = $^{(1)}/_{(2)}$ = water film thickness on external surface neglecting the interlayer water content

a) Effect of salt concentration

Plate 2 shows the S.E.M. picture for 1 M NaCl montmorillonite in 1 M NaCl solution at σ = 1.5. At this σ value, the size of the pores remains comparable to that of the same clay in a 10^{-3}M NaCl solution at the same σ and the network keeps its isotropic character. Plate 9 shows that there are 20 or more layers forming the quasi crystals. Therefore the number of layers increases with the concentration of the salt solution but the pore size on a µm scale is nearly constant.

b) Effect of suction

Plates 3 and 4 show the general fabric of the clays in 10^{-3}M and 1 M NaCl solution at σ = 3. For both samples the

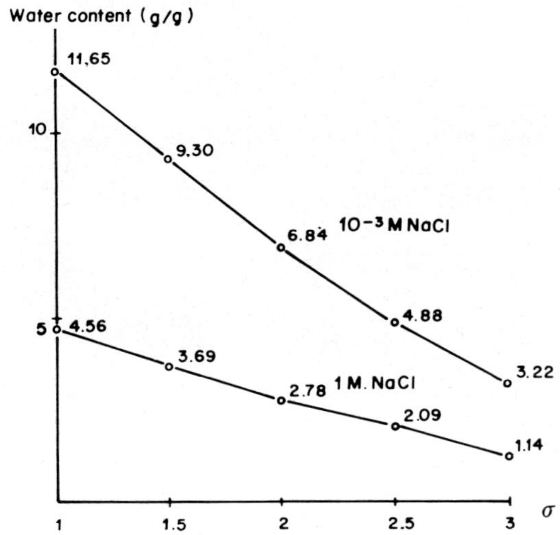

Fig. 1 - Water content as a function of σ for Na Greek montmorillonites in 10^{-3}M and 1 M NaCl solution

network becomes oriented. However the pore size in 1 M NaCl appears to be smaller than in 10^{-3}M NaCl solution.

Plate 8 shows the clay in 10^{-3}M NaCl at σ - 3. The number of layers in the quasi crystals is close to 15 and the dispersion of the number at layers becomes smaller than that in 10^{-3}M NaCl at σ = 1.5 (1 to 21 in this case).

We conclude that the decrease of water content in 10^{-3}M NaCl with increasing suction involves two mechanisms : first the pore size decreases slightly under the suction effect ; second, the layer number slightly increases and becomes more homogeneous due to regrouping of the thinnest particles.

III. DISCUSSION

Sample preparation for T.E.M. and S.E.M. involves some layer dehydration : freeze drying and substitution methods produce respectively a collapse of the layer distances to

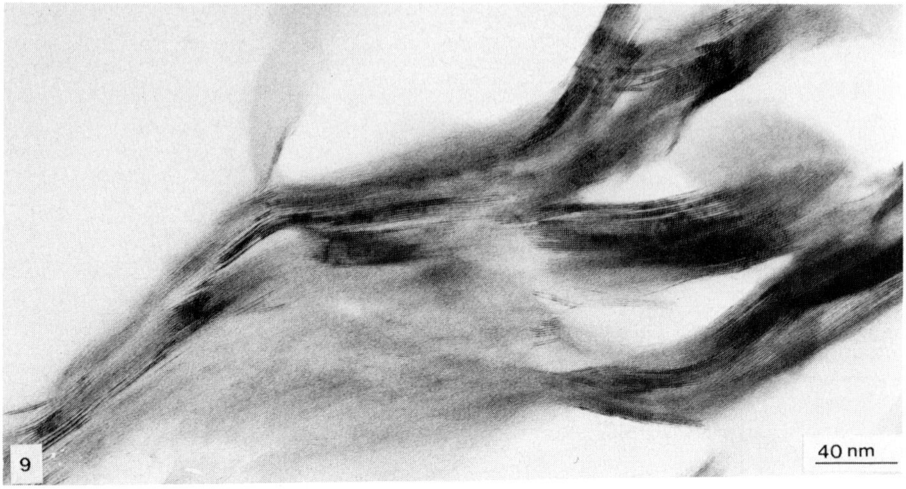

Plate 9 - L.I.T.E.M. for Na Greek montmorillonite, 1 M NaCl σ = 1.5

19 Å (Norrish and Rausell-Colom, 1962) and to 15.5 Å (Tessier, unpublished). S.E.M. results obtained by freeze drying and by critical point method (Tessier, unpublished) show comparable pore sizes for a montmorillonite in 1 M NaCl solution. As d(001) is about 19 Å in the wet state, the fabric may be considered to be representative for the clay water system.

For the clays in 10^{-3}M NaCl solution the freeze drying and displacement methods cause an increase in pore size on a μm scale compared with that for clay in 1 M NaCl. However data of Pons et al (1981) show that these methods do not affect the distribution of the number of layers in the quasi crystals.

*
* *

We can draw the following conclusions from our work :

(i). One of the main characteristics of the clay water systems in equilibrium at suction $\sigma < 3$ is the large inter quasicrystal pore size. This pore size sightly decreases and reaches about the same value at $\sigma = 3$ for all the samples studied. Thus, the pore size approaches a value estimated from Kelvin's law which is 1.4 µm at $\sigma = 3$. Consequently for different water contents obtained at a given suction, the smectite water systems mainly differ with respect to the number of pores.

(ii). At a given suction the amount of water in these clay water systems appears to be related to the external surface which depends on the average layer number for the quasicrystal. For Na smectites, the fundamental parameters are the layer charge (and probably its location) and the salt concentration.

We should mention, however, that suction may alter the external surface. This is the case when the diffuse double layer is not compressed, i.e. at low salt concentrations (cf. Pons et al, 1981).

REFERENCES

Aylmore L.A.G. and Quirk J.P., 1971. Domains and quasi crystalline regions in clay systems, Soil Sc. Soc. Amer. Proc. 35 : 652-654

Foster M.D., 1955. The relation between composition and swelling of clay minerals, ClaysClay Min. 3 : 205-220

Norrish and Rausell-Colom J.A., 1962. Effect of freezing on swelling of clay minerals, Clay Min. Bulletin, 9 : 15

Pons Ch.H., 1980. Mise en évidence des relations entre la texture et la structure dans les systèmes eau-smectites par diffusion aux petits angles du rayonnement synchrotron. Thèse Université d'Orléans, 175 p.

Pons Ch.H., Tessier D., Ben Rahiem H. and Tchoubar D.,1981. A comparison between X-ray studies and electron microscopy observations of smectite fabric (these proceedings)

Schultz L.G., 1969. Lithium and potassium absorption dehydroxylation temperature and structural water content of aluminous smectites, Clays Clay Min. 17 : 115-149

Tessier D. and Berrier J., 1979. Utilisation de la microscopie électronique à balayage dans l'étude des sols. Observation de sols humides soumis à différents pF, Science du Sol, 1 : 67-82

Tessier D. and Clinard C., 1981. Preparation method for lattice imaging observations of clays corresponding to a given wet state, Agronomie (to be published).

A COMPARISON BETWEEN X-RAY STUDIES AND ELECTRON MICROSCOPY OBSERVATIONS OF SMECTITE FABRIC

C.H. PONS*, D. TESSIER**, H. BEN RAHIEM*, D. TCHOUBAR*

*Laboratoire de Cristallographie, E.R.A. n°841
Université d'Orléans, 45046 Orléans Cedex (France)

**I.N.R.A., Station de Science du Sol,
78000 Versailles (France)

ABSTRACT.

This paper shows the value of using simultaneously small angle scattering (S.A.S.) and transmission electron microscopy (T.E.M.) to characterize quantitatively the micro and macrostructure of a water-clay system. These techniques were used to study the effects of salt concentration and of suction by air pressure on the structure of some clay-water samples prepared from a Greek montmorillonite.

INTRODUCTION.

The tridimensional organization of a water-clay system depends on the size of the elementary particles and their state of swelling and on their mutual arrangement which controls the pore structure and consequently the permeability of the system.

To describe the micro and macrostructure, it is essential to cover the full range of dimensions from Angström to micrometer. Two complementary methods are available to cover that range : electron microscopy, particularly transmission lattice imaging, and the small angle X-ray scattering technique. The latter method has been originally applied to the study of clay swelling by Norrish and Rausell Colom (1963) and later by Andrews et al. (1967) and others.

In this paper, both methods are reviewed and applied to the study of a paste of a Greek montmorillonite in salt solution which is submitted to a given hydrostatic suction. The particle organization in this system is described as a function of the suction parameter, the salt concentration, and the nature of the exchangeable cations.

DESCRIPTION OF THE METHODS.

Sample preparations.

The Greek montmorillonite has a layer charge of 0.4 per halt unit cell. The fraction of the clay particles smaller than 2 μm was first saturated with Ca or with Na. Pastes of the Na- and the Ca-clay were prepared with M or 10^{-3}M $NaC\ell$ or $CaC\ell_2$ solutions. The pastes were submitted to air pressure 3 200 Pa or 10^5 Pa. The corresponding values of the suction parameter P_F are respectively 1.5 and 3 ($P_F = \log p$ where p is expressed in cm water). The details of this suction method are given by Tessier and Berrier (1979).

A part of these pastes was used for small angle X-ray scattering measurements ; the other part was used to prepare thin sections for lattice imaging TEM observations.

Electron microscopy observations.

Observations were carried out with a microscope JEOL 100 CX on thin sections of about 500 Å thickness. To prepare these sections, a new method was used to avoid destruction of the structural organization of the paste (Tessier and Clinard, 1981). The liquid phase was first displaced by propylene oxide, which was in turn displaced by epoxy resin which stabilizes the sample structure by solidification, and allows the preparation of microtomic sections.

Small angle scattering experiments.

They were carried out with the X-rays emitted by the synchrotron storage ring D.C.I. at L.U.R.E., Orsay (France). The experimental conditions were described by Pons et al. (1981) ; we only recall that, with such an X-ray source, the diagrams are obtained without any instrumental deformations and can be directly compared with theoretical S.A.S. curves. We used computation models which are based on the notion that the particle are stackings of parallel layers. The internal structure of this particle is given by the degree of order in the succession of layers which is determined by a set of i translations $d_1, d_2 \ldots d_i$ and their respective probabilities $P_1, P_2 \ldots P_i$ (Pons et al. 1981). The small angle scattering intensity by extended layers is

$$I(s) = \frac{1}{2\pi\Omega}\,\frac{F_{00}^2(s)}{s^2}\left[\sum_M \alpha(M)G(S,M)\right] \qquad (1)$$

where

S is the scattering vector given by $2\sin\theta/\lambda$ when 2θ is the angle of scattering and λ the wave length.

Ω is the area of the plane unit cell.

$F_{00}^2(s)$ is the square of the layer structure factor along the 00 scattering rod (crossing the origin of the reciprocal space).

M is the number of layers for a particle or stack.

$\alpha(M)$ is the weight distribution of the number of layers for the set of stacks.

G(S,M) is the modulation function along the 00 rod which depends on the distribution $P_i d_i$.

To analyze the small angle scattering diagrams, we compare the shape of experimental curves with theoretical ones but it is necessary to consider two cases :

First case : The diagram shows a modulation at a relatively large angle, well separated from the central peak (fig.III.1). It correspond to the case where the strains due to the suction pressure of the salt concentration keep the elementary particles in a low state of hydration. The analysis of the diagram consists in comparing the experimental intensity curve with the theoretical one computed by expression (1).

Second case : The diagram shows a modulation very near the central peak. This situation corresponds to the case where the strains are weak and the particles are swollen. We can then be sure that the structure factor $F_{00}(S)$ does not become zero in the range of concern of S variation. The experimental modulation function can be determined from :

$$G_{exp} = \frac{I_{exp} \times 2\pi\Omega S^2}{F_{00}^2(S)} \qquad (2)$$

and directly compared with the term between brackets of the expression (1). This second method is more sensitive in showing the weak modulations in the S.A.S. curves (figure III.4).

EXPERIMENTAL RESULTS.

TABLE 1
List of samples and suction parameters applied.

N°	Nature of interlamellar cation	Salt concentration	Suction parameter P_F	Water contents after the application of the suction
1	Greek montmorillonite Ca	10^{-3}M $CaC\ell_2$	1.5	2.51 g of water per g of clay
2	Greek montmorillonite Ca	10^{-3}M $CaC\ell_2$	3	0.90
3	Greek montmorillonite Na	1 M $NaC\ell$	1.5	3.69
4	Greek montmorillonite Na	1 M $NaC\ell$	3	1.14
5	Greek montmorillonite Na	10^{-3}M $NaC\ell$	1.5	9.30
6	Greek montmorillonite Na	10^{-3}M $NaC\ell$	3	3.22

Three types of S.A.S. diagrams were obtained for the samples listed in Table I.

Figure III.1 shows the experimental scattering curves for samples 1 and 2 (black points and triangles) and the best fit with the theoretical curve (continuous line). From these curves an average number of layers per particle $M = 30$ is derived. The thickness distribution $\alpha(M)$ is A Gaussian between $M_{min} = 17$ and $M_{max} = 43$. The average distance between two adjacent layers is about 18.6 Å. The first neighbour distance distribution is indicated in figure III.1 by the diagram P_i versus d_i.

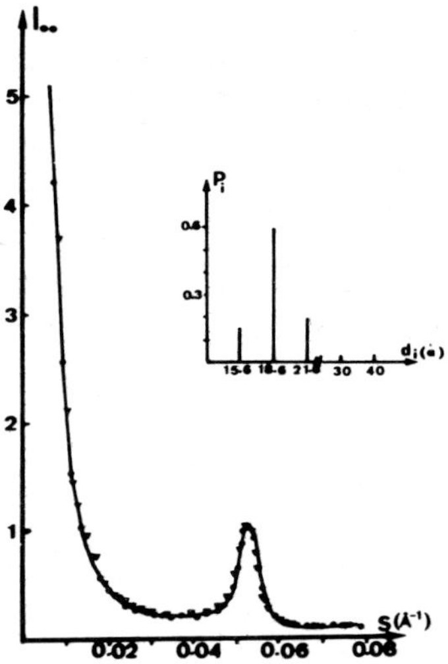

Fig.III.1. Experimental S.A.S. intensity from Greek montmorillonite-Ca with 10^{-3} Molar $CaC\ell_2$, $\sigma = 1.5$ (   ), 10^{-3} Molar $CaC\ell_2$, $\sigma = 3$ (   ) and theoretical curve (——).

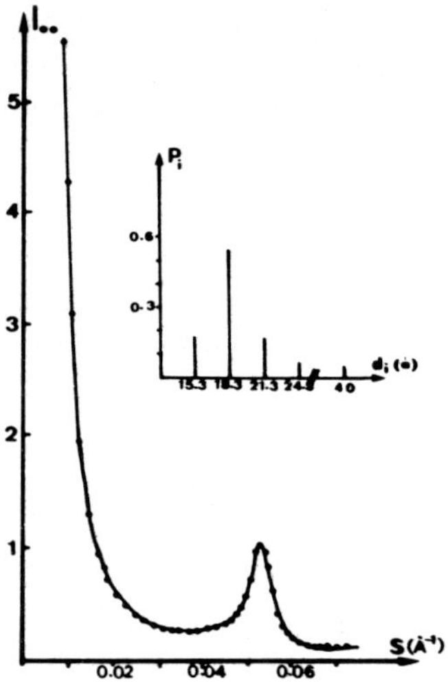

Fig.III.2. Experimental S.A.S. intensity from Na Greek montmorillonite with 1 M $NaC\ell$, $\sigma = 1.5$ and $\sigma = 3$ (   ) and theoretical curve (——).

From such parameter values, it is possible, to give a statistical picture of the particles by using a Monte Carlo method. It can be concluded that in these two samples, the elementary particles are disordered stacks of about 30 layers which include small ordered substacks of about 5 layers with a distance d corresponding to 3 monolayers of water molecules. These homogeneous substacks are separated from each other by distances which correspond to 2 or 4 monolayers of water or by some distances of about 30 and 40 Å.

Plate 1 shows these particles, for sample 1, which are organized as a porous system so that the average thickness of the pore walls is equal to the average thickness of the particles. One detects in these walls the homogeneous substacks derived from the S.A.S. curves. An important result in this case is that the method of section preparation for T.E.M. observation did not disturb the structure of the water clay system. The X-ray results show that the variation of the suction parameter P_F does not affect the particle organization in this sample. As shown by Tessier and Pédro (1981), the reduction of water contents by variation of the suction between $P_F = 1.5$ and $P_F = 3$ affects only the water in the pores. The walls of the pores are not changed.

Figure III.2 shows the S.A.S. diagram for the Na-samples 3 and 4 (1 M NaCℓ). At this concentration of NaCℓ, the behaviour of the systems is comparable to that of the Ca-samples, but the particle thicknesses are lower ($M_{min} = 7$, $M_{max} = 30$). And the distribution given in Figure III.2 shows a greater number of distances between 20 and 40 Å.

These characteristics are corroborated by the T.E.M. observations (plate II). As in the first cases, the suction strain modifies only the amount of water between particles.

Figure III.3 shows the S.A.S. diagram for the Na-clay with 10^{-3}M NaCℓ solution and $P_F = 3$ (sample 6). In this case, it was possible to determine the modulation function G_{exp} by using the expression (2), as illustrated in figure III.4 which shows the best fit with a computed

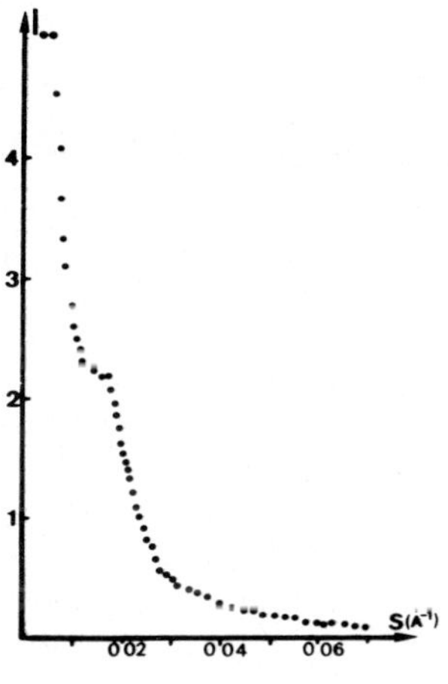

Fig.III.3. Experimental S.A.S. intensity from Na Greek montmorillonite with 10^{-3}M NaCℓ, $\sigma = 3$.

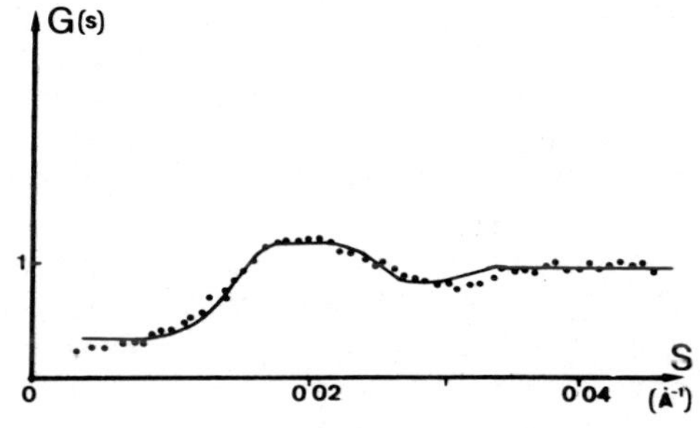

Fig.III.4. Experimental modulation function determined from the curve in fig.III.3 (   ) and theoretical function (——).

curve. Here we have a swollen system, in which the particles contain a smaller number of layers ($M_{min} = 1$, $M_{max} = 20$) and the distribution of interlamellar distances is very diffuse ($\bar{d} \simeq 50$ Å). Figure III.5 shows the two distance distributions Δ (di) corresponding respectively to $P_F = 1.5$ and 3. The high asymmetry of the curve for $P_F = 1.5$ agrees with the shape of the interaction potential distribution as

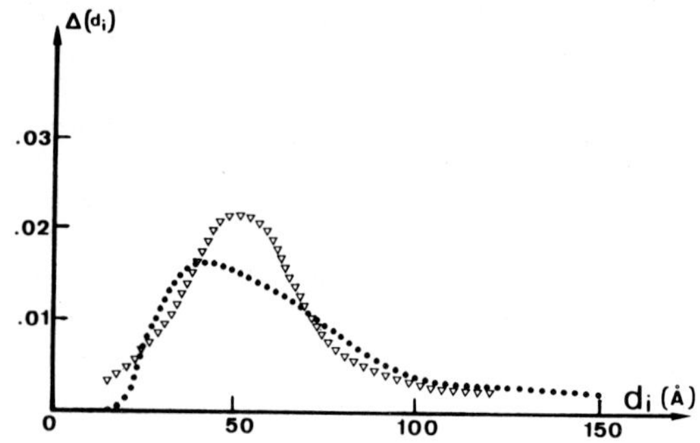

Fig.III.5. Distribution of interlamellar distances in the particles or pore walls which were determined from S.A.S. data ( ) Na Greek montmorillonite (10^{-3} M NaCℓ, $\sigma = 1.5$) and ( ) for Na Greek montmorillonite (10^{-3} M NaCℓ, $\sigma = 3$).

computed from the Gouy-Stern theory for diffuse electrical double layers (van Olphen, 1977). The effect of suction in this case is to regularize the particle structure (symmetrization of the distance distribution) and to increase slightly their thickness by regrouping the unit layers isolated by dispersion during swelling. In this case, the pore walls are slightly modified ($M_{min} = 6$, $M_{max} = 26$).

Plates III and IV for samples 5 and 6 confirm the X-ray results concerning the number of layers stacked in the pore walls. For these highly swollen samples, the preparation method of the T.E.M. sections causes a collapse of the layers in the particles but it does not modify the number of layers per particle. Therefore, in this case, the T.E.M. technique does not give a good picture of the internal structure of the particles. Nevertheless their macroscopic arrangement, that is to say the porosity at the micrometer scale, is not disturbed by the preparation method.

CONCLUSIONS.

Our results show that the S.A.S. and T.E.M. methods are complementary. Their simultaneous utilization enables one to describe quantitatively any clay water system, swollen or not, as well as particle thickness, their internal structure and mutual arrangement (porosity)

The application of both methods to the study of the effect of salt strain or pressure strain on some water-clay samples induces three main conclusions :

. The particles highly swollen or not are interconnected, generally face by face delimiting pores at micrometer scale.

. The salt strain affects the microstructure of the system (structure of the particles). A 1 M solution of NaCℓ prevents the swelling of Na-montmorillonite in such a way that its behaviour becom comparable to that of Ca-montmorillonite.

. The suction strain by air pressure between 3 000 and 10^5 Pa does not affect the internal structure of the slightly hydrated particles. Water removal by suction concerns only the water molecules located in the pores, with sizes up to about 1 µm. For the strongly swollen samples, the suction strain slightly modifies the structure of the pore walls by regrouping some unit layers and causes a homogenization of the interlamellar distances to a limit determined by the compressibility of the diffuse double layers.

ACKNOWLEDGMENTS.

The authors thank the people of Orsay linear accelerator and of L.U.R.E. Laboratory for their help and co-operation. They are also grateful to Dr. Van Olphen for his help in English and to M. Clinard for the T.E.M. photographs.

REFERENCES.

Andrews, D.E., Schmidt, P.W. and Van Olphen H., 1967. X-ray study of interactions between montmorillonite platelets. Clays Clay Min., 15: 321-330.

Norrish, K. and Rausell-Colom, J.A., 1963. Low angle diffraction studies of the swelling of montmorillonite and vermiculite. Clays Clay Min., 10: 123-149.

Pons, C.H., Rousseaux, F. et Tchoubar, D., 1981. Utilisation du rayonnement synchrotron en diffusion aux petits angles pour l'étude du gonflement des smectites. Partie I : Etude du système eau-montmorillonite-Na en fonction de la température. Clay Min., 16: 23-42.

Tessier, D. et Berrier, J., 1974. Utilisation de la microscopie électronique à balayage dans l'étude des sols. Observation des sols humides soumis à différents pF. Science du Sol. Bull. de l'AFES, 1: 67-82.

Tessier, D. et Clinard, C., 1981. Agronomie (to be published).

Tessier, D. et Pédro, G., 1981. Electron microscopy study of Na-smectite fabric. These proceedings.

Van Olphen, H., 1977. An introduction to clay colloïd chemistry. John Wiley and Sons (London, New York).

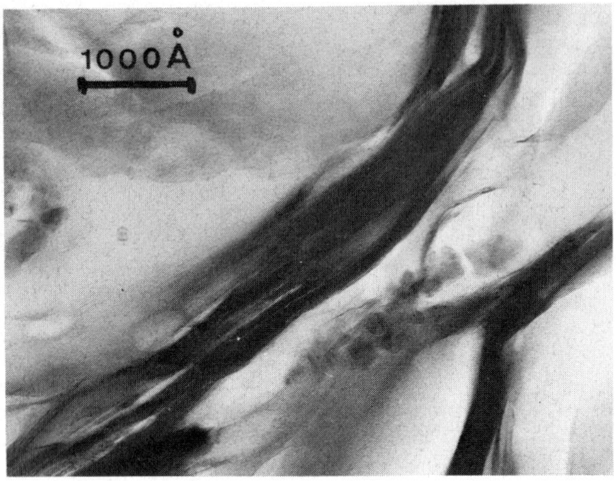

Plate I. L.I.T.E.M. for Ca montmorillonite, 10^{-3} M $CaC\ell_2$, $\sigma = 1.5$, showing the structure of pore walls and the size of micron scale porosity.

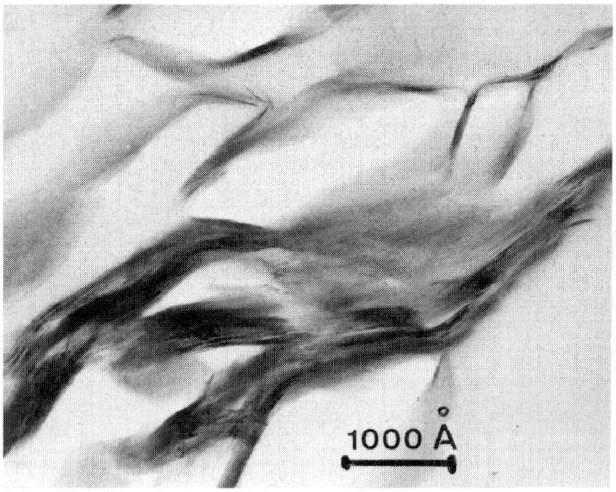

Plate II. L.I.T.E.M. for Na montmorillonite, 1 M $NaC\ell$, $\sigma = 1.5$. It shows a larger distribution of the pore walls (or particle thickness) than for the first sample.

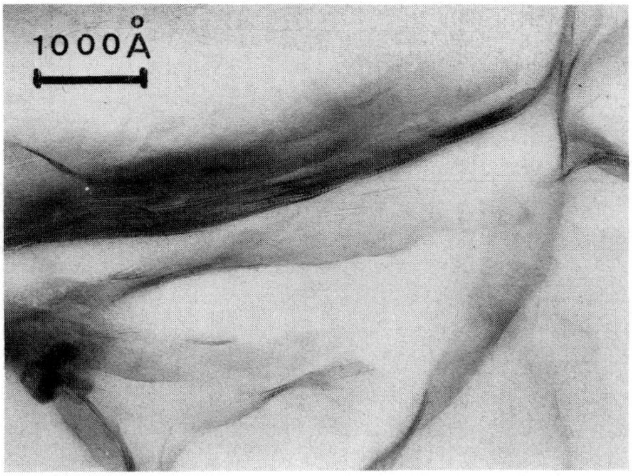

Plate III. L.I.T.E.M. for
Na montmorillonite,
10^{-3} M NaCℓ, $\sigma = 1.5$.

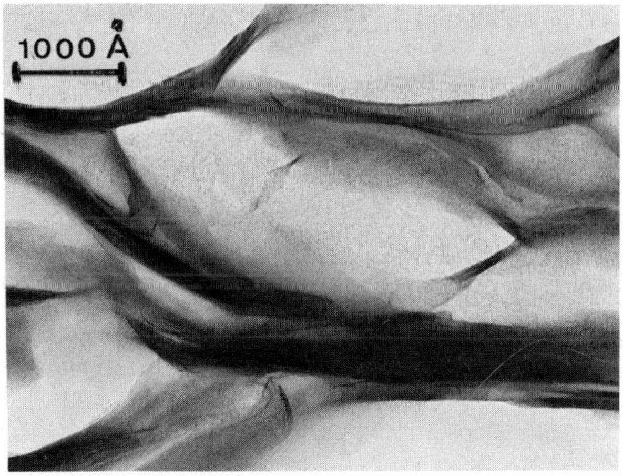

Plate IV. L.I.T.E.M. for
Na montmorillonite,
10^{-3} M NaCℓ, $\sigma = 3$.

NEAR INFRARED PROPERTIES OF WATER IN Na-HECTORITE PASTES

R. PROST - Institut National de la Recherche Agronomique,
Station de Science du Sol, Route de Saint-Cyr
78000 VERSAILLES - FRANCE.

ABSTRACT

It is shown that near-infrared spectroscopy gives important advantages, compared to the infrared region, in the study of clay-water systems. By using a method involving only differences between non-adjusted data it is shown that the near-infrared properties of water molecules are not perturbed by the clay surfaces beyond about two monolayers.

INTRODUCTION

The question of whether there is a long range effect of clay surfaces on the structure of water in clay-water systems has not yet received a satisfactory answer. It is a matter of fact that most researchers who claim that there is a long range effect are performing experiments in order to determine macroscopic properties of water, such as the density or other thermodynamic data (Low, 1979). As we know, these macroscopic properties are in fact related to both components of the clay-water system (Sposito and Prost, in press). For this reason we have performed experiments that reflect the molecular properties of water alone using near-infrared spectroscopy (Fornès and Chaussidon, 1976 ; Cariati et al., 1981).

MATERIALS AND METHODS

The sample used is Na-hectorite. The characteristics of this sample are described in a previous paper (Prost, 1976).

The spectra were obtained with a UV, visible, near-infrared spectrometer, the CARY 17D, connected to a computer. The measurements were performed with a quartz SUPRASIL cell of 0.2 mm pathlength.

To find out whether there is a long range effect of the clay surface on the near-infrared properties of water we have performed the following experiment.

A small quantity of water, $x_2 - x_1$, is added to a Na-hectorite sample whose water content is x_1 g water per g of clay to obtain a water content, x_2, slightly larger than the preceding one. We check whether this added water has the same spectroscopic properties as liquid water. In order to do this we subtract, for each wavelength, the optical density of the sample with water content x_1 from that with water content x_2. This subtraction is made on the non-adjusted data to avoid changes in the actual profile of the bands which can be produced when we take into account the base line.

Pastes were prepared with water contents decreasing from 10 to 0.4 g water per g of clay. They were squeezed between the plates of the quartz cell as shown in Fig. 1.

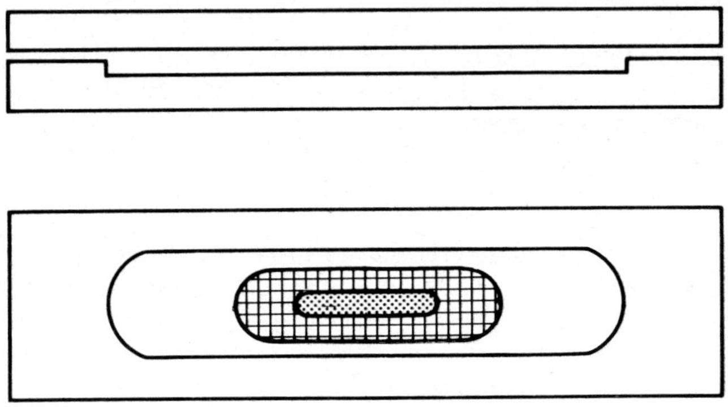

Fig. 1. Diagram representing the quartz cell. The clay paste does not fill the depression. The cross section of the near-infrared beam is less than that of the sample.

The difficulties encountered in squeezing the dryest paste in the quartz cell led us to use, for the lowest water contents, a deposit on a quartz window which was fixed in a cell under controlled relative humidity.

In order to compare the results, we have to standardize them with respect to the clay or to the water content. This was done by determining for each sample Z^W and Z^C, the thicknesses of the water and of the clay films, respectively, which are traversed by the near-infrared radiation. We assumed that the densities of water and clay are respectively 1 and 2.8 g/cm³. For convenience we chose as the standard thickness, 0.2 mm, the thickness of the cell. If x is the water content, we have :

$$Z^W = \frac{2.8 * 0.2\, x}{1 + 2.8\, x}$$

$$Z^C = \frac{0.2}{1 + 2.8\, x}$$

with Z^W and Z^C in mm and x in g water per g of clay.

We standardize the spectra as shown in fig. 2 with respect to :

1 - a water film 0.2 mm thick, by multiplying each optical density by a coefficient $K^W = \frac{1}{2.8\, x} + 1$

2 - a clay film 0.2 mm thick, by multiplying each optical density by a coefficient $K^C = 1 + 2.8\ x$.

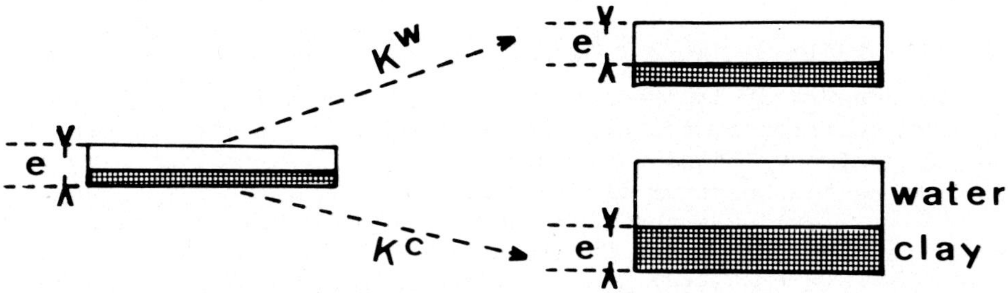

Fig. 2. Diagram showing how the standardizations are carried out.

To compare the spectroscopic properties of water added to a sample with water content x_1 (in order to obtain a sample with water content x_2) to the spectroscopic properties of liquid water, we must first of all standardize the spectra with respect to a clay thickness of 0.2 mm, as shown in Fig. 3.

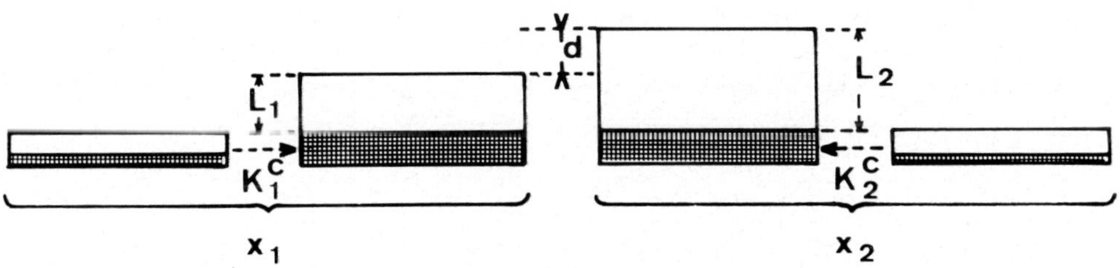

Fig. 3. Diagram showing how both steps of the standardization process are carried out in order to compare the near-infrared spectrum of water added, to a sample with water content x_1 to obtain a sample, with a larger water content of x_2 g/g clay.

After that we must standardize the difference, $d = L_2 - L_1$, to a water film 0.2 mm thick. Then it is easy to superpose the resulting spectrum with the spectrum of liquid water obtained in the same cell.

If L_1 and L_2 are the thickness of the water films after standardization with respect to a clay thickness of 0.2 mm, the coefficient used to standardize the difference, $d = L_2 - L_1$, is :

$$K^D_{21} = \frac{0.2}{L_2 - L_1} = \frac{0.2}{Z^W_2 K^C_2 - Z^W_1 K^C_1} = \frac{1}{2.8\ (x_2 - x_1)}$$

The spectra of samples whose water content increased from x_1 to x_n were subtracted step by step ($x_n - x_{n-1}$, $x_{n-1} - x_{n-2}$ etc). This was done also by other ways ; for example : $x_n - x_1$, $x_n - x_2$ etc. As long as the spectrum obtained by difference, $x_n - x_{n-1}$ or $x_{n-1} - x_{n-2}$ etc, has the same spectroscopic characteristics as the spectrum of liquid water obtained under the same conditions, we consider that the water added has the same near-infrared properties as liquid water.

RESULTS

Two sets of results were obtained. In figure 4 are the near-infrared spectra obtained with the Na-hectorite deposit. All correspond to the same thickness of the clay film.

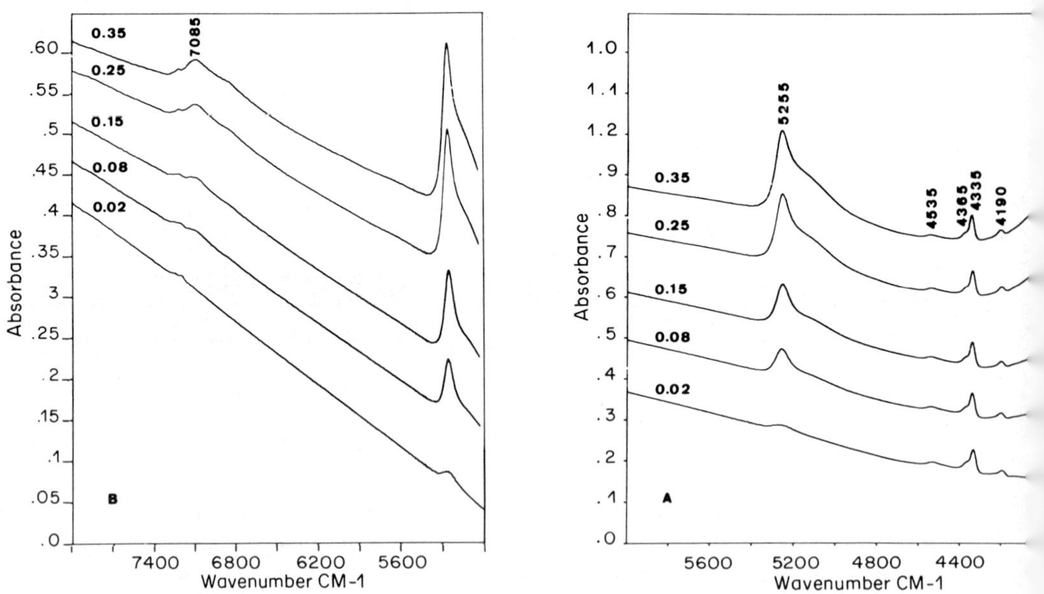

Fig. 4. A and B. Near-infrared spectra of a Na-hectorite paste deposited on a quartz window for different water contents expressed in g water per g of clay as indicated on each curve.

In figure 5 are the near-infrared spectra of water associated with pastes with water contents increasing from 0.5 to 5 g water per g of clay. These spectra correspond to the same 0.2 mm thick film of water.

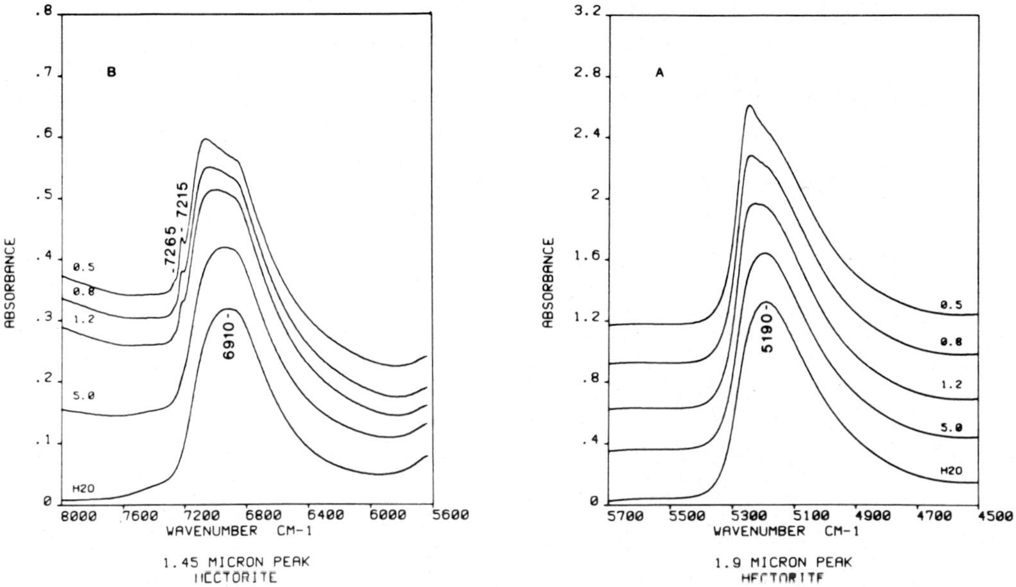

Fig. 5 A and B. Near-infrared spectra of Na-hectorite pastes with water contents in g/g clay as indicated on each curve. One of these spectra is for liquid water.

DISCUSSION

The near-infrared spectrum of the hydrated Na-hectorite consists of several bands. Two sets of 4 and 2 bands at 4 190, 4 335, 4 365, 4 535 cm^{-1} and at 7 215, 7 265 cm^{-1} are due to the structural OH groups of hectorite (Fig. 4 and 5). There are also two sets of 2 bands at 5 190, 5 255 cm^{-1} and at 6 910, 7 085 cm^{-1} the intensity of which increases when the water content increases. They are due to water.

As has been shown for the fundamental bands (Farmer, 1971 ; Frost, 1975, 1976), these two sets of bands differ also from the bands of bulk water or ice by the existence, on the high frequency side, of a well resolved band which is, respectively, at 5 255 and 7 085 cm^{-1}.

The 5 255, 7 085 cm^{-1} bands are due to water molecules which are involved in weak hydrogen bonding (Fornès and Chaussidon, 1976) and the 5 190, 6 910 cm^{-1} bands are due to water molecules which are involved in strong hydrogen bonding. It is remarkable that the 5 255, 7 085 cm^{-1} bands are more prominant, comparatively, than the 5 190, 6 910 cm^{-1} bands in the near-infrared range than the corresponding one in the infrared. This is probably due to the drastic decrease of the extinction coefficient of water molecules involved in strong hydrogen bonding. According to Luck (1973), in the overtone range, the extinction coefficient of OH involved in strong hydrogen bonding is 800 times less than in the fundamental range. This gives us better conditions to study the 5 255, 7 085 cm^{-1} bands, which correspond to water molecules which interact with the clay material. It is also remarkable that both components 5 190, 5 255 and 6 910, 7 085 cm^{-1} are better resolved than the equivalent 3 420, 3 630 cm^{-1} in the fundamental range. All these points added to the fact that we can work with thicker samples in the near-infrared range than in the infrared range, give important advantages to near-infrared spectroscopy.

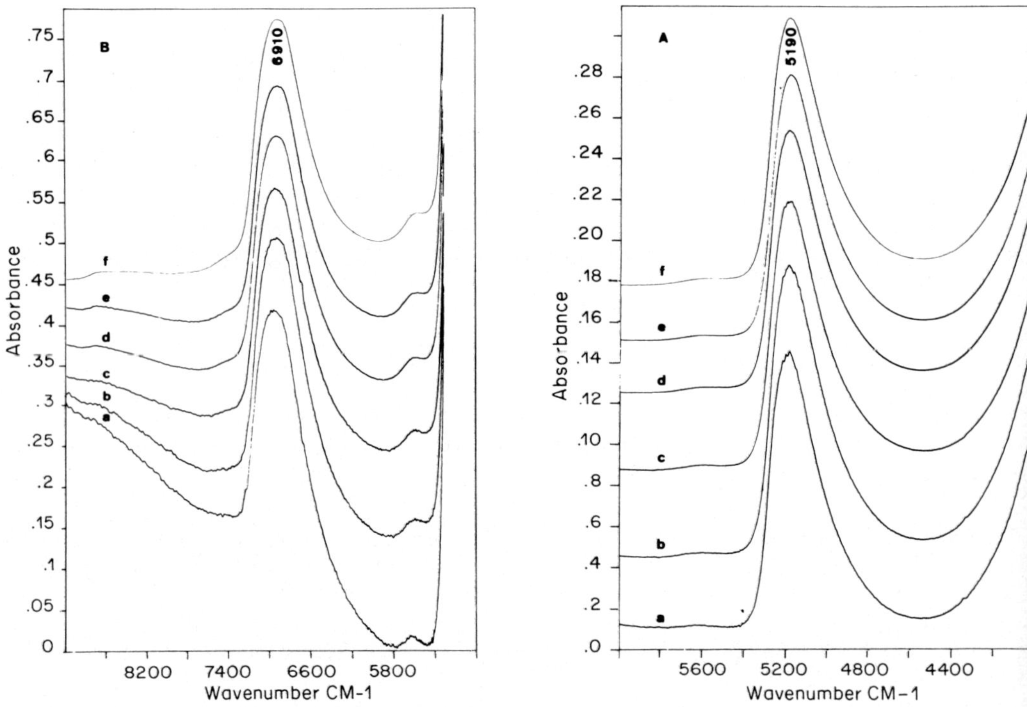

Fig. 6 A and B. These curves represent the difference of near-infrared spectra for Na-hectorite pastes with different water contents. Curve a corresponds to the difference (x_2 - x_1) between the spectrum of a paste at 0.8 g/g clay minus the spectrum of a paste at 0.5 g/g clay. The difference was standardized by comparison with the spectrum f for liquid water. b corresponds to 1.2 - 0.8 ; c to 2 - 1.2 ; d to 5 - 2 and e to 5 - 0.8 g/g clay.

From a qualitative comparison of the spectra in figures 4 and 5, it is clearly seen, that during the hydratation process, there is an increase first of all of the 5 255, 7 085 cm^{-1} bands. There is also a concomitent, but initially slow increase of the 5 190, 6 910 cm^{-1} bands. At high water contents the 5 190, 6 910 cm^{-1} bands increase in such a way that the 5 255, 7 085 cm^{-1} bands are no longer resolved. It is important to notice that the 5 255, 7 085 cm^{-1} bands remain at the same frequencies (fig. 4-a). This means that there is at least one kind of water molecule, which interacts strongly with the clay surface and which is not perturbed by the addition of water molecules during the first step of the hydration process. It means also that the clay system is essentially made of two "boxes" which are filled by water molecules with two different kinds of kinetics. One of these "boxes" could be the interlamellar spaces and the 5 255, 7 085 cm^{-1} bands could be due to water molecules which solvate the interlamellar cations. Indeed, it is well known, that the first step of the hydration process corresponds to the solvation of the cations. The 5 190, 6 910 cm^{-1} bands correspond to water molecules which are in micropores.

In order to determine whether or not there is a long range effect of the clay surfaces on the near-infrared properties of water associated with Na-hectorite, we have measured the difference between spectra as previously explained.

The differences obtained with Na-hectorite pastes are represented in figure 6. These curves, obtained after subtraction and standardization, have the same intensity and the same profile as the curve corresponding to liquid water. The change of the slope of the base line which is visible at low water content is due to the fact that the Na-hectorite pastes scatter the near-infrared radiation more when the samples have a low water content. The rigidity of the Na-hectorite paste at a water content less than 0.5 g water per g of clay makes the squeezing of the sample in the cell difficult. In this case the thickness of the sample was not known with enough accuracy with respect to the standardized process. Nevertheless these results show that there is no effect of the clay surface on the near-infrared properties of water added to the Na-hectorite paste with a water content of 0.5 g water per g of clay. This does not mean that this water content is the upper limit.

On the other hand, the spectra obtained by difference in the case of deposits (fig. 7) show that there is always an absorption at 5 255 and 7 085 cm^{-1} even at 0.3 g water per g of clay. It means that a significant portion of the water added to this sample has near-infrared properties different from those of liquid water. This water content does not correspond to a lower limit. The actual limit is therefore between 0.3 and 0.5 g water per g of clay.

To cover the surfaces of Na-hectorite, (specific surface area = 800 m^2/g), by one monolayer of water molecules requires 0.2 g of water. If the actual limit is, for example, 0.4 g water per g of clay, this corresponds to two monolayers of water molecules. This result is in good agreement with the calorimetric determinations of Rouquerol and Cases (to be published).

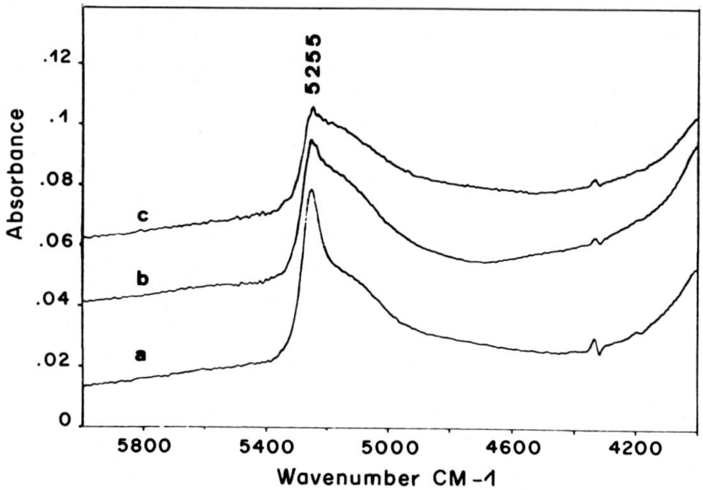

Fig. 7. These curves represent the difference of near-infrared spectra of Na-hectorite pastes deposited on a quartz window at different water contents. Curve a corresponds to the difference $(x_2 - x_1)$ between the spectrum of the sample containing 0.15 g/g clay minus the spectrum of the same sample containing 0.08 g/g clay ; b corresponds to 0.25 - 0.15 and c to 0.35 - 0.25 g/g clay.

CONCLUSION

This work shows the potential of near-infrared spectroscopy offers in the study of clay-water systems. Indeed we were able to show the existence of essentially two kinds of water molecules in hydrated Na-hectorite. One of them corresponds to water molecules which solvate the interlamellar cations. The other corresponds to water molecules which are located in the micropores of the clay material. In addition to this, it was possible, by using the particular method developed here, to show that there is no long range effect of the surface of Na-hectorite on the near-infrared properties of water beyond about two or three monolayers of water molecules.

ACKNOWLEDGMENT

This work was performed in part during a visit to PURDUE UNIVERSITY, WEST LAFAYETTE, IN. 47907 (U.S.A.). The author expresses gratitude to Professor P.F. LOW for the use of facilities and thanks Dr. C.B. ROTH and Mr. R.L. HUTCHINSON for their help to use the spectrometer and the computer.

REFERENCES

Cariati, F., Erre, L., Micera, G., Piu, P. and Gessa, C., 1981. Water molecules and hydroxyl groups in montmorillonites as studied by near-infrared spectroscopy. Clays Clay Min. 29 : 157-159.

Farmer, V.C. and Russell, J.D., 1971. Interlayer complexes in layer silicates : the structure of water in lamellar ionic solutions. Trans. Faraday Soc. 67 : 2737-2749.

Fornès, V. and Chaussidon, J., 1976. Near-infrared spectroscopic studies of clay-water systems. Proc. Internat. Clay Conf., Mexico, 383-390.

Low, P.F., 1979. Nature and properties of water in montmorillonite-water systems. Soil Sci. Soc. Am. Proc. J. 43 : 651-658.

Luck, W.A.P., 1973. Infrared studies of hydrogen bonding in pure liquids and solutions. In : F. Franks (editor), Water, A Comprehensive Treatise, Plenum Press, New York, Vol. 2, 235-321.

Prost, R., 1975. Etude de l'hydratation des argiles : interactions eau-minéral et mécanisme de la rétention de l'eau.
a - Etude d'une zéolite (natrolite) et de deux minéraux fibreux (attapulgite et sépiolite). Ann. agron. 26 : 400-462.
b - Etude d'une smectite (hectorite). Ann. agron. 26 : 463-535.

Prost, R., 1976. Interactions between adsorbed water molecules and the structure of clay minerals : hydration mechanism of smectites. Proc. Int. Clay Conf., Mexico, 351-359.

Sposito, G. and Prost, R. The structure of water adsorbed on smectites : a critical synthesis. Chem. Rev. (in press).

CLAY-ACTIVATED ORGANIC REACTIONS

B.K.G. THENG

Soil Bureau, Department of Scientific and Industrial Research
Lower Hutt, New Zealand

INTRODUCTION

Work on adsorption of organic compounds by clays has been reviewed by Mortland (1970) and by Theng (1974, 1979). Under certain conditions, clays are also capable of activating the decomposition, transformation, and polymerization of adsorbed organic species. In these processes, the mineral plays the dual role of a catalyst and that of an adsorbing or stabilizing surface for both reactants and reaction products; hence specificity is a characteristic of many clay-activated organic reactions. In many instances the clay surface becomes modified during the reaction; for this reason, the term 'activation' is preferred to 'catalysis'.

Compared with activation over zeolites (Poutsma, 1976; Spencer and Whittam, 1980), clay-activated organic reactions are less well understood. The earlier literature on clay-activated organic reactions has been summarized by Theng (1974). Reviews of a more specialized nature and restricted scope are those by Solomon (1968), Theng and Walker (1970), Theng (1971), Fripiat and Cruz-Cumplido (1974), Paecht-Horowitz (1977), and Thomas et al. (1977). This paper concentrates on the data which have accumulated over the past decade or so, in particular for reactions which take place on or at smectite surfaces. It also discusses some practical applications in relation to petroleum genesis, organic matter transformation in soils and sediments, and chemical evolution.

NATURE AND LOCATION OF ACTIVE SITES

The activity of clay minerals in converting, transforming, or polymerizing organic compounds under relatively mild conditions (e.g. at ambient temperature and humidity) may, almost invariably, be ascribed to the ability of clays to donate protons to, and/or accept electrons from, the adsorbed species. In other words, clay mineral surfaces can act as Brønsted and/or Lewis acids. Sometimes, clays can also donate electrons to adsorbed organics and thus behave as Lewis bases but this aspect of surface activity is only of minor concern here. For a given set of experimental conditions and clay pretreatments, either Brønsted- or Lewis-type acidity normally prevails. In some instances, however, both types of acidity may act sequentially or synergistically. When this occurs, it is difficult to make a clear distinction between Brønsted- and Lewis-site activation. Although the source and activity of Brønsted sites are fairly well understood, the nature and location of electron-accepting sites at clay mineral surfaces are not always clearly definable.

1. Brønsted Acidity

There is now ample and persuasive evidence to show that, for cation-exchanged clays, Brønsted sites essentially arise from the dissociation (or ionization) of adsorbed water. This dissociation is induced by the strong electric field or polarizing power of the exchangeable cations with which the water molecules are associated (Fripiat, 1968). Thus, the higher the charge of the cation and the smaller its radius, the greater is its polarizing effect on the surrounding water (Mortland, 1968). The proton-donating ability of clays is also greatly influenced by the water content of the system and, to some extent, by the origin of the layer charge, i.e. whether isomorphous substitution occurs chiefly in the octahedral or tetrahedral sheet (Mortland and Raman, 1968; Solomon et al., 1971; Frenkel, 1974). In the presence of much adsorbed water, e.g. in an aqueous suspension, the polarizing effect of the cation is presumably dissipated among the relatively large number of water molecules. However, as the water content of the clay is reduced to less than ∿ 5% w/w, the forces of polarization act on the few remaining molecules which are directly coordinated to the cation, forming its inner coordination sphere or hydration shell. There is then a marked increase in the dissociation of the residual water, $(H_2O)_x$, and hence, in the overall Brønsted acidity of the system. Such a situation may be represented by the following equilibrium where M is an exchangeable cation of valency n,

$$[M(H_2O)_x]^{n+} \rightleftharpoons [M(OH)(H_2O)_{x-1}]^{(n-1)+} + H^+ \qquad (1)$$

The degree of water dissociation as estimated from surface conductivity and nuclear magnetic resonance (NMR) measurements (Fripiat et al., 1965; Touillaux et al., 1968) on a monolayer hydrate of montmorillonite, ranges from 10^{-2} to 10^{-3} which is 5-6 orders of magnitude greater than for bulk (liquid) water at 25°C.

The protons (in reality hydronium ions) so derived are available for, and capable of, donating protons to a variety of adsorbed organic bases (B) according to the following equilibrium (Table 1),

$$H^+ + B \overset{K_e}{\rightleftharpoons} BH^+ \qquad (2)$$

The magnitude of K_e depends on the basicity of B. Such relatively strong bases as ammonia (pK_b 4.74), cyclohexylamine (pK_b 3.34), ethylamine (pK_b 3.27), and histidine (pK_b^3 4.67), adsorbed on divalent (Ca^{2+},Mg^{2+}) montmorillonite, are not only readily protonated by they also convert the exchangeable cations to their respective hydroxides (Mortland et al., 1963; Farmer and Mortland, 1965; Yariv and Heller, 1970; Heller-Kallai et al., 1973), rendering part of the cations non-exchangeable (cf. equilibrium (1)). Thus, 'dry' (water content <5% w/w) Ca^{2+}- montmorillonite could protonate ammonia to the extent of 0.8 meq/g, i.e. approaching the cation exchange capacity (CEC) of the clay. By comparison, only about half this quantity of pyridine (pK_b 8.85) was protonated under similar conditions (Mortland, 1968). By the same token, such extremely weak bases as acetamide (pK_b 13.34) and urea (pK_b 13.82) which, in homogeneous solutions, would require a pH <1 before being protonated, can only abstract protons from adsorbed water when the exchange sites of the clay are occupied by H^+ ions or such highly polarizing species as Al^{3+} and Fe^{3+} (Table 1). (When preparing hydrogen saturated clays by a cation exchange process, e.g. acid washing, part of the exchange positions become occupied by Al^{3+} ions or charged hydroxy-aluminium species (Barshad and Foscolos, 1970; Gerstl and Banin, 1980)).

TABLE 1

Protonation of some organic compounds on montmorillonite surfaces; the active protons either derive from dissociation of surface-adsorbed water under the polarizing influence of the exchangeable cations or are introduced as such into the exchange complex of the clay

Organic species protonated	Exchangeable cation on clay	References	Remarks
Amides (primary, secondary, tertiary)	H^+, Al^{3+}	Tahoun and Mortland (1966); Mortland (1968); Cloos (1972)	Protonation occurs on the oxygen atom ("O-protonation"); marked inductive effect of substituent groups, i.e. N,N'-diethylacetamide and chloro-N,N'-diethylacetamide fail to be protonated; hemisalt complexes may form
Amines (aliphatic, aromatic, cyclic)	various	Swoboda and Kunze (1968): Yariv et al. (1968); Yariv and Heller (1970); Harter and Ahlrichs (1969); Santos et al. (1970); Cloos and Laura (1972); Laura and Cloos (1975a, 1975b) Cloos et al. (1975, 1979); Tennakoon et al. (1974b)	Hemisalt complex formation and proton exchange may occur when excess amine is present or when ammonium is the exchangeable cation. With some amines protonation of the adsorbed species can be accompanied by a colour change of the complex. Metal hydroxides may form.
3-Aminotriazole	various	Russell et al. (1968a)	Extent of protonation generally increases with polarizing power of the exchangeable cation
Asulam (4-amino-benzene-sulphonyl-methyl-carbamate)	H^+, Al^{3+}	Fusi et al. (1980)	Protonation occurs on NH_2 group; hydrolysis may follow
Histidine	various	Heller-Kallai et al. (1973)	Concomitant formation of metal hydroxide; pronounced with Al^{3+} but also with less strongly polarizing cations
Porphyrins and Metalloporphyrins	various	Van Damme et al. (1978); Abdo et al. (1980)	With metalloporphyrins, demetallation as well as protonation may occur depending on the nature of the central metal ion in the porphin ring the kind of substituent groups around the ring, and the water content of the system
Pyridine	Mg^{2+}	Farmer and Mortland (1966); Swoboda and Kunze (1968)	Formation of $Mg(OH)_2$ and hemisalt complexes
s-Triazines	various	Cruz et al. (1968); Russell et al. (1968b); Brown and White (1969); White (1976)	When keto-enol tautomerism occurs as in hydroxy-triazines, the keto form predominates in the protonated species; hydrolysis often follows protonation, leading to loss of herbicidal activity; extent of protonation and hydrolysis depends on nature and location of substituent groups around the *s*-triazine ring
Urea	H^+, Al^{3+}	Mortland (1966)	see text for effect of basicity; may form hemisalt complex

A protonated base, B_1H, may transfer its proton to another adsorbed organic species according to the equilibrium

$$B_1H^+ + B_2 \rightleftharpoons B_2H^+ + B_1 \qquad (3)$$

which again is controlled by the relative basicity and concentration of B_1 and B_2. Proton interchange of this type has been observed to occur between pairs of ammonium and ethylammonium, ammonium and ethylenediammonium, pyridine and 3-aminotriazole, and triethylammonium and ethylenediammonium (Raman and Mortland, 1969; Cloos et al., 1975). Related to this process is the formation of 'hemisalt' complexes which may occur when the amount adsorbed of a given base, B, exceeds the number of available protons. The BH^+ species can then 'share' its protons with the neutral (excess) base to yield a cation of the type $B...H...B^+$. Hemisalt complex formation on montmorillonite has been observed with amides (Tahoun and Mortland, 1966), anilinium-aniline (Yariv et al., 1968), ethylammonium-ethylamine (Farmer and Mortland, 1965), ethylenediammonium-ethylenediamine (Cloos and Laura, 1972), pyridinium-pyridine (Farmer and Mortland, 1966), and urea (Mortland, 1966) (cf. Table 1).

That 'dry' (<5% w/w water) clay mineral systems can develop a high surface acidity has been confirmed by measurements of acid strength, based on colour changes of adsorbed Hammett indicators (Benesi, 1956; Fowkes et al., 1960; Solomon et al., 1971; Henmi and Wada, 1974). For some layer silicates and related minerals, the values derived in this way and expressed in terms of Hammett and Deyrup's (1932) H_0 scale, are listed in Table 2. Although H_0 values more negative than about -5.6 may be more apparent than real due to the physisorption and/or lack of desorption of the indicator (Frenkel, 1974; Take et al., 1975), surface acidities beyond $H_0 = -8.2$, i.e. equivalent to 90 weight per cent or more of sulphuric acid, can actually be attained under conditions of severe dehydration (Fig.1). The behaviour of some metalloporphyrins, adsorbed on smectite surfaces, serves to illustrate this point.

TABLE 2

Acid strength of some clay minerals and related alumino-silicates as measured using Hammett indicators

	H_o value (pK_a range of strongest acid sites)			
	A	B	C	D
Allophane				+1.5 to -5.6
Halloysite				-5.6 to -8.2
Imogolite				+1.5 to +3.3
Kaolinite	-3.0 to -5.6	-5.6 to -8.2	<-8.2	-5.6 to -8.2
Kaolinite treated with Na-hexametaphosphate			+1.5 to -3.0	
Montmorillonite	-3.0 to -5.6	-3.0 to -5.6	-5.6 to -8.2	-5.6 to -8.2
Palygorskite	+1.5 to -3.0	-5.6 (tr)	-8.2	
Talc	+3.3 (tr)		+4.0 to +3.3	

Values under A and B are taken from Benesi (1956) and Fowkes et al. (1960). They refer to samples as received and after drying at 120°C for 16 hours, respectively; tr = transition colour. Values under C are from Solomon et al. (1971) for samples with near-zero water content. Values under D are from Henmi and Wada (1974) for samples saturated with either Na^+ (halloysite and kaolinite) or H^+/Al^{3+} ions (allophane, imogolite, montmorillonite) after drying at 100°C.

Using a variety of spectroscopic techniques, Cruz and co-workers (Van Damme et al., 1978; Canesson et al., 1979; Abdo et al., 1980) have demonstrated that adsorbed metalloporphyrins can be protonated as well as demetallated, depending on the acid-stability of the compound and the water content of the clay. In the case of the acid-resistant tin-tetrapyridyl-porphyrin (Sn(IV)-TPyP) these two processes occurred when its complex (with hectorite or montmorillonite was dehydrated to near-zero water content. The acidity of the clay surface would then be at least equivalent to 100% sulphuric acid since tin porphyrin compounds generally fail to demetallate completely even under such extremely acid conditions (Buchler, 1975). Another interesting feature was that the demetallation process could be reversed by simply increasing the water content of the system, i.e. by reducing its proton-donating ability. This would indicate that the Sn^{4+} ion remained in close proximity to the free base formed by demetallation, enabling the metal to re-enter the porphin ring as surface acidity decreased. The protonation-demetallation of Sn(IV)-TPyP (on hectorite) may therefore be represented by the sequence of equilibria shown in Fig.2.

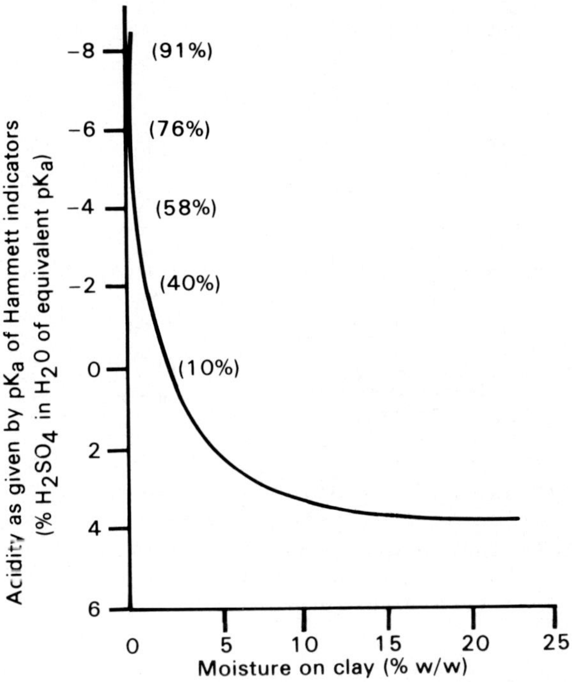

Fig.1. Surface acidity of kaolinite, as measured by the colour change of adsorbed Hammett indicators, as a function of clay moisture content. Values in brackets denote the equivalent weight per cent of sulphuric acid in water (after Solomon et al., 1971)

Hammett indicators when used in conjunction with amine titration of the 'dry' clay, suspended in a non-polar organic liquid (e.g. benzene, *n*-hexane), can also yield information about the distribution of acid sites (Benesi, 1967; Solomon et al., 1971; Frenkel, 1974). This assumes that all available acid sites on the mineral surface are accessible to the titrant. Even for kaolinites where amines appear to be preferentially adsorbed at crystal edges

(Conley and Lloyd, 1971; Sivalov et al., 1980), this assumption is not strictly valid because of orientation and steric effects (Fig.3). Steric effects would certainly be important for smectites where the active sites are mostly associated with interlayer surfaces, and charge density variations between silicate layers within a single crystal (Lagaly and Weiss, 1970; Stul and Mortier, 1974) may preclude complete and uniform penetration of the interlayer space by the amine molecules.

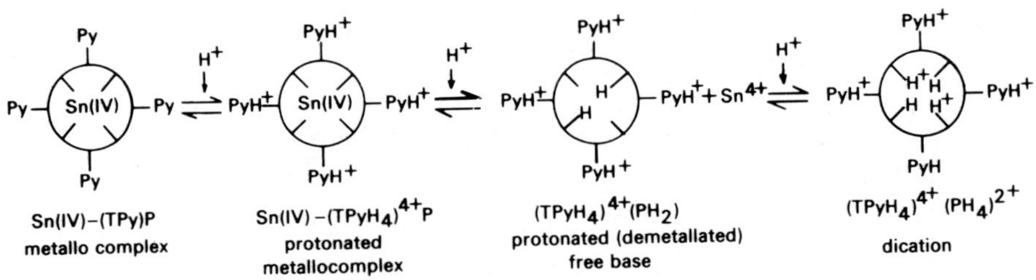

Fig.2. Sequence of protonation and demetallation of Sn(IV) - tetrapyridylporphyrin on hectorite as the system is dehydrated to near-zero water content. Note that protonation of the pyridyl substituents precedes demetallation and protonation of the pyrrole groups inside the porphin ring (after Abdo et al., 1980).

An alternative and perhaps less ambiguous means of assessing Brønsted site distributions is to adsorb ammonia or a suitable organic base, such as pyridine, and measure the amount of the corresponding protonated species formed using infrared (IR) spectroscopy (cf. equilibrium (1); Table 1). (Pyridine adsorption is also a useful method for estimating the number of Lewis acid sites - see Section 3). The values obtained are influenced by such factors as the basicity of the organic base, and clay pretreatment. As mentioned earlier, as much as 0.8 meq of ammonia per g clay can be protonated by 'dry' montmorillonite. Compared with smectites, kaolinite has relatively few acid sites (Fig.3) although the distributions of acid strengths are analogous.

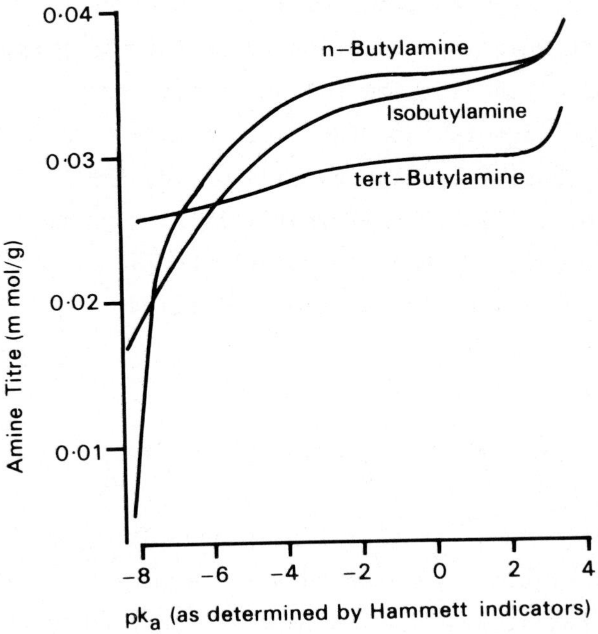

Fig.3. Distribution and relative strength of acid sites on kaolinite, as determined by amine titration in benzene using Hammett indicators. Differences in titration curves between the three isomeric butylamines reflect differences in relative site accessibility due to steric effects (after Solomon et al., 1971).

The question arises whether Brønsted sites still occur in cation-exchanged clays under anhydrous conditions. The available evidence suggests that as the water content of the system approaches zero, the proton-donating ability of the clay surface sharply increases (cf. Fig.1) and then declines as the last traces of water are removed (Sohn and Ozaki, 1980) (cf. Fig.9). The Brønsted sites which occur in the absence of surface-adsorbed water may be identified with structural hydroxyl groups on planar and edge surfaces (Davidtz, 1976). Their proton-donating ability may be greatly enhanced by the presence of adjacent electron-withdrawing aluminium ions (Sieskind et al., 1979) (Section 3). We should mention here that protons from surface-adsorbed water can and do exchange with those of structural hydroxyls (Russell and Fraser, 1971; Yariv and Heller-Kallai, 1973; Mortland

and Berkheiser, 1976).

A similar situation arises when ammonium clays are heated to temperatures of 200-300°C at which the NH_4^+ ions decompose to yield gaseous ammonia and protons. In dioctahedral smectites, the protons from this source can migrate into vacancies in the octahedral sheet where they associate with oxygens of adjacent aluminium ions (Wright et al., 1972). Alternatively, the protons may attach themselves to surface oxygens of the tetrahedral sheet, analogous to what has been suggested for X- and Y-zeolites (Uytterhoeven et al., 1965),

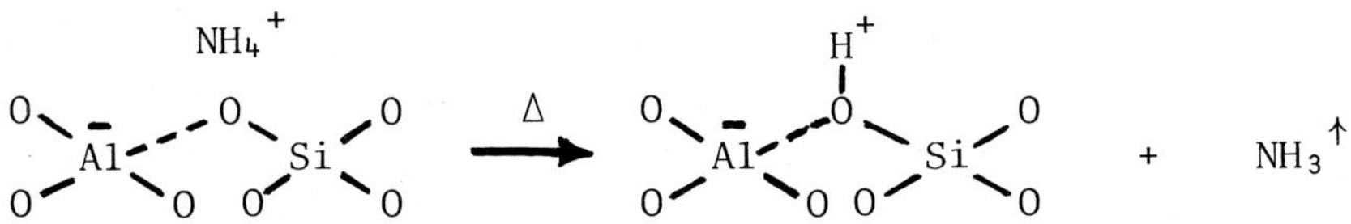

(4)

Whatever their ultimate fate, such structural protons — as opposed to those derived from dissociation of adsorbed water — play an important part in organic transformations which take place under anhydrous conditions and/or at elevated temperatures, e.g. in hydrocarbon cracking (Tamele, 1950; Swift and Black, 1974).

Besides acting as proton donors (Table 1), Brønsted sites arising from the dissociation of cation-coordinated water, can activate the decomposition and/or transformation of adsorbed organic molecules at clay mineral surfaces. Such reactions frequently involve the participation of carbonium ions as intermediates. An example is the decomposition of cobaltic hexammine cations on montmorillonite to yield $Co(OH)_2$, N_2, NH_3, and NH_4^+ ions (Fripiat and Helsen, 1966). Another reaction is the reversible transformation of adsorbed triphenylcarbinol to triphenylmethylcarbonium, also on montmorillonite (Fripiat et al., 1964) (Fig.4).

A Brønsted site-activated reaction which has attracted much attention is the decomposition of intercalated alkyl- or aryl-ammonium ions in montmorillonite (Weiss and Roloff, 1963; Chaussidon and Calvet, 1965; Fripiat, 1968; Durand et al., 1972;

Frenkel and Solomon, 1977). This occurs when the complexes are heated up to ∿200°C which is below the temperature at which amines and their hydrochlorides would normally decompose. Gas-liquid chromatographic (GLC) and IR spectroscopic analyses of the decomposition products indicated that the process involved C-N bond cleavage with formation of the corresponding alcohols or phenols, and ammonium. As was the case with the $Co(NH_3)_6{}^{3+}$-montmorillonite system, the NH_4^+ ions formed (reaction (5)) served to balance the negative surface charge of the clay. The phenols derived from arylammonium ions (e.g. anilinium and toluidinium) appeared to be acid-stable; on the other hand, the *n*-alcohols generated from *n*-alkylammonium ions were further transformed into a mixture of hydrocarbons. In an inert atmosphere (e.g. N_2), formation of alkenes seemed to be favoured over that of alkanes while the reverse was found when oxygen was present in the system. Transalkylation must also have occurred since di- and tri-ethylammonium ions, for example, were detected among the decomposition products of mono-ethylammonium complexes (reaction 10)).

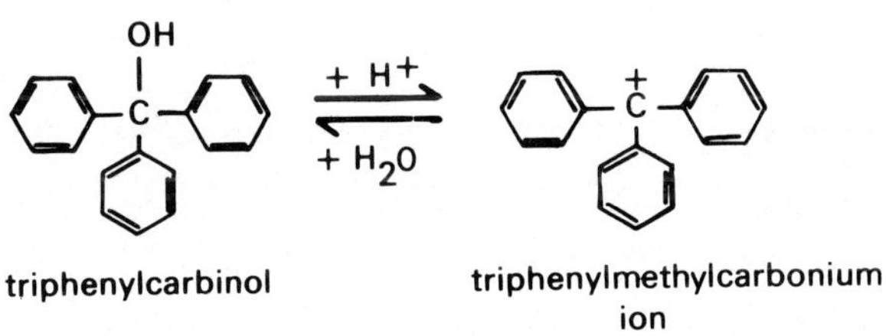

Fig.4. Brønsted site-activated transformation of triphenylcarbinol to triphenylmethylcarbonium on 'dry' montmorillonite. The reaction can be reversed by adding water to the system (after Fripiat et al., 1964).

These observations may be rationalized in terms of the following series of reactions where R represents a *n*-alkyl group,

$$RCH_2CH_2NH_3^+ \ [clay] \xrightarrow{H^+} NH_4^+[clay] + RCH_2\overset{+}{C}H_2 \quad \text{(carbonium ion)} \qquad (5)$$

$$RCH_2\overset{+}{C}H_2 \xrightarrow{+H_2O} RCH_2CH_2OH \ \text{(alcohol)} \qquad (6)$$

$$RCH_2\overset{+}{C}H_2 \xrightarrow{-H^+} RCH{=}CH_2 \ \text{(alkene)} \longrightarrow \text{dimers, telomers} \qquad (7)$$

$$RCH_2CH_2OH \xrightarrow{-H_2O} RCH{=}CH_2 \ \text{(alkene)} \longrightarrow \text{dimers, telomers} \qquad (8)$$

Reaction (8) seems to be the pathway involved in the clay-catalyzed synthesis of butadiene from ethanol (Aleixandre and Fernandez, 1960). In an inert atmosphere, the alkenes formed by reactions (7) and (8) could convert to alkanes by disproportionation,

$$2\ CH_3CH{=}CH_2 \longrightarrow CH_3CH_3 + CH_2{=}CHCH{=}CH_2 \qquad (9)$$

while transalkylation involving carbonium ions, may be represented by,

$$\underset{\text{monoalkylammonium}}{RCH_2CH_2NH_3^+ \ [clay]} \xrightarrow{+\ RCH_2\overset{+}{C}H_2} H^+ \ [clay] + \underset{\text{dialkylammonium}}{R_2(CH_2CH_2)_2NH_2^+} \qquad (10)$$

When air or oxygen is present in the system (additional) alkanes may form via fatty acids by decarboxylation,

$$RCH_2\overset{+}{C}H_2 \xrightarrow{+O_2} RCH_2COOH + H^+ \qquad (11)$$

and

$$RCH_2COOH \longrightarrow RCH_3 + CO_2 \qquad (12)$$

Some idea of the variety of hydrocarbons which are produced when alkylammonium-montmorillonite complexes are heated at 200°C (in air) is provided by the work of Weiss and Roloff (1963). For complexes

with *n*-hexylammonium, *n*-octylammonium, hexamethyldiammonium, and cyclohexylammonium ions, a series of alkanes up to C_8 and alkenes together with aromatic hydrocarbons were detected in the GLC traces. The latter group of compounds could derive from cyclohexene by a reaction of the type

$$3 \text{ cyclohexene} \longrightarrow 2 \text{ cyclohexane} + \text{benzene} \qquad (13)$$

That montmorillonite can activate the decomposition of *n*-alcohols to yield alkenes and alkanes has been reported by Galwey (1969a, 1969b, 1970). Using mild heat treatment (<250°C) with continuous product condensation at -195°C, *n*-hexanol, for example, yielded an equilibrium mixture of hexene isomers with no appreciable cracking. Both *n*-dodecanol and *n*-octadecanol, however, cracked to form nC_5-nC_8 alkenes together with isomeric hexenes; no significant yield of alkanes was found. These results were explained in terms of dehydration (cf. reaction (8)) and double bond migration during thermal desorption. Under conditions where the decomposition products and water vapour were allowed to be readsorbed, the same alcohols yielded mixtures consisting largely of iso-alkanes (C_4-C_8 at <200°C) while yields of alkenes were small. In addition, complex surface transformations occurred including polymerization, cracking, skeletal isomerization, and hydrogenation, strongly indicative of a Brønsted site-activated process involving carbonium ions as intermediates.

Although ethers may form from alkenes according to,

$$2 \text{ RCH=CH}_2 + \text{H}_2\text{O} \rightleftharpoons \text{RCH}_3\text{CHOCHCH}_3\text{R} \qquad (14)$$

they were not detected among the end products, probably because equilibrium (14) would lie far to the left as the water content of the clay was reduced on heating the alkylammonium complex under vacuum. However, secondary ethers with yields of up to 65% were obtained when 1-alkenes (in hexane) were refluxed in the presence of montmorillonite containing interlayer water (Adams et al., 1978). Similarly, Vu Moc Thuy and Maitte (1979) synthesized cyclic acetals in 40-70% yield by refluxing a mixture of diols and enol ethers (in benzene) to which montmorillonite had been added (Fig.5). Although no mechanism was suggested, it seems probable that the process involves a protonated aldehyde as an intermediate.

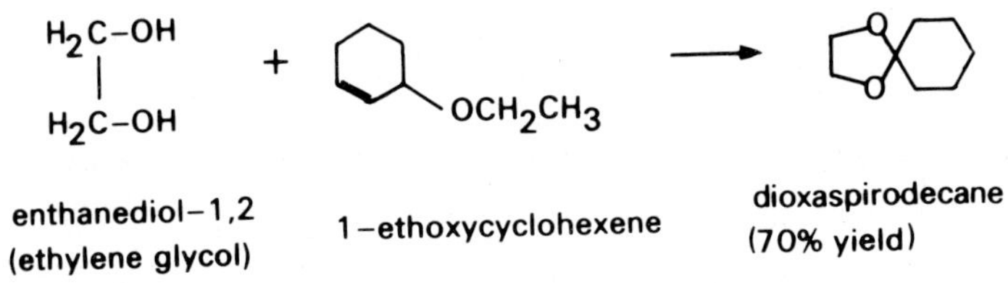

Fig.5. Synthesis of cyclic acetals (e.g. dioxaspirodecane) from diols (e.g. 1,2-ethanediol) and enol ethers (e.g. 1-ethoxycyclohexene) activated by montmorillonite (after Vu Moc Thuy and Maitte, 1979).

If, as indicated above, hydrocarbons can derive from clay-adsorbed amines through fatty acids while the amines themselves may form by decarboxylation of adsorbed amino acids, clay minerals might be expected to play an important part in petroleum genesis under natural conditions (Fripiat and Cruz-Cumplido, 1974). The clay-activated decarboxylation of long chain fatty acids followed by cracking of the *n*-alkanes produced (Almon and Johns, 1976), will be discussed later.

2. Lewis Acidity

For our purposes we shall define Lewis acids as electron-accepting sites. Their presence on clay surfaces has been deduced from, or demonstrated by, the ability of clays to oxidize various organic compounds, activate the polymerization of organic monomers, and form charge-transfer complexes with some organic molecules (Solomon, 1968; Theng and Walker, 1970; Theng, 1971). Unlike the Brønsted acidity of clays, the ability of layer silicates to accept electrons from adsorbed organic species is influenced by the hydration status of the system only in as much as water competes with the organic compound for Lewis sites. Nor is the nature of the exchangeable cations of great importance in Lewis site-activated reactions unless the cations themselves can act as electron acceptors (e.g. Fe^{3+}, Cu^{2+}). Thus, Lewis acidity tends to increase when all of the surface-adsorbed water has been removed, say, by evacuating the system at moderate temperatures (100-150°C) or by heating beyond 200°C. Indeed, Lewis acidity persists and may even increase on heating the clay at temperatures above 400°C when dehydroxylation occurs (Section 3).

In clays which have not been brought to the point of dehydroxylation and whose exchange sites are occupied by the more common cations (Na^{+}, Ca^{2+}, Mg^{2+}), the principal Lewis sites may be identified with ferric ions within the silicate structure and with octahedrally coordinated aluminium exposed at crystal edges. Electron spin resonance (ESR) spectroscopy indicates that structural iron may occupy up to three different octahedral sites in kaolinite and up to four different sites (two tetrahedral and two octahedral) in micas, smectites, and vermiculites (Hall, 1980a). That this iron is actively involved in the redox reactions of clays has long been recognized by the formation of coloured complexes when some aromatic amines and leuco dyes are brought into contact with clays (e.g. Theng, 1971). This principle has been widely used in the manufacture of pressure-sensitive, carbonless paper (e.g. Isaac and Oulton, 1977; McGibbon, 1980).

An outstanding example and one which has been extensively studied is the oxidation of benzidine to its blue monovalent radical cation (Fig.6) which can occur in an aqueous medium even under nitrogen (Dodd and Ray, 1960; Solomon et al., 1968; Theng, 1971; Furukawa and Brindley, 1973; Lahav and Anderson, 1973; Tennakoon et al., 1974a, 1974b; McBride, 1979). Either prior reduction with hydrazine or treatment of the clay with sodium hexametaphosphate, which specifically adsorbs on crystal edges, inhibits colour development or markedly reduces the intensity of the blue colour. These observations have been successfully rationalized in terms of electron transfer from the neutral, colourless diamine to either ferric ions at planar surfaces or aluminium at crystal edges or both.

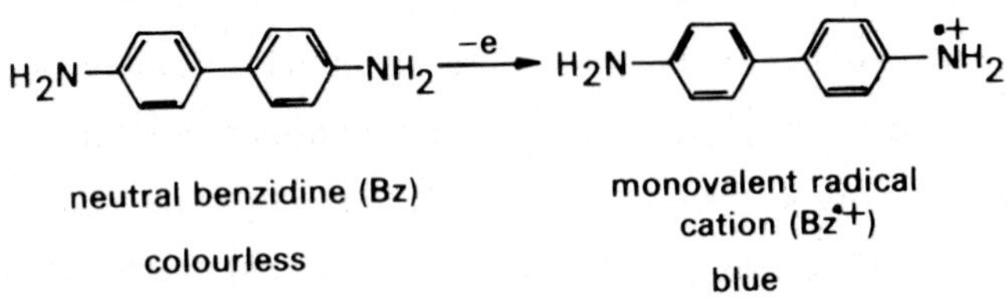

Fig.6. Oxidation of benzidine to the blue monovalent radical cation by electron transfer from the colourless diamine to structural ferric and/or aluminium ions at planar and edge surfaces of clays.

More direct evidence for the participation of structural ferric ions in the benzidine blue reaction has been provided by the adsorption and ^{57}Fe Mössbauer work of Tennakoon et al. (1974a, 1974b). They were able to show that at high adsorption levels (>0.6 meq/g) when intercalation occurred, the Fe^{2+}/Fe^{3+} ratio increased from 0.16 (for the parent montmorillonite) to 0.41 for the benzidine blue complex. A still larger increase in the Fe^{2+}/Fe^{3+} ratio was achieved by treating the clay with hydrazine, while hydrogen peroxide treatment reduced this ratio to zero. It seems clear that electron transfer can occur across the silicate layer boundary with relative ease, in accord with the findings of Rozenson and Heller-Kallai (1978).

However, when the amount of benzidine adsorbed was small and intercalation was absent, the formation of a blue complex was not accompanied by a measurable increase in the Fe^{2+}/Fe^{3+} ratio. This would indicate the presence of electron-accepting sites on external surfaces, including edge aluminium and adsorbed ferric ions, although dissolved oxygen would also contribute to colour development. Alternatively, only a small fraction of structural iron need be involved since the clay provides an adsorbing surface on which the blue species is stabilized by cation exchange. Thus, using hectorite and montmorillonite containing less than 0.05 meq benzidine per g clay, McBride (1979) failed to detect a significant change in the redox status of structural iron by ESR spectroscopy, although colour development was clearly visible, even in a nitrogen atmosphere.

It is not difficult to appreciate that dissolved oxygen would be equally effective in oxidizing adsorbed organic species and hence obscure the inherent electron-accepting activity of the clay surface. However, oxidation by dissolved oxygen in an aqueous clay slurry or suspension is generally a slow process, being limited by the rate at which oxygen can diffuse to the surface (McBride, 1979). Nevertheless, there are instances where electron transfer to oxygen seems to take precedence over that to structural Fe^{3+} or Al^{3+} ions, presumably because the former process has a favourable redox potential or the oxygen is already present at the mineral surface as a chemisorbed species. Thus, colour development in complexes of aniline with smectites (Furukawa and Brindley, 1973) and of dimethylaniline with laponite (a synthetic hectorite) (Vansant and Yariv, 1977) only occurred on exposing the system to atmospheric oxygen, the corresponding reaction products being stabilized by the presence of clays. These findings refer to clays saturated with alkali and alkaline earth cations. However, for clays containing exchangeable Fe^{3+} or Cu^{2+} ions, which themselves can act as electron acceptors (oxidizing agents), the presence of oxygen in the system is not essential for adsorbed aniline to be oxidized (Cloos et al., 1979). Similarly, the oxidation of hydroquinone to *p*-benzoquinone (in smectite slurries) appeared to require chemisorbed oxygen or its radical anion, no oxidation being observed in a nitrogen atmosphere unless Fe^{3+} or Cu^{2+} were the exchangeable cations (Thompson and Moll, 1973). On the other hand, tyrosine was altered in aqueous montmorillonite suspensions whether or not oxygen was present in the system, although the reaction products were different for each case (Thompson and Tsunashina, 1973).

Using IR spectroscopy, Eltantawy and Bavarez (1978) noted that acetaldehyde, adsorbed on montmorillonite, converted to acetic acid at room temperature and under nitrogen. To account for this observation, they postulated that residual oxygen in the system effected the following reactions,

$$H_2O + \tfrac{1}{2}O_2 \rightarrow 2(OH) \quad + \quad CH_3CHO \rightarrow CH_3COOH \quad + \quad H_2O \qquad (15)$$

or

$$H_2O + CH_3CHO \rightarrow CH_3CH(OH)_2 \quad + \quad \tfrac{1}{2}O_2 \rightarrow CH_3COOH \quad + \quad H_2O \qquad (16)$$

It seems unlikely, however, that oxygen was present under the specified conditions, nor is it necessary to invoke the participation of oxygen since the results could satisfactorily be explained in terms of a Cannizaro-type reaction, activated by Brønsted sites,

$$CH_3CHO \xrightarrow{H^+} CH_3\overset{+}{C}HOH \xrightarrow{CH_3CHO} CH_3CH_2OH + CH_3\overset{+}{C}{=}O \xrightarrow{H_2O} CH_3COOH + H^+ \qquad (17)$$

The involvement of carbonium ions in reaction (17) would also account for the observation that the acid formed was much more heat-stable compared with acetic acid which was added, as such, to the clay. At 180-230°C, the acetaldehyde-derived acid was partially converted to acetate; at still higher temperatures (200-300°C), both the acetate and the residual acid were found to interact with surface hydroxyls. This interaction appeared to promote clay dehydroxylation since such a process would not normally occur to any great extent until beyond 350°C. Dehydroxylation was followed by deposition of a carbonaceous material on the clay surface, indicative of polymer formation.

Although no mechanism was suggested by Eltantawy and Bavarez (1978), polymeric species might conceivably arise from dehydration of the ethanol formed in reaction (17), yielding ethylene by reaction (8). The latter would, in turn, undergo dimerization and polymerization through the 'normal' carbonium ion mechanism (Poutsma, 1976). This is not to say that a free radical pathway of polymerization could not be followed. Indeed, the latter alternative, involving electron transfer to the clay surface, may well be operative in the temperature range used by Eltantawy and Bavarez (1978). This is because Lewis acid sites would be exposed as water is removed from the surface (Solomon, 1968) while Brønsted acidity would diminish at the same time. In addition, defect structures and possibly 3-coordinated aluminium sites, acting as electron acceptors (reaction (20)) would be formed on partial dehydroxylation. That changes in coordination number of trivalent cations can occur in the octahedral sheet of layer silicates has recently been demonstrated by Rozenson and Heller-Kallai (1980).

In this context we should mention the work by Juppe and Rau (1969) on the thermal (452°C) decomposition of terphenyls. On the basis of the products formed and their relative abundance, these workers suggested that, in the presence of palygorskite ("attapulgite"), the process followed a carbonium ion pathway, whereas in the absence of clay a free radical mechanism was operative. This suggestion, however, may not be justified since steric and shape factors, related to the open-channel structure of palygorskite, would be expected to influence the decomposition process. By comparison, little doubt attaches to the involvement of structural and surface-adsorbed ferric ions in the oxidative degradation of organic compounds on palygorskite at room temperature and in aqueous media. Using ESR and UV spectroscopy, Cornejo et al. (1980, 1981), for example, have been able to relate the decomposition kinetics of hydrocortisone on palygorskite to the intensity of the signals at $g \sim 4.2$ and $g \sim 2$, due to structural and adsorbed Fe^{3+} ions, respectively. By the same token, sepiolite, which is structurally akin to palygorskite but contains less iron, was less active in decomposing hydrocortisone.

Carbon deposition and the attendant surface blackening are common phenomena when clay-organic complexes are subjected to a heating regime, even at temperatures which are not sufficiently high to cause decomposition of the free (unadsorbed) organic species (Table 3). Thus, with many organic liquids, carbonization could be induced by simply boiling or refluxing the liquid in the presence of clays (Walker, 1967; Eltantawy and Arnold, 1978). The underlying mechanism is not well understood because it is commonly difficult to extract and characterize the black deposit which is strongly held by the surface (Chou and McAtee, 1969). In the case of montmorillonite containing interlayer azobenzene, the deposit formed on heating the complex above 120°C, appeared to consist of a mixture of amines (Adams and Reid, 1977). For the complex with docosanoic or behenic acid, heating at 259°C gave rise to the deposition of a kerogen-like material on the montmorillonite surface (Shimoyama and Johns, 1971). Although both a free radical and an ionic mechanism may be involved in carbonization, the latter pathway seems to be favoured, at least under boiling/refluxing conditions. Thus, any prior treatment which causes a contraction or collapse of the interlayer space of expanding 2:1 type layer silicates, tends to promote surface blackening. Carbonization can be a serious drawback to using clays as catalysts for organic reactions since the deposited material brings about a reduction in catalytic activity, and ultimately a complete 'poisoning' of the catalyst.

Not all clay-organic systems turn visibly black or yield a carbonaceous surface deposit when heated to moderate temperatures (∿300°C). Rather, heating can induce some intercalated organic species to convert to well-defined products in remarkably high yield, thus providing a better means of synthesizing specific organic compounds than would otherwise be possible. A good example is the formation of (gaseous) aniline in 45% yield when an interlayer complex of montmorillonite with 4,4'-diamino-*trans*-stilbene is heated to ∿300°C under vacuum (Tennakoon et al., 1974c). Another Lewis acid-activated reaction is the almost quantitative conversion of triphenylamine to N,N,N',N'-tetraphenylbenzidine in the interlayer space of Na^{+}-montmorillonite (Tricker et al., 1975a). This conversion could follow one of two reaction pathways, (i) formation of a triphenylamine radical cation by electron transfer to the mineral surface, followed by dimerization, deprotonation, and the benzidine rearrangement,

$$2Ph_3N \xrightarrow{-2e} 2Ph_3N^{\cdot +} \longrightarrow (Ph_3\text{-}N\text{-}N\text{-}Ph_3)^{2+} \longrightarrow 2H^{+} + Ph_2N\text{-}Ph\text{-}Ph\text{-}NPh_2 \quad (18)$$

where Ph denotes a phenyl group, or (ii) direct coupling between the benzene rings in the *para* position of two radical cations, followed by proton elimination. Similarly, diphenylethylene gives the indan dimer in 30% yield, when refluxed for 30 min in the presence of Cu^{2+} -montmorillonite (Thomas et al., 1977).

TABLE 3
Clay-activated decomposition of some organic compounds leading to surface blackening and carbon deposition

Organic species	Mineral type	References	Remarks
glycerol	montmorillonite	Walker (1967)	Complex heated until dry
	vermiculite		Mineral immersed in boiling liquid
alkylammonium ions	hectorite and montmorillonite	Chou and McAtee (1969)	Complexes heated between 180 and 350°C. Reaction may involve dehydrogenation and polymerization
n-docosanoic acid (behenic acid)	montmorillonite	Shimoyama and Johns (1971)	Complex heated at 259°C. Reactions involved decarboxylation, cracking and polymerization
n-hexane, 1-butanol, 2-ethoxy ethanol (boiling point <140°C)	illite	Eltantawy and Arnold (1973)	Mineral boiled in organic liquid. Little or no blackening with air-dry montmorillonite but intense blackening occurred with montmorillonite first heated to 250°C which collapsed the interlayer space
ethylene glycol, glycerol 2,2'-oxydiethanol, *n*-dodecane (boiling point >200°C)	air-dry illite, montmorillonite, halloysite, kaolinite, synthetic goethite and gibbsite	Eltantawy and Arnold (1973)	Minerals boiled in organic liquid. At the high boiling point of these compounds, montmorillonite would have collapsed to a basal spacing of ∿1 nm.

The ability of exchangeable Cu^{2+} and Fe^{3+} ions in smectites to act as electron acceptors has already been referred to. In addition to their oxidizing properties, these cations can interact with a variety of aromatic compounds to yield interlayer complexes which show characteristic infrared spectra. Thus, benzene, when presented as a vapour to Cu^{2+}- montmorillonite (Cu-mont) under dehydrating conditions, forms two complexes which can be distinguished by IR spectroscopy: (i) type I (green) in which the molecule is edge-π-bonded to copper and both ring planarity and aromaticity are retained, and (ii) type II (red) where the benzene ring is distorted and aromaticity is lost (Doner and Mortland, 1969; Mortland and Pinnavaia, 1971; Van de Poel et al., 1973) (Fig.7). Type I complexes are obtained at a higher moisture content than type II, both of which are interconvertible by simply changing the hydration status of the system within a given, narrow limit. Symmetrical arenes, such as biphenyl, naphthalene, and anthracene (Rupert, 1973; Tricker et al., 1975b) as well as thiophene (Cloos et al., 1973) behave similarly to benzene. On the other hand, substituted benzenes such as toluene, mesitylene, the isomeric xylenes, and phenol only form type I complexes (Pinnavaia and Mortland, 1971; Fenn and Mortland, 1973); anisole, which forms both types of complexes, is a notable exception (Fenn et al., 1973); type II complexes are also formed with smectites containing Fe^{3+}, Al^{3+} and VO^{2+} ions (Pinnavaia et al., 1974; Cloos et al., 1979).

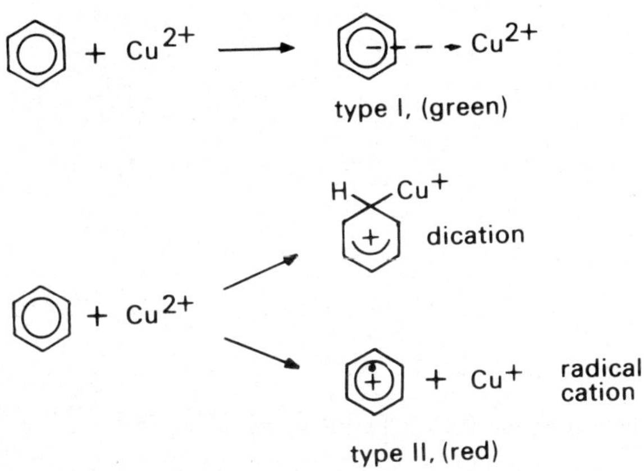

Fig.7. Formation of types I and II complexes between benzene and Cu^{2+}- montmorillonite under dehydrating conditions. Rapid electron interchange probably occurs between the dication and the monovalent radical cation in type II complexes.

On the basis of ESR measurements, Rupert (1973) has suggested that formation of type II complexes between Cu-mont and arenes involves the transfer of an electron from the arene, Ar, to the d^9 orbital of Cu^{2+} yielding a dication or a radical cation, $Ar^{\cdot +}$ (by electron interchange), plus a diamagnetic Cu^+ ion (Fig.7). The general reaction

$$Ar + M^{n+}[clay] \longrightarrow Ar^{\cdot +} + M^{(n-1)+}[clay] \quad (19)$$

is therefore analogous to the benzidine → benzidine blue transformation (cf.Fig.6) except that in reaction (19) it is the interlayer cations (rather than structural Fe^{3+} and/or Al^{3+} ions) which act as electron acceptors. ESR measurements by Pinnavaia et al. (1974) further indicate that the radical cation formed in type II complexes may be coupled or associated.

Whether singly or associated, such species are highly reactive and hence can subsequently interact, either mutually or with other adsorbed organics to yield oligomeric and polymeric products. Anisole on Cu-mont, for example, gave 4,4'-dimethoxybiphenyl (Fenn et al., 1973), presumably by coupling of two $Ph\text{-}OCH_3^{\cdot +}$ in the *para* position with the loss of protons (cf.reaction (18)). When benzene was also present in the interlayers, products consisting of anisole-benzene combinations, in varying proportions, together with anisole oligomers were formed (Mortland and Halloran, 1976). Similarly, on pyrolyzing benzene complexes with Cu^{2+} and Fe^{3+} -montmorillonite (between 180 and 400°C), a mixture of species with $m/e > 200$ were detected in the mass spectrograms. This contrasts with the findings by Tricker et al. (1975b) who obtained only negligible amounts of material having a molecular mass of 2-3 times that of benzene (by heating a type II Cu^{2+}-benzene montmorillonite complex at 50-250°C under vacuum).

The reason behind this discordance is not immediately apparent. However, Mortland and Halloran (1976) prepared their complexes under reflux rather than by the more conventional route of exposing the dry clay to the organic vapour. It is conceivable that the method of preparation and clay pretreatment would affect the orientation of the organic compound in the interlayer space (Theng, 1974). Subtle differences in orientation with respect to the metal cation and to the silicate surface may, in turn, influence the ease and extent of intermolecular coupling.

Stoessel et al. (1977), for example, were able to obtain interlayer polyparaphenylene (or *p*-polyphenyl) by subjecting a Cu^{2+}-benzene complex (type II) to several cycles consisting of dehydration-benzene adsorption-rehydration in air. They suggested that the process was initiated by either a radical cation or a carbonium ion.

The latter alternative was earlier postulated by Kovacic and Kyriakis (1963) for the polymerization of benzene to polyparaphenylene by aluminium chloride-cupric chloride (in the absence of clay). However, in the light of the ESR data (Hall, 1980b) and by analogy with the clay-activated polymerization of styrene (Solomon and Rosser, 1965; Theng and Walker, 1970), it seems more likely that polyparaphenylene formation is initiated by a benzene radical cation. This cation could rapidly dimerize to yield a carbonium ion which then acts as the propagating species. A similar pathway is probably followed during the polymerization of interlayer phenol, toluene, and the xylenes when their respective type I complexes with Cu^{2+}- and Fe^{3+}-smectites are heated at moderate temperatures (<250°C) (Pinnavaia et al., 1974; Tricker et al., 1975b; Mortland and Halloran, 1976). This tendency of substituted arenes to form oligomers and polymers, may explain why such compounds generally fail to yield type II complexes, but this does not mean that arene polymerization at clay surfaces cannot be initiated by a carbonium ion. Indeed, in the case of H^+ and H^+/Al^{3+} clay systems, this is undoubtedly the principal mechanism involved (Theng and Walker, 1970; Aragon de la Cruz and Viton Barbolla, 1975). Similarly, the formation of 'interlayer' polymethylene when diazomethane was added to an ethereal suspension of montmorillonite, saturated with H^+, Mn^{2+} and Cu^{2+} ions, at 0°C appeared to proceed by an ionic mechanism (Bart et al., 1979).

Mass spectrometry indicates that the oligomers formed when Cu-mont complexes with toluene and the xylenes are heated, essentially arise by intermolecular condensation with elimination of hydrogen (Tricker et al., 1975b). However, with aromatic molecules which also contain an ethylenic linkage, a different kind of oligomer is formed, in that no loss of hydrogen apparently occurs. Thus tetraphenylcyclobutane, a dimer of *trans*-stilbene, is the principal product formed on heating an interlayer Cu^{2+}-*trans*-stilbene montmorillonite complex at 50-250°C under vacuum, while the indene complex yielded the corresponding dimer and trimer. It would therefore appear that Cu-mont can discriminate between two kinds of C=C bonds, the ethylenic double bond being preferentially activated by copper compared with the benzene ring. The unique reactivity of Cu^{2+}- and Fe^{3+}-smectites towards aromatic compounds opens "...a vast territory of synthetic organic chemistry...." (Thomas et al., 1977) waiting to be exploited.

To conclude this section, we mention the reaction of aniline with Fe^{3+}-montmorillonite which Cloos et al. (1979) have shown to give a type II complex with formation of a radical cation (cf.reaction (19)). When an aqueous aniline solution was percolated through a sand column containing some Fe^{3+}-montmorillonite, coloured streaks and spots developed along the column indicative of complex formation between aniline and the clay (Cloos et al., 1981). Treating the separated complex with 0.1 M HCl or alkaline $Na_4P_2O_7$ failed to change its colour or its carbon content, a behaviour not unlike that of soil humin or kerogen. The IR and ESR spectra of the complex were consistent with formation of an interlayer type II species which subsequently reacted to yield a humin-like polymer. Although aniline itself is not a typical biogenic compound, it may not be unique in being able to form interlayer polymers in aqueous media through a type II smectite complex. If so (and this point needs to be tested) such a process would have important implications for organic matter transformations in soils and sediments under oxidising conditions, since iron is a ubiquitous element in natural systems.

3. Brønsted-Lewis Site Combination and Interaction

We saw earlier that Lewis acid sites, associated with structural ferric and aluminium ions at planar and edge surfaces, are capable of activating certain organic reactions even in an aqueous medium. As the water content of the clay falls below about 5% w/w, Brønsted acidity becomes very prominent, exerting an overriding influence on organic reactions, except for clays containing Cu^{2+} or Fe^{3+} ions. Thus by simply manipulating the hydration status of the system within quite narrow limits, we can either suppress or enhance the effect of one or the other type of active sites.

This 'interplay' between Lewis and Brønsted types of activity within a single system may be illustrated by the smectite-benzidine interaction in that the blue colour of the complex, in an aqueous medium, turns yellow on drying and back to blue when water is added to the system. Although structural iron is unlikely to be directly involved in the blue → yellow transformation, Tennakoon et al. (1974b) have observed an increase in the Fe^{3+} content of the clay (at the expense of Fe^{2+}). However, this may largely be a reflection of the ease with which structural iron in layer silicates can be oxidized, especially when the interlayers are highly acidic (Rozenson and Heller-Kallai, 1978). It is generally agreed that the yellow colour develops as a direct consequence of the large increase in Brønsted acidity on drying (cf. Fig.1) but some confusion still exists as to the mechanism involved. This arises because protonation of the blue monovalent radical cation may yield either a divalent 'semiquinone' radical cation or a non-radical 'quinodal' cation, both of which are yellow (Fig.8). Which particular divalent species would dominantly form appears to depend on clay type and the experimental conditions used (Tennakoon and Tricker, 1975; McBride, 1979).

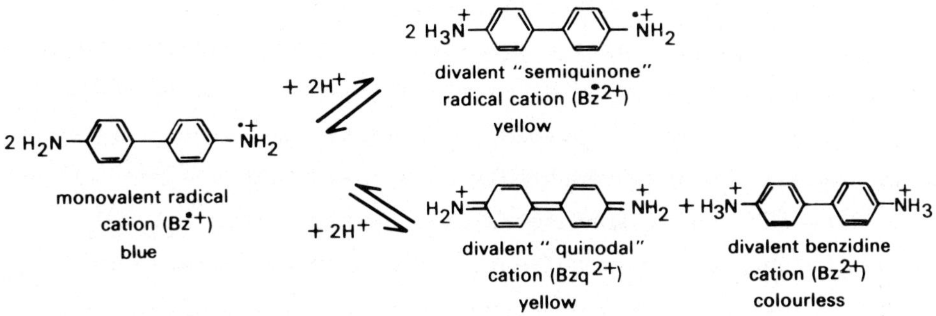

Fig.8. Brønsted acid induced transformation of benzidine blue to either (i) a divalent 'semiquinone' radical cation, or (ii) a 'quinodal' cation plus a divalent benzidine cation.

As mentioned earlier, Brønsted acidity derived from the dissociation of cation-coordinated water would peak as the water content of the clay approaches zero. This type of acidity would decline when all traces of surface-adsorbed water have been removed, e.g. by heating the mineral at 100-200°C in air or at lower temperatures under vacuum. The Brønsted acidity that remains under anhydrous conditions can be ascribed to structural hydroxyl groups. On further heating to 300-500°C, a proportion of such groups is removed in pair-wise fashion, giving rise to 3-coordinated aluminium with the loss of water (cf. reaction (4)),

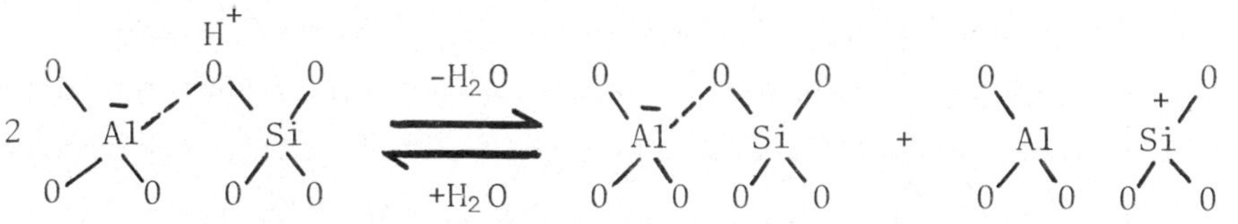

(20)

Brønsted site Lewis site

Thus, (partial) dehydroxylation might be expected to lead to the formation of additional Lewis acid sites at the expense of Brønsted sites, as suggested by Uytterhoeven et al. (1965) for deammoniated ('decationated') synthetic near-faujasites. These points may best be illustrated by the work of Sohn and Ozaki (1980) on the dimerization of ethylene and the isomerization of *n*-butene over a synthetic nickel-substituted montmorillonite. As shown in

Fig.9, they were able to establish that the dimerization and isomerization activity of the mineral ran closely parallel to its surface acidity as measured by *n*-butylamine titration in benzene (cf.Fig.3). The marked increase in acidity for samples previously heated at 50-100°C was undoubtedly associated with a similar rise in the dissociation of surface adsorbed water. Its removal between 100 and 300°C was reflected in the steep decrease in acidity in this temperature range. The parallel behaviour of activity clearly indicated that both dimerization and isomerization were induced by Brønsted sites, involving carbonium ions as intermediates. This view was further supported by the observation that addition of water to samples previously heated above 300°C could restore activity to its peak level (cf. reaction (21)). The slight increase in surface acidity beyond 300°C may be due to a synergistic interaction between Brønsted and Lewis acid sites, i.e. the presence of Lewis sites, formed by such a reaction as depicted in (20), may enhance the acidity of neighbouring Brønsted sites. This type of interaction will be further discussed below.

The occurrence in layer silicates of reaction (20) or a process akin to it has been demonstrated by Wright et al. (1972) for a synthetic (dioctahedral) ammonium mica-montmorillonite (SMM) with a unit cell formula of $[(Al)_4{}^{vi}(Al_xSi_{8-x})^{iv}O_{20}(OH,F)_4]^{x-}.xNH_4{}^+$. Using IR spectroscopy, these workers were able to show that pyridine adsorbed on deammoniated SMM, was completely converted to pyridinium ions (band at 1540 cm^{-1}). On the other hand, for SMM, which has previously been evacuated at 200°C, about equal amounts of pyridinium ions and Lewis acid site-coordinated pyridine (bands near 1463 and 1456 cm^{-1}) were formed. By exposing this complex to water vapour, the latter species was converted to pyridinium,

(21)

These observations and other IR data have led Wright et al. (1972) to conclude that the Lewis acidity of partially dehydroxylated SMM was due to 3-coordinated aluminium, located at the edge of the tetrahedral sheet. This view accords with the recent IR measurements of Chourabi and Fripiat (1981). They showed that the protons, produced on deammoniation of dioctahedral NH_4^+-smectites with tetrahedral substitutions, formed interlayer silanols which can act as Brønsted sites under anhydrous conditions. That both Brønsted and Lewis acid sites can also occur in kaolinite has been demonstrated by Solomon et al. (1971), using pyridine adsorption techniques. In using pyridine as a probe for Brønsted acidity in clays, however, we should be aware of the possibility of hemisalt complex formation which would effectively suppress the band near 1550 cm^{-1} (Farmer and Mortland, 1966; cf. Table 1).

Since Brønsted and Lewis acid sites can occur together under anhydrous conditions, it is not always easy to say which type of acidity might be responsible for activating a given reaction. Thus, the kaolinite-induced polymerization of styrene in various organic solvents displays the characteristics of both a Brønsted and Lewis acid activated reaction (Solomon et al., 1971). Similarly, the clay-activated decarboxylation of long chain fatty acids to their corresponding *n*-alkanes (cf. reaction (12)) and the subsequent cracking of the latter to yield a mixture of petroleum-like hydrocarbons appear to involve both Brønsted and Lewis acid sites (Johns and Shimoyama, 1972; Almon and Johns, 1976). In the case of docosanoic or behenic acid ($C_{21}H_{43}COOH$) the rate of decarboxylation at 250°C in the presence of different clay minerals decreased in the order nontronite > oxidized montmorillonite > montmorillonite > pyrophyllite > saponite > talc for 2:1 type clays, and kaolinite > chrysotile for 1:1 type layer silicates. For montmorillonite, this rate could be increased by ∿43% by adding H_2O_2 to the system, or decreased by ∿36% by pretreating the clay with Na-hexametaphosphate.

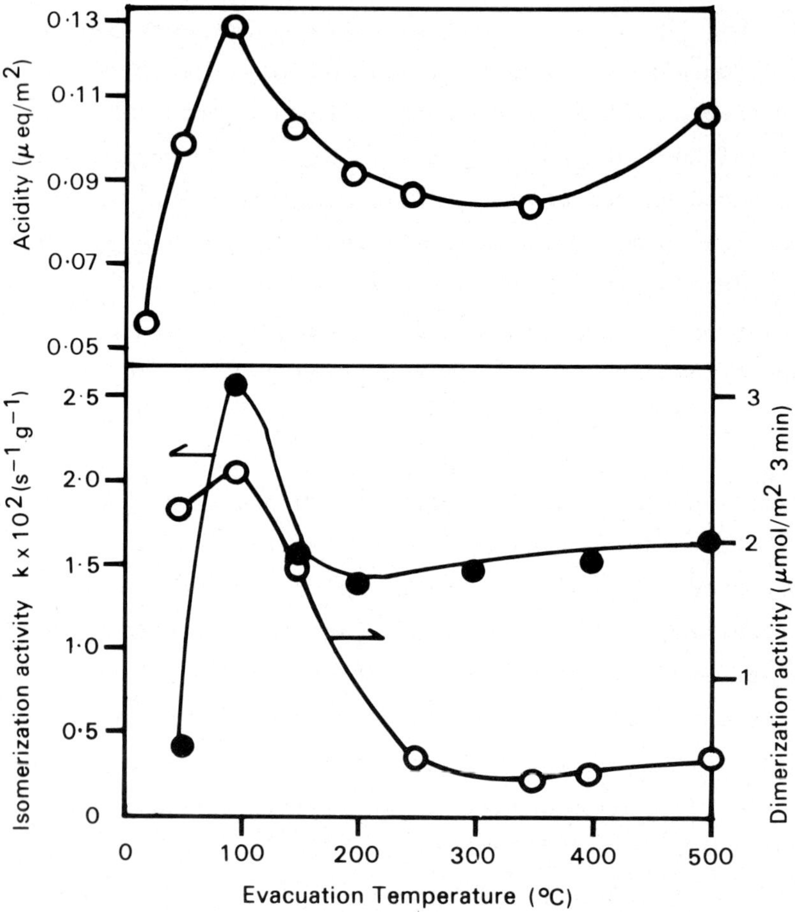

Fig.9. Surface acidity of a synthetic nickel-substituted montmorillonite and its corresponding activity for ethylene dimerization and *n*-butene isomerization, as a function of the temperature at which the mineral had previously been evacuated (after Sohn and Ozaki, 1980).

By analogy with the benzidine blue reaction, Almon and Johns (1976) have explained these observations in terms of Lewis acid site activation, involving the transfer of an electron from the fatty acid to structural ferric ions and edge aluminiums. The acid is converted to the corresponding acyl radical which decomposes to the alkyl radical and then to the normal alkane. Although

$C_{21}H_{44}$ was indeed the principal n-alkane produced in montmorillonite and kaolinite systems, a range of shorter chain, normal and branched, alkanes (C_{14}-C_{20}) were also detected in the GLC traces as well as minor amounts of C_{20}-C_{28} n-alkanes. In the presence of water the ratio of branched to normal alkanes was $\sim$0.1 whereas under anhydrous conditions, this ratio was of the order of 4.5. The small value found in hydrated clay systems would suggest that the cracking process was dominantly activated by Lewis acid sites, involving free radicals as Eisma and Jurg (1969) had proposed earlier. On the other hand, the alkane distribution observed in the absence of water indicated an ionic pathway of cracking. Although Almon and Johns (1976) made no suggestion as to how carbonium ions might be generated, it seems probable that such species are formed through direct attack of C_{21} n-alkane (RH) by Brønsted and Lewis sites,

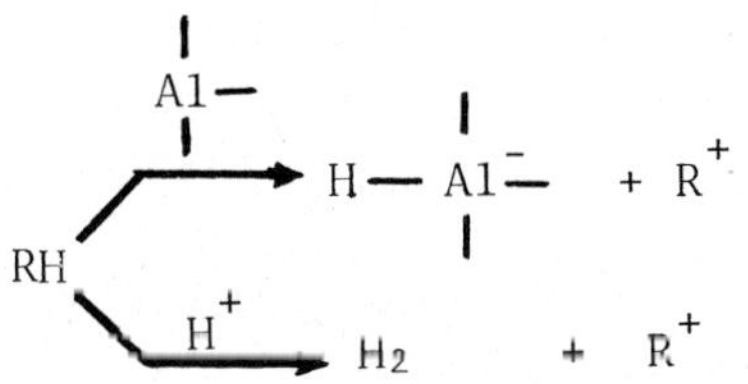

(22)

(23)

$$R^+ \xrightarrow{\text{cracking, isomerization}} \text{short chain alkanes} \qquad (24)$$

In a scheme which provides for the co-existence of Brønsted and Lewis acid sites (under anhydrous conditions) there is clearly scope for a synergistic interaction between the two types of active sites, in that the acidity of Brønsted sites, associated with structural hydroxyl groups (e.g. silanols) could be markedly enhanced through an inductive effect exerted by adjacent electron-deficient aluminiums, acting as Lewis acids (cf. reaction (20)),

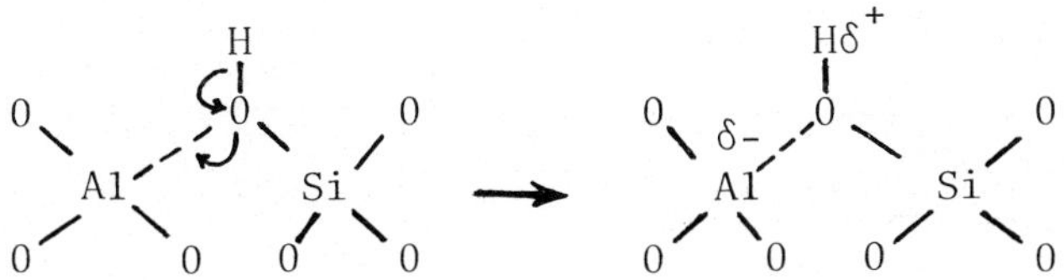

(25)

There is evidence to suggest that this type of Brønsted-Lewis site interaction occurs during the clay-activated cracking of *n*-alkanes (reaction 22-24). Thus, Swift and Black (1974) found that the activity of calcined nickel-substituted SMM (cf. Wright et al. 1972) for the cracking and isomerization of light hydrocarbons (e.g. *n*-hexane) increased with the nickel content of the sample (Table 4). Although they could offer no explanation for this observation at the time, it is clear from the subsequent work by Sohn and Ozaki (1980) that the increase in activity relates to the parallel rise in surface acidity with nickel substitution within the mineral structure. On this basis, it seems reasonable to suggest that calcination leads to dehydroxylation with formation of electron-deficient centres (Lewis sites). This in turn would have the effect of enhancing the acidity of Brønsted sites (associated with structural hydroxyl groups) and hence promote reaction (23).

TABLE 4

Activity of nickel-substituted synthetic mica-montmorillonite (SMM) for the isomerization and cracking of *n*-hexane (from Swift and Black, 1974)

Mineral sample [a]	Nickel content		*n*-hexane conversion			
	weight %	atoms/ unit cell	to isomers (at 260°C) weight %	activity [b]	to cracked products (at 315-343°C) weight %	activity [b]
A	0	0	3.5	0.035	2	0.031
B	1	6.75	45	0.356	28	0.325
C	2	14.3	71	0.636	69	0.918
D	2	15.0	76	0.730	82	1.113
E	3	21.6	80	0.638	100	1.176
F	4	26.4	84	0.588	n.d.	n.d.
G	5	30.5	86	0.599	100	1.021
H	6	35.7	85	0.596	100	1.152

[a] samples previously calcined in air at 537°C

[b] activity is expressed as moles of hexane converted per gram mineral per hour

n.d. = not determined

Likewise, the work of Davidtz (1976) on the conversion of tertiary butylalcohol to isobutylene and water,

$$(CH_3)_3COH \longrightarrow CH_3\overset{\displaystyle CH_3 \atop |}{C}=CH_2 \;+\; H_2O \tag{26}$$

over deammoniated smectites may best be explained in terms of a synergistic interaction between Brønsted and Lewis acid sites. Using a series of dioctahedral 2:1 type clays with varying amounts of isomorphous substitution in the octahedral and tetrahedral sheets, a constant rate of conversion was obtained in all instances (Fig.10a). Attempts at relating mineral activity, derived from the slope of the rate plots, to such parameters as total CEC, surface area, or amount of octahedral substitution, met with little or no success. However, when activity was plotted against tetrahedral surface charge density a good linear relationship was found (Fig.10b). Davidtz (1976) suggested that the protons which found their way into the octahedral sheet, were either shielded or lost during deammoniation. Chourabi and Fripiat's (1981) IR measurements indicate, however, that in these types of smectites, the Brønsted sites are identifiable with tetrahedral silanol groups. This being the case, only those aluminium present in the tetrahedral sheet might be expected to enhance the acidity of such groups by virtue of their mutual proximity. By the same token, this synergistic effect of aluminium on adjacent silanols would increase as the amount of tetrahedral substitution increases.

More direct evidence for Brønsted-Lewis site synergism in clay minerals has been provided by the studies of Sieskind et al. (1979) who found that when cholestanol was heated at 146°C for 16 hr in the presence of kaolinite, a mixture of steranes (e.g. 5α(H)-and 5β(H)-cholestanes) and sterenes (e.g. diacholestene) was formed in high (35-40%) yield. The pathway followed was rather complex, involving backbone rearrangement, formation of carbonium ions at positions 17 or 20 in the steroid structure, and possibly ring homologation with hydride abstraction. In the absence of clays, such a transformation could only occur under extremely acid ("superacidic") conditions, e.g. in toluenesulphonic-acetic acid or HF-SbF_5. Montmorillonite gave similar products but in smaller yields whereas talc, montmorillonite treated with Na-hexametaphosphate, and alumina yielded only trace amounts of Δ^2-cholestene. These observations indicate that Lewis acid sites (associated with aluminium at crystal edges) are not directly involved in activation; nor are Brønsted sites (surface silanols) active by themselves. However, the presence of Lewis sites is essential in order to enhance the protonic activity of adjacent silanol groups. We should note that since steranes and sterenes occur widely in ancient sediments and crude oil, clays are strongly implicated in the formation of these compounds under natural conditions.

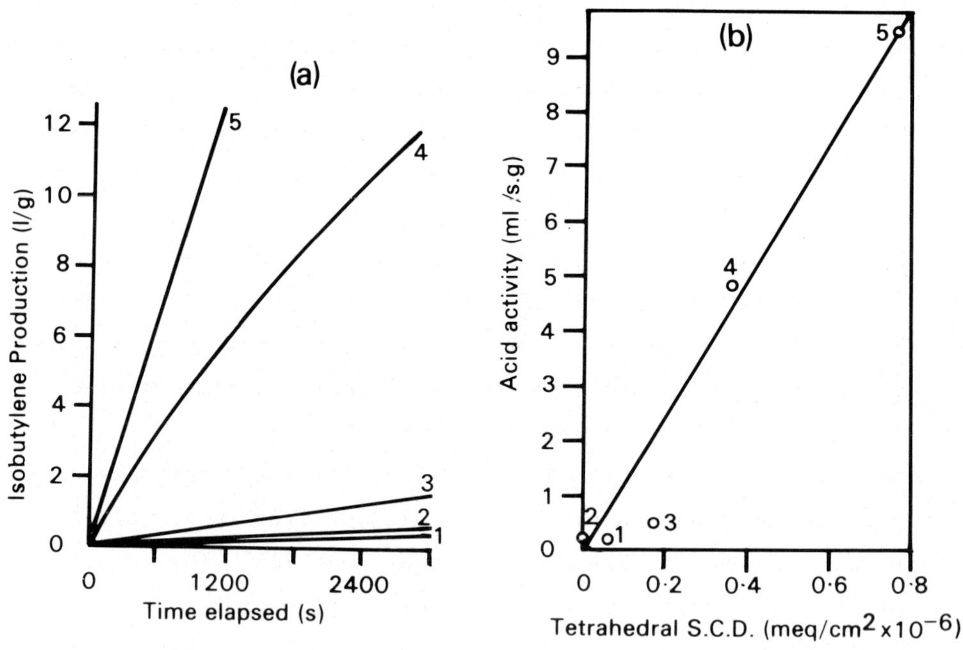

Fig.10. (a) rate plots for the production of isobutylene from tertiary butylalcohol activated by a series of smectites (samples 1-5) with different amounts of isomorphous substitution in both octahedral and tetrahedral positions; (b) relationship between clay activity (slope of rate plots) and tetrahedral surface charge density (S.C.D.); for each point, the respective sample number is given in brackets (after Davidtz, 1976).

REFERENCES

Abdo, S., Cruz, M.I. and Fripiat, J.J., 1980. Metallation-demetallation reaction of tin tetra(4-pyridyl)porphyrin in Na-hectorite. Clays Clay Miner., 28: 125-129.

Adams, J.M. and Reid, P.I., 1977. Azobenzene intercalates of montmorillonite. Clays Clay Miner., 25: 228-230.

Adams, J.M., Ballantine, J.A., Graham, S.H., Laub, R.J., Purnell, J.H., Reid, P.I., Shaman, W.Y.M. and Thomas, J.M., 1978. Organic synthesis using sheet silicate intercalates: Low-temperature conversion of olefins to secondary ethers. Angew. Chem. Int. Ed. Engl., 17: 282-283.

Aleixandre, V. and Fernandez, T., 1960. Influence catalytique des différents minéraux argileux sur la synthèse du butadiène a partir de l'éthanal. Silicates Inds., 25: 243-248.

Almon, W.R. and Johns, W.D., 1976. Petroleum forming reactions : clay catalyzed fatty acid decarboxylation. Proc. Int. Clay Conf. Mexico D.F., 399-409.

Aragon de la Cruz, F. and Viton Barbolla, C., 1975. Polymerization of styrene, indene and divinylbenzene in the interlamellar space of montmorillonite. Advan. Org. Geochem., Proc. 7th Int. Meet., 851-856.

Barshad, I. and Foscolos, A.E., 1970. Factors affecting the rate of the interchange reaction of adsorbed H^+ on the 2:1 clay minerals. Soil Sci., 110: 52-60.

Bart, J.C., Cariati, F., Erre, L., Gessa, C., Micera, G. and Piu, P., 1979. Formation of polymeric species in the interlayer of bentonite. Clays Clay Miner., 27: 429-432.

Benesi, H.A., 1956. Acidity of catalyst surfaces. I. Acid strength from colors of adsorbed indicators. J. Am. Chem. Soc., 78: 5490-5494.

Benesi, H.A., 1967. Acidity of catalyst surfaces. II. Amine titration using Hammett indicators. J. Phys. Chem., 61: 970-973.

Brown, C.B. and White, J.L., 1969. Reactions of 12 *s*-triazines with soil clays. Soil Sci. Soc. Am. Proc., 33: 863-867.

Buchler, J.W., 1975. Static coordination chemistry of metalloporphyrins. In: K.M. Smith (Editor), Porphyrins and Metalloporphyrins. Elsevier, New York, N.Y., p.147.

Canesson, P., Cruz, M.I. and Van Damme, H., 1979. X.P.S. study of the interaction of some porphyrins and metalloporphyrins with montmorillonite. Proc. Int. Clay Conf., Oxford, 217-225.

Chaussidon J. and Calvet, R., 1965. Evolution of amine cations adsorbed on montmorillonite with dehydration of the mineral. J. Phys. Chem., 69: 2265-2268.

Chou, C.C. and McAtee, J.L., Jr., 1969. Thermal decomposition of organo-ammonium compounds exchanged onto montmorillonite and hectorite. Clays Clay Miner., 17: 339-346.

Chourabi, B. and Fripiat, J.J., 1981. Determination of tetrahedral subsitutions and interlayer surface heterogeneity from vibrational spectra of ammonium in smectites. Clays Clay Miner., 29: 260-268.

Cloos, P., 1972. Interactions entre les pesticides et la fraction minérale du sol. Pédologie, 22: 148-173.

Cloos, P. and Laura, R.D., 1972. Adsorption of ethylenediamine (EDA) on montmorillonite saturated with different cations - II. Hydrogen- and ethylenediammonium-montmorillonite: protonation and hydrogen bonding. Clays Clay Miner., 20: 259-270.

Cloos, P., Laura, R.D. and Badot, C., 1975. Adsorption of ethylenediamine (EDA) on montmorillonite saturated with different cations - V. Ammonium- and triethylammonium-montmorillonite : ion exchange, protonation and hydrogen bonding. Clays Clay Miner., 25: 417-423.

Cloos, P., Van de Poel, D. and Camerlynck, J.P., 1973. Thiophene complexes on montmorillonite saturated with different cations. Nature Phys. Sci., 243: 54-55.

Cloos, P., Moreale, A., Broers, C. and Badot, C., 1979. Adsorption and oxidation of aniline and *p*-chloroaniline by montmorillonite. Clay Miner., 14: 307-321.

Cloos, P., Badot, C. and Herbillon, A., 1981. Interlayer formation of humin in smectites. Nature, 289: 391-393.

Conley, R.F. and Lloyd, M.K., 1971. Adsorption studies on kaolinites - II. Adsorption of amines. Clays Clay Miner., 19: 273-282.

Cornejo, J., Hermosin, M.C., White, J.L., Peck, G.E. and Hem, S.L., 1980. Oxidative degradation of hydrocortisone in presence of attapulgite. J. Pharm. Sci., 69: 945-948.

Cornejo, J., Hermosin, M.C., White, J.L. and Hem, S.L., 1981. Reactivity of structural and adsorbed iron in sepiolite and palygorskite for oxidation of hydrocortisone. Abstracts Int. Clay Conf., Bologna/Pavia, 64-65.

Cruz, M., White, J.L. and Russell, J.D. 1968. Montmorillonite-*s*-triazine interactions. Israel J. Chem., 6: 315-323.

Davidtz, J.C., 1976. The acid activity of 2:1 layer silicates. J. Catalysis, 43: 260-263.

Dodd, C.G. and Ray, S., 1960. Semiquinone cation adsorption on montmorillonite as a function of surface acidity. Clays Clay Miner., 8: 237-251.

Doner, H.B. and Mortland, M.M., 1969. Benzene complexes with Cu(II) montmorillonite. Science 166: 1406-1407.

Durand, B. Pelet, R. and Fripiat, J.J., 1972. Alkylammonium decomposition on montmorillonite surfaces in an inert atmosphere. Clays Clay Miner., 20: 21-35.

Eisma, E. and Jurg, J.W. 1969. Fundamental aspects of the generation of petroleum. In: G. Eglinton and M.T.J. Murphy (Editors), Organic Geochemistry. Springer, Berlin, pp. 676-698.

Eltantawy, I.M. and Arnold, P.W., 1973. Catalytic decomposition of organic molecules by clays. Nature Phys. Sci., 244: 144.

Eltantawy, I.M. and Bavarez, M., 1978. Clay-organic molecule interactions. Oxidation of acetaldehyde by montmorillonite in N_2 atmosphere at room temperature. Clays Clay Miner., 26: 285-290.

Farmer, V.C. and Mortland, M.M., 1965. An infrared study of complexes of ethylamine with ethylammonium and copper ions in montmorillonite. J. Phys. Chem.. 69: 683-686.

Farmer, V.C. and Mortland, M.M., 1966. An infrared study of the co-ordination of pyridine and water to exchangeable cations in montmorillonite and saponite. J. Chem. Soc. (A), 344-351.

Fenn, B.D. and Mortland, M.M., 1973. Interlamellar metal complexes on layer silicates. II. Phenol complexes in smectites. Proc. Int. Clay Conf., Madrid, 591-603.

Fenn, D.B., Mortland, M.M. and Pinnavaia, T.J., 1973. The chemisorption of anisole on Cu(II) hectorite. Clays Clay Miner., 21: 315-322.

Fowkes, F.M., Benesi, H.A., Ryland, R.B., Sawyer, W.M., Detling, K.D., Loeffler, E.S., Folckemer, F.B., Johnson, M.R. and Sun, Y.P., 1960. Clay-catalyzed decomposition of insecticides. J. Agr. Food Chem., 8: 203-210.

Frenkel, M., 1974. Surface acidity of montmorillonites. Clays Clay Miner., 22: 435-441.

Frenkel, M. and Solomon, D.H., 1977. The decomposition of organic amines on montmorillonites under ambient conditions. Clays Clay Miner., 25: 463-464.

Fripiat, J.J., 1968. Surface fields and transformation of adsorbed molecules in soil colloids. Trans. 9th Int. Congr. Soil Sci., Adelaide, 1: 691-699.

Fripiat, J.J. and Cruz-Cumplido, M.I., 1974. Clays as catalysts for natural processes. Ann. Rev. Earth Planet. Sci., 2: 239-256.

Fripiat, J.J. and Helsen, J., 1966. Kinetics of decomposition of cobalt coordination complexes on montmorillonite surfaces. Clays Clay Miner., 14: 163-179.

Fripiat, J.J., Helsen, J. and Vielvoye, L., 1964. Formation des radicaux libres sur la surface des montmorillonites. Bull.Groupe Fr. Argiles, 15: 3-10.

Fripiat, J.J., Jelli, A.N., Poncelet, G and André, J., 1965. Thermodynamic properties of adsorbed water molecules and electrical conduction in montmorillonites and silicas. J. Phys. Chem., 69: 2185-2197.

Furukawa, T. and Brindley, G.W., 1973. Adsorption and oxidation of benzidine and aniline by montmorillonite and hectorite. Clays Clay Miner., 21: 279-288.

Fusi, P., Ristori, G.G. and Malquori, A., 1980. Montmorillonite-asulam interactions: I. Catalytic decomposition of asulam adsorbed on H- and Al-clay. Clay Miner., 15: 147-155.

Gerstl, Z. and Banin, A., 1980. Fe^{2+}-Fe^{3+} transformations in clay and resin ion-exchange systems. Clays Clay Miner., 28: 335-345.

Galwey, A.K., 1969a. Reactions of alcohols adsorbed on montmorillonite and the role of minerals in petroleum genesis. J. Chem. Soc. (D), 577-578.

Galwey, A.K., 1969b. Heterogeneous reactions in petroleum genesis and maturation. Nature 223: 1257-1260.

Galwey, A.K., 1970. Reactions of alcohols and of hydrocarbons on montmorillonite surfaces. J. Catalysis, 19: 330-342.

Hall, P.L., 1980a. The application of electron spin resonance spectroscopy to studies of clay minerals: I. Isomorphous substitutions and external surface properties. Clay Miner., 15: 321-335.

Hall, P.L., 1980b. The application of electron spin resonance spectroscopy to studies of clay minerals. II. Interlamellar complexes - structure, dynamics and reactions. Clay Miner., 15: 337-349.
Hammett, L.P. and Deyrup, A.J., 1932. A series of simple basic indicators. I. The acidity functions of mixtures of sulfuric and perchloric acids with water. J. Am. Chem. Soc., 54: 2721-2739.
Harter, R.D. and Ahlrichs, J.L., 1969. Effect of acidity on reactions of organic acids and amines with montmorillonite clay surfaces. Soil Sci. Soc. Am. Proc., 33: 859-863.
Heller-Kallai, L., Yariv, S. and Riemer, M., 1973. Effect of acidity on the sorption of histidine by montmorillonite. Proc. Int. Clay Conf. Madrid, 651-662.
Henmi, T. and Wada, K., 1974. Surface acidity of imogolite and allophane. Clay Miner., 10: 231-245.
Isaac, P.A.H. and Oulton, T.D., 1977. Reduced charge montmorillonite pigment for sensitizing recording material. U.S. Patent 4,028,133.
Johns, W.D. and Shimoyama, A., 1972. Clay minerals and petroleum-forming reactions during burial and diagenesis. Am. Assoc. Petrol. Geol. Bull., 56: 2160-2167.
Juppe, G. and Rau, H., 1969. The attapulgus clay-catalysed thermal decomposition of terphenyls. J.Appl.Chem., 19: 120-124.
Kovacic, P. and Kyriakis, A., 1963. Polymerization of benzene to *p*-polyphenyl by aluminum chloride-cupric chloride. J. Am. Chem. Soc., 85: 454-458.
Lagaly, G. and Weiss, A., 1970. Inhomogeneous charge distributions in mica-type layer silicates. Réunion Hispano-Belga de Minerales de la Arcilla, Madrid, 179-187.
Lahav, N. and Anderson, D.M., 1973. Montmorillonite-benzidine reactions in the frozen and dry states. Clays Clay Miner., 21: 137-139.
Laura, R.D. and Cloos, P., 1975a. Adsorption of ethylenediamine (EDA) on montmorillonite saturated with different cations - III. Na,-K- and Li-montmorillonite : ion exchange, protonation, co-ordination and hydrogen bonding. Clays Clay Miner., 23: 61-69.
Laura, R.D. and Cloos, P., 1975b. Adsorption of ethylenediamine (EDA) on montmorillonite saturated with different cations - IV: Al-, Ca- and Mg-montmorillonite : protonation, ion-exchange, co-ordination and hydrogen bonding. Clays Clay Miner. 23: 343-348.
McBride, M.B., 1979. Reactivity of adsorbed and structural iron in hectorite as indicated by oxidation of benzidine. Clays Clay Miner. 27: 224-230.
McGibbon, G., 1980. Color developer materials. Ger.Offen.3,019,077.
Mortland, M.M., 1966. Urea complexes with montmorillonite : An infrared absorption study. Clay Miner., 6: 143-156.
Mortland, M.M. 1968. Protonation of compounds at clay mineral surfaces. Trans. 9th Int. Congr. Soil Sci., Adelaide 1: 691-699.
Mortland, M.M., 1970. Clay-organic complexes and interactions. Advan. Agron., 22: 75-117.
Mortland, M.M. and Berkheiser, V., 1976. Triethylene diamine-clay complexes as matrices for adsorption and catalytic reactions. Clays Clay Miner., 24: 60-63.

Mortland, M.M. and Halloran, L.J., 1976. Polymerization of aromatic molecules on smectite. Soil Sci. Soc. Am. J., 40: 367-370.
Mortland, M.M. and Pinnavaia, T.J., 1971. Formation of copper(II)-arene complexes on the interlammellar surfaces of montmorillonite. Nature Phys. Sci., 229: 75-77.
Mortland, M.M. and Raman, K.V., 1968. Surface acidity of smectites in relation to hydration, exchangeable cation, and structure. Clays Clay Miner., 16: 393-398.
Mortland, M.M., Fripiat, J.J., Chaussidon, J. and Uytterhoeven, J.B., 1963. Interaction between ammonia and the expanding lattices of montmorillonite and vermiculite. J. Phys. Chem., 67: 248-258.
Paecht-Horowitz, M., 1977. Polymerization on surfaces. Midland Macromol. Monogr., 89-113.
Pinnavaia, T.J. and Mortland, M.M., 1971. Interlamellar metal complexes of layer silicates. I. Copper(II)-arene complexes on montmorillonite. J. Phys. Chem., 75: 3957-3962.
Pinnavaia, T.J., Hall, P.L., Cady, S.S. and Mortland, M.M., 1974. Aromatic radical cation formation on the intracrystal surfaces of transition metal layer lattice silicates. J. Phys. Chem., 78: 994-999.
Poutsma, M.L., 1976. Mechanistic considerations of hydrocarbon transformations catalyzed by zeolites. In: J.A. Rabo (Editor), Zeolite Chemistry and Catalysis. ACS Monograph 171, American Chemical Society, Washington, D.C., pp.437-551.
Raman, K.V. and Mortland, M.M., 1969. Proton transfer reactions at clay mineral surfaces. Soil Sci. Soc. Am. Proc., 83: 313-317.
Rozenson, I. and Heller-Kallai, L., 1978. Reduction and oxidation of Fe^{3+} in dioctahedral smectities - III. Oxidation of octahedral iron in montmorillonite. Clays Clay Miner., 26: 88-92.
Rozenson, I. and Heller-Kallai, L., 1980. Dehydroxylation of dioctahedral phyllosilicates. Clays Clay Miner., 28: 355-368.
Rupert, J.P., 1973. Electron spin resonance spectra of interlamellar Cu(II)-arene complexes on montmorillonite. J. Phys. Chem., 77: 784-790.
Russell, J.D. and Fraser, A.R., 1971. I.R. Spectroscopic evidence for interaction between hydronium ions and lattice OH groups in montmorillonite. Clays Clay Miner., 19: 55-59.
Russell, J.D., Cruz, M. and White, J.L., 1968a. The adsorption of 3-aminotriazole by montmorillonite. J. Agr. Food Chem., 16: 21-24.
Russell, J.D., Cruz, M., White, J.L., Bailey, G.W., Payne, W.R., Pope, J.D. and Teasley, J.I., 1968b. Mode of chemical degradation of *s*-triazines by montmorillonite. Science, 160: 1340-1342.
Santos, A., Rodriquez, A., Gonzales Garmendia, J. and Barrios, J., 1970. Espectres infrarrojos de muestras homoionicas de montmorillonita y vermiculita y sus complejos interlaminares con amines. Réunion Hispano-Belga de Minerales de la Arcilla, Madrid, 87-97.
Shimoyama, A. and Johns, W.D., 1971. Catalytic conversion of fatty acids to petroleum-like paraffins and their maturation. Nature Phys. Sci., 232: 140-144.

Sieskind, O., Joly, G. and Albrecht, P., 1979. Simulation of the geochemical transformations of sterols : superacid effect of clay minerals. Geochim. Cosmochim. Acta, 43: 1675-1679.
Sivalov, E.G., Vasil'ev, N.G. and Ovcharenko, F.D., 1980. Nature of the acidity of kaolinite lateral faces. Effect of Group I exchange metals. Dokl. Akad, Nauk SSSR., 253: 1176-1179.
Sohn, J.R. and Ozaki, A., 1980. Acidity of nickel silicate and its bearing on the catalytic activity for ethylene dimerization and butene isomerization. J. Catalysis, 61: 29-38.
Solomon, D.H., 1968. Clay minerals as electron acceptors and/or donors in organic reactions. Clays Clay Miner., 16: 31-39.
Solomon, D.H. and Rosser, M.J., 1965. Reactions catalyzed by minerals. I. Polymerization of styrene. J. Appl. Polymer Sci., 9: 1261-1271.
Solomon, D.H., Loft, B.C. and Swift, J.D., 1968. Reactions catalysed by minerals. IV. The mechanism of the benzidine blue reaction on silicate minerals. Clay Miner., 7: 389-397.
Solomon, D.H., Swift, J.D. and Murphy, A.J., 1971. The acidity of clay minerals in polymerization and related reactions. J. Macromol. Sci. Chem., A5: 587-601.
Spencer, M.S. and Whittam, T.V., 1980. Catalysis on non-faujasitic zeolites and other strongly acidic oxides. In: C. Kemball and D.A. Dowden (Senior Reporters), Catalysis, Vol.3. The Chemical Society, London, pp. 189-227.
Stoessel, F., Guth, J.L. and Wey, R., 1977. Polymérisation de benzene en polyparaphenylene dans une montmorillonite cuivrique. Clay Miner., 12: 255-259.
Stul, M. and Mortier, W.J., 1974. The heterogeneity of the charge density in montmorillonites. Clays Clay Miner., 22: 391-396.
Swift, H.E., and Black, E.R., 1974. Superactive nickel-aluminosilicate catalysts for hydroisomerization and hydrocracking of light hydrocarbons. Ind. Eng. Chem., Prod. Res. Develop., 13: 106-110.
Swoboda, A.R. and Kunze, G.W., 1968. Reactivity of montmorillonite surfaces with weak organic bases. Soil Sci. Soc. Am. Proc., 32: 806-811.
Tahoun, S. and Mortland, M.M., 1966. Complexes of montmorillonite with primary, secondary, and tertiary amides: I. Protonation of amides on the surface of montmorillonite. Soil Sci., 102: 248-254.
Take, J., Kawai, H. and Yoneda, Y., 1975. Liquid-phase desorption kinetics of benzeneazodiphenylamine physisorbed on silica gel. J. Catalysis, 36: 356-360.
Tamele, M.W., 1950. Chemistry of the surface and the activity of alumina-silica cracking catalyst. Disc. Faraday Soc., No.8: 270-279.
Tennakoon, D.T.B. and Tricker, M.J., 1975. Surface and intercalate chemistry of layered silicates. Part V. Infrared, ultraviolet, and visible spectroscopic studies of benzidine-montmorillonite and related system. J.C.S. Dalton Trans., 1802-1806.

Tennakoon, D.T.B., Thomas, J.M. and Tricker, M.J., 1974b. Surface and intercalate chemistry of layered silicates. Part II. An iron-57 Mössbauer study of the role of lattice-substituted iron in the benzidine blue reaction of montmorillonite. J.C.S. Dalton Trans., 211-2215.

Tennakoon, D.T.B., Thomas, J.M., Tricker, M.J. and Graham, S.H., 1974c. Selective organic reactions in sheet-silicate intercalates. Conversion of 4,4'-diamino-*trans*-stilbene into aniline. J.C.S.Chem.Commun., 124-125.

Tennakoon, D.T.B., Thomas, J.M., Tricker, M.J. and Williams, J.O., 1974a. Surface and intercalate chemistry of layered silicates. Part I. General introduction and uptake of benzidine and related organic molecules by montmorillonite. J.C.S. Dalton Trans., 2207-2211.

Theng, B.K.G., 1971. Mechanisms of formation of colored clay-organic complexes. A review. Clays Clay Miner., 19: 383-390.

Theng, B.K.G., 1974. The Chemistry of Clay-Organic Reactions. Hilger, London, 343 pp.

Theng, B.K.G., 1979. Formation and Properties of Clay-Polymer Complexes. Elsevier, Amsterdam, 362 pp.

Theng, B.K.G. and Walker, G.F., 1970. Interactions of clay minerals with organic monomers. Israel J. Chem., 8: 417-424.

Thomas, J.M., Adams, J.M., Graham, S.H. and Tennakoon, D.T.B., 1977. Chemical conversion using sheet-silicate intercalates. In: J.B. Goodenough and M.S. Whittingham (Editors), Solid State Chemistry of Energy Conversion and Storage. Advan. Chem. Ser. 163, American Chemical Society, Washington, D.C., pp. 298-315.

Thompson, T.D. and Moll, W.F., Jr., 1973. Oxidative power of smectites measured by hydroquinone. Clays Clay Miner., 21: 337-350.

Thompson, T.D. and Tsunashina, A., 1973. The alteration of some aromatic amino acids and polyhydric phenols by clay minerals. Clays Clay Miner., 21: 351-361.

Touillaux, R., Salvador, P., Vandermeersche, C. and Fripiat, J.J., 1968. Study of water layers adsorbed on Na and Ca montmorillonite by the pulsed nuclear magnetic resonance technique. Israel J. Chem., 6: 337-349.

Tricker, M.J., Tennakoon, D.T.B., Thomas, J.M. and Graham, S.H., 1975b. Novel reactions of hydrocarbon complexes of metal-substituted sheet silicates: thermal dimerisation of *trans*-stilbene. Nature, 253: 110-111.

Tricker, M.J., Tennakoon, D.T.B., Thomas, J.M. and Heald, J., 1975a. Organic reactions in clay-mineral matrices: mass-spectrometric study of the conversion of triphenylamine to N,N,N',N'-tetraphenylbenzidine. Clays Clay Miner., 23: 77-82.

Uytterhoeven, J.B., Christner, L.G. and Hall, W.K., 1965. Studies of hydrogen held by solids. VIII - The decationated zeolites. J. Phys. Chem., 69: 2117-2126.

Van Damme, H., Crispin, M., Obrecht, F., Cruz, M.I. and Fripiat, J.J., 1978. Acid-base and complexation behaviour of porphyrins on the intracrystal surface of swelling clays: meso-tetraphenylporphyrin and meso-tetra (4-pyridyl) porphyrin on montmorillonites. J. Colloid Interface Sci., 66: 43-54.

Van de Poel, D., Cloos, P., Helsen, J. and Janini, E., 1973. Comportement particulier du benzene adsorbé sur la montmorillonite cuivrique. Bull. Groupe Fr. Argiles, 25: 115-126.

Vansant, E.F. and Yariv, S., 1977. Adsorption and oxidation of dimethylaniline by laponite. J.C.S. Faraday Trans. I., 73: 1815-1824.

Vu Moc Thuy and Maitte, P., 1979. Réaction des éthers d'énols sur diols en présence de montmorillonite : une synthèse commode d'acétals cycliques. Bull. Soc. Chim. Fr., 5-6: 264-265.

Walker, G.F., 1967. Catalytic decomposition of glycerol by layer silicates. Clay Miner., 7: 111-112.

Weiss, A. and Roloff, G., 1963. Die Rolle organischer Derivate von glimmerartigen Schichtsilikaten bei der Bildung von Erdöl. Proc. Int. Clay Conf., Stockholm, 2: 373-378.

White, J.L., 1976. Protonation and hydrolysis of *s*-triazines by Ca-montmorillonite as influenced by substitutions at the 2-, 3- and 6-positions. Proc. Int. Clay Conf., Mexico D.F., 391-398.

Wright, A.C., Granquist, W.T. and Kennedy, J.V., 1972. Catalysis by layer lattice silicates. I. The structure and thermal modification of a synthetic ammonium dioctahedral clay. J. Catalysis 25: 65-80.

Yariv, S. and Heller, L., 1970. Sorption of cyclohexylamine by montmorillonite. Israel J. Chem., 8: 935-945.

Yariv, S. and Heller-Kallai, L., 1973. I.R. evidence for migration of protons in H- and organo-montmorillonites. Clays Clay Miner., 21: 199-200.

Yariv, S., Heller, L., Sofer, Z. and Bodenheimer, W., 1963. Sorption of aniline by montmorillonite. Israel. J.Chem. 6: 741-756.

CLAY MINERALS AS CATALYSTS

F.D.OVCHARENKO

Institute of Colloid and Water Chemistry of the Ukrainian Academy of Sciences, Kiev, USSR

In their natural form clay minerals are scarcely used in direct catalysis due to the presence of different elements such as iron which poison catalysts, exchange alkaline earth ions, their abrasion resistance, low dispersivity and ion exchange ability (e.g. kaolinite), etc.

Applying methods of phase analysis, thermochemistry, adsorption, spectroscopy (especially radiospectroscopy), ion exchange, electrokinetics and others it became possible to clarify the chemical nature of the clay mineral surface as well as the nonhomogeneity of the energy and adsorption characteristics of its active centres. The mechanism of interaction between substances of different geometrical and electron structure with disperse phase surfaces has also been clarified (1 - 3). This knowledge laid foundations for developing methods of forming porous colloidal adsorbents and catalysts based on clay minerals.

Clay minerals acquire catalytic properties after acid treatment which substitutes alkaline and alkaline earth ions with hydrogene ions. The analysis of the crystallochemical structure of protonated forms of montmorillonite shows that the chemical action of acid causes a decrease of magnesium, iron and aluminium in isomorphous substitution of the cation-exchange capacity and of the negative surface charge. Previous studies on clays have not provided full information on the nature of their exchange acidity, chemical stability of H-forms of laminar silicates and the influence of different factors on the process of spontaneous transformation of H-forms, their ageing, etc.

Studies of these problems open up a wide application of acid forms of clay minerals as catalysts, stable phases in gas-adsorption chromatography, filling agents for polymers, etc.

Together with N.G.Vassilyev we carried out a broad research program on surface properties of clay minerals in the acid form and developed an understanding of the nature of exchange acidity (4,5), which proved to be useful for the study of chemical transformations by catalytic action. The results of this research are as follows :

1. We discovered a regularity of ion exchange for acid forms of montmorillonite, nontronite, vermiculite, beidelite, palygorskite and kaolinite. In the suggested model (Fig. 1) the three exchange protons compensate

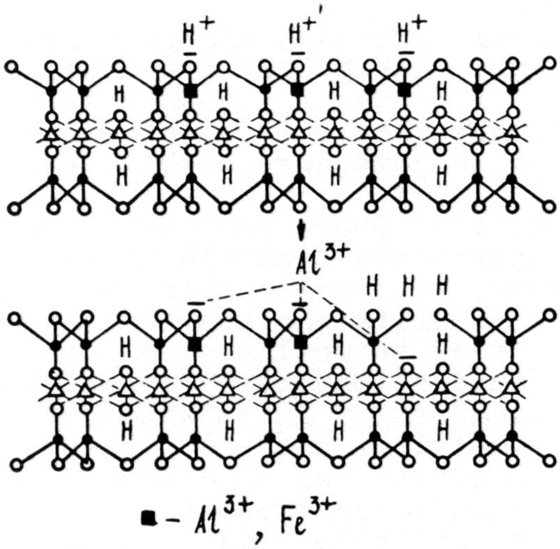

Fig.1. Destruction of tetrahedrons, in which Si is substituted by Al or Fe, by exchange protons.

the split bonds (i.e. the transformation of SiO^- into SiOH groups) and the single Al^{3+} (or Fe^{3+}) ion, liberated from the tetrahedron, neutralizing the charge of two adjacent stable tetrahedrons and the charge formed at the apex of the destroyed tetrahedron. At the same time the Al^{3+} (or Fe^{3+}) ions are capable of neutralizing charges of stable tetrahedrons in neighbouring basal planes. The transition of mineral charge from the tetrahedral to the octahedral sheet does not influence the stability of the clay minerals before and after the formation of H-forms. The destruction of defective tetrahedrons with hydrogen ions was also proved by other methods.

When the nonstoichiometric isomorphous substitutions were totally or partially located in the octahedral sheet, a hydrogen or mixed H-Al-Fe-mineral form with high acid properties was obtained. When the nonstoichiometric substitutions were located in the tetrahedral sheet the exchange protons quickly destroy the defective tetrahedrons and as a result Al^{3+} (or Fe^{3+}) ions appear on the aluminosilicate surface.

2. The H-form of clay minerals is known to be chemically instable and to transform spontaneously into the Al-, Fe- and Mg-form (ageing). Investigations (5-7) prove that the rate of ageing of clay minerals increases sharply with increase of temperature (Fig. 2).

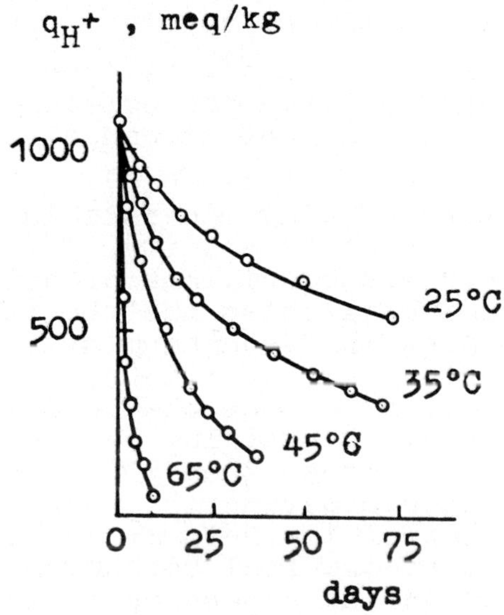

At the same time the rate of ageing decreases as the solid phase increases, especially in the form of a dense precipitate.

Water is one of the ageing process agents for H-clays as demonstrated by the fact that they never age in pure organic solvents, while in water mixtures the rate of ageing depends on the amount of the organic component in the mixture. We propose that the ageing process of clay minerals is a heterogeneous chemical reaction performed not by free protons but by hydronium ions.

The presence of water molecules in the system as well as the catalytic activity of the exchange H^+ and Al^{3+} ions causes a split of the Si-O-Si bonds and as a result two molecules of silicic acid are released to the aquilibrium solution :

$$2\left[\begin{matrix}\equiv Si-O \\ \equiv Si-O\end{matrix}\!\!>Si(OH)_2\right] + 4H_2O \longrightarrow 4\left[\equiv SiOH\right] + 2Si(OH)_4$$

The siloxane bonds are split according to the equation

$$2\left[\equiv Si\text{-}O\text{-}Si(OH)_3\right] + 2H_2O \longrightarrow 2\left[\equiv SiOH\right] + 2Si(OH)_4$$

After the total transformation of the hydrogen form of clay into the Al-form (ageing cycle) its exchange capacity decreases.

Water molecules are also fulfilling some additional functions. Thus, for example, due to a considerable energy of hydration of the Al^{3+}, $Fe^{3+}$and Mg^{2+} ions water assists in removing them from the octahedral sheet. Water promotes the activity of exchange protons due to the formation of hydronium ions (H_3O).

3. Catalytic activity of clay minerals is primarily connected with their surface acidity. That is why both clay composition and the conditions of acid ions in the surface of the aluminosilicates must be taken into account when determining the mechanism of chemical transformations under catalytic conditions.

For the aluminium form of the clay one must take into account the decrease in containing of exchanging Al^{3+} iones at temperatures higher than 500°C. Even greater caution is needed when studying the mechanism of catalytic transformations of the H-forms of minerals for which determination of the real mechanism of catalysis is vague due to the ageing processes of the H-form. In the absence of water the H-form is chemically stable up to 200°C. At higher temperatures the exchanging protons are spontaneously migrating into the interior of the mineral crystal, thus influencing its catalytic activity. After heating the H-forms of clay minerals at 300-400° the exchange capacity of the mineral irreversibly decreases within an hour by an amount equal to that of the hydrogen exchange ions.

4. When preparing clay catalysts by acid activation one must take into account that both the composition of the exchange cations and the exchange capacity of montmorillonite depend on the activation time Table 1). With prolongation of activation time the content of hydrogen ions decreases and that of aluminium ions increases on the mineral surface. After treatment of the mineral with

a hot 20% solution of H_2SO_4 a pactically homoionic Al-form may be created within 6 hours. However, weaker acid solutions are preferable for the acid treatment of clays.

TABLE 1.

Influence of the activation time on the amount of exchange H^+ and Al^{3+} ions of montmorillonite and the Al^{3+} ion concentration in the starting solution of Al^{3+} p-p. Activation temperature is 98°C.

Time of processing, hours	Exchange capacity, meq/kg	Amount of exchange ions, meq/kg		Al^{3+} p-p
		H^+	Al^{3+}	N eq/dm^3
1	960	500	465	0.06
2	880	390	490	0.07
3	790	310	485	0.10
4	780	180	610	0.13
5	700	100	605	0.17
6	650	40	660	0.21
7				
8	630	30	640	0.22

The decrease of the amount of H-ions during acid activation is due to the accumulation of Al-ions and polysilicic acid as a result of the partial dissolution of the mineral. This process depends on the content of the disperse phase in the system as well as on the composition of the intermicellar medium and the temperature.

Clay catalysts have found broad industrial application in petroleum cracking and in gasoline production. We shall discuss the results of investigations on clay catalysts in reactions of a large number of organic substances, carried out in the Soviet Union.

Ionic polymerization of monomers. Being solid active acids, clay minerals are of great interest as catalysts for ionic polymerization of monomers, capable of polymerization by cations.

Together with M.T.Bryck we recently carried out a broad investigation (8-12) of cation polymerization of cyclosiloxanes, in particular, octamethylcyclotetrasiloxane (OMCTS), catalyzed by clay minerals in different homoionic (Li-, Na-, K-, Ca-, Mg-, Cu-, Ni-, Al- and H) forms. The highest activity among them was shown by the hydrogen and aluminiun forms.

For polymerization of styrene from the vapour phase on Al-montmorillonite (6) and of OMCTS on each of the homoionic forms of different kaolinites (9-10) with different exchange capacities a symbatic dependence exists between the concentration of exchange ions on the mineral surface and their catalytic activity in the polymerization reaction. With a totally dehydrated H-form of dispersed minerals the OMCTS polymerization occurs under the influence of exchange protons while with air-dry and hydrated laminar silicates the polymerization is initiated by hydronium ions consisting of an exchange proton combined with several molecules of water.

The increased activity of the air-dry kaolinite samples in the Ca, Mg, Cu, Ni, and AL forms demonstrates that the acidity of the mineral surface necessary for the OMCTS polymerization is obtained by the polarization of water molecules combined with the exchange ions (11, 12). Polarization of water molecules increases with the decrease of the cation radius and increases with its charge. The yield of the polymer which is formed on the surface of various cation exchanged kaolinites is proportional to the ratio between the charge and the radius of the exchange ion (Ze/r) (Fig. 3) and proves the participation of nonexchange protons of polarized water molecules in the OMCTS polymerization.

The investigation of the OMCTS polymerization on partially dehydrated (at 200°C) samples of some ion forms of Glukhov Kaolinite has shown that the dehydration of the Cu- and Ni-forms of the mineral has practically no effect on the rate of the process and the yield of the polymers, while for the Mg-form the polymer yield is increased by a factor 1.5 and for the Ca-form by a factor 10 (Fig. 4).

The removal of weakly bound water with low acidity serving as an agent for transfering chains in OMCTS cation polymerization sharply increases the activity of the Mg- and Ca-form of Kaolinite and causes a considerable increase of molecular mass (from 30 000 - 50 000 to 300 000 - 380 000) of polydimethylsiloxane (PDMS) which are formed on their surfaces.

The data on OMCTS polymerization on different ion forms of kaolinite demonstrate that water, if strongly bound with the exchange ions, takes part in ion couple formation, thus initiating polymerization.

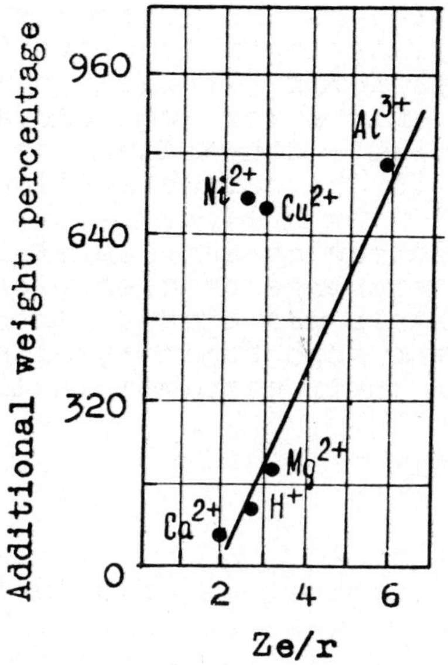

Fig. 3. Dependence of PDMS (as additional weight percentage) formed on the surface of different homoionic minerals on the ratio of ion charge and ion radius (Ze/r).

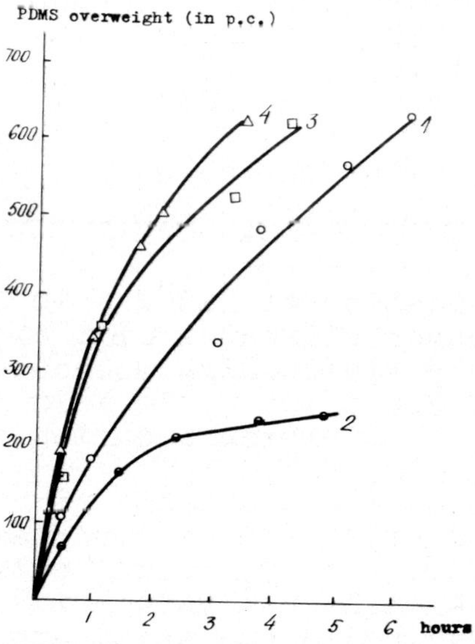

Fig.4.

Therefore, there is a definite difference between the nature of active centres on the surface and H-forms of ion exchangers for catalyzing polymerization. For dehydration of homoionic forms (Al-form included) polymerization increases only after the removal of weakly bound water. Upon complete dehydration of homoionic forms their activity in the polymerization reaction is only slight if any.

The OMCTS polymerization on samples of kaolinite and montmorillonite with exchange centres present only on the edges of the crystals was studied in order to clarify the influence of the topochemistry of active centres of layer silicates on their catalytic activity (Table 2).

TABLE 2.

Physical and chemical properties of acid forms of layer silicates

Mineral	Amount of exchange Al^{3} ions meq/kg	Amount of Si-OH groups mmol/kg	
Kaolinite			
from Glukhovets	12.0	22.0	10.0
from Prosyanov	15.0	35.0	17.0
from Glukhov	45.0	90.0	68.0
Montmorillonite			
from Pyzhev	70.0	165.0	55.0

The OMCTS polymerization at 263°C - 10°C in the presence of 1-10% of layer silicates with exchange centres on crystal edges only yields high molecular PDMS with a molecular mass of $3.0 \times 10^5 - 6.6 \times 10^5$. Under these conditions the OMCTS conversion amounted to 84-96% (Table 3).

The investigation of the synthesic processes on the monomer-mineral boundary has some important theoretical and practical aspects. A clay mineral catalyst in the H- and Al-form has a high catalytic activity in the ion polymerization of cyclosiloxane which are of great industrial interest for the synthesis of silicone rubbers. The cyclosiloxane conversion amounts to 92-95% (up to 85% in industry) within 1-3 hours of polymerization at relatively mild temperature conditions (60-120°C) and a molecular mass up to 300 - 500 thousand is obtained.

TABLE 3.

Degree of transformation of OMCTS, and molecular mass of POMS on layer silicate surfaces

Mineral	Concentration of disperse minerals in OMCTS mass %	Degree of transformation of OMCTS %	PDMS molecular mass
Kaolinite			
from Glukhovets	3.0	95.7	600 000
	5.0	95.6	730 000
	10.0	93.3	470 000
from Prosyanov	3.0	83.3	300 000
	5.0	94.0	660 000
	10.0	93.4	420 000
from Glukhov	3.0	94.0	360 000
	5.0	95.5	300 000
	10.0	92.4	370 000
Montmorillonite			
from Pyzhev	1.0	83.4	310 000
	3.0	91.4	340 000
	5.0	93.6	360 000
	10.0	95.4	270 000

At the same time the catalyst may play the part of a modified filler for a polymer composition with a high load capacity (up to 90-95%) and with homogenous distribution of the polymer in the product (the so called "polymerizing load).

Polymerization on the surface of such disperse catalysts-fillers promotes their homogeneity by filling the pores and cracks : it improves the hydrophobicity nature, corrosion resistance and strength of the composite polymer substance. We hope that in the near future a major success will be achieved both in the understanding of the mechanism of polymerization on the clay mineral surface and in the improvement of their properties as fillers for composite materials.

Alkylation. Effective catalysts used for alkylation, such as aluminium chloride, boron fluoride and acids are causing corrosion of equipment whereas cation exchange resins have a low thermal stability. H.I.Areshidze and his colleagues (13, 14) showed that these drawbacks may be eliminated by using montmorillonite (gumbrine) treated with a 0.15 N solution of aluminium chloride. Under optimal conditions the yield of cresols was 67.1% of the amount of carbolic acid in the reaction while the yield of xylenols from alkylation of meta- and para-cresols was respectively 66.8 and 57.3%.

The alkylation reaction of isobutane with butylenes on montmorillonite gave a high yield of the alkylate. When alkylating carbolic acid with hexene-1 the total yield of the product was 73.2%.

A kaolinite catalyst was also studied for the process of alkylating benzene with propylene (15).

Izomerization. Montmorillonite, kaolynite, vermiculite and hydrous mica after acid activation show good catalytic properties for the isomerization reaction of - and -pinenes to camphene and other substances of the terpene group (r-cymene, dipentene, tricyclene, 3-carene). Together with I.Dimov we showed that the reaction conditions in catalytic isomerization can be varied aiming at the production of camphene, r-cymene and dipentene, using H- and Al-montmorillonite. For H -vermiculite (15) the highest yield of the desired products (camphene and tricyclene) was 68%, for H - kaolinite 61% and for H-hydrous mica 62%.

Transformation of naphthene hydrocarbons in the presence of H-montmorillonite in gasoline fractions 60-150° was proved by isomerization of cyclopentane homologues to hydroaromatic and cyclohexane hydrocarbons (16) and of cyclic alcanol-1-methylcyclopentane-1 resulting in cyclohexane (17).

Amination. Amination of cyclohexanol on H-montmorillonite and montmorillonite with 8% of ferric oxyde was also investigated. Montmorillonite with 8% of ferric oxyde was also turned out to be the best catalyst with a yield of cyclohexamine of about 90.16% of the initial cyclohexanol (18). The same catalyst (containing 5% of ferric oxyde) has shown a high activity in the amination reaction of butanediol - 1.4 with a pyrrolidine yield comparable to that obtained with zeolitic modified catalysts (19).

Effective catalysts with oxydes of silver, vanadium, magnesium, iron and chromium on montmorillonite (Georgian gumbrine) were selected for oxidative ammonalysis of alkylaromatic hydrocarbons. In the reaction of oxidative ammonalysis of o- , m- and p-xylenes a catalyst with 10% vanadium oxyde on gumbrine produces a yield of dinitriles of phtalic acids of respectively 51.0, 51.9 and 54.6% (20).

Dehydration. H-montmorillonite is a good dehydrating catalyst (21, 22). Thus when dehydrating alcohol on an H-clay one can easily obtain ethylene or diethyl ether. In the butanediole-1.4 dehydration reaction on montmorillonite, activated with 15% hydrochloric acid the tetrahydrofurane yield equals 98% of theory (at 280°C and a volume rate of 0.3 h^{-1}). Dehydration of alkylcyclopentals with catalyzed by H-clay is even more successful than homogeneous acid dehydration.

Dehydrogenation. The catalyst formed by means of an ion exchange reaction between montmorillonite and 0.1N solution of $CrCl_3$ proved to be very effective for the synthesis of alkenylaromatic hydrocarbons in the dehydrogenation reactions of ethylenebenzene and isopropylenebenzene. The same catalyst gives higher yield of C_2-C_4 unsaturated hydrocarbons.

A Ni-bentonite catalyst is very effective in the hydrogenation reaction of benzene and fatty oils, and palladium (1% on clay) is especially active in the hydrogenation reaction of furane with a maximum yield of tetrahydrofurane of 80.4%.

Cracking. In the cracking reaction of a standard crude oil bentonite catalytic activity equals that of the available synthetic industrial catalysts. The conditions for augmenting the cracking ability of H-bentonites with water vapour are already well known as well as effects of rare-earth metals, molybdenum oxydes, aluminium and chromium on coking and regeneration of bentonite cracking catalysts.

Catalytic activity of H-bentonite on the vapours of the amylene-hexylene fraction of the benzene group is high based on yield of paraffins and cyclic hydrocarbons. Olefin hydrocarbons are especially effectively transformed at 250°C.

The same bentonite turned out to be an active catalyst for the olygomerization process of -olefins (C_9H_{14}) to lubricating oils.

Cracking catalyst (H-montmorillonite) is useful for treating both medium and heavy oil fractions. Natural bentonites of Azerbaidzhan are effective in cracking heavy fractions for diesel fuel even without any acid treatment. The yield of clear product is about 70% (23, 24).

REFERENCES

1. Ovcharenko, F.D., 1979. The Study of the Mechanism of Water Interaction with the Surface of Solids. In: Physical and Chemical Mechanics and Lyophility of Dispersed Systems. Naukova Dumka Publishers,Kiev, No.11,pp. 5-15.
2. Ovcharenko,F.D., Kruglitsky, N.N., Tarasevich,Yu.I.,

1980. Colloid and Chemical Criteria of the Industrial Application of Bentonites. In: Bentonites. Nauka Publishers,Moscow, pp. 147-155.

3. Tarasevich, Yu.I., 1976. Adsorption on Clay Minerals. Naukova Dumka Publishers, Kiev, 351 pp.

4. Ovcharenko, F.D., Vassilyev, N.G., 1973. In: Proceedings of the Academy of Sciences of the Ukrainian SSR, Kiev, No. 12, pp. 38-51.

5. Vassilyev, N.G., Ovcharenko, F.D., 1977. Surface Chemistry of the Acid Forms of the Natural Silicate Layers. In: Achievements in Chemistry, Moscow, vol. XVI, book 8, pp. 1488-1511.

6. Vassilyev, N.G., Ovcharenko, F.D., Buntova, M.A., 1974. Specific Aspects of the Arbitrary Transformation of the H-Montmorillonite into Al-Form. In: Colloid Magazine, Moscow, vol. 36, pp. 26-31.

7. Vassilyev, N.G., Buntova, M.A., Ovcharenko, F.D., 1975. Destruction Mechanism and Ageing in H-Form Clay Minerals. In: Colloid Magazine, Moscow, vol. 37, pp. 16-23.

8. Bryck, M.T., 1973. Polymerization on the Solid Surface of the Nonorganic Substances. In: A Summary in Science and Technology. Chemistry and Technology of Highly Molecular Compounds. Riga, vol. 4, pp. 142-184.

9. Savkin, A.G., Ovcharenko, F.D., Bryck, M.T., Vassilyev, N.G., 1979. The Study of the Polymerization Processes on H-Kaolinite. In: Proceedings of the USSR Academy of Sciences, vol. 247, No. 4, pp. 884-887.

10. Bryck, M.T., Bagley, N.N., Vassilyev, N.G., Ovcharenko, F.D., 1978. The Impact of the Solvent Adsorption on the Catalytic Transformation of OMCTS on the Surface of Ion Exchangers in H-Form. In: Proceedings of the USSR Academy of Sciences, vol. 238, No. 11, pp. 113-116.

11. Ovcharenko, F.D., Bryck, M.T., Bagley, N.N., Vassilyev, N.G., 1977. Absorption Water Impact on the Catalytic Transformation of the OMCTS on the Surface of Al-Kaolinite. In: Proceedings of the USSR Academy of Sciences, vol. 235, No. 4, pp. 865-868.

12. Scobets, I.Ye., Ovcharenko, F.D., Bryck, M.T., Vassilyev, N.G., 1979. OMCTS Catalytic Transformation on the Surface of the Al-Form of Kaolinite. In: Colloid Magazine, Moscow, vol. 41, No. 3, pp. 501-506.

13. Areshidze, Kh.I., Chivadze, G.O., Khakhnelidze, V. V.,Balakhishvily, L.I., 1977. Thermocatalytic Transformation of C_2-C_4 Hydrocarbons. In: Oil Chemistry, vol. 17, No. 6, pp. 852-857.

14. Areshidze, Kh.I., Chivadze, G.O., Cheidze, G.S., Mortikov, Ye. S., Zenkovsky, S.M., 1979. Gumbrine

Application for Alkylation. In: Proceedings of the Academy of Sciences of the Georgian SSR, vol. 5, No. 2, pp.111-117.
15.Batallova, Sh., Pak, N.D., Mukitayeva, T.R., 1977. The Study of the Impact of the Activity of Acid Centers of Catalysts on Their Selectivity in Izomerization Reaction. In: Proceedings of the Academy of Sciences of the Kazakh SSR, No. 2, pp. 33-38.
16.Areshidze, Kh.I., 1980. Research in Oil Chemistry. Metsnisreba Publishers, Tbilissy, 235 pp.
17.Areshidze, Kh.I., Tseretely, B.S., 1970. Isomerization of the 1-Methylcyclopentene-1 under the Conditions of the Expanding Cycles. In: Proceedings of the Academy of Sciences of the Georgian SSR, No. 2, pp. 325-328.
18.Ibid., 1971, vol. 61, No.4, pp.69-72.
19.Areshidze,Kh.I.,Chivadze, G.O., 1973. Synthesis of Pyrolidine and other Heterocycles from Butanediol-1.4 and Ethanediol-1.2. In: Chemistry of Heterocyclic Compounds. Riga, No. 7, pp. 937-941.
20.Areshidze, Kh.I., Chivadze, G.O., Naskidashvili, Ts. I., 1976. Oxidative Ammonalization of Xylenes. In: Oil Chemistry. Vol. 16, No. 2, pp. 269-274.
21.Areshidze, Kh.I., Tavartkiladze, Ye.K., Chivadze, G.O., 1970. The Study of the Catalytic Properties of both Natural and Activated Alumosilicates in the Dehydration of Butanediol-1.4. In: Journal of Applied Chemistry. Vol.13, pp.601-606.
22.Areshidze, Kh.I., Dolidze, A.V., Maisuradze, G.V., Balardjishvily, G.I.,1974. Catalytic Transformations of 1-Methylcyclopentanol-1 and 1.2-Dimethylcyclopentanols in the Presence of Gumbrine. In: Proceedings of the Academy of Sciences of the Georgian SSR, vol. 74, No. 2, pp. 341-344.
23.Zulfugarov, S.G., Rassulova, S.M., Shirinova, E.B., 1960. Cracking Properties of the Catalysts Made of Azerbaidzhan Bentonites. In: Works of the Chemistry Institute of the Academy of Sciences of the Azerbaidzhan SSR. Vol. 1 8, pp. 5-23.
24.Zulfugarov, S.G., Kuliev, R.Sh., Annagiev, M.Kh., Dzhavadov, M.N.,1978. Patent No 2501610/23-05.

THE CHEMISTRY OF RHODIUM ON THE SURFACE OF HECTORITE

R.A. SCHOONHEYDT, V. MARIEN and H. VAN BRABANT
Katholieke Universiteit Leuven
Centrum voor Oppervlaktescheikunde en Colloïdale Scheikunde
De Croylaan 42, B - 3030 Leuven (Heverlee), Belgium.

ABSTRACT

$[Rh(NH_3)_5Cl]^{2+}$ on the interlamellar surface of hectorite decomposes with the formation of mineral-bound NH_4^+. The resultant Rh(III) complex is transformed to a $Rh(I)(CO)_2$ moiety by gaseous CO but the reaction is incomplete. With H_2 the decomposed $[Rh(NH_3)_5Cl]^{2+}$ is reduced to Rh metal. Rh is adsorbed on hectorite from $RhCl_3.3H_2O$ solutions mainly by precipitation of chloro-hydroxyl complexes, while Na^+ is ion-exchanged by H^+.

INTRODUCTION

Several methods of incorporating cationic Rh into the interlamellar space of smectites have been presented in the literature. Pinnavaia and coworkers (1975, 1976, 1979) exchanged Rh_2^{4+} on H-hectorite in methanol to produce an active hydrogenation catalyst of end-standing double bonds after activation with H_2. An active catalyst is also obtained after reacting the mineral-bound Rh_2^{4+} with PPh_3 (triphenylphosphine) to mineral-bound $[Rh(PPh_3)_x]^+$. The exchange of Rh_2^{4+} in the presence of air and catalytic amounts of HF gives a Rh(III)-hectorite, which, when treated with PPh_3, produces a selective catalyst for the hydrogenation of acetylenes to olefines. Cationic Rh-complexes of the type $[Rh(COD)S_2]^+$ (COD = 1,5-cyclooctadiene, S = acetone or acetonitrile) and $[RhClH_2(PPh_3)_x(P-P^+)_{3-x}]$ (x = 1 or 2, $P-P^+ = Ph_2PCH_2CH_2P^+Ph_2CH_2Ph$) also can be intercalated (Pinnavaia et al., 1976; Quayle et al., 1979). The intercalated complex is active for the hydrogenation of 1-hexene. Finally, asymmetric hydrogenation of prochiral olefines has been reported with cationic Rh-complexes on several clay minerals; in some cases optical yields were strongly dependent on the type of clay mineral used as the support (Mazzei et al., 1980; Mazzei et al., 1979; Mazzei et al., 1979).

In this paper we extend the immobilization of Rh on clays to complexes with inorganic ligands. The complexes on the surface are characterized chemically and spectroscopically.

EXPERIMENTAL

Samples

After several washings of the raw powdered hectorite in 2 % suspension with 1N NaCl, the fraction < 2μm was separated by centrifugation, concentrated, washed until free of Cl^- in dialysis membranes, lyophilized and stored.

$[Rh(NH_3)_5Cl]^{2+}$-loaded hectorite ($[Rh(NH_3)_5Cl]^{2+}$-H) was prepared by exchanging 1 % Na-hectorite suspensions in dialysis membranes against 0.005N $[Rh(NH_3)_5Cl]^{2+}$-solutions overnight at room temperature. The samples were washed until free of Cl^- and dried in air at room temperature. The Rh-content was 0.352 mmol g^{-1} dry clay. The adsorption of Rh from $RhCl_3.3H_2O$ solutions was studied as a function of the [Rh]/([Rh] + [Na]) ratio in the equilibrium solutions and as a function of the pH. Batches of 40 cm^3 solutions of 0.01N total normality and of initial [Rh]/([Rh] + [Na]) ratios between 0.01 and 1 were shaken with dialysis membranes containing 10 cm^3 1 % Na-hectorite suspension. At high [Rh]/([Rh] + [Na]) ratios the pH was adjusted to 4 with NaOH. After the reaction, pH, Na^+ and Rh^{3+} were determined in the exchange solution,the latter two by atomic absorption (AA).

Materials

Hectorite was obtained from the Source Clays Repository of the Minerals Society. $[Rh(NH_3)_5Cl]Cl_2$, $RhCl_3.3H_2O$ were purchased from Strem Chemicals and used as such. The gases H_2, O_2, CO were ultra-high purity gases from L'Air Liquide and ethylene was purchased in lecture bottles from Matheson.

Physical and chemical characterization of the samples

The thermal decomposition of $[Rh(NH_3)_5Cl]^{2+}$ was followed by heating a known amount of sample in vacuo up to 673K in a spring balance with a sensitivity of .5 mg mm^{-1}. It was also followed by IR spectroscopy on a Beckman IR12 spectrometer in the range 1200 - 3800 cm^{-1} and by diffuse reflectance spectroscopy on a Cary-17 spectrometer with type I diffuse reflectance attachment in the range 700 - 220 nm. Na-hectorite was the reference.

The adsorption of CO was studied by IR spectroscopy in the range 1200 - 3800 cm^{-1}. After evacuation in the range 298 - 423 K 13.3 kPa CO was allowed to adsorb first at room temperature, then at the degassing temperature, and evacuated. In a closed circulating system with small volume the CO consumption and CO_2 production were followed up to 523K by measurement of the pressure changes with a Bell and Howell pressure transducer in the presence and absence of a solid CO_2-trap respectively. The reduction with H_2 in the presence of a liquid air trap and the reoxidation with O_2 in the presence of a solid CO_2-trap were studied with the same apparatus. Experiments were performed either with a linear temperature program of 0.0833K s^{-1} or isothermally. More details of the apparatus were published by Verdonck et al., 1980.

RESULTS

$[Rh(NH_3)_5CL]^{2+}$-H

The d001 spacing of $[Rh(NH_3)_5Cl]^{2+}$-H is 1.47 nm. It decreases to 1.311 nm after evacuation at 473K. In figures 1 and 2 the IR and reflectance spectra of the decomposition of $[Rh(NH_3)_5Cl]^{2+}$ on hectorite are summarized. Initially $[Rh(NH_3)_5Cl]^{2+}$ is present on the surface with its characteristic d-d transitions at 277 nm and 349 nm (Jørgensen, 1956). Upon degassing they broaden and shift to 305 nm and 366 nm respectively. After evacuation at 373K the 305 nm band has disappeared in the background and a shoulder around 425 nm is produced. The bathochromic shifts and the band broadening are indicative for the decomposition of $[Rh(NH_3)_5Cl]^{2+}$ and for the coordination of ligands weaker than NH_3 or Cl^-.

This instability of $[Rh(NH_3)_5Cl]^{2+}$ on hectorite is confirmed by IR (fig. 2). The characteristic bands of $[Rh(NH_3)_5Cl]^{2+}$-H are identical to those of $[Rh(NH_3)_5Cl]Cl_2$ (Kharitonov et al, 1971).

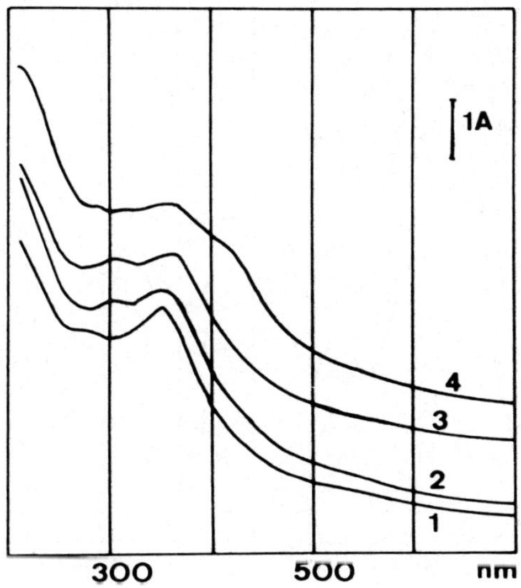

Figure 1

Reflectance spectra of $[Rh(NH3)5Cl]^{2+}$-H.
1, air-dry; 2, evacuated at room temperature;
3, evacuated at 323K; 4, evacuated at 373K.

Outgassing at 298K, but also standing in air, produces a band at 1430 cm^{-1}, characteristic of NH_4^+. It grows in intensity with increasing degassing temperature to attain its maximum intensity after evacuation at 573K. The band, initially at 1330 cm^{-1}, is the symmetrical deformation of coordinated NH_3 in $[Rh(NH_3)_5Cl]^{2+}$-H. Upon evacuation at high temperature this band shifts to 1300 cm^{-1}. This shift indicates changes in the coordination sphere of the complex on the surface.

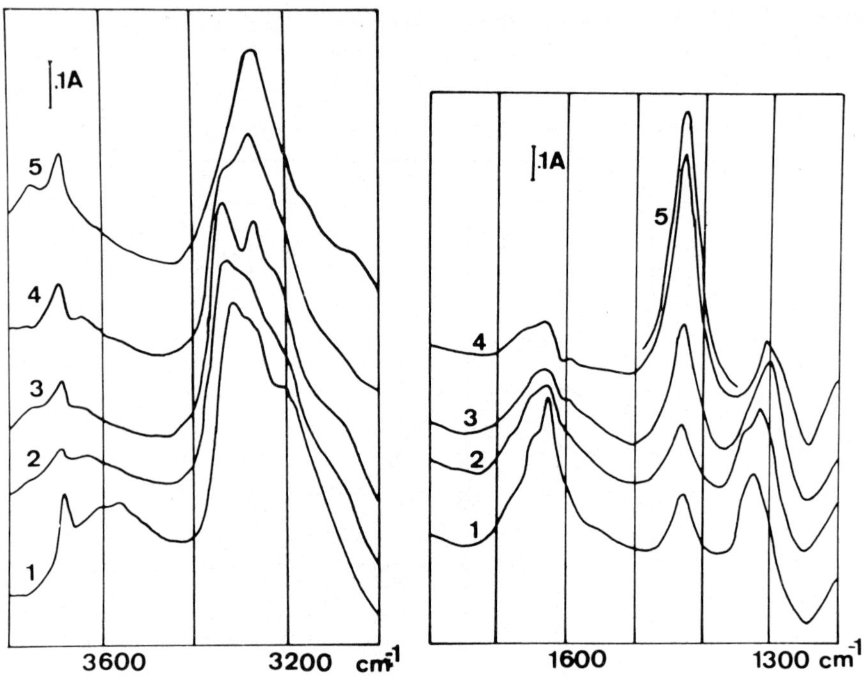

Figure 2

IR spectra of $[Rh(NH_3)_5Cl]^{2+}$-H.
1, evacuated at room temperature; 2, evacuated at 323K; 3, evacuated at 393K; 4, evacuated at 477K; 5, evacuated at 573K.

These spectral changes are accompanied by an almost continuous weight loss up to 473K, due to desorption of H_2O and NH_3. A second weight loss wave, observed above 533K involves the decomposition of NH_4^+.

The interaction with CO starts after evacuation at 373K. A dicarbonyl is formed as shown by the IR bands at 2115 cm^{-1} and 2045 cm^{-1} (Fig. 3). The weak band at 2160 cm^{-1} is due to gaseous CO.

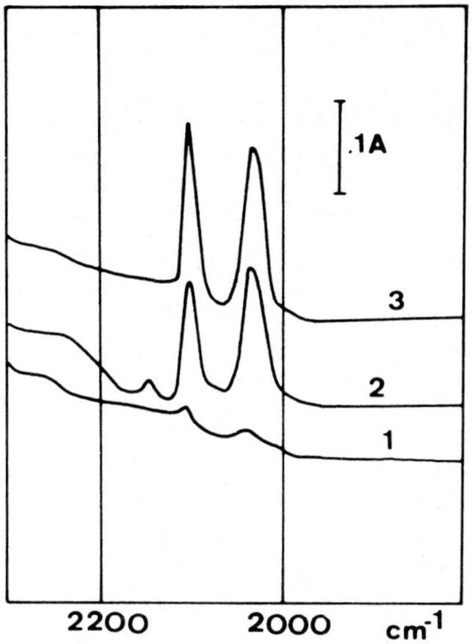

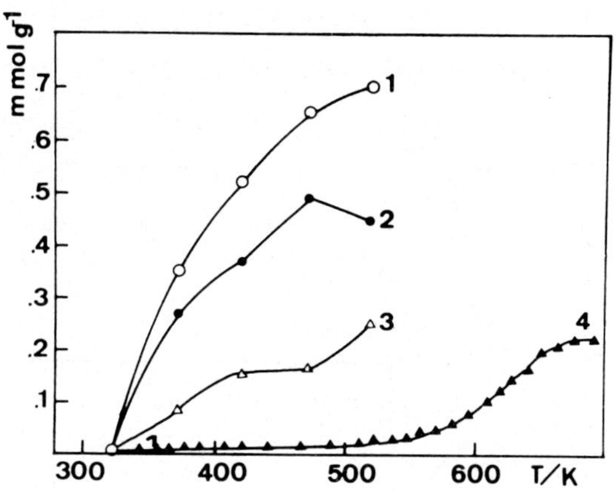

Figure 4

Total CO consumption (1), adsorbed CO (2), production of CO_2 (3) and desorption of CO (4) for $[Rh(NH_3)_5Cl]^{2+}$-H as a function of the temperature.

Figure 3

IR spectra of adsorption of CO on $[Rh(NH_3)_5Cl]^{2+}$-H. 1, at 373K; 2, at 423K; 3, at 298K after evacuation at 423K.

In fig. 4 the amounts of CO consumed, CO adsorbed, and CO_2 produced are plotted as a function of the pretreatment temperature. After evacuation at the given temperature, CO is allowed to adsorb at that temperature until constant pressure, gaseous CO is removed by evacuation and CO_2 is measured. Then, the evacuation temperature was increased and the procedure repeated. After the last adsorption step, gaseous CO is removed by evacuation,

CO_2 is measured, the system cooled to room temperature, and CO is desorbed with a linear temperature program of 0.0833K s^{-1} up to 700K. At 523K the ratio CO ads/CO_2 is 1.80 and the ratio CO (consumed)/Rh is 2. Desorption of CO starts above 500K and amounts to only half of the amount of CO adsorbed.

Reduction with H_2 starts at 373K. When the reduction is done in isothermal steps i.e. evacuation at a given temperature followed by reduction at that temperature with repetition of this procedure every 50K up to 550K, the amount of H_2 consumed is 0.6 mmol g^{-1}, in good agreement with the 0.528 mmol g^{-1} required for the reaction $Rh^{3+} + 3/2\ H_2 \rightarrow Rh(o)$ (Fig. 5).

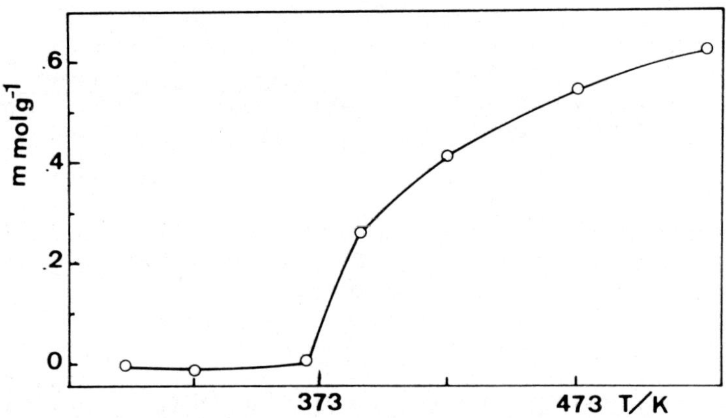

Figure 5

Isothermal H_2 consumption on $[Rh(NH_3)_5Cl]^{2+}$-H.

H_2-reduction during a linear temperature program up to 800K gives a continuous H_2-uptake between 373K and 800K. The total H_2-consumption is 0.665 mmol g^{-1}.

The subsequent O_2 uptake is 0.22 mmol g^{-1} and the second H_2-reduction needs 0.49 mmol H_2 g^{-1}. This amount is consumed between 340 and 440K.

$RhCl_3$-H

The spectra of the $RhCl_3.3H_2O$ solutions all have an intense charge transfer band at 225 nm (ε = 16000) and a d-d band, whose position is pH-dependent : pH = 7, λ = 475 nm; pH = 5.5, λ = 485 nm and pH = 2.1, λ = 497 nm. These values are to be compared to λ = 518 nm for $RhCl_6^{3-}$ and λ = 393 nm for $[Rh(H_2O)_6]^{3+}$ (Jørgensen, 1956). The positions of the d-d band in our solutions indicate the presence of complexes intermediate between $[RhCl_6]^{3-}$ and $[Rh(H_2O)_6]^{3+}$. A complicated adsorption behavior is expected. The results are summarized in Table 1. When the Rh content of the equilibrium solutions is very small, there is a large discrepancy between the amount Rh adsorbed and the amount of Na^+ released, due to proton uptake. In our experimental conditions the maximum Rh-uptake is around 0.40 mmol g^{-1}.

TABLE 1

Adsorption of Rh^{3+} from $RhCl_3.3H_2O$-solution

[Rh]/[Rh]+[Na]	initial pH	eq. pH	Rh-uptake	Na^+-release
7.3 x 10^{-4}	a	7.6	0.011[b]	0.373[b]
3.1 x 10^{-3}	a	7.3	0.014	a
8.1 x 10^{-1}	a	7.2	0.040	0.406
0.011	a	7.3	0.106	a
0.025	4.07*	6.7	0.221	0.566
0.150	4.03*	6.3	0.310	0.552
0.500	4.05*	5.8	0.390	0.587

* pH adjusted with NaOH; a : not measured; b : mmol g^{-1}.

DISCUSSION

Our data show that cationic complexes of Rh are easily exchanged to occupy positions in the interlamellar space of hectorite. The reaction is stoechiometric for $[Rh(NH_3)_5Cl]^{2+}$. For $RhCl_3 \cdot 3H_2O$ the spectroscopic data show that it is present in solution in a form intermediate between $RhCl_6^{3-}$ and $[Rh(H_2O)_6]^{3+}$. If the ligand field strengths of Cl^- and H_2O are simply added to predict the experimental d-d band position, not more than 1 to 2 water molecules are expected in the coordination sphere (Jørgensen, 1956). Thus, the complexes are anionic and cannot be exchanged. To explain the sorption data from $RhCl_3$-solution, extensive hydrolysis must be invoked, providing protons for exchange, and precipitation of chloro-hydroxyl complexes of Rh on the surface of hectorite.

$[Rh(NH_3)_5Cl]^{2+}$ is unstable on the surface of hectorite. It decomposes upon standing in the air at room temperature with formation of NH_4^+, but the reaction is accelerated upon heating. The primary reaction is

$$[Rh(NH_3)_5Cl]^{2+} + H_2O \longrightarrow NH_4^+ + [Rh(NH_3)_4(OH)(Cl)]^+ \quad (1)$$

The replacement of NH_3 in the coordination sphere by the weaker ligand OH explains the red shift of the d-d absorption bands. This process is continued upon heating as shown by the fact that the NH_4^+ - band at 1430 cm^{-1} attains its maximum intensity after evacuation at 573K.

Cl^--ions are removed by sublimation of NH_4Cl. The remaining NH_4^+ decomposes to NH_3 and gives an acid clay. This reaction scheme is entirely different from that of the thermal decomposition of $[Rh(NH_3)_5Cl]^{2+}$ on zeolites (Primet et al., 1978). In the latter case no NH_4^+ is formed, but the NH_3 liberated reduces Rh^{3+} to Rh(O) or escapes as NH_4Cl. The reduction of Rh^{3+} with NH_3 does not occur on hectorite. Indeed, H_2 is consumed according to the equation :

$$2Rh(III) + 3H_2 \longrightarrow 2Rh(O) + 6H^+ \quad (2)$$

In the subsequent oxidation-reduction cycle the $O_2:H_2$ ratio equals 2. This is indicative for the reactions :

$$Rh + O_2 \longrightarrow RhO_2 \quad (3)$$

$$RhO_2 + 2H_2 \longrightarrow Rh(o) + 2H_2O \quad (4)$$

$[Rh(NH_3)_5Cl]^{2+}$ as such does not coordinate CO. Partial decomposition is necessary as CO starts to interact from 373K on. The two carbonyl stretching frequencies at 2045 cm^{-1} and 2115 cm^{-1} are typical for $Rh(I)(CO)_2$ complexes (Primet et al., 1978).

From this evidence for a dicarbonyl and the quantities of CO consumed and CO_2 produced the equation of the reaction can be written as :

$$Rh(III) + 3CO + H_2O \longrightarrow Rh(I)(CO)_2 + CO_2 + 2H^+ \quad (5)$$

but the reaction is not complete, probably due to inaccessibility of Rh or to diffusion limitation.

ACKNOWLEDGMENT

Acknowledgment is made to the Donors of the Petroleum Research Foundation, administered by the American Chemical Society, for support of this research.

R.A.S. is indebted to the National Fund of Scientific Research (Belgium) for a grant as Senior Research Associate.

REFERENCES

Jørgensen, C.K., 1956. Complexes of 4d- and 5d-groups. I. Crystal fields spectra of Rhodium(III) and Iridium (III). Acta Chem. Scand., 10, 500-17.

Kharitonov, Y.Y., Knyazeva, N.A., Mazo, G.Y., Baronovskii, I.B. and Generalova, N.B., 1971. Normal vibrations of coordinated ammonia in pentamminerhodium complexes. Russian J. Inorg. Chem., 16, 1050-4.

Mazzei, M., Marconi, W. and Riocci, M., 1980. Asymmetric hydrogenation of substituted acrylic acids by Rh'-aminophosphine chiral complex supported on mineral clays. J. Molecular Catal., 9, 381-7.

Mazzei, M., Riocci, M. and Marconi, W., 1979. Procédé d'hydrogénation asymmétrique d'olefines. Brevet d'Invention (Belgique) N° 871.320, 8p.

Mazzei, M., Riocci, M., Monterotondo, R. and Marconi, W., 1979. Asymmetrische Hydrierung von prochiralen Olefinen durch Komplexe von Ubergangsmetallen, die in tonartigen Mineralien immobilisiert sind. Deutsche Offenlegungsschrift, 28 45 216, 10p.

Pinnavaia, T.J. and Welty, P.K., 1975. Catalytic hydrogenation of 1-hexene by Rhodium complexes in the intracrystal space of a swelling layer lattice silicate. J. Am. Chem. Soc., 97, 3819-20.

Pinnavaia, T.J., Welty, P.K. and Hoffman, J.F., 1976. Catalytic hydrogenation of unsaturated hydrocarbons by cationic rhodium complexes and rhodium metal intercalated in smectite. In S.W. Bailey, Ed., Proceedings Int. Clay Conference, 1975, 373-81. Applied Publishing Ltd.

Pinnavaia, T.J., Raythatha, R., Lee, J.G.-S., Halloran, L.J. and Hoffman, J.F., 1979. Intercalation of catalytically active metal complexes in mica-type silicates. Rhodium hydrogenation catalysts. J. Am. Chem. Soc., 101, 6891-6897.

Primet, M., Vedrine, J.C. and Naccache, C., 1978. Formation of rhodium carbonyl complexes in zeolite. J. Molecular Catalysis, 4, 411-21.

Quayle, W.H. and Pinnavaia, T.J., 1979. Utilization of cationic ligand for the intercalation of catalytically active rhodium complexes in swelling layer-lattice silicates. Inorg. Chem., 18, 2840-7.

Verdonck, J.J., Jacobs, P.A., Genet, M. and Poncelet, G., 1980. Redox behavior of transition metal ions in zeolites. 8. Characterization of a ruthenium metal phase in NaY zeolite. J.C.S., Faraday I, 76, 403-416.

THE INFLUENCE OF STRUCTURAL CHARACTERISTICS OF SMECTITES ON CATALYTIC DECOMPOSITION OF ASULAM.

G.G. RISTORI[(*)], P. FUSI[(**)] and S. CECCONI[(**)]
(*) Centro di Studio per i Collodi del Suolo del C.N.R., Firenze ItalY. (**) Istituto di Chimica Agraria e Forestale, Università di Firenze, Italy.

ABSTRACT

The nature of the decomposition products of Asulam herbicide on Mg-smectites, varied primarily with the location of isomorphous substitution affecting the polarising power of the Mg ion and, to a minor extent, with the total layer charge.

INTRODUCTION

Catalytic decomposition on moderate heating (75-90°C) of 4-aminobenzenesulphonylmethylcarbamate (Asulam herbicide) adsorbed on Wyoming Bentonite saturated with different cations has been described previously (Fusi et al., 1980, 1981; Ristori et al., 1981)

It was shown that the nature of the decomposition products is chiefly related to the polarising power of the exchangeable cations of the clay. For highly polarising cations e.g. H^+, Al^{3+}, Cu^{2+}, Mg^{2+} a large proportion of the adsorbed organic molecule is protonated, involving the NH_2 group. Upon heating the molecule is coordinated with the clay surface through the C=O group and decomposed to sulphanilic and carbamic acid. The latter stabilized by its coordination through the C=O group, decomposes when rehydratation of the clay, weakens the coordination bond. These reactions can be summarized as follows(Me is a metal cation of valency n):

Asulam is supplied by May & Baker, Ltd.

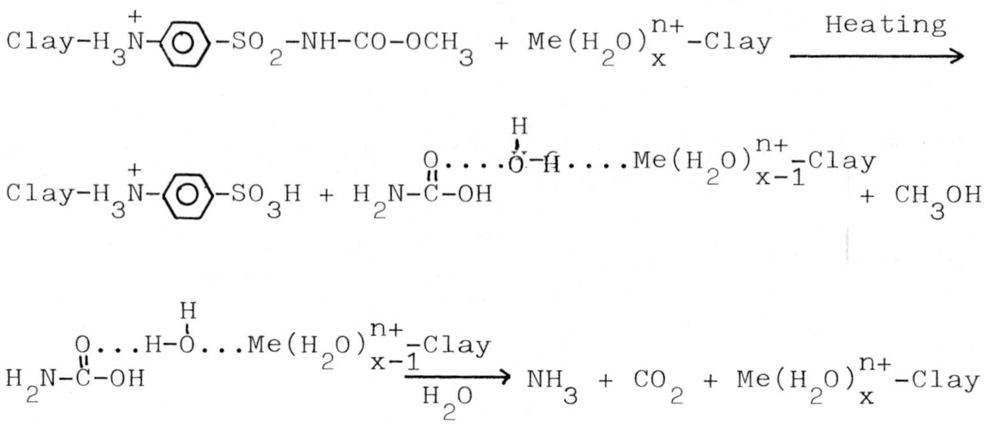

In Mg-clay systems, minor amounts of sulphanilamide were also produced, while in Ba- and Ca-complexes this compound becomes the prevailing reaction product; sulphanilamide results from the uncoordinated fraction of the molecule accordind to the following reaction:

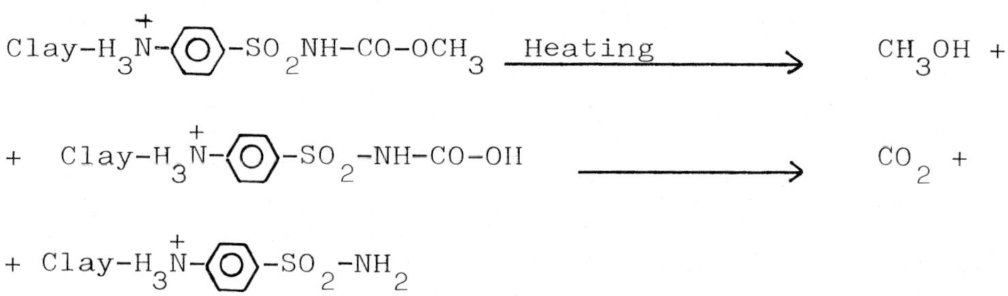

The alkali cations, due to their low solvation energy, areunable to coordinate Asulam and therefore only sulphanilamide and minor amounts of 4-aminobenzenesulphonylcarbamate anion were obtained.

The polarizing power of theexchange cations in smectites is also influenced by the structural characteristic of the clays (Mortland and Raman, 1968; Yariv and Heller, 1970; Swoboda and Kunze, 1968). Therefore, the type of smectite would influence the nature of the reaction products obtained by catalytic decomposition of adsorbed Asulam. The present study aimed to test the validity of this hypotesis, with several different smectites in the Mg form. Because of the intermediate polarizing power of the Mg-ion, the Mg-saturated clays may be expected to allow a good differentiation of the effect of various smectites on the nature of the reaction products.

MATERIALS AND METHODS

Dioctahedral and trioctahedral smectites, in which the charge is domintly located in either tetrahedral or octahedral sheet, and having a different total layer charge, were selected. The chemical composition of the 2 m fraction of the samples was determined by atomic absorption and X-ray fluorescence analysis; X-ray diffraction and IR spectroscopy confirmed that they were almost pure smectites. Their structural formulas, calculated on the basis of 10 O and 2 OH per half unit cell, are listed in Table 1.

Films of homoionic Mg-clays were immersed in a CCl_4 solution of Asulam. After washing with pure CCl_4, the samples were heated at 90°C, rehydrated by equilibration at 100% R.H., and analysed by IR spectroscopy and thin layer chromatography (TLC) at each stage of the thermal and hydration treatments. Details of these procedures have been described previously (Fusi et al., 1980, 1981; Ristori et al., 1981).

The relative amounts of the different decomposition products were estimated as follows: the films, used for TLC analysis, were weighed before extraction and elution; the spots formed on the chromatografic sheets were scraped off, exctracted in distilled water and the clear solution was analysed by UV spectrophotometry at the absorption maxima of the different products. The relative amounts of the different compounds were calculated as percentage of the total content of reaction products resulting from Asulam decomposition.

RESULTS AND DISCUSSION

From the IR spectra of the Mg-clay-Asulam complexes two group can be clearly distinguished: for those with predominant tetrahedral substitution (samples 6,7,8,9) the C=O stretching band at 1740 cm^{-1} of the adsorbed Asulam is remarcably reduced after heating at 90°C; by contrast for those with predominant octahedral substitution this band remains strong but is shifted to about 1695 cm^{-1}. This shift indicates the establishment of a coordination bond between the exchangeable cation and the C=O group of adsorbed Asulam, resulting in formation of sulphanilic and carbamic acids. The reduction in intensity of the 1740 cm^{-1} band in the absence of shift indicates the formation of sulphanilamide (Fusi et al., 1980; Ristori et al., 1981).

TABLE 1
Structural formulae of smectites calculated on the basis 10 O, 2 OH per formula unit for ignited material.

	Octahedral ions				Tetrahedal ions		Layer charge
	Al	Fe	Mg	Li	Si	Al	X^+
1-Bentonite Upton (Wyoming)	1.52	0.16	0.33	--	3.91	0.09	0.39
2-Hectorite (Hector,Cal.)	0.02	--	2.70	0.24	3.98	0.02	0.32
3-Bentonite Chambers(Arizona)	1.42	0.17	0.40	--	3.90	0.10	0.53
4-Bentonite Clay Spur(Wyoming)	1.52	0.16	0.34	--	3.88	0.12	0.40
5-Bentonite Otay (Calif.)	1.35	0.05	0.62	--	4.00	--	0.56
6-Beidellite(Drayton Australia)	1.12	0.42	0.71	--	3.48	0.52	0.48
7-Saponite (Mte Binjour Australia)	0.47	--	2.41	--	3.32	0.68	0.45
8-Nontronite (Manito,Wash.)	0.07	1.91	0.03	--	3.50	0.50	0.50
9-Bentonite Caffaro (Puglia, Italy)	1.38	0.55	0.13	--	3.55	0.45	0.40

Further and conclusive support for this interpretation is provided by the quantitative analysis of the reaction products (Table 2). The amount of sulphanilic acid formed on the primarily tetrahedrally substituted clays (samples 6,7,8,9) is low and principal product is sulphanilamide. By contrast, in the primarily octahedrally substituted clays (samples 1,2,3,4,5), Asulam decomposition leads mainly to sulphanilic and carbamic acids. Among the predominantly octahedrally substituted samples, Otay smectite (sample 5) was the least effective in producing these acids. This clay is also characterized by a very high total layer charge. Among the predominantly tetrahedrally substituted minerals, larger amounts of such compounds were produced on trioctahedral saponite. Minor amounts of 4-aminobenzenesulphanilcarbamate anion were also formed chiefly in the tetrahedrally substituted samples and this compound is a result of the neutralization of 4-aminobenzenesulphanilcarbamic acid due to the lower interlayer acidity of such clays, as discussed previously (Ristori et al., 1981).

These results clearly indicate that the solvation energy of the Mg ion largely depends on the location of layer charge arising from isomorphous substitutions. In tetrahedrally substituted smectites, the negative layer charge is relatively close to the interlayer space, and this would significantly decrease the solvating power of the cation; as a consequence, Asulam is only weakly coordinated and sulphanilamide is the prevailing decomposition product. By contrast, in octahedrally substituted smectites, the influence of layer charge is reduced and the exchangeable cation retains its solvation energy, the coordination bond through the C=O group can be established, and Asulam decomposes mainly to sulphanilic and carbamic acid.

A minor role might be played by other structural characteristics: (i) a high negative layer charge seems to have a greater effect on the polarizing power of the cation in octahedrally substituted smectites, leading to the formation of lower amounts of sulphanilic acid, (ii) the orientation of octahedral OH groups, normal to the silicate sheet in trioctahedral mineral such as saponite, seems to decrease the influence of tetrahedral negative layer charge on the polarizing power of the cation; this results in the formation of higher amounts of sulphanilic acid in comparison with other tetrahedrally substituted Mg-smectites.

TABLE 2

Relative amounts of degradation products of Asulam formed after heating Mg-clay complexes at 90°C expressed as percentage of total decomposition products.

Smectites	1	2	3	4	5	6	7	8	9
Sulphanilic acid (%)	77	71	70	69	55	25	38	26	5
Sulphanilamide (%)	15	21	20	21	36	60	47	54	65
4-aminobenzenesulphanilcarbamate anion (%)	8	8	10	10	9	15	15	20	30

REFERENCES

Fusi, P., Ristori, G.G. and Malquori, A., 1980. Montmorillonite-Asulam interactions: I° Catalytic decomposition of Asulam adsorbed on H-and Al-clay. Clay Miner.,15: 147-15!

Fusi, P., Ristori, G.G. and Franci, M., 1981.Oxidative processes related to degradation of Asulam herbicide adsorbed on smectites. Ann. Chim., 71: 443-451.

Mortland, M.M. and Raman, K.V., 1968. Surface acidity of smectites in relation to hydration, exchangeable cation and structure. Clays Clay Miner., 16: 393-398.

Ristori, G.G., Fusi, P. and Franci, M., 1981. Montmorillonite-Asulam interactions: II° Catalytic decomposition of Asulam adsorbed on Mg-, Ba-, Ca-, Li-, Na-, K-, and Cs-clay. Clay Miner., 16: 125-137.

Swoboda, A.R. and Kunze, G.W., 1968. Reactivity of montmorilonite surfaces with weak organic bases. Soil Sci. Soc. Am. Proc., 32: 806-811.

Yariv, S. and Heller, L., 1970. Sorption of cyclohexylamine by montmorillonites. Israel J. Chem., 8: 935-945.

GEOLOGY OF CLAYS

EMILIO GALAN
Dpto. Cristalografía y Mineralogía. Facultad de Ciencias. Universidad. Zaragoza, Spain.

Originally, clays were studied by chemists, agronomists, engineers and mineralogists (non-geologists). They proposed hypotheses about their origin based only on theoretical and laboratory data, ignoring field evidence and the role of geological agents in their formation and evolution.

Geologists only relatively recently began to take part in research on clay. This is probably due to the fact that clay must be studied by special techniques, with which the geologist, mainly trained in classical microscopy, was in general unfamiliar.

In 1951, on the occasion of a Meeting on "Geological Aspects of Clay Mineralogy", organized by the Clay Minerals Group of The Mineralogical Society, J.H. Taylor pointed out that "nearly 50% of all known sedimentary rocks belong to the clay grade and it is therefore a hard fact that our knowledge of the mineralogical composition of half the rocks of the sedimentary column is still in a rudimentary state". He also observed that "it is of the highest importance that adequate geological data should be included in papers dealing with clay mineralogy" (Taylor, 1952). However, at the same meeting, Boswell (1952) presented a general survey of the main British sediments in the fashion proposed by Taylor.

Before 1950, clays, shales and clayey materials were, in general, inadequately described, because their principal components, clay minerals, were hardly studied by geologists, with some notable exceptions (e.g. Lapparent and Correns). During the 50's, many interesting papers on geology of clays were written by the principal clay mineralogists. Among these pioneers we recognize: Grim, Keller, Murray, Hathaway, Weaver, Johns, Taylor, Millot, Kulbicki, Nicolas, Erhart, Konta, Correns, Galitelli, Martin Vivaldi, Chuckrov, Kossovskaya, Shutov, Sudo, and others. Many of them are still working for the benefit of our science.

In 1958, Weaver said:"Clay minerals can be used in interpreting and understanding such problems as tectonics, source, age, boundaries, facies, environments, zonation, correlation, and metamorphism...", and he also pointed out the necessity to integrate clay mineral information with other geological data: "Clay mineral information should be integrated with other geologic information and the geologic problems should be throughly understood if the clay data are to be used effectively".

With the generalized use of the mineralogical technique of X-ray diffraction in geological studies in the early 60's, the geologist definitely entered into the field of clay science, and a new field "geology of clays" was created, in which chemistry, crystal chemistry, physical-chemistry and the study of the geological environment of clay are brought together to obtain knowledge of the true role of clay in the sedimentary geochemical cycle and in the hydrothermal and weathering alteration processes of rocks.

The general aims of this new field were first formulated by Millot. His book titled "Geology of Clays", published in 1964, presented a methodology for the study of clays and gave a geological interpretation of various clay environments. Millot defines a new terminology, which seems normal to us today, and he incorporates it into the geological glossary. The originality and epistemological value of Millot's work have been recognized by every clay scientist and it has decisively contributed to the evolution of clay science.

Today, research in clay geology is comprised of basic and applied studies, as is usual in the geological sciences.

Basic studies are devoted to acquire information concerning a) sedimentary clays (enclosed shales), b) hydrothermal clays, and c) weathering horizons. Every study must start with material characterization. X-ray powder diffraction is the most useful technique to analyze quickly (qualitatively and quantitatively) a great number of samples. Routine chemical analyses of major components, usually carried out by atomic absorption, are also convenient for geochemical studies. Sometimes, DTA, TG, IR, electron microscopy, etc., can be used to solve some complicated cases. Nevertheless, up-to-now, no quick and accurate method for mineral quantification has been developed. Schultz's method (1964), and especially the method described by Biscaye (1965) (based on the method of Johns et al. 1954), have been widely used in sedimentology. Recently, Biscaye's method has been computer programmed for use in research of clay sediments (Karling, 1980). Other methods suitable for computer application have been developed by Pearson (1978), Heath and Pisias (1979), Hauff and Trump (1980), among others. Because, like in sedimentology, it is often sufficient to know the relative amounts of minerals rather than their exact percentages, many researchers still use a quantification method based on the ratios of the heights of the principal peaks of the minerals present (i.e. peaks at 7Å, 10Å, 14Å, etc.).

Comparison of the quantitative results presented in different papers is often quite impossible, because different and non-comparable methods were used. A standard method, universally accepted, must be developed to quantify clay (and non-clay) minerals in sedimentary series, to allow comparisons between sedimentary basins. In my opinion, the quantitative methods to be routinely used in geology are still inadequate.

Clay formation and evolution are geological events in the geologica history of a basin. Source area, transport, and depositional environme of clay must be studied, as well as clay transformation and diagenetic changes.

Clay minerals are particularly sensitive to pressure and temperature conditions and to variations in the chemical environment (sensibility expressed as crystallographical and mineralogical changes). Consequently, the study of clays can supply a great deal of information concerning the processes that occur in the evolution of the series of which they form a part (weathering, hydrothermal alteration, diagenesis, low-grade metamorphism,etc).

Authigenic and transformed clay minerals reveal the physical-chemical conditions, and the chemical activity of the environment in which they were formed, or in which they reached their equilibrium. Clays of weathering horizons furnish information about geomorphological and climatic conditions. The paleoclimatic interpretation of the clay minerals in a given horizon has been reviewed by Singer (1979/80). In short: many geological processes could be better understood if adequate clay data were available.

The principal investigations in basic clay geology are and/or will be devoted to:

a) present-day clay sedimentation in oceans, lakes, lagoons, etc. as related to chemical composition, salinity, pH and Eh, etc. of water, and to early diagenesis, submarine weathering, magmatic events, and presence of metals, organisms,etc. in these environments.
b) ancient sediments of continents in order to reconstruct the depositional environment and the evolution of sediments during diagenesis and low-grade metamorphism, and to the use of clay petrology as a lithostratigraphic criterium for correlation, whose absolute geochronology is also often derived from isotopic data (Rb/Sr, K/A methods).
c) the weathering crusts of any type of rocks in connection with climatic and tectonic features.
d) hydrothermal clays to discover the origin, characteristics and nature of fluids, and to study wall-rock alteration phenomena.
d) petrologic interpretation of clay mineral assemblages by means of experimental determination of clay mineral stabilities.

This last type of research appears to be very promising in the future. Genetical hypotheses are based on experimental thermodynamic models, and synthesis and alteration of clay minerals in conditions similar to those in nature. Thus, assumed geological features can be substantiated, or be modified according to the experimental data, as the case may be. So, it will be possible to establish typical mineral assemblage indicators for any climate, salinity, chemical equilibrium, etc. A very interesting attempt in this area was made by Velde (1977). His work represents a new way to look at clay mineralogy.

In applied clay geology, studies are very diverse. In the past, the most important ones were concerned with clay deposits (bentonite, kaolin, etc.), their origin and characteristics. But, at present, studies are directed more to clay mineralogy in relation to: a) petroleum generation, b) economical ore deposits, c) soil engineering and soil mechanics, d) nuclear waste disposal, e) environmental sciences, f) industrial clays.

Many of these topics are frequently treated in specialized literature rather than in the classical clay minerals journals. The most recent relevant papers on applied clay geology are devoted to industrial clay deposits (geology, genesis, exploration, evaluation, and so on) (Patterson and Murray, 1975), bentonites (Grim and Guven, 1978); problems of geothermal alteration of clays, with the goal of finding emplacement for radiactive wastes in argillaceous rocks (Weaver,1979); oil formation (generation and migration) related to clay diagenesis in oil field (Johns and Shimoyama, 1972; Foscolo and Powell, 1978; Sudo, 1978; Weaver 1979), diagenesis and low-grade metamorphism phenomena on clay (measured by Kubler's index) and its relation with the degree of coalification (measured by the vitrinite and volatile matter) (Weber, 1972; Wolf, 1975; Srodon, 1978).

Future research will be aimed at obtaining a better knowledge of the geology of clays of every country. Geological features will be better understood, and prospecting for resources (metals, coals, petroleum, clays and other industrial minerals) will be easier. The survey of Perrin (1971) on British sediments and "Clays of Japan" by Sudo and Shimoda (1978) are notable examples of adequate general studies.

In conclusion: The geology of clays is one of the modern branches of geology, whose importance grows daily, because it supplies data helpful in understanding various different geological processes in the Earth's surface, and also because it aids prospecting for natural resources.

REFERENCES

Biscaye, P.E., 1965. Mineralogy and sedimentation of recent deep-sea clays in the Atlantic and adjacent seas and oceans. Bull. Geol. Soc. Am., 76: 803-332.

Boswell, P.G.H., 1952. The contribution of clay mineralogy to the study of the diagenesis of sediments. Clay. Min. Bull., 1: 246-251.

Foscolo, A.E. and Powell, T.G., 1978. Mineralogical and geochemical transformation of clays during burial diagenesis (catagenesis): relation to oil generation. In: M.M. Mortland and V.C. Farmer (Eds) Proc. Int. Clay Conf. Oxford. Elsevier, Amsterdam, pp.261-270.

Grim, R.E. and Guven, N., 1978. Bentonites. Geology, Mineralogy, Properties and Uses. Elsevier, Amsterdam, 256 pp.

Hauff, P. and Trump, G van, Jr., 1980. XRD-CLAY-ID-A computerized system for clay mineral identification. 29th Annual Clay Minerals Conf. Waco. Texas. (Abstract).

Heath, J.R. and Pisias, N.G., 1979. A method for the quantitative estimation of clay minerals in North Pacific Deep-sea sediments. Clays Clay Min., 27: 175-184.

Johns, E.D.; Grim, R.E. and Bradley, W.F., 1954. Quantitative estimation of clay minerals by diffraction methods. J. Sediment. Petrol., 24: 242-251.

Johns, W.D. and Shimoyama, A., 1972. Clay minerals and petroleum forming reactions during burial and diagenesis. Am. Assoc. Petroleum, Geol. Bull., 56: 2160-2167.

Karling, R., 1980. Sediment sources and clay minerals distribution of the Oregon Coast. J. Sediment. Petrol., 50: 543-560.

Millot, G., 1964. Geologie des Argiles. Masson et cie. Paris, 320 pp.

Patterson, S. and Murray, H.H., 1975. Clays. In: S.J. Lefond (Ed). Industrial Minerals and Rocks (Non-metallics other than Fuels). 4th Ed. Am. Inst. Mining. Metall. Petroleum. Eng. Ins., N. Y, pp. 519-585.

Pearson, M.J., 1978. Quantitative clay mineralogical analysis from the bulk chemistry of sedimentary rocks. Clays Clay Min., 26: 423-433

Perrin, R.M.S., 1971. The clay mineralogy of British sediments. Mineral ogical Soc., London, 247 pp.

Schultz, L.G., 1964. Quantitative interpretation of mineralogical composition from X-ray and chemical data for the Pierre-Shales. Prof. Pap. U.S. Geol. Surv., 391-C, 31 pp.

Singer, A., 1979-80. The paleoclimatic interpretation of clay minerals in soils and weathering profiles. Earth-Sci. Rev., 15: 303-326.

Srodon, J., 1978. Correlation between coal and clay diagenesis in the Carboniferous of the Upper Silesian coal basin. In: M.M. Mortland and V.C. Farmer (Eds). Proc. Int. Clay Conf. Oxford. Elsevier, Amsterdam, pp. 251-260.

Sudo, T., 1978. Studies of clay minerals in sediments. A review. In: M.M. Mortland and V.C. Farmer (Eds). Proc. Int. Clay. Conf. Oxford. Elsevier, Amsterdam, pp. 241-249.

Sudo, T. and Shimoda, S. (Eds), 1978. Clays and Clay Minerals of Japan. Elsevier. Amsterdam, 326 pp.

Taylor, J.H., 1952. Clay minerals and the evolution of sedimentary rocks. Clay Min. Bull, 1: 238-243.

Velde, B., 1977. Clays and Clay Minerals in Natural and Synthetic Systems. Elsevier. Amsterdam, 218 pp.

Weaver, C.E., 1960. Possible uses of clay minerals in the search for oil. Clays Clay Min. (8th Nat. Conf., 1958), pp. 214-227

Weaver, C.E., 1979. Geothermal alteration of clay minerals and shales: Diagenesis. Technical Report ONWI-21. Battelle. U.S.A., 176 pp.

Weber, K., 1972. Kristalinitat des Illits in Tonschiefern und andere Kritereen Schwacher Metamorphose in nordostlichen Rheinischen Schiefer gebirge. Neues Jahrb. Geol. Palaontol. Abh., 141: 333-363.

Wolf, M., 1975. Uber die Beziehungen Zwischen Illit-Kristallinitat und Inkohlung. Neues Jahrb. Geol. Palaontol. Monatsh., 7: 437-447.

GENESIS OF RIO JARI KAOLIN

HAYDN H. MURRAY[1] AND PADRIAC PARTRIDGE[2]

ABSTRACT

A large sedimentary kaolin deposit is located on the east side of the Rio Jari in the territory of Amapa in Brazil. The kaolin occurs near the top of the Barreiras Series of Pliocene age. The kaolin reaches a thickness of 85 metres. The clays at and near the surface are laterized to a depth of 7 to 8 metres consisting of a yellowish-brown bauxitic kaolin. Below this laterized zone is a layer of iron oxide concretions ranging from 0.5 to 3 m in thickness. Below this is a mottled zone ranging in thickness between 8 and 16 m. A light colored kaolin layer about 30 m thick underlies the mottled zone. It is the upper 15 to 20 m of this zone that is presently being exploited. The kaolin deposit extends for approximately 12 km N-S along the Rio Jari and 5 km east. The authors suggest a lacustrine or deltaic fluvial depositional environment for the arkosic sands derived from the Guyana shield to the north. The complete mineralogy and chemistry of the kaolin profile is presented and discussed.

INTRODUCTION

In 1967 a kaolin deposit was discovered on the Rio Jari in the Amazonia region of Brazil near Monte Dourado (Fig. 1). The kaolin was discovered approximately 100 km upstream from the point at which the Rio Jari flows into the Amazon River. The kaolin deposit is located on a plateau on the east side of the Jari River in the territory of Amapa (Fig. 2). This plateau is dissected by many streams and is overgrown with a very dense tropical forest. The elevation of the plateau is 150 m above the river and the escarpment along the river valley is rather steep. Monte Dourado is a town established by Jari Florestal e Agropecuario to serve personnel for a major forestry project and a pulp mill installation. The kaolin deposit was drilled and tested in 1970 and '71, and in 1976 a kaolin processing plant was completed to produce coating grade kaolins which are shipped mainly to European markets under the trade name Amazon 88 kaolin.

[1]Department of Geology
Indiana University
Bloomington, Ind. 47405

[2]9947 Rudnick Avenue
Chatsworth, Calif. 91311

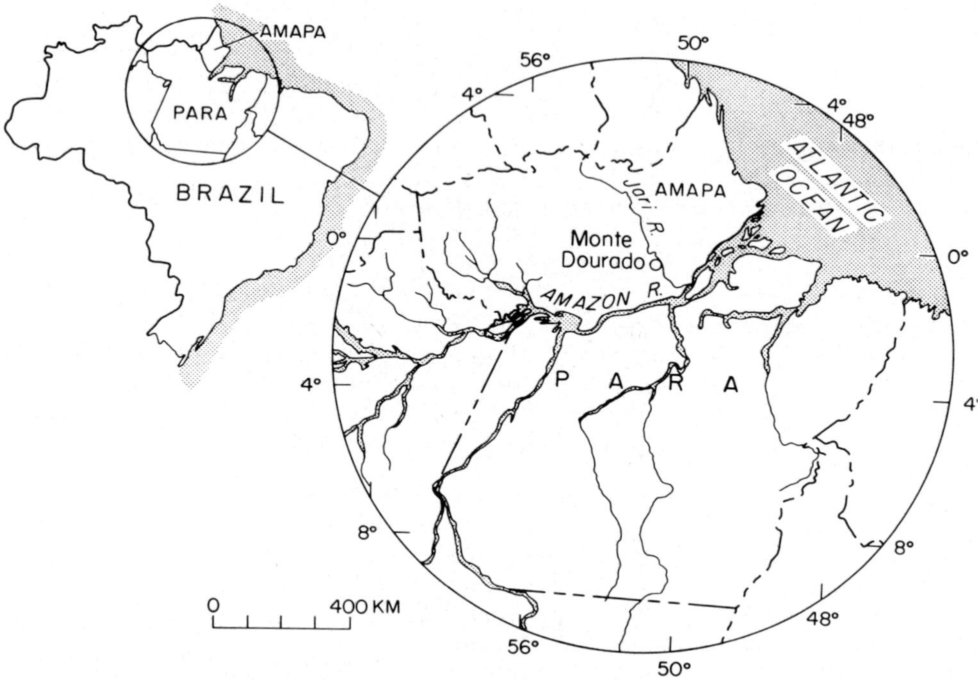
AMAPA
PARA
BRAZIL
0 400 KM
56°
50°
4°
48°
0°
8°
ATLANTIC OCEAN
AMAPA
Jari R.
Monte Dourado
AMAZON R.
P A R A

Fig. 1

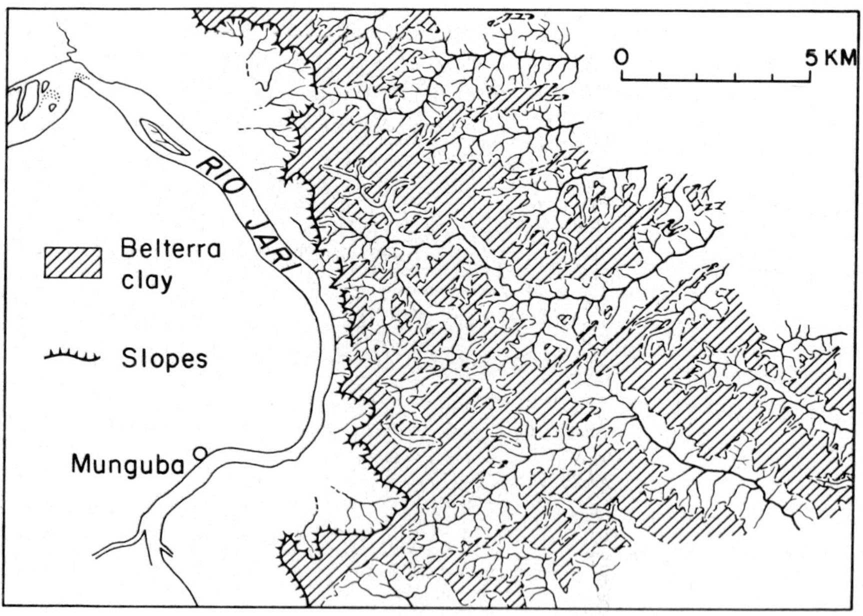
0 5 KM
RIO JARI
Belterra clay
Slopes
Munguba

Fig. 2

In 1979 after the open pit mine was well developed the deposit was completely sampled from the tropical soil zone at the top, through a lateritic zone which constitutes the overburden, and 30 m into the white kaolin for a mineralogical and chemical study. The first results from this study are reported in this paper. The purpose of the study is to understand better the genesis of the kaolin and the associated laterite overlaying the kaolin.

GEOLOGY, MINERALOGY, AND CHEMISTRY

The kaolin occurs near the top of the Barreiras Series of Pliocene age (Fig. 3). The Barreiras Series is a sequence of unconsolidated sands, sandy clays, kaolinitic clays, and conglomerates which reaches a thickness of up to 700 m in some areas (Sombroek, 1966). The uppermost 200 m are exposed in stream valleys which dissect the plateau. At the top of the sequence is the Belterra clay (Klammer, 1971).

The Barreiras Series of sediments were undoubtedly derived from the Guyana shield area to the north, and as shown on Figure 3 the sediments get finer upward in the section. The Belterra clay with the thick kaolin deposit extends some 12 km along the river and about 5 km in width east of the river. The deposit is thickest at the escarpment along the river and thins eastward and also to the north. It appears that the Belterra clay is either of lacustrine or deltaic fluvial origin but there is a question as yet as to whether the kaolin is detrital or is the result of alteration of arkosic sands that were dumped into the lake basin. Additional studies that will include extensive petrographic study and scanning electron micrography hopefully will provide additional information to settle this question.

The weathering in the Amazon area of Brazil is intense. The rainfall is very heavy and the temperature high. Jungle growth is abundant providing organic acids that intensify alteration at and near the surface. The Belterra clay is laterized to a depth of about 7 m consisting of a yellowish or orange-brown bauxitic kaolin (Fig. 4). At the base of this laterized zone is a layer of iron and alumina concretions ranging from 0.5 to 3 m in thickness. The concretions form a conglomeratic zone of pebble sized nodules embedded in a kaolin matrix. Underlying the nodular zone is a mottled zone in which there are white areas, red or brown areas, and streaks of each color along joints and channels in the kaolin. Liesegang rings are prominent on top of the kaolin surface which can be seen as the kaolin is mined. It appears that these rings are the result of water movement down solution channels. One such channel was traced in the high wall of the mine for a depth of 20 m. The mottled zone ranges from 8 to 16 m in thickness.

The light colored kaolin layer below the mottled zone averages about 30 m in thickness. Thin layers of sandy kaolin are interspersed through the kaolin. The upper 15 to 20 m of the white kaolin contains 5 percent or less quartz. A purple stained coarse sand layer about 1 m thick separates the upper kaolin with a low quartz content from a lower kaolin 10 to 20 m thick which has a much higher quartz content and in some areas the lower kaolin contains as much as 30 percent

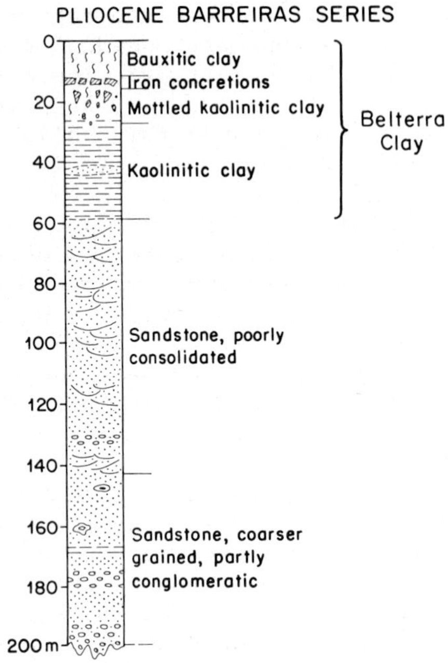

PLIOCENE BARREIRAS SERIES
0
20
40
60
80
100
120
140
160
180
200 m
Bauxitic clay
Iron concretions
Mottled kaolinitic clay
Kaolinitic clay
Belterra Clay
Sandstone, poorly consolidated
Sandstone, coarser grained, partly conglomeratic

Fig. 3

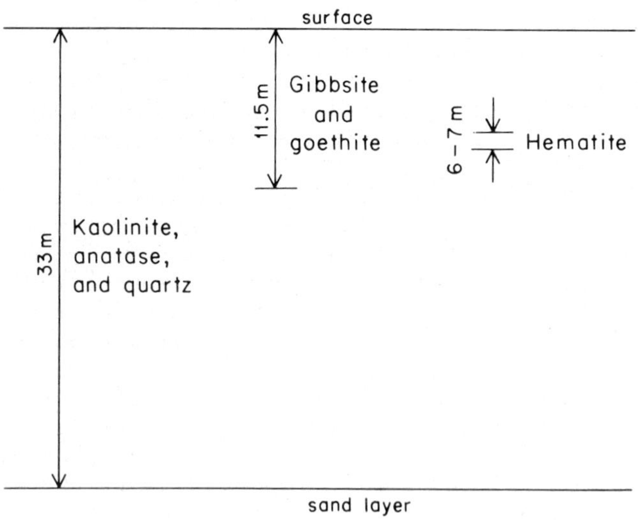

surface
11.5 m
Gibbsite and goethite
6 – 7 m
Hematite
33 m
Kaolinite, anatase, and quartz
sand layer

Fig. 4

quartz. This sand layer is shown at the base of Figure 4. Although analyses have been completed on the samples from the total section this paper only deals with the section above this sand layer. The lower kaolin is highly stained with red, brown, and purple coloration in fairly large areas extending several metres whereas in the upper white kaolin the staining appears to be restricted to joints, fractures, and solution channels. The quartz in the upper layer is etched and in some cases has been completely dissolved by the acidic weathering solutions. There are sharp angular voids in the kaolin where the quartz has been dissolved and in some of these voids there is a remaining small etched quartz grain. Below the purple sandy layer the quartz is sharp and angular and has not been etched or dissolved. It appears that the sand layer is a lateral channel through which the weathering solutions that descend vertically through the kaolin move because the quartz in the lower kaolin is fresh and angular indicating that the weathering solutions have not penetrated very far into the lower kaolin.

The minerals in this section from the soil zone to the sand layer are shown on Figure 4. Kaolinite, quartz, and anatase are present throughout the section with kaolinite being by far the dominant mineral. In the lateritic zone gibbsite and goethite are present and in the nodular zone hematite appears. Figure 5 shows an x-ray diffraction pattern from the lateritic zone, the nodular zone, and the kaolin layer. In a few x-ray diffraction patterns of kaolin from core samples to the north of the present mine are traces of smectite and an occasional trace of illite but these are not present in the section sampled for this study.

The samples were analyzed chemically. The major oxides and trace elements except for rubidium were determined using Inductively Coupled Argon Plasma (ICAP) with NBS sample 98 as the standard for the major oxides and GSP-1 and ACV-1 as standards for the trace elements. The values for rubidium were determined using a flame emission spectrophotometer.

Table 1 shows the chemical analyses data for the bauxitic kaolin and the kaolin layer. Table 2 gives the trace element analyses data for the bauxitic kaolin and the kaolin.

In comparing the values shown in Table 1 and 2 the elements that are depleted in the laterite are Si, Ba, Cu, Ni, and Pb. The elements that are enriched in the laterite are Al, Fe, Ti, Mn, Mg, Ca, K, P, Cr, Ga, Li, Mo, Rb, Th, V, Y, and Zn. Elements that are neither enriched or depleted are Na, B, and Sr.

Figures 6-10 show graphically the concentration of these elements with depth which reveals some interesting information which will be discussed in the following section.

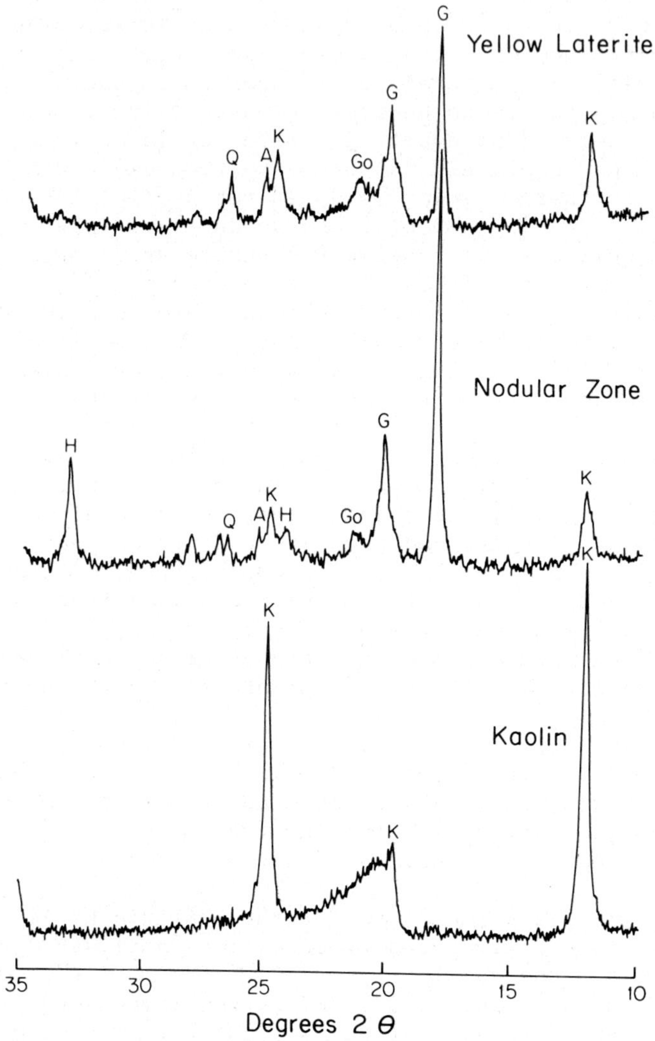
Yellow Laterite
G
G
K
Q
A
K
Go
Nodular Zone
G
H
K
K
Q
A
H
Go
K
K
Kaolin
K
35
30
25
20
15
10
Degrees 2 θ

Fig. 5

TABLE 1 Chemical Analyses Data (%)

	SiO_2	Al_2O_3	Fe_2O_3	TiO_2	MnO	MgO
Bauxitic Kaolin	24.54	43.79	5.10	2.78	0.012	0.023
Kaolin	44.45	37.37	1.95	1.39	0.009	0.015
	CaO	Na_2O	K_2O	P_2O_5	LOI	TOTAL
Bauxitic Kaolin	0.021	0.011	0.204	0.127	23.08	99.69
Kaolin	0.013	0.011	0.123	0.102	14.45	99.89

TABLE 2 Trace Element Analyses Data (PPM)

	B	Ba	Cr	Cu	Ga	Li	Mo	Ni
Bauxitic Kaolin	10	54	162	13	32	4	5	21
Kaolin	9	68	82	17	26	2	4	23
	Pb	Rb	Sr	Th	V	Y	Zn	
Bauxitic Kaolin	40	1.6	31	33	147	11	85	
Kaolin	45	1.0	30	13	69	6	63	

DISCUSSION OF RESULTS

The kaolin in the Belterra clay member of the Barreiras Series was either deposited as a detrital kaolinite derived from intense weathering of granitic rocks on the Guyana Shield or resulted from the alteration of a thick arkosic sand. At the present time based on field evidence and only partial analytical results the latter concept is favored. The lower kaolin below the thin sand layer contains up to 30 percent fresh, angular quartz grains whereas in the upper kaolin above the sand layer most of the quartz is etched and dissolved. A problem exists as to how to explain that there is no evidence of feldspar in the lower kaolin and the quartz is fresh and angular, but hopefully this can be resolved when the study is completed and all the chemical analyses, mineralogical, petrographical and scan electron micrograph data have been analyzed.

This study is restricted primarily to the upper kaolin and the associated laterite which has been derived from the kaolin by intense tropical weathering. The minerals in the kaolin are kaolinite, quartz, and anatase and in the laterite or bauxitic kaolin are kaolinite, gibbsite, goethite, hematite, and anatase. Note on Figure 5 that the basal reflection for kaolinite is relatively broad and has a much reduced intensity in the laterite x-ray diffraction pattern as compared to the kaolin x-ray pattern. This would suggest that the kaolinite is being altered to the aluminum and iron minerals, gibbsite, goethite, and hematite. The chemical analyses of a relatively pure processed kaolinite from this deposit shows 1.90 percent iron and the brightness of the kaolin is 88 percent (Murray, 1981). This would indicate that iron is probably substituting for aluminum in the kaolinite crystal and when the kaolinite breaks down during weathering gibbsite, goethite, and hematite are formed. The iron in the kaolinite crystal would be a source for the iron in the laterite. Additional iron from another lateral source is necessary because the aluminum and iron oxides have not accumulated in the laterite in the same proportion as their relative content in kaolinite.

As shown in Figure 6 and 7 SiO_2, Al_2O_3, Fe_2O_3 and TiO_2 and the loss on ignition have predictable values relative to depth. SiO_2 is depleted in the lateritic zone and Al_2O_3, Fe_2O_3, and TiO_2 are enriched, and the loss on ignition is considerably higher in the laterite zone. The Fe content shows a high concentration near the base of the laterite which corresponds to the nodular zone in which hematite is present (Fig. 5).

Figure 7 indicates that K_2O is concentrated along with the Fe. Additional study is needed to find where K_2O is concentrated. Na_2O shows a difficult to explain unusual concentration in the middle of the laterite zone. As shown on Figures 6 and 7 MgO, CaO, P_2O_5, and MnO all show a slight enrichment in the laterite.

Figure 8 shows that Ba is depleted in the soil zone and increases continuously with depth through the laterite and is erratic in concentration through the kaolin. Cr shows an enrichment in the laterite similar to iron. B shows slight enrichment in the laterite and the depth profile is roughly parallel to that of iron. Cu is enriched in the soil zone and in the nodular layer but is depleted in the rest of the laterite compared to the amount in the kaolin. The Rb profile (Fig. 9) runs parallel with the K_2O profile. The lead profile is similar to that of Ba. Lead is concentrated in the nodular zone along with Fe_2O_3, Cr, and Cu. Ni is concentrated at the top of the laterite and again in the nodular zone but depleted in most of the laterite relative to the kaolin. Li and Mo show a slight but definite enrichment in the laterite as does Ga although the distribution of Ga is more erratic. Th, Y, and Zn (Fig. 10) are all enriched in the laterite zone and the distribution pattern is similar to the profile of TiO_2. V shows a very large enrichment in the laterite and the profile is similar to iron and Cr. Sr shows a general increase through the laterite and the highest values are at the base of the laterite a distribution pattern somewhat similar to Ba and Pb.

Goldschmidt (1937) suggested that the behavior of the various ions in weathering reactions can be related directly to their ionic potential. According to this concept Al_2O_3, Fe_2O_3, TiO_2, Cr, Mo, Ga, Th, V, Y, and Zn should be concentrated in the laterite which they are. SiO_2 is a soluble complex anion and should be depleted in the laterite as indeed it is. Phosphate and boron should also be depleted but they are not. This observation means that other factors such as pH and Eh or parent mineral structure may have an equally important influence on controlling the loss or retention of these elements. The elements Rb, K_2O, Na_2O, Li, Ba, Pb, Ni, Sr, Ca, MnO, Cu, and MgO should be depleted according to their ionic potential but only Ba, Pb, Ni and Cu follow the theory. The other elements are enriched except for Na_2O which remains essentially constant, again indicating that the other factors mentioned above have an equally or more important control over the loss or retention of these particular elements.

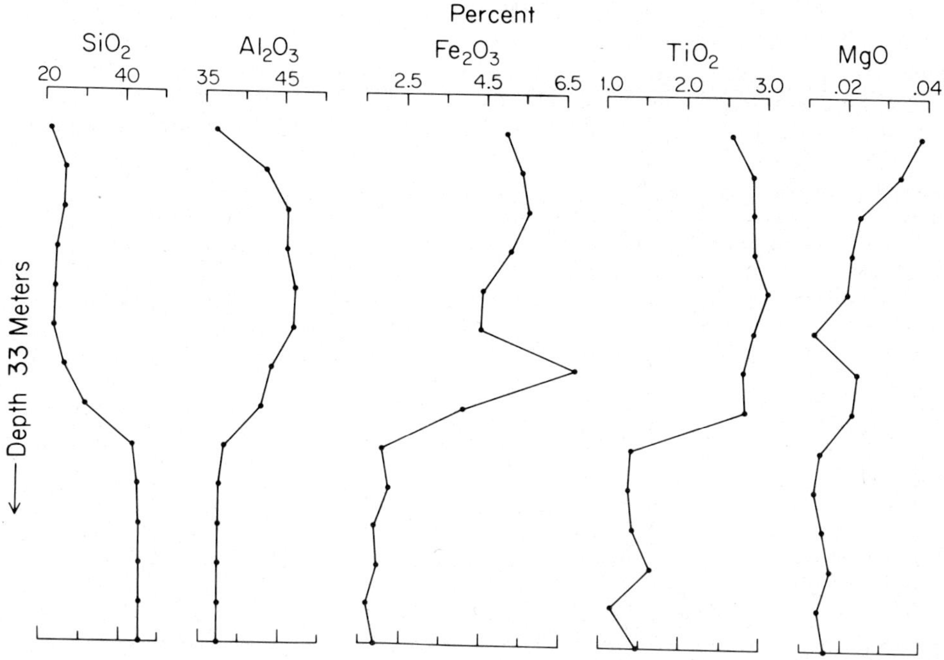
Percent
SiO2
Al2O3
Fe2O3
TiO2
MgO
20
40
35
45
2.5
4.5
6.5
1.0
2.0
3.0
.02
.04
Depth 33 Meters

Fig. 6

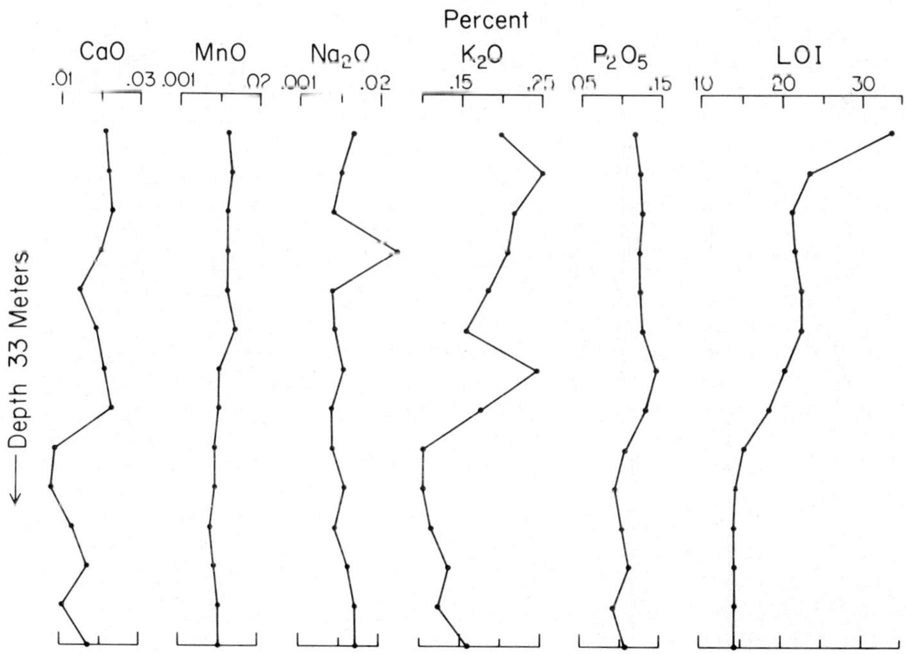
Percent
CaO
MnO
Na2O
K2O
P2O5
LOI
.01
.03
.001
.02
.001
.02
.15
.25
.05
.15
10
20
30
Depth 33 Meters

Fig. 7

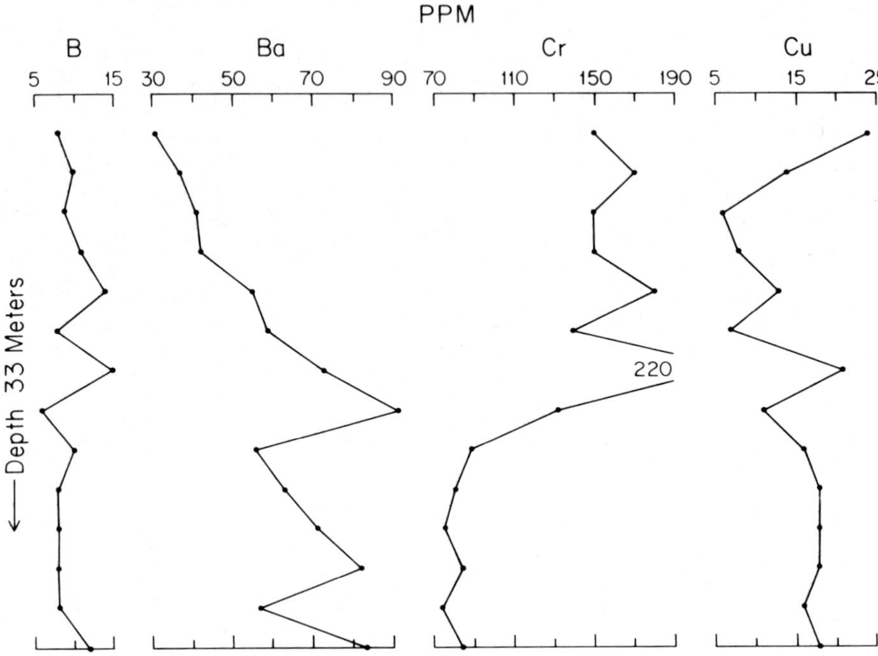
PPM
B
Ba
Cr
Cu
5
15
30
50
70
90
70
110
150
190
5
15
25
220
← Depth 33 Meters

Fig. 8

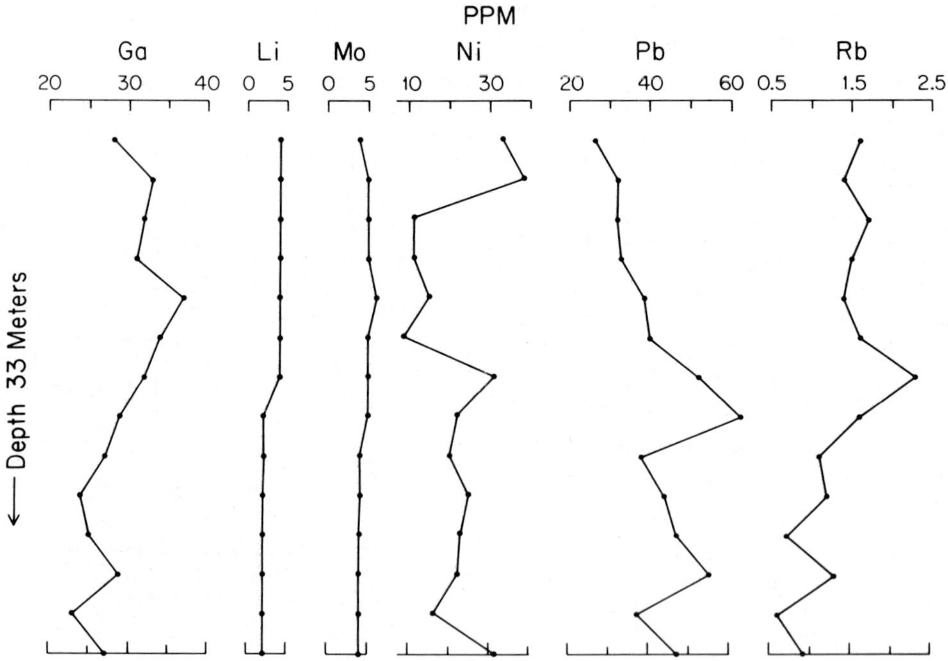
PPM
Ga
Li
Mo
Ni
Pb
Rb
20
30
40
0
5
0
5
10
30
20
40
60
0.5
1.5
2.5
← Depth 33 Meters

Fig. 9

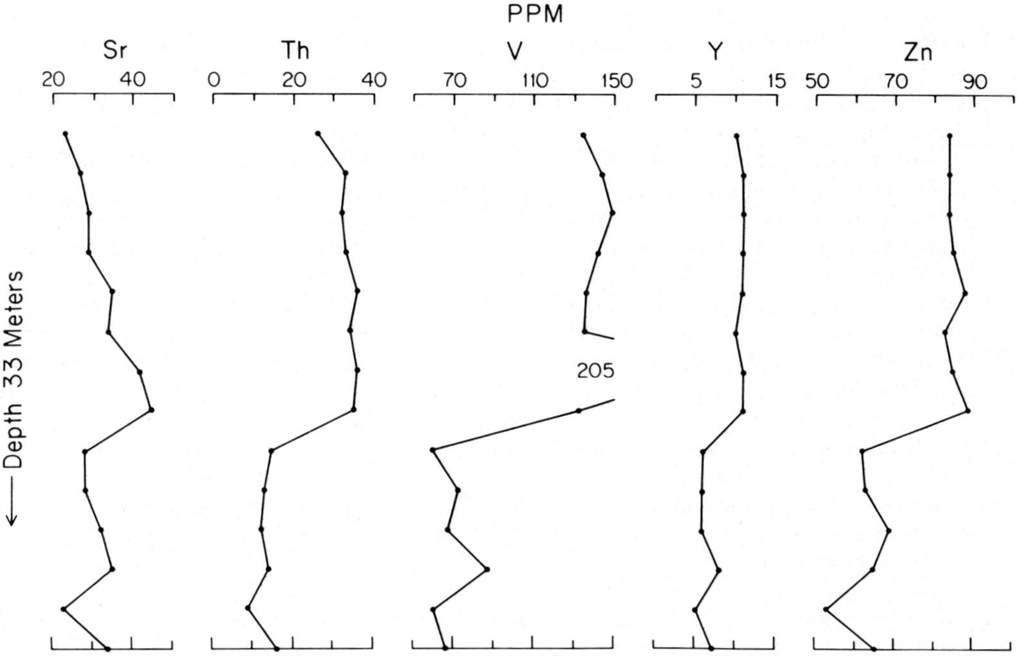
PPM
Sr
20 40
Th
0 20 40
V
70 110 150
205
Y
5 15
Zn
50 70 90
Depth 33 Meters

Fig. 10

Organic acids certainly have a major effect on the solubility of minerals in the weathering process (Sapozhnikov, 1963). As pointed out by Valeton (1972) the liberation of elements from a mineral or rock is governed by (1) bonds in the crystal structure of the minerals to be disintegrated; (2) solubility of secondary mineral phases; (3) pH and Eh of the solution; (4) charge of the elements; (5) temperature and concentration of the weathering solution; and (6) other ions present in the weathering solution. Therefore, the process of dissolution, transportation, and precipitation of gels or crystals is exceedingly complex. From the concentrations of the various elements and the minerals present it would appear that the weathering solutions were probably on the alkaline side because the normal case is that in the presence of alkalis the gibbsite structure forms. However Schoen and Robertson (1970) have shown that gibbsite can nucleate at the expense of acidic aluminum solutions. The fact that calcium and sodium are concentrated in the laterite would substantiate that the laterite was probably formed under alkaline conditions. It may be that the aluminum in the kaolinite was first dissolved by acid solutions (carbon dioxide bearing and organic) in the tropical rainy season. In the subsequent dry period the aluminum was carried upward together with the alkaline ground water by capillarity and gibbsite precipitated.

The formation of the iron minerals goethite and hematite strongly depends on pH. Schwertmann (1966) showed that from altered gels at pH 10 only goethite crystallized whereas below pH 10 hematite also occurred, but transformation of goethite to hematite is accelerated by temperature and is influenced by other inorganic anions and by organic components (Valeton, 1972). Schwertmann and Taylor (1977) reported that goethite may also form in acidic environments.

Titanium dissolves and precipitates under conditions somewhat similar to those for aluminum so that concentration of titania in the laterite would be expected and that the most common titanium mineral in laterites is anatase. Lelong (1969) pointed out that titanium does not dissolve and precipitate under conditions exactly similar to aluminum and for this reason isotitanium values may be very useful in the study of laterite genesis.

CONCLUSIONS

1. The Rio Jari kaolin deposit is sedimentary and was probably formed by the alteration of arkosic sands although the deposition of detrital kaolinite cannot yet be ruled out.
2. Tropical weathering of the kaolin produced a laterite or bauxitic kaolin on the upper seven to ten metres.
3. Minerals present in the kaolin are kaolinite, quartz, and anatase.
4. Minerals present in the laterite are kaolinite, gibbsite, goethite, hematite, anatase, and quartz.
5. Elements enriched in the laterite are Al, Fe Ti, Mn, Mg, Ca, K, P, Cr, Ga, Li, Rb, Th, V, Y, Zn.
6. Elements depleted in the laterite are Si, Ba, Cu, Ni, Pb.

REFERENCES

Goldschmidt, V. M., 1937. The principles of distribution of chemical elements in minerals and rocks. J. Chem. Soc. 139: 655-675.
Klammer, G., 1971. Uber Pro-pleistocene terrassen and ihre sedimente im unteren Amazonas Gebeit. Zeit. F. Geomorphologie 15, Heft 1, 62-106.
Lelong, F., 1969. Nature et genese des produits d´alteration de roches cristallines sous climat tropical humide (Guyana Francaise). Sciences de la Terre, Mem. 14: 188 pp.
Murray, H. H., 1981. Kaolin project in the Amazonia area-Brazil. Soc. of Min. Engrs. of AIME, Preprint 81-57, 7 pp.
Sapozhnikov, D. G., 1963. On the subtraction of aluminum by organic acids from minerals and rocks in the course of weathering. Symp. Bauxites, Zagreb, 1: 107-113.
Schoen, R. and Robertson, C. E., 1970. The structures of aluminum hydroxide and geochemical implications. Amer. Min. 55: 43-77.
Schwertmann, V., 1966. Die Bildung von Goethit and Hümatit in Böden und Sedimenten, Proc. INt. Clay Conf., Jerusalem: 159-165.
Schwertmann, V. and Taylor, R. M., 1977. Iron oxides: Chap. 5 in Minerals in Soil Environments, Dixon, J. B. and Weed, S. B., eds., Soil Sci. Soc. Amer.: 145-176.
Sombroek, W. G., 1966. Amazon soils: Ph.D. thesis Agric. Univ. Wageningen, The Netherlands.
Valeton, I., 1972, Bauxites: Elsevier, Amsterdam: 226 pp.

EOLIAN INFLUENCE ON TERRA ROSSA SOILS OF ITALY TRACED BY QUARTZ OXYGEN ISOTOPIC RATIO

M. L. JACKSON(*), R. N. CLAYTON(**), A. VIOLANTE(***), and P. VIOLANTE(***)

(*)Soil Science, College of Agriculture, University of Wisconsin, Madison, USA - (**)Chemistry-Geophysics, University of Chicago, Illinois, USA - (***)Istituto di Chimica Agraria, Università di Napoli, Portici, ITALY.

ABSTRACT

Analysis of the oxygen isotopic ratios of quartz from main regions of terra rossa soils lying over and under carbonate rocks of Italy provided definitive evidence of provenance or source. The oxygen isotopic ratio, expressed as $\delta^{18}O$, is high (22 to 34‰) for quartz crystallized at sea temperatures in some of the rocks. The $\delta^{18}O$ values of the long range aerosol size quartz (1-10 μm diameter) from the terra rossa soils ranged from 17.4 to 20.2‰ (mean, 18.5±0.9‰), the intermediate, mixed-provenance range common for the Northern Hemisphere. The $\delta^{18}O$ values for the whole quartz ranged from 11.9 to 17.1‰ (mean, 15.5±1.6‰), reflecting the usual trend for a higher proportion of high temperature quartz, in coarser fractions of soils. The whole quartz content was highest (43 to 56%) at lower altitudes (20-30 m, low $\delta^{18}O$ values, 11.9 to 13.4‰, characteristic of quartz in beach sands) and decreased with altitude (r=-0.7) to 2% at 640 m. With increasing altitudes to 640 m, the proportion of 1-10 μm size fraction of the quartz increased (r=0.8) as a percentage of total quartz (1.5 to 27%). The percentage of total quartz and 1-10 μm size quartz generally decreased markedly with soil horizon depth. One atypical $\delta^{18}O$ value of 34.3‰ was found in organogenic calcarenite rock which had an unusually high quartz content of 20.8% with large quartz overgrowths (as shown by scanning microscopy). The quartz is believed to have formed under a warm, arid climate during the late Miocene when the Mediterranean Sea evaporated to dryness. The data indicate that only a small part of the quartz in the terra rossa soils originated as residuum from the carbonates, and therefore that a major portion of the now-weathered terra rossa material may also be of eolian (pyroclastic and Saharan) provenance.

INTRODUCTION

Reddish clay soils ("terra rossa" term of Italian origin) overlying hard carbonate rocks occur commonly in the countries bordering on the Mediterranean Sea and a few other regions with Mediterranean type of climate (Bridges, 1970). They are widespread in Italy. Their origin has long been debated by pedologists and geologists and contradictory hypotheses have been put forth (Blanck, 1930). They were considered to be residue left after the dissolution of limestone, but also from wind-borne Saharan materials and (or) some detritus from the neighboring volcanoes (Galdieri, 1913; Lippi-Boncampi et al., 1955; Yaalon and Ganor, 1973). Because of the strong contrast with the underlying, pure, hard, compact, whitish limestone, terra rossa was also believed to originate from residues of older sediments, weathered volcanic ashes, and residues of overlying shale or sandstone (Kubiena, 1970). Some role of pyrites was seen in the biochemical weathering of carbonate rock residue as a factor of terra rossa genesis (Glazovskaya and Parfenova, 1974).

The oxygen isotopic ratio of monomineralic quartz has been used to determine quartz origin, from igneous-metamorphic rock material (low isotopic ratio, which reflects high-temperature origin) or from marine sediments including limestones and dolomites (high isotopic ratio, which reflects low-temperature origin)(Degens and Epstein, 1962). Quartz of the two types, i.e. of mixed provenance, was identified from the isotopic ratio after eolian transport and deposition in Hawaii during soil formation (Jackson et al., 1968, 1971) and in Pacific pelagic sediments (Clayton et al., 1968, 1972). The purpose of this paper is to detect, through the determination of the oxygen isotopic ratios and of the morphology of soil quartz by scanning electron microscopy, the significance of additions of eolian materials to the residue of dissolved carbonate rocks during the genesis of terra rossa soils in Italy.

MATERIALS

Sampling sites (Figure 1) were chosen to represent typical terra rossa soils covering large areas in Italy and overlying carbonate rocks of different geological ages. Many of the soils occupy areas which are almost plains and permit cultivation. The soil profiles are generally less than 1 m in depth; some are lithosols. Erosion is widespread. Colluviation may have been responsible for the presence of a considerable amount of clay in profiles not situated on plateau sites or hilltops.

METHODS

Quartz was isolated from the soils and carbonate rocks by treatment with 6$\underline{N}$ HCl and concentrated HNO_3 at 100°C to remove carbonates, iron oxides and some clay. The quartz from the 1-10 μm elutriated fraction from the soils and rocks was isolated by sodium pyrosulfate fusion and hexafluorosilicic acid treatment (Jackson, 1975; Jackson et al., 1976). Cristobalite, volcanic ash glass or phytoliths, if

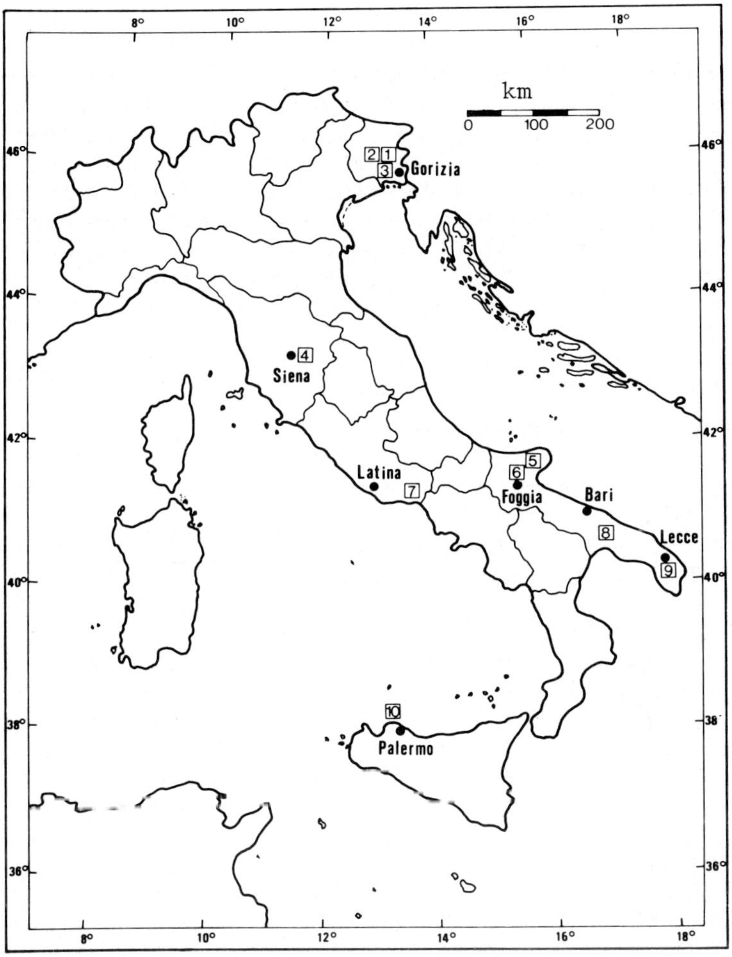

Fig. 1. Outline map of Italy with terra rossa sampling locations.

present, were removed by heavy liquid separation at specific gravity 2.33 in tetrabromoethane-nitrobenzene. Each sample, free of contaminating minerals as checked by means of X-ray diffraction, was treated in saturated H_3BO_3 solution to remove any fluorate and washed with H_2O. The oxygen was isolated from the quartz fractions by bromine pentafluoride and analyzed for isotopes as CO_2 by means of a 15-cm 60° sector double-collecting mass-spectrometer (Clayton and Mayeda, 1963). The $^{18}O/^{16}O$ ratios are expressed as $\delta^{18}O$, in parts per thousand (‰) relative to Standard Mean Ocean Water (SMOW). The SEM observations of morphological properties of quartz were done with a Cambridge S4 stereoscan.

RESULTS AND DISCUSSION

Quartz distribution downward in the profile

The soil quartz content tends to decrease with depth in profile (Table 1), especially relative to the B2 horizon just above the calcareous rock. This observation confirms a fairly limited downward movement of eolian quartz through mixing by faunal and floral activities (Jackson et al., 1971). The amount of the 1-10 µm fraction, as percent of whole quartz shows a consistent decrease of the aerosol size fraction (1-10 µm) with depth in the profile (Table 1). The "Quartz (%)" is of the 6N HCl-HNO_3 residue.

TABLE 1.

Quartz content of Terra Rossa soils of Italy.

Altitude m	Site (No.)	Horizon	Quartz Total (%)	Quartz 1-10 µm / Total (%)
25	Palermo	B	43	1.5
	(10)	R	21	--
30	Lecce	A	51	2.5
	(9)	B	56	2.1
50	Latina	A	53	2.0
	(7)	C	49	0.2
128	Gorizia	A1	28	10.3
	(2)	B1	3	1.9
220	Gorizia	A1	34	13.1
	(1)	B1	41	10.3
		B2	15	2.0
240	Foggia	A	21	14.1
	(5)	B1	20	14.5
		B2	3	4.4
260	Bari	Ap	23	13.8
	(8)	B1	22	13.1
		B2	12	3.4
520	Siena	B1	32	27.7
	(4)	B2	6	25.9
640	Foggia	B2	2	11.5[b]
	(6)		r=-0.75[a]	r=0.79[a]

[a]Against altitude ($n=19$; $p<0.01$).
[b]Remnant of erosion; r=0.99, if this value is omitted, and r=0.9965 if all B2 horizons are omitted.

Variation of quartz content with landscape elevation

Total quartz content decreases with altitude (correlation coefficient, r=-0.75; Table 1), a relationship attributed to eolian quartz winnowed from the beaches (low-elevation), as in Australia (Jackson et al., 1972). A general relation of quartz content to elevation and rainfall has been demonstrated for surface soils of the Hawaiian Islands (Jackson et al., 1971).

The long range eolian (1-10 μm) quartz content relative to total quartz increases with landscape elevation (r=0.79), not only for the surface horizons but also for the lower ones (Table 1). This could indicate that long range eolian (finer) silt fraction accretion is the dominant process at high altitude and that the accretion has continued during the whole formation time of terra rossa.

Oxygen isotopic ratios of the quartz

The oxygen isotopic ratios are high (23 to 34‰; typical of Phanerozoic chert) in the rocks (R) from the Palermo, Gorizia, and higher elevation Foggia, S. Marco in Lamis area (6) (Table 2). Both calcarenite (calcite) in Gorizia and Palermo and dolomite rocks in the Foggia area (6) are represented.

Intermediate $\delta^{18}O$ values of about 16 to 20‰ in the soils (Table 2) characterize 1-10 μm quartz as being of mixed provenance (a mixture from the high temperature and low temperature sources). The values of 18.3 to 19.1‰ for quartz in the Bari, Siena, and Latina calcarenite and fossiliferous micrite rocks (R) indicate that quartz of mixed provenance was incorporated into the rocks during the period of their fossil sedimentation and dolomitization. The rocks (R) from the Lecce, Marina di Castro area (9) and Foggia, Sannicandro Garganico area (5) were so low in quartz that no quartz could be isolated. The irregular, broken particle shapes for all of the mixed provenance samples are represented in Figure 2. For comparison, the $\delta^{18}O$ values for a paleosol "terra rossa" soil of southwestern Wisconsin ranged from 26.4‰ for 1-10 μm quartz from Sinnipee (Galena) dolomite to 26.9‰ for the 1-10 μm quartz from red clay close above this rock; but quartz from the red soil and loess further up in the profile had $\delta^{18}O$ values of 19.7 and 18.6‰, respectively, (Frolking and Jackson, 1981), indicative of eolian quartz of mixed provenance.

The $\delta^{18}O$ values for total quartz from the terra rossa soils of Italy (Table 2) range from 11.9 to 17.1‰ (mean, 15.5±1.6‰). The lower end of this range reflects the usual trend for a higher proportion of high temperature quartz in coarser fractions of soils. The lowest two values of 11.9 and 13.4‰ (Table 2) occur for the low elevations, where fine eolian material is winnowed from beach sand (Jackson et al., 1972).

While the quartz contents of the rocks are typically <1 to 3%, an atypically high quartz content of 20.8% and a high $\delta^{18}O$ value of

TABLE 2.

Quartz oxygen isotopes ($\delta^{18}O$) for terra rossa soils[a].

Location (No.)	Alt. m	Quartz Soil (%)	Quartz R (%)	Quartz ($\delta^{18}O$) for selected horizons and calcite or dolomite, R (‰): Total B1	Total B2/C	1-10 μm B1	1-10 μm B2/C	R
Palermo (10)	25	43	21	13.4	--	18.3	--	34.2
Gorizia (1)	220	41/15[b]	1.3	17.1	16.7	19.6	18.8	27.2
Gorizia (2)	128	28/3	2.2	16.8[c]	16.6	18.3[c]	17.4	24.2
Foggia (6)	640	2	<1	16.1	--	17.7	--	22.7
Latina (7)	50	51	2.1	--	14.1	--	20.2	19.1
Siena (4)	520	20/3	2.8	16.6	16.8	19.4	19.4	18.4
Bari (8)	260	22/12	<1	15.0	14.7	18.3	18.1	18.3
Foggia (5)	240	20	--	16.1	--	17.7	--	--
Lecce (9)	30	56	--	11.9	--	17.9	--	--
Mean±Std. Dev.				15.5±1.6‰		18.5±0.9‰		
Wisconsin (1-10 μm)				Loess 18.6	U.R.C.[d] 19.7	L.R.C.[d] 26.9		Dolo.[d] 26.4

[a]Listed in decreasing order of $\delta^{18}O$ in R.
[b]Ratio for upper/lower horizon
[c]A1 horizon.
[d]Upper Red Clay and Lower Red Clay and Dolomite.

34.3‰ was found for the organic calcarenite from Monte Pellegrino in Palermo (10), Sicily (Table 2). Prolific quartz overgrowths in the Cretaceous organogenic calcarenite, observed by scanning electron microscopy (Fig. 3), are believed to have formed during multistage quartz-overgrowths under a very warm, arid climate during the late Miocene. At that time, the Mediterranean Sea evaporated to dryness (Hsü, 1972) and the $H_2^{18}O$ content of the sea water would have become highly enriched by selective evaporation of $H_2^{16}O$. This carbonate rock was later uplifted and emerged as Monte Pellegrino during middle to late Pleistocene.

Fig. 2. Scanning electron micrograph of 1-10 µm quartz from the Palermo terra rossa soil B horizon, typical of the irregular particles from detrital pyroclastic and Saharan dust.

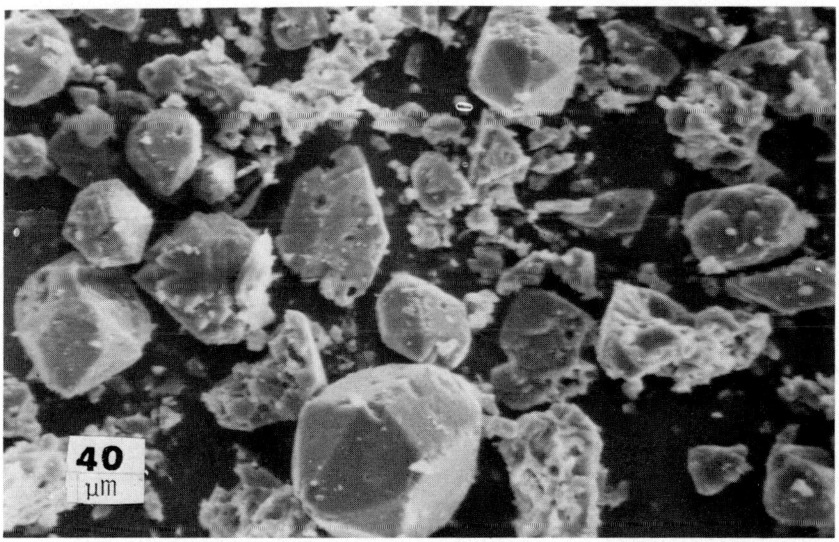

Fig. 3. Scanning electron micrograph of large, unusually abundant (20.8%), euhedral quartz crystal overgrowths from Palermo organogenic calcarenite, now uplifted as Monte Pellegrino and underlying the terra rossa soil (Fig. 2).

ACKNOWLEDGEMENTS

This research was supported in part by the College of Agricultural and Life Sciences of the University of Wisconsin, in part by the National Science Foundation EAR 76-19783-Jackson, EAR 74-19038-Clayton; in part by an exchange visitor's grant (for A.V.) from NATO; and in part by the Istituto di Chimica Agraria, Universita di Napoli, Portici; through an International Consortium for Interinstitutional Cooperation in the Advancement of Learning (ICICAL). Special gratitude is extended to Ms. Cynthia Sodini and Mrs. T. K. Mayeda for help with the oxygen isotopic ratio analyses, and Mrs. Chrystie Jackson for her meticulous care in editing and typing the manuscript.

REFERENCES

Blanck, E., 1930. Die Mediterran-Roterde (Terra Rossa). In: E. Blanck (Editor), Handbuch der Bodenlehre. Springer, Berlin, pp. 194-257.

Bridges, E. M., 1970. World Soils. Cambridge University Press, Cambridge, 89 pp.

Clayton, R. N. and Mayeda, T. K., 1963. The use of bromine pentafluoride in the extraction of oxygen from oxides and silicates for isotopic analyses. Geochim. Cosmochim. Acta, 27:43-52.

Clayton, R. N., Rex, R. W., Syers, J. K. and Jackson, M. L., 1968. Oxygen isotope abundance in quartz from Pacific pelagic sediments. Abstr., Geol. Soc. Am. Spec. Paper 121, p. 56.

Clayton, R. N., Rex, R. W., Syers, J. K. and Jackson, M. L. 1972. Oxygen isotope abundance in quartz from Pacific pelagic sediments. J. Geophys. Res., 77:3907-3915.

Degens, E. T. and Epstein, S., 1962. Relationship between $^{18}O/^{16}O$ ratios in coexisting carbonates, cherts and diatomites. Bull. Am. Assoc. Petrol. Geol., 46:534-542.

Frolking, T. A. and Jackson, M. L., 1981. Origin of red clay over dolomite in Wisconsin driftless uplands. Agronomy Abstr., v. 1981, p. 241.

Galdieri, A., 1913. L'origine della terra rossa. Ann. R. Scuola Sup. Agric. Portici, 11, p.46.

Glazovskaya, M. A. and Parfenova, E. I., 1974. Biogeochemical factors in the formation of terra rossa in the Southern Crimea. Geoderma, 12:57-82.

Hsü, K., 1972. When the Mediterranean dried up. Scientific American, 227(6), 26-36.

Jackson, M. L., 1975. Soil Chemical Analysis Advanced Course. Second Edition, 10th printing. Publ. by the author, Madison, WI, 895 pp.

Jackson, M. L., Gibbons, F. R., Syers, J. K. and Mokma, D. L., 1972. Eolian influence on soils developed in a chronosequence of basalts of Victoria, Australia. Geoderma, 8:147-163.

Jackson, M. L., Levelt, T. W. M., Rex, R. W., Clayton, R. N. and Sherman, G. D., 1968. Geomorphological relationships of tropospherically-derived quartz in soils of the Hawaiian Islands. Agronomy Abstr., v. 1981, p. 153.

Jackson, M. L., Levelt, T. W. M., Syers, J. K., Rex, R. W., Clayton, R. N., Sherman, G. D. and Uehara, G., 1971. Geomorphological relationships of tropospherically derived quartz in the soils of the Hawaiian Islands. Soil Sci. Soc. Am. Proc., 35:515-525.

Jackson, M. L., Sayin, M. and Clayton, R. N., 1976. Hexafluorosilicic acid reagent modification for quartz isolation. Soil Sci. Soc. Am. J., 40:958-960.

Kubiena, W. L., 1970. Micromorphological Features of Soil Geography. Rutgers University Press, New Brunswick, NJ, 254 pp.

Lippi-Boncampi, C., Mackenzie, R. C. and Mitchell, W. A., 1955. The mineralogy of some soils from Central Italy. Clay Miner. Bull., 2:281-288.

Yaalon, D. H. and Ganor, E., 1973. The influence of dust on soils during the Quaternary. Soil Sci., 116:146-155.

THE SIGNIFICANCE OF FLUORINE IN SOME ARGILLACEOUS ROCKS OF ENGLAND

R.K. HARRISON, M.E. STUART and G.E. STRONG
Institute of Geological Sciences, London SW7 2DE, UK.

ABSTRACT

F commonly occurs in volcanic eruptions. The stability of F even under high P-T conditions, together with its wide dispersal and reactiveness (as HF) make it a potential geochemical 'fingerprint' of penecontemporaneous volcanic contributions to sediments.

66 mudstone/marl/micrite samples were analysed for F, and ranged from Devonian to Early Tertiary; certain horizons relating to suspected tectonothermal or volcanogenic events were included. F was separated by pyrohydrolysis and determined by means of a specific ion electrode method. Duplicate runs were made using a higher flux-to-sample ratio to ensure complete decomposition of the samples. Bulk mineralogy and clay fractions (<2μm) were determined by X-ray diffractometry.

Palaeozoic tonsteins gave sub-average (<500 ppm) F values (320-480 ppm F); the High Main-Rider Coal tonsteins (Westphalian C) gave 640 and 600 ppm F. Kaolinite-mudstones, however, do not appear generally to be enriched in F.

Mesozoic smectite-rich fuller's earths, known to be volcanogenic, contain up to 2580 ppm F, but some Aptian earths contain only 530 ppm F. Illite-rich micrites from the Bathonian fuller's earth contain 1340, 1450 ppm F. Since phosphate is sparse in many samples, the F substitutes mainly for $(OH)^-$ in smectite, illite or mixed-layer clay. High-F also occurs in certain Rhaetic, Liassic and Upper Jurassic samples. In the Cretaceous, the Wealden and Gault clays have only average tenors, but a smectite-rich insoluble residue from the Lower Chalk contains 1410 ppm F. In the Tertiary, tuffaceous Thanet Beds contain up to 750 ppm F. Some relation emerges between high-F and volcanogenic input; no detrital F-minerals per se were detected.

Historic records of fluorosis in England and Wales relate to F-enriched groundwaters from aquifers and aquicludes some containing volcanogenic fluorine. Volcanogenic rocks elsewhere may indicate possible hazards of fluorine-enriched groundwaters.

INTRODUCTION

Except where fluoride-mineralisation has occurred, fluorine is a minor element in many magmatic rocks, particularly mica-granites and pneumatolysed rocks. Apart from fluoride-minerals per se (e.g. fluorite), F is a minor constituent of phosphates, biotite, phlogopite, lepidolite and hornblende; concentrations of detrital F-minerals might occur near F-enriched source rocks. Fluorine is often voluminous in the gaseous products of volcanism. Zies (1929) calculated the annual discharge of F from the Valley of Ten Thousand Smokes (Alaska) to be 135,000 tonnes, with 75% F being added directly to the sea through atmospheric agencies. Hydrothermal sources too may produce F as in the hot springs of Foochow (China), attributed by Tiao-Hsin Wang and Yuan-Tsun Lin (1940) to deep volcanic sources; many other examples occur. Through denudation, F may be released to the hydrosphere, and detrital F-bearing minerals incorporated in sediments. The possible sources of F in sediments are thus complex. Because of the apparently ready substitution of F^- for OH^- (with the same ionic radius of 1.33Å) it may be difficult to discriminate between detrital F-bearing mineral and those in which F was incorporated from pore waters during, or subsequent to, deposition.

Global values of F in sedimentary rocks (Turekian and Wedepohl, 1961) indicate the highest F (1300 ppm) in deep-sea clays, comparable with the highest for igneous rocks (1200 ppm) in syenites, and these values may support a connection between volcanogenic, submarine-exhalative F and fixation in the oceanic sediments.

Correns (1956) collated data on F in argillaceous rocks (mainly from Germany) on recent deep-sea sediments, and on individual clays and clay-minerals. Mansfield (1940) stressed the relation between volcanism and phosphate accumulation, with volcanogenic F acting as an indirect (though determinative) factor in this process. Spada (1969) described fluorite- and baryte-bearing sandstones and shales associated with tuffs around Castel Giuliano, Rome Province, with hydrothermal fluoritisation. Shishkina et al., (1972) found high F (up to 5400 ppm) in oceanic productive zones off Peru, Chile and SW Africa, and noted the correlation with glauconite and phosphorite enrichment. Akaiwa and Aizawa (1973) found that in sedimentary rocks from Tomioka, Japan, F unlike Cl is not easily removed by groundwaters. Montmorillonite tuffs are more enriched in F (up to 630 ppm) than those composed of clinoptilolite. Schulze (1977) also found that fluoride compounds are stable and F is only released under high P-T conditions. Baker (1958) determined up to 1827 ppm F in lakes of the Gregory Rift Valley, Kenya. Schneider et al., (1977) deduced that submarine exhalative sources were probably responsible for F enrichment (3000 to 8000 ppm) in the insoluble clay residues of the East Alpine mid-Triassic carbonates.

Bloxam and Thomas (1970) concluded that the highest F of Carboniferous shales from the South Wales Coalfield occurred in offshore marine facies and decreased shorewards. F increases in the clay fraction (1627 ppm) and in collophane. Edmunds (1971) found high F anomalies in groundwaters from mineralised areas in the Derbyshire Dome, both with fluorite deposits and low in fluorite.

Within the Mesozoic sediments of the British Isles only sporadic evidence of volcanism has so far been found, except for volcanogenic Bathonian and Aptian fuller's earths (Jeans et al., 1977). Petrographical evidence (e.g. vitroclastic textures) may, however, have been removed through argillisation, and so a potential geochemical indicator (such as F) of volcanic or other thermal events could provide useful corroboratory evidence.

SAMPLING: TECHNIQUES

66 samples* were analysed for F, from a stratigraphically and lithologically wide range. Most were taken from deep boreholes to avoid superficial alteration. Stratigraphical horizons were chosen broadly on the basis of known or inferred volcanogenic or other thermal events (Harrison et al., 1979a) in the Mesozoic and early Tertiary. Some Westphalian tonsteins and a tonstein from the Lower Old Red Sandstone were also included. Samples were air-dried and ground in a Tema mill; splits were reserved for X-ray diffractometry, P analysis, and thin sections were prepared of many samples.

Samples were analysed for fluorine using the pyrohydrolysis method described by Jeffery (1970) modified for use in the Analytical Chemistry Unit of the Institute of Geological Sciences (Thomas, 1970; Branch and Hutchison, 1979/80). The samples were ground in an agate mortar with a weighed excess of vanadium (V) oxide, which was used to accelerate the decomposition of the sample by its action as an oxygen transfer catalyst. The mixture was transferred to a specimen tube and shaken in a mixer mill. A convenient weight (approximately 1g) was then transferred to a silica boat which was placed in a silica tube in a tube furnace at a temperature of 650°-700°C. Steam was passed over the sample at a fixed rate and condensed into a polythene bottle containing sodium hydroxide solution. The furnace temperature was then raised to 750°-800°C and maintained there for 35 minutes. The fluorine was released, probably as hydrofluoric acid which attacked the walls of the tube to form volatile hydrofluorosilicic acid which was carried over into the distillate. The distillate was cooled to room temperature, the pH adjusted to between 5.5 and 6.5 (Haynes, 1978) using dilute nitric acid and the solution made up to volume.

The fluoride content was measured using a specific-ion electrode, calibrated against standards from Analar sodium fluoride and containing amounts of sodium hydroxide and acid similar to those in the sample solution. The amount of fluorine in the original sample was then calculated. The determination was repeated using a higher proportion of flux to sample to ensure that the more refractory materials were decomposed. The use of vanadium (V) oxide prepared by heating Analar ammonium metavanadate reduced the blank determination to a negligible level.

RESULTS: INTERPRETATION

New fluorine analyses are listed in Tables 1 and 2. Also published whole-rock (classical) analyses, which reported F as a routine, have been abstracted and incorporated in these tables (denoted by 'A').

---

*Sample details are filed at the Library of the Institute of Geological Sciences, Exhibition Road, London SW7 2DE and are available on request.

Palaeozoic rocks, Devonian: A tonstein, described by Mr R.W. Elliot (in Welch and Trotter, 1961) as tuffaceous, from the Raglan Marl Group (Lower Old Red Sandstone) of the Welsh Border, contains biotite, devitrified shards, traces of zircon and apatite. This sample contains 500 ppm F, which is about the average for shales.
Carboniferous: Nine tonsteins from the Upper Carboniferous (Westphalian A, B, C) contain an average of 460 ppm F similar to that of fireclays and other clays in the Upper Carboniferous. The highest F values (600 to 4200 ppm) relate to samples from the Westphalian 'C'. A tonstein (with 640 ppm F) from the High Main Coal (Nottinghamshire and North Derbyshire Coalfield) consists of major kaolinite, minor illite, carbonate and Fe-oxides; its lateral equivalent in North Staffordshire, the Rowhurst Tonstein, is enriched in 4200 ppm F (Wilson et al., 1966) attributed to about 78% crandallite as a phosphate host.

A tuffaceous sample (Westphalian A) from the Black Rake horizon (Francis et al., 1968) gave only 480 ppm F whereas turbiditic and interturbiditic (pelagic) mudstones beneath tuffs in the Viséan of Derbyshire contain 840 to 900 ppm F. The predominantly alkali basaltic character of Carboniferous volcanism in the Central Province perhaps resulted in little or no addition of F to the sedimentary basins, since F is more commonly associated with alkaline or acid volcanism.

Mesozoic Rocks

Known volcanogenic horizons. Fuller's earths (i.e. Ca-montmorillonite rich clays) contain abundant pyroclasts (pumice, crystals, shards, accessory minerals) (Jeans et al., 1977). Chemical analyses of Bajocian and Aptian fuller's earth samples (Lab. Nos. 2752 and 2751 respectively : Table 2) (Inst. geol. Sci. 1978; Merriman, 1978) show abnormally high F (2580 ppm and 1230 ppm F). Two Bajocian samples from the IGS Stowell Park borehole, Gloucestershire (Green and Melville, 1956) and, 140 km to the ESE, two from the Warlingham borehole, Surrey (Worssam and Ivimey-Cook), 1971) gave high F (1020-1560 ppm). XRD analyses confirmed that the Stowell Park borehole samples are smectite-rich, but those from the Warlingham borehole are impure with illite in excess of mixed-layer illite-smectite, subordinate kaolinite and chlorite. Only 0.11% P_2O_5 occurs in the fuller's earths (Lab. Nos. 2752, 2751) and from microscopic examination traces only of apatite occur. No F-bearing detrital minerals per se were detected. Though F is enriched in smectite-dominant fuller's earth it apparently remains where primary smectite has altered to illitic mixed-layer clay. Smectite-rich Aptian fuller's earths from two boreholes near Faringdon, Oxfordshire, some 100 km from the high-F sample (1230 F) from Redhill, (Surrey) contain average and subaverage F (530 and 310 ppm). No obvious pyroclasts have, so far, been found in the Oxfordshire sample unlike the Redhill earths. Assuming that the Faringdon fuller's earths are volcanogenic, then, tentatively, the low F may be due to a fluorine-deficient volcanic source (e.g. basaltic) compared with that contributing to the fuller's earths in SE England.

The insoluble clay residue of the Lower Chalk is rich in smectite and this together with other evidence of pyroclasts (Jeans et al.,

1982) indicates volcanism during its deposition. This residue contains abnormally high F (1410 ppm). In all these samples, phosphate may occur only as sparse biotic scraps accounting for only a minor proportion of the F which is mainly held in the clay structures. Again, no F-bearing detrital minerals were detected.

Other Mesozoic horizons. Clay, marl and other samples were analysed from the two deep IGS boreholes (noted above) in Gloucestershire and Surrey, from the Keuper to the Lower Cretaceous.

Triassic. A major mineralising event occurred in the British Isles at about 220 My (mid-Triassic; in Harrison et al., 1979a). This event might relate to the widespread mid-Triassic mineralisation in Europe noted above. A sample of Keuper Red Marl taken near its contact with the underlying Bunter gave, however, only 520 ppm F. Values increase in the Keuper Tea Green Marl. A sample from close to the phosphate-rich Rhaetic Bone Bed contains only 530 ppm F.

Jurassic: (Lower) : Lias. In 9 samples of shale and mudstone with variable calcite the principal clay mineral is illite with subordinate kaolinite, mixed-layer smectite-illite, chlorite and vermiculite. Apatite occurs in only trace amounts. High-F occurs nevertheless in 4 samples (900-1090 ppm F), with the highest value in the *Schlotheimia angulata* zone (1090 ppm F), and is mainly held, presumably, in illite. A Middle Lias oolite contains chamosite, calcite, siderite, and in places, minor goyazite, to which less than half of the F (930 ppm) can be attributed. An Upper Lias mudstone contains 1% apatite (2000 ppm P), which perhaps accounts for about 400 ppm F of the total F (1020 ppm).

Middle Jurassic. Two micrites from the Inferior Oolite gave above-average F (650, 840 ppm). Dissolution of the carbonate concentrated the F (2100 ppm) in the insoluble clay residue (19% wt). Above the F-rich fuller's earths in the Great Oolite Series, two samples of clayey micrite and micritic clay also gave high F values (1170, 1060 ppm).

Upper Jurassic. A sample of Lower Oxford Clay (930 ppm F) contains dominant illite with subordinate interlayered illite-smectite, kaolinite, and minor components; carbonaceous material is important. Less than 1% apatite (720 ppm P) occurs in this sample. Ovey and Gay (1960) identified discrete fluorite in the Upper Oxford Clay; the unabraded euhedral forms suggest an authigenic rather than detrital origin.

A micritic Kimmeridge Clay sample gave an above average F value of 880 ppm; the insoluble clay residue (49% wt) contained 1100 ppm F. Illite is the main clay mineral, well in excess of mixed-layer illite-smectite and kaolinite; only 710 ppm P are present.

Cretaceous. Samples of Lower Cretaceous Wealden Beds gave low F (370-420 ppm). The Gault Clay shows above-average F (660 ppm); volcanogenic components have been identified in this clay elsewhere (Jeans et al., 1982). Volcanicity during sedimentation of the Lower Chalk (Jean et al., 1982) is supported, perhaps, by the high-F (1410 ppm) of the insoluble, smectite-rich residue.

Lower Tertiary. Volcanogenic activity occurred widely during the early Tertiary. (Harrison et al., 1979b; Knox, 1979; Knox and Ellison, 1979). There is F-enrichment in tuffaceous smectite clays (750 ppm) of Thanetian age from a North Sea Well; the Basal Eocene 'ash-marker' gave however only average F.

CONCLUSIONS

Assessment of results

Of the Palaeozoic samples only two Lower Carboniferous mudstones are enriched in F and this might be attributed to some volcanogenic input. The Upper Carboniferous tonsteins contain average F except in the laterally widespread High Main tonstein and the phosphate-rich equivalent Rowhurst tonstein. The kaolinitic mineralogy of tonsteins, however, might restrict fluorine tenors.

In the Mesozoic, there is positive correlation between high-F and known volcanogenic sediments, and in other samples the high-F reported might support some volcanogenic/hydrothermal input. Phosphate probably contributes only a minor part of the F in the values determined. There is no correlation (except in phosphate-rich samples) between F and P. Fifteen samples with 280 to 2000 ppm P showed no overall correlation with their F contents. Some correlation is shown between F-enrichment and smectite, but carbonate-rich samples with mainly illitic clay (for example, from the fuller's earths) also contain high F, and the F is probably held in illite or mixed-layer illite-smectite. In the Cretaceous rocks the high-F of some samples of Aptian volcanogenic fuller's earth and Lower Chalk might correlate with volcanic episodes, but is less marked in the Tertiary samples. Correlations must, however, be tentative pending more extensive analyses. While it is recognised that detrital F-minerals might be widespread, none have been identified with certainty in the present samples.

Fluorosis and F in Mesozoic-Lower Tertiary sediments

Bromehead (1943) discussed the geographical distribution of historic cases of known fluorosis in the 1940's. In areas of obvious contaminants of groundwater (e.g. flooded F-Pb-Zn mines of Weardale, Co. Durham), severe fluorosis was only of minor occurrence due to the low solubility of CaF_2. Of the other areas, some with severe fluorosis, however, the aquifers or aquicludes relate to sediments (including those confirmed from the present study) to be enriched in F. Bromehead related the high-F to phosphate in these beds but no analytical data were given. As indicated above, high-F relates (in general) to the bulk clay mineralogy rather than to phosphate or carbonate. Interestingly, Mo causing teart in cattle (from contaminated herbage) was also attributed to groundwater circulating in the Lower Lias. Some sediments containing volcanogenic/exhalative components may thus contain high F (and other poisons) and so be suspected as possible sources of contaminants to groundwater, but F appears to be relatively fixed in clay-mineral structures even through diagenesis and compaction.

TABLE 1
Fluorine analyses[1] of some Palaeozoic rocks

Stratigraphical horizon	F (ppm)	IGS Lab No/ Source
Upper Carboniferous		
-285 My-		
Upper Coal Measures : Stephanian		
Pennant Series Fireclay	300	(A) 609
Middle Coal Measures:		
Westphalian C		
Rowhurst Rider Tonstein	460+	(A) 3006
Rowhurst Tonstein	4200	(A) 1951
Rider Coal above 4-ft seam	600+	3004
" " "	370+	3005
Tonstein in High Main Coal	640+	2975
Below High Main Coal	320+	3009
Tonstein in Sharlston Muck Seam	420+	3007
Tonstein P27	400+	2997
" P26	320+	3010
Tonstein in Upper Hollywell Wood Seam	440+	3008+
Fireclay, Half-Yard	490	(A) 615
Westphalian B		
Tonstein below Clowne Coal	420+	3011
Clay below Top Main Coal	520	(A) 624
Fireclay, horizon not stated	800	(A) 632
" " " "	160	(A) 612
Westphalian A		
Stanllyd (9 ft) seam	700	(A) -
'Best Pot Clay'	420	(A) 687
Black Rake tuffaceous siltstone	480+	3015
Namurian		
Passage Group seatclay	300	(A) 2049
Fireclay	160	(A) 465
"	290	(A) 314
"	330	(A) 317
Interturbidite pelagic mudstone	810	(A) 2186
Turbidite silty mudstone	650	(A) 2187
Lower Carboniferous		
Viséan P.		
(Tuffs 12m (Turbidite silty mudstone	840	(A) 2184
above) (Interturbidite pelagic mudstone	900	(A) 2185
-345 My-		
Devonian		
Lower Old Red Sandstone: tonstein in Raglan Marl Group	500+	2976

[1] Analyst: (+) Robertson Research International Ltd.

[A] Analyses abstracted from the Chemical Analysis Memoirs of the Institute of Geological Sciences.

TABLE 2
Fluorine analyses of some Mesozoic-Lower Tertiary rocks

			F (ppm)	Lab No/ Source
TERTIARY	LOWER	Oligocene: Bovey Beds	300	(A)
		Eocene: London Clay	480*	2962
		Palaeocene: Thanetian	580+	2978
		"	460+	3012
		" (Tuffaceous)	750+	2979
		"	450+	3013
		- 65 My -	440+	3014
MESOZOIC	CRETACEOUS	Upper Cretaceous: Lower Chalk (insol. residue)	1410*	2964
		Lower Cretaceous: Albian : Gault Clay	660*	2965
		Aptian (Fuller's Earth)	530*	2961
		" "	1230	(A) 2751
		" "	310+	3018
		Wealden: Weald Clay	370*	2955
		" "	390*	2954
		" "	280	(A)
		-140 My - Ashdown Beds	420*	2966
	JURASSIC UPPER	Purbeckian	440*	2956
		Kimmeridgian Kimmeridge Clay	880*	2957
		" Hartwell Clay	330*	2967
		Oxfordian Oxford Clay	520	(A)
		" Lower Oxford Clay	930*	2958
	MIDDLE	Great Oolite Series : Stonesfield 'Slate'	1170*	2980
		" " " " "	1060*	2981
		" " " Fuller's Earth	1020*	2946
		" " " " "	1560*	2947
		" " " " "	1340*	2959
		" " " " "	1460*	2960
		" " " (Tuffaceous)	2580	(A) 2752
		" " " " "	270*	2982
		" " " " "	1240*	2983
			1270*	2984
		Bajocian: Inferior Oolite Series	840*	2985
		" " "	650*	2986
	LOWER	Upper Lias	630*	2987
		Middle Lias Marlstone Rock Bed	930*	3016
		Lower Lias *ibex* zone	530*	2948
		" " *jamesoni* zone	440*	2988
		" " *semicostatum* zone	1020*	2989
		" " *bucklandi* zone	480*	2990
		" " *angulata* zone	1090*	2949
		" " *planorbis* zone	550*	2991
		" "	1020*	2999
		" "	630*	3000

		F (ppm)	Lab No/ Source
MT	-195 My - Blue Lias	480+	3001
	Upper Rhaetic: Cotham Beds	730*	2950
ER	" " " "	1000+	3002
	" " " "	1000+	3003
SI	Lower Rhaetic: <u>contorta</u> shales	660*	2951
	" " " "	530*	2992
OA	Keuper Marl: Tea Green Marls	860*	2952
	" " " " "	860*	2993
ZS	" " " " "	380*	2994
	Red Keuper Marls	470*	2995
OS	" " "	400*	2996
	" " "	340*	2997
II	" " "	430*	2998
	" " "	520*	2953
CC	Muschelkalk, Germany	1780	(A)

New analyses for this paper:- *(1) M.E. Stuart

+(2) Robertson Research International Ltd.

A. Other analyses abstracted from the Chemical Analysis Memoirs of the Institute of Geological Sciences and other references quoted.

ACKNOWLEDGEMENTS

This paper is published with permission of the Director of the Institute of Geological Sciences (N.E.R.C.) The authors wish to thank their colleagues Messrs B.R. Young and R.J. Merriman for assistance with X-ray diffraction work, Miss E. Waine, Mr. J.A. Bain and Dr W.M. Edmunds for helpful discussions, Mr. C.W. Wheatley for thin sections of some very fragile samples, and Mrs. O.M. Burke for typing.

REFERENCES

Akaiwa, H. and Aizawa, S., 1973. Behaviour of fluorine during sedimentary processes - fluorine contents of sedimentary rocks from Tomioka, southern Gunma Prefecture, Japan, Geochem.J, 7: 1-9.

Baker, B.H., 1958. Geology of the Magadi Area. Rep.Geol.Surv. Kenya No. 42 81 pp.

Bloxam, T.W. and Thomas, R.L., 1970. Fluorine in Carboniferous shales from the South Wales coalfield. Chem.Geol. 5: 179-193.

Branch, C.H. and Hutchison, D.G., 1979/80. A.C.U. Internal Report No. 50 Inst.geol.Sci. (unpub).

Bromehead, C.E.N., 1943. Presidential address, Geology and Health. Trans. S.E. Union Sci.Soc., Papers contributed to Congress, 1943.

Correns, C.W., 1956. The Geochemistry of the Halogens. In: Ahrens, L.H., Rankama, K. and Runcorn, S.K., Ed., Physics and Chemistry of the Earth, 1,pp. 181-234.

Edmunds, W.M., 1971. Hydrochemistry of groundwaters in the Derbyshire Dome with special reference to trace constituents. Report 71/7, Inst.Geol.Sci.

Francis, E.H., Smart, J.G.O. and Raisbeck, D.E., 1968. Westphalian volcanism at the horizon of the Black Rake in Derbyshire and Nottingham, Proc.Yorks Geol.Soc., 36: 395-416.

Green, G.W. and Melville, R.V., 1956. The Stratigraphy of the Stowell Park Borehole (1949-51). Bull.geol.Surv.Gr.Bt. No. 11: 1-33.

Harrison, R.K., Jeans, C.V. and Merriman, R.J., 1979(a). Mesozoic igneous rocks, hydrothermal mineralisation and volcanogenic sediments in Britain and adjacent regions. Bull.geol.Surv.Gr.Bt. No. 70: 57-69.

Harrison, R.K., Knox, R.W.O'B. and Morton, A.C., 1979(b). Petrography and mineralogy of volcanogenic sediments from DSDP Leg 48, Southwest Rockall Plateau, Sites 403 and 404. Init.Rep. DSDP 48, Washington: 771-785.

Haynes, S.J., 1978. Joint determination of fluorine and chlorine in granitic rocks with ion-selective electrodes. Talanta, 25: 85-89.

Institute of Geological Sciences 1978 Annual Report for 1977, 111.

Jeans, C.V., Merriman, R.J. and Mitchell, J.G., 1977. Origin of Middle Jurassic and Lower Cretaceous fuller's earths in England. Clay Miner. 12: 11-44.

Jeans, C.V., Merriman, R.J., Mitchell, J.G. and Bland, D., 1982. Volcanic clays in the Cretaceous of Southern England and Northern Ireland. Clay Miner. Vol. 17; in press.

Jeffrey, P.G., 1970. Chemical methods of rock analysis. Pergamon, Oxford, pp. 234-237.

Knox, R.W.O'B., 1979. Igneous grains associated with zeolites in the Thanet Beds of Pegwell Bay, north-east Kent. Proc.geol.Assoc. 90: 55-59.

Knox, R.W.O'B. and Ellison, R.A., 1979. A Lower Eocene ash sequence in SE England. J.geol.Soc. 136: 251-253.

Mansfield, G.R., 1940. The role of fluorine in phosphate deposition. Am.J.Sci. 238: 863-879.

Merriman, R.J., 1978. The fuller's earth deposits of the Redhill area. in Guide book to South East England, Sixth Int.Clay Conf. Oxford, p. 306.

Ovey, C.D. and Gay, P., 1960. The discovery of fluorite and perovskite in the Upper Oxford Clay at Warboys, Huntingdonshire. Proc.Geol.Assoc. 71: 340.

Schneider, H.J. Möller, P., Parekh, P.P. and Zimmer, E., 1977. Fluorine contents in carbonate sequences and rare earths distribution in fluorites of Pb-Zn deposits in East-Alpine mid-Triassic. Mineral.Depos. 12: 22-36.

Schulze, H., 1977. Geochemistry of fluorine in the exogenous range. Z.Angew.Geol. 23: 65-70.

Shishkina, O.V., Baturin, G.N. and Bukova, V.S., 1972. Fluorine in sediments and ooze waters of high-productive zones of the ocean. Geokhimya, No. 8: 988-996.

Spada, A., 1969. Il giacimento di fluorite e baritina esalativo-sedimentario in 'facies' lacustra, intercalata nei sedimenti piroclastici della zona di Castel Giuliano, in prov. di Roma. Ind. Mineraria, 20: 501, 504-510, 515-518.

Thomas, C.H., 1970. A.C.U. Internal Report. Inst.geol.Sci. (unpub.).

Tiao-Hsin Wang and Yuan-Tsun Lin., 1940. The analysis and study of hot-spring water in Foochow. Am.J.Sci., 238: 799-804.

Turekian, K.K. and Wedepohl, K.H., 1961. Distribution of the elements in some major units of the Earth's crust. Bull.geol.Soc.Am. 72: 175-192.

Welch, F.B.A. and Trotter, F.M., 1961. Geology of the country around Monmouth and Chepstow. Mem.geol.Surv.Gr.Bt. 164 pp.

Wilson, A.A., Sergeant, G.A., Young, B.R. and Harrison, R.K., 1966. The Rowhurst Tonstein, North Staffordshire and the occurrence of crandallite. Proc.Yorks.Geol.Soc. 35: 421-427.

Worssam, B.C. and Ivimey-Cook, H.C., 1971. The stratigraphy of the Geological Survey Borehole at Warlingham, Surrey. Bull.geol.Surv. Gr.Bt. No. 36: 1-146.

Zies, E.G., 1929. The Valley of Ten Thousand Smokes. Nat.Geog.Soc. Conf.Tech.Papers. 1, Katmai Ser., No. 41: 66.

HYDROGEN AND OXYGEN ISOTOPIC COMPOSITIONS OF KAOLIN MINERALS IN JAPAN

K.MARUMO, Y.MATSUHISA and K.NAGASAWA

ABSTRACT

Hydrogen and oxygen isotopic compositions were determined for kaolin minerals of various origins in Japan. The hydrogen isotopic composition depends on the latitude of sample locality; this suggests that the kaolins were formed by the action of the meteoric waters. The oxygen isotope fractionation factor varies considerably with temperature, and therefore the δD-$\delta^{18}O$ diagram is useful to distinguish hydrothermal kaolins from kaolins of weathering and sedimentary origins. The δD value of halloysite is higher than that of kaolinite of similar origin and locality; This is probably due to fractional loss of H on sample preparation.

1 INTRODUCTION

Hydrogen and oxygen isotopic compositions of minerals provide a useful clue for elucidating the origin of water contained in them and the conditions of their formation. Several isotopic studies have been published on kaolin minerals of either weathering or hydrothermal origin.

The writers have studied hydrogen and oxygen isotopic compositions of kaolin minerals of various origins in Japan. A part of the result was reported by Marumo et al.(1980), who concluded that the hydrogen isotope fractionation factor between kaolin and water is not sensitive to the temperature in the range below 250°C based on hydrogen isotopic data of kaolins in geothermal areas.

In this paper, the hydrogen and oxygen isotopic compositions of kaolin minerals of various origins are given and they are discussed with regard to the kind and genesis of the minerals.

2 SAMPLES

The origins of the samples examined are:

1) Hydrothermal, relating to Quarternary volcanic or geothermal activities...Ohnuma, Matsukawa, Fukenoyu, Oyasu, Itaya, Ohtake, Iriki.

2) Hydrothermal, Neogene or older...Ohtoge, Akatani, Kampaku, Bodai, Hiraki, Shokozan, Kodachi.

3) Alteration product of Pliocene pyroclastics...Sendai.

4) Weathered Quarternary pumice...Hachinohe, Ina, Yame.

5) Weathered Pliocene pumice...Naegi A, Naegi B (Naegi A is an alteration product of phenocrystic biotite, while Naegi B is that of volcanic glass of the same pumice bed).

6) Weathered Cretaceous pyroclastics unconformably overlain by Oligocene flint clay...Iwaizumi.

7) Sedimentary product...Seto, Hara.

The geographical distribution of sample localities and the mineralogy of samples are shown in Figs.1 and 2. All samples were purified by sedimentation.

3 EXPERIMENTAL PROCEDURE

The experimental procedure for the determination of the hydrogen isotopic composition was described by Marumo et al.(1980). The hydrogen isotopic composition is reported in per-mil deviations from a standard: $\delta D = (R_{sample}/R_{standard} - 1)1000$, where R=D/H and a standard is Standard Mean Ocean Water (SMOW).

The oxygen from the kaolin samples was extracted by reaction with bromine pentafluoride after removing adsorbed and interlayer water by heating at 200°C in vacuo. The oxygen was then quantitatively converted to carbon dioxide which was used for mass spectroscopic analysis. The results are expressed in terms of $\delta^{18}O$ per-mil relative to Standard Mean Ocean Water.

4 RESULTS AND DISCUSSION

4.1 Hydrogen isotopic composition

Figs.1 and 2 show the δD value of kaolin samples of various origins. Since the hydrogen isotope fractionation factor between kaolin and water is not sensitive to the temperature in the range below 250°C (Marumo et al.,1980), the hydrogen isotopic composition of kaolin minerals is expected to depend mainly on that of water with which the respective kaolin mineral was in equilibrium. As shown in Fig.1, most kaolinite and dickite samples from northeastern and central Japan show δD values between -85 and -100 per-mil, while those from southwestern Japan show δD values around -70 per-mil. Because the δD value of meteoric water largely depends on the latitude (Sheppard et al.,1969), this difference in δD of kaolinite and dickite samples from different regions of Japan shows that the water in equilibrium with the kaolin minerals was of meteoric origin. For some localities, the correlation between δD values of kaolin and meteoric water was confirmed by measuring the δD of meteoric water near the mineral localities.

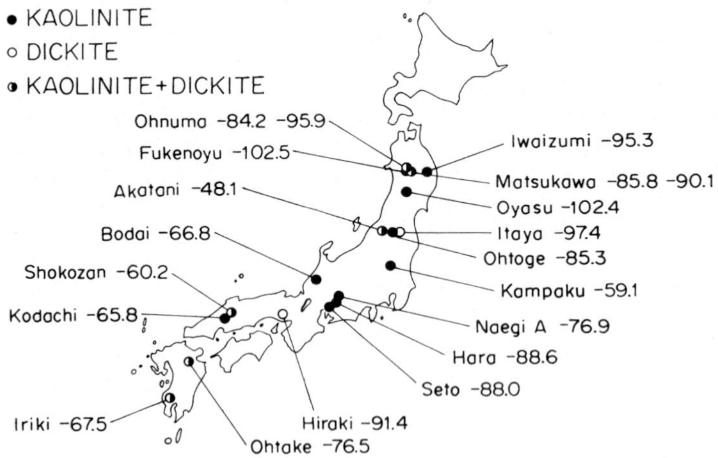

Fig.1 δD values of kaolinite and/or dickite samples.

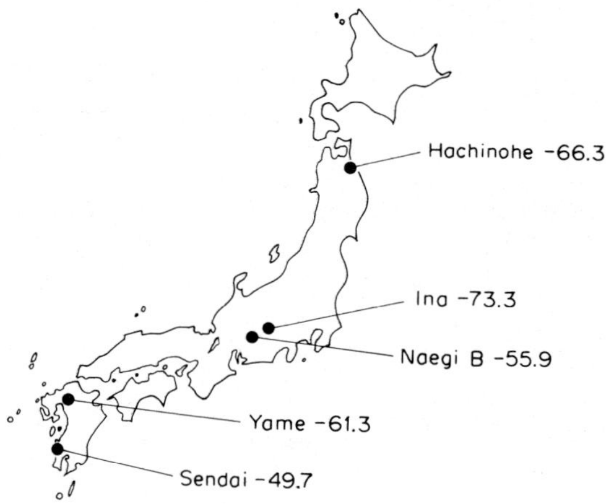

Fig.2 δD values of halloysite samples.

Some samples from older hydrothermal kaolins, e.g. Kampaku and Akatani, show higher δD values than expected from the latitude. The cause of these high δD values cannot be known, since we have no information about the δD of meteoric water at the time of mineral formation. The high δD of the hydrothermal solution responsible for these older kaolins may be due to a contribution from sea water.

Hydrogen isotopic compositions of halloysite samples shown in Fig.2 have no clear relation to the latitude of the sample localities. They will be discussed later, using δD-δ^{18}O diagram.

4.2 Temperature dependence of oxygen isotope fractionation

In Fig.3, oxygen isotope fractionation factors calculated for borehole samples from Ohnuma and Ohtake geothermal areas are plotted against the actual temperature. A similar plot for Yellowstone kaolinite by Sheppard et al.(1969) is also shown. These plots lie near the extention of the fractionation line experimentally determined by Kulla and Anderson(1978). The figure shows that the oxygen isotope fractionation factor varies considerably with temperature and therefore suggests that oxygen isotopic composition of kaolin minerals provides a useful clue for distinguishing whether they are of hydrothermal or of weathering origin.

4.3 δD-δ^{18}O relation

Fig.4 is the δD-δ^{18}O diagram for kaolin samples of weathering and sedimentary origins. The δD-δ^{18}O plots for meteoric waters near the mineral localities are also shown. The plots for kaolinite samples are near the kaolinite line of Savin and Epstein(1970), whereas the plots for the halloysite samples are on the higher δD side of that line. Halloysite from Naegi shows 21 per-mil higher δD than the coexisting

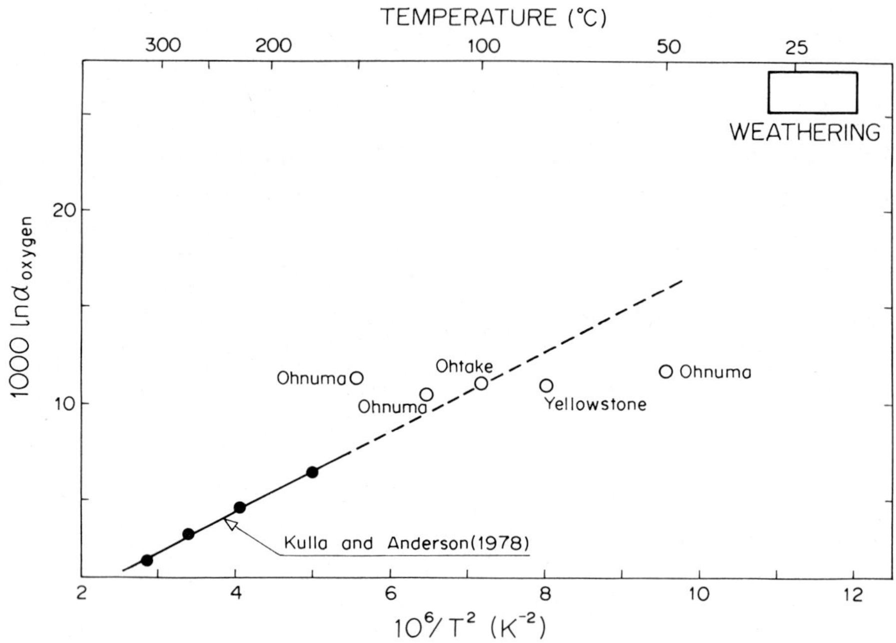

Fig.3 Temperature dependence of the oxygen isotope fractionation.

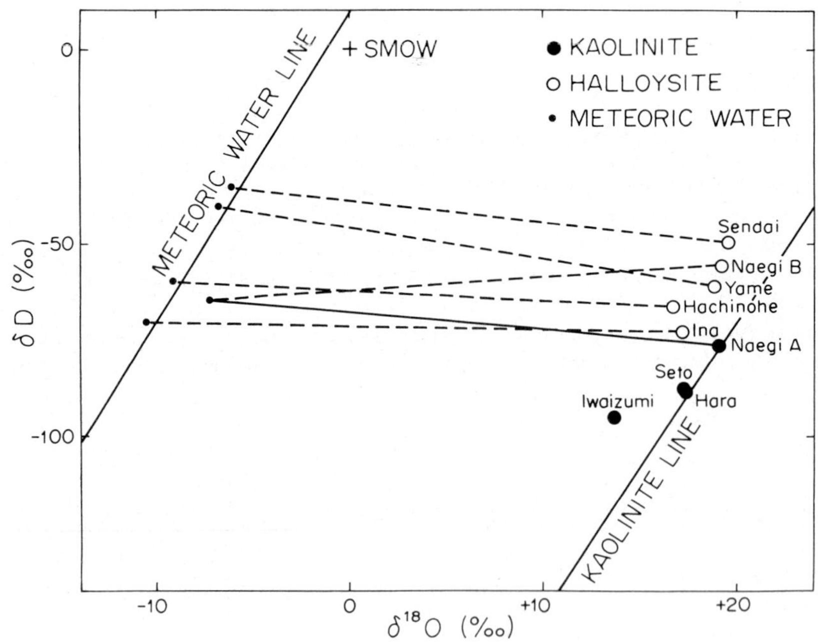

Fig.4 δD-δ^{18}O diagram for kaolins of weathering and sedimentary origir

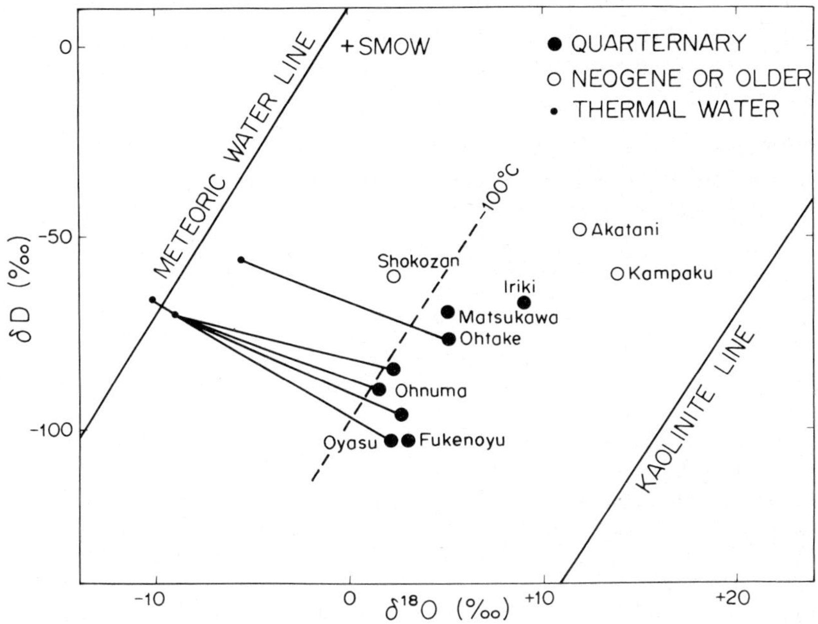

Fig.5 δD–δ^{18}O diagram for hydrothermal kaolins.

kaolinite, although they show nearly identical δ^{18}O values. All of the plots for the halloysite samples would be near the kaolinite line, if their δD's were reduced by 21 per-mil. This relation can be reasonably explained by the change in hydrogen isotopic composition during the preparation of the samples. Fractional release of H on air-drying of the sample and on removal of adsorbed and interlayer waters at 270°C in vacuo may have occurred by an exchange of H and D between interlayer water and inner-surface hydroxyl groups(Ledoux and White,1964; Wada, 1967; Lawrence and Taylor,1971).

Data for halloysite from Sendai are plotted in Fig.4, although lack of detailed geological information does not permit to reach firm conclusion on its genesis. The writers' field observations and works by Minato(1976) and Fujii(1962) on a similar nearby deposit at Okuchi suggest that the deposit was formed under near-surface conditions by ground water or thermal water related to volcanic activity.

Fig.5 is the δD-δ^{18}O diagram for kaolinite and/or dickite samples of hydrothermal origin. The δD-δ^{18}O plots of thermal waters near the mineral localities are also shown. As is expected from the large temperature dependence of oxygen isotope fractionation, hydrothermal kaolins are clearly distinguished from kaolins of weathering and sedimentary origins in this diagram. The dashed line labeled 100°C shows the δD-δ^{18}O relation of kaolinite which is in equilibrium with water on the meteoric water line at 100°C. All data for kaolins from active geothermal areas (marked as Quaternary in the figure) fall on the higher $\delta^{18}O$ side of this line. This does not necessarily mean that they were formed below 100°C, because, in many cases, plots of δD and $\delta^{18}O$ values of thermal waters are on the higher $\delta^{18}O$ side of the meteoric water line.

ACKNOWLEDGEMENTS

The writers wish to thank Professor S. Matsuo of Tokyo Institute of Technology, Professor Y. Kuroda of Shinshu University and Professor N. Nakai of Nagoya University for instruction and for permission to use facilities of their laboratories.

REFERENCES

Craig,H.,1961. Isotopic variations in meteoric waters. Science, 133: 1702-1703.

Fujii,N.,1962. On Okuchi clay deposits, Kagoshima Prefecture. Bull. Geol.Surv.Japan, 13: 231-238.

Kulla,J.B. and Anderson,T.F.,1978. Experimental oxygen isotope fractionation between kaolinite and water. Short Papers 4th Intern. Conf.Geochron.Cosmochron.Isotope Geol.,1978, U.S.G.S. Open-file Rept.,78-701: 234-235.

Lawrence,J.R. and Taylor,H.P.Jr.,1971. Deuterium and oxygen-18 correlation: Clay minerals and hydroxides in Quarternary soils compared to meteoric waters. Geochi.Cosmochi.Acta, 35: 993-1003.

Ledoux,R.L. and White,J.L.,1964. Infrared study of the OH groups in expanded kaolinite. Science, 143: 244-246.

Marumo,K., Nagasawa,K. and Kuroda,Y.,1980. Mineralogy and hydrogen isotope geochemistry of clay minerals in the Ohnuma geothermal area, northeastern Japan. Earth Planet.Sci.Lett., 47: 255-262.

Minato,H.,1976. Mineralogy and genesis of hydrothermal kaolin in Japan. The 7th Symposium on Genesis of Kaolin, Tokyo, 17-31.

Savin,S.M. and Epstein,S.,1970. The oxygen and hydrogen isotope geochemistry of clay minerals. Geochi.Cosmochi.Acta, 34: 25-42.

Sheppard,S.M.F., Nielsen,R.L. and Taylor,H.P.Jr.,1969. Oxygen and hydrogen isotope ratios of clay minerals from porphyry copper deposits. Econ.Geol., 64: 755-777.

Wada,K.,1967. A study of hydroxyl groups in kaolin minerals utilizing selective deuteration and infrared spectroscopy. Clay Miner., 7: 51-61.

CLAY MINERAL DISTRIBUTION IN THE NIGERIAN TAR SAND SEQUENCE

E. I. ENU[1], O. S. ADEGOKE[1], C. ROBERT[2], B. D. AKO[1], T. R. AJAYI[1], S. A. ADEDIRAN[1], and A. ORIYOMI[1]

1. Department of Geology, University of Ife, Ile-Ife, Nigeria
2. Laboratoire de Geologie Marine, Centre Universitaire de Luminy-Marseille

ABSTRACT

The clay mineral suite of the vast tar sand sequence of Ondo State, Nigeria was studied by X-ray diffraction analysis. The studied samples were obtained from among the forty-four core-holes drilled in a 17 km^2 area in the tar sand belt near Agbabu.

Kaolinite is the dominant and most frequently encountered clay mineral. Mixed layer minerals are next in abundance, markedly increasing downdip. Illite, chlorite and smectites occur generally in small amounts.

The abundant occurrence of kaolinite is attributable to the effects of weathering and leaching, aided by the pronounced relief in the source areas. This also confirms the prevalence of humid tropical conditions throughout the period of deposition. The mixed layer minerals are considered as products of incomplete diagenetic alterations, formed as a result of variations in some environmental factors, particularly gradient, drainage and salinity, in the depositional basin.

The average clay content of the tar sand pay zones is lower (2 - 5%) than that in equivalent horizons in

Athabasca, Canada (15 - 20%). Therefore, environmental problems accompanying mining and oil extraction from the tar sands, particularly those related to high oil retention in the tailings pond, will be easier to manage in the Nigerian situation.

INTRODUCTION

The Nigerian tar sands occur within a belt of about 120km, stretching from Ogun through Ondo to Bendel States of Nigeria (Fig. 1). Previous works on the tar sand were aimed at determining the suitability of the material as raw material for producing asphaltic mixes for road surfacing (Coursey, Hubbard and Hitch, 1963; Crocket and Wescott,

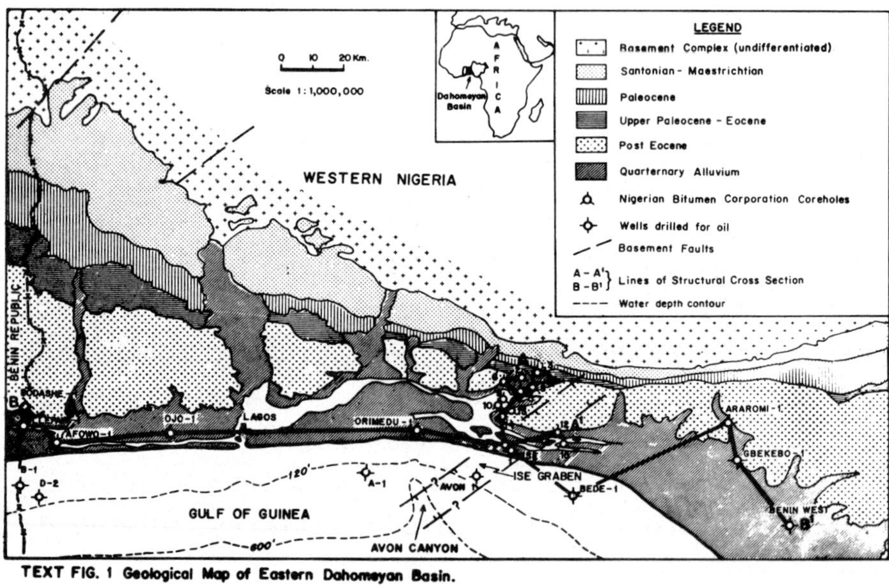

TEXT FIG. 1 Geological Map of Eastern Dahomeyan Basin.

1954) as well as the production of heavy oil by the application of insitu recovery techniques (Nigerian Bitumen Corporation 1907 - 1914; Tennessee Nigerian Inc. in: Durham and Pickett, 1966).

The present detailed study of part of the tar sand belt by the Geological Consultancy Unit, Department of Geology, University of Ife, started in 1974 and involved field mapping, drilling of over 40 boreholes, geophysical and geochemical studies. Subsurface data from the study area shows that the stratigraphic sequence consists of oil rich sands at the base, overlain by clays, shales and interbeds of limestones and siltstones. The impregnated sands occur as two distinct stratigraphic bands, separated by a uniformly thick organic shale. The upper clays and shales generally act as the cap rock preventing further updip migration of oil.

The tar sands and the overlying sediments belong to the upper part of the Araromi Formation and are thus Upper Maastrichtian - Paleocene in age (Omatsola and Adegoke, 1980).

The purpose of this paper is to present data on the clay mineral suite and their distribution in the tar sand sequence with a veiw to elucidating the paleo-environmental conditions in the depositional medium. The results obtained to date using the two most promising technologies for extraction viz. hot water extraction and the Lurgi-Ruhrgas processes were appraised. The most important constraint in the application of the hot water extraction process is the environmental problem created by the need for extensive tailings ponds and the high oil retention by the clay particles. This study of the clay mineral content of the tar bearing horizons will form the basis of further critical evaluation of the possible influence of the clays in the extraction of oil from the tar sands.

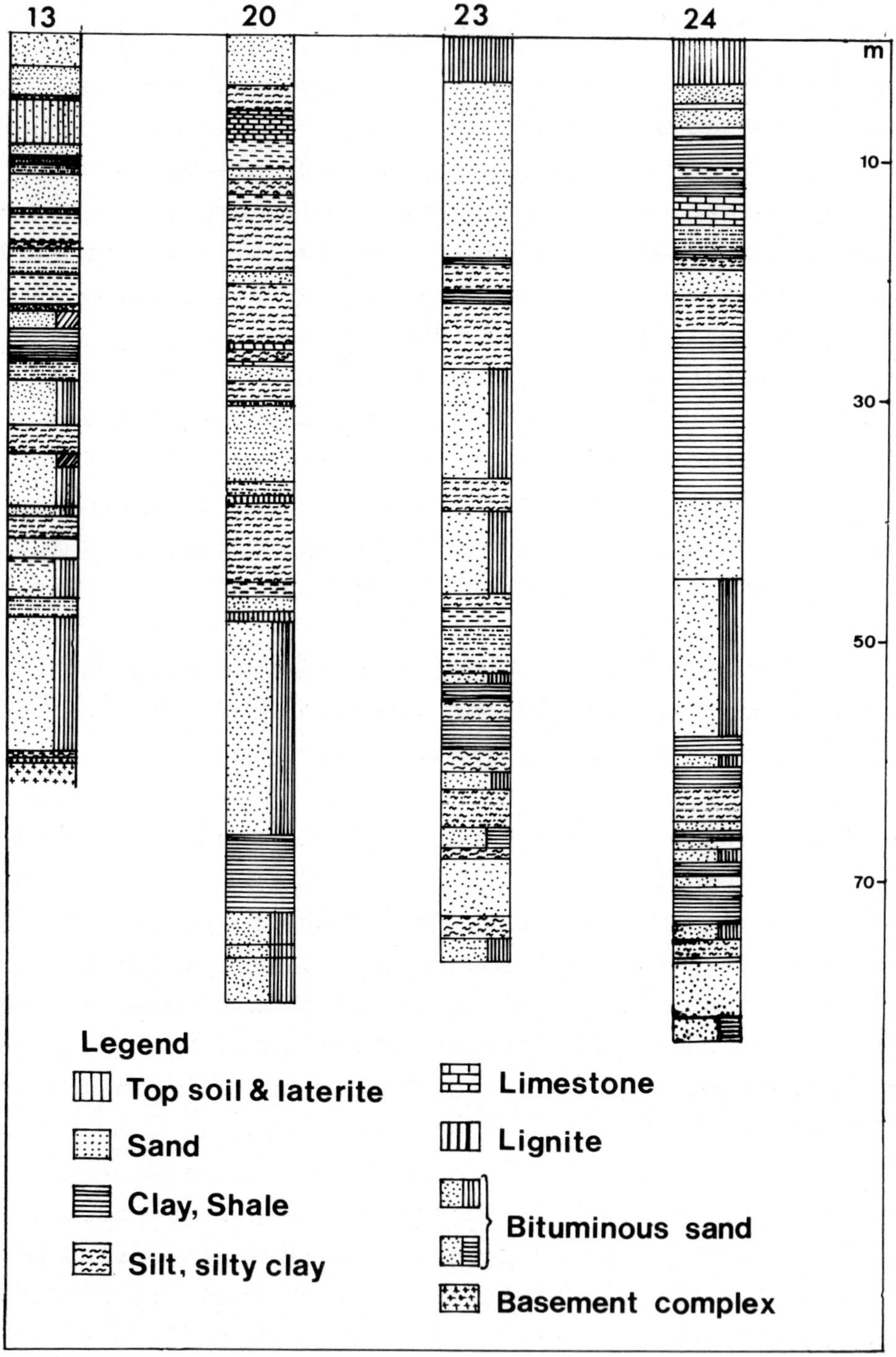

Fig. 2 Columnar sections of the boreholes sampled.

LITHOLOGY AND STRATIGRAPHY

The tar sand sequence occurs within the eastern Dahomey basin. The stratigraphy of the Nigerian sector of the basin was recently reviewed by Omatsola and Adegoke (1980). The oldest, post drift, coarse and conglomeratic continental sediments which overlie the Basement Complex are referred to as the Ise Formation. This is overlain by the transitional to marine sands and sandstones with variable interbedded shales and siltstones which now constitute the Afowo Formation (Abeokuta Formation of Reyment, 1965; Adegoke, 1969; and Billman 1980). The uppermost and youngest sequence composed dominantly of shales and siltstones with thin interbeds of sands, limestones and lignite are referred to as the Araromi Formation (Araromi Shale of Reyment, 1965).

In the study area, the block-faulted Basement Complex is uncomformably overlain by basal conglomeratic sands and weathered basement materials, followed by a 15 meter thick bitumen-impregnated layer (Horizon Y). An oil saturated, uniformly thick (8 meter) shale separates Horizon Y from the upper tar sands (Horizon X). A variable sequence of upper shales, sandy clay with interbedded limestone, lateritic clays and shales and top soil overlie Horizon X (Fig. 2). The beds show a consistent dip of about 10-12° to the south.

Faunal studies have shown that the tar sands and the overlaying sediments are of Upper Maastrichtian to Paleocene age.

ANALYTICAL METHODS

X-ray quantitative analyses were carried out on samples collected from updip, central and downdip boreholes drilled in the study area. The samples were washed, crushed and sieved through 63μm mesh sieve. The fine fraction was decarbonized in 0.2M hydrochloric acid and excess acid was removed by subsequent centrifugation. The < 2μm fraction was obtained by standard sedimentation processes. Oriented aggregates were prepared on glass slides from a clay suspension. Three X-ray diffraction patterns were recorded for each oriented specimen, viz in the untreated state, after glycolation, and after heating to 490°C for 2 hours. The analyses were carried out with a CGR Θ 60 diffractometer (copper radiation focused by a quartz crystal monochromator), at a scanning speed of 1° 2Θ/min. A receiving slit of 1.25 mm was used to allow for a better determination of mixed layer minerals. A semi-quantitative evaluation of the clay minerals is based on peak heights and areas (Chamley 1971). The 001 illite and chlorite peak heights from the glycolated diffraction patterns were used as a reference.

DISTRIBUTION OF CLAY MINERALS

Some typical X-ray diffraction patterns are shown in Figure 3. Kaolinite, irregular mixed layers, chlorite, illite and smectite are the clay minerals identified from the X-ray diffractograms of the clay fraction in the tar sand sequence. Kaolinite is the dominant (17 - 100%) and most frequently occurring clay mineral. The high kaolinite contents of the clay fractions are not restricted to tar impregnated sands only but also to silt, clay and shaly horizons (Table 1). Irregular mixed layer clays are next in abundance to kaolinite and ranging from 3% in updip areas (BH. 13) to 85% in the downdip (BH. 23 & 24) boreholes.

The mixed layer clays identified include illite-chlorite (1-C); illite-vermiculite (I-V); illite-smectite (I-S) and chlorite-smectite (C-S). Of these, I-S is more common. Chlorite and illite are present in small amount (14% - 26%) especially in downdip areas. Smectite is absent updip but occurs in small quantities (12-22%) in a few horizons downdip.

The clay mineral suite of the tar sand sequence does not exhibit any distinct lateral or vertical variations. Though individual clay mineral abundances may vary slightly, the assemblage remains virtually identical. Associated non-clay minerals are quartz, calcite, feldspar, goethite and gypsum.

DISCUSSION

The clay mineral distribution in the Upper Cretaceous - Paleocene Nigerian tar sand sequence show that kaolinite is the dominant clay mineral. It is known that kaolinite can be produced under suitable acid leaching conditions of alkali earth elements. Heavy mineral studies of sediments from the tar sand sequence indicate that Al-silicate rocks (granites) of the Precambrian Basement complex constitute the major source rock (Coker, 1976; Adediran, 1980; Olawoki, 1980; Adesida, 1980). These rocks occur five to six kilometers north of the study area. They can produce kaolinite under suitable environmental conditions.

Kaolinite is not restricted to a particular lithofacies. It is found in both sandy and clay-rich horizons (Table 1). Its abundant occurrence throughout the sequence reflects the intensity of weathering and leaching conditions aided by pronounced relief in the source areas. Furthermore, a warm and humid climatic condition which generally favours the formation of kaolinite must have prevailed during the Upper Cretaceous - Paleocene times. These environmental conditions agree with those put forward for the Upper Cretaceous sediments in the Lower and Upper Benue Trough respectively where an identical clay mineral suite has been recorded by Agagu (1979), Enu (1978, 1979) and Chamley _et_ _al_. (1979).

Kaolinite is relatively more abundant in the oldest

sediments updip. Irregular mixed layer clays, chlorite and illite on the other hand increase downdip at the expense of kaolinite. The commonest mixed layer clays identified are I-S, I-C and I-V.

The primary clay minerals (illite and chlorite) observed downdip could have been either detrital in origin or products of diagenetic alterations or both. There is a second group of sediments possibly derived from the nearby source different from the main and distant source rock of kaolinite (Adegoke et al. 1980). We suspect that these are the older sediments within the same basin, elevated as a result of the tectonic events associated with continental rifting. The presence of pyroxene, iron and magnesium hydroxides together with the observed textural immaturity of some of the sediments confirm the existence of this nearby sediment source. Chlorite and illite could have been derived from this source. However, the observed decrease in kaolinite, the abundant occurrence of mixed layer clays and the relatively deeper levels of occurrence of the horizons downdip tend to favour derivation of the illites, chlorites and mixed layer clays from kaolinite and smectite by diagenetic transformation. It is known that in the presence of K^+, smectite and/or kaolinite can easily be transformed to illite (Weaver and Pollard, 1973). Apart from the basal continental sands, the overlaying sediments in the tar sand sequence are essentially marine. These sediments have the K^+ concentration required to produce the diagenetic transformation proposed here (Garrel and Christ 1965). The mixed layers probably resulted from the incomplete adsorption of K^+ on these clay minerals.

PALEOGEOGRAPHICAL CONCLUSION

Tectonic activities which accompanied the opening of the Atlantic Ocean and the separation of West Africa from

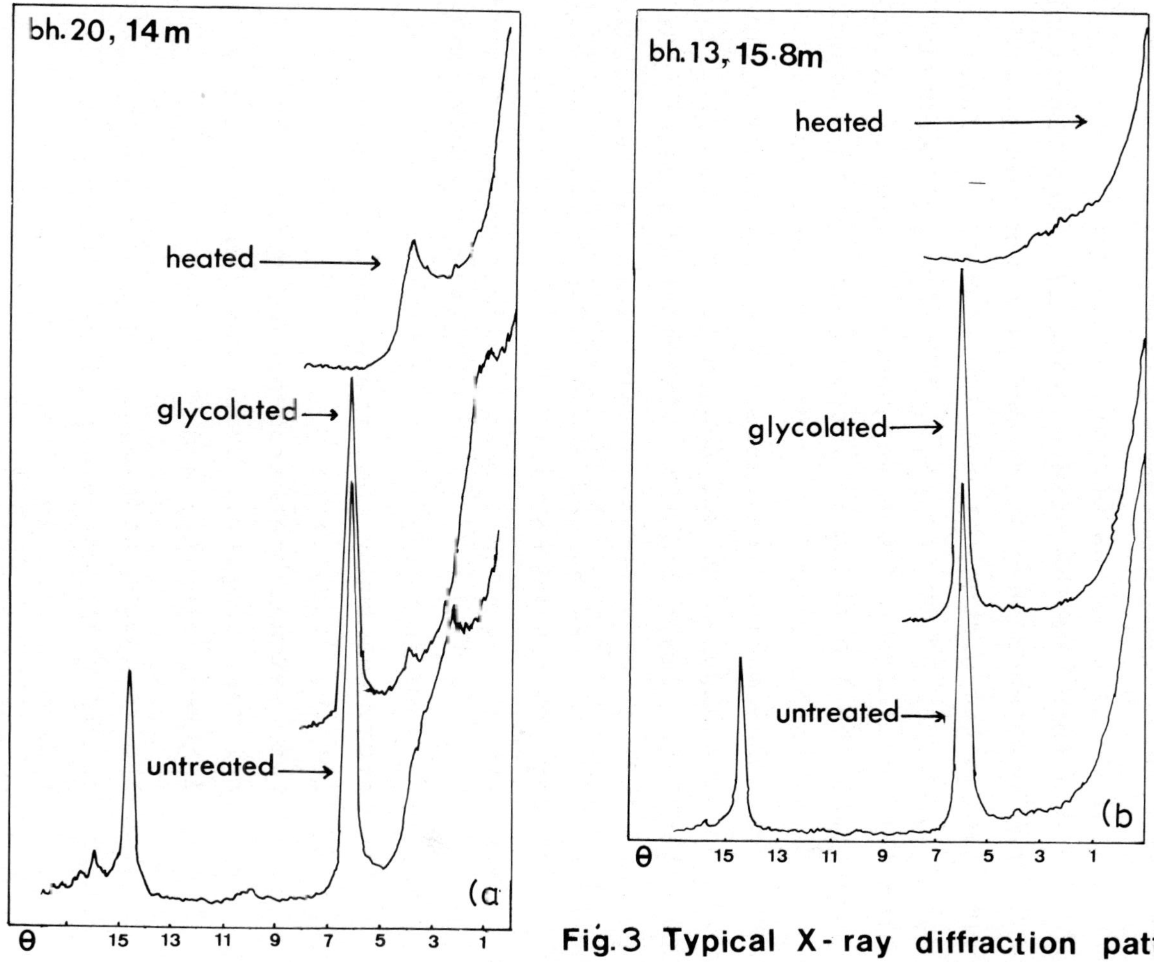

Fig. 3 Typical X-ray diffraction patterns.

Brazil in Cretaceous times are believed to have triggered off the development of the Dahomey basin (Omatsola & Adegoke, 1980). The Late Cretaceous tectonics resulted in the formation of a series of horsts and grabens. The study area lies within the Ise graben (Fig. 1). The continental sediments were subsequently eroded from the horsts and deposited in the grabens as the Ise Formation. These deposits correspond to the kaolinite rich sediments in which the kaolinite content may become as high as 100%.

During the Maastrichtian, the basin subsided gently and deposition occurred in a rapidly changing environment with occasional marine influence. The clay mineral suite still shows a high value for kaolinite (60% to 80%), but the mixed layer clays and the illites show a remarkable increase (17% to 85%) (Figs. 3a,b). This can be attributed to occasional transgressions which accentuated transformation of clays deposited earlier.

The upper horizons are essentially composed of marine sediments and the mixed layer clays and the primary clays are becoming more abundant.

This study has shown that there are two sources of sediments for the study area:

(a) the residual weathered products from the basement complex. These contribute largely to the kaolinite; and

(b) the weathered sediments within the basin contributing clays deposited earlier and aiding diagenetic alteration of kaolinite by adsorption of K^+.

REFERENCES

Adediran, S.A., 1980. Aspects of the sedimentary geology of bituminous sands (Maastrichtian) Ondo State, Nigeri Thesis, Department of Geology, University of Ife, Nigeria.

TABLE 1

Semiquantitative distribution of clay minerals in the tar sand sequence.

Bh. no.	Depth (m)	Lithology	CLAY MINERAL %				
			C	I	M.L.	S	K
13	11.0	Fine Sand			7		93
"	13.5	Clay			10		90
"	15.8	Silty Clay			6		94
"	25.4	Shale					100
"	30.0	Fine Sand (Bituminous)					100
"	37.0	Fine grained bituminous sand		3			
20	9.0	Fossiliferous Limestone	23		23		54
"	14.0	Silt	35				65
"	15.0	Silty Clay			17		83
"	21.0	Silty Clay	21		15		64
"	35.0	Fine Sand			14		86
"	45.0	Silty Clay			6		94
"	53.0	Fine grained bituminous sand	26				74
23	22.4	Clayey sand		70	30		
"	23.9	Sandy clay		15	85		
"	26.0	Sandy clay			32	22	46
"	52.0	Shale			20	12	68
"	56.2	Silt	26		42		26
"	60.0	Silt	14		35		51
"	65.0	Silt	20		60		20
24	8.8	Shale	8		85	7	
"	11.5	Silt		5	27	5	
"	13.0	Shale		5	27	5	63
"	21.4	Pyritiferous Sand		14	51		35
"	31.0	Shale		7	85		8
"	35.2	Fossiliferous shale			30		70
"	78.5	Medium grained sand	23				77

C - Chlorite, I - Illite, M.L. - Mixed Layers,

S - Smectite K - Kaolinite

Adegoke, O.S., 1969. Eocene stratigraphy of Southern Nigeria. Memo. Bur. Rech. Geol. Min., 69: 23-49.

Adegoke, O.S., (Coordinator) 1980. Geotechnical investigations of the Ondo State bituminous sands. Report, Geol. Consultancy Unit, Dept. of Geology, University of Ife, Nigeria. 257 pp., 33 folios.

Adesida, A., 1980. Geology of the Ore tar sands (Maastrichtian - Paleocene), Ondo State, Nigeria. M.Sc. Thesis, Dept. of Geology, University of Ife, Nigeria.

Agagu, O.K., 1979. Interpretative value of clay mineral species in Upper Cretaceous shales in the Lower Benue Trough, Eastern Nigeria. Abstract, Journ. Min. Geol. 16, No. 1, 145.

Billman, H.G., 1980. Offshore stratigraphy and paleontology of the Dahomey embayment, West Africa. Proc. 7th Afr. Micropal. Coll. (in press).

Chamley, H. 1971. Recherches sur la sédimentation argileuse en Méditerranée. Sci. Geol., Strasbourg, Mem. 35, 225 p.

Chamley, H., Enu, E., Moullade, M., Robert, C., 1979. La sédimentation argileuse du bassin de la Bénoué au Nigeria, reflêt de la tectonique du Crétacé superieur; C. R. Acad. Sci., Paris, Série D, 288, 1143-1146.

Coker, S.J.L. 1976. Sedimentology and Petroleum Geology of the Okitipupa tar sand deposits. M.Sc. Thesis, University of Ife, Nigeria.

Coursey, D.G., Hubbard, F.H. and Hitch, L.S. 1963. Preliminary observations on the tar sand deposits of Western Nigeria. Research Report No. 18, Federal Ministry of Commerce & Industry, Lagos, Nigeria.

Crockett, L.O. and Wescott, B.B., 1954. Asphalt samples from West Africa. Extracts from Report 17617, The geology and oil prospects of Nigeria by J. Irving and D. C. Ion.

Durham, K.N. and Pickett, C.R., 1966. Oil mining lease 47, Lekki Corehole programme (February - April 1966). Report Tennessee Nigeria, Inc.

Enu, E.I. 1978. Contribution a l'étude sédimentologique des formations crétacées dans le Bénoué (région de Cham, Nigeria Nord Oriental). Ph.D. Thesis, Univ. of Nice, France.

Enu, E.I. Lithostratigraphy and sedimentology of the Upper Cretaceous Formations in the Cham area. (Upper Benue Trough Nigeria). Annales du Museum d'histoire Naturelle de Nice, 5, 12 pp. 10 Figs. (In press).

Garrels, R.M. and Christ, C.L., 1965. Solutions, Minerals and Equilibria. Harper & Row, New York, N.Y., 450 pp.

Olawoki, O.A., 1980. The geology of the bituminous sands south of Foriku, Ondo State. M.Sc. Thesis, Univ. of Ife, Nigeria.

Omatsola, M.E. and Adegoke, O.S., 1980. Tectonic evolution of the Dahomeyan Basin, (West Africa) and its implications on the opening of the North and South Atlantic. Proc. 26th Int. Geol. Cong. Paris (in press).

Reyment, R.A., 1965. Aspects of the geology of Nigeria. Ibadan Univ. Press, 133 pp. 18 pl.

Weaver, C.E. and Pollard, L.D., 1973. The chemistry of clay minerals. Developments in Sedimentology 15, Elsevier, New York, 1973, 213 p.

HIGH-CHARGE SMECTITE FROM LIGNITE OVERBURDEN OF EAST TEXAS

K. EGASHIRA
Faculty of Agriculture, Kyushu University, Fukuoka 812, Japan
J.B. DIXON and L.R. HOSSNER
Department of Soil & Crop Sciences, Texas A&M University, College Station, Texas, 77843, U.S.A.

ABSTRACT

A "high-charge" smectite separated from a lignite overburden in East Texas, which has a 001 spacing of 18 Å when Mg-, Ca-, Ba-, Na-, or Li-saturated and glycerol-solvated, and of 10 Å when K-saturated and air dried, was characterized. Due to the consistent expansion to 18 Å the mineral name, smectite, was chosen over vermiculite even though more than 70% of the exchange sites of this "high-charge" smectite fixed potassium on heating at 110°C, a criterion often used to quantify soil vermiculites. The layer charge which was estimated from the 23 Å basal spacing of the dodecylammonium-clay complex was nearly equal to that of vermiculite.

Muscovite of the $2M_1$ polytype (5-2 μm fraction) was separated from the same sediments. The mica was reacted with molten lithium nitrate and the product had the same basal spacings as the "high-charge" smectite. Thus, the "high-charge" smectite may have formed from muscovite by charge reduction and K-depletion.

INTRODUCTION

Vermiculite and smectite differ from each other in charge per formula unit (designated as X). For minerals of the vermiculite group $X \sim 0.6$-0.9 and for those of the smectite group $X \sim 0.2$-0.6 (Bailey, 1980). Differentiation of vermiculites and smectites is made by X-ray diffraction rather than by layer charge determination, since vermiculites and smectites are members of a continuous series with varying chemical composition and layer charge (Walker, 1957, 1958; Barshad and Kishk, 1969). Expansion of the basal spacing by glycerol-solvation after Mg-saturation has been employed as a differentiating criterion (Walker, 1957, 1958; Brindley, 1966; Harward et al., 1969; Douglas, 1977). Vermiculites expand to 14.2 Å (monolayer) and smectites expand to 17.7 Å (double layer). Harward et al (1969) proposed as another differentiating criterion the degree of collapse of the basal spacing by K-saturation and air-drying at RH 54%. Vermiculites collapse to 10.4 Å. and smectites collapse to 12.2 Å. Both expansion and collapse currently are widely used to differentiate vermiculites and smectites.

A mineral which shows a basal spacing of 18 Å on Mg-saturation and glycerol-solvation and of 10 Å upon K-saturation and air-drying at RH 45-65% was found in a lignite overburden from Troup in East Texas. This mineral may be an intermediate mineral which behaves as a smectite in one test and as a vermiculite in another. A mineral which shows the same behavior has been reported by van Breemen (1976) and Stoch and Sikora (1976) in acid sulfate soils in Thailand and in weathering crusts of gneiss in Poland, respectively. They used the name of "high-charge" smectite but did not characterize these minerals further. This paper presents mineralogical and chemical properties of the mineral. The designation "high-charge" smectite for layer silicates with these properties is proposed.

MATERIALS AND METHODS

A "high-charge" smectite sample was separated from the 3rd layer at 2.3 m depth in lignite overburden from Troup, in East Texas. An air-dry sample (<2 mm) was successively treated with 1M $NaCH_3COO$ (pH 5) to remove carbonates, with 6% H_2O_2 to remove organic matter and with 0.3M Na citrate-1M $NaHCO_3$-Na dithionite to remove free iron compounds (Mehra and Jackson, 1960). After an ultrasonic wave treatment, a coarse clay fraction (0.2-2 μm) was collected by repeated centrifugation and decantation. High gradient magnetic separation (Schulze and Dixon, 1979) was applied to concentrate "high-charge" smectite, because the 0.2-2 μm size fraction contained large amounts of kaolinite and quartz. The 0.2-2 μm fraction as a <0.5% by weight suspension in pH 10.5 Na_2CO_3 solution was flowed through a stainless steel wool filter at a flow rate of 300 cm^3/min (flow velocity of 2.5 X 10^{-2} m/s in the filter). The magnetic fraction separated at 1.2 Tesla was Ca-saturated and freeze-dried. This material was used in our work as the "high-charge" smectite sample.

To prepare different mono-ionic forms, 25 mg of the sample was saturated with various cations by washing several times with solutions of the chlorides after an ultrasonic wave treatment for 5-10 seconds. Excess salt was removed by washing with water. The suspension (0.5 ml) was spread on a glass slide, and air-dried for X-ray determination. Relative humidity during air-drying and X-raying of specimens was in the range of 45-65%. The specimen was then solvated with glycerol by spraying with a 10% solution in water and X-rayed.

Potassium fixation in the "high-charge" smectite was assessed by the determination of CEC(Ca/Mg) and CEC(K//NH_4) according to Alexiades and Jackson (1965), using the CEC determination method of Wada and Harada (1969) as follows: The sample was first washed with 1M $NaCH_3COO$ (pH 7) and saturated with Ca by washing with 0.5M $CaCl_2$ and subsequently with 0.025 $CaCl_2$. The content of interstitial Ca salt which was measured by weighing was subtracted from the total Ca content obtained by extraction with 0.5M $MgCl_2$. The sample was again washed with 1M $NaCH_3COO$ (pH 7) and saturated with K by washing with 1M KCl and subsequently with 0.05M KCl. The sample was heated at 110°C overnight and then washed with 1M NH_4Cl to extract K. Exchangeable K was obtained by subtracting interstitial K from total extractable K. The potassium fixation capacity was expressed as
{CEC(Ca/Mg) - CEC(K//NH_4)} X 100/CEC(Ca/Mg).

The layer charge of the "high-charge" smectite was estimated by comparing the basal spacing of its dodecylammonium complex with those of the dodecylammonium complexes of vermiculites and smectites with known cation exchange capacities (Lagaly and Weiss, 1969):

Five ml of 0.1M dodecylamine·HCl solution was added to 30 mg of clay. The suspension was heated at 65-70°C in a water bath for 24 hours. Excess dodecylamine·HCl was removed by washing with 5 ml of an ethanol-water mixture (1:1) and subsequently with ethanol. The paste was spread on a glass slide, air-dried and X-rayed.

In addition to the above sample, for a study of the genesis of the "high-charge" smectite, a mica sample (2-5 μm, muscovite of $2M_1$ polytype) was separated from the 18.9 m depth of the same core which the "high-charge" smectite sample was obtained, by particle-size fractionation and high gradient magnetic separation following successive treatments with 1M $NaCH_3COO$ (pH 5), 6% H_2O_2 and 0.3M Na citrate-1M $NaHCO_3$-Na dithionite. This mica sample was reacted with molten $LiNO_3$ (White, 1956, 1958). Fifty mg of sample and 3 g of $LiNO_3$ were heated at 300°C in an electric furnace. After cooling, water was added to dissolve $LiNO_3$. The suspension was divided into two portions. They were saturated respectively with Mg and K. The K-saturation was performed by reacting with 1M KCl at 25°C for 3 days during which period the KCl solution was renewed four times. X-ray diffraction patterns were obtained after air-drying and after glycerol-solvation.

RESULTS AND DISCUSSION

X-ray diffraction patterns show that the "high-charge" smectite sample contains considerable amounts of other minerals, i.e. smectite, kaolinite, mica and quartz in addition to the "high-charge" smectite, characterized by a 14.7 Å peak for the Mg-saturated and air-dried specimen, which expanded to 18.0 Å upon glycerol-solvation (Fig. 1). No peak or shoulder was observed around 14.2 Å in the glycerol-solvated specimen. The 14.7 Å peak collapsed to 12.7 Å and 10.0 Å after K-saturation and air-drying. The collapse to 10.0 Å is shown by the remarkable increase of the height of the 10.0 Å peak. A tailing of the 10.0 Å peak on the low angle side suggests that a few layers contain interlayer water. The expansion to 18.0 Å by Mg-saturation and glycerol-solvation and the collapse to 10.0 Å by

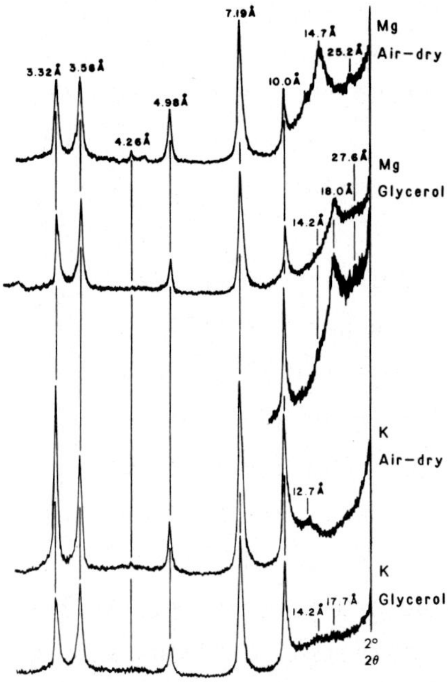

Fig. 1. X-ray diffraction patterns of the "high-charge" smectite sample.

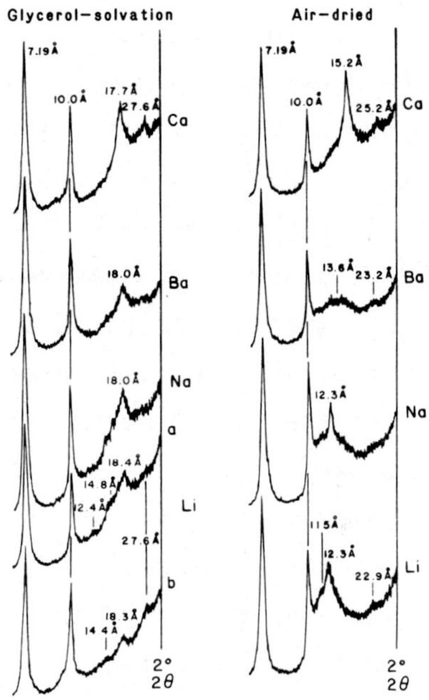

Fig. 2. X-ray diffraction patterns of the "high-charge" smectite sample saturated with different cations (b, Greene-Kelly's test).

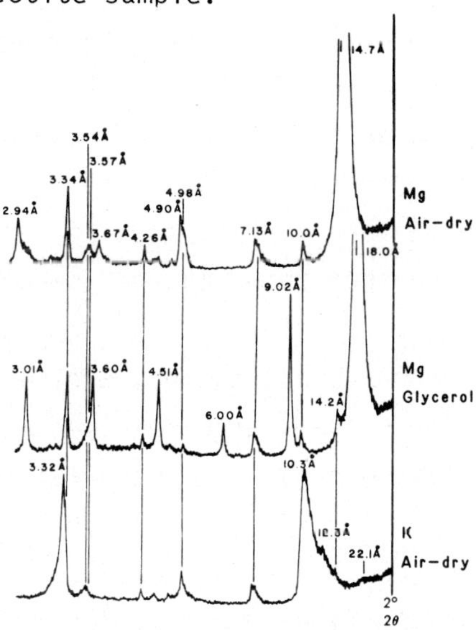

Fig. 3. X-ray diffraction patterns of the Troup mica reacted with molten $LiNO_3$ for 24 hours.

K-saturation and air-drying are indicative of "high-charge" smectite. The peak at 25.2 Å (with a second order peak at 12.6 Å) of the Mg-saturated and air-dried specimen which expands to 27.6 Å on glycerol-solvation and collapses to 10.0 Å on K-saturation and air-drying reflects a regular interstratification of mica and "high-charge" smectite.

First order basal spacings of the air-dried specimens of "high-charge" smectite saturated with various ions were as follows: Ca, 15.2 Å; Ba, 13.6 Å (very broad, random interstratification of "high-charge" smectites having one and two interlayers of water); Na, 12.3 Å; Li, 12.3 Å and 10.0 Å (Fig. 2). These values for monovalent ions agree better with those of vermiculites than those of smectites (Douglas, 1977; MacEwan and Wilson, 1980). The divalent ions gave similar spacings for both mineral types. But the first basal spacings of the glycerol-solvated specimens all showed expansion to 17.7-18.4 Å. Also, the peak width for the glycerol complexes is larger and more like that for smectite than that for vermiculite. These are the reasons why we consider this mineral as a smectite rather than a vermiculite. A small shoulder at 14.8 Å was observed for the Li-saturated specimen. Figure 2 also showed the presence of regularly-interstratified "high-charge" smectite and mica. This was most clear for the Ca-saturated specimen. A powder X-ray diffraction pattern of this specimen shows a 060 spacing at 1.50 Å indicating that "high-charge" smectite is a dioctahedral mineral.

To determine the location of isomorphous substitution, Greene-Kelly's Li-test (1953) was applied. The Li-saturated specimen was air-dried and solvated with glycerol (pattern a) and subsequently heated at 250°C for 7 hours (pattern b). According to Greene-Kelly, octahedrally substituted smectites (montmorillonites) collapse to 9.5 Å, while tetrahedrally substituted smectites (beidellites and nontronites) retain their 18 Å spacing. As shown in pattern b, "high-charge" smectite had characteristics of the tetrahedrally substituted minerals.

In order to compare "high-charge" smectite with other smectites a K-saturated specimen was glycerated in wet condition to differentiate between high and low charge smectite according to Barshad (1960). In this experiment, smectites with low exchange capacities (92-95 meq/100 g) give a 001 basal spacing of 18.5 Å, and those with high exchange capacities (115-135 meq/100 g) spacing of 14.5 Å. The K-saturated and glycerol-solvated specimen of "high-charge" smectite gave a very broad peak ranging from 14.2 Å to 17.7 Å in which some expanded smectite is evident (Fig. 1). But the 10.0 Å peak of "high-charge" smectite still remained and did not expand to 14 Å. Hence, "high-charge" smectite can fix potassium even in wet condition and apparently it has a layer charge higher than that of the smectites reported by Barshad.

Cation exchange capacity and potassium fixation capacity determinations, of the "high-charge" smectite sample indicate that its K-fixation capacity is much greater than that of other smectites. Its cation exchange capacity ($K//NH_4$) of 12.8 meq/100 g was less than half of its CEC(Ca/Mg) of 29.6 meq/100 g; therefore the potassium fixation capacity was 57%. The low CEC values of the "high-charge" smectite sample indicate the presence of a large amount of mineral with a low CEC. If we assume that the contribution of mica and kaolinite to the total CEC is 5 meq/100 g (Alexiades and Jackson, 1965) and that these minerals essentially do not fix potassium, the corrected potassium fixation capacity of "high-charge" smectite plus smectite amounts to 68%. The potassium fixation capacity of "high-charge" smectite itself may be more than 70%.

Potassium fixation of smectites has been well studied (Alexiades and Jackson, 1965; Weir, 1965; van Olphen, 1966; Ross and Mortland, 1966; Horváth and Novák, 1976). A high total layer charge and high degree of tetrahedral substitution are two factors promoting potassium fixation by smectites. Smectites (including montmorillonite, beidellite and nontronite were found to fix potassium by 0-40% of their total layer charge after heating at 105-110°C and no value exceeded 40%. On the other hand, it has been observed that in soil vermiculites more than 70% of total layer charge was occupied by fixed potassium after heating at 110°C (Ross and Mortland, 1966; Barshad and Kishk, 1970). The estimated potassium fixation capacity of the "high-charge" smectite sample of more than 70% is as high as that of soil vermiculites and remarkably higher than that of smectites. These observations suggest that "high-charge" smectite has a charge nearly equal to that of vermiculites and also that it has a high degree of tetrahedral substitution.

Since the "high-charge" smectite sample is a mixture of several clay minerals the determination of layer charge by chemical analysis and/or CEC measurement is difficult. Therefore, the method of Lagaly and Weiss (1969, 1976) was applied which permits the determination of layer charge even in mixtures containing very small amounts of swelling layer silicates.

Table 1 shows that the basal spacing of dodecylammonium complexes increase with increasing layer charge. The basal spacing of "high-charge" smectite is the same as that of the two vermiculites, hence its layer charge may be estimated to be of the order of 170 meq/100 g or more, although the exact value is not available because of thc small number of reference clays. In conclusion, the data for potassium fixation capacity and layer charge show that "high-charge" smectite is indeed a high layer charge mineral similar to vermiculite.

Finally, since "high-charge" smectite seemed to be a likely weathering product of mica its artificial formation from mica was attempted. Various inorganic salts, molten salts, organic cations and biological agents have been employed in the artificial alteration of mica (Norrish, 1973; Fanning and Keramidas, 1977). We have chosen molten $LiNO_3$ because the lithium ion exchanges with interlayer potassium and migrates into vacant octahedral sites of muscovite (White, 1956, 1958; Bronson et al., 1960) and sericite (Tomita and Sudo, 1971). And by this reaction, the total layer charge decreases but the tetrahedral sheet still retains a high degree of substitution of aluminum for silicon. We considered that this would be a prerequisite for the formation of "high-charge" smectite.

The original mica sample contains small amounts of smectite, vermiculite, chlorite, kaolinite, quartz and feldspars in addition to muscovite. After reacting with $LiNO_3$ for 6-24 hours, the height of the mica peaks considerably decreased and an increasingly strong peak appeared at 14.7 Å in the Mg-saturated and air-dried specimen (Fig. 3). This peak expanded to 18.0 Å with a rational series of its high orders upon glycerol-solvation. Therefore, this peak was considered to be due not to vermiculite but to smectite. This smectite peak collapsed to 10.0 Å (6 h) to 10.3 Å (24 h) upon K-saturation and air-drying. The collapse to 10.0 Å indicates that the newly-formed smectite is a "high-charge" smectite and the broad and asymmetric peak at 10.3 Å with a tailing toward low angles was considered to be due to a random interstratification of "high-charge" smectite (10.0 Å) and smectite (12.2 Å).

The "high-charge" smectite-like product obtained by White (1958) by reacting Delamica muscovite (1 to 2 μm) with molten $LiNO_3$ had a layer charge of 155 meq/100 g. However, this value is in the range of the layer charge of vermiculites rather than of smectites. Muscovite was expected to change into "high-charge" smectite through a vermiculite phase but no vermiculite was observed not even in the sample reacted for only 30 minutes.

CONCLUSION

A "high-charge" smectite, which shows its first basal spacing at 18 Å on the Mg-saturation and glycerol-solvation and at 10 Å on K-saturation and air-drying, was found in a lignite overburden from East Texas. The first basal spacing of its glycerol-solvated specimen also expanded to 18 Å after saturation with Ca, Ba, Na and Li. Determination of the potassium fixation capacity and layer charge showed that this mineral has a high layer charge in the range of vermiculites. Preparation of a mineral from mica separated from the same sediment and with the properties of "high-charge" smectite suggests that it may be a weathering product of mica.

TABLE 1

First basal spacing of dodecylammonium-clay complexes.

Sample	Cation-exchange capacity (meq/100 g)	DDA-first basal spacing (Å)
Montmorillonite-Gonzales, TX	85*	16.5*
Nontronite (<2 μm)-Trinidad Hill, WA	104	18.2
Montmorillonite (<1 μm)-Otay, CA	131*	18.0
Vermiculite (<2 μm)-Transvaal, S. Africa	170*	23.2
Vermiculite (<2 μm)-Llano, TX	210*	23.2
High-charge smectite (0.2-2 μm)	-	23.4

*Measured by Brian E. Viani, unpublished data.

REFERENCES

Alexiades, C.A. and Jackson, M.L., 1965. Quantitative determination of vermiculite in soils. Soil Sci. Soc. Am. Proc., 29: 522-527.

Bailey, S.W., 1980. Summary of recommendations of AIPEA Nomenclature committee, Clay Miner., 15: 85-93.

Barshad, I., 1960. X-ray analysis of soil colloids by a modified salted paste method. Clays Clay Miner., 7: 350-364.

Barshad, I. and Kishk, F.M., 1969. Chemical composition of soil vermiculite clays as related to their genesis. Contributions to Mineralogy and Petrology, 24: 136-155.

Barshad, I. and Kishk, F.M., 1970. Factors affecting potassium fixation and cation exchange capacities of soil vermiculite clays. Clays Clay Miner., 18: 127-137.

Brindley, G.W., 1966. Ethylene glycol and glycerol complexes of smectite and vermiculite. Clay Miner., 6: 237-259.

Bronson, R.D., Spain, J.M. and White, J.L., 1960. Potassium-depleted muscovite: Part I. Preparation using filtration process for treatment with molten lithium nitrate. Clays Clay Miner., 8: 39-43.

Douglas, L.A., 1977. Vermiculites. In: J.B. Dixon and S.B. Weed (Editors), Minerals in Soil Environments. Soil Sci. Soc. Am., Madison, Wis., pp. 259-292.

Fanning, D.S. and Keramidas, V.Z., 1977. Micas. In: J.B. Dixon and S.B. Weed (Editors), Minerals in Soil Environments. Soil Sci. Soc. Am., Madison, Wis., pp. 195-258.

Greene-Kelly, R., 1953. Identification of montmorillonoids. J. Soil Sci.,4: 233-237.
Harward, M.E., Carstea, D.D., and Sayegh, A.H., 1969. Properties of vermiculites and smectites: expansion and collapse. Clays Clay Miner., 16: 437-447.
Horváth, I. and Novák, I., 1976. Potassium fixation and the charge of the montmorillonite layer. Proc. Int. Clay Conf., Mexico, pp. 185-189.
Lagaly, G. and Weiss, A., 1969. Determination of the layer charge in mica-type layer silicates. Proc. Int. Clay Conf., Tokyo, pp. 61-80.
Lagaly, G. and Weiss, A., 1976. The layer charge of smectitic layer silicates. Proc. Int. Clay Conf., Mexico, pp. 157-172.
MacEwan, D.M.C., and Wilson, M.J., 1980. Interlayer and intercalation complexes of clay minerals. In: G.W. Brindley and G. Brown (Editors), Crystal Structures of Clay Minerals and Their X-ray Identification. Mineralog. Soc., London, pp. 197-248.
Mehra, O.P. and Jackson, M.L., 1960. Iron oxide removal from soils and clays by a dithionite-citrate system buffered with sodium bicarbonate. Clays Clay Miner., 7: 317-327.
Norrish, K., 1973. Factors in the weathering of mica to vermiculite. Proc. Int. Clay Conf., Madrid, pp. 417-432.
Ross, G.J. and Mortland, M.M., 1966. A soil beidellite. Soil Sci. Soc. Am. Proc., 30: 337-343.
Schulze, D.G. and Dixon, J.B., 1979. High gradient magnetic separation of iron oxides and other magnetic minerals from soil clays. Soil Sci. Soc. Am. Proc., 43: 793-799.
Stoch, L. and Sikora, W., 1976. Transformations of micas in the process of kaolinization of granites and gneisses. Clays Clay Miner., 24: 156-162.
Tomita, K. and Sudo, T., 1971. Transformation of sericite into an interstratified mineral. Clays Clay Miner., 19: 263-270.
van Breemen, N., 1976. Genesis and solution chemistry of acid sulfate soils in Thailand. Centre for Agricultural Publishing and Documentation, Wageningen.
van Olphen, H., 1966. Collapse of potassium montmorillonite clays upon heating - "potassium fixation". Clays Clay Miner., 14: 393-405.
Wada, K. and Harada, Y., 1969. Effects of salt concentration and cation species on the measured cation-exchange capacity of soils and clays. Proc. Int. Clay Conf., Tokyo, pp. 561-571.
Walker, G.F., 1957. On the differentiation of vermiculites and smectites in clays. Clay Miner. Bull., 3: 154-163.
Walker, G.F., 1958. Reactions of expanding-lattice clay minerals with glycerol and ethylene glycol. Clay Miner. Bull., 3: 302-313.
Weir, A.H., 1965. Potassium retention in montmorillonites. Clay Miner., 6: 17-22.
White, J.L., 1956. Reactions of molten salts with layer-lattice silicates. Clays Clay Miner., 4: 133-146.
White, J.L., 1958. Layer charge and interlamellar expansion in a muscovite. Clays Clay Miner., 5: 289-294.

THE COMPOSITION OF CHLORITES IN A PROPYLITIZED SYENITE NEAR CEYSSAT, PUY-DE-DOME, FRANCE

A. MEUNIER and D. PROUST
E.R.A. 220 - Laboratoire de Pédologie
Faculté des Sciences
40, avenue du Recteur Pineau

86022 POITIERS CEDEX

ABSTRACT

A petrographic study using an electron microprobe has been made of a propylitized partially mylonitized syenite near Ceyssat (Puy-de-Dôme, France). Particular attention was devoted to the destabilization of biotites in relation with rock texture and crystallization sites.

The main phyllosilicate minerals in the undeformed rock were :

1. Magnesian chlorite, growing at quartz-(001) biotite interfacial planes.
2. Corrensite, replacing biotite.
3. Mg-Fe chlorite, replacing corrensite which is associated with Fe-rich epidote and pyrite.

The main minerals in the mylonitized rocks were :

1. Fe-Mg chlorites, replacing deformed biotites
2. Vermicular Fe-rich chlorite growing in large veins.

Simultaneous presence in small rock volumes of corrensite and Fe rich epidote shows that the temperature of crystallization was between 250°C and 280°C. It appears that chloritic phases tend to become more and more silicon-magnesium depleted and richer in iron as the fissural pattern develops. The composition of chlorites appears to be dependent on whether alteration process occurs in partially closed or in open micro-systems. Significant amounts of materials are transported out of the reaction area, and are precipitated at various distances in the vein systems of the rock.

INTRODUCTION

Propylitized igneous rocks occuring in the outer alteration fringe of porphyry copper deposits (Lowell and Guilbert, 1970) are commonly defined by calcite-chlorite-epidote assemblages (Creasey, 1959). Primary biotites and amphiboles are generally replaced along cleavages by chlorite pseudomorphs (Lanier <u>et. al.</u>, 1978 ; Carson and Jambor, 1974) which have constant chemical compositions throughout the propylitic zone, except when thermal gradients or successive hydrothermal-fluid intrusions occur.

The study of the propylitized syenite, which exhibits an alternating sequence of coarse grained and mylonitized facies, shows a wide range of chlorite compositions depending on their crystallization sites.

The scope of this study is to provide a better understanding of the extent to which rock texture and particularly crack patterns govern the chemical behavior of chloritic phases.

PROCEDURE

Thin sections were made on each sampled drill core after epoxy resin impregnation. Primary minerals were first separated from crushed, dry-sieved material using a Frantz magnetic separator and heavy liquids (bromoform + alcohol). Mineral purification was checked by optical and X-ray diffraction, then the material was ultrasonically treated in order to release alteration products. The latter were identified by X-ray diffraction using a Philips diffractometer (40 Kv-40 mA) with iron filtered COKα radiation. Samples were prepared as oriented aggregates or random powders with various cationic saturations.

Secondary phases were first located on polished thin sections using a petrographic microscope and then analyzed in situ using a CAMECA MS 46 electron microprobe and a TRACOR energy dispersive analyser. Data were refined with a Z.A.F. type program. A low-energy electron beam (1,5 nA-15 kV) was used in order to minimize loss of elements through volatilization during the count-time (120 seconds). Bulk chemical analyses were obtained using a Perkin-Elmer (107) atomic absorption spectrophotometer.

MATERIAL

The "INAG 1" drilling hole is located near Ceyssat (Puy-de-Dôme, France) ; it was drilled through 90 m of basalt and 160 m of highly fractured and tectonized monzonitic syenite. Tectonization is contemporaneous with a large faulted area east of the hole. Several shear planes developed with crushed minerals at various depths in the crystalline body. The syenite basalt interface, at the top of the hole, is marked by a reddish zone (90 to 96.30 m) associated with numerous green-coated cracks with occur down to 105 m.

Thirteen samples were studied outside the reddish zone, from 97.25 to 121 m. Calcite occurs either as straight crossing veins in coarse grained rock or pervasive veinlets in mylonitized zones. Geologically, the deformation of the syenite is related to a tectonic event followed by a comprehensive propylitic alteration ; the late basaltic flow overlying the syenite has indiced the rubefaction zone at the top of the hole by heating the hydrous zones.

The least altered syenite samples contain potassic feldspar (3.2 %), andesine (43.9 %), biotite and chlorite (8.2 %), quartz (3.2 %), hornblende (2.3 %), calcite veins (1.9 %) and accessory minerals (sphene, allanite, rutile, pyrite, magnetite) 2,8 %.

ROCK DESCRIPTION

Coarse grained and mylonitized zones are easily recognizable, either in hand-specimen or on a microscopical scale. Thin sections of the coarse grained rock exhibit straight microcracks (50 μm to 100 μm wide) underlining shear planes, while the mylonitized rock has preserved underformed rock relics. All samples studied were cut by calcite veins showing no specific alteration along their walls.

Coarse-grained rock

Biotites are replaced by two mineralogical assemblages which can coexist in the same sample.

(i) Biotite relics-corrensite-chlorite : Mineral reactions generally occur along mineral joints. Fresh potassic feldspars are partly converted into albite along andesine grain contacts. Hornblendes and sphenes are replaced respectively by calcite-chlorite and calcite-rutile assemblages growing from the rims of the minerals towards the core. Andesine alteration patterns, i.e. sericitization twinning, bear no textural evidence of being affected by reactions at intermineral joints.

TABLE 1

Microprobe analyses of scarcely altered biotites and corrensite-residual biotite zones. (11 oxygen basis ; Fe total = Fe^{3+}).

	Corrensites + résidual biotites											Biotites		
	1	2	3	4	5	6	7	8	9	10	11	1	2	3
Si	2.44	2.56	2.48	2.49	2.42	2.47	2.48	2.57	2.41	2.49	2.39	2.74	2.74	2.72
Al	1.51	1.46	1.46	1.36	1.49	1.47	1.50	1.53	1.39	1.45	1.53	1.16	1.23	1.32
Ti	0.25	0.26	0.33	0.39	0.27	0.13	0.12	0.15	0.39	0.36	0.07	0.26	0.23	0.22
Fe^{2+}	1.34	1.34	1.28	1.19	1.33	1.35	1.43	1.23	1.29	1.13	1.60	1.38	1.37	1.21
Mg	1.77	1.55	1.52	1.49	1.74	1.97	1.96	1.86	1.62	1.59	2.03	1.42	1.39	1.30
Mn	0.04	0.03	0.03	0.04	0.03	0.04	0.06	0.05	0.04	0.04	0.07	0.06	0.06	0.05
Ca	0.02	-	0.18	0.25	0.05	0.02	0.01	0.02	0.31	0.30	0.04	0.04	0.05	0.04
Na	0.01	0.01	-	0.05	0.17	-	-	-	-	-	-	-	-	0.01
K	0.30	0.49	0.35	0.41	0.30	0.33	0.16	0.26	0.14	0.15	0.11	0.72	0.71	0.74

Biotites show two alteration patterns which are respectively : reaction growths on (001) planes in contact with corroded quartz (Plate 1) and internal alteration in biotite cores. Biotites, ranging from light to dark brown pleochroic flakes show exfoliated bent layers which are separated by lenticular chlorite zones. X-ray diffraction patterns indicate the presence of three phases : remnants of biotites with a weak, enlarged 10 Å reflection, chlorite, an interstratification of expandable and chlorite-like layers. The interstratification is close to that in Lippmann's corrensite description (1954) with no enhancement of the 10 Å reflection or appearance of a 12 Å reflection when heated to 550°C, which could be assigned to a chlorite-saponite or chlorite-vermiculite interstratification. The natural occurences of corrensite were summarized in Kubler's paper (1973).

Microprobe analyses (Table 1) show that the biotite relics are associated with various amounts of corrensite intergrowths. Potassium contents of brown pleochroic zones range from 0.10 to 0.70 atom per O_{10} $(OH)_2$; grey polarizing zones have chlorite compositions similar to chlorite pseudomorphs in biotites. Chemical analyses with potassium contents higher than 0.70 atom are close to primary biotite compositions ; they have no excess magnesium in the octahedral layer (occupancy 3 atoms). Si/Al ratios range from 2-2.3 in primary biotites to 1.5-2 in relic biotite-corrensite assemblages. The Mg/(Mg + Fe) ratio remains unchanged. Corrensite replacement of biotite proceeds then through silicon and potassium release, calcium-sodium enrichment with constant Fe-Mg contents.

TABLE 2

Microprobe analyses of chlorite growths on (001) biotite face-quartz joint (14 oxygen basis ; Fe total = Fe^{2+}).

	Chlorite growths							
	1	2	3	4	5	6	7	8
Si	3.29	3.33	3.39	3.33	3.33	3.39	3.43	3.37
Al	2.33	2.46	2.21	2.05	2.36	2.06	2.20	2.29
Ti	0.01	-	0.01	-	0.02	0.03	0.02	0.01
Fe^{2+}	0.77	0.74	0.59	0.43	0.64	0.81	0.70	0.88
Mg	3.07	2.83	3.26	3.81	3.03	3.19	3.05	2.73
Mn	0.02	0.03	0.01	0.01	0.01	0.01	0.02	0.01
Ca	0.03	0.03	0.01	0.03	0.05	0.02	0.03	0.03
Na	-	-	-	-	-	-	-	-
K	0.02	0.05	0.03	0.01	0.07	0.05	0.02	0.06

Intermineral joints between (001) biotite planes and quartz exhibit corroded zones, up to 50 μm thick which are formed by colourless to ligh green pleochroic "podlike" chlorites growing into corroded quartz (Plate 1). "Podlike" chlorites are rich in magnesium (Table 2), and belong to the penninite group according to Foster's classification (1962).

(ii) Chlorite-epidote-pyrite assemblage : Biotite-corrensite assemblages are partly or entirely replaced by light to dark green pleochroic chlorites ; the origianl shape, cleavage pattern and inclusions of primary biotites are still easily recognizable. Chlorites crystallize together with lenticular leucoxene patches which occur either as layer intergrowths or crystal rims with pyrite and magnetite. Numerous chloritic veinlets, emerging from coarse chlorites invade neighbouring minerals, especially andesines (Plate 2) ; these sinuous veinlets exhibit straight outlines when joining tectonic rock cracks. Coarse chlorites belong to the brunsvigite group (Table 3) with higher iron contents than the chlorites in the podlike growth zones.

Epidotes occur either at grain contact between chlorites and andesines or as underlining of chloritic veinlets in andesines ; their chemical compositions are close to pistacites which were described in the "Salton sea geothermic system" (Keith <u>et al.</u>, 1968 ; Mac Dowell and Elders, 1980), without any chemical zonation.

TABLE 3

Microprobe analyses of chlorites pseudomorphosing corrensites-residual biotites from the undeformed rock (14 oxygen basis, Fe total = Fe^{2+}).

	Chlorites without epidote							Chlorites with epidotes							
	1	2	3	4	5	6	7	1	2	3	4	5	6	7	8
Si	2.81	2.68	2.77	2.93	2.90	2.73	2.78	2.79	2.85	2.86	2.72	2.81	2.68	2.91	2.91
Al	2.48	2.65	2.60	2.44	2.36	2.64	2.54	2.45	2.36	2.33	2.39	2.28	2.66	2.36	2.35
Ti	0.01	0.01	0.03	0.02	0.02	-	0.02	0.01	0.01	0.01	0.01	-	-	-	0.01
Fe^{2+}	2.39	2.18	2.26	2.41	2.58	2.48	2.63	2.61	2.43	2.43	2.64	2.19	2.33	2.44	2.54
Mg	2.78	2.20	2.39	2.34	1.91	1.99	1.79	2.04	2.21	2.23	2.15	2.67	2.20	2.18	2.06
Mn	0.05	0.08	0.04	0.04	0.09	0.10	0.18	0.08	0.08	0.09	0.10	0.08	0.09	0.04	0.05
Ca	0.01	0.01	0.01	-	-	-	-	0.01	0.01	0.02	-	-	-	-	-
Na	-	-	-	-	-	-	-	-	-	-	-	-	-	-	-
K	-	-	0.01	0.07	0.05	0.01	-	-	-	0.01	-	-	-	-	-

Andesines appear to be strongly sericitized and converted into albite. Sericites which were extracted from these andesines by ultrasonic treatment are 2M polytype micas according to X-ray diffraction patterns. Their chemical compositions are close to muscovite with a high Al content and a layer charge ranging from 0.85 to 0.95 per formula unit according to microprobe analyses.

Potassic feldspars remain unaltered, except for the occurrence of albitic rims which are found next to andesines. Sphenes and hornblendes are entirely replaced respectively by calcite-rutile and calcite-chlorite pseudomorphs which preserve the original shape of the mineral.

Mylonitized rock

Mylonitized rock exhibits newly crystallized, fine grained zones surrounding crushed or deformed primary minerals. Samples are cut by calcite veins branching out on shear planes. Some open fractures are fringed with small sized vermicular chlorites (50-80 μm).

(i) Mylonitized zones : Chlorites occur as flexible flakes, occasionally distributed in more or less regular alignments (Plate 3) ; pleochroism is much weaker than for chlorites in undeformed rock ; primary inclusions such as apatites or zircons have disappeared. Their chemical compositions are close to those of an iron-rich brunswigite (Table 4). Chlorite outlines or cleavage planes are fringed with lenticular patches of leucoxene which is sometimes associated with pyrite and secondary apatite. Numerous chloritic veinlets, joining tectonic fractures, emerge from the flexible chlorite flakes and invade neighbouring minerals.

Strongly albitized and sericitized andesines are easily identified by their deformed multiple lamellar twinnings. Quartz and potassic feldspars occur as a fine grained groundmass which can be ascribed to pressure effects. Amphibole and sphene have disappeared from the mylonitized rock.

TABLE 4

Microprobe analyses of chlorites from tectonized rock and vermiculous fissural chlorites (14 oxygen basis ; Fe total = Fe^{2+}).

	Chlorites from tectonized rock							Vermiculous fissural chlorites				
	1	2	3	4	5	6	7	1	2	3	4	5
Si	2.80	2.77	2.86	2.72	2.76	2.74	2.74	2.63	2.74	2.68	2.77	2.69
Al	2.42	2.46	2.50	2.50	2.42	2.45	2.46	2.76	2.71	2.76	2.53	2.58
Ti	0.01	0.01	0.01	0.01	0.01	0.02	0.01	0.01	0.01	0.01	0.01	0.01
Fe^{2+}	2.84	3.13	3.08	3.15	2.85	2.86	2.87	3.00	3.11	3.29	3.09	3.07
Mg	1.82	1.53	1.36	1.53	1.93	1.89	1.85	1.53	1.26	1.11	1.46	1.58
Mn	0.08	0.08	0.07	0.01	0.05	0.05	0.10	0.05	0.06	0.09	0.07	0.08
Ca	-	0.02	-	0.01	0.03	-	-	-	0.01	0.02	-	-
Na	-	-	-	-	-	-	-	-	-	-	-	-
K	-	-	-	-	-	-	-	-	0.01	0.01	0.04	-

(ii) Vermicular chlorite bearing veins : Tectonic joints are fringed with small sized vermicular chlorites (50 μm) which are weakly pleochroic, blue to purple under crossed nicols and which invade or entirely mask cross minerals (Plate 4). No primary or secondary inclusions (zircon, apatite, leucoxene, pyrite) have been observed in vermicular chlorites which have iron rich compositions (Table 4). Similar phases have been described as ripidolites (Helminthe variety) by Lacroix (1896).

DISCUSSION

Calcitic and chloritic veins cutting the drilling hole exhibit undistorted outlines. Since calcite and chlorite are related to the hydrothermal phases, we can conclude that syenite alteration has developed subsequently to tectonization. Alteration proceeded simultaneously in coarse grained and mylonitized rock, resulting in specific mineral assemblages.

Formation fields of corrensite and iron rich epidote, ranging respectively from 150-200°C to 280°C (Velde, 1977) and from 220°C 50°C to 350°C (Tomasson and Kristmandottir, 1972 ; Seki, 1972 ; Keith et al., 1968) indicate that hydrothermal alteration proceeded in a 250-280°C temperature range. Assuming the existence of a normal gradient in the hole seems unrealistic since corrensite and epidote occur simultaneously in all samples studied and locally become increasingly important as tectonically related phases.

The chemical composition of chlorites appear then to be governed by their crystallization sites. Chlorite growth occurs in confined microsystems, along intermineral joints between quartz and (001) biotite planes. They are enriched in the most mobile elements, according to temperature, i.e. silicon and magnesium. Chlorite growths have been previously observed in propylitized granite and microgranite (Meunier et. al., 1982) where they occur prior to biotite replacement. This first stage does not occur in the syenite where biotites are already partly replaced by corrensite pseudomorphs. Corrensite crystallization occurs through silicon leaching with a constant Fe/(Fe + Mg) ratio. Biotite replacement creates an increase of volume inducing a pattern of radiating microcracks around the biotite. This process is enhanced by chlorite replacement of corrensite. Microcracks are filled by chloritic phases which are brought about by leaching of elements from biotitic sites. Silicon and magnesium are partly leached out, while iron tends to concentrate in chlorite and pyrite. Sporadic epidote (pistacite) occurrences do not seem to change the chemical composition of chlorite.

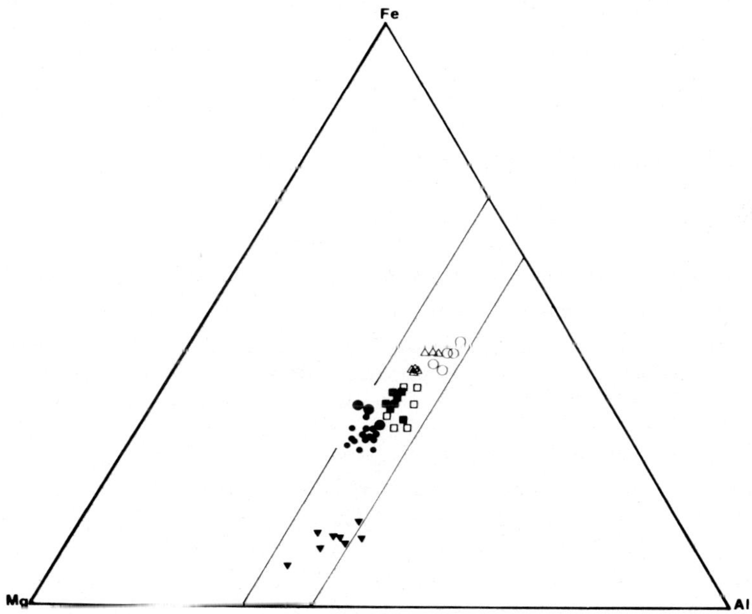

Fig. 1 - Chemiographic representation of biotite-corrensite-residual biotite and chlorites in the Fe-Mg-Al system. Lines delimite the composition area of metamorphic chlorites (Velde, 1977).
Biotites (large dots) ; corrensite-residual biotites (small dots) ; chlorites growths (full triangle) ; chlorite from undeformed rock (without epidote : empty square ; with epidote : full square) ; chlorites from deformed rock (large crystals : empty triangle ; vermiculous fissural chlorites : circles).

Chlorites from deeply fractured and mylonitized zones are richer in iron and poorer in silicon than chlorites from undeformed zones, reaching the ripidolite field for vermicular chlorites in enlarged fractures. It appears that chloritic phases tend to be progressively depleted of silicon-magnesium and enriched in iron as the fissural pattern develops. Chemical compositions, when plotted in Al-Fe-Mg coordinates, fall into the metamorphic chlorite field (Velde et. al., 1979), with a constant Al content (Fig. 1). The molar fraction of clinochlore decreases from 80 % in the pod-like chlorites to 35 % in the vermicular chlorites (Bayliss, 1975).

One can consider that the nature and chemistry of new replacement phases are governed by the rate of exchange between biotites and hydrothermal fluids. Various rates of leaching of elements create first silicon and magnesium-rich microsystems where chlorites crystallize. Exchange between biotites and hydrothermal fluids is then enhanced by radiating microcracks which create secondary, silicon-magnesium leached microsystems. Aluminium remains constant while iron concentrates in vermicular chlorite which crystallizes in microcracks.

It appears that rock texture strictly governs the chlorite compositions in propylitized zones. They become progressively richer in aluminium-iron (less mobile elements) with increased rock-fluid exchange.

REFERENCES

BAYLISS, 1975. Nomenclature of the trioctahedral chlorites. Can. Min., 13, 178-180.

CARSON, D.J.T. and JAMBOR, J.L., 1974. Mineralogy, zonal relationships and economic significance of hydrothermal alteration at porphyry copper deposits, Babine Lake Area, British Columbia. Can. Min. Metall. Bull., 1-24.

CREASEY, S.C., 1959. Some phase relations in hydrothermally altered rock of porphyry copper deposits. Econ. Geol., 54, 351-373.

FOSTER, M.D., 1962. Interpretation of the composition and a classification of the chlorites. Professional Paper U.S. Geological Survey, 414 A, 1-33.

KUBLER, B., 1973. La corrensite, indicateur possible de milieux de sédimentation et du degré de transformation d'un sédiment. Bull. Centre Rech., Pau, S.N.P.A., 7, 543-556.

KEITH, T.E.C., MUFFLER, L.J.P. and CREMER, M., 1968 - Hydrothermal epidote formed in the Salton Sea geothermal system, California. Amer. Min., 53, 1635-1644.

LACROIX, A., 1896. Minéralogie de la France et des ses anciens Territoires d'Outre-Mer. Réed. 1962-1964. Libr. Scient. Tech., Paris, 6 t.

LANIER, G., RAAB, W.J., FOLSOM, R.B. and CONE, S., 1978. Alteration of equigranular monzonite, Bingham mining district, Utah. Econ. Géol., 73, 7, 1270-1286.

LIPPMANN, F., 1954. On a Keuper clay from Zaisersweiher, near Maulbronn. Heidelb. Beitr. Miner. Petrog., 4, 130-134.

LOWELL, J.D. and GUILBERT, J.M., 1970. Lateral and vertical alteration-mineralization zoning in porphyry ore deposits. Economic geol., 65, 373-408.

MEUNIER, A., BEAUFORT, D. and PARNEIX, J.C., 1982. Chloritization facies of biotites in natural propylitized rocks. Submitted (Min. Mag.).

MC DOWELL, S.D. and ELDERS, W.A., 1980 . Authigenic layer silicate minerals in Borehole Elmore 1; Salton Sea geothermal field, California, U.S.A. Contrib. Mineral. Petrol., 74, 293-319.

SEKI, Y., 1972 - Lower grade stability limit of epidote in the ligh of natural occurrences. J. Geol. Soc. Japan, 78, 405-413.

TOMASSON, J. and KRISTMANSDOTTIR, H., 1972. High temperature alteration minerals and thermal brines, Regkjanes, Iceland. Contrib. Mineral. Petrol., 36, 123-134.

VELDE, B., PROUST, D. and MEUNIER, A., 1979. Chlorite compositions during sedimentatin, diagenesis and low grade metamorphism. Sci. Géol., 53, 71-73.

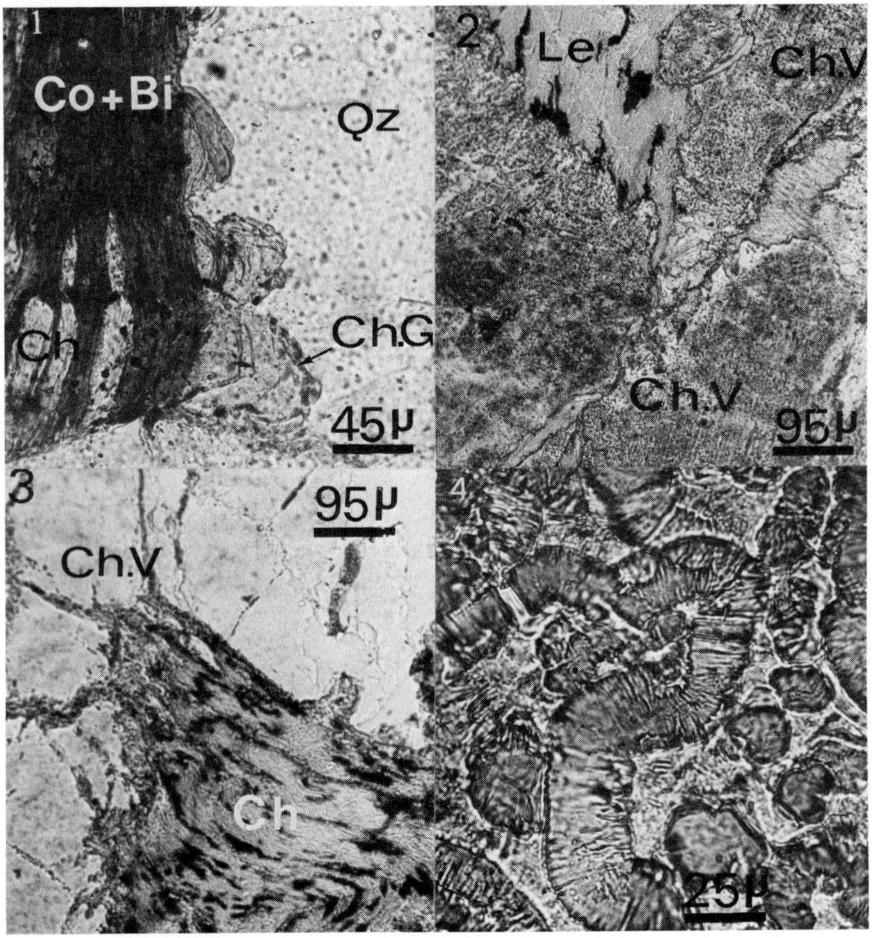

Plate 1 - Chlorite growth (Ch.G) on the biotite-quartz interface. Original biotite is replaced by corrensite-residual biotite (Co + Bi) and lenticular chlorite (Ch).

Plate 2 - Chloritic veinlets (Ch.V) issued from chlorite pseudomorph of biotite (Ch). Leucoxene (Le).

Plate 3 - Chlorite pseudomorph (Ch) of a deformed biotite with radiating chloritic veinlets (Ch.V).

Plate 4 - Vermicular ripidolites in large veins inside mylonitized zones.

SUPERGENE ALTERATION OF HORNBLENDE IN AN AMPHIBOLITE FROM MASSIF CENTRAL, FRANCE

D. PROUST
E.R.A. 220 - Laboratoire de Pédologie
Faculté des Sciences
40, avenue du Recteur Pineau
86022 POITIERS CEDEX, France

ABSTRACT

Progressive mineralogical and chemical changes of a Tschermakitic hornblende in different horizons of weathering amphibolite have been studied. Microscopic and microprobe analyses of new phases produced in or around hornblendes allow one to classify the different alteration stages relative to their position in the profile. In the unaltered rock, metamorphic processes have occurred at the interfaces between primary minerals with subsequent talc-chlorite replacement after hornblende. In the early supergene alteration stages, when rock structure is still preserved, saponite occurs either as an internal weathering product in hornblende or as a reaction phase between primary chlorites and hornblendes.

Further weathering leads to the replacement of saponite by trioctahedral vermiculite at hornblende surfaces and dioctahedral vermiculite within intramineral microcracks. The last stage of weathering is characterized by the destabilization of trioctahedral vermiculite, giving way to a dioctahedral vermiculite-beidellite assemblage.

INTRODUCTION

Previous studies on the alteration of amphiboles demonstrate the great variety of their weathering products : iron hydroxides such as goethite (Leneuf, 1959), talc (Jayaraman, 1940 ; Delvigne, 1965 ; Nicolas and Sagon, 1966) and nontronite (Stringham and Taylor, 1950). Mixed layered minerals also have been observed frequently : chlorite-saponite (Wilson and Farmer, 1970) or chlorite-vermiculite giving way to vermiculite (Stephen, 1952). Nontronite, talc, trioctahedral vermiculite and ferric beidellite have been observed recently in an altered gabbro (Ildefonse, 1978). The objectives of the present study are to define, on a microscopic scale, the exact mineralogical and chemical composition of the new phases, in order to provide a better understanding of the influence of primary minerals and fissural pattern upon the newly crystallized minerals.

SAMPLING AND ANALYTICAL METHODS

The weathering profile is developed on an amphibolite from **Limousin**, previously described by Proust (1981). Amphiboles were extracted respectively from fresh rock, saprock, saprolite and overlying prismatic horizon, and separated from crushed, dried and sieved material using a magnetic separator and heavy liquids (bromoform + alcohol). Samples were prepared, after ultrasonic treatment, as oriented K or Mg saturated aggregates or random powders and examined by X-ray diffraction using a Philips diffractometer (40 kV-40 mA) with iron filtered CoKα radiation. Chemical analyses were obtained from polished thin sections, using a CAMECA MS 46 electron microprobe and a TRACOR solid state detector. Low beam current (1.5 nano-ampère) was used in order to minimize the loss through volatilization. Count-time was 120 sec.

RESULTS

- Fresh rock

The X-ray powder patterns yielded by amphiboles are those of hornblende with strongest reflections at 9.03, 8.45, 8.38, 3.09 and 2.70 Å. Additional reflections of oriented samples can be assigned to five secondary phases, respectively, talc, chlorite, an expandable phase, chlorite-smectite mixed layered mineral and goethite.

The microscopical study of the fresh rock shows that hornblende have two distinct alteration pattern (Plate 1). Talc and chlorite replacements appear first after hornblende and are located along (110) cleavage planes. These replacements are cut by secondary intramineral microcracks (Stoops _et al._, 1979) filled by blue to light brown clay material and iron oxides.

The microprobe analyses (Table 1) show that the hornblendes have a Tschermakitic type composition according to amphibole nomenclature (Leake, 1978) and that clay materials from microcracks have trioctahedral smectite (saponite) compositions (Fig. 1.B). They have distinct iron, aluminium and silicon contents according to their crystallization sites in hornblendes. Saponites which crystallize at hornblende surfaces are clearly differentiated from those occuring in microcracks through higher silicon, aluminium and lower iron-magnesium contents.

Plate 1

Altered hornblende from the fresh rock

t.c. : talc and chlorite replacements
s. : saponites
h. : hornblende

TABLE 1

Electron microprobe analyses from the fresh rock

* Total iron expressed as Fe^{2+}.

	A : Unweathered hornblendes							B : Chlorites	
Si	6.34	6.39	6.38	6.34	6.35	6.44		2.82	2.97
Al	1.66	1.61	1.62	1.66	1.65	1.56		1.18	1.03
Al	1.07	1.24	1.04	1.07	0.99	0.94		1.28	1.39
* Fe^{2+}	0.95	0.95	0.97	1.03	1.02	0.97		0.60	0.57
Mg	3.30	3.22	3.25	3.18	3.27	3.31		4.02	3.79
Mn	0.02	0.03	0.02	0.01	0.04	-		0.01	0.02
Ti	0.01	0.01	0.003	0.01	0.02	0.01		0.001	0.01
Ca	1.67	1.41	1.78	1.77	1.71	1.79		0.001	0.04
Na	0.54	0.64	0.48	0.42	0.44	0.54		0.001	0.001
K	0.001	0.01	0.001	0.01	0.02	0.01		0.01	0.001
	C : Saponites at hornblende surface								
Si	3.53	3.62	3.60	3.54	3.50	3.52	3.46	3.54	3.51
Al	0.47	0.38	0.40	0.46	0.50	0.48	0.54	0.46	0.49
Al	0.53	0.76	0.69	0.65	0.61	0.70	0.63	0.89	0.77
* Fe^{2+}	0.63	0.59	0.70	0.71	0.68	0.70	0.74	0.69	0.84
Mg	1.66	1.31	1.31	1.30	1.50	1.31	1.33	1.01	1.11
Mn	0.002	0.01	0.002	0.002	0.01	0.002	0.01	0.004	0.002
Ti	0.002	0.01	0.004	0.01	0.01	0.02	0.01	0.01	0.01
Ca	0.13	0.11	0.13	0.15	0.14	0.13	0.12	0.12	0.11
Na	0.001	0.001	0.001	0.13	0.002	0.001	0.15	0.10	0.001
K	0.02	0.02	0.03	0.02	0.02	0.02	0.02	0.01	0.04
			D : Saponites from microcracks						
Si			3.46	3.48	3.52	3.35	3.37	3.28	3.26
Al			0.54	0.52	0.48	0.65	0.63	0.72	0.74
Al			0.41	0.43	0.54	0.34	0.40	0.30	0.31
* Fe^{2+}			0.87	0.90	0.89	0.88	0.98	1.08	1.20
Mg			1.60	1.45	1.33	1.75	1.46	1.50	1.48
Mn			0.01	0.002	0.01	0.01	0.004	0.004	0.03
Ti			0.01	0.01	0.01	0.01	0.03	0.01	0.01
Ca			0.17	0.15	0.15	0.16	0.21	0.29	0.18
Na			0.002	0.15	0.001	0.002	0.002	0.002	0.002
K			0.01	0.01	0.03	0.01	0.001	0.03	0.02

- Saprock

X-ray diffraction patterns of hornblendes from the saprock indicate a newly formed non expandable 14 Å phase with a vermiculitic behaviour (Robert and Tessier, 1974). This mineral is associated with chlorite, talc, goethite and saponite, previously observed in the fresh rock.

The hornblende weathering microsites from the saprock (Plate 2) exhibit three alteration patterns which can be classified according to their location in hornblende and the mineralogy of their secondary phases. The massive hornblende cores contain lamellar intergrowths, parallel to the cleavage planes, with optical properties similar to those of talc and chlorite observed in the fresh rock. In more strongly weathered zones, hornblendes show irregular linear alteration through intramineral microcracks filled by distinct secondary clay material and iron oxides. Hornblende residues have non accordant surfaces, including dissolution cavities where brown to yellow pleochroic, sheaf-like minerals crystallize. Another secondary mineral occurs within microcracks as fine grained material with random orientation and yellow birefringence colours. This mineral is associated with iron oxides. Interfaces between hornblendes and chlorites are often rimmed by clay minerals growing from the chlorite surface and invading corroded hornblendes. Their optical properties are similar to those of saponites from the fresh rock.

The microprobe analyses of the weathering microsites (Table 2) have been plotted in Si-R^2-R^3 coordinates (Fig. 1.C). Three compositional fields appear which can be related to the crystallization sites described. The chemical compositions of sheaf-like minerals may be placed in the trioctahedral vermiculite field (Foster, 1963). The compositions of the fine grained material may be placed in the dioctahedral vermiculite field (Proust, 1981). They are distinct from sheaf-like minerals because of the higher silicon and aluminium contents. The compositions of the reaction phases between chlorites and hornblendes may be placed in the trioctahedral smectite (saponite) field (Velde, 1977). The layer charge is located in tetrahedral site while the magnesium rich octahedral layer has an occupancy from 2.83 to 2.89 atoms.

- **Saprolite and new prismatic horizon**

X-ray diffraction data of hornblendes from the saprolite and the overlying prismatic horizon can be assigned to four secondary phases, respectively, talc, minor amounts of chlorite, a vermiculitic phase and a smectitic phase which appears to be of a beidellite type according to the Hofmann-Klemen treatment (1950).

Enlarged intramineral microcracks in hornblendes shows different alteration patterns according to their crystallographic orientation. (001) planes show cross banded alteration, parallel to (110) cleavages with organized hornblende residues : the original hornblende shape is preserved. (h k l) planes exhibit an irregular banded alteration with random residues : the original hornblende shape is destroyed. Intramineral microcracks are filled by clay material showing grey polarization colours and associated with iron oxides ; talc occurs as yellow-flakes in coherent hornblende cores.

The chemical compositions of grey materials from microcracks (Table 3) plotted in Si-R^2-R^3 coordinates (Fig. 1.D), may be placed in the dioctahedral vermiculite field with low silicon contents, aluminium rich dioctahedral layers and high interlayer charges. They appear to grade towards the dioctahedral smectite pole through vermiculite-smectite mixtures, close to those described by Ildefonse (1978).

DISCUSSION and CONCLUSION

Each secondary phase observed in the course of hornblende weathering results from specific chemical microsystems which seem to be governed either by host mineral chemistry when rock structure is preserved or by weathering solution compositions when rock structure is destroyed. The talc-chlorite-hornblende assemblage (Fig. 1.A) can be considered as a primary assemblage for the weathering profile. Several

TABLE 2

Electron microprobe analyses from the saprock.
* Total iron expressed as Fe^{2+} or Fe^{3+}.

	A : Hornblendes	B : Saponites réactionnelles			C : Vermiculites trioctaédriques									
Si	6.83	3.30	3.39	3.54	2.93	2.89	2.93	2.95	3.00	2.92	2.98	2.90	2.95	2.80
Al	1.17	0.70	0.61	0.46	1.07	1.11	1.07	1.05	1.00	1.08	1.02	1.10	1.05	1.20
Al	0.79	0.49	0.58	0.36	0.43	0.54	0.57	0.45	0.56	0.55	0.43	0.71	0.48	0.46
* Fe^{2+}	1.00	0.50	0.51	0.48	1.16	1.09	0.71	0.81	0.61	1.15	1.14	0.85	0.92	0.91
Mg	3.39	1.84	1.76	2.04	1.39	1.36	1.71	1.74	1.82	1.29	1.41	1.42	1.59	1.62
Mn	0.01	0.002	0.002	0.001	0.01	0.01	0.01	0.003	0.01	0.001	0.01	0.01	0.01	0.01
Ti	0.02	-	0.003	0.01	0.01	0.002	0.004	0.001	0.003	0.01	0.01	0.01	0.002	0.004
Ca	1.78	0.14	0.06	0.06	0.10	0.11	0.10	0.10	0.11	0.11	0.17	0.06	0.16	0.11
Na	0.38	0.10	0.09	0.14	0.002	0.001	0.001	0.001	0.001	0.001	0.001	0.001	0.04	0.03
K	0.01	0.04	0.06	0.04	0.08	0.04	0.05	0.08	0.06	0.12	0.07	0.09	0.10	0.06
Mg	-	-	-	-	0.17	0.16	0.11	0.15	0.08	0.10	0.09	0.09	0.04	0.20

	D : Vermiculites dioctaédriques														
Si	3.09	2.96	3.13	3.00	3.22	3.14	3.01	3.14	3.07	3.19	3.18	3.12	3.17	2.98	3.05
Al	0.91	1.04	0.87	1.00	0.78	0.86	0.99	0.86	0.93	0.81	0.82	0.88	0.83	1.02	0.95
Al	1.84	1.79	1.84	1.83	1.64	1.49	1.50	1.69	1.33	1.57	1.40	1.48	1.48	1.58	1.48
* Fe^{3+}	0.27	0.43	0.28	0.38	0.44	0.52	0.52	0.41	0.71	0.41	0.49	0.45	0.43	0.55	0.48
Mg	-	-	-	-	-	-	-	-	-	0.01	0.10	0.06	0.08	-	0.02
Mn	-	0.01	0.002	0.002	0.002	0.002	-	-	0.002	0.002	0.002	0.01	0.003	0.002	0.01
Ti	0.01	0.004	-	0.004	0.01	0.01	0.001	0.01	0.01	0.003	0.004	-	0.01	0.01	0.01
Ca	0.05	0.02	0.03	0.03	0.03	0.07	0.10	0.03	0.04	0.04	0.02	0.05	0.02	0.04	0.03
Na	0.001	0.07	0.07	0.04	0.08	0.09	0.15	0.03	0.07	0.13	0.06	0.10	0.03	0.001	0.03
K	0.03	0.02	0.02	0.01	0.01	0.06	-	0.004	0.04	0.05	0.02	0.01	0.02	0.01	0.01
Mg	0.21	0.12	0.18	0.12	0.17	0.26	0.28	0.22	0.27	0.27	0.40	0.36	0.41	0.26	0.45

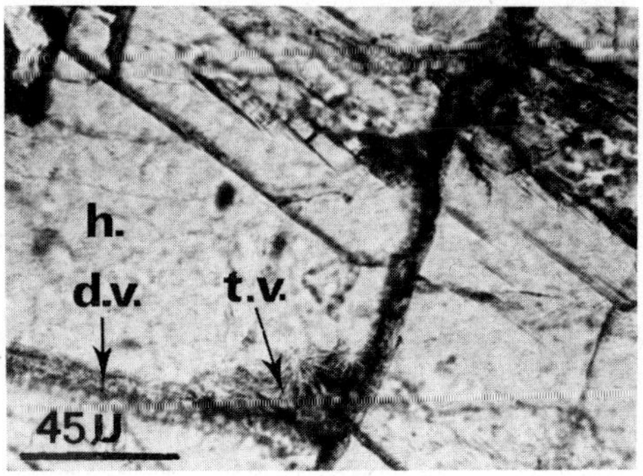

Plate 2

Altered hornblende from the saprock

t.v. : trioctahedral vermiculite
d.v. : dioctahedral vermiculite
h. : hornblende

TABLE 3

Electron microprobe analyses from the saprolite and the prismatic horizon. * Total iron expressed as Fe^{3+}.

	A : Dioctahedral vermiculites							B : Mixture of beidellite and dioctahedral vermiculite												
Si	2.87	2.88	2.93	2.96	2.97	2.98	3.03	3.05	3.09	3.11	3.13	3.14	3.16	3.17	3.18	3.18	3.18	3.21	3.21	3.25
Al	1.13	1.12	1.07	1.04	1.03	1.02	0.97	0.95	0.91	0.89	0.87	0.86	0.84	0.83	0.82	0.82	0.82	0.79	0.79	0.75
Al	0.97	1.08	1.75	1.72	1.61	1.23	1.77	1.85	1.63	1.76	1.77	1.81	1.78	1.76	1.69	1.76	1.88	1.37	1.79	1.76
* Fe^{3+}	1.05	0.94	0.29	0.27	0.34	0.73	0.26	0.21	0.30	0.28	0.17	0.22	0.28	0.18	0.31	0.21	0.19	0.64	0.14	0.26
Mg	-	-	-	-	0.04	-	-	-	0.05	-	0.05	-	-	-	-	0.01	-	-	0.06	-
Mn	0.01	0.01	0.01	0.01	0.003	0.02	0.01	0.01	0.01	0.01	0.002	0.01	0.01	-	0.01	0.01	0.01	0.002	0.002	0.002
Ti	0.03	0.03	0.02	0.01	0.01	0.02	0.004	0.002	0.01	0.003	0.01	0.01	0.01	0.01	0.02	0.01	0.004	0.02	0.01	0.004
Ca	0.21	0.16	0.07	0.04	0.03	0.05	0.02	0.08	0.12	0.07	0.03	0.04	0.07	0.03	0.09	0.03	0.02	0.09	0.04	0.04
Na	0.002	0.002	0.06	0.06	0.05	0.14	0.05	0.05	0.22	0.08	0.07	0.001	0.001	0.04	0.05	0.001	0.001	0.001	0.001	0.07
K	0.10	0.08	0.05	0.02	0.03	0.06	0.03	0.002	0.02	0.04	0.02	0.003	0.03	0.01	0.05	0.02	0.02	0.08	0.03	0.03
Mg	0.20	0.27	0.30	0.45	0.47	0.34	0.36	0.26	0.26	0.24	0.39	0.32	0.20	0.38	0.22	0.38	0.27	0.19	0.37	0.25

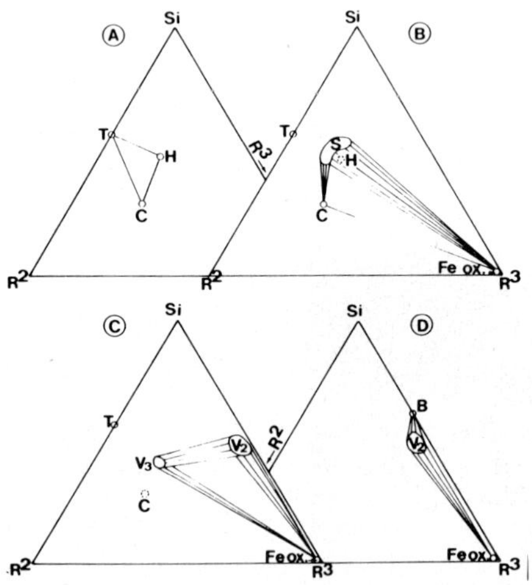

Fig. 1. Electron microprobe analyses and general phase relations in $Si-R^2-R^3$ coordinates.
T : talc ; C : chlorite ; H : hornblende ; S : saponite ; Fe.ox. : iron oxides ; V_3 : trioctahedral vermiculite ; V_2 : dioctahedral vermiculite ; B : Beidellite.
Lines refer to two-phase fields.

microscopical observations indicate that talc has a metamorphic origin and pre-date the saponite crystallization. The chemistry of saponites from the fresh rock is governed either by the host mineral compositions at hornblende - chlorite interfaces with high magnesium enrichment, or by solution compositions within hornblende microcracks with two differently concentrated zones : a zone of relative aluminium enrichment and magnesium depletion at hornblende surfaces and a zone of relative magnesium enrichment within microcracks created by moderate magnesium leaching. Iron appears to concentrate in microcracks as iron oxides.

The chemistry of vermiculitic phases from the saprock tend to be more and more aluminium rich as the fissural pattern develops, i.e. weathering microsystems become more and more permeable to leaching solutions. These solutions induce then a compositional trend for vermiculitic phases similar to the one described for the saponites, except that the magnesium leaching is more pronounced in microcracks. The vermiculite-beidellite mixtures from the saprolite and prismatic horizon occur in a strongly porous weathering plasma, highly permeable to leaching solutions. The leaching process is then the main factor which controls the chemistry of the secondary phases, obliterating the host mineral influence. The most mobile elements, i.e. magnesium, calcium, sodium, are strongly leached out from the weathering microsites which are relatively enriched in silicon and aluminium, this being clearly demonstrated by the crystallizations of aluminous vermiculite and beidellite.

It appears then that the rock structure is the main factor which controls the chemistry of the secondary phases in the weathering profile. These phases become more and more aluminium rich as weathering increases and as the weathering microsystems become more permeable to leaching solutions.

REFERENCES

DELVIGNE, J., 1965. Pédogenèse en zone tropicale. La formation des minéraux secondaires en milieu ferrallitique. O.R.S.T.O.M., Dunod éd., Paris, 177 p.

FOSTER, M.D., 1963. Interpretation of the composition of vermiculites and hydrobiotites. Clays Clay Min., Nat. Conf., 10th, 1961, 70-89.

HOFMANN, U. and KLEMEN, R., 1950. Verlust der Austauschfähigkeit von Lithiumionen an Bentonit durch Erhitzung. Z. anorg. chem., 262, 95.

ILDEFONSE, Ph., 1978. Mécanismes de l'altération d'une roche gabbroïque du Massif du Pallet (Loire-Atlantique). Thèse 3ème cycle, Fac. Sci. Poitiers, 142 p.

JAYARAMAN, N., 1940. Alteration of tremolite to talc in the dolomitic marbles of Yellandu Warangal District (Hyderebad, Dn.). Proc. Indian Acad. Sci., sect. A, 12, 65-71.

LEAKE, B.C., 1978. Nomenclature of amphiboles. Amer. Miner., 63, 1023-1052.

LENEUF, N., 1959. L'altération des granites calcoalcalins et des granodiorites en Côte d'Ivoire forestière et les sols qui en sont dérivés. Thèse Sciences, Paris, O.R.S.T.O.M., 210 p.

NICOLAS, J. and SAGON, J.P., 1966. Alteration en talc d'une diabase au Sud de Quintin (Côtes-du-Nord). Bull. Gr. Fr. Arg., XVIII, 14, 47-50.

PROUST, D., 1981. Supergene alteration of metamorphic chlorite in an amphibolite from Massif Central, France. Clay Min. (in press).

ROBERT, M. and TESSIER, D., 1974. Méthode de préparation des argiles et des sols pour les études minéralogiques. Ann. Agron., 25, (6), 859-882.

STEPHEN, I., 1952. A study of rock weathering with reference to the soils of the Malvern Hills. II : Weathering of appinite and "Ivy Scar rock". J. Soil Sci., 3, 219-237.

STOOPS, G., ALTEMULLER, H.L., BISDOM, E.B.A., DELVIGNE, J., DOBROVOLSKY, V.V., FITZPATRICK, E.A., PANEQUE, G. and SLEEMAN, J., 1979. Guidelines for the description of mineral alterations in soil micromorphology. Pedologie, XXIX, 1, 121-135.

STRINGHAM, B.F. and TAYLOR, A., 1950. Nontronite at Bingham, Utah. Am. Miner., 35, 1060-1066.

VELDE, B., 1967. Clays and clay minerals in natural and synthetic systems. Developments in Sedimentology, 21, Elsevier Scientific Publishing Company, 1 vol., 218 p.

WILSON, M.J. and FARMER, V.C., 1970. A study of weathering in a soil derived from a biotite-hornblende rock. II : Weathering of hornblende. Clay Min., 8, 435.

DETERMINATION OF THE COMPLETE MINERAL COMPOSITION OF CLAYS

WIEGMANN, J., HORTE, C.H. and KRANZ, G.
Zentralinstitut für Anorganische Chemie der AdW, Berlin, DDR.

ABSTRACT

A method is developed for the determination of the complete composition of mineral samples instead of determining the content of each mineral in the mixture separately. For the calculation of the composition any quantitatively interpretable information may be used, such as chemical analysis, X-ray, and thermoanalytical data. The method is illustrated by some examples.

PRINCIPLE OF THE METHOD

Geological, genetical, chemical, physical and technical properties of clays are mainly correlated with their mineralogical composition. Therefore, the quantitative evaluation of the mineral composition of clay samples is a central problem in nearly all types of clay investigations.

By combining various methods it is usually possible to detect all mineral components in a clay sample. Also,numerous methods have been developed for quantitative determination of the contents of specific minerals, but for more complicated mineral mixtures the quantitative determination of the complete mineral composition of the clay remains a problem. Difficulties arise particularly when the samples contain minerals of variable composition or structure such as smectites, micaceous or mixed-layer minerals. Causes of these difficulties are primarily (a) the lack of suitable standards for calibration of the analytical method, and (b) interference of determination criteria due to overlapping effects of similar minerals present in the mixture.

To eliminate these problems, several methods have been developed for the evaluation of the mineral composition of clays by combining the results of mineralogical and chemical analysis (Pearson, 1978;

Tchernjak et al., 1979; Urban, 1978).

The following method focuses on the complete composition determination using quantitative information that is available from generally applied methods in the investigation of clays. If a sample comprises the mineral components 1,2,...n with contents $C_1, C_2, \ldots C_n$, then a value M_m characterizing the behavior of the sample when measured by a method (m), usually may be written as a linear combination of the mineral contents as follows:

$$M_m = a_{m1}C_1 + a_{m2}C_2 + \ldots a_{mn}C_n \quad \text{or} \quad M_m = \sum_n a_{mn} C_n$$

in which a_{mx} is a calibration coefficient describing the contribution of mineral content C_x to the measured value of M_m. For example, if M_m is the weight loss of the sample as recorded thermographically between two temperatures, a_{mx} is the weight loss of the pure mineral under identical conditions. Naturally, some a_{mx} values may be zero. Treating all measured values of the sample by various available methods in this way, a set of linear equations is obtained of the type:

$$M_1 = \sum_n a_{1n} C_n = 1 = \sum_n C_n \; ; \; M_2 = \sum_n a_{2n} C_n \; ; \; \ldots \; M_m = \sum_n a_{mn} C_n$$

The first equation refers to the results of a qualitative examination of the sample and states that the sum of the contents of all of the minerals present in the sample should be equal to unity (conditional equation). The number m of equations defined for such a system usually will be larger than the number of components n ($m > n$). When solving the equations to obtain the contents $C_1 \ldots C_n$ one must keep in mind that each equation is based on physical results and does not have a purely mathematical meaning since the measured values M_x as well as the calibration coefficients a_{xy} are subject to experimental errors. Therefore, to obtain a realistic result all equations should be used but each equation representing a certain measurement should be weighted in accordance with actual error estimates. The larger the estimated errors of a given measurement, the lower the weight the corresponding equation gets. The following rule was chosen for the calculation of the weights W'_x :

$$W'_x = 1 : (\Delta M_x/M_x + \Delta a_{x1}/a_{x1} + \dots \Delta a_{xn}/a_{xn})$$

where ΔM_x is the standard deviation of the measured value M_x and Δa_{xy} is the standard deviation of a_{xy} or the probable span in which a_{xy} lies. For the conditional equation ($M_1 = 1$) $\Delta a_{1y} = 0$ is applied and ΔM_1 is estimated from the detection limits of qualitative analysis. Usually W'_1 is between 25 and 100.

For the application of statistical tests to the final result the weights may be standardized so that $\sum_m W_m = m$, and $W_x = W'_x.m/(\sum_m W'_m)$

Using the system of weighted equations the probable mineral contents are calculated by a least squares method, minimizing the following expression: $\sum_m W_m(M_m - \sum_n a_{mn} C'_n)^2$

Partial differentiation of this expression with respect to all C'_n and equating the partial differential coefficients to zero lead to a system of equations which can be solved for C'_1, C'_2 ... C'_n.

APPLICATIONS OF THE METHOD

Three examples taken from investigations of raw clay materials used in ceramics are chosen to demonstrate the application of the method.

The main components of the ceramic clays were kaolinite, quartz, feldspar (primarily potassium feldspars) and dioctahedral 2:1 layer minerals of variable types. The conditional equation therefore reads:

$$K + Q + F + T = 1 - i$$

Small amounts of impurities such as anatase, amorphous silica, and iron oxides present in the samples were determined separately by chemical methods and in the calculations they were only considered by the correction term i. Data obtained by the following methods were used in the calculations:

q_x : quartz content determined by X-ray method

q_{th} : quartz content determined by DTA

q_s : quartz content determined by chemical solution (H_3PO_4)

k_x : kaolinite content determined by X-ray method

f_x : feldspar content determined by X-ray method

OH_{th} : thermographically determined weight loss accompanying the dehydroxylation of clay minerals

t_x : approximate content of 2:1 layer minerals estimated from the intensity of the 15 Å diffraction peak of potassium saturated and dried samples.

In addition, the results of a complete chemical analysis of the K-saturated sample or of the analysis corrected to the K form based on the cation exchange capacity, may be included in the calculations.

The best results are obtained by carrying out the calculations in several runs. The first run was usually done without taking into account the results of chemical analysis. This run primarily served to obtain an approximate content of 2:1 layer minerals. In further runs the results of chemical analysis were included in the calculation by using the following equations:

$$0{,}465\ K + Q + 0{,}65\ F + f_s\ T = s - i_s$$
$$0{,}396\ K + 0{,}186 F + f_a\ T = a - i_a$$
$$0{,}16\ F + f_k\ T = k$$
$$0.139\ K + f_g\ T = g$$

where

$s - i_s$ is the SiO_2 content corrected for impurities (i_s)

$a - i_a$ is the Al_2O_3 content corrected for impurities (i_a)

k is the K_2O content of the K-saturated form

g is the loss on ignition of the K-saturated form.

The calibration factors f_s, f_a, f_k, f_g for the 2:1 layer mineral content in the corresponding equations depend on the special structure and composition of these minerals. Approximate values for these factors were chosen in the following way. It was assumed that for the dioctahedral 2:1 layer minerals present in the samples the following formula applies for the K-saturated form:

$$K_{\alpha+\mu}(Si_{8-\alpha}Al_{\alpha})(Al_{4-\varphi-\mu}Fe_{\varphi}Mg_{\mu})O_{20}(OH)_4$$

From the swelling behavior of these minerals observed by X-ray diffraction methods (Brown, 1961) the layer charge ($\alpha+\mu$) was estimated. At least four groups with layer charges 0,6–1,1; 1,1–1,5; 1,5–1,8 and 1,8–2 may be distinguished. Assuming that all Fe (not belonging

	Kemmlitz	f_s	.557	.545	Frohnsdorf		.523	.509	Mittelherwigs-		.566
	"Glück auf"	f_a	.262	.273	Emaillierton		.209	.279	dorf		.208
	Kaolin	f_k	.0727	.0784			.0838	.0917	"Bentonit"		.048
	Z 8	f_g	.0463	.0461			.0458	.0452			.046
	Meas. val. %	Calc. 1 ap.	Calc. 2 ap.	Calc. 3 ap.	Meas. val.%	Calc. 1 ap.	Calc. 2 ap.	Calc. 3 ap.	Meas. val.%	Calc. 1 ap.	Calc. 2 ap.
Σ	99,89	99,51	99,80	99,87	98,0	98,2	97,9	97,9	92,57	92,41	92,3
q_x	12	13,2	11,6	11,9	25	27,5	27,5	27,6	7	6,9	6,6
q_{th}	12	13,2	11,6	11,9	27	27,5	27,5	27,6	8	6,9	6,6
q_s					28,9	27,5	27,5	27,6	6,7	6,9	6,6
k_x	7	9,7	7,3	7,1	49	48,5	49,1	49,1	3	4,7	4,1
f_x	20	22,2	19,8	19,9	3	-1,6	2,3	2,2	11	12,6	10,2
OH_{th}	3,9	3,7	3,7	3,7	7,0	7,5	7,3	7,3	4,0	3,6	3,7
t_x	47	54,0	-	-	30	23,8	-	-	65	68,2	-
$s-i_s$	61,3	-	61,9	61,4	60,4	-	61,8	61,5	53,7	-	55,5
$a-i_a$	23,2	-	22,6	23,2	25,3	-	25,4	25,2	18,4	-	18,4
k	8,3	-	7,6	8,0	2,11	-	1,96	2,11	4,1	-	5,0
g	4,2	-	3,8	3,8	8,6	-	7,7	7,7	5,2	-	3,9
Min. comp.	K'	9,7	7,3	7,1		48,5	49,1	49,1		4,7	4,1
	Q'	13,2	11,6	11,9		27,5	27,5	27,6		6,9	6,6
	F'	22,6	19,8	19,9		-1,6	2,3	2,2		12,6	10,2
	T'	54,0	61,1	60,9		23,8	19,0	19,1		68,2	71,4
	$10^5 \tilde{s}^2$		2,08	0,57	$10^4 \tilde{s}^2$		1,71	1,66	$10^4 \tilde{s}^2$		0,91
Comput. T best appr.		$\alpha + \mu = 1,3$ $\mu = 0,37$ $\varphi = 0,38$			2:1 min.: mica 50 mix.l. 25 smect. 25	$\alpha + \mu = 1,55$ $\mu = 0,3$ $\varphi = 0,6$				$\alpha + \mu = 0,8$ $\mu = 0,2$ $\varphi = 1,2$	

Table 1. Measured values and calculated mineral composition of three ceramic clays.

to impurities) and Mg belong to the 2:1 layer component estimates of φ and μ were made. Then it was possible to calculate first approximation values for f_s, f_a, f_k and f_g as the silica, alumina, potassium oxide and water contents respectively, according to the compositional formula. Study of the swelling behavior showed that different 2:1 layer components, both swelling and non-swelling, were present in the samples. No attempts were made to distinguish these different forms by the calculations, but for an estimate of the layer charge the proportions of these forms were derived from the intensities of the corresponding 001 reflections to arrive at an average layer charge

Table 1 shows the example of a kaolin from the Kemmlitz pit "Glück auf" (DDR). Measured values are entered in the first column; the secon column shows the approximate results of the first calculation run for the composition, including the numbers calculated from these results for the measured values. At the top of the third column the calibration factors are shown which were derived from the first approximation T' and the analytical results, as described before. Below these the results of the second approximation are tabulated. By comparing the measured values with the corresponding numbers of the second approximation it is possible to correct the calibration factors for the chemical contents due to the 2:1 layer components, as shown at the top of the fourth column. The calculation for a third run using these corrected factors yields the results shown in the fourth column. When comparing the data for the mineral composition with those of the second approximation there appears to be only a slight change and nearly all of the numbers calculated for the measured values in the third approximation agree with the actually measured values within the limits of experimental errors. Therefore, it may be concluded that the mineral composition given by the third approximation is the most probable one on the basis of the experimental information used in the calculations. The variance $\tilde{s}^2$ (shown below columns 3 and 4) between measured values and the calculated ones can be used to characterize the quality of the derived fit. As shown in the theory of statistics (Benett and Franklin, 1954)

$\tilde{s}$ is proportional to the confidence span of the results for the mineral contents. The value of $\tilde{s}^2$ may also be applied to check whether the calculation of further approximations will be useful.

The results for the Kemmlitz Z 8 sample show that it is a rather unusual kaolin with a low kaolinite content and a large proportion of 2:1 layer mineral. Study of the swelling behavior indicated that the latter is a random mica-montmorillonite mixed layer mineral with approximately equal parts of swelling and non-swelling layers. This agrees well with the derived layer charge $\alpha + \mu = 1,3$.

The second sample shown in the Table, Frohnsdorf Emaillierton, was chosen since it contains according to qualitative examination three types of 2:1 layer minerals in minor amounts: mica, montmorillonite, and a random mica-montmorillonite mixed-layer clay with about 40% swelling layers, in the proportion 2:1:1. Therefore, it would seem to be very difficult to determine the mineral contents by means of separate methods of mineral determination. A comparison between the measured value t_x = 30% for the 2:1 layer mineral content by the direct X-ray method and the final calculated result T = 19,1% demonstrates the difficulties. The stable results for the second and third calculation runs shows that also in this case a satisfactory accuracy is obtained. The relatively high $\tilde{s}^2$ value is mainly due to the differences in the quartz contents determined by different methods.

The third sample in the Table shows the application of the method to a hematite bearing bentonite from Mittelherwigsdorf. The hematite content of 5,3%, determined by chemical and X-ray methods was treated as an impurity in the calculation. In this case, the smectite content t_x estimated from X-ray measurements was only slightly different from the final calculated result. Therefore, already the second approximation converges sufficiently. In agreement with the swelling behavior it follows from the results that the Mittelherwigsdorf sample contains a highly iron bearing smectite with a layer charge of 0.8 per formula unit.

The applicability of the described method is not restricted to clays with compositions similar to those presented here. Weighting of the measured values and the corresponding equations on the basis of their accuracy opens the possibility to use any information which can be expressed in numbers for the calculation.

REFERENCES

Pearson, M.J., 1978. Quantitative clay mineralogical analysis from the bulk chemistry of sedimentary rocks. Clays Clay Min., 26: 423-43

Tchernjak, L.P., Nesterenko, I.P. and Soroka, A.C., 1979. Determinatio of the mineralogical composition of clay raw materials using compute techniques. Steklo i ceramica (Moscow), 3: 17-19

Urban, H., 1978. Eine Methode zur rechnerischen Bestimmung von Illit und Kaolinit in Tonen und keramischen Massen. Sprechsaal 111: 373-37

Brown, G., 1961. The X-ray identification and crystal structures of clay minerals. Mineralogical Society, London.

Benett, C.A. and Franklin, N.L., 1954. Statistical analysis in chemistry and the chemical industry. Wiley, New York.

INTERLAYER HYDRATES AND COMPLEXES OF CLAY MINERALS OBSERVED BY ELECTRON MICROSCOPY USING AN ENVIRONMENTAL CELL

Norihiko KOHYAMA, Kurio FUKUSHIMA(*) and Akira FUKAMI(*)

National Institute of Industrial Health, Ministry of Labor,
6-21-1 Nagao, Tamaku, Kawasaki 213, Japan.
* Department of Physics, College of Humanities and Sciences,
Nihon University, Sakurajosui, Setgayaku, Tokyo 156, Japan.

ABSTRACT

To observe interlayer hydrates and complexes of clay minerals in the electron microscope, a film-sealed high resolution environmental cell has been developed. By observation in both wet air and in vacuum, the structural changes were revealed that occur when hydrazine complexes of kaolin minerals and hydrated smectites are evacuated. The structure and morphology of halloysite particles intercalated with hydrazine are quite similar to those of halloysite(10Å) in both tubular and spherical form in a wet air environment. The transformation of a platy kaolin mineral to a tubular one or the reverse transformation does not easily take place by hydrazine treatment alone. Morphological changes of the hydrated form of smectite, especially the curling of the particle, was observed to accompany removal of wet air in the cell by evacuation. Since the environmental cell has a high resolution power, the layer lattice images of hydrated smectite and tubular halloysite(10Å) could be observed.

This environmental cell technique provides some important information in the investigation of clay minerals and in related fields.

INTRODUCTION

To observe electron microscopically the structures of specimens in the presence of water or which contain other volatile components, it is necessary to place the specimens in a special cell which contains water vapor or gas. For this purpose, a film-sealed high resolution environmental cell (abbreviated to E.C.) was developed by one of the authors(A.F.)(Fukami et al., 1972).

Fully hydrated particles of halloysite were first observed using the E.C. by Kohyama et al.(1978a, 1978b). The E.C. enables continuous electron microscopic observation of specimen in different hydration and complexing states under controlled gas and water vapor environments.

In this study, we describe the changes in morphology and layer lattice images revealed by observation in wet air and in vacuum which occur when hydrazine complexes of kaolin minerals as well as hydrated smectite and tubular halloysite are evacuated.

MATERIALS AND METHODS

The following specimens were studied: Dehydrated halloysites of tubular and spherical forms (abbreviated to halloysite(7Å)) from Kusatsu, Gumma Prefecture, and Utsunomiya, Tochigi Prefecture, respectively; kaolinite from the Kanpaku mine, Tochigi Prefecture; dickite from the Tooho mine, Aichi Prefecture; and nacrite from the Shokozan mine, Hiroshima Prefecture, Japan. Halloysite(7Å) and the other kaolin minerals have no interlayer water and are basically non-expandable in water, but they can form interlamellar complexes with some organic compounds such as formamide, urea, hydrazine and others. The kaolin minerals were separated by a centrifugation technique to prepare the <2μm size fraction and treated in 40% of hydrazine water solution at 60°C for 6 days to prepare interlamellar complexes(Weiss et al., 1963) By this treatment, the basal spacing of about 7.2Å increased to about 10.4Å. The spacing was reduced to 7.4Å by vacuum drying. Hydrated halloysite (abbreviated to halloysite(10Å)) of tubular form from Algeria, Africa was also used for the observation of the layer lattice image in the E.C.. Smectites came from Wyoming, U.S.A. and from the Seta mine, Hokkaido, Japan.

The film-sealed E.C. is constructed by a layering of gaskets, coppe disks with apertures and a specially prepared sealing film and spacer. The principal specification of the E.C. are as follows: 1) Acceleratin voltage: 100 to 200kV, 2) gas in the E.C.: saturated water vapor with carrier gas(air) of 50-80 Torr, 3)thickness of the gas layer: 50μm, 4) sealing film: evaporated carbon film 20nm thick with plastic micro-grid', 5) transmittance of electron beam: 60% at 100kV, and 6) resolvin power: 1 nm. Specimens to be observed are placed on the underside of the sealing film. In order to observe wet specimen in a wet gas enviro ment, a water saturated carrier gas (air) from a circulation system outside the microscope column is supplied into the E.C. under controll pressure about 50-60 Torr. The gas in the E.C. can be evacuated through piping of the gas circulation system to observe the specimen under vacuum. Accordingly, in-situ observation in various environments from a wet gas atmosphere to vacuum can be achieved in the cell.

The electron microscope employed in this study was the JEM 7A model but the lattice images of the hydrated smectite and halloysite(10Å) were taken with the JEM 200CX model.

RESULTS AND DISCUSSION

Tubular and spherical halloysite intercalated with hydrazine

The results of the present study for tubular halloysite are shown in Fig. 1. A and B in Fig.1 show the structural change of halloysite particles by dehydration which was reported previously(Kohyama et al., 1978a). C shows the interlamellar complex of tubular halloysite with hydrazine in wet air. After hydrazine treatment, the stripes observed in the tubes of halloysite(7Å) disappeared and changed into smooth structure in the wet air environment. The structure resembles that of halloysite(10Å). But peculiar contrasts such as stripes appeared again in the tubes in vacuum (Fig. 1D). The corresponding selected area electron diffraction (SAD) patterns shows that the halloysite treated with hydrazine is fully intercalated in wet air and completely de-

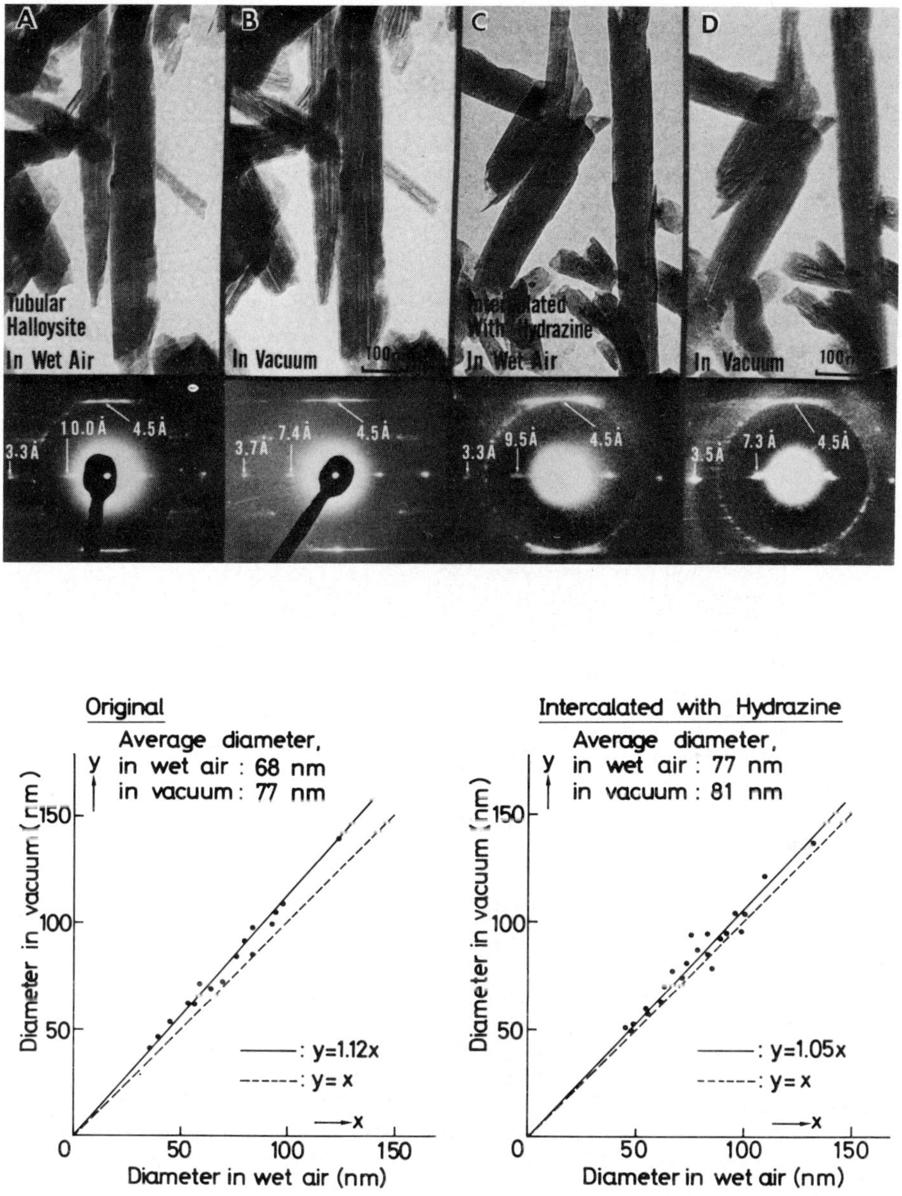

Fig. 1. Observation of tubular halloysite in the E.C.
A: Hydrated state(10Å phase) observed in water saturated air at 60 Torr, B: Dehydrated state(7Å phase) observed in vacuum, C: Intercalated state with hydrazine(10Å phase) observed in water saturated air at 60 Torr, D: De-intercalated state(7Å phase) observed in vacuum. The lower two figures show the changes of halloysite(10Å) tube diameters and those of lamellar complex with hydrazine in both environments. Accelerated voltage: 100kV.

intercalated in vacuum. The average diameter of halloysite(7Å) tubes (77nm) was not affected by hydrazine treatment but it increased slightly to 81nm when de-intercalated in vacuum as shown on the lower right of Fig. 1.

Fig. 2 summarizes the structural change of tubular halloysite as shown by dynamic observation in the E.C. The bottom illustrates the

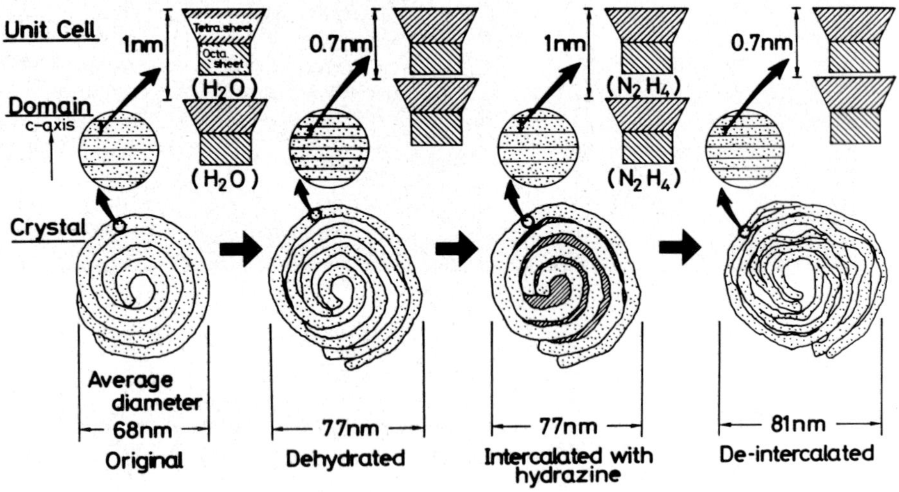

Fig. 2. Summaries of structural and morphological changes of tubular halloysite observed in the E.C.

cross-sectional view of halloysite tubes. Each spiral-like plane means a packet of layers equal to a domain. Each domain is enlarged in the circle. One domain consists of several kaolinite layers. Accompanying the de-intercalation of hydrazine, the thickness of the layers (and also of the domain) decreases about 30%. The average diameter, on the contrary, increases slightly as shown in the pair at the right hand side in Fig. 2. This behavior is quite similar to the trend observed for the untreated halloysite particles and may be explained by the appearance of many gaps in the tubes by removal of hydrazine and the expansion of the rolled configuration caused by relaxation of inherent strain accompanying the segregation of each domain.

Fig. 3 shows analogous observations for spherical halloysite. A and B in Fig. 3 represent the structural changes of spherical halloysite particles by dehydration reported previously (Kohyama et al., 1978b). C and D in Fig. 3 show the particles after hydrazine treatment As was the case for tubular halloysite, a smooth contrast without

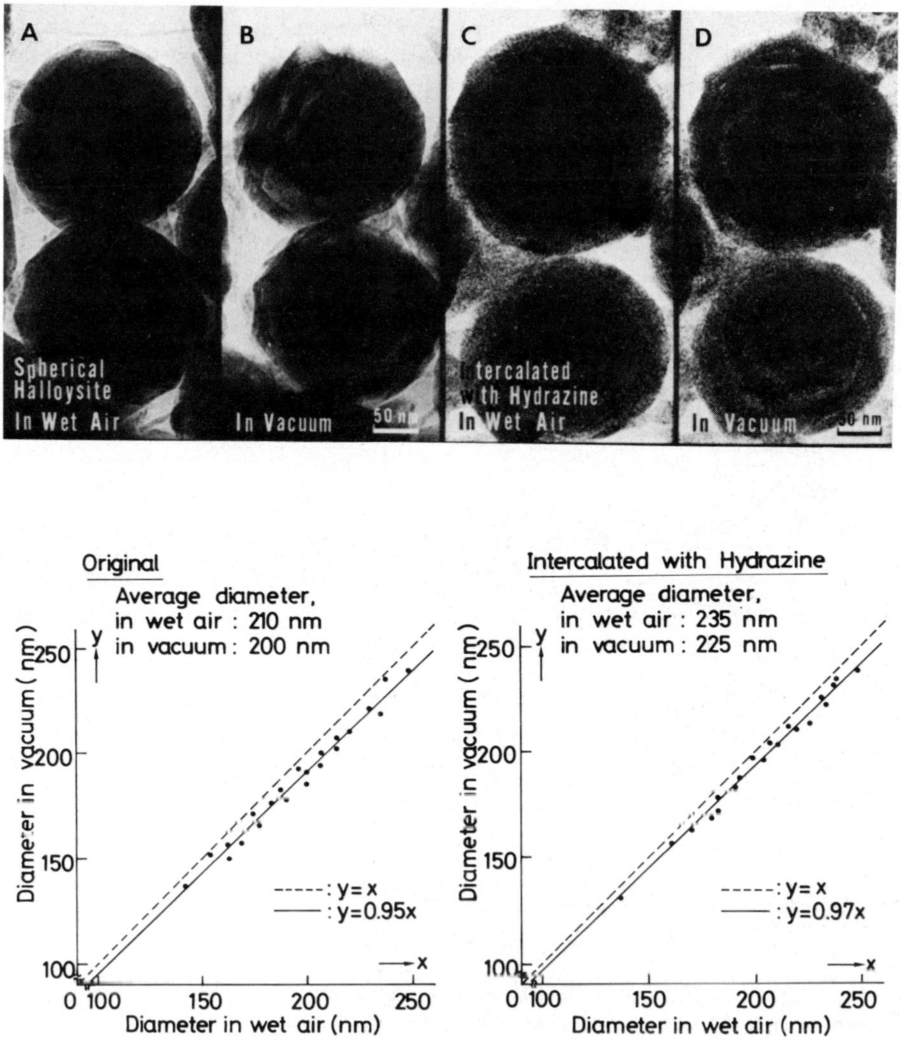

Fig. 3. Observation of spherical halloysite in the E.C.
A: Hydrated state(10Å phase) observed in water saturated air at 60 Torr,
B: Dehydrated state(7Å phase) observed in vacuum, C: Intercalated state with hydrazine(10Å phase) observed in water saturated air at 60 Torr,
D: De-intercalated state of C(7Å phase) observed in vacuum. The lower two figures show the changes of halloysite(10Å) sphere diameter and those of the interlamellar complex with hydrazine in both environments. Accelerated voltage:100kV.

structure was observed in the spherical halloysite intercalated with hydrazine and in the hydrated state when observed in a wet air environment(A and C), but a peculiar contrast appeared in the particles showing a zonal structure when they were observed in vacuum(B andD). After hydrazine treatment, the average diameter in vacuum was slightly smaller than that in wet air as shown at the lower right of Fig. 3.

The structural changes of sperical halloysite are similarly summarized in Fig. 4. The features of the layer and domain level are the

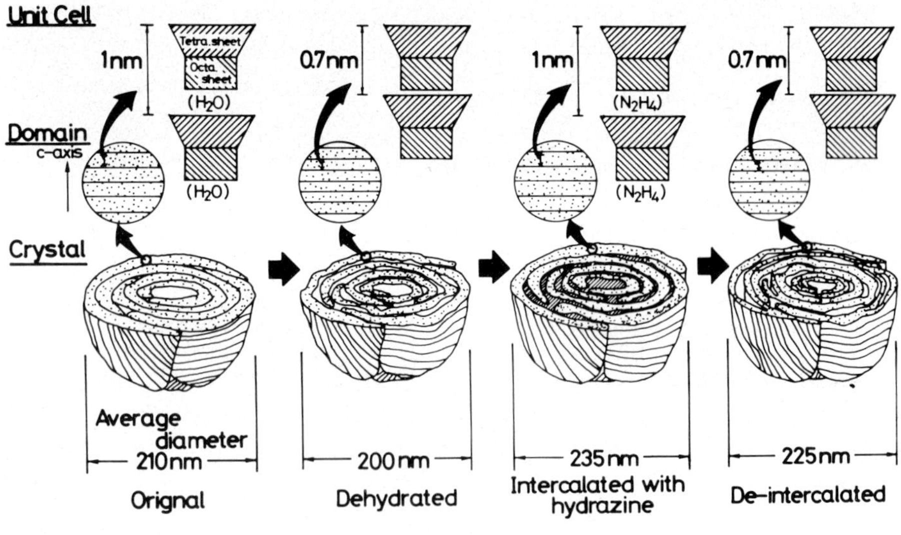

Fig. 4. Summaries of structural and morphological changes of spherical halloysite observed in the E.C.

same as that of tubular halloysite. The particles of spherical halloysite consist of many domains and the structure looks like that of a cabbage or an onion according to previous studies(Dixon et al., 1974; Kirhman, 1977; and Sudo et al, 1977) as well as present observation. The cross-sectional view of the particles is illustrated at the bottom of Fig. 4. The layers and also the domains are shrunk by dehydration or removal of intercalated hydrazine by about 30%, but the average diameter decreases only by about 5% in both cases. This discrepancy can again be explained by the appearance of gaps or interior-spaces in the particles accompanying dehydration or de-intercalation. The reductuion of the diameter of the spherical particles after dehydration or de-intercalation and the opposite effect for tubular halloysite may be the result of the structural differendes between tubular and spherical particles.

Platy kaolin minerals intercalated with hydrazine

The E.C. observation shows that the particle of kaolinite, dickite and nacrite complexes with hydrazine remain in the original platy form in a wet air environment. The preservation of the hydrazine complex in a wet air environment was confirmed by occasional observation of the basal spacing of 10A in the SAD pattern at the curled edge of a platy particle. The transformation of kaolin minerals from platy to tubular forms or the reverse is an interesting problem. From the present study, it appears to be difficult, if at all possible, to transform a platy kaolin mineral to a tubular one by hydrazine treatment alone.

Hydrated smectite

Hitherto the crystalline structure and morphology of smectites in the hydrated state have not been observed electron microscopically because the interlayer water is easily dehydrated upon evacuation. Wyoming montmorillonite shows a typical aggregate structure under a conventional electron microscope. In these aggregates, thin but wide montmorillonite particles are mostly irregularly folded and have a foliated appearance.

When a smectite specimen with a thin water layer is carefully transfered to the E.C., flat and unfolded lamellar particles are observed in a wet air environment (Fig. 5A). Upon evacuation, the flat edges of the lamellar particles gradually curled up and any curly edges that existed in a wet environment curled more and more as shown in Fi. 5B.

Fig. 6 shows basal spacings in wet air and in vacuum in SAD patterns of the smectite observed at the curled edges together with the corresponding patterns of X-ray powder diffraction. The basal spacing observed in the SAD patterns changed from 17Å to 10Å during removal of interlayer water by evacuation. As the interlayer water is sensitive to the environment, the partial or full dehydration easily occurs through evacuation of wet air in the E.C.. SAD patterns of (hk0) in Fig. 6, so-called "net patterns", were not significantly affected by the difference of environmental conditions. The dehydrated smectite could be hydrated again within 30 min. by introducing air saturated with water vapor in the E.C. as shown by the increase of the basal spacing to 16Å in the SAD pattern and by the increase of the thickness of curled edges.

Direct observation of layer lattice images of hydrated smectite and halloysite

The basal diffraction spot of about 15Å and that of the higher order were detected for the hydrated smectite. Since the E.C. has high resolution power, we tried direct observation of the lattice images using the E.C. at an accelerating voltage of 100kV.

The smectite specimen studied was a Ca-beidellite from the Seta mine, Hokkaido, Japan. The layer lattice image of the clay in the hydrated state was observed at the curled edges in a wet air environment in the E.C. as shown in Fig. 7A. The optical diffraction pattern clearly

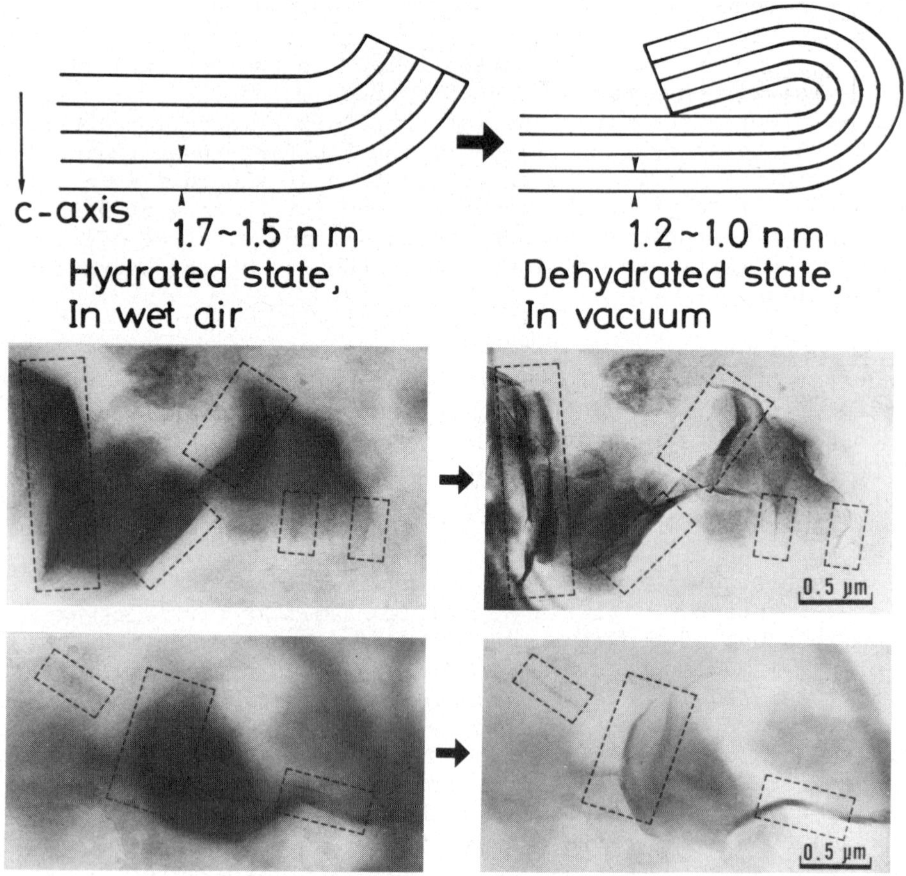

Fig. 5. Curling of smectite particles due to dehydration.

shows a basal spacing of 15.6Å. In vacuum, partial dehydration occurred resulting in a basal spacing of 12.7Å as shown in Fig. 7B. Moreover, a defect of lattice such as a dislocation is seen in the both environments (arrowhead). This means the crystallinity was not so affected by the dehydration process.

The layer lattice image of tubular halloysite(10Å) was also observed in a wet air environment in the E.C.(Fig. 8). The lattice image of 10Å was clearly seen parallel to the tube direction. The area around the centre of the capillary is somewhat dark in contrast. This might reflect the presence of adsorbed water in the capillary in a wet air environment. In vacuum, the dark contrast disappeated and the central capillary was clearly seen in the tube.

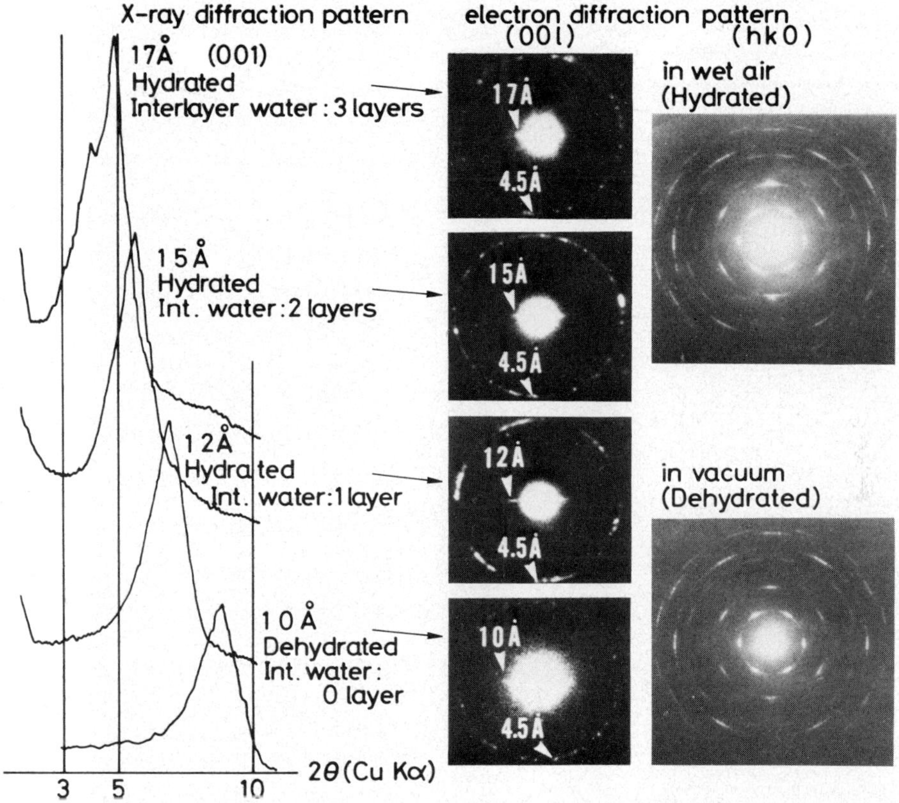

Fig. 6. SAD patterns of various stages of hydration of smectite together with the corresponding X-ray diffraction patterns.

CONCLUSIONS

Through the dynamic observation using the E.C. of halloysite particles, the morphological similarity between the hydrated and the hydrazine intercalated states was clearly revealed as well as that between the dehydrated and the de-intercalated states. Curling of smectite particles accompanying dehydration could be shown. Direct observation of layer lattice images was achieved in hydrated smectite and halloysite(10Å).

The application of the E.C. to the study of kaolin minerals complexed with organic compounds is the subject of further work.

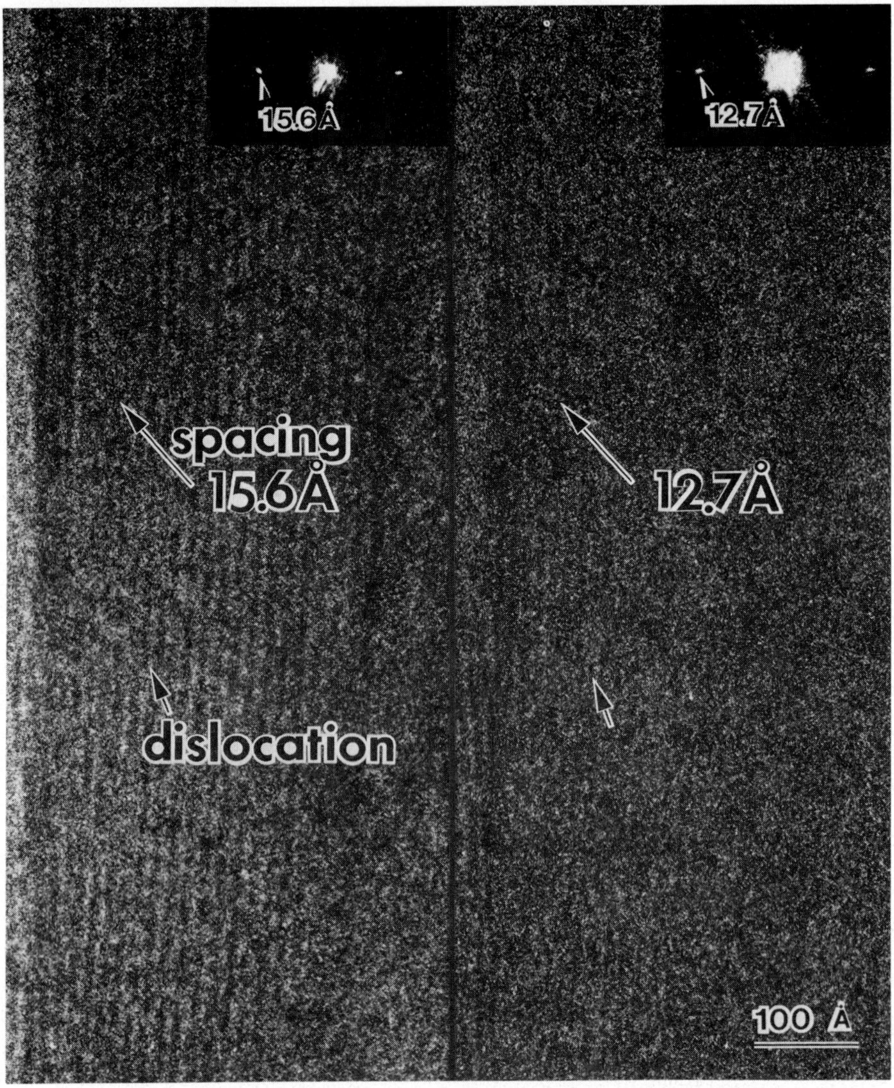

Fig. 7. Observation of layer lattice images of smectite in the E.C. A: In water saturated air at 60 Torr, B: In vacuum after evacuation for 30 minutes. Accelerated voltage: 100kV.

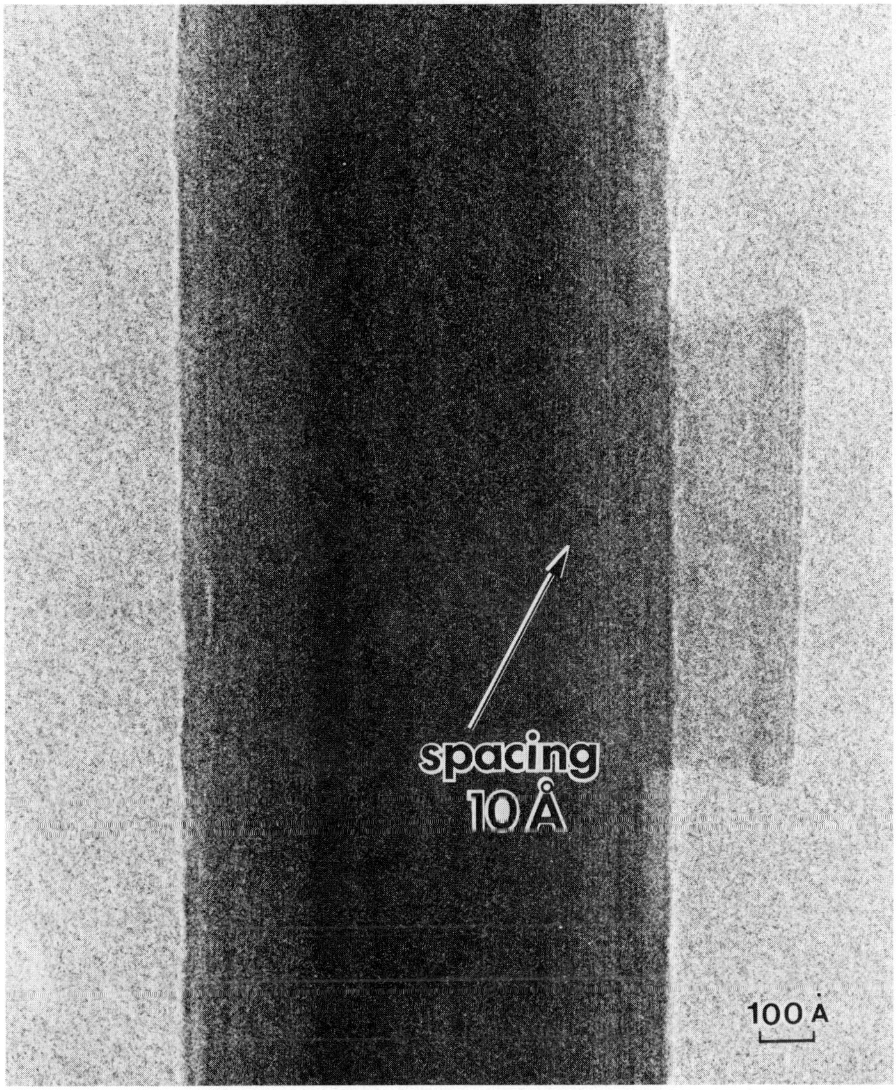

Fig. 8. Observation of layer lattice images of halloysite(10Å) in the E.C. of water saturated air at 60 Torr. Accelerated voltage: 100kV.

ACKNOWLEDGEMENT

We wish thank Dr. H. Yotsumoto and Mr. T. Etoh of JEOL LTD. for permitting us to use the JEM 200CX and for their helpful discussions.

REFERENCES

1 Dixon, J.B. and Mckee, T.R., 1974. Internal and external morphology of tubular and spherical halloysite particles: Clays Clay Min. 22: 127-137.
2 Fukami, A. and Katoh, M., 1972. Construction and application of environmental cell: Proc. 30th EMSA Meeting, 614-615.
3 Kirkman, J.H., 1977. Possible structure of halloysite discs and cylinders observed in some New Zealand rhyolitic tephras: Clay Miner. 12: 199-216.
4 Kohyama, N., Fukushima, K. and Fukami, A., 1978a. Observation of the hydrated form of tubular halloysite by an electron microscope equiped with an environmental cell: Clay Clay Min. 26: 25-40.
5 Kohyama, N., Fukushima, K. and Fukami, A., 1978b. Hydrated form of some clay minerals observed using a film-sealed environmental cell Ninth Int. Congr. Electron Microscopy, Toronto, 72-73.
6 Sudo, T. and Yotsumoto, H., 1977. The formation of halloysite tubes from spherulitic halloysite: Clays Clay Min. 25: 155-159.
7 Weiss, A., Thielepape, W., Ritter, W., Schafer, H. and Goring, G., 1963. Hydrazine kaolinite: Z. anorg. allg. Chem. 320: 183-204.

AMORPHOUS CLAY MINERALS — CHEMICAL COMPOSITION, CRYSTALLINE STATE, SYNTHESIS, AND SURFACE PROPERTIES

KOJI WADA
Faculty of Agriculture, Kyushu University 46, Fukuoka 812, Japan

INTRODUCTION

I thank the Organizing Committee of the VII International Clay Conference for the invitation to present the introductory lecture to the session on "Amorphous and ill-crystallized components of soils and sediments". Amorphous clay minerals have a large specific surface area and develop a variable charge on their surfaces. The presence of these minerals markedly affect physical, chemical, and biological properties of soils and sediments. The proper management and use of those soils and sediments therefore require sound knowledge and understanding on the nature and properties of amorphous clay minerals. During the past decade, there has been a considerable progress in studies on the subject, particularly in relation to soils with variable charge. We can see this in the relevant chapters in "Minerals in Soil Environments" (Editors, J.B. Dixon and S.B. Weed), and in "Soils with Variable Charge" (Editor, B.K.G. Theng). Here I shall attempt to present a summary of the existing information on amorphous clay minerals along four major lines: chemical composition, crystalline state, synthesis, and surface properties, which are heavily relied on the chapters quoted above. For brevity only those references are cited here, which were not included in those chapters.

CHEMICAL COMPOSITION AND NOMENCLATURE

The AIPEA Nomenclature Committee recommended the use of the term "noncrystalline" rather than the commonly used term "amorphous"

(Bailey, 1979). The term amorphous is, however, retained in this lecture as a conventional and inclusive term.

The Committee also recommended that specific names not be given to newly discovered amorphous clay minerals, unless it becomes apparent that particular ranges of chemical composition exist for them (Bailey, 1979). Brown (1955) proposed a scheme of classification and nomenclature of amorphous clay minerals. A renewal of classification and nomenclature in this scheme may be considered in the light of recent progress in characterization of amorphous clay minerals.

Opaline silica — allophane — kliachite (Table 1) Clay size opaline silica has been found in young soils derived from volcanic ash. Silt and clay size, biogenic opal is a minor but ubiquitous constituent of most soils, and opaline microfossils have been recognized in many geological materials. Opal is a hydrated silica (SiO_2 = 85 to 95 %) but contains some Al (Al_2O_3 < 5 %) and Fe (Fe_2O_3 < 2 %).

TABLE 1.
Classification and nomenclature of amorphous clay minerals (I)

SiO_2/Al_2O_3 ratio	Brown (1955)	Present	
∞	Opaline silica	Opaline silica	
2	Allophane	----------	
		↑ Allophane ↓	
1		----------	Imogolite
		↑ Allophane-like constituents ↓	
0.25		----------	
0	Kliachite	?	

Brown (1955) proposed the name allophane as an end member of aluminosilicate with the molar SiO_2/Al_2O_3 ratio of 2.0. Allophanes occur most commonly in soils derived from volcanic ash. Possibly, there are two end members of allophanes having SiO_2/Al_2O_3 molar ratio 1 and 2 respectively and different charge characteristics. Whether an allophane with an intermediate SiO_2/Al_2O_3 ratio represents a mixture of two end members, or a member in a series of minerals with composition of continuous variations is still debated. Imogolite is a newly discovered aluminosilicate with an ideal composition $SiO_2 \cdot Al_2O_3 \cdot 2.5H_2O(+)$. Imogolite commonly has been found in association with allophane, but the two can be most easily distinguished by their morphological characteristics. Aluminosilicates termed allophane-like constituents with SiO_2/Al_2O_3 ratios lower than 1.0 have been distinguished from allophane and imogolite by dissolution upon treatment with dithionite-citrate and hot 2 % Na_2CO_3 aq. Whether they represent a phase separate from, or part of allophane is debated.

Hsu stated in his chapter in "Minerals in Soil Environments" that large amounts of noncrystalline aluminum hydroxides and oxyhydroxides are present in nature, but gave no particular references. So far as I know, however, the presence of a mineral that can be called kliachite has not been confirmed in soils and sediments, presumably due to the high reactivity of noncrystalline aluminum hydroxide. On the other hand, polymeric hydroxy-Al ions are often present in interlayers of expansible layer silicates and in the forms of complexes with humus.

Opaline silica — hisingerite — limonite (Table 2) Brown (1955) used the name hisingerite as a Fe analogue of allophane. However, minerals hitherto called hisingerites may be classified into two groups; one includes poorly ordered and finely crystallized 2:1 layer silicates with high Fe contents (e.g. Kohyama and Sudo, 1976), and another includes Si-bearing ferrihydrite, with SiO_2/Al_2O_3 ratios lower than 0.5 (e.g. Carlson and Schwertmann, 1981). A hisingerite-like mineral reported by Ossaka et al. (1979) in spring sediments may come in this category.

The name limonite was abolished (Rooksby, 1961), because mineral samples hitherto called limonites were found to be finely divided and poorly crystallized goethite, lepidocrocite or mixtures of them.

Ferrihydrite is a new iron oxide mineral, formerly called amorphous ferric hydroxide, with chemical composition $5Fe_2O_3 \cdot 9H_2O$ and occurs in ochreous precipitates from Fe-bearing waters.

TABLE 2.
Classification and nomenclature of amorphous clay minerals (II)

SiO_2/Al_2O_3 ratio	Brown (1955)	Present
∞	Opaline silica	Opaline silica
4		↑
2	Hisingerite	Poorly ordered nontronite, Fe-saponite etc.
1.5		↓
1		↑
∿ 0.5		Si-containing ferrihydrite ↓
0	Limonite	Poorly ordered goethite, lepidocrocite etc.
		Ferrihydrite

Intermediates between kliachite and limonite were not considered in Brown's scheme, but the substitution of Fe by Al has been known very common in iron oxides. This substitution can occur to the extent of 1/3 of the octahedral position in goethite and to the extent of 1/6 in hematite. Crystallinity is improved at low substitution and at high substitution crystal growth is retarded along directions specific to the mineral (Schwertmann, 1979). Fibrous particles of Al-substituted goethite 20 - 50 Å wide and several μm long have been found in soils (Nakai and Yoshinaga, 1980).

CRYSTALLINE STATES

The amorphous state is usually defined as one which lacks order and the crystalline state as one which has order. The two states of matter are also referred to as those having short-range order (non-crystalline state) and those with long-range order (crystalline state). Diffraction characteristics and morphology of amorphous clay minerals has shown that an intermediatory, paracrystalline state should also be included to describe the whole range of crystalline states that exists. Four crystalline states can be distinguished: only short-range order (non-crystalline) is present, or long-range order in one, or two (paracrystalline) or three dimensions (crystalline), combined with the presence or absence of defects and disorders in the existing long-range order. The following are descriptions of the structure of several minerals which exemplify the respective crystalline states.

Noncrystalline state — e.g. opaline silica Opaline silica appears as 0.2 to 4 μm plates of circular, elliptical or other shapes, or 0.1 to 0.8 μm spheres. They consist of a three dimensional framework of SiO tetrahedra with only a short-range order due to atomic coordination.

Paracyrstalline state with defects — e.g. allophane Allophane has been called a noncrystalline aluminosilicate. However, allophane consists of 30 - 50 Å hollow spheres with diameters of 30 to 50 Å and with walls which are probably made up of one Al-octahedral and one Si(Al)-tetrahedral sheet (Wada, 1979). Evidently, an order different from that of opaline silica developes in allophane. The wall structure implies the existence of long-range order in two dimensions, though diffraction effects due to this ordering do not appear because of the form and fineness of allophane particles. It contains defects which allow for water molecules to pass through the wall. Whether or not the wall of allophane spheres is composed of a defect kaolin or imogolite structure is debated.

Paracrystalline state without defect — e.g. imogolite A long-range order without defects exists in imogolite, which enables the formation of tubes a micrometer long with an external diameter of about 20 Å. The basic structure of imogolite is a single gibbsite sheet curved in the form of a tube, with orthosilicate groups attached to the inside of the tube, each group replacing three OH groups around an empty octahedral site. The long and flexible tubes combine to form a thread with somewhat parallel yet imperfect tube alignments.

Crystalline state with defects and disorders — e.g. ferrihydrite Ferrihydrite consists of spheres with diameter of 3 to 10 nm and has a hematite-like structure with a hexagonal unit cell. In its structure, parts of the Fe sites are vacant and O is replaced by OH_2. Ferrihydrite with good crystallinity gives five diffraction bands with hkl indices, while samples of poor crystallinity show only two prismatic diffraction bands (Carlson and Schwertmann, 1981). Chemisorbed silicate suppresses ordering of ferrihydrite and inhibits its transformation to more stable forms, i.e. hematite or goethite. Most ill-crystallized layer silicates and iron oxides also come in this category.

Very finely crystallized layer silicates (e.g. nontronite in Table 2) and iron oxides (e.g. goethite in Table 2) form another category of amorphous minerals. An embryonic form of halloysite formed in paddy soils derived from volcanic ash (Wada, K., Gondo, H. and Wada, S., Unpublished) can be placed in the same category. Its X-ray pattern shows that it is nearly amorphous and DTA shows no well-defined dehydroxylation peak characteristic of halloysite. The infrared spectrum, however, suggests bonding features similar to those in halloysite, and electron micrographs show scrolls with a diameter of 20 to 50 nm and with a lattice image of about 7 Å spacing.

SYNTHESIS AND NATURAL OCCURRENCE

Synthesis of allophane, imogolite and ferrihydrite, and their precursors have thrown light on the genesis and structure of amorphous clay minerals. An important finding on the reaction of aluminum ion and $Si(OH)_4$ in acid solution is the formation of water soluble polymeric hydroxy-Al silicate complexes or cations (Farmer and Fraser, 1979; Wada and Wada, 1980); e.g. $\left[{}^{Al(VI)}_{Al(VI)} > O - Si(OH)_3\right]^+$, where Al(VI) stands for 6-fold coordinated Al. Heating of a solution containing hydroxy-Al silicate ions at 95 to 100°C for 3 to 6 days results in formation of imogolite (Farmer et al., 1977) and allophane (Wada et al., 1979). These products are similar to their natural counterparts in morphology, SiO_2/Al_2O_3 ratio, and infrared spectrum (Fig. 1). Their precursor, the hydroxy-Al silicate ion, has an infrared spectrum similar to that of "allophane-like constituents" and imogolite (Fig. 1).

In alkaline solution, the reaction between aluminate and orthosilicate ions first yields silicatoaluminate ion (Wada and Wada, 1981); e.g. $[(OH)_3Al(IV) - O - SiO(OH)_2]^{2-}$, where Al(IV) stands for 4-fold coordinated Al. Acidifying the solution results in the formation of a product which is similar to natural allophane in both morphology and SiO_2/Al_2O_3 ratio but their infrared spectra are slightly different (Farmer et al., 1979; Wada and Wada, 1981) (Fig. 1). The reaction is interesting in that it may explain why part of Al in allophanes, particularly with high SiO_2/Al_2O_3 ratio, is in 4-fold coordination.

The infrared spectra of all synthetic precursors and products and their natural counterparts show a common absorption band in the region from 800 to 350 cm^{-1} (Fig. 1). A similar, though less well-defined, band also appears in polymeric hydroxy-Al ions. The basic structure common to allophane, imogolite and allophane-like constituents is probably an octahedral Al OH sheet derived from the polymeric hydroxy-Al ions.

Addition of alkali to the solution containing Fe^{3+} and $Si(OH)_4$ yields Si-bearing ferrihydrite (e.g. Carlson and Schwertmann, 1981; Wada, 1981). Orthosilicic acid retards hydrolytic polymerization of Fe^{3+}, but addition of alkali results in formation of polymeric "hydroxy-Fe silicate" complexes or cations (Wada, 1981). X-ray diffraction and infrared spectroscopy of all these products indicated that the common basic structural elements are hydroxy-Fe units and that Si is incorporated by formation of Si-O-Fe bonds.

The occurrence of allophane and imogolite as pedogenic products have been reported not only in Ando Soils derived from volcanic ash and pumice but also in those derived from massive basalts (Hetier et al., 1977; Mizota, 1981), Podzols (Tait et al., 1978; Ross and Kodama, 1979; Farmer et al., 1980; McKeague and Kodama, 1981), and

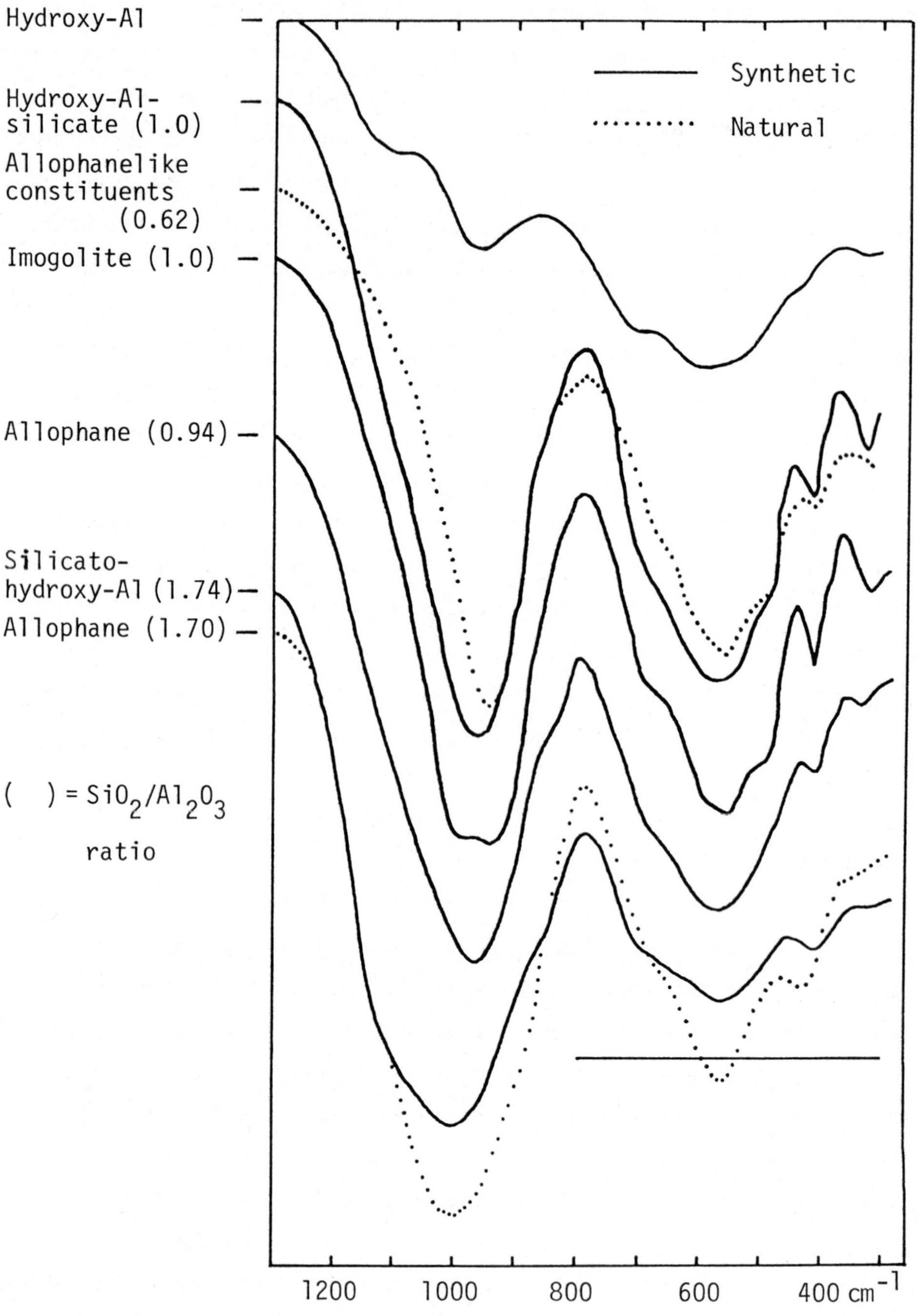

Fig. 1. Infrared spectra of hydroxy-Al, hydroxy-Al silicate, and silicato-hydroxy-Al ions, and allophane-like constituents, imogolite, and allophane (Wada et al., 1979; Wada and Wada, 1980; Wada and Wada, 1981).

Brown Forest Soils derived from glacial till (Tait et al., 1978), all under cool to temperate humid climates. It has been argued that a soluble aluminum silicate complex is the predominant mobile form in which aluminum is transported to B_2 and lower horizons of Podzols (Farmer et al., 1980). Allophane similar to that synthesized from alkaline solution was found in river sediments in a hot spring region (Wells et al., 1977). Well-preserved plant remains, completely replaced by allophane in a scoriaceous volcanic ash deposit have been reported (Tachibana, 1980).

SURFACE PROPERTIES

Amorphous clay minerals are characterized by a large specific surface area with strong chemical reactivity. The specific surface area of allophane spheres with external and internal diameters of 5 and 3.5 nm respectively is 1050 m^2/g and that of imogolite tubes with external and internal diameters of 2.1 and 0.65 nm respectively is 1025 m^2/g, based on a particle density of 2.6 g/cm^3. The calculated areas are comparable with those measured by adsorption of ethylene glycol monoethyl ether, i.e. 700 to 1100 m^2/g for allophane and 1000 to 1100 m^2/g for imogolite (Egashira and Aomine, 1974). A specific surface area of 220 to 560 m^2/g has been reported for natural ferrihydrites (Carlson and Schwertmann, 1981).

Hydroxyl or hydronium groups bonded to only one Si, Al, or Fe atom are abundant on the surfaces of amorphous minerals. They represent broken bonds and give rise to variable electric charge which depends on pH and on the electrolyte concentration in solution. Exchanging cations or anions, if they are heavy metal ions or phosphate, fluoride, arsenate and some organic anions, may become specifically adsorbed (chemisorbed). An intensified interest in soils of variable charge has stimulated studies on charging phenomena and specific adsorption, which have covered a range of soils for a range of conditions.

In those studies, commonly the development of net charge has been measured by potentiometric titration, and the result was compared with the theory. Available information on the development of negative and positive charge has been limited, which is practically more important but requires the direct measurement of adsorption of counter ions. For allophane, however, we can illustrate the development of negative and positive charge per sphere at different pHs and electrolyte concentrations, based on empirically determined CEC and AEC data and their functional relationships with pH and electrolyte concentration (Fig. 2) and surmise probable charging reactions (Table 3).

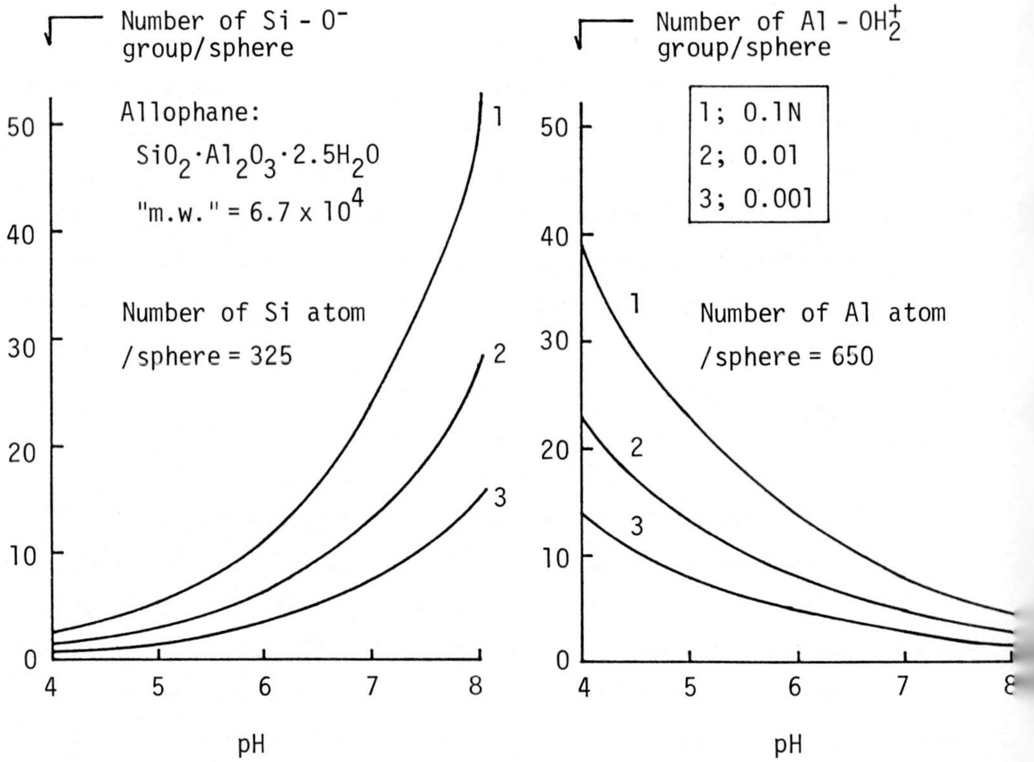

Fig. 2. Development of negative and positive charge per allophane sphere: based on CEC (pH 7.0, $C = 0.01$ N) = 20 me/100g, AEC (pH 5.0, $C = 0.01$ N) = 20 me/100g, $\log \mathrm{CEC} = a\,\mathrm{pH} + b \log C + c$ and $\log \mathrm{AEC} = a'\mathrm{pH} + b'\log C + c'$, where $a = 0.325$, $b = 0.26$, $a' = -0.23$, $b' = 0.225$ (Wada and Okamura, 1977).

TABLE 3

Probable charging reactions for allophane and imogolite

Mineral	Charging reactions at pH 5 - 8
Imogolite & Allophane ($SiO_2/Al_2O_3 = 1$)	$\equiv Si-OH + OH^- \rightarrow \equiv Si-O^- + H_2O$ $\equiv Al\langle{}^{OH}_{H_2O} + H^+ \rightarrow \left[\equiv Al\langle{}^{H_2O}_{H_2O}\right]^+$
Allophane ($SiO_2/Al_2O_3 = 2$)	$\equiv Si-OH + OH^- \rightarrow \equiv Si-O^- + H_2O$ $\equiv Al(IV)-OH_2 + OH^- \rightarrow \equiv Al(IV)-OH^- + H_2O$

Specific adsorption shows a marked hysteresis. The adsorption of heavy metal cations may (iron oxides) or may not (layer silicates, allophane) exceed the CEC, whereas that of phosphate can far exceed the AEC, depending on the phosphate concentration and pH. The adsorption affinity of heavy metal cations varies depending on the nature of adsorption sites (broken bonds > isomorphous replacement), pH (neutral > acid), and on cation species (Cu, Pb > Zn > Co, Cd). Thus, the formation of coordination complexes at broken bond sites, e.g.

$$\begin{matrix} Si(Al,\ Fe) - O \\ Si(Al,\ Fe) - O \end{matrix} \rangle Cu$$

has been suggested as a possible mechanism of the specific adsorption of heavy metal cations. On the other hand, phosphates may react with surface Al or Fe to form complex anions or insoluble salts. In the former case, they will reverse the originally positive surface charge into negative. The reactivity of (Al + Fe) in allophane and imogolite toward phosphate at pH 5.0, as expressed in terms of P/(Al + Fe) ratio was estimated to increase from 0.14 to 0.42 with increasing P concentration from 5×10^{-3} to 1.5 M (Gunjigake and Wada, 1981). The importance of specific adsorption of phosphate by allophane and imogolite has been recognized in relation to phosphate fixation, particularly in Ando soils. Little attention has been paid to the correlation between the specific adsorption and the deficiency and excess of heavy metal cations in soils containing amorphous clay minerals. This together with the possible use of amorphous clay minerals for removal of heavy metals deserve further study from environmental standpoints.

For allophane and imogolite, the location of cation-exchange sites inside the sphere and tube suggests possible exclusion of large organic cations which has been observed for cetylpyridinium, orthophenanthroline, and alkylammonium ions. The phenomenon has practical significance in predicting the effect, toxicity and persistence of herbicide (e.g. paraquat and diquat) and insecticide (caltap) which are applied to the soil in cationic forms.

REFERENCES

Baily, S. W., 1979. Summary of recommendations of AIPEA Nomenclature Committee. Clay Sci., 5: 209-220.

Brown, G., 1955. Report of the clay minerals groups sub-committee on nomenclature of clay minerals. Clay Minerals Bull., 2: 294-302.

Carlson, L. and Schwertmann, U., 1981. Natural ferrihydrites in surface deposits from Finland and their association with silica. Geochim. et Cosmochim. Acta. 45: 421-429.

Dixon, J. B. and Weed, S. B. (Editors), 1977. Minerals in Soil Environments. Soil Sci. Soc. Amer. Madison, Wisconsin.

Hsu, P. H., Aluminum oxides and oxyhydroxides; Schwertmann, U. and Taylor, R. M., Iron oxides; Wada, K., Allophane and imogolite; Wilding, L. P., Smeck, N. E. and Drees, L. E., Silica in soils: Quartz, Cristobalite, Tridymite, and Opal.

Egashira, K. and Aomine, S. 1974. Effects of drying and heating on the surface area of allophane and imogolite. Clay Sci., 4: 231-242.

Farmer V. C. and Fraser, A. R., 1979. Synthetic imogolite, a tubular hydroxyaluminum silicate. Proc. Int. Clay Conf. Oxford, 547-553.

Farmer, V. C., Fraser, A. R. and Tait, J. M., 1977. Synthesis of imogolite. J. Chem. Soc., Chem. Comm., 13: 462-463.

Farmer, V. C., Fraser, A. R. and Tait, J. J., 1979. Characterization of the chemical structure of natural and synthetic aluminosilicate gels and sols by infrared spectroscopy. Geochim. Cosmochim. Acta, 43: 1417-1420.

Farmer, V. C., Russell, J. D. and Berrow, M. C., 1980. Imogolite and proto-imogolite allophane in spodic horizons: Evidence for a mobile aluminum silicate complex in Podzol formation. J. Soil Sci., 31: 673-684.

Gunjigake, N. and Wada, K., 1981. Effects of phosphorus concentration and pH on phosphate retention by active aluminum and iron of Ando soils. Soil Sci., 132: 347-352.

Hetier, J. M., Yoshinaga, N., and Weber, F., 1977. Formation of clay minerals in Andosols under temperate climate. Clay Min., 12: 299-307.

Kohyama, N. and Sudo, T., 1976. Hisingerite occurring as a weathering product of iron-rich saponite. Clays Clay Min., 23: 215-218.

McKeague, J. A. and Kodama, H., 1981. Imogolite in cemented horizons of some British Columbia soils. Geoderma, 25: 189-197.

Mizota, C., 1981. Clay mineralogy of seven Dystrandepts developed from basalts, Northland, French Massif Central and Western Oregon. Soil Sci. Plant Nutr., In press.

Nakai, M. and Yoshinaga, N., 1980. Fibrous goethite in some soils from Japan and Scotland. Geoderma, 24: 143-158.

Ossaka, J., Hirabayashi, J. and Imaeda, M., 1979. Synthesis of hisingerite-like minerals. Abst. Pap. Annu. Meet. Clay Sci. Soc. Japan., pp.30 (in Japanese).

Rooksby, H. P., 1961. Oxides and hydroxides of aluminum and iron. In: G. Brown (Editor), "The X-ray Identification and Crystal Structures of Clay Minerals", pp.354-492. Mineral Soc. London.

Ross, G. J. and Kodama, H., 1979. Evidence for imogolite in Canadian soils. Clays Clay Min., 27: 297-300.

Schwertmann, U., 1979. Noncrystalline and accessory minerals. Proc. Int. Clay Conf. Oxford, 491-499.

Tachibana, K., 1980. Plant remains replaced by clay mineral (allophane) in the volcnaic ash beds — allophanization —. Chigaku Kenkyu 31:

481-490. (in Japanese, with English abstract).

Tait, J. M., Yoshinaga, N., and Mitchell, B. D., 1978. The occurrence of imogolite in some Scottish soils. Soil Sci. Plant Nutr., 24: 141-151.

Theng, B. K. G., (Editor), 1980. Soils with Variable Charge. New Zealand Soc. Soil Sci., Lower Hutt.

Bowden, J. W., Posner, A. M. and Quirk, J. P., Adsorption and charging phenomena in variable charge soils; Parfitt, R. L., Chemical properties of variable charge soils; Wada, K. Mineralogical characteristics of Andisols.

Wada, K., 1979. Structural formulas of allophanes. Proc. Int. Clay Conf. Oxford, 537-545.

Wada, K. and Okamura, Y., 1977. Measurement of exchange capacities and hydrolysis as means of characterizing cation and anion retentions by soils. Proc. Int. Seminar on Soil Environment and Fertility Management in Intensive Agriculture, 811-815.

Wada, S.-I., 1981. Effects of orthosilicic acid on hydrolysis of ferric ion. Abst. Pap. 1981 Meeting. Soc. Sci. Soil Manure Japan. 27: 24.

Wada, S.-I. and Wada, K., 1980. Formation, Composition and structure of hydroxy-aluminosilicate ions. J. Soil Sci., 31: 457-467.

Wada, S.-I. and Wada, K., 1981. Reactions between aluminate ions and orthosilicic acid in dilute alkaline to neutral solutions. Soil Sci., 132: 267-273.

Wada, S.-I., Eto, A. and Wada, K., 1979. Synthetic allophane and imogolite, J. Soil Sci., 30: 347-355

Wells, N., Childs, C. W. and Downes, C. J., 1977. Silica springs, Tongariro National Park, New Zealand — analysis of the spring water and characterization of the alumino-silicate deposit. Geochim. Cosmochim. Acta, 41: 1497-1506.

A COMPARATIVE STUDY OF SELECTIVE CHEMICAL DISSOLUTION METHODS FOR SEPARATING NON-CRYSTALLINE COMPONENTS PRODUCED BY GRINDING OF SILICATES

H. KODAMA and M. JAAKKIMAINEN

Chemistry and Biology Research Institute,
Agriculture Canada, Ottawa, Ontario K1A 0C6

ABSTRACT

Four different chemical dissolution methods were examined for removing non-crystalline components from crystalline silicates. The methods examined included treatments with 0.5 N NaOH to extract Si and Al, with dithionite-citrate-bicarbonate to extract Fe and Al, with alkaline Tiron solution to remove Si, Al, Fe and Mg, and with 5% Na_2CO_3 to extract Si and Al. Extraction experiments were carried out on partially non-crystalline specimens prepared by mechanical grinding of six different silicates.

For ground kaolinite, the NaOH, Alkali-Tiron, and Na_2CO_3 treatments gave extracts with Al/Si ratios close to unity, indicating little selectivity with regard to the two elements by the three methods. The extraction capacity decreased in the following order: NaOH > Alkali-Tiron > Na_2CO_3. For ground mafic minerals such as Fe-clinochlore, biotite and hornblende, the Alkali-Tiron treatment was superior to the other three treatments for extracting Si, Al, Fe and Mg, except for extracting Si from biotite. For biotite, the Alkali-Tiron method yielded selective extractions of Al, Fe and Mg. A strong selectivity for Fe was always observed with the dithionite-citrate-bicarbonate treatment. For ground alkali feldspars, the NaOH treatment showed no selectivity with regard to Si and Al, but had a superior capacity for extracting the two elements compared to the Alkali-Tiron treatment. The latter treatment exhibited a preference for Al over Si.

From these analyses, it is evident that the chemical compositions of non-crystalline components affect the extraction capacity and the cation selectivity of each of these chemical dissolution methods. This should be borne in mind when one derives compositions of non-crystalline components from extraction data.

INTRODUCTION

Non-crystalline hydrous oxides and silicates are important components that influence the physical, chemical and biological activities of soils. Selective chemical dissolution methods have been widely used in the study of non-crystalline components in soils and sediments (see for example Schwertmann 1959; Hashimoto and Jackson 1960; Mehra and Jackson 1960; Mitchell et al. 1964; Schwertmann 1964; Segalen 1968 and many others). Little is still known about the relation between the composition of the substances and the effectiveness and selectivity of extraction procedures, therefore we examined four different chemical dissolution methods for six compositionally different silicates which were mechanically ground to prepare partially non-crystalline specimens. We emphasize that these mechanically ground silicates generally cannot be equated with many of the non-crystalline components that influence the physical, chemical and biological activities of soils. However, we believe that with the ground silicates valuable information can be obtained on the relation between chemical composition and effectiveness of extraction methods.

MATERIALS AND METHODS

We selected the following six minerals; kaolinite, Fe-clinochlore, biotite, hornblende, albite and microcline. The locality, particle fraction and composition of these minerals are listed in Table 1. Fractionations were done by sedimentation for kaolinite; by breaking into small particles with a blender after cutting with scissors and then wet-sieving for biotite; by crushing, breaking into small particles with a blender and wet-sieving for Fe-clinochlore, and by crusing and wet-sieving for hornblende, albite and microcline. Subsequently the samples were ground in a mechanical mortar for various periods to produce specimens containing various amounts of non-crystalline components. The amounts of non-crystalline components were expressed in terms of "relative crystallinity" defined by the ratio of peak intensities of a characteristic diffraction line before and after grinding recorded under identical operational conditions. For these specimens Table 2 shows grinding time, "relative degree of crystallinity" and sample codes. The specimens were dried overnight in an oven at 110°C before use.

The following four dissolution methods were employed: treatments with 0.5 N NaOH (Hashimoto & Jackson 1960) to extract Si and Al, with dithionite-citrate-bicarbonate (DCB method, Mehra & Jackson 1960) to remove Fe and some Al and Si, with alkaline Tiron solution (Biermans and Baert 1977) to extract Si, Al, Fe and Mg, and with 5% Na_2CO_3 (Follett et al, 1965) to extract Si and Al. Extraction is greatly affected by the actual procedures for these methods and since some of these methods were modified, a detailed description of each procedure will be provided by the authors on request.

Elemental analyses were based on reference standards prepared in solution obtained under conditions equivalent to those of the corresponding dissolution method. Si was determined colorimetrically (Weaver et al. 1968), Al, Fe and Mg by atomic absorption spectrophotometric method.

RESULTS

Kaolinite

The NaOH, Alkali-Tiron and Na_2CO_3 methods were examined on this composition. For the K-0 sample, the NaOH and Na_2CO_3 methods showed slight selectivity for Si, while the Alkali-Tiron method showed no selectivity. For all ground specimens, the NaOH method gave extracts with an Al/Si ratio close to unity as is expected from the original kaolinite composition (Table 1 and Fig. 1). The extraction capacity of the NaOH method was always higher than that of the other two methods. As compared with the extraction capacity of the NaOH method on the K-144 sample, the extracting efficiency was reduced to 75% with the Alkali-Tiron method and to 50% with the Na_2CO_3 method. Also, no specific selectivity was indicated in the latter two methods.

Fe-Clinochlore

The Alkali-Tiron method gave the most effective extraction performance of all methods with regard to its capacity and selectivity. With the C-132 sample, the capacity reached more than 80% for all four elements. Since one mg of the C-0 sample contains 140.89 μg Si, 88.60 μg Al, 72.40 μg Fe and 185.94 μg Mg, the Al/Si, Mg/Si and Fe/Si ratios are 0.63, 1.32 and 0.51, respectively. The ratios obtained for the C-132 sample were 0.71, 1.49 and 0.55 respectively, and are close to the theoretical ones. These ratios were higher for the C-60 sample and much higher for the C-24 sample, indicating an extraction selectivity for Al, Mg and Fe over Si. The NaOH method showed a much lower extraction capacity for Si and Al (Fig. 1), although the Al/Si ratio showed a relatively small change from 0.84 in the C-24 sample to 0.72 in the C-132 sample. The extraction capacity of the Na_2CO_3 method for Si was similar to that of the NaOH method. The Al/Si ratio of the Na_2CO_3 method, however was 0.40 with the C-132 sample and lower than the theoretical value of 0.63, indicating selectivity for Si with the Na_2CO_3 method. Surprisingly, the Fe extraction capacity of the DCB method was lower than that of the Alkali-Tiron method, although the DCB method showed a strong selectivity for Fe (Fig. 1). With the C-0 sample again, there were no consistencies in extraction ratios, suggesting a strong extrinsic effect on the ratios.

Biotite

The largest extraction capacity was obtained by the Alkali-Tiron method. The method, however, showed a strong selectivity for Al, Mg, Fe over Si which is illustrated by the ratios obtained with the B-180 sample (Fig. 1 and Table 1). The Al/Si, Mg/Si and Fe/Si ratios were 0.72, 0.92 and 1.69, respectively, and these were nearly twice as high as the respective ratios in the unground biotite. With respect to Si extraction, the NaOH and Na_2CO_3 methods showed as much efficiency as the Alkali-Tiron method. Furthermore, both methods showed no selectivity for Al over Si. Contrary to the experience with chlorite, the Fe extraction capacity of the DCB method was almost the same as that of the Alkali-Tiron method. Regardless of length of grinding time, Al/Si, Mg/Si and Fe/Si ratios changed little with the Alkali-Tiron method. Similarly Al/Si ratios remained unchanged with the NaOH and Na_2CO_3 methods.

Hornblende

In view of the overall extraction performance of the four elements in hornblende (Fig. 2), the Alkali-Tiron treatment extracted the largest amounts. With the H-150 sample, the method gave ratios of 0.26 for Al/Si, 0.41 for Mg/Si and 0.66 for Fe/Si. These ratios were close to the theoretical values of 0.27, 0.38 and 0.61, respectively. For other ground samples, the Alkali-Tiron method also gave similar ratios. The extractions of Fe by the DCB method was nearly identical to that of the Alkali-Tiron method. The NaOH method was rather ineffective for Al and Si extraction of the hornblende sample.

Albite and Microcline

The Alkali-Tiron and NaOH methods were applied to extract Al and Si from these two minerals. With either mineral, the NaOH method was superior in extraction capacity for both elements. Little selectivity was observed in NaOH extraction as can be seen from the Al/Si ratios of 0.35 for the A-152 sample and the ratio of 0.37 for the M-160 sample. Theoretical ratios for the A-0 and M-0 samples, are 0.31 and 0.32, respectively. Unlike the NaOH method, the Alkali-Tiron method showed a strong selectivity for Al over Si in both mineral compositions, as indicated by Al/Si ratios of 1.29 for the A-152 sample and of 1.39 for the M-160 sample.

DISCUSSION

For the extraction of Si, the NaOH and Alkali-Tiron methods were effective. The NaOH method was especially good when it was applied to minerals of Si-Al and Si-Al-K(Na) compositions. The Alkali-Tiron method appeared to be effective in extracting Si, Al, Mg and Fe from mafic minerals containing little alkali elements such as chlorite and hornblende. For minerals containing substantial amounts of alkali elements, however, the method showed selectivity for elements other than Si. The Na_2CO_3 method was also useful for extracting Si and Al, but its extracting capacity was generally moderate compared to that of the NaOH and Alkali-Tiron methods. Concerning extraction of Fe, the Alakli-Tiron method showed its superiority over the DCB method in extraction experiments on chlorite, biotite and hornblende. This may indicate that Fe in the non-crystalline components is still bound to residual silicate structures from which the Fe is not readily extractable by the DCB method. The effective simultaneous extraction of other elements, such as Al, Mg and in some cases Si, may also contribute to the extraction efficiency of Alkali-Tiron for iron in these minerals.

Although in some cases two or more methods gave similar values for extraction capacity, there is no basis for distinguishing between crystalline and non-crystalline phases. The chemical dissolution methods are basically operational for substances in solid state. From the standpoint of quantification of non-crystalline fractions, therefore, the methods showing high extraction capacity are not always good for this purpose. They may be too drastic and destroy some of the crystalline fraction. We can also view this problem in relation to the definition of the non-crystalline phase. Variations in extraction capacity of different methods may be considered in terms of the concept of a continuum between non-crystalline and crystalline phases. Furthermore, partial structures, such as tetrahedral sheets may persist in the "non-crystalline" portion which influences the extraction selectivity and capacity.

There seems to be no simple solution to the separation and determination of non-crystalline components. Based on the results of the present investigation, however, we may draw the following conclusions:

1) The proportion of elements extracted is influenced by chemical composition of non-crystalline components. In other words, the chemical composition affects the cation selectivity of chemical dissolution methods. These observations should be borne in mind when one derives compositions of non-crystalline components from extraction data.

2) Depending on the type of dissolution method and chemical composition of non-crystalline components, both the extraction capacity and the selectivity vary.

3) It is advisable to compare the results obtained by two or more separate dissolution methods unless the nature of a sequential application of two or more dissolution methods would be known; i.e., the effect of the first method on the second and so on.

4) Extraction experiments on unground minerals gave very inconsistent data on the compositions of the fractions extracted, indicating strong effects of the nature of their external surfaces. Therefore, it is meaningless to use these unground minerals for testing the efficiency of dissolution methods for removing non-crystalline fractions from minerals.

REFERENCES

Biermans, V. and Baert, L., 1977. Selective extraction of the amorphous Al, Fe and Si oxides using an alkaline Tiron solution. Clay Miner. 12: 127-135.

Follett, E.A.C., McHardy, W.J., Mitchell, B.D. and Smith, B.F.L., 1965. Chemical dissolution techniques in the study of soil clays: Part 1 Clay Miner. 6: 23-34.

Hashimoto, I. and Jackson, M.L., 1960. Rapid dissolution of allophane and kaolinite - halloysite after dehydration. Clay Clay Miner. 7: 102-113.

Mehra, P.O. and Jackson, M.L. 1960. Iron oxide removal from soils and clays by a dithionite - citrate system buffered with sodium bicarbonate. Clays Clay Miner. 7: 317-327.

Mitchell, B.D., Farmer, V.C. and McHardy, W.J., 1964. Amorphous inorganic materials in soils. Adv. Agron. 16: 327-373.

Schwertmann, U., 1959. Mineralogische und chemische Unterschungen an Eisenoxyden in Böden und Sedimenten. Neues Jahrbuch für Mineralogie, Abhandlungen 93: 67-86.

Schwertmann, U. 1964. Differenzierung der Eisenoxide des Bödens durch Extraktion mit Ammoniumoxalat - Lösung Z. pfl. Ernähr. Düng. Bodenk. 105: 194-202.

Segalen, P., 1968. Note sur une méthode de détermination des produits minéraux amorphes dans certains sols a hydroxydes tropicaux. Cah. ORSTOM sér Pédol. vol. VI no 1: 105-126.

Weaver, R.M., Syers, J.K. and Jackson, M.L., 1968. Determination of silica in citrate-bicarbonate-dithionite extracts of soils. Soil Sci. Soc. Amer. Proc. 32: 497-501.

Table 1. Chemical composition of six minerals used and their major elemental ratios

	KAOLINITE	Fe-CLINOCHLORE	BIOTITE	HORNBLENDE	ALBITE	MICROCLINE
	MACON GEORGIA	CHESTER VERMONT	BANCROFT ONTARIO	FARADAY ONTARIO	KEYSTONE S. DAKOTA	PERTH ONTARIO
	2-5 μm	<44 μm	2-44 μm	<74 μm	<74 μm	<74 μm
SiO_2	45.08%	30.14%	39.32%	42.10%	69.36%	63.65%
Al_2O_3	39.49	16.74	11.58	9.92	19.28	17.87
Fe_2O_3	0.35	0.56	4.73	2.74	0.43	1.03
FeO		8.81	17.42	13.08		
MnO	tr.	0.03	1.13	0.63	0.01	0.01
MgO	0.03	30.83	14.08	12.44	--	0.01
CaO	0.07	0.01	0.01	9.98	0.26	0.16
Na_2O	0.03	--	0.62	3.32	10.71	2.19
K_2O	0.18	--	9.50	1.23	0.24	13.39
Ign Loss	13.74	11.88	--	0.52	0.39	0.06
		Amounts of major elements (μg/mg sample) and their ratios				
Si	210.72	140.89	183.80	196.79	324.22	297.53
Al	209.00	88.60	61.29	57.50	102.04	94.58
Fe	2.45	72.40	168.49	120.84	3.01	7.20
Mg	0.18	185.94	84.92	75.03	--	0.06
Al/Si	0.99	0.63	0.33	0.27	0.31	0.32
Fe/Si	--	0.51	0.92	0.61	--	--
Mg/Si	--	1.32	0.46	0.38	--	--

Table 2. Mineral specimens prepared for the extraction experiments

Mineral	Duration of grinding (hrs)	Relative degree of crystallinity*	Sample codes
Kaolinite	0	1.00	K-0
	24	0.27	K-24
	72	0.08	K-72
	144	0.00	K-144
Fe-Clinochlore	0	1.00	C-0
	24	0.20	C-24
	60	0.02	C-60
	132	0.00	C-132
Biotite	0	1.00	B-0
	24	0.51	B-24
	84	0.10	B-84
	180	0.05	B-180
Hornblende	0	1.00	H-0
	24	0.53	H-24
	72	0.30	H-72
	150	0.12	H-150
Albite	0	1.00	A-0
	24	0.94	A-24
	80	0.72	A-80
	152	0.53	A-152
Microcline	0	1.00	M-0
	30	0.61	M-30
	70	0.60	M-70
	160	0.36	M-160

* see text

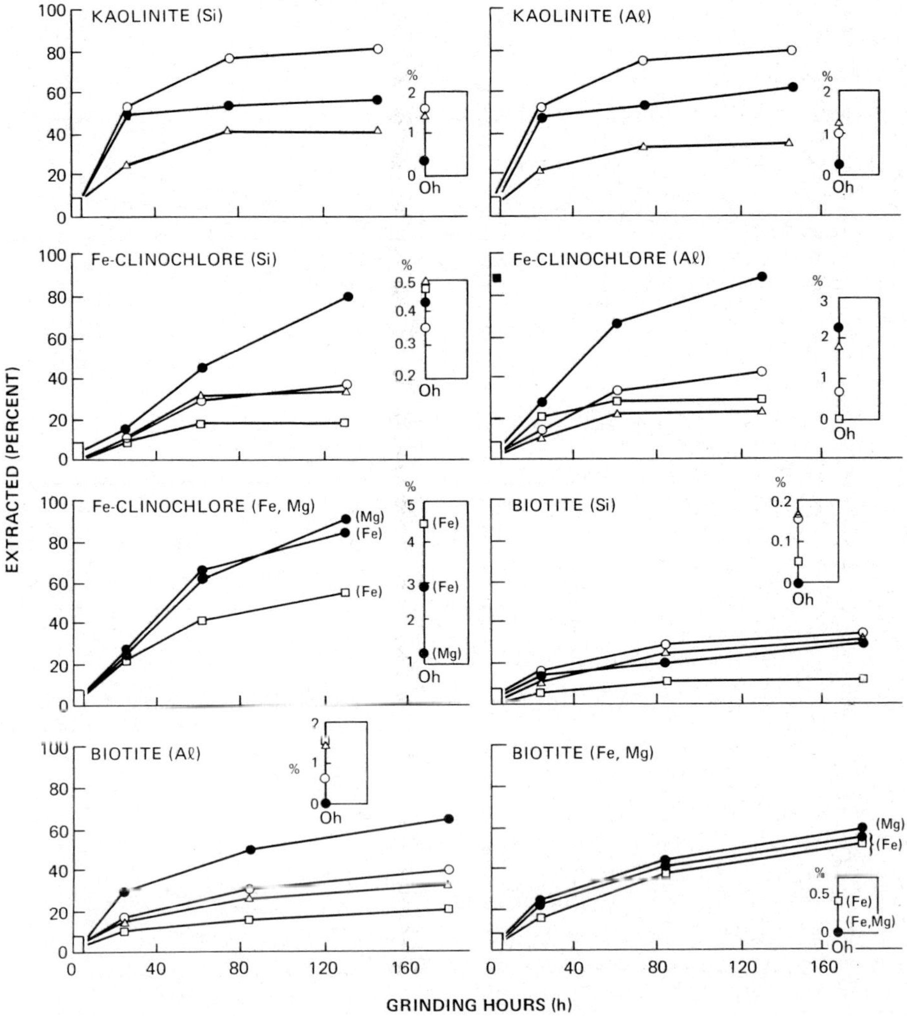

Figure 1. Elements extracted from kaolinite, Fe-clinochlore and biotite by various dissolution methods.
Note: Expressed by (amount of element extracted/total amount of element in original sample) x 100.
○ NaOH method; ● Tiron method; △ Na_2CO_3 method;
□ DCB method.

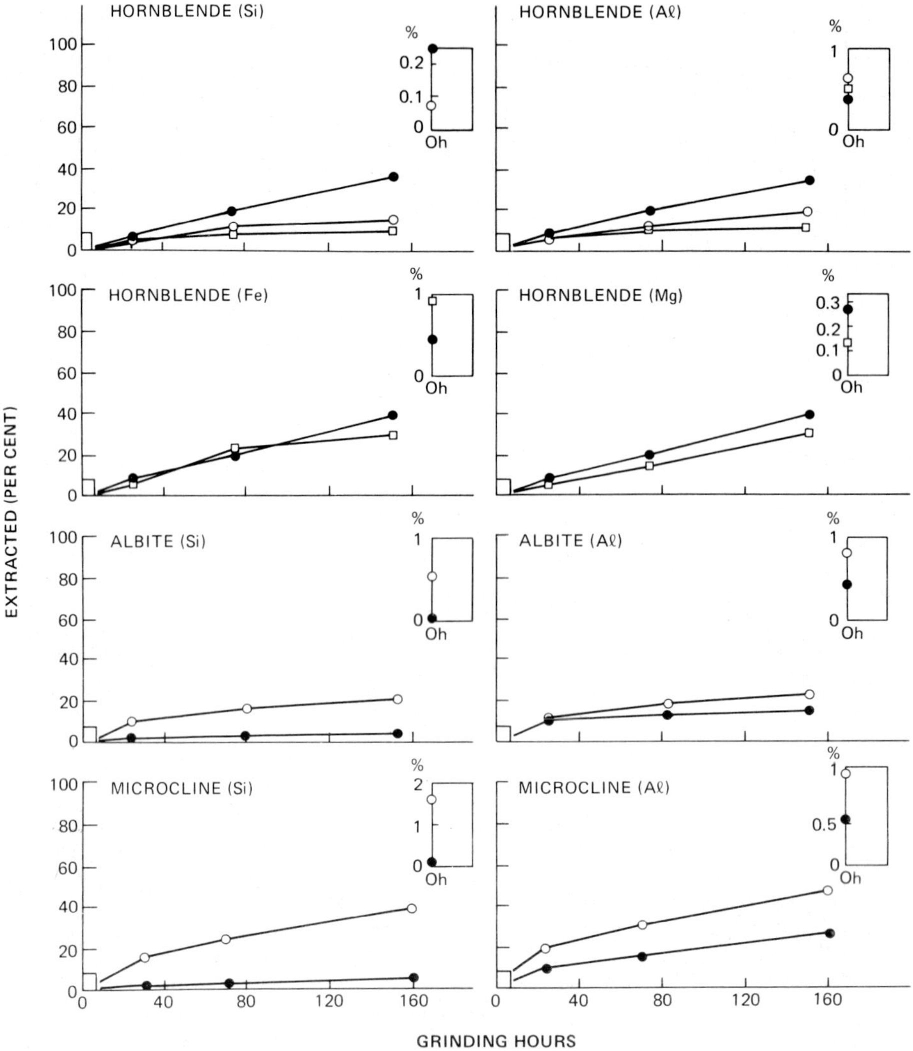

Figure 2. Elements extracted from hornblende, albite and microcline by various dissolution methods.
Note: Expressed by (amount of element extracted/total amount of element in original sample) x 100.
○ NaOH method; ● Tiron method; △ Na_2CO_3 method;
□ DCB method.

ACTION OF AMORPHOUS COMPOUNDS ON CLAY PARTICLE ASSOCIATIONS

M.ROBERT*, J.BERRIER*, G.VENEAU* and M.A.VICENTE**

* I.N.R.A., Station de Science du Sol, 78000 Versailles (France)

**Centro de Edafologia, C.S.I.S., Salamanca (Espana)

ABSTRACT

Using the S.E.M. with a special service and a special sample preparation technique, it can be shown that amorphous Al and Fe compounds in the wet state form a highly hydrated gel with a threedimensional structure which can link rigid clay crystallites (kaolinite).

Both hydration and binding properties are decreased by drying or by increasing the crystallinity of the Al and Fe hydrate.

For smectite, the threedimensional network of the gel is superimposed on that of the smectite and changes its properties more than its appearance. Soluble Al or Fe hydroxy polymers penetrate in the interlayer space inducing the formation of thicker particles resulting in breaking up of the network into micro-aggregates.

INTRODUCTION

Iron or aluminum compounds play a major role in the formation of the microstructure of certain soils where they generally appear to be responsible for clay particle association which may lead to the formation of pseudosands or pseudosilts (Chauvel et al, 1976). However, it is still rather difficult to define the different roles of simple more or less hydroxylated ions, of short-range crystallized compounds (ferrihydrite and cliachite) - usually called non-crystalline or amorphous compounds - and of long-range crystallized compounds such a gibbsite, goethite or hematite (Blackmore, 1973).

Taking advantage of recent development in scanning electron microscopy we studied the bonding of clays in the wet state in the presence of various iron or aluminum compounds synthesized in vitro.

MATERIAL AND METHODS

1. Clay selection

As a representative rigid clay particle we selected Saint-Austell kaolinite particles with a diameter < 1 μm an thickness of a few hundred Å.

Smectite (Greek montmorillonite) of 0.40 charge used mostly in the calcium form saturation, was selected to represent clays which are strongly hydrated and give a three-dimensional structure resulting from edge-to-edge and edge-to-face associations between the particles.

2. Preparation of iron and aluminum compounds

The following compounds were prepared :

- Hydroxylated ferrous ions in the presence of kaolinite or montmorillonite at pH 2.7 (Nalovic, 1977).
- Hydroxylated Al ions in the interlayer spaces of smectite (Shawney, 1968).
- "Amorphous" aluminum compounds, by Kee Kwong and Huang's method (1978) using 10^{-3}N malic acid as a crystallizatic inhibitor.
- "Amorphous" iron compounds using 10^{-3}N citric acid (Cornell and Schwertmann, 1979).
- Gibbsite, goethite and others which were crystallized without organic inhibitor.

The amorphous and crystallized compounds were prepared both in the absence and the presence of clays. The synthesized materials were identified using traditional chemical and mineralogical methods (X-ray diffraction, thermal analyses, extraction by ammonium oxalate) (Landa and Gast, 1973).

3. S.E.M. observation in the wet state

The sample is never dried prior to observation. Its state of hydration is fixed by submitting the wet sample to a given stress in an ultra-filtration device (between pF 1 and pF 4) (Tessier and Berrier, 1979). The sample is then rapidly frozen in freon at the temperature of liquid nitrogen, fractured, mettalized, and observed with the scanning electron microscope using a special device (Cryoscan from Jeol). Traditional observations were also made in the dry state using the S.E.M.

EXPERIMENTAL RESULTS

1. Amorphous Fe and Al compounds in the wet state

The amorphous iron and aluminum compounds in the wet state show a threedimensional network flexible with many ramifications and a high porosity (plate I.1). As soon as larger crystals form, the network becomes coherent and rigid. This is already true for aluminum gels in which in spite of the presence of organic acid gibbsite crystals larger than one micron develop from gel ramifications at some spots (plate I.2 and 3, 4). Developing goethite in the experiment without organic acid forms much smaller crystals (500 Å), but again some collapse is observed (plate I.5,6). Such a modification of amorphous compounds also occurs during desiccation as is clearly seen at pF 4 where the three-dimensional network disappears to become bidimensional (plate II.1).

2. Kaolinite particle association by hydrates

Crystals in Saint-Austell kaolinite are usually separated and in the wet state (pF 1), and this mineral has a homogenous and oriented appearance (Tessier, 1979). The addition of 1 to 5 % iron or aluminum gels has a cementing effect that causes some isotropy. Edge-to-edge, face-to-face or edge-to-face bonds are formed between kaolinite crystallites. The resulting structure is composed of particle packets or threedimensional card-houses with a porosity of several microns (plate II.2, 3). Jones and Uehara, 1973 using transmission microscopy with a special technique could obtain a few illustrations on such associations. However, it is much easier to show the bonding effect of amorphous compounds using S.E.M. on wet state samples. Since in the dry state, these structures collapse completely.

When goethite or gibbsite crystals grow in the systems the bonds between kaolinite particles are more difficult to observe (plate II.4). Two distinct phases usually join and depending on the relative particle size one may observe kaolinite crystals around gibbsite crystals (plate II.5), or small gibbsite crystals on kaolinite crystals (plate II.6

It is slightly more difficult to determine goethite or hematite/kaolinite associations due to the small size of iron hydroxide and oxide crystals. Some micrograph suggest also the two possibilities : deposits on kaolinite, (plate III.1) or independent crystallization.

We note that addition of more or less hydroxylated Fe^{3+} and Al^{3+} cations which induce a flocculation of kaolinite does not lead to the formation of particle association or microstructures that are sufficiently stable to be chara terized under our experimental conditions. Therefore, it is indeed the presence of gel (proportion 1 to 5 %) that cause cementation in microstructures.

3. Action of iron and aluminum compounds on smectites

In the wet state Na or Ca smectite particles are linked by edge-to-edge and edge-to-face bonds in a threedimensional network with high porosity (pore size ≠ 1 μm)(Tessier, 1979)

We first consider treatment with hydroxylated Fe and Al cations :these cations are usually fixed in interlayer spaces and hydroxyferrous or hydroxyaluminous smectites are formed. A major morphological change can be observed : the apparently more rigid network is distorted and breaks (plate III.2, 3). Therefore, there is a tendency to indivualization of particles or particle stacks (micro-aggregation). This is even more marked when the clay is previously subjected to desiccation before observation.

When amorphous or crystallized compounds are prepared in the presence of clay, it is certainly difficult to avoid Fe and Al cation penetration between layers (anti-gibbsite effect) and the above mentioned morphological changes will thus occur. When gels are formed or mixed with smectite, the gel threedimensional network merges with the smectite network and it is very difficult to distinguish between them. Anyway, amorphous compounds will have no effect on particle associations as these particles are already associated within the network. However, in the presence of gel (plate II. 4) the aggregate has a different appearance.

When oxide or hydroxide crystals are formed, these crystals are tightly locked in by the clay network. Hydroxide crystal genesis could actually occur in the pores at the smectite frame, which would account for the round shape of resulting products (plate III.5 and 6).

DISCUSSION

Under wet conditions, amorphous compounds form flexible highly porous threedimensional network which becomes rigid and compact upon desiccation. This change is irreversible : the original structure is not reestablished by rehydration.

This characteristic gel network can establish various bonds between "rigid" clay particles (kaolinite, micas...) by creating microstructures (table 1).

For smectites which associate in a threedimensional network, this network will be superimposed on that of the amorphous gel and the effects are less easily distinguishable. The main effect seems to be a change in properties of the network which tends to break up due to the increased thickness of the smectite particles in the mixed system. This effect can already be observed upon treatment of smectites with Al and Fe soluble compounds penetrating into interlayer spaces, resulting in more rigid micro-aggregates (table 1).

These morphological effects are correlated with measurements of water contents in products formed at different pF. Amorphous compounds have a high hydration capacity and they can fix 10 to 16 times their weight of water at pF 1. Of course, this is not true for crystallized hydrates (1 or 2 times their weight of water) in which gels are necessaril mixed. The presence of gel increases hydration in the kaoli nite system but decreases that in a montmorillonite system. In all cases, gel hydration is irreversible as for allophanes.

"Amorphous compounds" or compounds crystallized at short distance seem to be made up of basic units of 10 to 40 Å. In the wet state these units are relatively free to associate in a loose and flexible threedimensional network. During desiccation (pF 4 and above) or during crystallization these basic units collapse or arrange themselves in an irregular pattern to form larger crystals.

In this respect, a parallel can be drawn between "mineral gels" and organic compounds of high molecular weight such as polysaccharides, fulvic and humic acids, which have a comparable morphology in the wet state as well as identical properties with regard to hydration or clay particle association.

CONCLUSIONS (Table 1)

Our experiments show that the so called amorphous minerals, like organic compounds (Vicente and Robert, 1981), play a major role in hydration and microstructure of soils. One to five per cent amorphous compounds in a soil can totally change its properties.

Most of the effects of these compounds can only be observed in the wet state. Desiccation which is certainly accompanied by an increasing crystallinity and a larger crystal size will induce an irreversible change of the properties. These observations may account for the disagreements regarding the role of such compounds in the literature (Deshpande and al, 1964).

REFERENCES

Blackmore A.V., 1973. Aggregation of clay by the products of iron (III) hydrolysis. Aust. J. Soil Res. 77 : 75-82

Chauvel A., Pédor G., Tessier D., 1976. Rôle du fer dans l'organisation des matériaux kaoliniques. Science du Sol, 2 : 101-113

Cornell R.M. and Schwertmann U., 1979. Influence of organic anions on the crystallization of ferrihydrite. Clays Clay Min., 27 : 402-410

Deshpande T.L., Greenland D.J. and Quirk J.P., 1964. Role of iron oxides in the bonding of soil particles. Nature, 4 : 107-108

El Swaify S.A. and Emerson W.W., 1975. Changes in the physical properties of soil clays due to precipited aluminum and iron hydroxides. Soil Sci. Soc. Amer. Proc., 39 1056-1063

Jones R.C. and Uehara G., 1973. Amorphous coating mineral surfaces. Soil Sci. Soc. Amer. Proc., 37, 792-798

Kee Kwong K.F. and Huang P.M., 1978. Nature of hydrolytic precipitation products of aluminum as influenced by low molecular weight complexing organic acids. Proc. Int. Clay Conf. Oxford, 527-536.

Kaolinite (INDIVIDUAL PARTICULES)		Al or Fe COMPOUNDS		SMECTITE (TACTOIDS)
NO STABLE BONDS FLOCCULATION	←	SOLUBLE Al Fe	→	INTERLAYER Fe and Al RUPTURE OF THE NETWORK MICRO AGGREGATION
ASSOCIATION AND BONDING INCREASE OF THE MICRO-POROSITY AND HYDRATION	←	AMORPHOUS Fe/Al FORMING THREEDIMENSIONAL NETWORK	→	SUPERPOSITION ON THE NETWORK
- KAOLINITE AROUND THE LARGE CRYSTALS - GROWTH OF Fe/Al CRYSTALS ON KAOLINITE PARTICLES	←	CRYSTALLIZED COMPOUNDS = RIGID COMPONENTS	→	LOCALIZATION IN THE MICROPOROSITY

TABLE 1 - ROLE OF THE DIFFERENT IRON AND ALUMINUM COMPOUNDS ON KAOLINITE AND SMECTITE

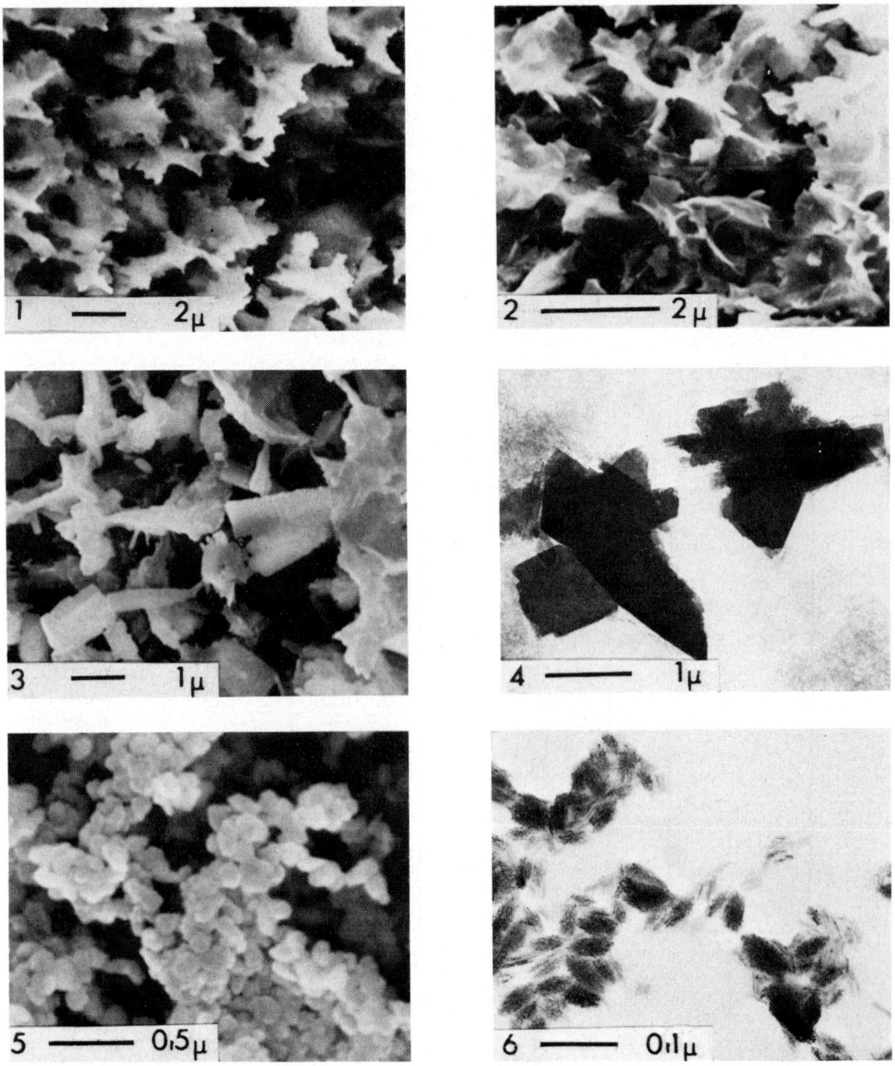

Plate I - Scanning and Transmission micrographs. 1) amorphous Fe pH1 ; 2) amorphous Al (pF1 ; 3) $Al(OH)_3$ crystals formed from the gel ; 4) the same in TEM ; 5) goethite crystals ; 6) the same in TEM.

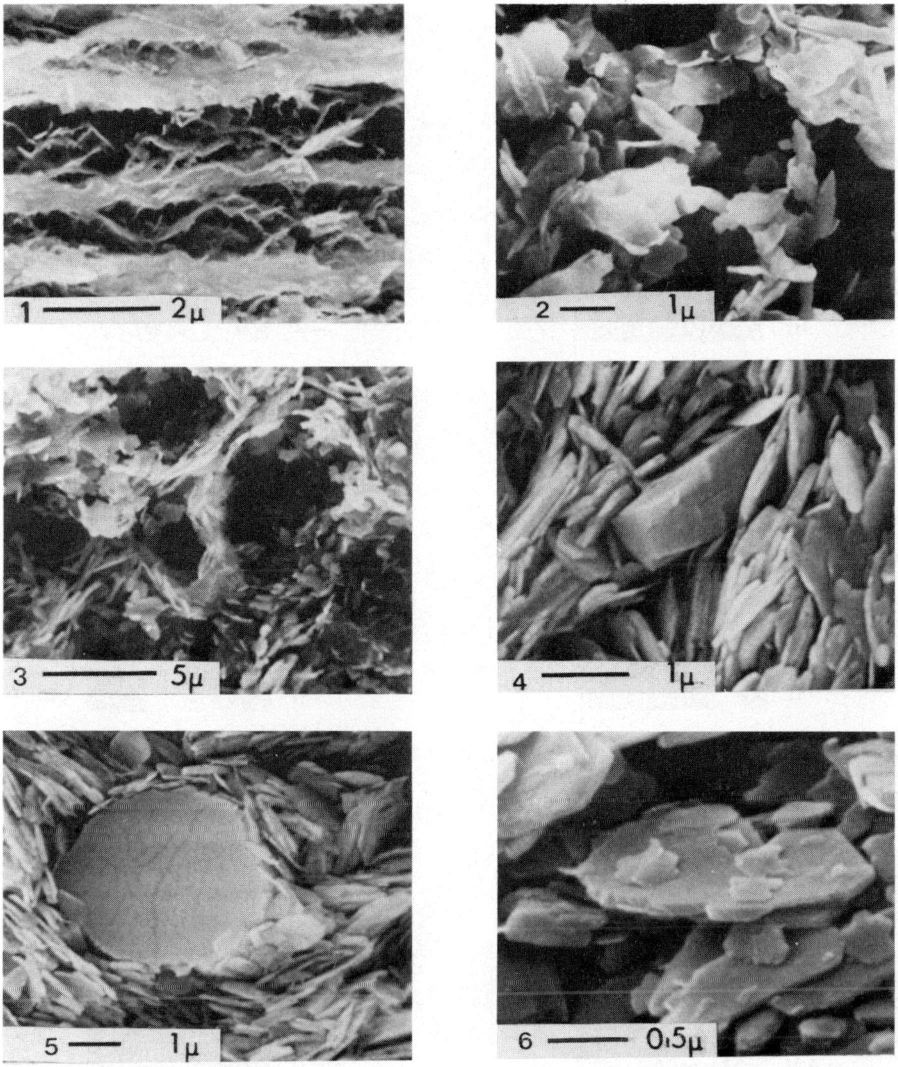

<u>Plate II</u> - Scanning electron micrographs. 1) collapse of the gel at pF4 ; 2) kaolinite + Al gel ; 3) kaolinite + Fe gel ; 4) kaolinite + $Al(OH)_3$ crystal ; 5) kaolinite crystals around kaolinite ; 6) small $Al(OH)_3$ crystals on kaolinite.

Plate III - Scanning electron micrographs. 1) Iron crystals on kaolinite ; 2 and 3) effect of interlayer Al compounds on smectite tactoïds ; 4) tactoïd + gel ; 5 and 6) iron crystal in the smectite porosity.

BORN-HABER CYCLES FOR INTERLAYER CATIONS OF MICAS

R.M. GARRELS (*) and Y. TARDY (**)

(*) Dept. of Marine Science, Univ. of South Florida, St. Petersburg, U.S.A.

(**)Laboratoire de Minéralogie, Univ. Paul Sabatier, Toulouse, France.

ABSTRACT

The Born-Haber Cycle below was constructed by Cs, Rb, K, Na, Li, Ba, Sr, Ca, and Mg. The table gives the Gibbs energy values in kilojoules per equivalent for the various legs of the cycle. The energies to place the gaseous ions in dry interlayers are proportional to those of hydrating the gaseous ions in water; the proportionality constant for alkali cations differs from that for earth alkali cations. A similar relation exists between the energy of hydration of the dry interlayer sites (per equivalent) and the hydration numbers of the cations. The calculated anionic radius of the dry interlayer sites, for all the alkali metals, is about 2.1 Å. The radius was calculated from Coulomb's law using ΔG from leg 2→5 and the radius of the cation.

INTRODUCTION

Tardy and Garrels (1974) described a method for estimating the Gibbs energies of formation of layer silicates. By assuming that the matrix of a layer silicate behaves as a solvent in which constituent oxides (or cations) dissolve ideally, they were able to solve sets of simultaneous equations to obtain values for the standard Gibbs energies of formation per equivalent of the exchange cations of layer silicates. In particular, they obtained values describing standard molar Gibbs energies of formation of K_2O and Na_2O in the interlayers of mica. The numbers obtained are equivalent to twice the standard Gibbs energies of formation of K^+ and Na^+ in the mica interlayers according to the reactions:

$$K^{o}_{element} \longrightarrow K^{+} \text{ silica interlayer} \qquad \Delta G^{o}_{f\ sil} = -393 \text{ kJ}$$

$$Na^{o}_{element} \longrightarrow Na^{+}_{mica\ interlayer} \qquad \Delta G^{o}_{f\ sil} = -343 \text{ kJ}$$

These reactions are analogs of the familiar reactions of the elements to form hydrated ions in aqueous solution:

$$K^{o}_{element} \longrightarrow K^{+}_{aq} \qquad \Delta G^{o}_{f\ aq} = -285 \text{ kJ}$$

$$Na^{o}_{element} \longrightarrow Na^{+}_{aq} \qquad \Delta G^{o}_{f\ aq} = -264 \text{ kJ}$$

The availability of these numbers for the standard Gibbs energy of formation of Na^+ and of K^+ in the dry interlayers of micas suggested the possibility of generating Born-Haber Cycles for micas. Conceptually, one can run around the periphery of the cycle of Fig. 1, starting with the element, ionizing it to the gas (1→2), plunging the gaseous ions into water (2→3), removing them from water and putting them into hydrated mica interlayers (3→4), removing the water from the interlayers (4→5), leaving the dry ions in the contracted interlayers, and, finally, removing the ions from the interlayers and discharging them to produce the element again (5→1). If these peripheral values are known, values for putting the gaseous ion into the dry interlayer (2→5), or putting the hydrated ions into the dry interlayer (3→5) are automatically provided.

The numbers obtained for the cycles are shown in Table 1.

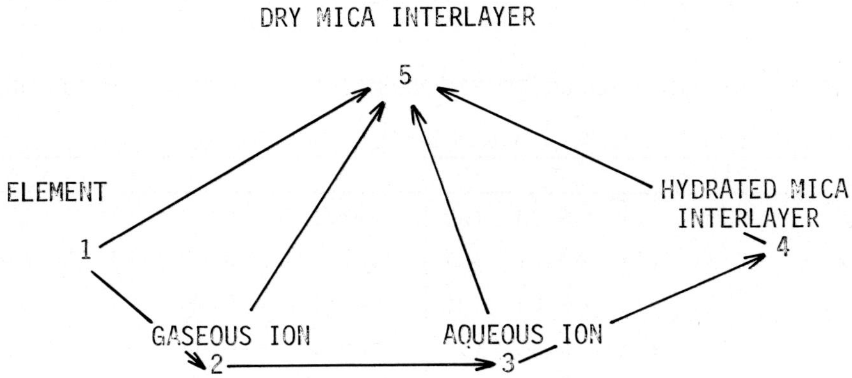

Figure 1. The Seven Legs of the Born-Haber Mica Cycle

TABLE 1. GIBBS ENERGIES PER EQUIVALENT (kJ) FOR VARIOUS LEGS OF THE BORN-HABER MICA CYCLE

ELEMENT	1 → 2	2 → 3	1 → 3	3 → 4	4 → 5	3 → 5	2 → 5	1 → 5
Cs	-42	-243	-285	-117	-8	-126	-368	-410
Rb	-17	-276	-293	-109	+8	-117	-393	-410
K	+13	-297	-285	-105	-4	-109	-406	-393
Na	+109	-372	-264	-100	+21	-79	-452	-343
Li	+180	-473	-293	-100	+46	-54	-527	-347
Ba	+339	-619	-280	-100	+46	-54	-674	-335
Sr	+406	-686	-280	-100	+50	-50	-736	-331
Ca	+481	-757	-276	-100	+54	-46	-803	-322
Mg	+690	-916	-226	-100	+59	-42	-958	-268

CALCULATION OF GIBBS ENERGIES FOR THE LEGS OF THE BORN-HABER CYCLE

Cycle legs 1→2, 2→3, 1→3: Gibbs energies of formation of aqueous and gaseous ions

The Gibbs energy of formation of the aqueous ion from the element, Leg 1→3, was taken from Robie et al. (1978), and the hydration energy of the gaseous ion from Hunt (1963). The step from element to gaseous ion was obtained by difference (first three columns of Table 1).

TABLE 2. EXCHANGE ENERGIES FOR VERMICULITES AND MONTMORILLONITES (kJ)

	A	B	C	D		A	B	C	D
Mont.	-0.35	(1)	+0.33	(a)	Verm.	-0.49	(6)	-0.10	(h)
Bent.	-0.35	(1)	+0.21	(a)	Verm	-0.49	(7)	+0.10	(h)
Verm.	-0.94	(1)	+6.04	(a)	Mont.	-0.46	(7)	-1.38	(g)
Mont.	-0.35	(1)	+0.30	(b)	Mont.	-0.34	(9)	-0.22	(i)
Bent.	-0.38	(2)	-2.56	(c)	Mont.	-0.46	(9)	-2.07	(f)
Beid.	-	(2)	-9.12	(d)	Verm.	-0.49	(9)	+2.05	(h)
Mont.	-0.40	(2)	-4.02	(b)	Mont.	-0.46	(10)	-12.7	(j)
Mont.	-0.35	(2)	-2.80	(b)	Verm.	-	(11)	+8.99	(k)
Mont.	-0.46	(2)	-5.65	(b)	Verm.	-	(11)	+8.62	(1)
Bent.	-0.38	(3)	-5.31	(c)	Mont.	-	(11)	+6.02	(1)
Bent.	-	(3)	-19.4	(e)	Mont.	-	(11)	+10.4	(1)
Bent.	-0.38	(4)	-9.04	(c)	Mont.	-0.34	(11)	+6.07	(i)
Mont.	-0.46	(4)	-18.0	(f)	Mont.	-0.46	(12)	-4.60	(c)
Mont.	-0.46	(4)	-19.1	(g)	Verm.	-0.48	(13)	-0.68	(m)
Verm.	-0.49	(5)	+1.10	(h)	Bent.	-0.45	(13)	-0.64	(m)

Columns: (A): interlayer charge; (B): exchange reaction; (C): ΔG exchange (kJ/2 equiv.); (D): references.

Mont. - Montmorillonite; Bent. - Bentonite; Verm. - Vermiculite; Beid. - Beidellite.

ΔG for exchange reactions, using Na_2X_2 -Cs_2X_2 as an example, is for the reaction: $Na_2X_2 + 2\,Cs^+ (aq) = Cs_2X_2 + 2Na^+(aq)$.

Column (B): (1): Na_2X_2-Li_2X_2; (2) Na_2X_2-K_2X_2; (3): Na_2X_2-Rb_2X_2; (4) Na_2X_2-Cs_2X_2; (5): Na_2X_2-MgX_2; (6) Na_2X_2-CaX_2; (7): Na_2X_2-SrX_2; (9): Na_2X_2-BaX_2; (10): K_2X_2-Cs_2X_2; (11): K_2X_2-CaX_2; (12) Rb_2X_2-Cs_2X_2; (13): CaX_2-SrX_2.

References: (a): Gast and Klobe (1971); (b) Tabikh et al. (1960); (c) Gast (1969); (d): Truesdell and Christ (1968);

(e): Kunishi and Heald (1968); (f): Lewis and Thomas (1963); (g):Eliason(1966); (h): Wild and Keay (1964); (i): Laudelout et al. (1968); (j): Faucher and Thomas (1954); (k): Dolcater et al. (1968); (l) Rich and Black (1964); (m): Heald (1960).

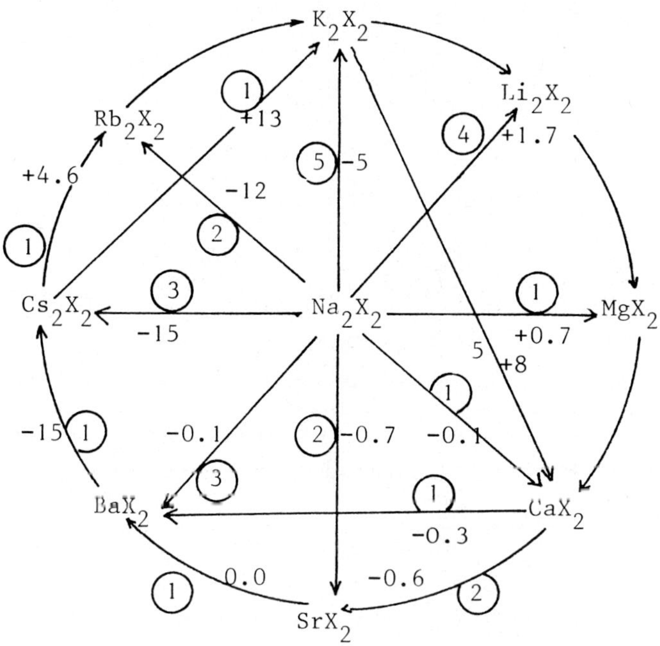

Figure 2. Consistency net for ion exchange reaction such as: $Cs_2X_2 + 2Na^+(aq) = Na_2X_2 + 2\ Cs^+\ (aq)$. (2): number of determinations averaged. -0.6: average value of exchange energy (kJ) from data of Table 2.

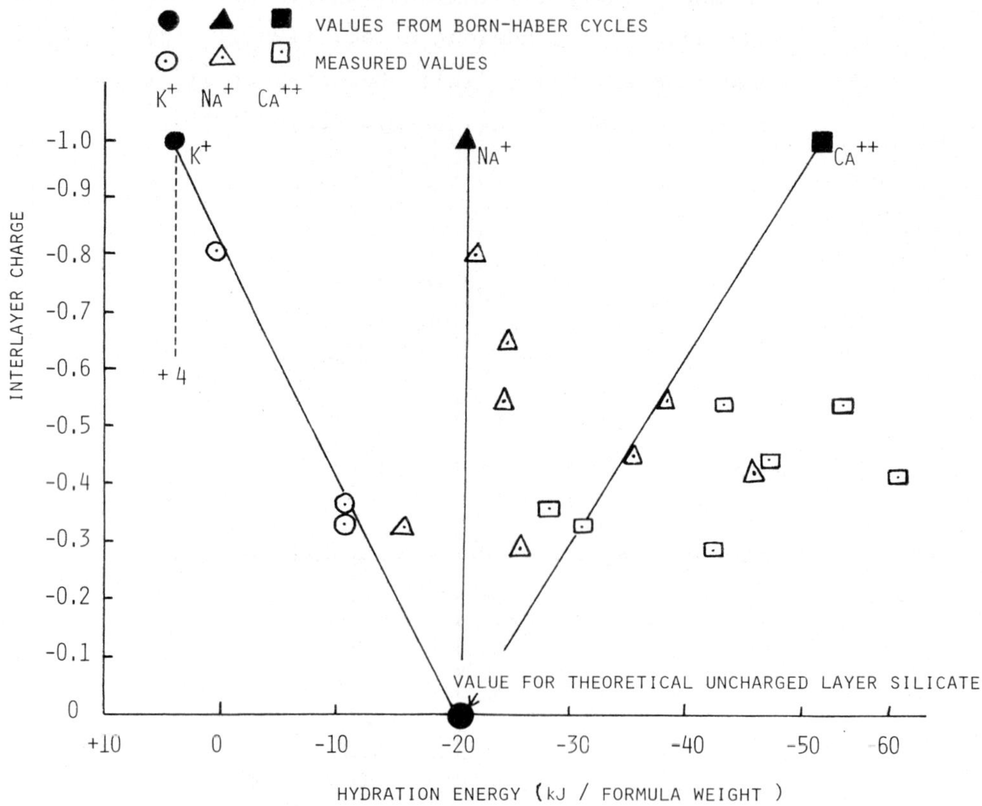

Figure 3. Comparison of measured hydration energies of K, Na, and Ca layer silicates with those selected for Born-Haber Cycles (data from Table 3).

<u>Cycle Leg 3→4: Gibbs energy to place aqueous ion in hydrated interlayer</u>

The specific numbers used for the Gibbs energy to place the aqueous ion in hydrated interlayers have been derived by consideration of Legs 1→5 and 5→4. The differences of the numbers for various cations are based on a compilation of exchange energies for the cations considered (Table 2). Figure 2 is a diagram we call a "consistency net". It shows the cation exchange energies listed in Table 2.

TABLE 3. HYDRATION ENERGIES OF SOME CLAYS, OBTAINED BY INTEGRATING DESORPTION ISOTHERMS (kJ/FORMULA WEIGHT)

		A	B	C	D			A	B	C	D
Na^+	Verm.	0.80	22	-21.8	(a)	Cs^+	Mont.	0.33	20	-12.1	(h)
Rb^+	Verm.	0.66	22	-22.2	(b)	Rb^+	Mont.	0.33	20	-11.1	(h)
Na^+	Verm.	0.66	22	-24.6	(b)	K^+	Mont.	0.33	20	-11.0	(h)
Li^+	Verm.	0.66	22	-38.2	(b)	Na^+	Mont.	0.33	20	-16.6	(h)
Ba^{++}	Verm.	0.66	22	-27.4	(b)	Li^+	Mont.	0.33	20	-23.4	(h)
Mg^{++}	Verm.	0.66	22	-46.5	(b)	H^+	Mont.	0.33	20	-27.3	(h)
Na^+	Beid.	0.54	20	-24.0	(c)	Ba^{++}	Mont.	0.33	20	-24.9	(h)
Ca^{++}	Beid.	0.54	20	-43.7	(c)	Sr^{++}	Mont.	0.33	20	-25.1	(h)
Na^+	Beid.	0.54	20	-38.4	(d)	Ca^{++}	Mont.	0.33	20	-31.1	(h)
Ca^{++}	Beid.	0.54	20	-53.5	(d)	Mg^{++}	Mont.	0.33	20	-27.6	(h)
Na^+	Sap.	0.45	20	-35.4	(d)	Cs^+	Hect.	0.29	20	-20.8	(1)
Ca^{++}	Sap.	0.45	20	-47.1	(d)	K^+	Hect.	0.29	20	-21.0	(1)
Na^+	Mont.	0.42	20	-40.2	(e)	Na^+	Hect.	0.29	20	-26.0	(1)
Ca^{++}	Mont.	0.42	20	-58.5	(e)	Li^+	Hect.	0.29	20	-36.0	(1)
K^+	Mont.	0.36	20	-10.9	(f)	Ba^{++}	Hect.	0.29	20	-31.4	(1)
Li^+	Mont.	0.36	20	-23.4	(g)	Sr^{++}	Hect.	0.29	20	-36.8	(1)
Ba^{++}	Mont.	0.36	20	-28.1	(g)	Ca^{++}	Hect.	0.29	20	-42.6	(1)
Ca^{++}	Mont.	0.36	20	-28.0	(f)	Mg^{++}	Hect.	0.29	20	-45.7	(1)

Columns: (A): layer charge; (B): temperature of the experiment (^{o}C); (C): ΔG, hydration energy; (D): references.

Verm. - Vermiculite; Beid. - Beidellite; Sap. - Saponite; Mont. - Montmorillonite; Hect. - Hectorite.

References: Van Olphen (1963); (b): Ovcharenko et al. (1963); (c): Glaeser and Mering (1968); (d): Suquet (1978); (e): Glaeser et al. (1968); (f): Tarasevich et al. (1967); (g): Tarasevich et al. (1968); (h): Mooney et al. (1952); (1): Prost (1975) and Tardy et al. (1980).

Furthermore, because the data compiled represent a whole series of montmorillonites and vermiculites of various interlayer charges and exchange capacities, we conclude that the exchange energies for various cation pairs are the same for all hydrated interlayers, irrespective of charge. In other words, the values for exchange energies generated from Figure 2 are presumed to be valid for micas with hydrated interlayers, as well as for vermiculites and montmorillonites. Examination of Table 1 shows the remarkable conclusion that the exchange energies for occupations of hydrated interlayers are the same to within one kJ for 2 Na^+, 2 Li^+, Ba^{2+}, Sr^{2+}, Ca^{2+}, and Mg^{2+}. Only the large ions, which are less hydrated, even in a hydrated interlayer, show significant differences in bonding energy from the cations listed above.

The individual numbers assigned to Leg 3→4 will be derived in the following evaluation for Leg 4→5.

<u>Leg 4→5: Energy of dehydration of the mica interlayers</u>

All of the reliable hydration energy values for montmorillonites and vermiculites that we could find are compiled in Table 3. They show a large spread in values for a given cation and a given interlayer charge. Figure 3 shows a plot of data for K^+, Na^+, and Ca^{2+} from the Table.

The minimum values for Na^+ are nearly constant at about -21 kJ per formula weight, independent of charge. The various values for hydration energies much larger than those that lie on the 21 kJ line we explain as belonging to montmorillonites with a large surface area in addition to that of the interlayers. In other words, we attribute the 4-25 kJ per formula weight increase of hydration energy to grain size effects. If it is assumed that the 21 kJ per formula weight represents the energy of hydrating some 370 m^2 per gram of silicate surface, we obtain 0.140 J/m^2 as the energy involved. This number compares with Van Olphen's (1965) values of 0.150J/m2 for Na-vermiculite; with Keenan's (1951) values of 0.125J/m2 for Li-kaolinite (that he interpreted as having an uncharged surface); and Iler's

(1955) number of 0.130 J/m^2 for heat of wetting of silicate surfaces.

There are not many data to support the line drawn for the hydration energy of K^+ layer silicates. The point of zero hydration energy at an interlayer charge of -0.8 is well established (Barshad, 1954). If the line is accepted, it extrapolates to the same value as that for the interlayers of our theoretical uncharged Na^+ layer silicate; furthermore, it yields a value of +4 kJ per formula weight for the hydration of a K-mica. A positive ΔG^o for this reaction is quite reasonable since K^+ micas do not spontaneously hydrate in water at room temperature.

The very tentative line shown for Ca^{2+} in Figure 3 is used here to suggest the behavior of hydrated Li^+ as well as that of all the alkaline earths. The extrapolated value for Ca-mica at unit charge, clearly based on poor data, is between 50 and 60 kJ per formula weight. In support of the surface area explanation for the spread of hydration energy values at a given interlayer charge, we note that those clays that show values lying above the line chosen for Na^+, also show higher values for Ca^{2+}.

The hydration energies decrease for K^+, remain constant for Na^+, and increase for Ca^{2+}, Mg^{2+}, and Li^+, as the interlayer charge increases. Our interpretation is that the K^+ ion is presumably not hydrated in the interlayer sites. At low charge, the hydration energy is dependent upon the energy of the uncharged surfaces. As the K^+ occupancy increases, the area of hydratable uncharged surface diminishes as does the hydration energy. At an interlayer charge of -0.8 the interlayer completely collapses to 10 Å and no hydrated surface is available. Increase of interlayer charge to -1.0 (that of mica), increases the bonding of the interlayers so that the hydration energy becomes +4 kJ per formula weight. On the contrary, as the hydration energy of an interlayer Ca^{2+} ion is considerably greater than that of intervening neutral surfaces with increasing interlayer charge, hydration energy per formula weight also increases.

In summary, with respect to Leg 4→5, numbers for K^+ and Na^+ have been derived from Figure 3 as well as values for Ca^{2+}. Because all the water of hydration in the interlayers occupied by alkaline earths and lithium are associated with a 15 Å spacing, it seems likely that energies for Leg 4→5 will be similar for lithium and alkaline earths.

Eisenman (1962) made theoretical calculations of the energies involved in removing (or adding) gaseous cations from (or to) various silicate sites. His numbers apply to Leg 2→5 of our Born-Haber Cycles. Our data for Leg 2→5 are derived from the values we have assigned to Leg 1→5, plus standard values from the chemical literature for Leg 1→2. Eisenman made calculations for the energy of placing a gaseous ion on an Al-O-Si site, exactly equivalent to an interlayer mica site. Table 4 lists his calculated values, our estimated values, and his values and our values for the energies relative to that of Ca^+. Our numbers assigned to specific ions differ markedly from his, but agree, except for Li^+, with differences in energy referred to Cs^+.

The agreement of our estimated values for Leg 2→5 with those of Eisenman for several alkali-metals bolsters confidence in the estimate (Tardy and Garrels, 1974) we have assigned to Leg 1→5 for those elements and for the alkaline earths as well.

TABLE 4. COMPARISON VALUES FROM EISENMAN (1962) FOR ELECTROSTATIC ENERGY TO PLACE CATIONS ON Al-O-Si SITES WITH VALUES DERIVED HERE FOR PLACING CATIONS ON DRY INTERLAYER SITES OF MICAS (VALUES IN kJ/MOL)

	Eisenman's electrostatic sites	Cations to dry mica interlayer	Differences relative to Cs (Eisenman)	Differences relative to Cs (this study)
Li^+	-515	-527	-119	-159
Na^+	-482	-452	- 86	- 84
K^+	-436	-406	- 40	- 38
Rb^+	-419	-393	- 23	- 25
Cs^+	-396	-368	0	0

Adjustment of the cycles by consistency tests

The numbers we have assigned in Table 1 to the previously unknown legs of the Born-Haber Cycles have been derived, in part as already discussed, but they have been refined by plotting the various numbers against each other, and using relations so discovered to give a set of numbers consistent with apparently trustworthy systematic relations among the variables.

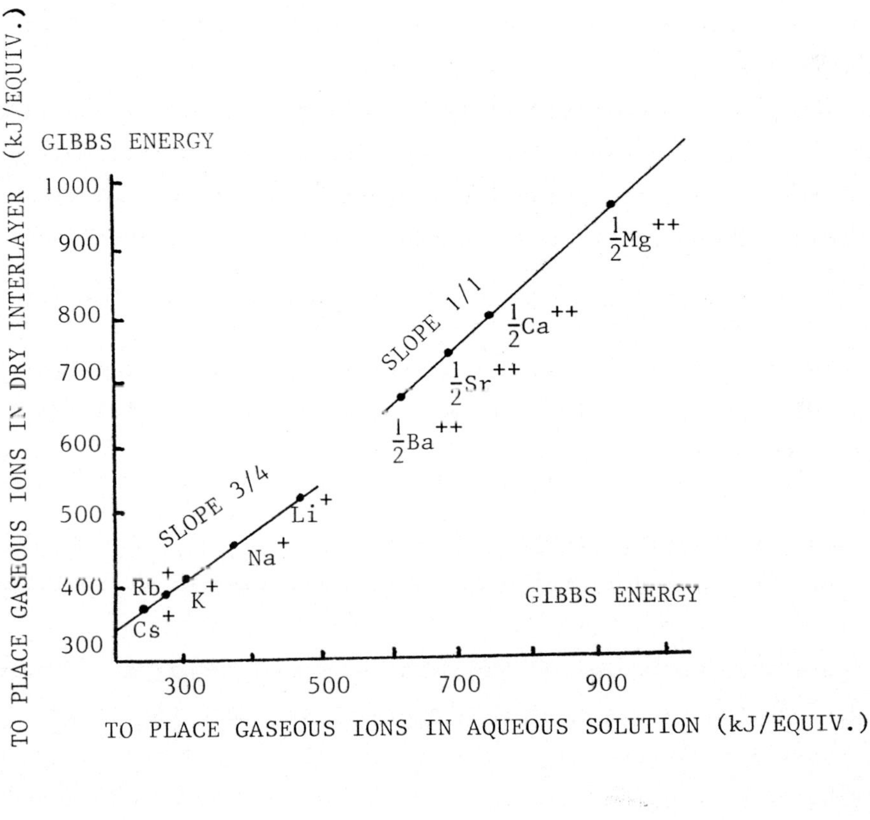

Figure 4. Gibbs energy of hydration of gaseous ions versus Gibbs energy to place gaseous ions in dry mica interlayers

To return to the Gibbs energy values for Leg 3→4, the differences in values were obtained from exchange energies. The absolute number base was derived chiefly from Legs 1→5 and 4→5. Values for K^+ and

Na^+ for these two legs served to give values for these cations for Leg 3→4. Then, knowledge of differences provided the remaining required numbers.

Incidentally, the energy to remove Na^+, for example, from a hydrated interlayer to aqueous solution is so great that the concept of "dissociated clay acids" can have no reality. Negative sites on clays are never deserted!

Legs 2→5 and 3→5; gaseous ion to dry interlayer and aqueous ion to dry interlayer

Legs 2→5 and 3→5 were obtained by difference from the other legs of the cycles.

We imitated Eisenman (1962) in calculating an "equivalent anionic radius" for the negative site of the interlayers by using our energies for removing the cation from the site to make the gaseous cation, the radius of the cation and Coulomb's law. Following Eisenman, we wrote:

$$\Delta G\ 2\rightarrow 5 = -1390\ \text{kJ}/\ [(\text{radius cation}) + (\text{radius anion})]$$

and solved for radius anion. The values obtained for the equivalent radius of the dry negative site are: 2.08 Å (Cs^+); 2.05 Å (Rb^+); 2.09 (K^+); 2.12 (Na^+); 2.03 Å (Li^+).

Additional consistency tests

Figures 4, 5, and 6 show plots of the numbers for various Legs against each other.

Figure 4 shows a plot of the Gibbs energies required to hydrate the gaseous ion versus the Gibbs energies to place the gaseous ion into a dry interlayer. One inference is that the layer silicate interlayers indeed behave as a uniform solvent for cations, and that the energies involved are proportional to those of hydration of gaseous cations in water.

Figures 5 and 6 show the relations of the hydration numbers of cations (Robinson and Stokes, 1959) to the hydration energies of the cations. For hydration in an aqueous system, the number of water

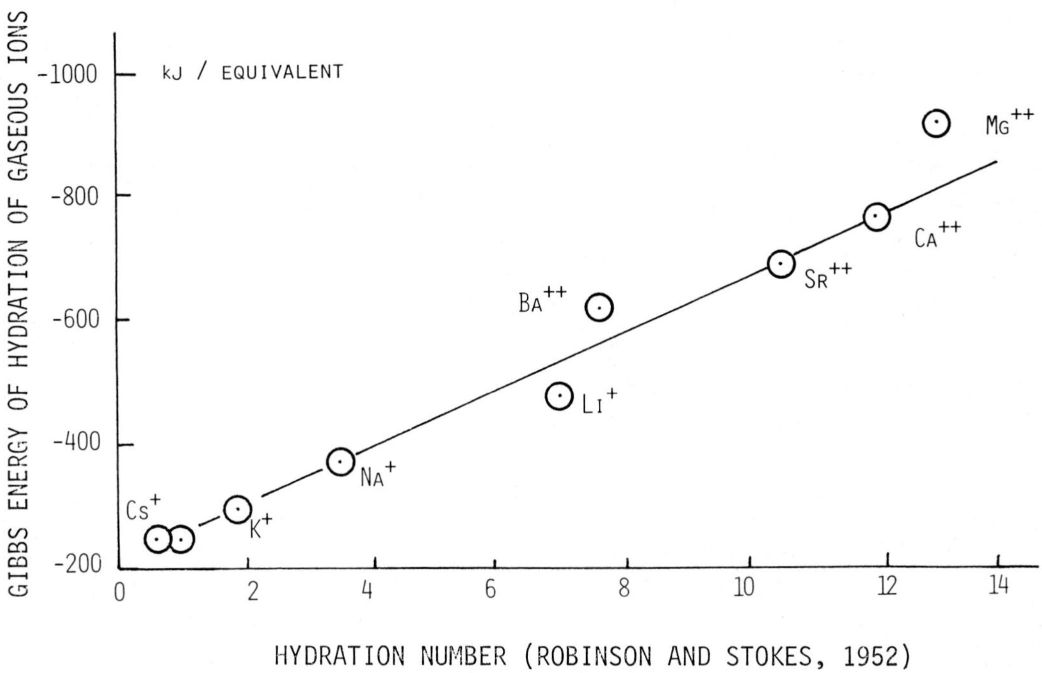

Figure 5. Gibbs energy of hydration of gaseous ions as a function of hydration number

molecules presumably associated with the cation is proportional to the hydration energy (per equivalent) of the cation. For the derived values of hydration energies of interlayer sites, the alkali cations and alkaline earths show linear relationships.

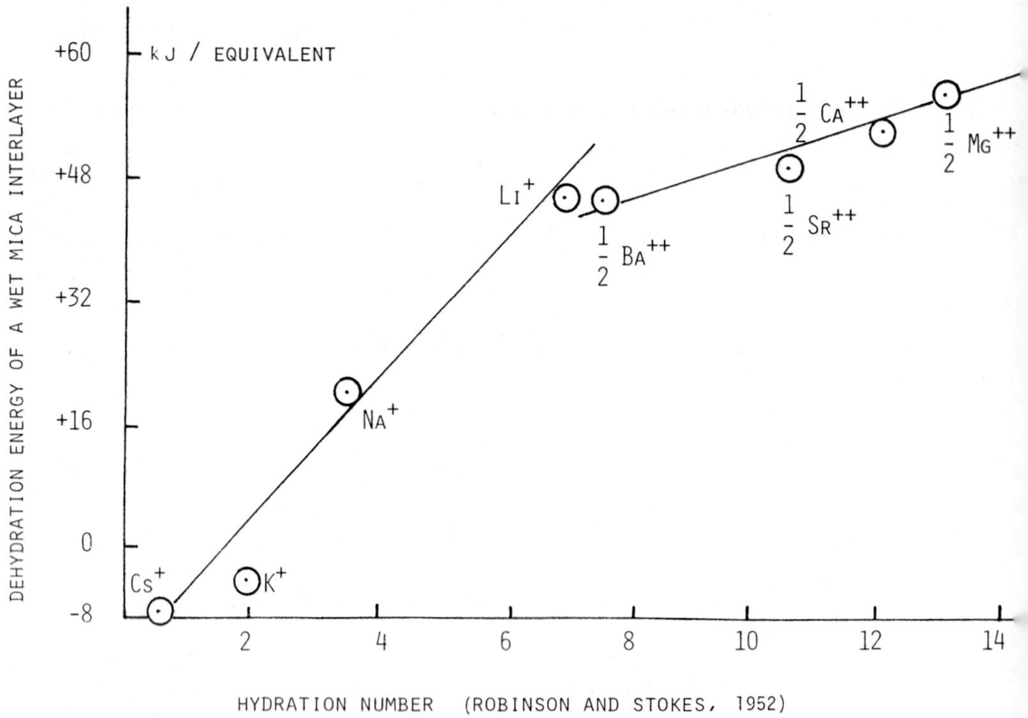

Figure 6. Gibbs energy of hydration of wet mica interlayers as a function of hydration number of aqueous cations

SUMMARY

In summary, we have prepared a series of diagrams showing the Born-Haber Cycles of cations in gas, water, and in the interlayers of micas. We conclude that the numbers we have derived are within a few kilojoules of the correct values for the differences among the cations, but may be many kilojoules (perhaps 10 to 20) from absolute values for individual species. The values derived show an interesting analogy between the interlayers of silicates as a "solvent" and water itself. At this stage it appears that the matrix of layer silicates can be thought of as a solvent and that the concept of nearly ideal solution of the components of layer silicates in their matrix is a satisfactory first approximation. We suggest that differences of the hydration energies of clays of the same interlayer charge are measures of grain size effects that can amount to tens of kilojoules per formula weight. Studies need to be made of hydration energies versus surface areas of selected clays. If these differences in hydration energies are indeed a function of surface area in excess of interlayer surface area it will be necessary to have an estimate of this area before calculating the standard Gibbs energy of formation of a clay mineral.

ACKNOWLEDGEMENTS

Over many years we have discussed problems related to the present manuscript with various colleagues. We thank especially K. Beck, the late C. L. Christ, J. I. Drever, D. Eberl, R. Eddy, W. Gunter, J. Hower, the late E. A. Perry, Jr., R. Robie, the late D. Waldbaum, and C. Weaver.

The work was supported by National Science Foundation Grant GA 29207 and by Délégation Générale à la Recherche Scientifique et Technique, "Récupération Assistée du Pétrole", contrat n° 78 7 2206.

REFERENCES

Barshad, L., 1954. Cation exchange in micaceous minerals: II Replaceability of ammonium and potassium from vermiculite, biotite, and montmorillonite. J. Soil Sci., 78: 57-76.

Dolcater, D.L., Loste, E.G., Syers, J.R., and Jackson, M., 1968. Cation exchange selectivity of some clay-sized minerals and soil materials. Soil Sci. Soc. Amer. Proc., 32: 795-798.

Eisenman, G., 1962. Cation selective glass electrodes and their mode of operation. Biophys. Journ. 2: 259-323.

Eliason, J.R., 1966. Montmorillonite exchange equilibria with strontium-sodium-cesium. Am. Min. 51: 324-335.

Faucher, J.A., and Thomas, H.C., 1954. Adsorption studies on clay minerals. IV. The system montmorillonite-cesium-potassium. J. Chem. Phys. 22: 258-261.

Gast, R.G., 1969. Standard free energies of exchange for alkali metal cations on Wyoming bentonite. Soil Sci. Soc. Amer. Proc. 33: 37-41.

Gast, R.G., and Klobe, W.D., 1971. Sodium-lithium exchange equilibria on vermiculite at 25° and 50°C. Clays Clay Min. 19: 311-319.

Glaeser, R., Mering, J., 1968. Domaines d'hydratation homogène des smectites. C.R. Acad. Sci. Paris, 267(D): 463-466.

Glaeser, R., Mantine, I., and Mering, J., 1967. Observation sur la beidellite. Bull. Groupe Français Argiles 19: 125-130.

Heald, W.R., 1960. Characterization of exchange reactions of strontium or calcium on four clays. Soil Sci. Soc. Amer. Proc. 24: 103-106.

Hunt, J.P., 1963. Metal Ions in Aqueous Solution. W.A. Benjamin, Inc. 124 pp.

Iler, R.K., 1955. The Colloid Chemistry of Silica and Silicates. Cornell Univ. Press, Ithica, NY, 324 pp.

Keenan, A.G., Mooney, R.W., and Wood, L.A., 1951. The relation between exchangeable ions and water adsorption on kaolinite. J. Phys. Colloid. Chem. 55: 1462-1474.

Kunishi, H.M., and Heald, W.R., 1968. Rubidium-sodium exchange on kaolinite and bentonite. Soil Sci. Amer. Proc. 32: 201-204.

Laudelout, H., Van Bladel, R., Bolt, G.H., and Page, A.L., 1968. Thermodynamics of heterovalent cation exchange reactions in a montmorillonite clay. Trans. Faraday Soc., London, 67: 1477-1488.

Lewis, R.J., and Thomas, H.C., 1963. Adsorption studies on clay minerals. VIII. A consistency test of exchange sorption in the systems sodium-cesium-barium montmorillonite. J. Phys. Chem. 67: 1781-1783.

Mooney, R.W., Keenan, A.G., and Wood, L.A., 1952. Adsorption of water vapor by montmorillonite. I. Heat of desorption and application of BET theory. J.A.C.S. 74: 1367-1371.

Mooney, R.W., Keenan, A.G., and Wood, L.A., 1952. Adsorption of water vapor by montmorillonite. II. Effect of exchangeable ions and lattice swelling as measured by x-ray diffraction. J.A.C.S. 74: 1371-1374.

Orchiston, H.D., 1955. Adsorption of water vapor: III. Homoionic montmorillonites at 25°C. Soil Sci. 79: 1, 71-78.

Ovcharenko, F.D., Tarasevich, Y.S., Valitskaya, U.M., and Polyakov, V.E., 1967. Adsorption of water vapor on cation-substituted vermiculite, translated from Kolloidnyi zh. 29: 565-568.

Prost, R., 1975. Etude de l'hydratation des argiles: interactions eau-minéral et mécanisme de la rétention de l'eau. Annales agronomiques 26: 4, 401-461.

Quinson, J.F., Escoubes, M., and Blanc, R., 1972. Profil énergétique de l'adsorption d'eau à 25°C sur montmorillonites homoioniques (K^+, Na^+, H^+, Ca^+). Bull. Groupe Français Argiles. 24: 49-67.

Rich, C.D. and Black, W.R., 1964. Potassium exchange as affected by cation size, pH and mineral structure. Soil Sci. 97: 384-390.

Robie, R.A., Hemingway, B.S., and Fisher, J.S., 1978. Thermodynamic Properties of Minerals and Related Substances at 298.15K and 1 bar (10^5 Pa) Pressure and At Higher Temperatures. U.S. Geol. Sur. Bull., 1452: 456 pp.

Robinson, R.A., and Stokes, R.H., 1959. Electrolyte Solutions. Butterworths, 571 pp.

Suquet, H., 1978. Propriétés de Gonflement et Structure de la Saponite. Comparaison Avec la Vermiculite. Thèse Sci. Univ. Paris VI, 281 pp.

Tabikh, A.A., Barshad, I., and Overstreet, R., 1960. Cation exchange hysteresis in clay minerals. Soil Sci. 90: 219-226.

Tarasevich, Yu.S. and Ovcharenko, T.D., 1967. Nature of the reaction between water and a montmorillonite surface, translated from Ukr. Khim. Zh. 33: 165-172.

Tarasevich, Yu. S., Valitskaya, V.M., and Ovcharenko, T.D., 1968. Adsorption of water vapor on cation-substituted montmorillonite, translated from Kolloidnyi Zh. 6: 889-896.

Tarasevich, Yu.S., Orazmuradov, A.O., and Ovcharenko, T.D., 1971. Heat of adsorption of water on cation-substituted vermiculite, translated from Kolloidnyi Zh. 33: 597-602.

Tardy, Y., and Garrels, R.M., 1974. A method of estimating the Gibbs energies of formation of layer silicates. Geochim. Cosmochim. Acta, 38: 1101-1116.

Tardy, Y., Lesniak, P., Duplay, J., and Prost, R., 1980. Energie d'hydratation de l'hectorite. Bull. Minéral, 103: 217-223.

Truesdell, A.H., and Christ, C.L., 1968. Cation exchange in clays interpreted by regular solution theory. Amer. Jour. Sci. 266: 402-412.

Van Olphen, H., 1963. Compaction of clay sediments in the range of molecular particle distances. Clays Clay Min., 178-187.

Van Olphen, H., 1965. Thermodynamics of interlayer adsorption of water in clays. J. Colloid Sci. 20: 822-837.

Wild, A. and Keay, J., 1964. Cation exchange equilibria with vermiculite. J. Soil Sci. 15: 136-144.

CHEMICAL COMPOSITION OF INDIVIDUAL CLAY PARTICLES: AN IDEAL SOLID SOLUTION MODEL

Y. TARDY (*), J. DUPLAY (**) and B. FRITZ (**)
(*) Laboratoire de Minéralogie, Université Paul Sabatier, Toulouse, France
(**) Institut de Géologie et Centre de Sédimentologie de la Surface, Strasbourg, France

ABSTRACT

Chemical analyses of individual particles of different clay minerals yield a large range of variation in composition. Correlations, which were found among chemical constituents, can be interpreted in terms of combinations of several end members. An ideal solid solution model is proposed as a way to calculate the contribution of each of the end members involved.

CLAY PARTICLES ANALYSES

Five samples are described here. They are:

- A standard glauconite, called Odin's glauconite, from thanetian sediments of the Paris Basin.
- An illite from Orzilhac, in sediments of the intracratonic basin of Le-Puy, France.
- A beidellite from vertisoils of Godola, Cameroun.
- A nontronite from Pensylvania, U.S.A.
- A stevensite from the sediments of Jebel Ghassoul, Marocco.

Each of these five samples was treated to remove the major impurities, saturated subsequently with calcium and with potassium, and examined in a transmission electron microscope. The individual particles were chemically analysed with a microprobe, according to the method of Duplay (1982). About 30 particle analyses were made for each sample. The chemical compositions of the individual particles were averaged and the calculated structural formulae are given in Table 1.The five samples selected represent a large variety of compositions: from dioctahedral to trioctahedral and from Al to Fe to Mg clay types.

CORRELATIONS AMONG CHEMICAL CONSTITUENTS

For each of the five sets of chemical analyses of individual clay particle, correlation coefficients were calculated. Table 2 presents some of the most significant ones among the principal variables. It is shown that:

- The tetrahedral charge (CHT) is negatively correlated with octahedral aluminium (Al^{VI}) in all cases, except for stevensite, for which the correlation coefficient is positive.

- The tetrahedral charge (CHT) is positively correlated with octahedral iron in all cases.
- The tetrahedral charge (CHT) is almost not correlated with magnesium except for stevensite for which the correlation coefficient is negative.
- The octahedral charge (CHO) is positively correlated with Al^{VI} except for the two micas, negatively correlated with Fe^{VI} in all cases and negatively correlated with Mg^{VI} in three of the five cases.
- Octahedral charge (CHO) and tetrahedral charge (CHT) are not correlated for beidellite and nontronite, are negatively correlated in the two micas and positively correlated for stevensite.

The two major correlations,i.e. CHT - Al^{VI} and CHT - Fe^{3+VI} are shown in Fig. 1 for the Odin's glauconite. These correlations are somewhat similar to those presented for example by Shutov et al. (1973), Weaver and Pollard (1973), or Velde (1977), for glauconite, illite or various smectites. Our data and those from the literature suggest for example that in clay minerals, the contribution of Fe^{3+}-muscovite is generally greater than that of Fe^{3+}-pyrophyllite and also, that the contribution of talc is greater than the contribution of phlogopite as suggested by the fact that the correlation coefficients CHT - Fe^{3+VI} are almost always positive and CHT - Mg^{2+VI} always negative.

However, the correlation coefficients must be used with precaution to interpret contributions of the different mineral constituents. First of all, ferrous iron cannot be determined by microprobe so that some of the chemical formulae may be incorrect. Therefore, the percentage of Fe^{2+} in the type of clays presented is generally small, so that the error will not be important. Let us consider, for example, the correlations involving the negative octahedral charge (CHO):

$$CHO = 6 - 3p - 3q - 2r, \text{ per 12 oxygens in formula}$$

in which p, q and r respectively stand for the number of atoms of Al^{3+},Fe^{3+} or Mg^{2+} in octahedra. The higher the number of the ions present in octahedra, the smaller is the octahedral negative charge (CHO), so that the variable CHO tends to be negatively with octahedral Al^{3+}, Fe^{3+} or Mg^{2+}. Consequently, a positive correlation coefficient should be interpreted as significant for a phenomenon different from that of the mass balance constraint.

It seems clear that the tetrahedral negative charge (CHT) as well as the octahedral negative (CHO) are not determined by mass balance constraints but are also determined by a selected contribution of some mineral constituents. The problem is now how to distribute or how to weight the contribution of each of the possible mineral constituents.

TABLE I

Averaged structural formulae of the five clays studied.

GLAUCONITE: $K_{0.72}Ca_{0.03}[Si_{3.70}Al_{0.30}]^{IV}[Al_{0.41}Fe^{3+}_{1.17}Mg_{0.37}]^{VI}O_{10}(OH)_2$
ILLITE: $K_{0.65}Ca_{0.016}[Si_{3.43}Al_{0.57}]^{IV}[Al_{1.21}Fe^{3+}_{0.28}Mg_{0.60}]^{VI}O_{10}(OH)_2$
BEIDELLITE: $K_{0.08}Ca_{0.005}[Si_{3.40}Al_{0.60}]^{IV}[Al_{1.32}Fe^{3+}_{0.65}Mg_{0.22}]^{VI}O_{10}(OH)_2$
NONTRONITE: $K_{0.01}Ca_{0.01}[Si_{3.56}Al_{0.44}]^{IV}[Al_{0.28}Fe^{3+}_{1.73}Mg_{0.18}]^{VI}O_{10}(OH)_2$
STEVENSITE: $K_{0.13}Ca_{0.009}[Si_{3.89}Al_{0.11}]^{IV}[Al_{0.31}Fe^{3+}_{0.09}Mg_{2.37}]^{VI}O_{10}(OH)_2$

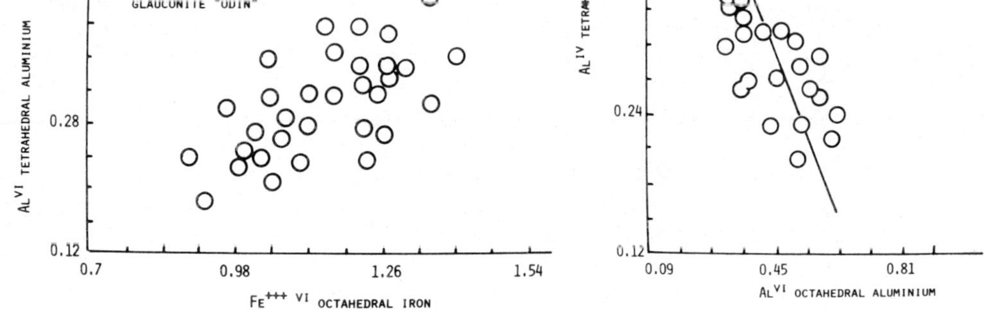

Fig. 1. Correlation diagram, showing variations of tetrahedral aluminium (Al^{IV}) as a function of octahedral aluminium (Al^{VI}) or octahedral iron (Fe^{3+} VI).

TABLE 2.

Correlation coefficients between major chemical variables of the five clay minerals studied.

GLAUCONITE

	C H T	C H O
Al tot.	-0.41	-0.26
Al^{VI}	-0.80	-0.01
Fe^{3+VI}	+0.75	-0.05
Mg^{2+VI}	+0.10	-0.45
C H T	1	-0.34

ILLITE

	C H T	C H O
Al tot.	+0.65	-0.59
Al^{VI}	-0.13	-0.07
Fe^{3+VI}	+0.23	-0.18
Mg^{2+VI}	+0.32	-0.38
C H T	1	-0.86

BEIDELLITE

	C H T	C H O
Al tot.	-0.06	-0.10
Al^{VI}	-0.60	+0.43
Fe^{3+VI}	+0.52	-0.49
Mg^{2+VI}	-0.00	+0.26
C H T	1	-0.78

NONTRONITE

	C H T	C H O
Al tot.	—	0.00
Al^{VI}	-0.54	+0.55
Fe^{3+VI}	+0.64	-0.74
Mg^{2+VI}	0.00	+0.65
C H T	1	-0.71

STEVENSITE

	C H T	C H O
Al tot.	+0.76	+0.47
Al^{VI}	+0.53	+0.50
Fe^{3+VI}	+0.52	-0.21
Mg^{2+VI}	-0.54	-0.50
C H T	1	-0.07

For a confidence level of 99% the correlation coefficient should be smaller than -0,45 or greater than +0,45.

CALCULATION OF THE MINERAL COMPONENT CONTRIBUTION

Clay minerals may be considered as mixtures of a good number of mineral components. They are, for four octahedral cations (Al^{3+}, Fe^{3+}, Fe^{2+}, Mg^{2+}):

4 potassium micas (muscovite, phlogovite etc...)
4 potassium celadonites (negative octahedral charge)
4 potassium vermiculites (positive octahedral charge)
4 pyrophyllites (Al^{3+} or Fe^{3+}), talc or minnesotaite
12 sets of corresponding clays, Na, Mg or Ca exchanged.

The total number of these components is 52; but only a few are needed to describe a given clay composition. An example of four mineral components is given below. Let us consider an entirely hypothetic clay fomula such as:

$$K_{0.4}(Al_{0.4}Si_{3.6})^{IV}(Al_{1.33}Mg_{1.0})^{VI}O_{10}(OH)_2$$

The tetrahedral negative charge is: CHT = 0.4
The octahedral negative charge is zero: CHO = 6 - 3x1.33 - 2x 1.0 = 0

The point, representing this mineral is shown in the composition diagram of Fig. 2. This diagram has four corners which represent the compositions of the four mineral constituents indicated in the figure.

For the clay formula given in the example, only three mineral components are actually needed to describe the composition. In this example, the mass balance constraints require one to retain muscovite and pyrophyllite but leave the choice to select either phlogopite (solution 1) or talc (solution 2) as the third component. If X_i are the mole fractions of the i=4 mineral components, (with $\Sigma X_i=1$), the distribution of the X_i in the two different situations is the following:

Solution I:

$X_{muscovite}$ = 0.4 - 1/3 = 0.066
$X_{pyrophyllite}$ = 3/3 - 0.4 = 0.600
$X_{phlogopite}$ = 1/3 = 0.333
with $\Sigma X_i = 1$

Solution II

$X_{muscovite}$ = 0.40 = 0.400
$X_{pyrophyllite}$ = 2/3 - 0.4 = 0.266
X_{talc} = 1/3 = 0.333
with $\Sigma X_i = 1$

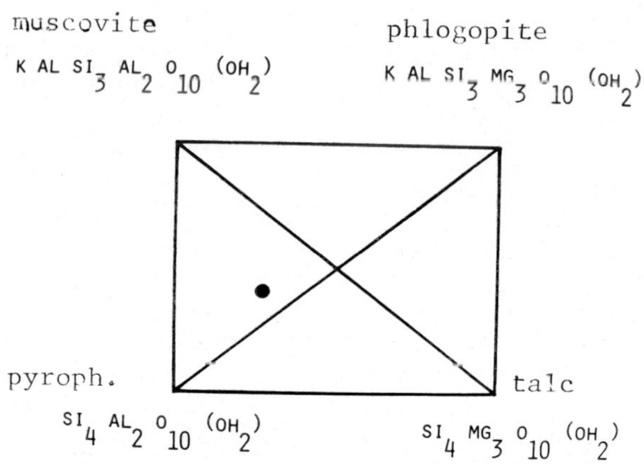

Fig. 2. Composition diagram showing the point representative of the clay: $K_{0.4}\ Al_{0.4}Si_{3.6}Al_{1.33}\ Mg_{1.0}\ O_{10}(OH)_2$ in the system of four mineral components: muscovite, pyrophyllite, phlogopite and

talc.

The two distributions are different and there are no evident reasons to select one or the other. Furthermore, there is potentially an infinite number of possibilities to combine four mineral components when only three are required. In other words, there is an infinite number of possibilities to combine solution I and solution 2. However, the examination of variations of chemical composition of individual clay particles within a given population contains informatio allowing to make a choice. In our example, if solution I is working, the clay population would show a strong positive correlation between the tetrahedral negative charge and octahedral magnesium, simply because the mole fraction of muscovite is small compared to the contribution of phlogopite. On the other hand, if solution 2 would apply a negative correlation between CHT and Mg^{VI} should be excepted. This analysis simply means that correlations among chemical constituents of an individual clay particles should reflect the weighted contribution of each of the mineral components which describe a given composition. As discussed in the next section, the weighted contribution of each mineral component can also be computed by the use of solid solution models. Then mineral components will be called end members.

AN IDEAL SOLID SOLUTION MODEL

An ideal solid solution model of end members in the example cited above would require that only three end members be considered. Furthermore, solubility products of the four end members considered should be so distributed that solution I and solution 2 are equivalent and give the same solubility for a given composition.

The model presented here is not, properly speaking, a model of solid solution of end members, because the four end members are all taken into consideration, and weighted as a function of their solubility products as follows:

$$\frac{X_{phlogo.}}{X_{talc}} = \frac{X_{talc}\, K_{musco.}\, X_{musco.}}{K_{phlogo.}\, K_{pyro.}\, X_{pyro.}},$$

in which X_i and K_i stand respectively for the mole fractions and the solubility products of the end members involved. The contribution (X_i) of each of the end members (i) is weighted as a function of K_i, the corresponding solubility product: the higher K, the greater the solubility, and the smaller the contribution (X).

This method was proposed by Tardy and Fritz (1981) and extensively described by Fritz (1981) for a large number of components. Relationships concerning partial equilibria, the type of which were described by Garrels and Christ (1965) for binary mixtures, involve activities of aqueous species [] and can be written as:

$$\frac{[Mg^{2+}]^3}{[Al^{3+}]^2} = \frac{K_{phlogo.}\ X_{phlogo.}}{X_{musco.}\ X_{musco.}} = \frac{K_{talc}\ X_{talc}}{X_{pyro.}\ X_{pyro.}} = etc..$$

Among the infinite number of ways to weight the contribution of the different components in an ideal way, the solution presented here is the one which requires the minimum energy.

By a method of trial and error, Fritz (1981) has determined the end member solubility products (K_i) required to form clay minerals of observed composition. They are generaly higher than those of well crystallized phyllosilicates (Table 3).

Application to the clay mineral studied

In our model, there is no provision for dealing with octahedral minerals with more than 6 positive charges in octahedra, so that beidelllite and nontronite are not considered here.

The results of calculated contributions of each of the 9 end members according to our ideal solid solution model, are listed in Table 4 and illustrated for glauconite in Fig. 3.

It is shown that for illite, glauconite and stevensite, the contribution of Fe^{3+}-muscovite is important compared to that of Fe^{3+}-pyrophyllite. It is shown also that the contribution of phlogopite is very low compared to the contribution of Mg-celadonite and talc. Finally, in glauconite the contribution of Al-celadonite is greater than the calculated contribution of muscovite or pyrophyllite.

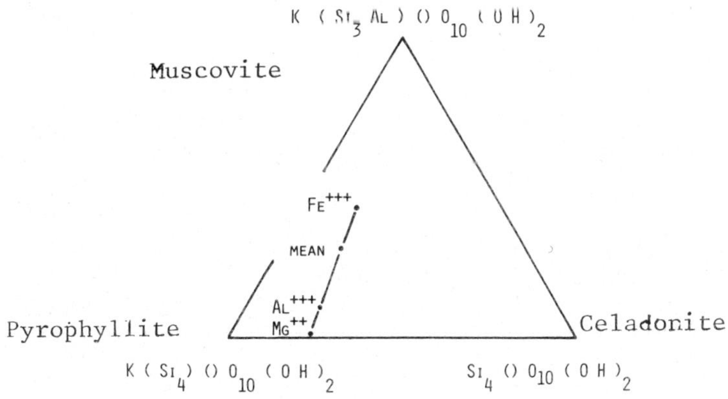

Fig. 3. Calculated contributions of the three types of layer silicates: pyrophyllite, muscovite and celadonite, for the average composition (mean) and for Mg, Al and Fe^{3+} octahedral occupation, in the Odin's glauconite.

TABLE 3.

Log K solubility products of K^+-exchanged clays. Column 1: solubility product of the well crystallized minerals. Column 2: solubility products of the corresponding clays entering in solid solutions.

	1	2		1	2		1	2
Phlogopite	38.3	41.9	Talc	21.6	25.1	Mg-Celadon.	-	25.4
Annite	23.3	23.3	Minnesotaite	7.2	7.2	Fe^{2+}-Celad.	-	10.5
Muscovite	14.6	16.6	Pyrophyllite	1.1	1.1	Al-Celadon.	-	5.3
Fe^{3+}-Musco.	-	1.1	Fe^{3+}-Pyroph.	-	-13.7	Fe^{3+}-Celad.	-	-5.9

TABLE 4.

Weighted contribution of the nine end members in the clays analyzed.

	Si_4 () O_{10} $(OH)_2$	K Si_4 () O_{10} $(OH)_2$	K Al Si_3 () O_{10} $(OH)_2$
Glauconite			
Al	0.048	0.160	0.029
Fe	0.087	0.270	0.270
Mg	0.029	0.106	0.001
Illite			
Al	0.053	0.166	0.415
Fe	0.002	0.011	0.125
Mg	0.047	0.151	0.030
Stevensite			
Al	0.082	0.027	0.048
Fe	0.008	0.004	0.035
Mg	0.622	0.139	0.027

The solid solution model seems to be good enough to support the conclusions drawn from the correlation coefficients:

- the Fe^{3+}-mica is the predominant form, so that the higher the iron content, the higher is the tetrahedral charge;
- Mg-mica (phlogopite) makes a very small contribution, so that the higher the magnesium content, the lower is the tetrahedral charge;
- conclusions concerning the respective contributions of Al or Fe or Mg-celadonites are less clear.

DISCUSSION AND CONCLUDING REMARKS

Velde (1977) has extensively reviewed the compositional variations of solid solutions of a large number of clay minerals. Furthermore, Lippmann (1977, 1979) has discussed the conditions of formation of solid solutions and concluded that it is unlikely that clay minerals form in nature under the conditions of thermodynamic equilibrium. Truesdell and Christ (1968), Helgeson and Mackenzie (1970), Stoessel (1979) and Aagaard et al.(1981) have proposed regular solid solution models or regular site-mixing models for calculating smectite, illite or mixed layer clay stabilities. These model studies were performed for a relatively small number of end members which do not include celadonites and Fe^{3+}-clays. Furthermore, nothing is said about the way to select only four or five components among many possibilities. Details and discussion are found in Fritz (1981).

Tardy and Garrels (1974) have presented a model which allows calculations of a large number of individual solubilities but this model does not allow the simulation of a continuous variation of clay composition as functions of continuous chemical changes in aqueous solutions.

Our ideal solid solution model has the advantage to present a relatively simple expression (compared to the regular one) and to use a large number of components which cover all possible compositions of natural clays. It also permits to simulate continuous variations in composition and to explain correlations among chemical constituents within clay particle populations, understood in terms of contributions of mineral components.

This model should be improved further. Additional observations are required to reduce the uncertanities. Both ideal and regular solid solution models should also be tested in a large variety of cases before deciding which model would be the most appropriate.

ACKNOWLEDGEMENTS

This work was supported by the grants CCE (European Community) contract n° 076 76 EGF, DGRST (Délégation à la Recherche Scientifique et Technique) and contract 78 7 2206 (récupération assistée du pétrole); and Société Nationale Elf Aquitaine, contract n° 3065.

REFERENCES

Aagaard, P.,Helgeson, H.C and Benson, L.V, 1981. Chemical and thermodynamic consequences of preferential site occupancy in montmorillonite and illites. (in print).

Duplay, J., 1982. Analyses chimiques de populations de particules argileuses. Thèse Univ. Poitiers, (in preparation).

Fritz, B., 1981. Thermodynamique et modélisation des réactions hydro-thermales et diagénétiques. Thèse Univ. Strasbourg (in press).

Garrels, R.M. and Christ, C.L., 1965.Solutions, Minerals and Equilibria. Harper and Row, 450 pp.

Helgeson, H.C. and Mackenzie, F.T., 1970. Silicate sea weater equilibria in the ocean system. Deep Sea Research, 17: 877-892.
Lippmann, F., 1977. The solubility products of complex minerals, mixed crystals and three-layer clay minerals. N. Jb. Miner. Abh., 130, 3: 243-263.
Lippmann, F., 1979. Stabilitätsbeziehungen der Tonminerale. N. Jb. Miner. Abh., 136, 3: 287-309.
Shutov, V.D., Katz, M.Ya., Dritz, V.A., Sokolova, A.L. and Kazakov, G.A., 1973. Crystallo-chemical heterogeneity of glauconite as depending on the conditions of its formation and postsedimentary changes. Proc. Int. Clay Conf. Madrid, 269-279.
Stoessel, R.K., 1979. A regular solution site-mixing model for illites. Geochim. Cosmochim. Acta, 43: 1151-1159.
Tardy, Y. and Garrels, R.M., 1974. A method for estimating the Gibbs energies of formation of layer silicates. Geochim. Cosmochim. Acta, 38: 1101-1116.
Tardy, Y. and Fritz, B., 1981. An ideal solid solution model for calculating solubility of clay minerals. Clay Miner. (in press).
Truesdell, A. and Christ, C.L., 1968. Cation exchange in clays interpreted by regular solution theory. Amer. Jour. Sci., 266: 402-412.
Velde, B., 1977. Clays and clay minerals in natural and synthetic systems. Elsevier, 218 pp.
Weaver, C.E. and Pollard, L.D., 1975. The chemistry of clay minerals. Elsevier, 213 pp.

DEMIXING PHENOMENON IN LOW WATER CONTENT BIIONIC K-Na MONTMORILLONITES

Jean-Pierre GAULTIER and Jean MAMY
I.N.R.A. - Station de Science du Sol - Route de Saint-Cyr,
78000 VERSAILLES - FRANCE

ABSTRACT

Experiments were carried out to determine whether cation distribution in interlamellar spaces of biionic K-Na montmorillonite is influenced by the mode of preparation of the sample. In all cases it appears that demixing of the cations occurs in the range 30-100 % Na during the drying process.

INTRODUCTION

In previous studies of cation demixing in biionic K-Na and K-Ca montmorillonites, it was shown that cations segregation occurs in the interlamellar spaces allowing potassium fixation (Gaultier and Mamy, 1978, 1980). This demixing phenomenon is a necessary condition for potassium fixation in biionic systems and the nature of the second cation, Na or Ca is not a determining factor.

Additional experiments were carried out to answer two questions concerning this phenomenon of demixing :

1) is the phenomenon influenced by the mode of preparation of the samples, i.e. is the initial saturation of the mineral important ?

2) at what time during dehydration does demixing occur ?

MATERIALS AND METHODS

The two cations used in the experiments are potassium and sodium. The mineral is a Wyoming montmorillonite.

Study of ool X-ray diffraction peaks is one of the best methods for showing cation segregation in clay interlamellar spaces (Chaussidon, 1963 ; Gruner, Le Dred and Wey, 1979). The only condition required is that the d_{001} spacing i.e. hydration state should not be the same for the two end-member phases. Assuming that the repartition of the cations is homogeneous in the interlamellar spaces, mixing of the two cations gives rise to one of the two d_{001} values depending on the relative proportions of the two cations (Mering and Glaeser, 1954) ; in demixing we can observe either regular variation in the mean value of d_{001} between the two end-member values in the case of random-interstratification, or simultaneous presence of two d_{001} values when two distinct phases are present.

Four methods were used to prepare biionic K-Na samples :

In the first method the starting samples were pure Na-montmorillonite to which potassium chloride solutions at various concentrations were added to obtain different K/Na ratios in the samples. Then the biionic mixtures were stirred for 24 hours and washed. The proportions of exchangeable potassium and sodium were determined by extraction with ammonium acetate. The percentages of sodium in the different samples that were prepared in this way were 10, 17, 26, 40, 50, 65, 81 and 91 %.

In the second method the starting samples were pure K-montmorillonites to which sodium chloride solutions at various concentrations were added to obtain different K/Na ratios in the samples. After stirring and washing the proportions of sodium determined in the resulting samples were 9, 16, 29, 34, 41, 56, 71 and 80 %.

In the third method the calcium initially present in pure Ca-montmorillonite was completely exchanged by sodium - potassium chloride solutions of different K/Na-ratios. After stirring and washing the proportions of sodium determined in the resulting samples were 8, 10, 24, 31, 39, 47, 57 and 66 %.

The fourth method consisted of mixing various amounts of pure Na-montmorillonite and pure K-montmorillonite suspensions in water. After stirring the proportion of sodium in the samples were 15, 30, 40, 50, 60 and 80 %.

Oriented films of all these different samples were gently dried in a dessiccator in which the relative humidity was maintained at 80 %. Drying under these conditions keep potassium montmorillonite from collapsing irreversibly. In taking the X-ray data on the samples, the relative humidity was subsequently fixed at 80 %, at 40 % and after heating of the sample at 150°C at 0 %.

EXPERIMENTAL RESULTS

The d_{001} values of the pure K and Na montmorillonites at 80, 40 and 0 % RH are presented in table 1.

TABLE 1. d_{001} values of pure K and Na-montmorillonites.

	K-montmorillonite (Å)	Na-montmorillonite (Å)
80 % R H	12,40	15,50
40 % R H	12,27	12,60
0 % R H	10,28	9,60

At 80 % R H, because the d_{001} values of pure K and Na montmorillonites are very different, mixing or demixing of the cations in the interlamellar spaces can be easily observed by observing the evolution of the maxima and the profile of the diffraction lines. This evolution of the 001 diffraction line is represented in figure 1. The diagrams are those of biionic K-Na montmorillonite samples prepared according to the first method i.e. Na-montmorillonite + potassium chloride. The evolution of the 001 diffraction line of samples prepared according to the other methods one similar to that represented in figure 1. The maximum position varies from 12.40 Å to 15.50 Å in a non-linear way because of the presence of shoulders which are characteristic of the presence of two distinct phases, due to demixing of the cations.

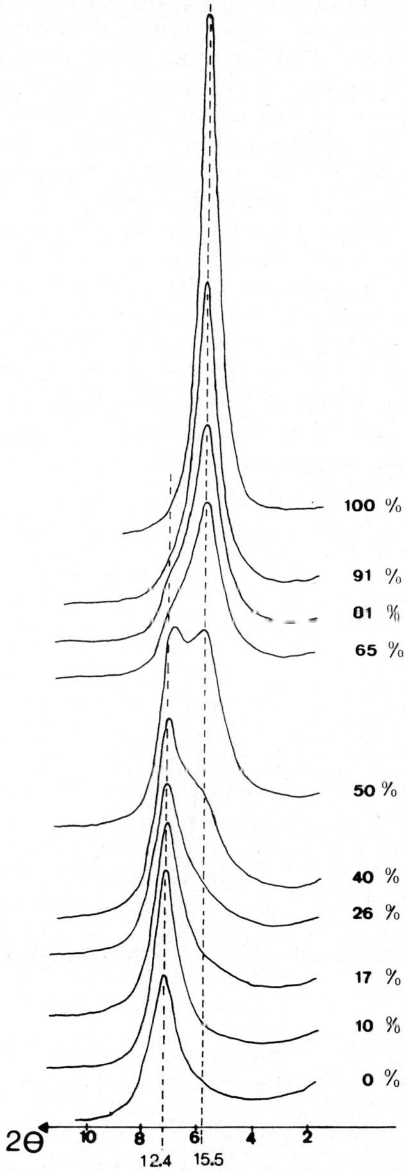

Fig. 1. Evolution of the 001 diffraction line profile according to the variation of the percentage of exchangeable sodium.

A semi-quantitative attempt was made to determine whether demixing occurs, i.e. whether there are two distinct phases both for small and for hight proportions of sodium in the samples. Therefore, 001 diffraction line profiles were calculated for pure K-montmorillonite and for pure Na-montmorillonite assuming that they have a Gaussian form in the reciprocal space. If there is demixing of the two cations in biionic samples and a formation of two phases the experimental profile will be the sum of two calculated end-member curves. On the other hand it will be composed of only one curve if there is mixing of the two cations in the interlamellar spaces.

The best fit obtained for three samples prepared according to the first method, and which contain 10, 50 and 81 % Na respectively is represented in figure 2. For the sample containing 10 % sodium only one curve is sufficient to obtain a good fit ; the difference between observed and calculated curve is very small. In the other two cases (50 and 81 % Na) two curves are required to obtain a good fit. It was found that the 001 diffraction line of all the biionic samples can be represented by either one or two calculated curves depending on the proportion of sodium in the samples. Only one curve is sufficient for a good fit as long as the sodium percentage is under 25-30 ; two curves are required in the range 30 - 100 % Na. These results are independent of the method of preparation of the various samples.

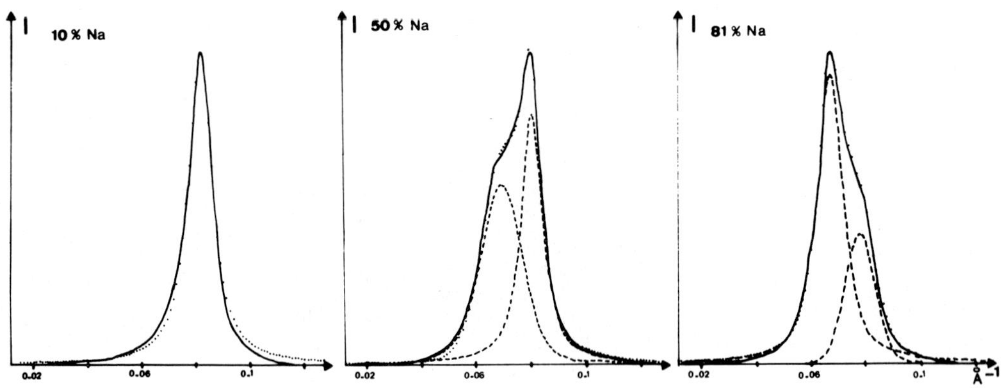

Fig. 2. Calculated (...), (---) and experimental (——) line profiles for K-Na montmorillonites with various proportions of exchangeable sodium.

These results indicate that beyond 30 % Na two phases exist and that at lower percentage the mineral is composed of only one phase. With regard to the latter it must be noticed that if there were two phases in samples with low sodium content they would show in the diffraction patterns : at a relative humidity of 80 % the structure factors for K-montmorillonite and Na-montmorillonite are respectively 220 and 310. Because intensities of X-ray diffraction peaks are proportionnal to the square of these structure factors, a sodic phase should be seen if it would exist even at only 10 % Na. For the 81 % Na sample such a phase can be seen (in this case a potassic phase) and the corresponding value of the structure factor is the lower.

The above conclusion is confirmed by the variations in the d_{001} mean values when the samples are submitted to relative humidities of 80 %, 40 % and 0 %. In figure 3, the d_{001} values are plotted versus the percentage of sodium for all samples.

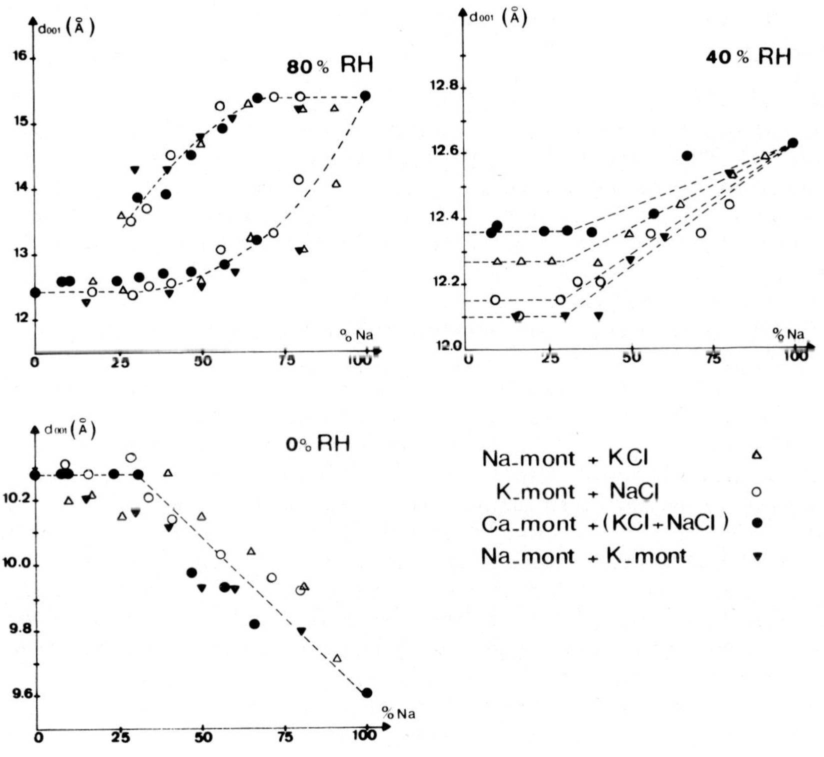

Fig. 3. d_{001} mean values of K-Na-montmorillonites versus the proportion of exchangeable sodium.

At 80 % RH when two phases exist, the two corresponding values of d_{001} appear. In all cases, at 80, 40 and 0 % RH, there is only one phase when the sodium percentage is lower than about 30, the value of d_{001} beeing always the same.

In the range 30-100 % Na, d_{001} has two maximum values at 80 % RH and at 40 and 0 % RH it varies linearly with the percentage of sodium, characteristic of the presence of two phases. Higher orders of diffraction were observed. At 80 % RH only one other order appears at about 3.15 Å but it is weak and diffuse. It is the superposition of the fourth order and of the fifth order of the 12.4 and 15.5 Å. At 40 and 0 % RH the maximum values of the other orders of diffraction have the same type of variation that the first order. They all indicate the presence of only one phase when the sodium percentage is lower than about 30 and they vary linearly with the percentage of Na in the range 30-100 % Na. At 40 % RH, the history of the sample seems to be important. The d_{001} value appear to be higher when the original sample is sodic or calcic than when it is potassic. Perhaps the kinetics of dehydration and the arrangement of interlayer water molecules slightly vary depending on the mode of preparation of the sample.

DISCUSSION

(i) Demixing phenomena are independent of the mode of preparation of biionic K-Na samples of montmorillonite.

(ii) Cation demixing phenomena exist in biionic K-Na samples of montmorillonite not only after wetting and drying cycles which permit potassium fixation, but also as soon as the first drying cycle is performed and it is clear that this phenomenon occurs before the relative humidity of 80 % is reached during the drying process. At this relative humidity K and Na-montmorillonites have different water contents as shown by their d_{001} values. Such a difference in hydration properties seems to be significant for segregation phenomena in interlamellar spaces. In the present case cation demixing occurs as soon as this difference in hydration properties between K and Na-montmorillonite appears.

(iii) The demixing phenomenon is an asymmetric phenomenon, i.e. the behaviour of biionic clays is not a regular one in the whole range 0-100 % Na. What occurs in the 0-30 % Na range does not occur in the 0-30 % range : from the hydration point of view, a biionic sample with less than 30 % Na has a potassic behaviour whereas biionic samples with less than 30 % K have a behaviour intermediate between the behaviour of Na and K-clays.

Apparently a small quantity of potassium added to a sodic sample is sufficient to produce a potassic phase i.e. interlamellar spaces which are fully occupied by potassium. Such a phenomenon is new evidence for the great affinity of potassium for cation interlamellar positions of smectites. On the other hand the introduction of a small quantity of sodium in a potassic sample does not induce such a behaviour. Because of the great affinity of potassium for interlamellar spaces, sodium is probably first located on the external surfaces, but considering that external surfaces represent less than 30 % of the exchange capacity, there may be mixing of potassium and sodium ions in the interlamellar spaces in the range 0-30 % Na.

These results demonstrate the specific behaviour of potassium in clays. In the case of biionic K-Na-montmorillonite, potassium seems to produce the demixing phenomenon rather than sodium : as soon as potassium is introduced in sodic samples it modifies their physico-chemical properties whereas the structure and the physico-chemical properties of K-montmorillonite are not changed when sodium is introduced. This fact implies a very high stability of the monoionic potassic interlayers. This stability is so high that it allows the introduction of another cation like sodium up to 25-30 % of the exchange capacity of the mineral without producing any difference in its X ray diffraction behaviour.

REFERENCES

Chaussidon, J., 1963. Evolution des caractéristiques chimiques et cristallographiques des montmorillonites biioniques K-Ca au cours d'alternances répétées d'humectations et de dessiccations. Proc. Int. Clay. Conf., Stockholm 195-201.

Gaultier, J.P., Mamy, J., 1978. Evolution of exchange properties and crystallographic characteristics of biionic K-Ca montmorillonite submitted to alternate wetting and drying. Proc. Int. Clay. Conf., Oxford, 167-175.

Gaultier, J.P., Mamy, J., 1980. Study of exchange properties of potassium in biionic K-Na montmorillonites. Fourth European Clay Conference, Abstracts p. 56.

Gruner, L., Le Dred, R., Wey, R., 1979. Obtention de minéraux interstratifiés sodium-alkylammonium de type 1/1 réguliers à partir de vermiculite. C.R. Acad. Sci. D 288, 661-663.

Mering, J., Glaeser, R., 1954. Sur le rôle de la valence des cations échangeables dans la montmorillonite. Bull. Soc. Franc. Miner. Crist., 77, 519-530.

GEOCHEMISTRY OF SPRING WATERS . SEASONAL VARIATIONS AND ALUMINUM CONTROL

G. BOURRIE

Laboratoire de Sciences du Sol. I.N.R.A., 65, Rue de Saint-Brieuc, 35042 RENNES CEDEX (FRANCE)

ABSTRACT

Spring waters issuing from acidic brown earth soils, under Beech-Oak forest in Fougères (Armorican Massif, France) bear wide seasonal variations. Total Al ranges from 56 to less than 0,1 μM and Si from 470 to 60 μM ; pH ranges from 4.8 to 7.4, and temperature from 5 to 14°C. As temperature increases, in April and early June, alkalinity, Ca, Si and pH increase, sulfate decreases ; in summer, most samples are deeply depleted either in Al (<0,2 μM) or Si (<7 μM) or both ; in autumn, Si, Al and sulfate increase, pH decreases. If one considers only monomeric forms of aluminum hydroxocomplexes, and using Debye-Hückel extended law to calculate activity coefficient, the waters are much supersaturated with respect both to gibbsite (log Q/K = + 1 to + 4.2) and kaolinite (log Q/K = + 4.5 to + 17). When considering polynuclear hydroxocomplex along with mononuclear forms, log Q/K for gibbsite does not exceed 0.36, if corrected for log K variation with temperature. Silica concentration is not correlated with Al activity and pH. Al is more soluble than if it were controlled by imogolite, kaolinite or various 2-1 clay minerals tested.

INTRODUCTION

Aluminum content in spring waters is usually low, except at low or high pH, but in waters issuing from forested soils, it was found to be unusually high. It can thus no longer be considered as an inert element controlled by a solid phase and subject to few variations. The following results are part of a study undertaken since 1978 in order to assess variation of water quality in relation to weathering of granite and pedogenesis.

SITUATION AND METHODS

The forested massif of Fougères (Armorican Massif) is situated on a Vire type 2- micas granite. Acidic brown earth soils are developed on a partly allochtonic silt loam material covering weathered granite, with a slight surface podzolization in some places, under a Beech-oak forest in a humid temperate climate. Spring waters were collected about twice a month, pH and temperature were measured in situ and an aliquot for Al filtered at 0.025 μm pore diameter and acidified in situ. Al was analyzed by flameless atomic absorption. Na, K, Mg, Ca, Cl^-, alkalinity (Alk), NO_3^-, SO_4^{2-} and silica were analyzed after 0.45 μm filtration.

SEASONAL VARIATIONS OF MAJOR ELEMENTS

Results presented here concern 70 samples collected from Dec. 1977 to April 1979. Spring water temperature is about 7 to 8°C in winter and spring, increases sharply to about 12°C in early June and decreases slowly in autumn. Alkalinity is correlated with temperature and Ca, and negatively with SO_4^{2-}. On the other hand, Na, Cl, and to a lesser degree Mg, are linked due to the proximity of the sea. As a result, when the influence of input from the sea is dominant, Na, Cl and Mg are high, Si, Alk. and Ca are low, whereas as temperature increases and organic matter is oxidized, Alk., pH and Si, Ca increase, whereas Na, Cl, Mg and SO_4^{2-} are low. This results in some slightly negative correlation between Mg and Si, that is not apparently a solubility relationship.

In summer some samples are depleted in either silica or aluminum or both. Samples depleted in silica (less than limit of detection i.e. 5-7 10^{-6}M), correspond always to peat-bogs. Therefore this depletion is likely to be a biological effect. In autumn, SO_4^{2-} increases, Alk. decreases, pH decreases, Al and Si increase. K and NO_3^- are always minor elements.

ALUMINUM AQUEOUS AND SOLID SPECIES

Nature of aqueous species considered and thermodynamic data at standard conditions

Reliable thermodynamic data exist at either low or high pH, when monomeric species $Al(H_2O)_6^{3+}$ or $Al(OH)_4^-$ predominate. Between pH 4 and pH 8 a variety of monomeric species, such as $AlOH^{2+}$, $Al(OH)_2^+$, $Al(OH)_3^\circ$, AlO_2^-, dimeric species such as $Al_2(OH)_2^{4+}$ or polynuclear forms were considered. Hemingway et al. (1978) discouraged use of AlO_2^-. Bottero et al. (1980) considered dimer forms $Al_2(OH)_2^{4+}$, $Al_2(OH)_x^{6-x}$ (x undefined) and polynuclear ion $Al_{13}O_4(OH)_{28}^{3+}$ from RMN and potentiometric measurements. The dimer is formed by two octahedra sharing an edge, consisting of OH^- groups (Hem, 1970). However, the dimer forms are always minor species and Hsu (1977) considers them as unstable due to electrostatic repulsion between two Al^{3+} ions. The polynuclear ion is formed by a central tetrahedrally coordinated Al surrounded by two hexagons of octahedrally coordinated Al, thus leading to the formula $Al_{13}O_4(OH)_{28}^{3+}$. Single hexagon and double hexagons are symmetrically stabilized and are considered by Hsu (1977) and Hem (1970) as intermediate forms in the nucleation of gibbsite.

The values used here are listed in Table 1. The basic data are Gibbs free energies of formation at 298.15 K, 10^5 Pa for Al^{3+}, $Al(OH)_2^+$, $Al(OH)_4^-$, from Hemingway and Robie (1977), and Hemingway et al. (1978). Data for $AlOH^{2+}$ are derived from log K = 5.00 (Sarazin,1979) for the reaction :

$$AlOH^{2+} + H^+ = Al^{3+} + H_2O$$

Data for $Al(OH)_3^\circ$ are from log K = 5.86 (Naumov et al., 1971) for the reaction :

$$Al(OH)_3^\circ + H^+ = Al(OH)_2^+ + H_2O$$

Bottero et al. (1980) proposed log K = 105 for the reaction :

TABLE 1
Thermochemical data for aqueous species at 298.15 K, 10^5 Pa

Species	ΔG°_f (kJ mol^{-1})	Reaction	ΔG°_R (kJ mol^{-1})	log K[a]	Ref
H_2O (1)	-237.141				1
$OH^-_{aq.}$	-157.328	$H_2O = H^+ + OH^-$	79.813	-13.983	1
$H_4SiO_4^\circ$ aq.	-1308.00				1
$Al^{3+}_{aq.}$	-489.4				2
$AlOH^{2+}_{aq.}$	-698.	$AlOH^{2+} + H^+ = Al^{3+} + H_2O$	-28.54	5.00	3
$Al(OH)_2^+$ aq.	-914.2	$Al(OH)_2^+ + 2 H^+ = Al^{3+} + 2 H_2O$	-49.482	8.669	4
$Al(OH)_3^\circ$ aq.	-1117.9	$Al(OH)_3^\circ + 3 H^{\pm} + Al^{3+} + 3 H_2O$	-82.931	14.529	5
$Al(OH)_4^-$ aq.	-1305.	$Al(OH)_4^- + 4 H^+ = Al^{3+} + 4 H_2O$	-132.964	23.294	4
$Al_{13}O_4(OH)_{28}^{3+}{}_{aq}$	-13351.	$Al_{13}O_4(OH)_{28}^{3+} + 36 H^+ = 13 Al^{3+} + 32 H_2O$	-599.336	105.	6

a. log $K = -\Delta G^\circ_R / (\ln 10)RT$; R = 8.31440 J mol^{-1} K^{-1} ; T = 298.15 K. 1. Robie, Hemingway and Fisher (1978). 2. Hemingway and Robie (1977). 3. Sarazin (1979), cf. Text. 4. Hemingway, Robie and Kittrick (1978). 5. Naumov et al. (1971). 6. Bottero et al (1980). 7. Mattigod and Kittrick (1980) Farmer et al. (1979)

$$Al_{13}O_4(OH)_{28}^{3+} + 36\ H^+ = 13\ Al^{3+} + 32\ H_2O$$

These data allow computation of Gibbs free energies of formation and of log K for dissociation reactions, using data for water and OH^- aq. from Robie et al. (1978). Reliability of thermodynamic data depends on the nature of aqueous species considered - or ignored - in the computations. Most uncertain date are those concerning $Al(OH)_3^{\circ}$ and polynuclear species.

Complex formation and activity corrections

The mass balance for Al can be written :

$$m_{Al} = \Sigma m_{Al(OH)_n^{3-n}} + 13\ m_{Al'}$$

if Al' designates the polynuclear species $Al_{13}O_4(OH)_{28}^{3+}$

This mass balance can be combined with the mass action equations :

$$\log m_{Al(OH)_n^{3-n}} = \log m_{Al^{3+}} + n\ pH - \log K'_n$$

$$\log m_{Al'} = 13 \log m_{Al^{3+}} + 36\ pH - \log K'_{13}$$

where $\log K'_n = \log K_n + \log\gamma_{Al(OH)_n^{3-n}} - \log\gamma_{Al^{3+}}$

and $\log K'_{13} = \log K_{13} + \log \gamma_{Al'} - 13 \log \gamma_{Al^{3+}}$

Activity coefficients are computed from the Debye-Hückel extended law :

$$\log \gamma_i = - A z_i^2 I^{1/2} / (1 + a_i B I^{1/2}) + C I$$

with at 25°C and 10^5 Pa, A = 0.5092, B = 0.3284, C = 0.041 ;

a_i = 9 for Al^{3+}, 6 for $AlOH^{2+}$, 4 for $Al(OH)_2^+$ and $Al(OH)_4^-$ (Fritz, 1975). Bottero et al. (1980) proposed a_i = 19.7 for $Al_{13}O_4(OH)_{28}^{3+}$, the Debye-Hückel extended law giving a better fitting to their experimental data than the Güntelberg law. These equations allow calculation of ionic distributions, activities and molalities of every aluminum species, the ionic strength I being computed from the complete analysis and ionic distribution, by the definition :

$I = (\Sigma z_i^2 m_i)/2$ where z_i is the electrostatic charge of the species.

The contribution of aluminum species to the acid neutralizing capacity (A.N.C.) of the solution can be computed from the equation :

$$A.N.C._{Al} = \Sigma\ n\ m_{Al(OH)_n^{3-n}} + 36\ m_{Al'}$$

which is a term in the equation of alkalinity :

$$Alk = A.N.C._{Al} + m_{HCO_3^-} + 2\ m_{CO_3^{2-}} + m_{OH^-} - m_{H^+}$$

This gives an independent check of the validity of the calculations,

$$A.N.C._{Al} + m_{H^+} - m_{OH^-} < Alk$$

TABLE 2

Thermochemical data for aluminosilicate minerals at 298.15 K, 10^5 Pa

$Al(OH)_3$	$S° = 68.44$	$\Delta G_f^\circ = -1154.889$	Ref. 1
Gibbsite	Gibbsite + 3 H^+ = Al^{3+} + 3 H_2O		
	$\Delta G°_R = -45.934$	log K = 8.047	
$Si_2Al_2O_5(OH)_4$	$S° = 203.05$	$\Delta G_f^\circ = -3799.364$	do.
Kaolinite + 6 H^+ = 2 Al^{3+} + 2 H_4SiO_4 + H_2O			
	$\Delta G°_R = -32.577$	log K = 5.71	
$Si_2Al_2O_5(OH)_4$	$S° = 203.00$	$\Delta G_f^\circ = -3780.713$	do.
Halloysite + 6 H^+ = 2 Al^{3+} + 2 H_4SiO_4 + H_2O			
	$\Delta G°_R = -51.228$	log K = 8.97	
$SiAl_2O_3(OH)_4$	S° unknown	$\Delta G_f^\circ = -2926.7$	Ref.8
Imogolite + 6 H^+ = 2 Al^{3+} + H_4SiO_4 + 3 H_2O			
	$\Delta G°_R = -71.523$	log K = 12.53	

S° J mol^{-1} K^{-1} ; ΔG_f° and ΔG_R° kJ mol^{-1}

same references as Table 1

Nature of secondary aluminosilicate minerals considered

Along with gibbsite, kaolinite and imogolite (Farmer et al., 1979), 2-1 clay minerals having a Si/Al ratio from 1.1 to 1.25 were considered (Tables 2 and 5). These latter minerals were found in soils and weathered granitic arene in Armorican Massif (Curmi, 1979). Al is largely dominant and iron was not considered.

Variation of gibbsite solubility with temperature

The heat capacity of gibbsite between 13 and 480 K was measured by Hemingway et al.(1977). Their values between 273.15 and 298.15 K are C_p° = 84.77 ; 86.73 ; 89.51 and 91.7 J mol^{-1} K^{-1} at respectively 273.15 ; 280 ; 290 ; 298.15 K. This allows to fit in this range a polynom in the form of a Maier-Kelley function :

$$C_p^{\circ} = a + b\ 10^{-3}\ T + c\ 10^{5}\ T^{-2}$$

The values are listed in Table 3, with values for aqueous species Al^{3+} from Naumov et al. (1971), and liquid water from Mattigod and Kittrick (1980).
The integration was performed using the form proposed by Mattigod and Kittrick (1980) :

$$\ln K_T = \frac{1}{R}\ (A \ln T + B\ T + (C/T^2) + (D/T) + E)$$

where $A = \Delta a \quad B = \Delta b/2 \quad C = \Delta c/2$

$$D = \Delta a\ T_r + (\Delta b/2)\ T_r^{2} - (\Delta c/T_r) - \Delta S^{\circ}_{T_r}\ T_r + RT_r \ln K_{T_r}$$

$$E = -\ \Delta a - \Delta a \ln T_r - \Delta b\ T_r + (\Delta c/2T_r^{2}) + \Delta S^{\circ}_{T_r}$$

$$\Delta S^{\circ}_{T_r} = S^{\circ}_{Al^{3+}} + 3\ S^{\circ}_{H_2O} - S^{\circ}_{gibbsite}$$

T_r = reference temperature

The value chosen for $S^{\circ}_{Al^{3+}}$ is − 308.67 J mol^{-1} K^{-1} instead of −308 ± 1 given by Hemingway and Robie (1977), because the chosen value is based on ΔS°_R = − 141 for reaction Al + 3 H^+ = Al^{3+} + 3/2 H_2. With S°_{Al} = 28. and $S^{\circ}_{H_2}$ = 130.68 (Robie et al., 1978), one obtains − 308.67 J mol^{-1} K^{-1} Hence, $\Delta S^{\circ}_{T_r}$ = − 167.26 J mol^{-1} K^{-1}. The solubility of gibbsite increases from log K = 8.64 at 15°C to 9.27 at 5°C (Table 4).

TABLE 3

Heat capacity data for solubility variation with temperature

Species	C_p° function (J mol^{-1} K^{-1}) $C_p^\circ = a + b\ 10^{-3}\ T + c\ 10^5\ T^{-2}$			Ref.
	a	b	c	
H_2O (l)	49.66	54.31	8.44	7
$H_4SiO_4{}^\circ{}_{aq.}$	0	719.65	0	5
$Al^{3+}_{aq.}$	0	128.03	0	5
Gibbsite	48.198	185.898	- 10.600	This study
	$Al(OH)_3$ +	$3 H^+$ =	$Al^3 + 3 H_2O$	
ΔS°_R	Δa	Δb	Δc	
-167.26	100.782	105.062	19.040	

Kaolinite

$$C_p^\circ = 1430.26 - 0.78850\ T + 3.0340\ 10^{-4} T^2 - 1.85158\ 10^4 T^{-1/2} + 8.3341\ 10^6\ T^{-2}$$ (Ref. 2)

$$\text{Kaolinite} + 6\ H^+ = 2\ Al^{3+} + 2\ H_4SiO_4 + H_2O$$

ΔS°_R = - 390.44

$$\Delta C_p^\circ = -\ 1380.6 + 2.53817\ T - 3.0340\ 10^{-4} T^2 + 1.85158\ 10^4\ T^{-1/2} - 7.4901\ 10^6 T^{-2}$$

TABLE 4

Variation of solubilities of gibbsite and kaolinite from 0°C to 25°C

t°C	0	5	10	15	20	25
log K Gibbsite	9.615	9.270	8.947	8.635	8.335	8.047
log K Kaolinite	8.16	7.63	7.11	6.62	6.155	5.71
log $a_{H_4SiO_4}$	-5.53	-5.46	-5.39	-5.32	-5.26	-5.19

Variation of kaolinite solubility with temperature and crystallinity

(i) Crystallinity variations at 25°C

Hemingway et al. (1978) consider that "the thermodynamic properties of several gibbsite samples of varying crystal size, in the range from 50 to 0.05 μm, are identical". However, amorphous $Al(OH)_3$ may form (Sarazin, 1979). Unlike gibbsite, kaolinite is known to present variations in solubility, ranging from log K = 8.02 to log K = 7.02 for "England" and "Georgia 2" kaolinites (Kittrick, 1980), for the reaction

$$Si_2Al_2O_5(OH)_4 + 6\ H^+ = 2\ Al^{3+} + 2\ H_4SiO_4 + H_2O$$

Natural samples of kaolinite can be considered as ranging from well crystallized kaolinite (log K = 5.71) to halloysite (log K = 8.97). As a consequence, equilibrium between gibbsite and kaolinite occurs at activities of aqueous silica ranging from $\log a_{H_4SiO_4} = -5.2$ to -3.6.

(ii) Temperature variation

Hemingway et al. (1978) proposed a function for C_p°, which combined with C_p° of Al^{3+}, aqueous silica and water give a ΔC_p° function (Table 3), hence by integration :

$$RT \ln K_T = RT_r \ln K_{T_r} + \Delta S^\circ_{R,T_r} (T - T_r) - (L(T) - L(T_r)) + T(K(T) - K(T_r))$$

where $L(T) = \int \Delta C_p^\circ\, dT$ and $K(T) = \int \Delta C_p^\circ\, d\ln T$

$$L(T) = aT + b\,T^2/2 + c\,T^3/3 + 2d\,T^{1/2} - e\,T^{-1}$$

$$K(T) = a \ln T + b\,T + c\,T^2/2 - 2\,d\,T^{-1/2} - e/2\,T^2$$

The variation with temperature is practically the same as obtained by Fritz and Tardy (1973) and Mattigod and Kittrick (1980) using Maier Kelley type functions. Results are given in Table 4. Activity of aqueous H_4SiO_4 at equilibrium between gibbsite and kaolinite ranges from $10^{-5.46}$ to $10^{-5.32}$ from 5 to 15°C.

RESULTS AND DISCUSSION

The temperature dependance of polynuclear species is at present unknown. In a first step, polynuclear species were ignored and a computation was made at the actual temperature of the sample, using procedure THERMA, designed by Fritz (1981) (Fig. 1). In a second step, polynuclear species were considered, though calculations were restricted to 25°C., on the same 70 samples (Fig. 2). Ionic strength computed in the first step was used in the second. In all cases, aluminum contribution to alkalinity was minor, but not negligible. Though polynuclear species are not always dominant, $\log a_{Al^{3+}}$ decreases dramatically from -7.3 to -9.8 on average. Ignoring polynuclear species leads to considerable apparent supersaturation with respect to both minerals, gibbsite (log Q/K = + 1 to + 4.2) and kaolinite (log Q/K = + 4.5 to + 17). The slope of the regression line of $\log a_{Al^{3+}}$ versus pH is about 2, diffe-

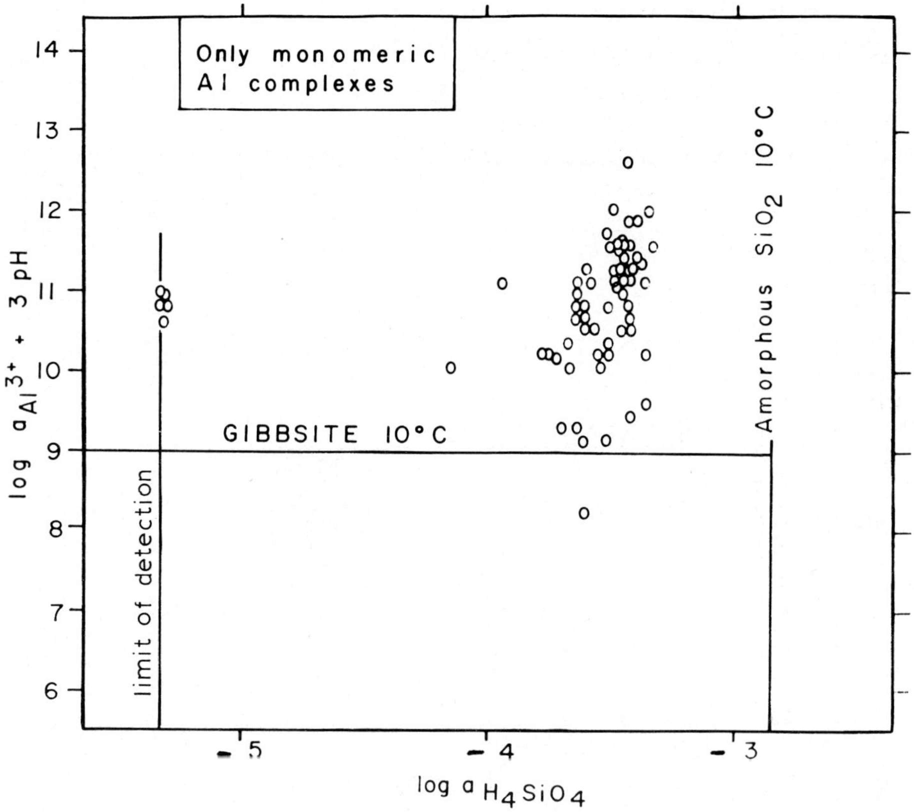

Fig. 1. Al - Si activities diagram considering only monomeric Al hydroxocomplexes

rent from the theoretical value 3. As pH increases, the discrepancy increases, i.e. Al dissolved is much greater than predicted.

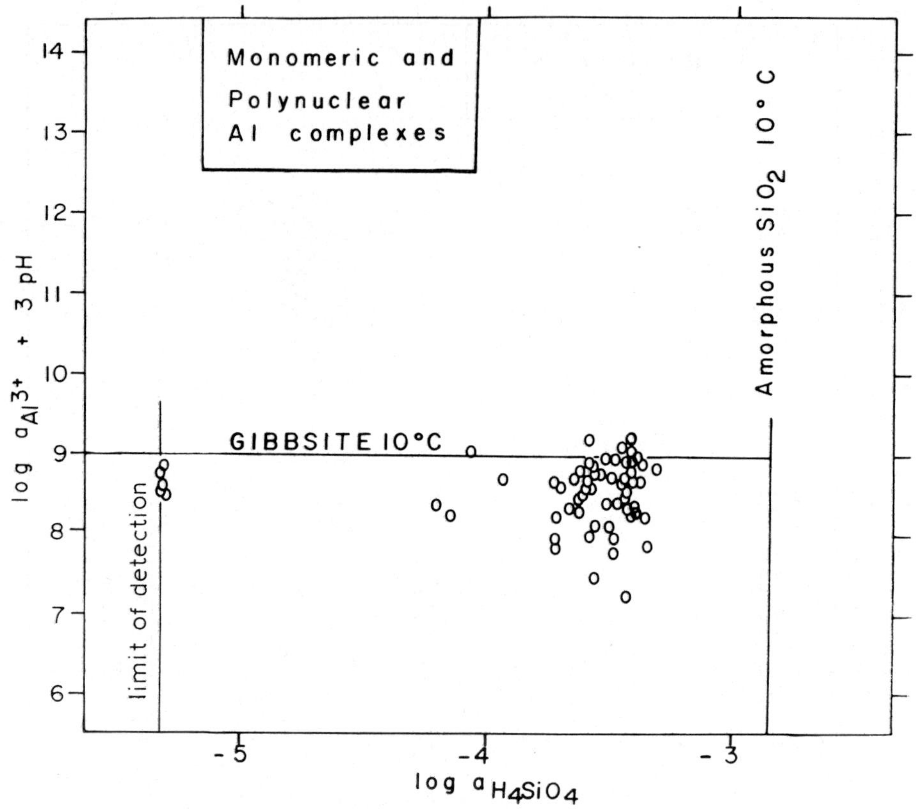

Fig. 2. Al - Si activities diagram considering Al monomeric and polynuclear species $Al_{13}O_4\ (OH)_{28}^{3+}$

With polynuclear and monomeric species (Fig. 2), log $a_{Al}3+$ + 3 pH remains between 7 and 9.2, i.e. the variation is two logarithmic units lower, with an average value 8.3.

Samples that are depleted in aqueous silica (Si<7 10^{-6}M) bear log Q = 8.4 to 8.9. This value seems thus to correspond to a control. This is higher than gibbsite solubility at 25°C (log K = 8.047), but the temperature of the samples ranges from 5 to 15°C, with an average value of 10°C. It seems thus necessary to take into account the variation of solubility with temperature (Fig. 3). When corrected for temperature variation log Q/K does not exceed + 0.36. Uncertainty of gibbsite

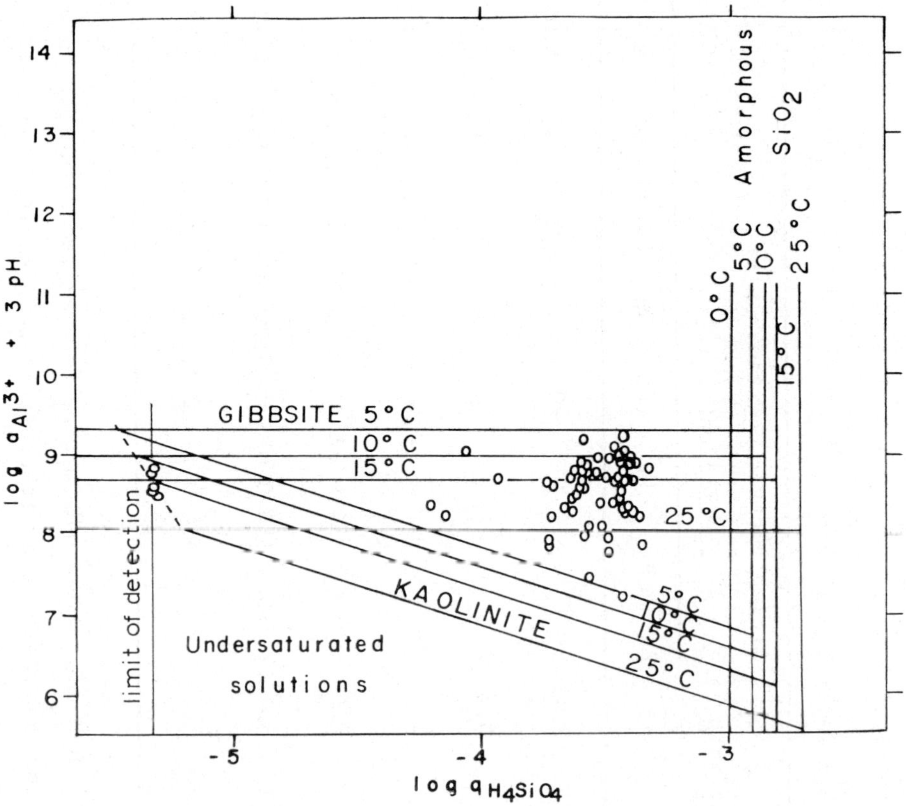

Fig. 3. Variation of solubilities of gibbsite, kaolinite and amorphous silica as a function of temperature

solubility is about 0.12 (Hemingway and Robie, 1977). Uncertainties of total aluminum (<30 %), and of pH (0.02) lead to about 0.2 variation of log Q. It seems thus that 0.2 to 0.3 log. unit is a minimum. Samples for which log Q/K < - 0.4 will be considered as undersaturated with respect to gibbsite. Half of the samples are undersaturated with respect to

TABLE 5

Assessment of aluminosilicate - water equilibria

Name	Structural formula	$\frac{Si}{Al}$	log K	Stoichiometric coefficient b_i log a_i coefficient							Linear Slope		regression coefficient[a]	
				Al^{3+}	K^+	Na^+	Ca^{++}	Mg^{++}	$Si(OH)_4$	pH	m_1	m_2	r_1	r_2
Imogolite	Si Al_2O_3 $(OH)_4$	0.5	12.53[b]	2	0	0	0	0	1	6	6.05	5.61	-0.99	-0.98
Kaolinite	$Si_2Al_2O_5$ $(OH)_4$	1	5.71	2	0	0	0	0	2	6	5.98	5.50	-0.99	-0.98
Hydroxy Al vermiculite		1.1	18.1[c]	2.9	0.05	0.05	0.05	0.25	3.15	9.4	8.63	7.91	-0.99	-0.98
Weathered biotite (BA - 7)		1.1	17.1[c]	2.876	0.134	0.075	0.075	0.298	3.132	9.583	8.55	7.80	-0.99	-0.975
Weathered biotite (BA -10)		1.15	16.4[c]	2.765	0.134	0.075	0.075	0.308	3.195	9.27	8.21	7.48	-0.99	-0.97
Vermiculite (VERM)		1.25	15.7[c]	2.6	0.4	0.1	0.10	0.25	3.70	9	7.70	6.89	-0.99	-0.97

a. $\log Q'_i = \Sigma b_i \cdot \log a_i$, without pH, versus pH. m_1, r_1 for samples saturated (n = 36) with respect to gibbsite ; m_2, r_2 for samples undersaturated with respect to gibbsite. 2/1 clay minerals analyses from Curmi (1979). b. Farmer et al. (1979). c. estimated by Aurousseau (personal communication) following Tardy and Garrels method (1974).

gibbsite and only 15 % give log Q/K < - 0.75.

The problem of aluminosilicate-water equilibria

Undersaturation may result from two situations : equilibrium not yet attained or control of activities by another mineral more insoluble than gibbsite. All samples are supersaturated with respect to kaolinite, silica concentration ranging from $10^{-4.2}$ to $10^{-3.3}$, except samples depleted in silica. Computing ionic products for aluminosilicate minerals leads always to excellent correlation with pH (Table 5) due to aluminum activity being strongly pH dependent, but as the Si/Al ratio increases, the slope of the regression line departs more and more from the theoretical (stoichiometric) slope. Al is thus more soluble than would be predicted, if it were controlled. Average values of log Q/K are equal to + 1 for imogolite, + 4.3 for kaolinite, - 0.3 for weathered biotite BA7, - 1 for weathered biotite BA10, - 1.3 for vermiculite and - 2 for hydroxy-Al vermiculite, when compared to log K values at 25°C (Table 5). However, no control on aluminum is apparent, since there is no correlation between $\log a_{Al^{3+}} + 3\ pH$ and $\log a_{H_4SiO_4}$ (r = 0.06), even if one considers separately waters saturated and undersaturated with respect to gibbsite, though some waters are supersaturated or close to saturation with respect to clay minerals.

CONCLUSION

Aluminum variation in spring waters from Fougères is governed by gibbsite solubility and its variations with temperature. This field evidence agrees with recent experimental data on the stability and heat capacity of gibbsite determined independently. There is no need to consider more soluble forms, at least on slowly weathering rocks such as granite. Dissolved aluminum is probably present partly as polynuclear complexes in near neutral environment. More data are needed to precise the stoichiometry, electrostatic charge and temperature dependence of equilibrium constants for these species. Ignoring these forms leads to a large apparent supersaturation with respect to gibbsite and to higher solubility of $Al(OH)_3$.

Waters are supersaturated with respect to kaolinite, which is present, but not effective enough to control aluminum. These two facts result in rather large dissolved aluminum contents in natural waters.

Silica variations are rather narrow, except in some summer periods probably due to its biological removal. Silica concentration is not predictable from aluminum and pH and the strong pH dependence of Al on pH may be misleading when computing ionic products for complex composition minerals.

Acknowledgments

I wish sincerely thank G. Michard and G. Sarazin (Paris) for assistance to analyze Al in waters, and B. Fritz (Strasbourg) for help to compute ionic activities.

TABLE 6

Chemical composition of some spring waters, representative of the whole range of variation observed

Range of Log Si	<- 5.3	- 4.8 - 4.30	- 3.72 - 3.96	- 3.60 - 3.72	- 3.48 - 3.60	- 3.36 - 3.48	- 3.24 - 3.36
n[(a)]	5	3	5	12	16	27	2
Sample No.	213	209	149	194	191	230	112
Date	June 78	June 78	Feb.78	Ap.78	Feb.80	Sept.78	Dec.7
Temperature °C	11.0	12.9	6.3	8.7	6.0	11.40	5.5
pH	6.22	5.56	5.20	5.52	4.96	5.80	6.90
I[b]	1.20	1.00	0.85	1.20	1.06	1.63	1.18
log Si	<- 5.3	- 4.15	- 3.74	- 3.62	- 3.54	- 3.37	- 3.33
total Al[(c)]	1.50	1.5	7.4	10.7	25.6	5.6	1.9
log a_o [(d)]	- 7.67	- 6.65	- 5.45	- 5.59	- 4.83	- 6.37	- 8.66
log a'_o [(e)]	-10.1	- 8.44	- 6.96	- 7.72	- 6.14	- 8.54	-11.9
log m_{Al} [(f)]	-12.2	-14.5	- 8.19	- 6.49	- 6.13	- 7.07	-11.4
log Q/K[(g)] gibbsite	- 0.32	- 0.53	- 0.30	- 0.19	- 0.46	0.00	- 0.45

a. n = number of samples in this range of variation of log Si .
b. I = ionic strength, mM. c. total Aluminum, μM. d. log a_o is logarithm of activity of Al^{3+} with only monomeric species, computed by using program THERMA (Fritz, 1981). e. log a_o' is logarithm of activity of Al^{3+} computed considering polynuclear species. f. $m_{Al'}$ is computed molality of the polynuclear species. g. log Q/K = log a'_o + 3 pH − log K (T) .

Supplementary material

Complete analyses and ionic distribution of aqueous species are available on request.

REFERENCES

1. J.Y. Bottero, J.M. Cases, F. Fiessinger and J.E. Poirier, J. Phys. Chem., 84 (1980) 2933-2939.
2. P. Curmi, Altération et différenciation pédologique sur granite en Bretagne, étude d'une toposéquence, Thesis, Rennes, 1979, 176 p.
3. V.C. Farmer, B.F.L. Smith and J.M. Tait, Clay Minerals, 14 (1979) 103-107.
4. B. Fritz, Etude thermodynamique et simulation des réactions entre minéraux et solutions. Application à la géochimie des altérations et des eaux continentales. Thesis, Strasbourg, 1975 and Sci. Geol. Mém. 41, 152 p.
5. B. Fritz, Etude thermodynamique et modélisation des réactions hydrothermales et diagénétiques, Thesis, Strasbourg, 1981.
6. B. Fritz and Y. Tardy, Sci. Geol. Bull., Strasbourg, 26 (1973) 339-367.
7. J.D. Hem, in K.H. Wedepohl (Ed.), Handbook of Geochemistry (1970), pp. 13G-13L.
8. B.S. Hemingway and R.A. Robie, Geochim. et Cosmochim. Acta, 41 (1977) 1402-1404.
9. B.S. Hemingway, R.A. Robie and J.A. Kittrick, Geochim. et Cosmochim. Acta, 42 (1978) 1533-1543.
10. P.H. Hsu, in J.B. Dixon and S.B. Weed (Eds.) Minerals in soil environments, Soil Sci. Soc. Am. (1977), pp. 93-98.
11. J.A. Kittrick, Soil Sci. Soc. Am. J., 44 (1980) 139-142.
12. S.V. Mattigod and J.A. Kittrick, Soil Sci. Soc. Am. J., 44 (1980) 149-154.
13. G.B. Naumov, B.N. Ryzhenko and I.L. Khodakovsky, Handbook of thermodynamic data (in Russian), Atomizdat, Moscow, 1971, 239 p.
14. R.A. Robie, B.S. Hemingway and J.R. Fisher, U.S. Geol. Surv. Bull., 1452 (1978), 456 p.
15. G. Sarazin, Géochimie de l'aluminium au cours de l'altération des granites et des basaltes sous climat tempéré, Thesis, Paris, 1979, 169 p.
16. Y. Tardy and R.M. Garrels, Geochim. et Cosmochim. Acta, 38 (1974) 1101-1116.
17. J.V. Walther and H.C. Helgeson, Am.J. Sci., 277 (1977) 1315-1351.

THE THERMODYNAMIC STATUS OF CLAY MINERALS

F. LIPPMANN, Mineralogisches Institut der Universität, Tübingen, WEST GERMANY

INTRODUCTION

In modern petrology, rock-forming minerals are regarded as thermodynamic phases which form under conditions near chemical equilibrium. The stability fields of most minerals in terms of temperature and pressure are now well known for the temperature range above, say, 300°C. Closer to temperatures prevailing at the surface of the earth, the reactivity of silicates, especially of alumosilicates, decreases to the extent that establishment of equilibrium remains in doubt for many experiments carried out below 300°C. Similarly, many natural silicate assemblages that originated near earth-surface temperatures, especially clays, may be expected not to have attained thermodynamic equilibrium. There are many obvious indications supporting this view. Some strictly thermodynamic reasoning will be presented in this paper. However, since establishment of thermodynamic equilibrium is the final stage of a system at any temperature, calculation of silicate equilibria and their representation as phase diagrams for ordinary temperature remains extremely interesting, because we can at least learn in which direction a certain system would react, in case reactivity conditions become more favourable due to some change in environmental parameters.

Variability of temperature and pressure is rather restricted in sedimentary environments in comparison to the wide ranges covered by metamorphism. In calculating stability relations for clays and related minerals from thermochemical data obtained from calorimetric measurements and by extrapolation of high-temperature equilibria, we are thus not so much concerned with variations of temperature and pressure. In view of the errors affecting thermochemical data of silicates, it is sufficient to start the discussion for standard conditions, i.e. 25°C and 1 bar.

In contrast with metamorphic systems for which the composition of the reacting aqueous phase is determined largely by the minerals present, the water compositions near the surface of the earth show wide variations in different geological environments. It is therefore more important to determine mineral stabilities as functions of the composition of aqueous solutions. This problem is related to determining aqueous solubilities of clay minerals. Solubilities would represent adequate and, at the same time, very practical parameters of stability, because all clays are known to have formed in or from aqueous solutions. However, the measurement of their solubilities is impaired by sluggish reactivity in the same way as is the establishment of other silicate equilibria at ordinary temperature. Furthermore, there is no agreement so far as to how the aqueous solubility of more complicated clay minerals, such as illite and montmorillonite, should be defined, although the present writer proposed such a measure in the form of the total solubility product (Lippmann, 1977a, 1980).

In order to see these problems more clearly and, more important, to answer the question which clay minerals, if any, are endowed with fields of thermodynamic stability, we must study them under the aspect of thermodynamic phases, i.e. enquire into their thermodynamic status.

METASTABILITY IN OXIDE AND HYDROXIDE COMPONENTS OF CLAY MINERALS

The chemical component species of clay minerals are already characterized by slow reaction rates from ordinary temperature well into the range of hydrothermal reactions. This experience is reflected in the number of polymorphs which exist for the oxides and hydroxides of silicon, iron, and aluminium. Of the SiO_2 minerals, quartz, several varieties of cristobalite and of opal are well known as neoformations in clays. Of these, quartz is the most stable mineral, because it is least soluble in water. The other SiO_2 minerals, in spite of their lower stabilities compared with that quartz, require geological periods of time to recrystallize to quartz.

Of the system Fe_2O_3 - H_2O, the minerals hematite, maghemite, goethite, lepidocrocite, amorphous $Fe(OH)_3$ and some more recently described crystalline minerals are found in soils and sediments. Goethite and, more probably, hematite are the candidates eligible for stable phases. Obviously, the rate of formation of the most stable phase can be so slow that it is easy to intercept the other unstable minerals for determination.

All of the several hydroxides of of aluminium are capable of forming in soils and sediments, especially bauxites. These are often composed of two or more Al hydroxides. Some certainly non-metamorphic bauxites contain in addition fine-grained authigenic corundum. According to current thermochemical data (Robie et al., 1978), diaspore is the stable phase in the system Al_2O_3 - H_2O at ordinary temperature. The other Al hydroxides, boehmite, gibbsite, nordstrandite, and bayerite are less stable and must be regarded as metastable precursors of diaspore. Their persistence in natural materials must be due to the extremely slow rate of formation of diaspore. Corundum becomes more stable than diaspore above 300°C.

APPLICATION OF THE PHASE RULE

In addition to the persistence of individual unstable polymorphs, the coexistence of two or more hydroxides in a bauxite is already indicative of thermodynamic non-equilibrium. According to the *mineralogical phase rule* (cf. Barth, 1962, p. 398), a bauxite characterized by the system Al_2O_3 - H_2O should not contain more than one hydroxide mineral under equilibrium conditions, because the number of phases (one hydroxide phase + aqueous solution) should not exceed the number of components (Al_2O_3 and H_2O).

At least four different minerals with the composition $Al_2Si_2O_5(OH)_4$ are known: dickite, nacrite, ordered (triclinic) kaolinite and disordered (monoclinic) kaolinite. Since these minerals are known to form from aqueous solutions, it is realistic to consider the system $Al_2Si_2O_5(OH)_4$ - H_2O. Consequently, hydrated halloysite can be regarded as an additional polymorph of kaolinite. Again, the coexistence of two or more kaolinite minerals (e.g. halloysite + kaolinite, or ordered kaolinite + disordered kaolinite) in a clay or a soil represents a violation of the phase rule and indicates non-equilibrium. An assemblage of several kaolinite polymorphs can at most be regarded as a system on its way toward some equilibrium, e.g. complete conversion to ordered kaolinite. However, according to lattice-energy calculations by Giese (1973), dickite is more stable than kaolinite. Consequently, it should be less soluble in water and thus crystallize at the expense of kaolinite. The experience that kaolinite persists in all surface deposits shows that this reaction is practically frozen under earth-surface conditions. The formation of dickite by burial diagenesis does not mean that it is a high-temperature phase, because the lattice energy indicates stability at low temperatures. The mode of occurrence of dickite can merely be explained by recognizing that temperature, or rather thermal activation, promotes the formation of the most stable phase. Only nacrite remains as a possible candidate for a high-temperature polymorph.

Halloysite, disordered kaolinite, more and more completely ordered kaolinites, and finally dickite form a hierarchy characterized by increasing thermodynamic stability. For most practical purposes, it is justifiable to consider fields for kaolinite and gibbsite in activity diagrams encompassing these compositions. However, in order to interpret correctly the rare but indisputable occurrences of dickite and diaspore in unmetamorphosed sediments, it must be borne in mind that the fields of kaolinite and gibbsite are not, strictly speaking, stability fields but rather fields of metastable existence delimited by metastable equilibria. It is interesting that dickite is one of the few minerals that have not been synthesized so far.

The phase rule may also be applied to coexisting illite polymorphs. The assemblage of $2M_1$ and 1Md illites reported by Velde and Hower (1963) is certainly more wide-spread in clays but may have escaped detection in many cursory determinations of illite. The two-phase assemblage indicates non-equilibrium. Most likely, the $2M_1$ polymorph is more stable than 1Md.

On the basis of the argument "too many phases", one might conclude that many zeolites and assemblages of zeolites are perhaps not absolutely stable but that they are metastable precursors of assemblages of feldspars plus quartz.

CONCLUSIONS FROM THE COMPOSITION OF THE THREE-LAYER CLAY MINERALS

Most of the minerals discussed so far show simple chemical compositions characterized by whole-number subscripts in their chemical formulae. These features they have in common with many other minerals forming near ordinary temperature, especially evaporite minerals. By contrast, the three-layer clay minerals, notably illites and montmorillonites, show more complicated chemical formulae characterized by fractional subscripts. These are normally explained as the result of various crystal-chemical substitutions based on similar dimensions

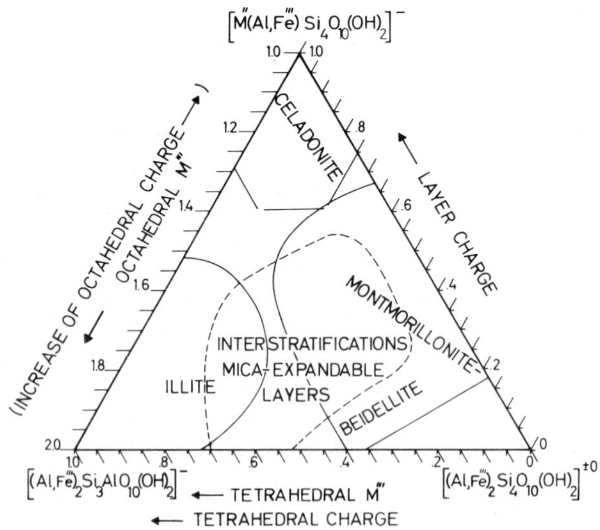

Fig. 1. The chemical compositions of three-layer clay minerals in the system mica-pyrophyllite-celadonite (after Weaver and Pollard, 1973)

of the ions involved. For chemical thermodynamics, it is essential to express the compositions in terms of end members. If we disregard different interlayer cations and their hydration, illites, montmorillonites and related interstratifications may be represented with fair approximation in a triangular diagram with mica, pyrophyllite and "celadonite" as the end members (Fig. 1) . It turns out that the minerals in question show solid solubility to an extent which is unusual at ordinary temperature. For example, in the system paragonite-muscovite (Fig. 2), phases with comparable degrees of substitution would only be in equilibrium above about 600°C. Furthermore, the end members mica and pyrophyllite of Fig. 1 show no appreciable miscibility in metamorphic rocks up to their upper limits of stability.

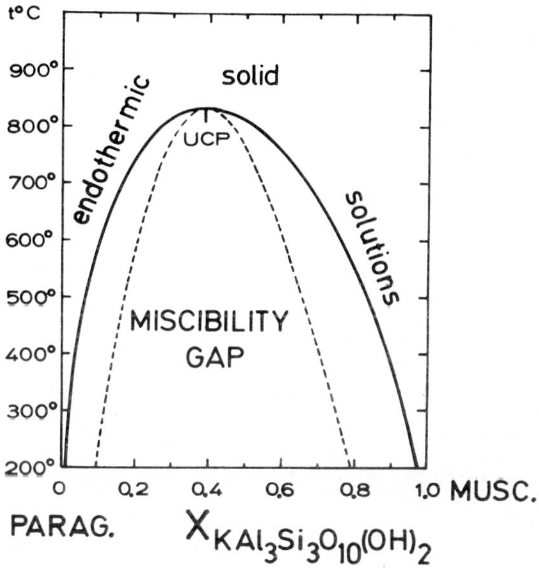

Fig. 2. The system paragonite-muscovite (after Eugster et al., 1972) showing a miscibility gap characterized by an upper critical point (UCP). Solid solution outside the continuous trace are thermodynamically stable. Inside the dashed trace (spinode), solid solutions are completely unstable. Mixed crystals between the continuous and the dashed lines are metastable.

This observation is explained by the fact that their crystal structures differ still more than those of paragonite and muscovite. Solid solubility in the latter system is characterized by an upper consolute temperature or upper critical point which is located above 800°C and is marked UCP in Fig. 2. Above this temperature, the two end members would form a complete series of stable mixed crystals. Toward lower temperatures, the equilibrium changes more and more in favour of the pure end members, and mutual solubility is negligible at ordinary temperature. This type of diagram, showing an upper critical point, is common to many mixed-crystal systems characterized by incomplete miscibility, e.g. alkali feldspars and NaCl - KCl. It explains why so many low-temperature minerals have rather simple chemical formulae and contain only trace impurities.

For the wide range of solid solubility of illites and montmorillonites to represent thermodynamically stable phases, one might consider a solubility diagram characterized by a lower critical point as known for binary systems of certain organic compounds with water (Fig. 3).

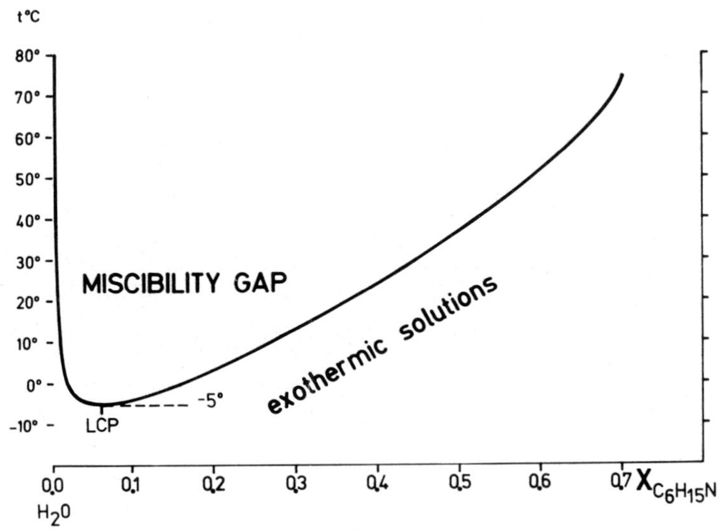

Fig. 3. The system di-n-propylamine - water showing a miscibility gap characterized by a lower critical point (LCP). In a system of this type, the mixing process is exothermic.

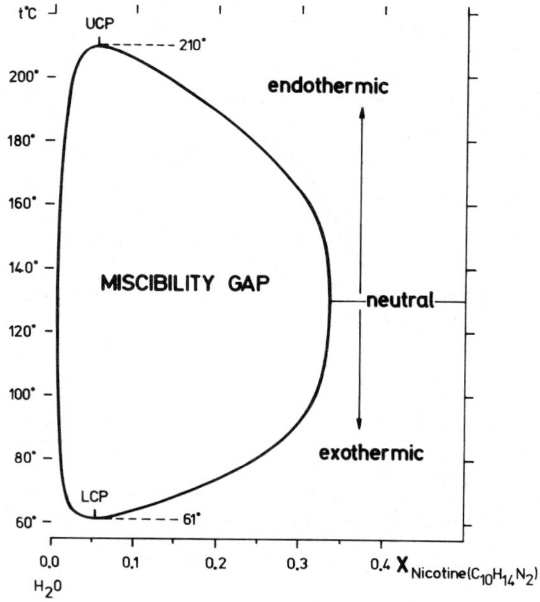

Fig. 4. The system nicotine - water showing a closed miscibility gap with an upper (UCP) and a lower critical point (LCP); endothermic, neutral, and exothermic indicate the sign of the heat of mixing.

In order to explain the limited miscibility in low-grade metamorphism and to allow, at the same time, for an increase of solid miscibility toward higher temperatures, one might even consider a closed miscibility gap with a lower *and* an upper critical point as known e.g. for the system nicotine-water (Fig. 4). However, for a lower critical point to exist, chemical thermodynamics require that the mixing in the pertinent temperature range be exothermic (Prigogine and Defay, 1954, p. 287). By contrast, in all mixed crystals for which the heat of mixing is known, the sign is endothermic. Calculations of the electrostatic excess lattice energies of predominantly ionic mixed crystals have also yielded an endothermic sign throughout (Urusov, 1975; Lippmann, 1980; H.D.B. Jenkins, private communication), for example, the excess lattice energies in the system paragonite- pyrophyllite as shown in Fig. 5. The value of the maximum excess energy of 17.9 kJ/mol (4.29 kcal/mol) is remarkable. After recalculation according to the theory of regular solutions (see Lippmann, 1980), equilibrium miscibility is found to be negligible between paragonite

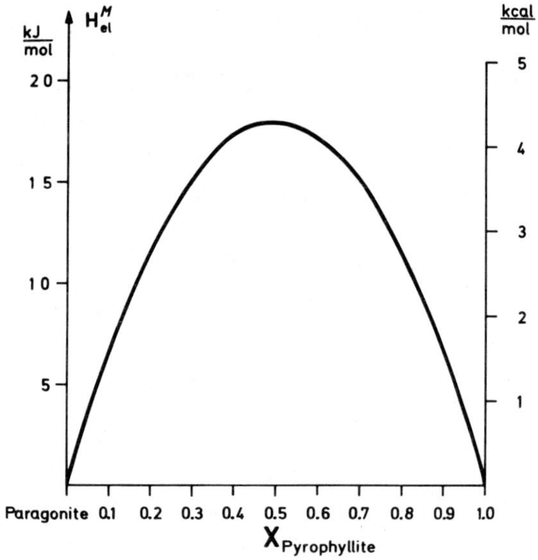

Fig. 5. First estimate of the electrostatic excess lattice energies H^M_{el} of mixed crystals between paragonite and pyrophyllite by H.D.B. Jenkins. For regular solutions, H^M_{el} is equal to the excess Gibbs energy of mixing G^E .

and pyrophyllite from ordinary temperature up to and into the range of metamorphism. One might object that the excess lattice energy could be compensated by the heat of hydration, e.g. in a beidellite composition. The experimental heat of hydration determined by Keren and Shainberg (1975) for Na montmorillonite is 18.4 kJ/mol (4.4 kcal/mol). This value is probably high, because it contains the heat evolved by water adsorption on the external surfaces. By contrast, electrostatic excess energies tend to be low. Hydration thus appears not to be sufficient to render the mixing of paragonite and pyrophyllite outright exothermic and to lead to a distinct increase of equilibrium miscibility toward low temperature, although hydration explains why expandable minerals form at all.

The stability of montmorillonites is impaired also by substantial substitution of pyrophyllite by end members that are likely to be unstable themselves, if they exist at all, i.e. sodium and calcium "celadonites". Even the potassium-celadonite end members contained in illites and phengites are probably less stable than an assemblage of biotite + K-feldspar + quartz (Lippmann, 1979, 1981).

The general conclusion from the above analysis of three-layer clay minerals in terms of their end members is that illites and montmorillonites must be less stable thermodynamically than the end members (or equivalent stable assemblages) of which they are composed. Similarly, interstratifications of illite and expandable layers should also be unstable, and this may apply as well to regularly interstratified minerals, such as rectorite, corrensite, and tosudite.

Of the fibrous minerals, sepiolite is certainly less stable than talc by virtue of aqueous solubility relations. It appears to be a metastable precursor of talc, especially in evaporite deposits, and palygorskite (attapulgite) is likely to be unstable as well.

The stability of all clay minerals is certainly decreased by their fine grain size. The fact that clay minerals do not grow into coarser crystals through solution-precipitation cycles (Ostwald ripening), not even in the course of geological periods of time, is characteristic of the conditions near the surface of the earth, where crystal growth in phyllosilicates is largely inhibited. On the other hand, the fine grain size is indicative of abundant nucleation and attests to the formation of clay minerals from highly supersaturated solutions.

We have seen that the three-layer clay minerals must be less stable than mechanical mixtures of minerals like pyrophyllite, muscovite, K-feldspar, quartz and biotite. These are formed at the elevated temperatures of metamorphism and in hydrothermal bombs. From such experiments and from calorimetric data, the Gibbs energies have been recalculated for 25°C and 1 bar for the above mentioned "metamorphic" minerals and for typically low-temperature minerals (Robie et al., 1978). From these values, solution equilibria can be calculated for ordinary temperature, e.g. for the system K_2O-Al_2O_3-SiO_2-H_2O, and represented in a diagram with the concentrations (activities) of K^+, OH^-, and H_4SiO_4 as the variables (Fig. 6).

The "metamorphic" minerals muscovite and pyrophyllite have true stability fields in this diagram, although they do not form at 25°C or even higher temperatures. The stability fields of these and other minerals characterized by simple chemical formulae cover the entire plain of the diagram so that no space is left for any fields for three-layer clay minerals.

Accepting these facts, we may explain the conditions of formation of illites and montmorillonites as follows. From the fine grain size of clays, we conclude that they precipitate at considerable supersaturation. In this way, component species are adsorbed from solution not only at "correct" sites compatible with the ideal crystal structures of muscovite or pyrophyllite but also at "wrong" sites where they fit nevertheless by virtue of similar size. The same happens to species extraneous to ideal muscovite or pyrophyllite, such as ions of magnesium, iron, calcium, and water molecules. Because of supersaturation, rejection from wrong sites and of wrong species is impossible. They are trapped or "carried down" and give rise to the substitutions that distinguish illite from muscovite, and montmorillonite from pyrophyllite. Supersaturation implies that the clay minerals form well within the stability field of their most abundant end

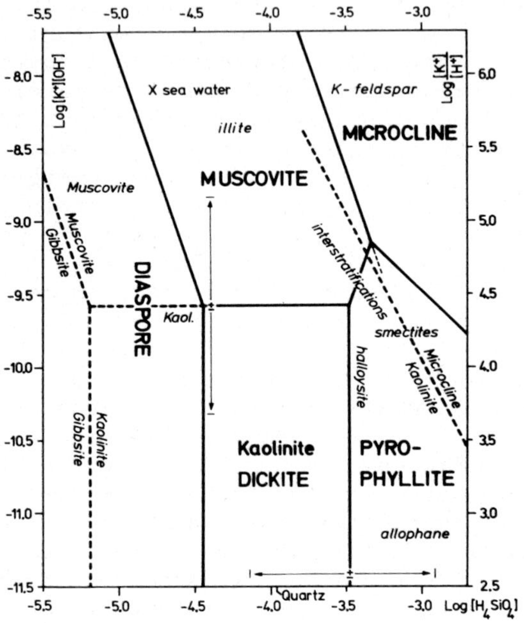

Fig. 6. The system K_2O-Al_2O_3-SiO_2-H_2O at 25°C calculated from data of Robie et al. (1978) with Al_2O_3 as the immobile component. Explanations: CAPITALS: stable minerals; *Italics* with first letter capital: metastable minerals reacting in metastable equilibrium; *italics*: unstable minerals; continuous lines: stable phase boundaries; dashed lines: metastable phase boundaries. Limits of uncertainty are indicated by arrows.

member: illite in the field of muscovite, and montmorillonite in that of pyrophyllite. Irregular interstratifications may be expected near the boundary separating the stability fields of muscovite and pyrophyllite.

The result of the present enquiry is that virtually all clay minerals are more or less metastable or even completely unstable . Nevertheless, they persist through geological times. They owe their existence and their many-varied properties not to thermodynamic equilibrium but to the kinetic inhibitions inherent in ordinary-temperature conditions. Thermodynamic instability must not be regarded as a shortcoming. Let us remember that the majority of organic compounds and virtually all biochemical substances are metastable as well.

Only metastable systems are capable of carrying information. By virtue of their metastability, clay minerals are useful as stratigraphic markers, as indicators of incipient metamorphism (Kübler, 1966) or of diagenesis as different from metamorphism (Lippmann, 1977b). If silicates would readily equilibrate at ordinary temperature, neither soils nor clays would exist and the earth would be covered by a shell of barren rocks to be classified near the greenschist facies.

REFERENCES

Barth, T.F.W., 1962. Theoretical Petrology. 2nd ed. Wiley, New York, 416 pp.

Eugster, H.P., Albee, A.L., Bence, A.E., Thompson, Jr., J.B. and Waldbaum, D.R., 1972. The two-phase region and excess mixing properties of paragonite-muscovite solid solutions. J. Petrology, 13: 147-179.

Giese, Jr., R.F., 1973. Interlayer bonding in kaolinite. Clays Clay Min., 21: 145-149.

Keren, R. and Shainberg, I., 1975. Water vapor isotherms and heat of immersion of Na/Ca-montmorillonite systems - I: Homoionic clay. Clays Clay Min., 23: 193-200.

Kübler, B., 1966. La cristallinité de l'illite et les zones tout à fait supérieures de métamorphisme. Etages tectoniques, 105-122. A la Baconnière, Neuchâtel.

Lippmann, F., 1977a. The solubility products of complex minerals, mixed crystals, and three-layer clay minerals. Neues Jahrb. Mineral. Abh., 130: 243-263.

Lippmann, F., 1977b. Diagenese und beginnende Metamorphose bei Sedimenten. Bull. Acad. serbe Sci. Arts; Classe Sci. nat. math.; Sciences nat., 56: 49-67.

Lippmann, F., 1979. Stabilitätsbeziehungen der Tonminerale. Neues Jahrb. Mineral. Abh., 136: 287-309.

Lippmann, F., 1980. Phase diagrams depicting aqueous solubility of binary carbonate systems. Neues Jahrb. Mineral. Abh., 139: 1-25.

Lippmann, F., 1981. Stability diagrams involving clay minerals. Eighth Conf. Clay Min. Petrology, Teplice 1979, 153-171.

Prigogine, I. and Defay, R., 1954. Chemical Thermodynamics; transl. by Everett, D.H.. Longman, London, 543 pp.

Robie, R.A., Hemingway, B.S. and Fisher, J.R., 1978. Thermodynamic properties of minerals and related substances at 298.25 K and 1 bar (10^5 Pa). U.S. Geol. Survey Bull. 1452.

Urusov, V.S., 1975. Energetic theory of miscibility gaps in mineral solid solutions. Fortschritte Mineral. (Special issue) 52: 141-150.

Velde, B. and Hower, J., 1963. Petrological significance of illite polymorphism in Paleozoic sedimentary rocks. Am. Miner. 48: 1239-1254.

Weaver, C.E. and Pollard, L.D., 1973. The Chemistry of Clay Minerals. Elsevier, Amsterdam, 213 pp.

GENESIS AND SYNTHESIS: KAOLINS IN SEDIMENTS

F.C. LOUGHNAN

School of Applied Geology, University of New South Wales, Kensington, 2033, Australia

INTRODUCTION

In thanking the organising committee for the invitation to introduce the symposium on genesis and synthesis of clay minerals, I am indeed cognisant of the honour bestowed upon me, particularly when one reflects on the list of distinguished speakers, who have officiated in this capacity at previous conferences.

The mechanisms by which clay minerals develop are generally considered under the headings of chemical weathering, diagenesis and hydrothermal activity but to assign a particular deposit to one or other such categories may prove difficult. The reasons being that the paths taken by weathering and hydrothermal solutions are not too dissimilar and moreover, weathering is not invariably a process of continuous congruent or incongruent dissolution of pre-existing minerals as sometimes envisaged. Rather, at times, due to variations in climate, water table level or other factors, the reactions may cease or be reversed leading to development of authigenic or neoformed minerals within the zone of weathering. Such processes, which Grim (1968) termed "reversions of the weathering cycle", really fall within the realm of diagenesis.

WEATHERING

Our understanding of the mechanisms of weathering of primary aluminium silicate minerals and the genesis of specific clay minerals has been greatly enhanced over the past decade or so mainly through the widespread application of the scanning electron microscope and energy-dispersive X-ray analyser and, more recently, by development of new techniques in the use of the high-resolution transmission electron microscope. Earlier held concepts based on data from bulk chemical and mineralogical changes have been modified and extended and it is now apparent that attack on the parent mineral commences at dislocation defects and other sites of energy excess in the crystal structure, both at the surface and within the crystal, with development of solution pits (Seifert, 1967; Parham, 1969; Berner and Holdren, 1977; Keller, 1978a) and that alteration proceeds in a sequential manner irrespective of the environmental conditions (Parham, 1969; Eggleton and Buseck, 1980; Eggleton and Boland, 1981). Thus, from a study by high resolution transmission electron microscopy of the weathering of K-feldspar under humid conditions, Eggleton and Buseck (1980) concluded that "it is not appropriate to consider reactions occurring simply at the crystal surface with feldspar inside and reaction product plus water outside. Rather, there is an initial reaction between feldspar and diffusing water beginning at crystal defects and gradually expanding within the crystal.........the beginning of feldspar weathering is, in this instance, an equilibrium reaction process leading to the formation of an illite/montmorillonite mixed layer". Similar 2:1 layer structures have been recorded as initial weathering products of other primary minerals including plagioclase (Page and Wenk, 1979) and enstatite (Eggleton and Boland, 1981).

The development of 2:1 layer minerals from feldspars involves partial loss of silica in addition to alkalis (or calcium) and consequently, where such depletion is inhibited structural breakdown of the parent mineral ceases. Pertinent in this respect, Meunier and Velde (1979) have stressed that the chemical forces producing new minerals "are often constrained to small volumes, frequently on the scale of a mineral grain or contact between grains". These authors have shown that in the weathering of granite near Parthenay, France, new phases generally develop along the junctions of grain boundaries but, where quartz is in contact with other minerals, no new phases are evident. Apparently, at this incipient stage of weathering, solutions in close proximity to quartz grains are approaching saturation with respect to silica and are therefore, ineffective in the dissolution of silica from minerals in the immediate vicinity.

For development of kaolins, representing the next stage in the sequential breakdown of feldspars, the specific requirements are generally believed to be retention of alumina and the presence of at least sufficient silica to combine with the alumina, and that the aluminium be in 6-fold co-ordination. The latter, in turn, necessitates not only depletion of alkalis but also concentration of hydrogen ions, for as Milliken et al. (1950) have pointed out, whereas aluminium in the absence of silica assumes octahedral co-ordination at pH values below 10-12, in silica-alumina gel complexes the change from 4- to 6- fold co-ordination does not occur until a pH of about 4 is reached, i.e. at a value where aluminium is becoming increasingly soluble. However, despite the fact that kaolins are relatively abundant as alteration products of feldspars and other aluminium silicate minerals, the mechanism by which they form and the factors controlling development of the various species are far from understood.

The fact that aluminium in mixed silica-alumina gels is generally in tetrahedral co-ordination and that low-temperature synthesis of kaolinite has only been achieved from highly diluted silica and alumina solutions, led Siffert (1978) to the conclusion that "aluminous phyllites are not formed by crystallisation of mixed gels but following dissolution (dispersion) and recombination of the dissolved elements (neoformed)". It would appear therefore, that kaolinite development in the weathering and diagenetic environments is probably by the same process and furthermore, that pseudomorphous kaolinisation of minerals such as biotite, must occur in very localised solutions, thus emphasising the point made by Meunier and Velde (1979).

In a survey of the literature pertaining to genesis of kaolins in recent weathering environments, Parham (1969) found that "halloysite is commonly the alteration product formed from parent rocks of widely different compositions under varying climatic conditions". Keller (1978a), on the other hand, concluded from examination by scanning electron microscopy of weathered products throughout the world, that the kaolin minerals have two distinct morphologies (i) an elongated form comprising halloysite and possibly imogolite that develops in near surface or highland areas where the geochemical systems are rather transitory and (ii) a platy, "book" type comprising mainly ordered kaolinite, that characterises thick saprolitic zones or weathered crusts derived from feldspathic crystalline rocks under geochemical conditions which persisted essentially at equilibrium for protracted periods.

Neither author mentioned the b-axis disordered form of kaolinite, or dickite as a weathering product although both minerals are found in sedimentary rocks, the former quite abundantly as fine grained, anhedral flakes particularly in fireclays, ball clays and underclays, and the latter of much more restricted distribution in certain sandstones (Dick, 1908; Smithson, 1954; Bayliss et al., 1965) and limestones (Hayes, 1967).

The ultimate stage in the weathering of aluminium silicate minerals is marked by development of bauxite minerals usually but not invariably at the expense of the kaolins. Gardner (1970), who examined the mechanism of alteration of kaolinite to gibbsite, considered the silica and hydrogen ion concentrations of the infiltrating waters and the presence or absence of an intermediate aluminous gel phase, as the critical factors. In the absence of an intermediate gel phase the essential requirements are a silica concentration less than $10^{-4.6}$ mol/dm^3 (1-2 ppm) and a pH above 4.2, but, if transformation is by way of amorphous $Al(OH)_3$ the specifications are much more stringent with the maximum silica concentration $10^{-6.3}$ mol/dm^3 and the minimum pH 5.6. From these findings he concluded that "bauxites do not readily form on quartz-bearing rocks or in cool regions where pH of the infiltrating water is low due to slow decay or organic matter and the production of organic acids". Presumably, the low silica tolerance of gibbsite accounts for the virtual absence of the mineral in older rocks although partial dehydroxylation to form the monohydrate phases boehmite and diaspore, could be a contributing factor in this respect.

Nevertheless, gibbsite is found in bauxites dating back to the Cretaceous and also has been recorded in present-day oceanic sediments, notably of the Atlantic equatorial region (Biscaye, 1965), as well as in alluvial, colluvial and near-shore marine sediments derived from Cretaceous and younger bauxites (Gordon et al., 1958; Nicolas, 1968; Grubb, 1970). Since the silica concentration of natural waters is rarely less than $10^{-4.6}$ mol/dm^3, the persistence of gibbsite in these sedimentary rocks is probably attributable to low diffusion rates of the groundwaters.

DIAGENESIS

If, as Parham (1969) has contended, halloysite is the common kaolin resulting from weathering of aluminium silicate minerals whereas kaolinite (ordered and disordered) predominates in sedimentary rocks, presumably much of the newly formed halloysite is subsequently transformed into kaolinite either in the thick saprolitic zones described by Keller (1978a) or within the depositional environment following erosion and transport. The reaction is certainly possible on thermodynamic grounds (Hem et al., 1971) but what is not certain is the manner in which it is achieved.

Possibly halloysite-kaolinite conversion is brought about through recrystallisation in the solid state by rising pressures and temperatures due to increasing depth of burial. However, it would be difficult to attribute all kaolinite deposits to this process and furthermore, from the study by Weather and Beck (1971) of clay mineral genesis with increasing depth, illite and chlorite appear the more likely products of such a mechanism.

Alternatively, aluminium of the halloysite may be complexed with fulvic, humic and other organic acids and, with destruction or separation of these acids, silica and alumina may recombine yielding kaolinite. Linares and Huertas (1971) synthesised a poorly ordered form of kaolinite in this manner and believed that the mechanism could be important in the low-temperature development of kaolinite in natural environments. Subsequently, La Iglesia and Galan (1975) realised the reaction sequence halloysite → disordered kaolinite → ordered kaolinite by means of an intense "weathering" action involving comminution of the halloysite to the point of breakdown of the crystal structure and treatment of the product with EDTA and oxalic acid. These authors considered that interlocking grains of ordered kaolinite that characterise flint clays, could originate by a somewhat similar process and, in this respect, their views are in accord with those of Keller (1978b). Nevertheless, although solution appears essential as the initial step in the conversion of halloysite to kaolinite, a sedimentary environment that could promote such rigorous destruction of clay material is difficult to envisage.

Not all kaolins in sedimentary rocks however, are necessarily detrital or derived from detrital clay material. In more permeable sandstones, conglomerates and possibly some limestones, primary aluminium silicate minerals such as feldspars and micas, may be altered *in situ* by circulating groundwaters and, pertinent in this respect, Smithson (1954) invoked such a mechanism to account for the presence of dickite in sandstones of Jurassic and Carboniferous age in England. But, alteration by groundwater is not dissimilar in principle, to that of surface weathering and thus, one would expect the two processes to yield the same product. Consequently, if the conclusions of Parham (1969) are substantially correct halloysite should be the common kaolin formed not only by surface weathering but also by circulating groundwaters.

A further mechanism by which kaolinite may form in sedimentary rocks and one that has been invoked by many authors, is through silicification of gibbsite. Thus, Wilson (1922), in describing the "bauxitic" or flint clay of Ayrshire, Scotland, drew attention to the unusual density of the material and attributed the phenomenon to authigenic development of kaolinite at the expense of bauxite minerals. And, more recently, Curtis and Spears (1971) concluded from published solubility data "that free alumina compounds formed in lateritic soils will tend to silicify spontaneously in all sedimentary environments" and that this "has been an important mechanism of kaolinite formation in ancient sediments".

De Kimpe et al. (1964) attempted synthesis of the gibbsite → kaolinite reaction and, although unsuccessful, they considered that "six fold co-ordination of Al in the starting material is not alone sufficient. Experiment proved that the gibbsite lattice is too stable to be attacked by depolymerised silica solutions unless the alumina phase is in a 'dynamic' stage where the lattice is on the point of reorganisation to boehmite, though boehmite itself also is unsuitable". These findings may explain not only the ability of gibbsite to persist in some sedimentary environments but also, the frequent absence of boehmite from such deposits. The gibbsite-kaolinite association in parts of the Nanafalia Formation in the southeastern United States is a good example in this respect.

BAUXITIC CLAYS OF THE SOUTHEASTERN U.S.A.

The Nanafalia Formation forms part of the Eocene Wilcox Group, which on the eastern side of the Mississippi embayment, crops out continuously over the Gulf coastal plain in a broad arc extending from the southern tip of Illinois to central Georgia and, on the western side, intermittently across Arkansas and Texas to the Mexico border. In the Georgia-Alabama area the Nanafalia comprises mainly quartz-mica sands with some kaolinite but in places, notably near Andersonville and Springvale, Georgia, and Eufaula, Alabama, discrete ovate to lenticular clay bodies measuring from a few metres to several kilometres across, are evident within these sands. The smaller bodies are composed mostly of kaolinite whereas the larger ones frequently contain a gibbsite-rich core that grades outwardly through successive envelopes of kaolinite, sandy kaolinite, kaolinitic sand into the enclosing quartz-mica sand. Neither boehmite nor diaspore has been recorded in these deposits.

The origin of the gibbsite-kaolinite bodies has been discussed by many authors including Shearer (1917), Rettger (1925), Burst (1965) and Clarke (1972) and, although some dissension in detail exists, the consensus of opinion is that they formed by a two-stage process involving (i) initial erosion of weathered zones in the Piedmont to the north, and deposition of the detritus on an emerging coastal plain where much of the clay was segregated into discrete bodies; and (ii) subsequent intense leaching of these bodies resulting in partial conversion of kaolinite to gibbsite. However, in a recent study of a vertical section through one of these gibbsite-kaolinite bodies, Lukas et al. (1981) found that whereas immediately below the core the kaolinite has a disordered structure, the kaolinite within and above the core is predominantly well-ordered. These authors have questioned the concept of development of gibbsite *in situ* on the grounds that (a) the leaching solution would need to be extraordinarily selective to bring about depletion of silica from the relatively impermeable clay bodies and yet leave the associated, much more permeable quartz-mica sands virtually unaffected and (b) since gibbsite is restricted to the core, leaching would of necessity be most active in this central zone and decrease in intensity in all directions radiating outward from such.

As an alternative proposal, these authors have suggested that the original detritus for the clay bodies contained appreciable amounts of gibbsite and that alteration of this mineral to kaolinite has been and probably is being effected by slow diffusion of silica-charged waters from the surrounding sands. This mechanism could account not only for the unusual form of the gibbsite-kaolinite bodies but also for the absence of boehmite from the deposits for, if as De Kimpe et al. (1964) have contended, silicification of gibbsite is possible only when the mineral is on the verge of re-organising to boehmite, formation of the latter mineral would be inhibited by all available alumina combining with silica to yield kaolinite.

CONCLUSIONS

If one may be permitted to speculate, it would appear from this brief resume, that low-temperature development of kaolin minerals in natural environments is possible only from dilute, molecularly dispersed systems and that the critical factor is not so much the composition or co-ordination state of the starting materials but rather, the pH of the localised solutions; indeed, the optimum pH for the reaction seems to be within a very narrow range. Moreover, since halloysite has the higher ΔG^{o} value and evidently is the common product of "aggressive" weathering activity whereas kaolinite seems more characteristic of those zones where the alteration processes are somewhat slower, presumably the principal factor governing development of the various kaolin species is the rate of precipitation from localised solutions.

REFERENCES

Bayliss, P., Loughnan, F.C., and Standard, J.C., 1965. Dickite in the Hawkesbury Sandstone of the Sydney Basin, Australia. Amer. Miner., 50: 418-426.

Berner, R.A., and Holdren, C.R. 1977. Mechanism of feldspar weathering: some observational evidence. Geology, 5: 369-372.

Biscaye, P.E., 1965. Mineralogy and sedimentation of recent deep-sea clay in the Atlantic ocean and adjacent seas and oceans. Bull. Geol. Soc. Amer., 76: 803-832.

Burst, J.F., 1965. Genetic relationship of the Andersonville, Georgia, and Eufaula, Alabama, bauxite-kaolin areas. Amer. Inst. Min. Engin., Transactions, 256: 137-143.

Clarke, O.M., 1972. Bauxite and kaolin in the Eufaula district. Geol. Surv. Ala., Bull. 100: 89 pp.

Curtis, C.D., and Spears, D.A. 1971. Diagenetic development of kaolinite. Clays Clay Min., 19: 219-217.

De Kimpe, C., Gastuche, M.C., and Brindley,G.W. 1964. Low temperature synthesis of kaolin minerals. Amer. Miner., 49; 1-16.

Dick, A.B., 1908. Supplementary note on the mineral kaolinite. Mineral. Mag., 15: 124-128.

Eggleton, R.A., and Boland, J.N., 1981. The weathering of enstatite to talc through a sequence of transitional phases. Clays Clay Min. (in press).

Eggleton, R.A., and Buseck, P.R., 1980. High resolution electron microscopy of feldspar weathering. Clays Clay Min., 28:173-178.

Gardner, L.R., 1970. A chemical model for the origin of gibbsite from kaolinite. Amer. Miner., 55: 1380-1389.

Gordon, M., Tracey, J.I. and Ellis, M.W., 1958. Geology of the Arkansas bauxite region. U.S. Geol. Surv., Prof. Paper, 299, 117 pp.

Grim, R.E., 1968. Clay Mineralogy. 2nd Ed. McGraw Hill, N.Y. 596 pp.

Grubb, P.L.C., 1970. Mineralogy, geochemistry and genesis of the bauxite deposits on the Gove and Mitchel plateaux, Northern Australia. Miner. Deposita, 5: 248-272.

Hayes, J.B., 1967. Dickite in Lansing Group (Pennsylvanian) limestones Wilson and Montgomery Counties, Kansas. Amer. Miner., 52: 890-896.

Hem, J.D., Roberson, C.E., Lind, C.J., and Polzer, W.L., 1971. Chemical interactions of aluminium with aqueous silica at 25°C. U.S. Geol. Surv., Water-Supply Paper, 1827-E, 57 pp.

Keller, W.D., 1978a. Kaolinisation of feldspars as displayed in scanning electron micrographs. Geology, 6: 184-188.

Keller, W.D., 1978b. Flint clay facies illustrated within one deposit of refractory clay. Clays Clay Min., 26: 237-243.

La Iglesia, A. and Galan, E., 1975. Halloysite-kaolinite transformation at room temperature. Clays Clay Min., 23: 109-113.

Linares, J. and Huertas, F., 1971. Kaolinite synthesis at room temperature. Science, 171: 896-897.
Meunier, A.F. and Velde, B., 1979. Weathering mineral facies in altered granite. The importance of small scale local equilibria. Mineral. Mag., 43: 261-268.
Lukas, T.C., Loughnan, F.C. and Eades, J.L., 1981. Origin of bauxite, Eufaula district, Alabama (in preparation).
Milliken, T.H., Mills, G.A., and Oblad, A.G., 1950. The chemical characteristics and structure of cracking catalysts. Disc. Faraday Soc., 8, 279-290.
Nicholas, J., 1968. Nouvelles donnés sur la genèse des bauxites à mur karstique du sud-ést de la France. Mineral. Deposita, 3: 18-33.
Page, R., and Wenk, H.R., 1979. Phyllosilicate alteration of plagioclase studied by transmission electron microscope. Geology, 7: 393-397.
Parham, W.E., 1969. Formation of halloysite from feldspar: low temperature artificial weathering versus natural weathering. Clays Clay Min., 17: 13-22.
Rettger, R.E., 1925. The bauxite deposits of southeastern Alabama. Econ. Geol., 20: 671-681.
Seifert, K.E., 1967. Electron microscopy of etched plagioclase feldspar. Amer. Ceram. Soc. Journ., 50: 660-661.
Shearer, H.K., 1917. A report on the bauxite and fuller's earth of the coastal plain of Georgia. Georgia Geol. Surv., Bull. 31, 340 pp.
Siffert, B., 1978. Genesis and synthesis of clays and clay minerals: Recent developments and future prospects. Proc. 6th Intn. Clay Conf., Oxford, 337-348.
Smithson, F., 1954. The petrology of dickitic sandstones in North Wales and Northern England. Geol. Mag., 16, 177-178.
Weaver, C.E., and Beck, K.C., 1971. Clay mineral diagenesis during burial: how mud becomes gneiss. Geol. Soc. Amer., Spec. Paper 124.
Wilson, G.V., 1922, The Ayrshire bauxitic clay. Mem. Geol. Surv., U.K.

EXPERIMENTAL HYDROTHERMAL ALTERATION OF ALBITE TO CLAYS: PRELIMINARY RESULTS

N. GUVEN(*), G. M. LAFON (**) AND L. J. LEE (*)

*Department of Geosciences, Texas Tech University, Lubbock, Texas 79415, U.S.A.

**Exxon Production Research Co., P. O. Box 2189, Houston, Texas 77001, U.S.A.

ABSTRACT

Alteration of albite by KCl solutions was conducted at 200°C and 500 bar in Dickson-type reaction vessels. Changes in the compositions of reaction fluids with time were monitored. The reactant and product solids were examined with electron microscopy (SEM, TEM, STEM, EDS, and diffraction modes). Special attention was given to the roles of pH and dissolved SiO_2.

Incongruent dissolution of albite appears to be primarily the result of reactions of the albite surface with dissolved species, especially $Si(OH)_4$, H^+, and Al. Under initially acid conditions, alteration products consist of boehmite, kaolinite, and gels. Production of boehmite is favored by low concentrations of SiO_2, while higher concentrations favor production of kaolinite. Under neutral conditions, small amounts of illite and very few crystals of kaolinite are formed on albite surfaces. We report the first experimental growth of "fibrous illite", a particular habit commonly found as diagenetic clay in reservoir sandstones. Under basic conditions, no alteration products are formed.

INTRODUCTION

The reservoir quality of sandstone formations strongly depends on their diagenetic histories. Growth of authigenic clay minerals in pore spaces can greatly reduce porosity and especially permeability. The ensuing degradation of reservoir quality varies considerably, depending not only on the amount of diagenetic clays but also on their types and their morphologies. In contrast, selective dissolution of unstable feldspars can result in significant development of secondary porosity in deeply buried rocks (e.g. Land and Milliken, 1981). Practical detailed mapping and prediction of the reservoir quality of sandstones would greatly benefit from a thorough understanding of the kinetics and mechanisms of diagenetic reactions, which is not yet available. As a first step toward achieving such an understanding, we have reacted albite with KCl-bearing solutions at hydrothermal conditions. We paid particular attention to changes of solution composition with time and to precise characterization of alteration products in relation to reaction kinetics. We report here preliminary results of these experiments, emphasizing the solid alteration products. More complete accounts of the alteration sequences and of the reaction kinetics will be given elsewhere.

PREVIOUS WORK

Previous experimental studies of reactions between feldspars and aqueous solutions have proceeded along two main distinct tracks, with minor amounts of overlap: 1) Examination of "feldspar dissolution kinetics", with relatively little or no effort to identify the precise nature of alteration products and their parageneses; occasionally the solids were characterized by bulk x-ray diffraction. 2) Detailed examination and characterization of alteration products, mainly by electron microscopic techniques, leading to inferences regarding their growth mechanisms; in these experiments, variation with time of the compositions of aqueous solutions generally was not recorded.

In all dissolution studies the ratio of Al to Si in solution was observed to be much lower than that in the original feldspars. This "incongruent dissolution" has been attributed to the precipitation of an aluminum-rich phase, possibly both within the aqueous solution and as a permeable alteration rind coating feldspar grains (Wollast, 1967; Busenberg and Clemency, 1976). The presence of such a hypothetical alteration layer has also been invoked to explain the general observation that the dissolution rate of feldspar decreases rapidly with time during the course of an experiment (Correns and von Engelhardt, 1938; Wollast, 1967; Busenberg and Clemency, 1976). In contrast, Lagache (1965, 1976) convincingly showed that the dissolution rate of previously altered feldspar is identical to that of freshly prepared material. Her results strongly suggest that the dissolution rate of feldspar is primarily controlled by reactions between the solid aluminosilicate surfaces and dissolved species of aluminum, silicon and cations. All direct observations by electron microscopy of experimentally altered feldspar surfaces (Wyart et al., 1963; Tchoubar, 1965; Berner and Holdren, 1977; Holdren and Berner, 1979) and by x-ray photoelectron spectroscopy (Petrovic et al, 1976; Holdren and Berner, 1979) show <u>no</u> evidence of an alteration layer, whether as a distinct rind coating feldspar grains or as a near-surface thin zone of atomic disorder and significant compositional changes.

There is general agreement that a number of experimental variables are likely to affect significantly the dissolution kinetics of feldspars and the nature, abundance and crystallinity of the solid alteration products. Dissolution rates appear to be markedly higher in acid solutions than in neutral or basic solutions (Wollast, 1967; Lagache, 1965). Increasing the dissolved silica concentration of altering solutions tends to favor production of kaolinite or smectite, rather than boehmite (Tchoubar, 1965). The ratio of the surface area of the feldspar to the volume of the solution, the concentrations of alkali cations and their ratios to that of hydrogen ion, have also been mentioned as possibly important parameters. The precise roles of these variables still remain generally unclear.

Feldspar alteration can lead to precipitation of a variety of aluminous and alumino-silicate compounds: most commonly found are boehmite, kaolinite, allophane and, more rarely, illite and smectite. Observations of these alteration products by electron microscopic techniques (SEM, TEM, electron diffraction) have led investigators to form hypotheses as to their growth mechanisms. The best documented alteration sequence is probably that observed by Tchoubar (1965). After reacting (001) cleavage plates of albite with pure H_2O at 200°C, he first observed the development of corrosion figures, followed by growth of boehmite fibers with significant epitaxial orientation. With increasing reaction time, these fibers evolved into well-formed pseudo-hexagonal crystals. When 150 ppm of dissolved silica was added to the altering solution, kaolinite was formed, apparently by reaction between poorly crystalline boehmite fibers and $Si(OH)_4$. When 700 ppm of colloidal silica was added to the altering solution, Tchoubar observed the formation of what probably was smectite.

EXPERIMENTAL METHODS

The experiments were conducted in modified Dickson-type reaction vessels (Seyfried et al., 1979) at 200°C and 500 bar. This choice of apparatus allowed the use of large solution to solid ratios, thus more closely simulating natural alteration conditions. In addition, use of relatively large solution volumes permitted accurate monitoring of chemical changes with time in the aqueous phase.

In all experiments, the reactant feldspar was British Chemical Standard No. 375 albite ground to a size range of 10 to 44 μm, washed repeatedly with distilled water until the supernatant was clear, with a specific surface area of 0.23 $m^2 g^{-1}$.

The aqueous solution usually was a 1.0 M KCl brine whose pH and dissolved silica content were adjusted to the desired values at the start of each experiment. Three preliminary runs were conducted with 0.0, 0.01 and 0.1 M KCl. One gram of feldspar was reacted with 160 to 180 cm^3 of KCl solution. Each experiment was terminated when inspection of the fluid composition suggested that the desired alteration products had been formed in sufficient quantity. Runs lasted from 9 to 84 days.

Upon termination of each experiment, the solids were separated, dialyzed to remove the KCl and examined by x-ray diffraction and by electron microscope techniques. The feldspar grains were examined by scanning electron microscopy (SEM) to determine the morphological effects of dissolution and the spatial relations between growing alteration products and the feldspar surfaces. In addition, a part of the solids was further dispersed and disaggregated and the smaller particles were mounted for examination by a high-resolution Analytical Electron Microscope (JEOL-100 CX). We used the transmission (TEM), scanning transmission (STEM), diffraction (SAD for the particles larger than 0.5 μm), microdiffraction (for particles less than 0.5 μm) and analytical (EDS) modes of this instrument to identify unambiguously the very small particles which commonly constituted most of our reaction products.

Thus far the effects of two of the important reaction variables listed above have been investigated: pH, and the concentration of dissolved silica. Four main acidity regimes were covered: acid (quench pH 2 to 3); initially acidic evolving toward neutral (quench pH varying from 2 to 6); neutral; and initially basic (quench pH evolving from 10 to 8). This report is primarily focused on a description of the solid alteration products, although changes in the composition of the fluid phase are also described briefly. Experimental formation of illite was paid special attention. Authigenic illite is commonly formed during the diagenesis of sandstones and is a major factor in reducing their quality as reservoirs. The illite crystals often appear as thin, flat, very elongated laths which suggest a "fibrous" habit in SEM images at magnifications lower than 5000x (Gray and Rex, 1966; Wilson and Pittman, 1977; Guven et al, 1980). Several runs were designed to produce experimentally illite from feldspar alteration, preferably "fibrous" illite.

SOLUTION CHEMISTRY DURING FELDSPAR ALTERATION

Three important new observations were made regarding the incongruent character of dissolution, the role of pH, and that of dissolved silica.

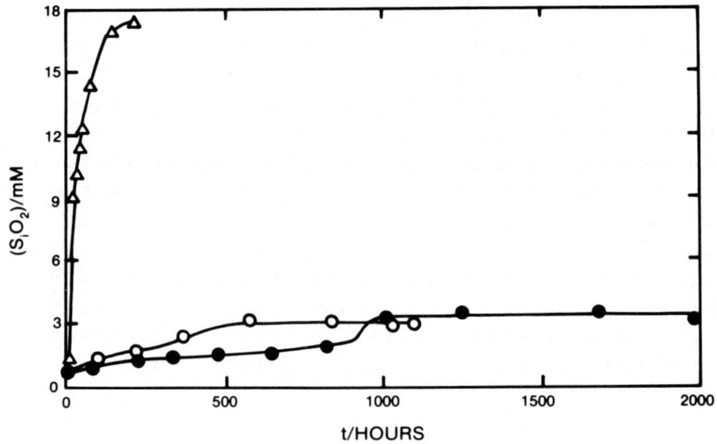

Fig. 1. Concentration of SiO_2 during the dissolution of albite at 200°C and 500 bar. (Δ) quench pH varying from 2.0 to 3.2; (.) quench pH approximately neutral; (o) quench pH varying from 9.9 to 9.1.

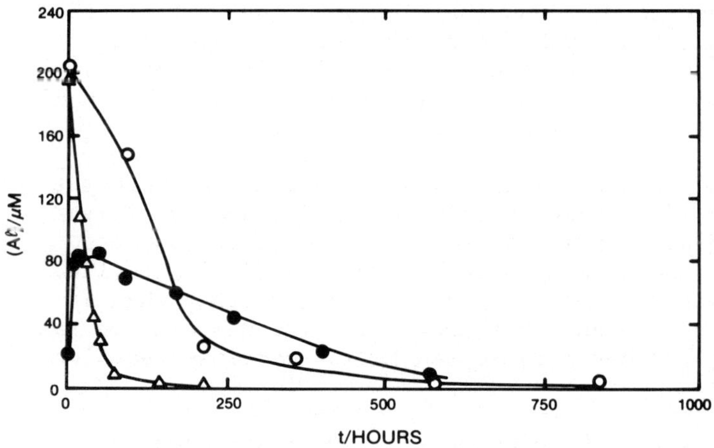

Fig. 2. Concentration of total dissolved Al at 200°C and 500 bar. (Δ) quench pH varying from 2.0 to 3.2; (.) quench pH approximately neutral; (o) quench pH varying from 9.9 to 9.1.

First, albite dissolution appears to be intrinsically incongruent, irrespective of the possible precipitation of aluminous compounds. The amounts of Al-bearing alteration products formed were far too small to account for the extremely low Al to Si ratios in solution, which are orders of magnitude smaller than that in albite. Furthermore, the low Na to Si ratios present in solution cannot be explained by precipitation of the observed alteration products since they do not contain sodium. Incongruence of the dissolution is best demonstrated by two experimental runs in which no solid alteration products were formed according to the detailed SEM and TEM examination. In these experiments, Al/Si in solution dropped very rapidly to 1/1500 and 1/2100 respectively, and Na/Si in solution stabilized at 1/6 and 1/8 (for comparison, in albite, Al/Si = Na/Si = 1/3).

Thus, the concept of a single dissolution rate for albite appears to be largely meaningless, at least in the initial stages of the dissolution reaction. Rates for the release of Si, Al and Na from albite to aqueous solution should be considered separately. These rates are probably linked indirectly but they are not governed by the stoichiometric ratios of the elements in the feldspar. Therefore, initial dissolution of albite must result in a change of the composition of this mineral. Additional work on this dissolution process is in progress. These conclusions apply only to the initial stages of albite dissolution, where the amount of Si released represents no more than a few percent of Si in the mineral. When dissolution is pursued to completion, it must eventually become congruent and stoichiometric.

Secondly, pH plays a complex role, affecting in different ways the rates of release to solution of Si, Na and Al, and the nature and precipitation rates of alteration products. In agreement with the data of Lagache (1965) and Wollast (1967), release of Si is much more rapid in acid solutions than in neutral and basic solutions (Figure 1). The dissolution rate of Na^+ is also further complicated by ion-exchange reactions with K^+ and H^+.

Reactions between albite and dissolved Al are complex (Figures 2 and 3). Under basic conditions, the Al initially dissolved during experimental run-up is adsorbed on the albite surface at run conditions, and there is no measurable release of Al by the feldspar. Under acid conditions, Al release from albite is marked by more rapid precipitation of aluminous compounds (boehmite, allophane, kaolinite). At near-neutral pH, the initially rapid release of Al is promptly slowed down by accumulation of Al and Si in solution. Later, over time spans of hundreds of hours, dissolved Al, Si and K react with one another to form clays, thus steadily lowering the concentration of Al.

An important new observation is the poisoning by $Si(OH)_4$ of Al release from albite at near-neutral pH (Figure 3). With an initial silica concentration of 150 ppm, no release of Al to solution is observed and the small amount of Al dissolved during run-up is slowly adsorbed by the albite surface. This behavior is in marked contrast to the pattern of Al release described above. Thus, moderate concentrations of dissolved silica can qualitatively alter the dissolution rate of Al from albite, and are probably a major cause of the incongruent dissolution of feldspar. Poisoning of Al dissolution by $Si(OH)_4$ has not been documented previously and must be considered as one of the more important controlling factors for reactions between alumino-silicates and aqueous solutions.

Precipitation of illite is observed only at near neutral pH, and requires the presence in solution of relatively large concentrations of Al (about 100 μM) for periods of time of $\sim 10^2$ hours at 200°C.

DESCRIPTION OF ALTERATION PRODUCTS

Reactions between albite and KCl solutions produce three main distinct alteration patterns depending on whether the solution pH is initially acidic (pH at 25°C from 2 to 3), neutral, or initially basic (pH at 25°C about 10).

Initially Acidic Conditions

In reactions with initially acidic solutions, albite is altered to boehmite, kaolinite, and allophane. The extent of alteration can vary greatly from grain to grain. Some particles are densely covered by a large number of overlapping alteration products which form a highly porous and permeable "fur" (not a continuous alteration layer). Others exhibit corrosion features and etch pits but little or no growth of secondary minerals.

Upon reaction with a dilute HCl solution (pH at 25°C of 2) for 9 days, we observed abundant formation of boehmite, formation of allophane, and of a small amount of kaolinite. Boehmite crystals display two habits (Fig. 4): dense bundles of fibers, and subhedral platelets. Fluorescent x-ray spectra (EDS) obtained from these fibers consistently indicate the presence of minor but significant amounts of Si which suggests that $Si(OH)_4$ is probably adsorbed on boehmite surfaces. The SAD patterns of well-formed boehmite platelets display orthogonal symmetry with parameters: $\underline{a}$ = 0.287 ± 0.002 nm, $\underline{c}$ = 0.382 ± 0.001 nm. Discrepancy between this value of $\underline{c}$ and the one generally accepted for boehmite (0.369 nm) is attributed to

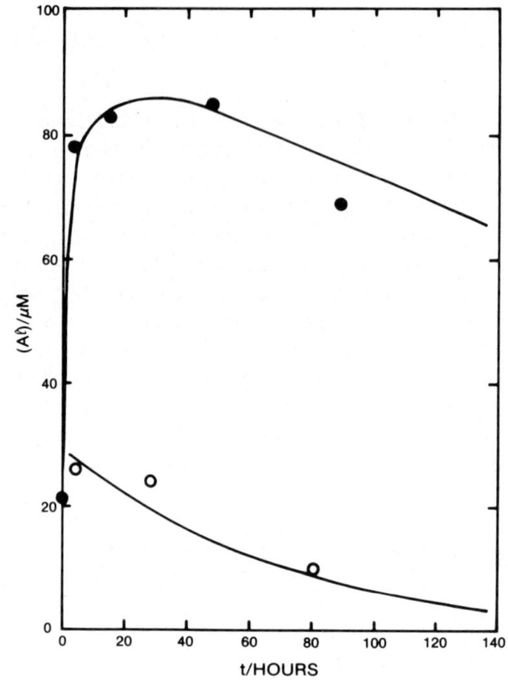

Fig. 3. Concentration of total dissolved Al at 200°C and 500 bar at neutral pH in the initial stages of reaction. (.) no initial dissolved SiO_2; (o) initial concentration of SiO_2 of 150 ppm.

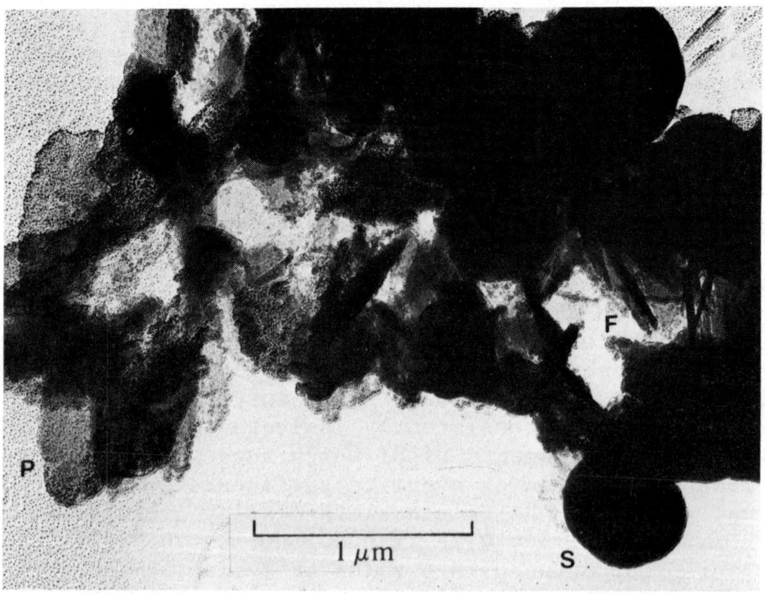

Fig. 4. Alteration products of albite reacted with dilute HCl. Bundles of boehmite fibers (F), boehmite platelets (P) and allophane spherules (S).

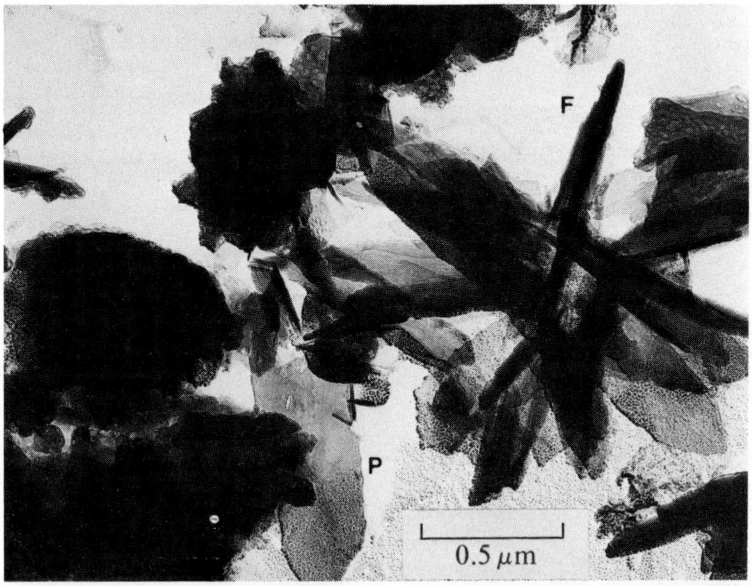

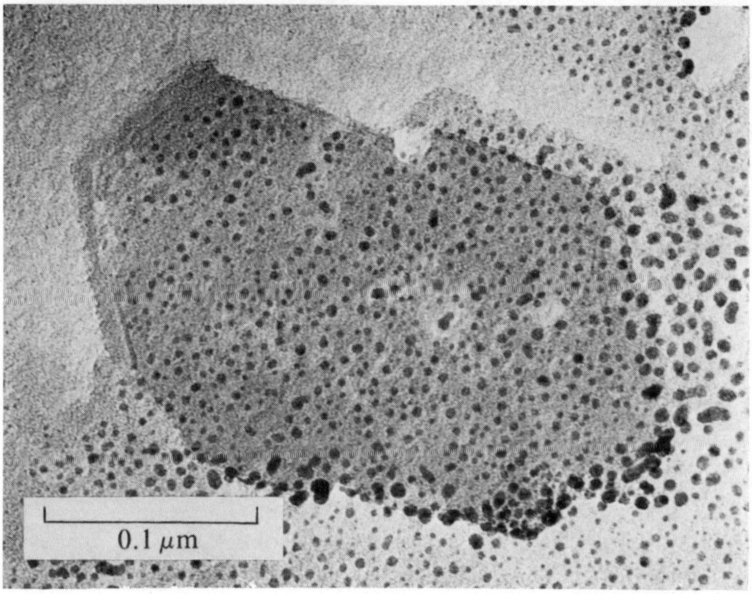

Fig. 5. Common forms of boehmite produced by reaction of albite with acidic KCl solution: boehmite fibers (F) aggregated in curved bundles, and elongated platelets (P) with triangular terminations.

Fig. 6 Kaolinite platelet produced by reaction of albite with acidic KCl solution.

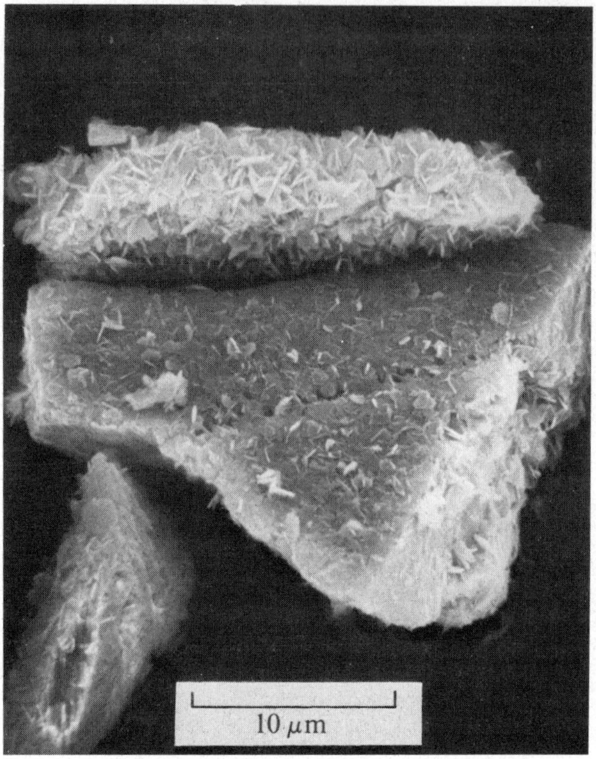

Fig. 7. Albite grains after reaction with acidic solution containing 1.0 M KCl and 150 ppm silica. Note the abundant pseudohexagonal platelets of kaolinite covering the surface of one albite particle, whereas the next grain has very little reaction products on it.

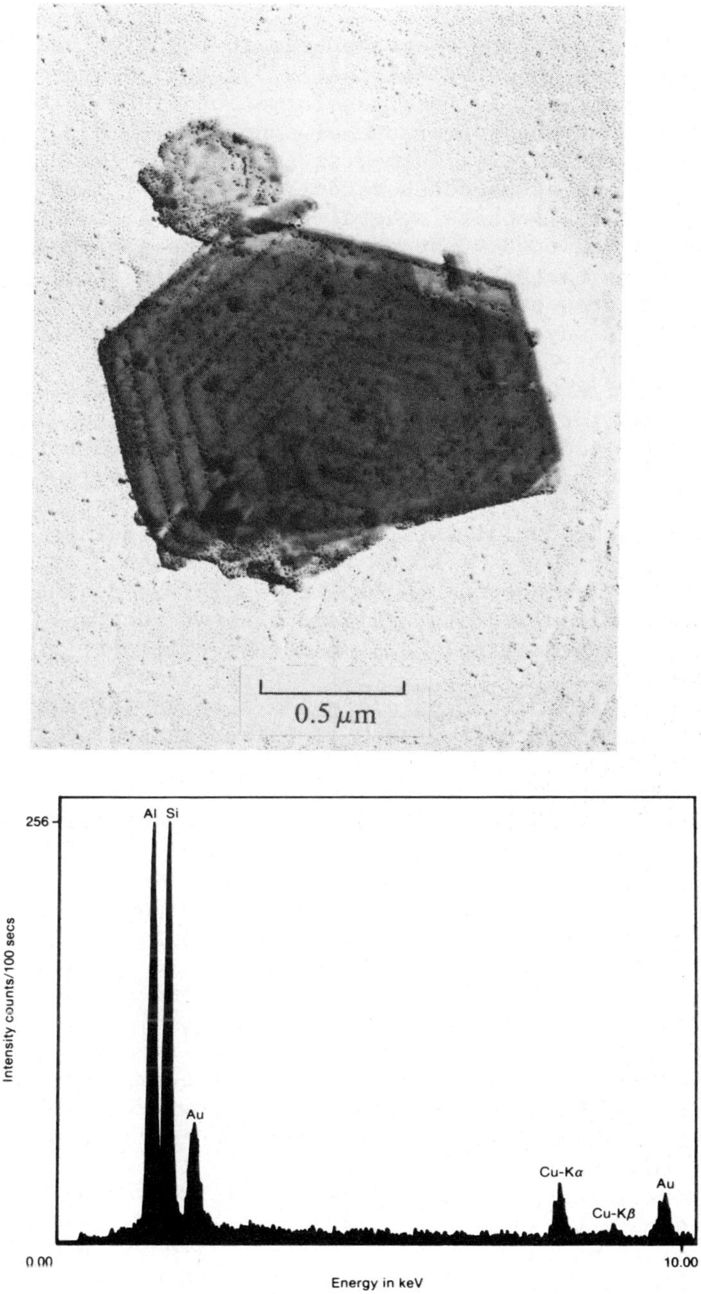

Fig. 8. a) An idiomorphic platelet of kaolinite displaying spiral growth steps decorated by gold. Note also the very small, electron-dense particles of amorphous alumina sprinkled over the kaolinite. b) EDS spectrum obtained from this kaolinite crystal; the Al/Si intensity ratio is 1.04.

partial dehydration under the electron beam in the high vacuum of the microscope. As shown by Wilson (1979), this dehydration can eventually result in the transformation of boehmite into γ-Al_2O_3 and cause the appearance of superlattice diffractions. Allophane typically is formed as small spherules (Fig. 4) and kaolinite as pseudo-hexagonal platelets. EDS spectra obtained from allophane spherules yield an Al to Si intensity ratio of 0.8 to 0.9, suggesting that their composition is very similar to that of kaolinite. All alteration products appear to grow outward from the albite surface, but we have not observed any epitaxial relations to this substrate.

Reaction of albite with 0.1 M KCl acid solution (initial pH at 25°C of 2.15) for 25 days produced in order of abundance: boehmite, kaolinite and minor amounts of allophane. Boehmite displayed fabrics similar to those observed in the previous experiment (Fig. 5): bundles of fibers often showing marked curvature, and elongated platelets with triangular endings. EDS spectra from both forms of boehmite showed evidence of small amounts of Si, as above. Kaolinite was formed as small, thin, pseudo-hexagonal platelets. Figure 6 shows such a kaolinite crystal, approximately 0.3 µm across, without evidence of any obvious precursor. The typical growth mechanism of these kaolinite crystals appears to be through surface nucleation and lateral growth of island-shaped nuclei.

To further elucidate the growth mechanisms of kaolinite, albite was reacted for 43 days with an acid 1.0 M KCl solution (initial pH at 25°C of 2.03) containing 150 ppm of dissolved SiO_2. In this experiment large amounts of kaolinite were produced as well as a moderate quantity of an amorphous aluminous phase and minor amounts of boehmite fibers. Figure 7 shows the spatial relations between kaolinite crystals and their albite substrate. Figure 8a is a TEM photograph of a typical kaolinite crystal from this experiment, clearly showing its idiomorphic character. In addition, the obvious spiral steps on the top surface of this crystal unambiguously demonstrate screw-dislocation control of its growth mechanism. Thus, kaolinite crystals formed in our experiments can grow by two distinct mechanisms: surface nucleation and lateral layer growth, and spiral growth through operation of screw-dislocations. The EDS spectrum of the kaolinite platelet in Fig. 8a is shown in Fig. 8b, indicating that its Al to Si intensity ratio is 1.04. This ratio is somewhat higher than Al to Si intensity ratios typically obtained under the same analytical conditions for Georgia kaolinite, with values ranging from 0.7 to 0.8. We attribute this discrepancy to the presence of very small aluminous particles intimately associated with the kaolinite crystals. These numerous small but dense particles can be seen sprinkled over the kaolinite platelet in Fig. 8a.

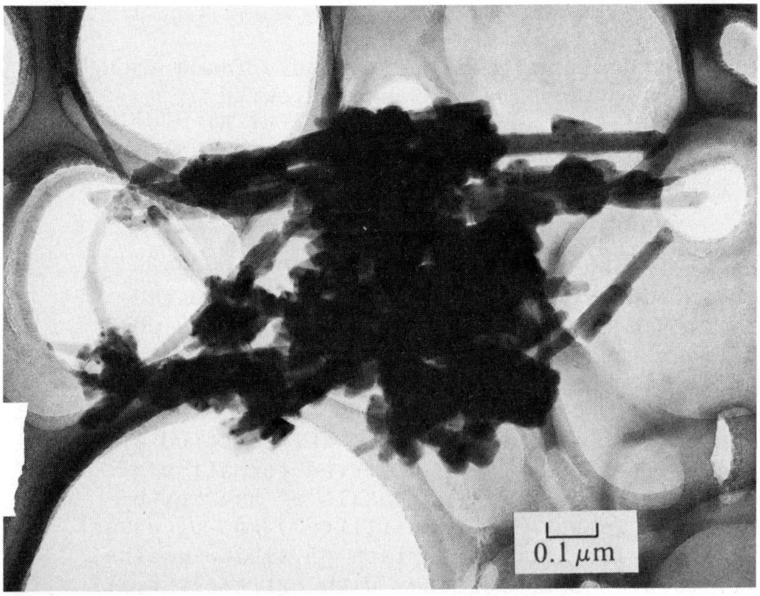

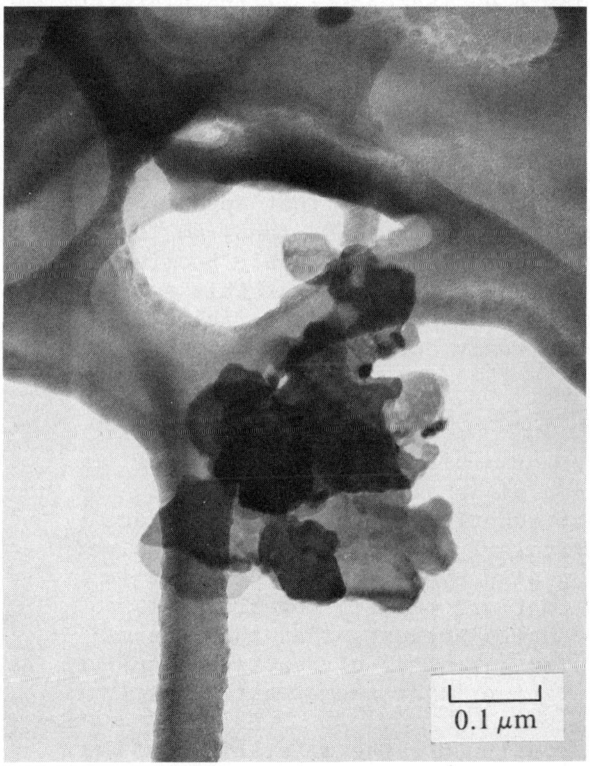

Fig. 9. Laths and platelets of illite produced by reaction of albite with near-neutral 1.0 M KCl solutions.

Fig. 10. Subhedral to euhedral illite platelets formed under the same conditions as in Fig. 9.

Aggregates of very similar particles are commonly found among the alteration products. They are disk-shaped, with a diameter of about 0.1 μm, and appear to consist of Al oxy-hydroxides. They do not yield electron diffraction patterns and are probably amorphous.

Neutral and basic conditions

Alteration of albite by KCl solutions at 200°C under neutral conditions produces illite and minor kaolinite and yields relatively large concentrations of Al (about 100 μM). Under basic conditions (quench pH from 9.9 to 9.1), no formation of alteration products was observed.

After reaction between albite and 1.0 M KCl solution at near neutral pH for 24 to 86 days, we observed formation of small quantities of illite and a few crystals of kaolinite. Sparse growth of illite laths ("fibrous illite") and of equant illite platelets takes place at the surface of albite grains. Diamond-shaped illite platelets together with extremely small illite laths are shown in Figure 9. Subhedral illite platelets with a tendency to develop euhedral terminations are shown in Figure 10. EDS spectra obtained from both laths and platelets indicate that their major elements are Si, Al and K, in proportions closely similar to those of standard illites. Illite laths similar to the natural materials examined by Guven et al. (1980) were formed in a few instances. We show in Figure 11a a bundle of such laths ("fibers") that appear to grow from a dense core. The corresponding EDS spectrum is given in Figure 11b. A few well-formed idiomorphic kaolinite platelets were found among alteration products of this series of experiments. In addition, minor amounts of Al-rich amorphous precipitates were commonly encountered (e.g., background of Figure 11a).

CONCLUSIONS

1) The incongruent nature of albite dissolution is not primarily due to precipitation of an aluminum-rich phase. Rather, it appears to reflect reaction of the albite surface with dissolved species, particularly $Si(OH)_4$, H^+ and dissolved Al. The coverage of albite grains by alteration products is highly variable. On grains that are densely covered, the arrangement of alteration products suggests that their permeability is extremely high. Thus, control of reaction kinetics by diffusion through an alteration layer seems most unlikely.

2) Under initially acid conditions, the alteration products of albite consist primarily of boehmite, kaolinite and amorphous gels. Boehmite appears to precipitate massively and very rapidly in the initial stages of the reaction where the concentration of dissolved silica is low. At first, it grows

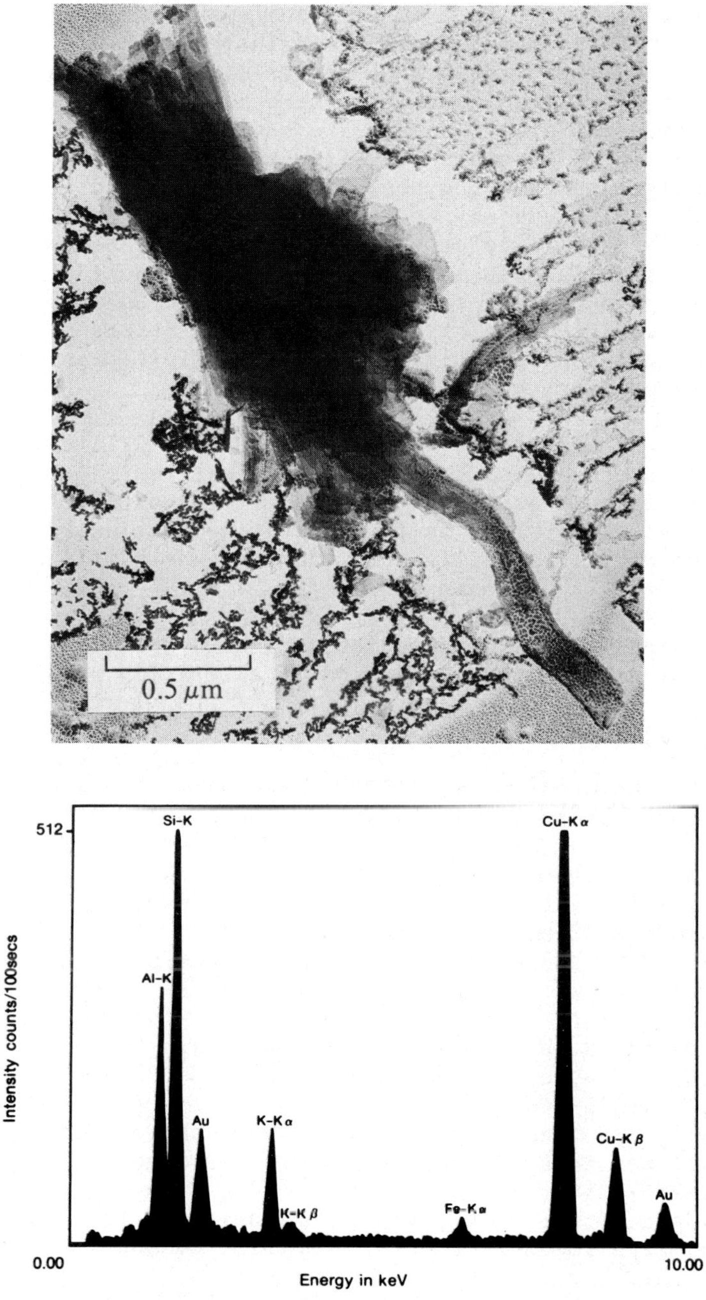

Fig. 11. a) Illite laths ("fibrous illite") formed at near-neutral pH in 1.0 M KCl solutions. b) EDS spectrum of the illite laths.

as numerous fibers originating from the albite surface and showing no particular epitaxy. This stage is followed by growth of boehmite plates with better crystalline order. As the concentration of silica increases, there is later precipitation of kaolinite and of allophane. We have no clear evidence that kaolinite crystals grow from boehmite precursors, although spatial relations among coexisting boehmite and kaolinite suggest that this may be the case in a few instances. The morphologies of kaolinite crystals suggest simultaneous action of two distinct growth mechanisms: surface nucleation followed by layer growth, and spiral growth associated with screw dislocations. Presence of moderate concentrations of silica in solution (~150 ppm) clearly favors precipitation of kaolinite and suppresses that of boehmite.

3) Under neutral conditions, small quantities of illite and a few crystals of kaolinite grow on the albite surfaces. These illites occur both as elongated laths ("fibrous illite") and as equant platelets. This is the first time that experimental formation of illites similar to the diagenetic illites of sandstone reservoirs has been achieved.

Under basic conditions, dissolution of albite does not release enough Al to cause precipitation of alteration products.

REFERENCES

Berner, R. A. and Holdren, G. R., 1977. Mechanism of feldspar weathering: some observational evidence. Geology, 5:369-372.

Busenberg, E. and Clemency, C. V. , 1976. The dissolution kinetics of feldspar at 25°C and at 1 atm CO_2 partial pressure. Geochim. Cosmochim. Acta, 40:41-49.

Correns, C. W. and Von Engelhardt, W., 1938. Neue Untersuchungen uber die Verwitterung des Kalifedlspates. Chem. Erde, 12:1-22.

Gray, D. H. and Rex, R. W., 1966. Formation damage in sandstones caused by clay dispersion and migration. Clays Clay Miner., 14:335-366.

Guven, N., Hower, W.F., and Davies, D.K., 1980. Nature of authigenic illites in sandstone reservoirs. J. Sed. Pet., 50:762-766.

Holdren, G.R., and Berner, R.A., 1979. Mechanism of feldspar weathering-I. Experimental studies. Geochim. Cosmochim. Acta, 43:1161-1171.

Lagache, M., 1965. Contribution à l'étude de l'altération des feldpaths dans l'eau entre 100°C and 200°C sous diverses pressions de CO_2 , et application à la synthèse des minéraux argileux. Bull. Soc. Franc. Minéral. Cristallogr., 88:223-253.

Lagache, M., 1976. New data on the kinetics of the dissolution of alkali feldspar at 200°C in CO_2 charged water. Geochim Cosmochim. Acta, 40:157-161.

La Iglesia, A., Martin-Caballero, J.L., and Martin-Vivaldi, J.L., 1974. Formation de kaolinite par précipitation homogène à température ambiante. Emploi de feldspaths potassiques. Compt. Rend., 279:1143-1145.

Land, L.S. and Milliken, K.L., 1981. Feldspar diagenesis in the Frio Formation, Brazoria County, Texas Gulf Coast. Geology, 9:314-318.

Parham. W.E., 1969. Formation of halloysite from feldspar: low temperature, artificial weathering versus natural weathering. Clays Clay Miner., 17:13-22.

Petrovic, R., Berner, R.A., and Goldhaber, M.S., 1976. Rate control in dissolution of alkali feldspar-I. Study of residual feldspar grains by x-ray photoelectron spectroscopy. Geochim. Cosmochim. Acta., 40:537-548.

Seyfried, W. E., Jr., Gordon, P. C., and Dickson, F. W., 1979. A new reaction cell for hydrothermal solution equipment. Amer. Mineralogist, 64:646-649.

Tchoubar, C., 1965. Formation de la kaolinite à partir d'albite altérée par l'eau à 200°C. Etude en microscopie et diffraction électronique, Bull. Soc. France. Minéral. Crist., 88:483-518.

Wilson, M.D., and Pittman, E.D., 1977. Authigenic clays in sandstones: Recognition and influence on reservoir properties and paleoenvironmental analysis. J. Sed. Pet., 47:3-31.

Wilson, S.J., 1979. The dehydration of boehmite, γ-AlOOH to γ-Al_2O_3. J. Solid State Chemistry, 30:247-255.

Wollast, R., 1967. Kinetics of the alteration of K-feldspar in buffered solutions at low temperature. Geochim. Cosmochim. Acta, 31:635-648.

Wyart, J., Oberlin, A., and Tchoubar, C., 1963. Etude en microscopie et microdiffraction électroniques de la boehmite formée lors de l'altération de l'albite: Compt. Rend., 256:554-555.

SCAN ELECTRON MICROGRAPHS OF CLAY DEPOSITS OF BUENOS AIRES PROVINCE, ARGENTINA

P. E. ZALBA

Centro de tecnología de Recursos Minerales y Cerámica (CETMIC) - CIC

Calle 47 y 115. Departamento Tecnología
1900 - La Plata, Argentina

ABSTRACT

Scan electron micrographs are shown of (1) residual or primary kaolinite, pyrophyllite and muscovite; (2) sedimentary kaolinite, illite, pyrophyllite and (3) diagenetic halloysite, dickite, montmorillinite, illite and related diaspore and alunite, collected from clay deposits of different marine sedimentary sequences in the Buenos Aires Province, Argentina.

SEMs show textures of plastic and semiplastic "fire clays" which reflect the environment of deposition and diagenetic processes involved. Textures with contorted, "swirl" pattern, as in typical ball clays, are shown.

Matted clay fibers of halloysite (7Å and 10Å) forming interesting hexagonal plates and "wet grass" structures, suggest the possible transformation from kaolinite into halloysite.

Late diagenesis in limestones is exemplified by crystallographycally oriented pits, due to chemical etching on authigenic pyrite crystals. Furthermore, crystals of alunite, associated with diaspore in kaolinitic clays, have formed diagenetically by desilicification processes.

INTRODUCTION

Clay mineral admixtures from different marine sedimentary sequences of two depositional basins, which overlap in the zone of Barker in the

Province of Buenos Aires (Fig. 1), were studied by scanning electron microscopy.

These sequences are integrated by the Sierras Bayas Formation, Upper Precambrian in age (Amos *et al.*, 1972; Bonhomme *et al.*, 1978; Dalla Salda *et al.*, 1979); the Cerro Negro Formation, Upper Precambrian ? (Iñiguez *et al.*, 1974) and the Balcarce Formation, Cambro-Ordovician in age (Dalla Salda *et al.*, 1979).

The Sierras Bayas Formation overlies granite and related migmatite rocks (Precambrian) and consists of orthoquartzites, dolomites, claystones and limestones, outcropping in the zone of Olavarría (González Bonorino, 1954) and in the Barker area.

The Cerro Negro Formation is represented typically in the locality of that name, near Olavarría, by an alternating succession of shales and sandstones and it overlies the Sierras Bayas Formation (Iñiguez *et al.*, 1974).

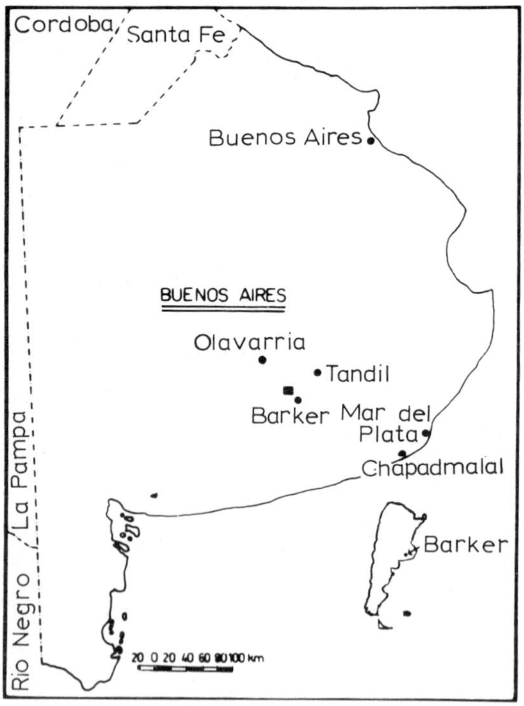

Fig. 1. Map shows main localities related to clay deposits.

The Balcarce Formation consists of a succession of clastic strata, for the most part sandstones, with a minor amount of claystones, siltstones and orthoconglomerates. It is found from Chillar, in the northwest of Barker to Chapadmalal, on the Atlantic coast.

Clay deposits occurring in the above mentioned formations are spread along 310 km of the Cordón de Tandilia, in a NW-SE direction, from Olavarría to Chapadmalal (Fig. 1).

Geological and mineralogical investigations (Iñiguez, 1965; Caillère _et al._, 1967) led to the conclusion that these materials follow a zone of lateritic alteration formed on the granitic rocks. Iñiguez _et al._ (1974) confirmed the detrital origin of these deposits derived from weathered granites and related migmatite rocks.

CLASSIFICATION OF CLAYS

Scan electron micrographs permit classification of clays geologically, mineralogically, morphologically and texturally. A tentative classification was made taking into account industrial uses of the different clays and technological test results.

Geologically, they are (1) residual or primary, which comprise (a) argillization of granitic basement rocks, producing kaolinite (Fig. 2); (b) hydrothermally altered granitic basement rocks, leading to the formation of kaolinite, pyrophyllite and muscovite (2) transported or sedimentary with different clay mineral associations; and (3) diagenetic clay minerals, such as dickite, halloysite and related diaspore and alunite; (Zalba, 1978; 1979 a, 1981 b), smectite and illite.

Residual or primary clays

Residual or primary clays comprise (a) in situ kaolinized granitic basement rocks and (b) hydrothermally altered granitic basement rocks.

Mineralogically classified, there are kaolinite deposits with illite and quartz as impurities (Fig. 2), kaolinite with feldspar associations (Fig. 3); and kaolinite-pyrophyllite, with muscovite (sericite) admixtures (Fig. 4).

Morphologically, kaolinite forms "stacks," in coarse, loosely packed plates (Fig. 2), and in some cases one sees "books" forming from feldspars (Fig. 3), pyrophyllite forms platy stacks with straight edges (Fig. 4) and muscovite also forms stacks of very large platy crystals (Fig. 5).

Texturally these primary clays consist of euhedral to subhedral platy aggregates of flakes, showing different compaction. Texture of residual kaolinite, formed by argillization (kaolinization) of primary silicate rocks, is shown in Fig. 2. Note "book" aggregates (not floccules) and scalloped kaolinite flakes (original crystallization or later differential dissolution ? (Keller, 1977)). The texture of hydrothermally altered feldspar (pitted feldspar, presumably due to incongruent dissolution) is seen in Fig. 6. A typical hydrothermal texture is seen in Fig. 7, with pyrophyllite in tight face-to-face packing, but the orientation changes within a short distance. The overall texture resembles that of typical "ball clays" described by Keller (1976) which will be discussed later.

Technologically these kaolinitic clays are classified as plastic "fire clays."

Transported or sedimentary clays

Geologically, Upper Precambrian and Cambro-Ordovician clay deposits are sedimentary or transported (Iñiguez, 1965, Zalba, 1978, Dalla Salda *et al.*, 1979). These clays show different grain size distribution, as in typical sedimentary deposits (Fig. 8). They are found in the Sierras Bayas Formation, the Cerro Negro Formation and the Balcarce Formation.

Mineralogically, from SEM evidence, as well as from X-ray powder diffraction, chemical analyses and DTA, the sedimentary clays are mechanical admixtures of several clay minerals: the Sierras Bayas Formation is illitic (Dalla Salda *et al.*, 1979), the Cerro Negro Formation is illitic-chloritic, with irregular mixed layer clays of chlorite-montmorillonite (Iñiguez *et al.*, 1974), whereas the Balcarce Formation has two different clay mineral associations, (a) kaolinite-pyrophyllite-illite with related halloysite, diaspore and alunite (Zalba, 1981 b*) and (b) kaolinite muscovite (sericite).**

Morphologically, clay minerals may occur as irregular plates of ill-defined forms, as in illitic-chloritic-montmorillonitic clays; as individual hexagonal plates or "books" of hexagonal flakes, as in kaolinite; as large plates with straight edges as in detrital pyrophyllite (Fig. 9), and large, platy flakes of muscovite.

Texturally, euhedral to subhedral individual flakes or stacks, show different compaction. SEMs provide evidence of kaolinitic clays which have been deposited in the flocculated state, and settled out in face-to-face relationship, which Murray (1976) reported as having been deposited under marine conditions. Williamson (1980) created the term "card-pack-flocs" for a face-to-face association of platelets, which have been deposited as flocs. Some sedimentary kaolinitic clays show a typical "swirl" pattern, (Fig. 10), probably resulting from "movements during packing, settling, or other soft rock deformation" (Keller, 1978), a texture that has been considered typical of "ball clays." Their texture is very similar to that of the designated ball clays. Although ball clays are essentially kaolinitic, they usually contain accessory illite, as in the case reported in this paper. Geologically, they are young clays in other parts of the world, occurring mainly in Tertiary-age rocks, (Keller, 1976). However, some of our Cambro-Ordovician kaolinitic clays exhibit a distinctive "ball clay" texture in the scan electron micrographs (Figs. 10 and 11).

Most of the sedimentary clays from the Balcarce Formation, are plastic and semi-plastic "fire clays" (Fig. 12). We also observed textures resembling "alkali" semi-flint clays, a term used by Keller (1968) in his proposal on flint clay facies. This last term takes into account

* Diagenetic processes involving these deposits were responsible for the formation of the halloysite (7Å and 10Å), diaspore, alunite and dickite in the Lower Paleozoic deposits and smectite, illite, alunite and pyrite in the Upper Precambrian sequences.

** Paleoenvironmental and stratigraphic studies which the author is carrying out will probably lead to the subdivision of the Balcarce Formation into different members.

the amount of "alkali" minerals present in the clays, which are in this case, for the most part illite and less alunite (Zalba, 1981 b). Kaolinitic clays bearing authigenic dickite, have characteristic semi-flint textures (Fig. 13) of interlocking clay mineral crystals, i.e., a hard and dense, highly siliceous semi-flint, which breaks with a conchoidal fracture resembling "flint" (Fig. 14).

Diagenesis in sedimentary clays

Diagenetic processes were responsible for the formation of different minerals in the sedimentary clays of the Cordón de Tandilia.

As regards the age of diagenesis, it is evident that several stages of diagenesis have occurred, but their dating is difficult. On the one hand, Rb/Sr and K/Ar radiometric data obtained from illitic clays of the Sierras Bayas Formation, Olavarría, by Bonhomme et al. (1978) show that no diagenetic processes have occurred after the late Precambrian for these deposits. On the other hand, there are no available data related to diagenesis in the Balcarce Formation.

Early diagenetic pyrite, with superficial hematite, crystallized in perfect cubes, can be seen in silty clays associated with limestones in the Sierras Bayas Formation (Figs. 15 and 16).

Illitic-chloritic-montmorillonite clays of the Cerro Negro Formation show, from SEM evidence, diagenetic illite, oriented perpendicular to the grain surface, showing a characteristic box-work texture, (Fig. 17) (Waugh, 1978). This honeycomb texture shows platelets arranged in an edge-to-edge cellular pattern. As Walker et al. (1978) have suggested, they may have formed "as a result of replacement and dissolution of some or all elements of the detrital silicate minerals."

Other evidence of diagenesis is authigenic smectite which has filled cracks or holes (syneresis cracks?) and shows a "flower" pattern. Higher magnification shows that the "flowers" are formed of tiny shreds of smectite which appear to have grown in spirals. The shreds are lying flat (Figs. 18 and 19). (Perhaps algal remains were replaced by the smectite to give this unusual texture). Bright border hollows, common on muscovite crystals (Fig. 20), may be related to previous mineral inclusions (Morrison-Smith, 1976). Also, the crystallographically oriented pits on authigenic pyrite crystals (Figs. 15 and 16), suggest chemical etching of crystal defects (Bull, 1978), during late diagenesis. Furthermore, pseudo-cubes of alunite formed diagenetically, are seen as restricted deposits in illitic-chloritic-montmorillonitic clays. Presumably its formation is related to the presence of pyrite (Fig. 21) (Goldbery, 1980).

Semi-plastic "fire clays" of the Balcarce Formation with diaspore-halloysite (Zalba, 1978 and 1979, a and b) and diaspore-halloysite-alunite (Zalba, 1981 b) suggest depotassification and desilicification processes, through dissolution, hydrolysis and dialysis of clay minerals, as stated by Reesman (1966); Reesman _et al_. (1968), and Keller (1968). These three minerals have a related diagenetic origin (Zalba, 1981 b). Halloysite (7Å and 10Å), confirmed by X-ray patterns and thermal treatments, shows typical elongate forms and a "wet grass" structure (Fig. 22), as described by Keller (1977). Matted fibers of elongated particles in a hexagonal pattern indicates a possible transformation from platy kaolinte to halloysite (Fig. 23). Note the curved borders on the platy kaolinite crystals. Nevertheless, Murray and Keller (personal communication), have seen repeated evidence that elongated halloysite particles may have changed to platy kaolinite. Perhaps the same process took place in this case, or the elongated particles were formed later from kaolinite flakes, because of the presence of sulphate in an acidic environment. The simultaneous presence of sulphate and halloysite suggests that sulphur compounds may influence halloysite crystallization: alunite has been shown to occur associated with halloysite and more commonly with diaspore (Fig. 24), Hendricks (1937); Dana _et al_. (1957) and Weaver _et al_. (1971). The presence of sulphate, in addition to acidity, is important in the argillation process, as suggested by Keller (1964), who further stated that "high-alumina minerals have been observed to originate in abundance, _secondarily_, in an acidic environment."

Allen (1952), cited evidence for the process that "The desilicication of feldspars and parent rocks is usually a two-stage process in which clay minerals such as kaolinite, halloysite, and nontronite are formed first and then silica is removed from them to form gibbsite, boehmite, and diaspore," Kennedy (1959) investigated the Al_2O_3-H_2O system and derived from thermodynamic data, that "diaspore in bauxitic and clay deposits may have formed at atmospheric pressures and temperatures, a familiar conclusion from field observations." Eberl and Hower (1975) could convert halloysite to kaolinite at 150°C "indicating that halloysite is a low temperature polymorph."

The clay minerals, which have been deposited in the flocculated state (resembling typical "ball clay" textures), may have been geochemically reorganized. Silica was removed in an acidic environment, as a result of the presence of sulphate in evaporitic deposits. Unconsolidated, porous, plastic mud permitted downward migration of silica. Leaching and desilicication of clay minerals yielded the formation of high-alumina minerals: halloysite, diaspore and alunite. Pseudo-cubic crystals of alunite (Fig. 25) associated with diaspore and halloysite in kaolinitic clays, are interpreted as formed by desilicication processes during early diagenesis (Zalba, 1981 b).

ACKNOWLEDGMENTS

Research for this project was supported by the Centro de Tecnologia de Recursos Minerales y Cerámica from La Plata, Argentina.

I wish to thank Dr. W. D. Keller for valuable suggestions and for taking some halloysite micrographs. Thanks are also due to Dr. H. H. Murray for suggestions and for his kind criticisms of this manuscript, and to Dr. G. W. Brindley for valuable comments.

Finally, I want to thank the Instituto de Neurobiología from the CONICET, especially Mr. Jorge Hohmann, for the scan electron micrographs.

REFERENCES

Allen, V. T., 1952. Petrographic relations in some typical bauxite and diaspore deposits. Bull. Geol. Soc. Amer., 63, 649-688.

Amos, A. J., Quartino, B. J. and Zardini, R. A., 1972. El "Grupo La Tinta" (Prov. de Buenos Aires), Paleozoico o Precámbrico?. XXV Congr. Bras. Geol. San Paulo. 211-221.

Bonhomme, M. and Cingolani, C., 1978. Rb/Sr and K/Ar dating of Upper Precambrian sediments of the "Serie de La Tinta," Buenos Aires Province, Argentina. Short papers of the Fourth International Conference, Geochronology, Cosmochronology, Isotope Geology, U.S. Geol. Survey, Open File Report 78-701, 45-46.

Bull, P. A., 1978. A quantitative approach to scanning electron microscope analyses of cave sediments. In Whalley, W. B., Editor, Scanning Electron Microscopy in the Study of Sediments. A symposium. 201-226, Geo Abstracts, Norwich, England.

Caillière, S. and Iñiguez, A. M., 1967. Etude minéralogique de "La Tinta" formation argileuse de la Province de Buenos Aires, République Argentine: Bull. Soc. Fr. Cristallogr., 90, 246-251.

Dalla Salda, L. and Iñiguez, A. M., 1979. "La Tinta," Precámbrico y Paleozoico de Buenos Aires. Actas VII Congr. Geol. Arg., I, 539-550.

Dana, J. D. and Dana, E. D., 1957. The system of mineralogy. 7th Edition. J. Wiley and Sons. New York, II, 556-559.

Eberl, D. and J. Hower, 1975. Kaolinite Synthesis: The role of the SiAl and (Alkali)/(H+) ratio in hydrothermal systems. Clays Clay Min., 23, 301-309.

Goldbery, R., 1980. Early diagenetic Na-alunites in Miocene algal mat intertidal facies, Ras. Sudar, Sinai. Sedimentology, 27, 189-198.

González Bonorino, F. 1954. Geología de las Sierras Bayas. Rev. Anales LEMIT, Serie II, 55, La Plata, Argentina.

Hendricks, S. B., 1937. The crystal structure of alunite and the jarosites. Amer. Min. 22, 773-787.

Iñiguez, A. M., 1965. Mineralogía de algunos yacimientos de arcillas de la Prov. de Buenos Aires. Doc. Thes., Fac. Cs. Nat. y Museo. La Plata, Arg., 1 130.

Iñiguez, A. M. and Zalba, P. E., 1974. Nuevo nivel de arcilitas en la zona de Cerro Negro, Partido de Olavarría, Provincia de Buenos Aires. Rev. Anales LEMIT, II, 264, 95-100, Argentina.

Keller, W. D., 1964. The origin of high-alumina clay minerals-A review. Clays Clay Min., 19, 129-151.

Keller, W. D., 1968. Flint clay and a flint clay facies. Clays Clay Min., 16, 113-128.

Keller, W. D., 1976. Scan electron micrographs of kaolins collected from diverse environments of origin, II. Clays Clay Min., 24, 114-117.

Keller, W. D., 1977. Scan electron micrographs of kaolins collected from diverse environments of origin, IV. Georgia kaolin and kaolinizing source rocks. Clays Clay Min., 25, 311-345.

Keller, W. D., 1978. Classification of kaolins exemplified by their textures in scan electron micrographs. Clays Clay Min., 26, 1-20.

Kennedy, G. C., 1959. Phase relations in the system Al_2O_3-H_2O at high temperatures and pressures. Am. Jour. Sci., 257, 563-573.

Morrison-Smith, D. J., 1976. Dislocation structures in synthetic quartz. In Wenk, H. R., Editor, Electron Microscopy in Mineralogy. Springer, New York, 410-418.

Murray, H. H., 1976. The Georgia sedimentary kaolins. 7th Symp. Genesis of Kaolin, Int. Geol. Correlation Program, Committee on Correlation of Age and Genesis of kaolin, Tokyo, 114-125.

Reesman, A. L., 1966. A study of clay mineral dissolution. Ph.D. Thesis, Univ. of Missouri at Columbia, 163. In W. D. Keller, 1968. Clays Clay Min. 16, 113-128.

Reesman, A. L. and Keller, W. D., 1968. Aqueous solubility studies of high-alumina and clay minerals. Am. Miner. 53, 929-942.

Walker, T. R., Waugh, B., and Crone, A. J., 1978. Diagenesis in first cycle desert alluvium of Cenozoic age, Southwestern United States and northwestern Mexico. Geol. Soc. Am. Bull., 89, 19-32.

Waugh, B., 1978. Diagenesis in continental Red Beds as revealed by scanning electron microscopy. A review. Scanning Electron Microscopy in the Study of Sediments, A symposium. In W. B. Whalley, Editor, Geo. Abstracts. Norwich, England.

Weaver, Ch. E. and Beck, K. C., 1971. Clay water diagenesis during burial. How mud becomes gneiss. Geol. Soc. Amer. Special Paper 134, 1-96.

Williamson, W. O., 1980. Experiments relevant to the genesis of clay mineral orientation in natural sediments. Clay Miner., 15, 95-97.

Zalba, P. E., 1978. Estudio geológico-mineralógico de los yacimientos de arcillas de la zona de Barker, Partido de Juárez, Pcia. de Buenos Aires y su importancia económica. Doc. thes., Fac. Cs. Nat. y Museo, La Plata, Arg., 1-175.

Zalba, P. E., 1979 a. Sobre la presencia de diasporo y halloysita en las arcilitas de la zona de Barker, Prov. de Buenos Aires. Actas, VII Congr. Geol. Arg. II, 337-349.

Zalba, P. E., 1979 b. Clay deposits of Las Aguilas Formation, Barker, Buenos Aires Province, Argentina. Clays Clay Min., 27, 433-439.

Zalba, P. E., 1981 a. Nuevo nivel de arcilitas sobre las calizas de la zona de Barker, Prov. de Buenos Aires. Rev. Asoc. Geol. Arg. XXXVI, 99-102.

Zalba, P. E., 1981 b. Alunite in association with diaspore and halloysite in the Las Aguilas Formation, Argentina and its significance, 7th Int. Clay. Conf. Italy, Abstract 346.

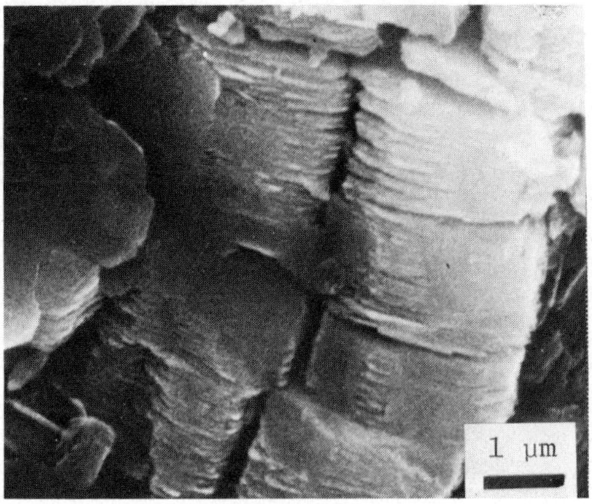

Fig. 2. Residual kaolinite formed by argillization processes on granite basement rocks, "books" in face-to-face orientation. Note scalloped kaolinite flakes. Mina Constante 10, Juárez.

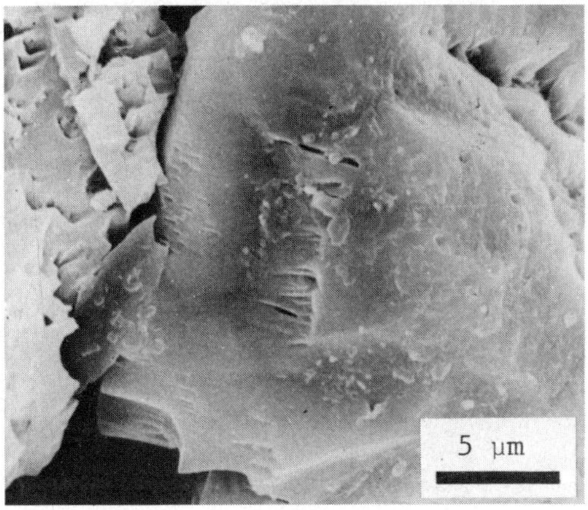

Fig. 3. Residual kaolinite formed by hydrothermal processes. Note "books" forming from feldspars. Cerro Segundo, Balcarce.

Fig. 4. Residual pyrophyllite formed by hydrothermalism. Accessory sericite and kaolinite. San Manuel, Lobería.

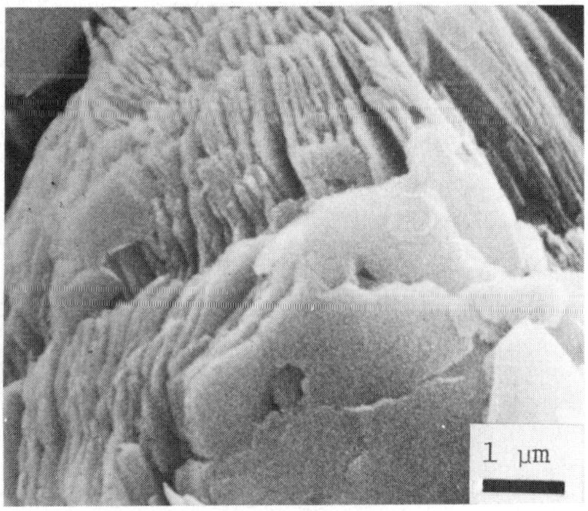

Fig. 5. Muscovite crystals forming "books." Cerro del Medio, Lobería.

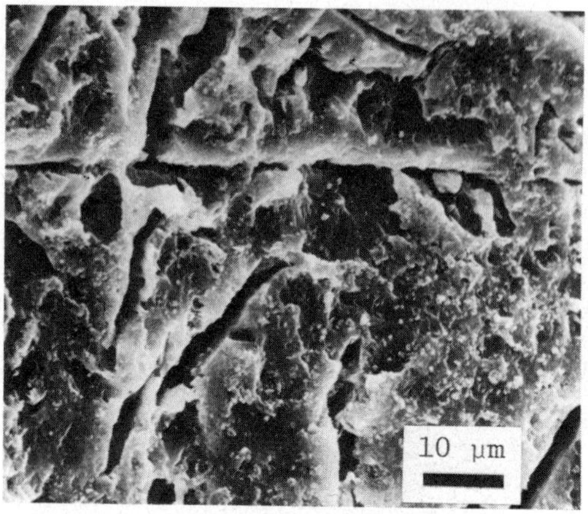

Fig. 6. Hydrothermally altered granite. Pitted feldspar, presumably incongruent dissolution. Residual kaolinite. Cerro Segundo, Balcarce.

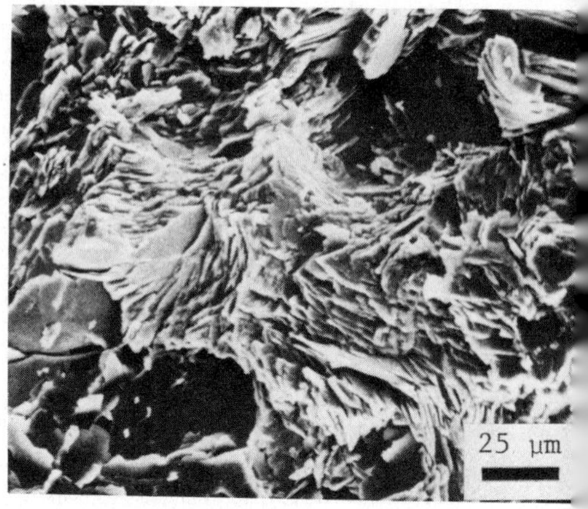

Fig. 7. Pyrophyllite, tight face-to-face packing. Some "swirl" patterns. San Manuel, Lobería.

Fig. 8. Pseudo-hexagonal plates of detrital kaolinite. Apparent bi-modal size distribution. La Barrosa, Balcarce.

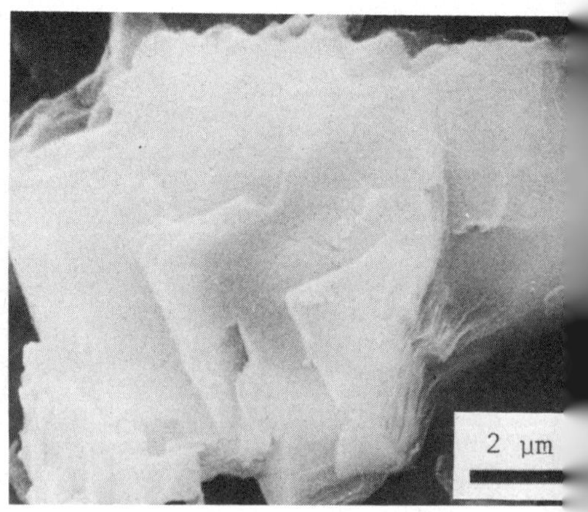

Fig. 9. Detrital pyrophyllite with straight borders. Barker, Juárez.

Fig. 10. Plastic "fire clay." Kaolinite in face-to-face orientation which curves into circles producing the typical "swirl pattern of ball clays." San Antonio, Juárez.

Fig. 11. Ilitic-kaolinitic plastic "fire clay." Flakes in face-to-face orientation. Note "swirl" pattern. typical of "ball clays." Barker, Juárez.

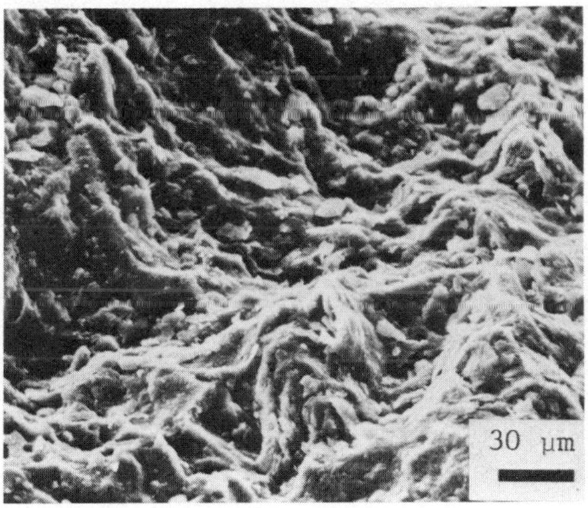

Fig. 12. Plastic "fire clay." Similar to "ball clay" textures. Kaolinite and pyrophyllite flakes in face-to-face orientation. Barker, Juárez.

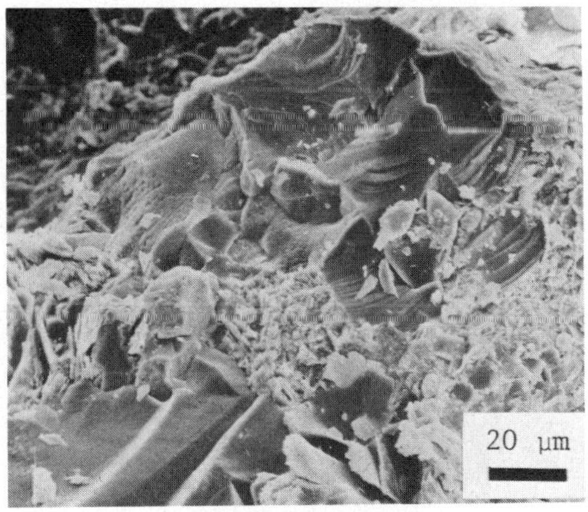

Fig. 13. "Alkali-semiflint" clay (Keller, 1968). Characteristic conchoidal fracture in kaolinitic clays bearing dickite. Constante 10, Juárez.

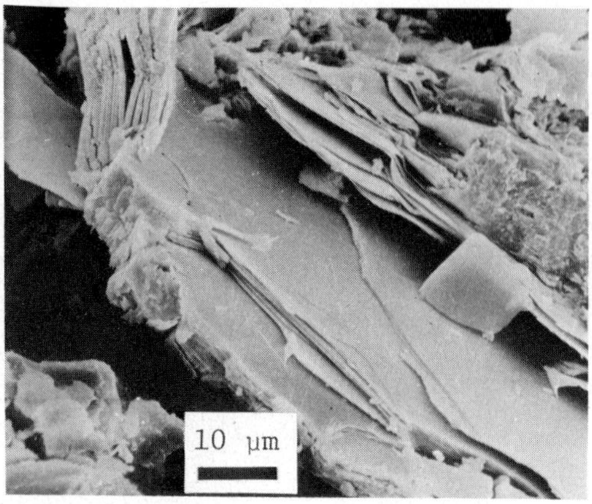

Fig. 14. Same specimen as Fig. 13. Large platlets of authigenic dickite. Note some "swirl" patterns. Constante 10. Juárez.

Fig. 15. Early diagenetic pyrite crystals. Crystallographically oriented pits due to chemical etching. Loma Negra, Juárez.

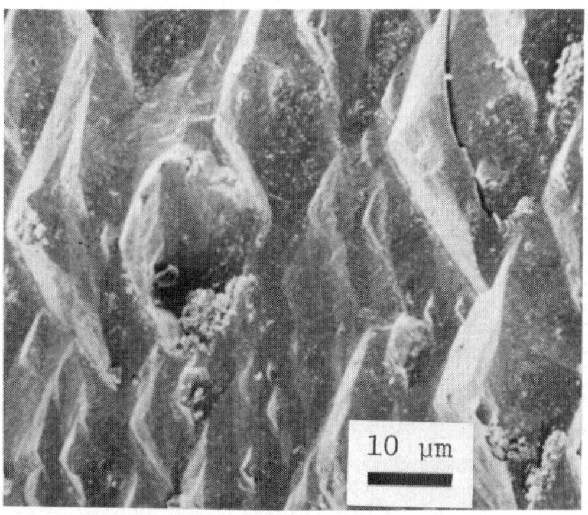

Fig. 16. View of part of Fig. 16. Oriented solution V-etches.

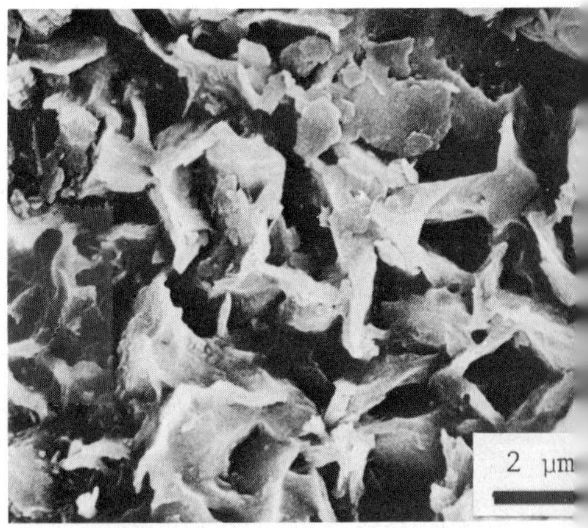

Fig. 17. Illitic-chloritic-montmorillonitic clay. Authigenic illite forming characteristic boxwork texture. Loma Negra, Juárez.

Fig. 18. Authigenic smectite filling cracks which show a "flower" pattern. Shreds are lying flat. Syneresis cracks or algal remains replaced by smectite? Loma Negra, Juárez.

Fig. 19. Smectite "flowers" enlarged. Flat-lying shreds forming spirals to form the "flower" pattern. Loma Negra, Juárez.

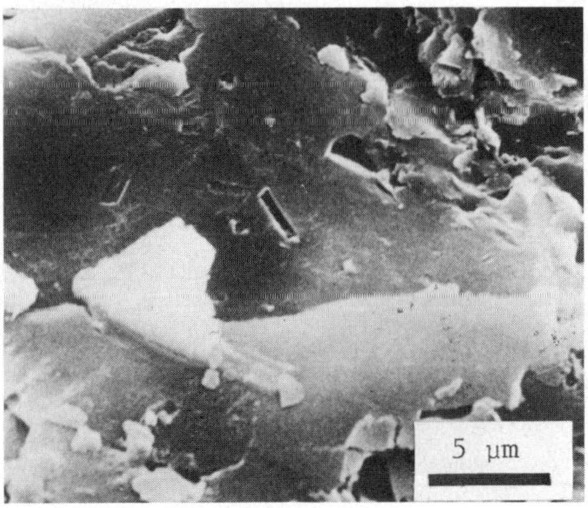

Fig. 20. Bright border hollows on muscovite crystals. Previous mineral inclusions? Loma Negra, Juárez.

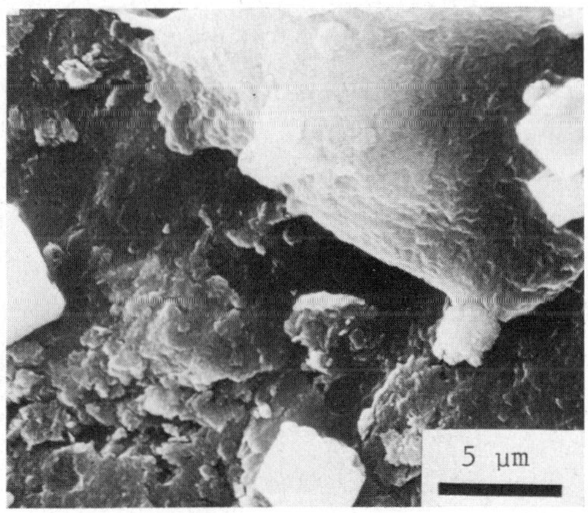

Fig. 21. Pseudo-cubes of alunite diagenetically formed, presumably related to pyrite presence. Loma Negra, Juárez.

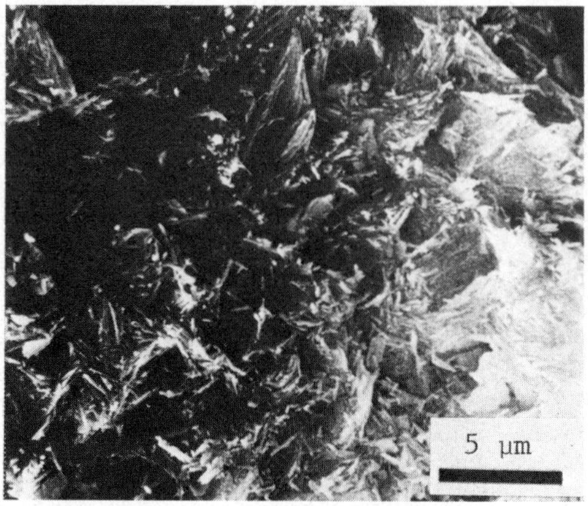

Fig. 22. Matted fibers of halloysite (7 and 10 Å) forming "wet grass" structure. Barker, Juárez.

Fig. 23. Pseudo-hexagonal kaolinite crystals with curled borders; transformation from kaolinite to halloysite (7 and 10 Å)? Barker, Juárez.

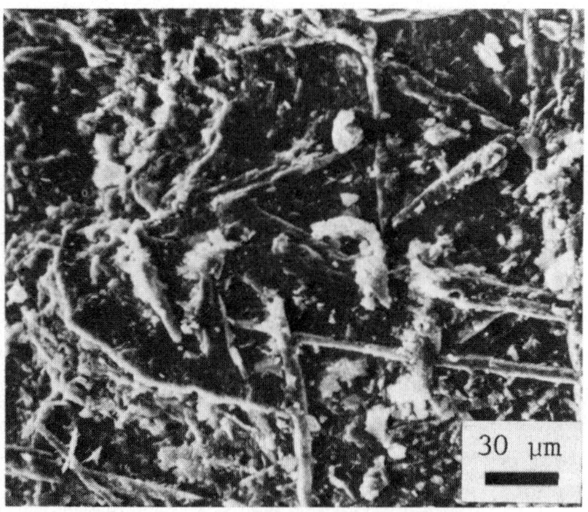

Fig. 24. Geometric disposition of authigenic diaspore crystals in kaolinitic-halloysitic clays with related alunite. Barker, Juárez.

Fig. 25. Pseudo-cubes of diagenetic alunite crystals in kaolinitic-halloysitic clays wit related diaspore. Barker, Juárez.

THE INFLUENCE OF COMPLEX-FORMING AGENTS ON THE STRUCTURE OF PRECIPITATED ALUMINIUM HYDROXIDE

O. LAHODNY-ŠARC* and Z. DRAGČEVIĆ**

University of Zagreb, *Mining, Geology and Petrol Eng. Dept., **Technological Department.

ABSTRACT

To stimulate the precipitation of aluminium hydroxides in the presence of organic ligands under natural conditions, aluminium ions were slowly released from amalgamated aluminium to ligand solutions, and the structure of the precipitated hydroxide was determined by X-ray diffraction after various periods of aging.

In pure water the initially amorphous precipitate becomes bayerite with gibbsite on aging. In the presence of sodium borogluconate, EDTA or citric acid aging yields pseudoboehmite or an amorphous gel depending on nature and concentration of the complexing agent and the pH of the solution.

INTRODUCTION

The influence of complex-forming organic ligands on the rate of formation and the nature of aluminium hydrolytic products has been extensively studied. Of particular note are the recent papers by Violante and Jackson (1978), Violante and Violante (1980), Lind and Hem (1975), Kwong and Huang (1975, 1978), and Serna et al. (1978).

The primary interest of these authors was to demonstrate the influence of organic ligands on the precipitation of dissolved aluminium species arising from weathering of aluminium -bearing rocks. Their results provide data for a better understanding of the mechanisms of formation and alteration of noncrystalline and crystalline aluminium hydroxides in soils and sediments. Many low molecular weight organic acids as well as large organic molecules occur in hydrous natural systems, usually associated with natural coloring material (Hem, 1975). These compounds may react with aluminium by forming complexes which control the rate of hydrolysis nucleation and crystal growth, and thus affect the nature of the precipitation products. An increase of the stability of the complex results in a decrease in the rate of crystallization and promotes the formation of an amorphous or poorly-ordered phase, generally recognized as pseudoboehmite. Since an acid environment favours the crystallization of gibbsite and/or norstrandite relative to bayerite, the complexing acids apparently promote formation of gibbsite and norstrandite, (Violante and Violante, 1980), and Volska (1977). Kwong and Huang (1978) attributed the inhibition and alteration of the crystallization of aluminium hydroxide by organic ligands to coordination of aluminium with carboxyl groups of organic ligands. Because of steric factors, the arrangement of the unit layers normally occuring in crystalline aluminium hydroxide is distorted by organic ligands, resulting in the formation of amorphous or poorly crystalline precipitates. Kwong and Huang (1978) have shown that the molar ratio of organic acid to aluminium has a decisive effect on the nature of the aluminium hydroxide formed.

EXPERIMENTAL

Most of the above mentioned experiments were carried out in solutions of selected well defined concentrations of both aluminium ions and organic ligands. In our work we used amalgamated aluminium as a source of aluminium ions to study the effect of several organic ligands on aluminium species initially present in small concentrations, as would be the case in natural systems where aluminium ions are slowly released by weathering, and also in order to obtain pure precipitates free of foreign ions.

Sheets of very pure aluminium were exposed for several seconds to a solution of mercury chloride and then immersed in a solution of the organic ligands, citric acid, EDTA and sodium borogluconate. After the formation of aluminium hydroxide the precipitate was allowed to age for various periods of time in the solutions, then washed with distilled water and air-dried at room temperature. The samples thus obtained were studied by X-ray diffraction and thermal analysis.

Details of the experimental techniques are presented in a previous paper on effects of dioxane, acetone and inorganic salts (Lahodny-Šarc et al.,1978).

RESULTS AND DISCUSSION

Fig. 1 shows X-ray diffractograms of samples obtained in citric acid of concentrations ranging from 10^{-3} to 10^{-1} mol/dm^3. All samples were aged three days in the equilibrium solution. The pH before and after every experiment is marked on the diagram; the initial pH is that of the solution of ligand and the final pH is that for the supernatant equilibrium liquid just prior to separation of the precipitates. Pattern 1 is for the sample prepared in pure water and shows formation of bayerite and of a small quantity of gibbsite. Patterns 2-5 are for the precipitates obtained from solutions with

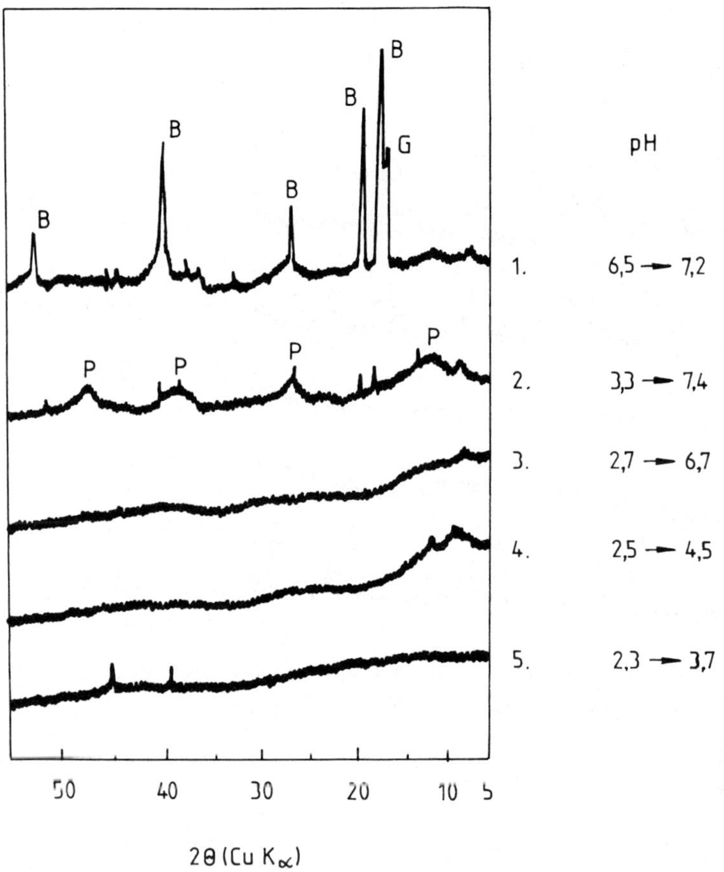

Fig. 1. X-ray diffraction patterns of aluminium hydroxides obtained from amalgamated aluminium after aging of 3 days: 1, in pure water, 2 to 5 in citric acid of 2.8×10^{-3}, 2.8×10^{-2}, 5.7×10^{-2} and 1.1×10^{-1} mol/dm^3, respectively.

concentrations of citric acid from 2.8×10^{-3} to 1.1×10^{-1} mol/dm^3. At concentrations below 10^{-3} mol/dm^3 the patterns cannot be distinguished from those for samples obtained in pure water. Starting at a concentration of about 10^{-3} mol/dm^3, pseudoboehmite is formed with small amounts of bayerite. At higher concentrations of citric acid only amorphous precipitates are produced. Two weak, but sharp peaks at 2.024 A and 2.338 A, correspond to the traces of aluminium metal in the sample. Fig. 2 shows patterns for the same system after aging of 15 days. In pure water (1) gibbsite is formed along with bayerite and a small amount of an amorphous phase. With a low concentration of citric acid pseudoboehmite

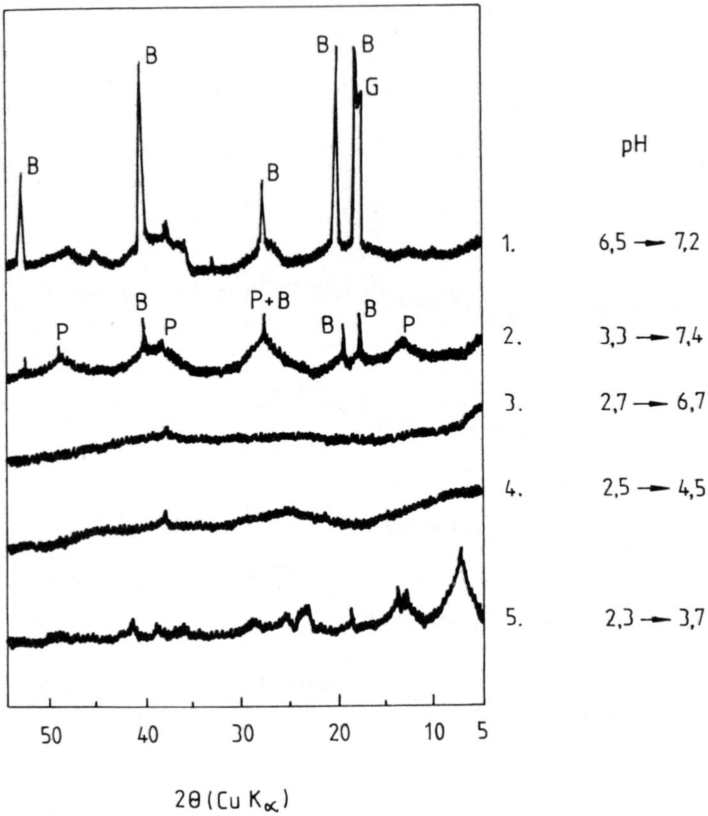

Fig. 2. X-ray diffraction patterns of aluminium hydroxides obtained from amalgamated aluminium after aging of 15 days 1, in pure water, 2 to 5 in citric acid of 2.8×10^{-3}, 2.8×10^{-2}, 5.7×10^{-2} and 1.1×10^{-1} mol/dm^3, respectively.

appears, but at higher concentrations the precipitate is completely amorphous. At the highest concentration (1.1×10^{-1} mol/dm^3)diffuse diffraction peaks are observed which are not due to aluminium hydroxide but probably to solid citric acid. Fig. 3. shows diffractograms of samples obtained in solutions of ethylenediaminetetraacetic acid (EDTA) which is a strong chelating agent. The results are similar to those in the experiments with citric acid. Bayerite is formed along with pseudoboehmite in pure water and up to a concentration of about 10^{-3} mol/dm^3 EDTA. With a higher concentration, pseudoboehmite is formed and at the highest concentrations studied an amorphous precipitate.

Fig. 4 shows the results of experiments carried out in solutions of sodium borogluconate which has a more pronounced effect in hindering the forma-

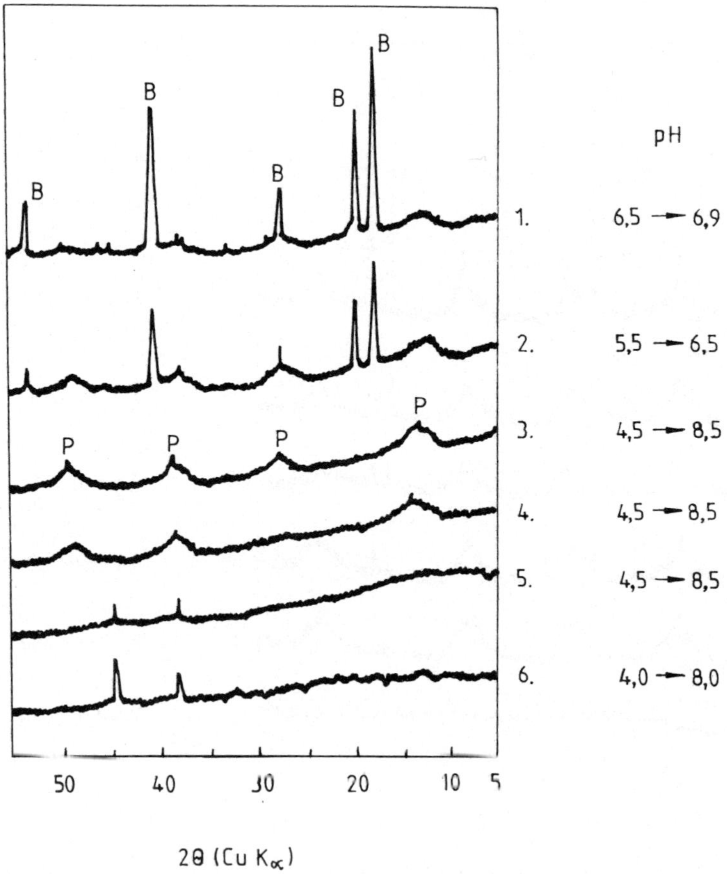

Fig. 3. X-ray diffraction pattems of aluminium hydroxides obtained from amalgamated aluminium after aging of 5 days : 1, in pure water, 2 to 5 in EDTA of 1.6×10^{-3}, 1.6×10^{-2}, 3.2×10^{-2}, 3.2×10^{-3}, 1.6×10^{-1} mol/dm^3 and in a saturated solution, respectively.

tion of bayerite and promoting the sequence pseudoboehmite-amorphous hydroxide. Small concentrations 7.1×10^{-6} mol/dm^{3}cause the formation of pseudo boehmite along with bayerite, which disappears at a concentration of 10^{-4} mol/dm^3, an d in solutions of 1.5×10^{-2} mol/dm^3 completely amorphous precipitate si formed. The pH in this series of experiments is

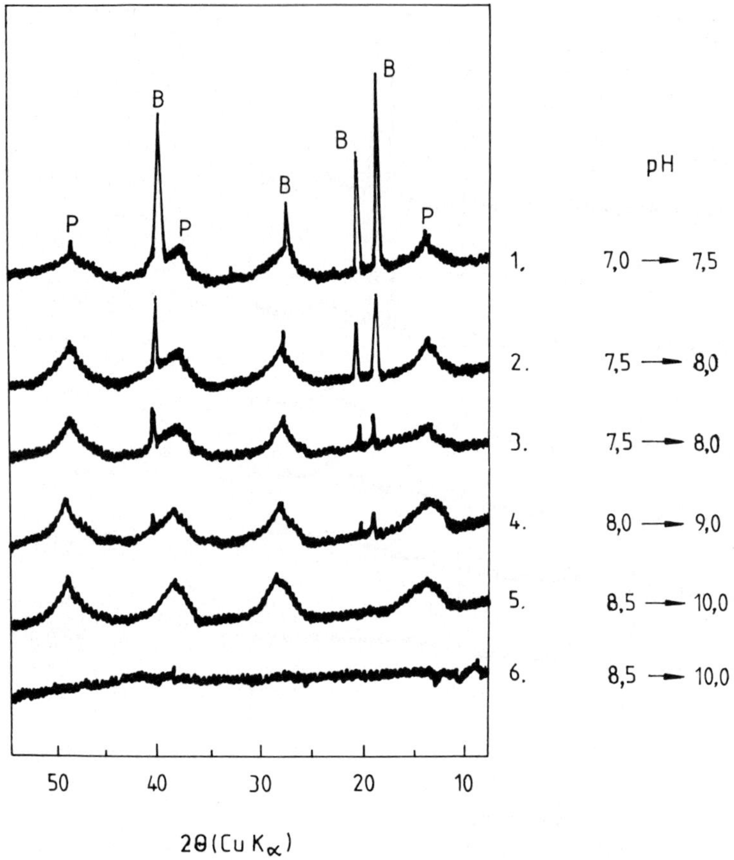

Fig. 4. X-ray patterns of aluminium hydroxides obtained from amalgamated aluminium in solutions of sodium borogluconate of: 1, 7.1×10^{-6}, 2, 1.07×10^{-4}, 3, 2.14×10^{-4}, 4, 4.8×10^{-4}, 5, 7.8×10^{-4} and 6, 1.5×10^{-2} mol/dm^3, after aging of 5 days.

higher than in previous studies (from 7.5 to pH 10, increasing with concentration), which also promotes formation of pseudoboehmite. The pronounced effect of the inhibition of crystallization by borogluconate must be related to the steric effect of the large molecule with a molecular weight of 700, (Telzinskaja et al., 1971.).

CONCLUSIONS

The difference between the results of our experiments in which very small quantities of aluminium ions are initially released in an organic ligand solution of a given concentration, and the results reported in the literature is that in the slow release process a higher concentration of organic ligand (about 10^{-3} to 10^{-1} mol/dm^3 of citric acid and of EDTA) is necessary to inhibit the formation of bayerite and to promote the formation of pseudoboehmite and amorphous hydroxide. The inhibiting concentration of 10^{-3} mol/dm^3 represents a relatively high concentration. The procedure with amalgamated aluminium is not only reproducible, but it better represents the natural conditions under which aluminium ion is released from aluminium-bearing minerals in the solution phase.

REFERENCES

Kwong Ng Kee, K.F. and Huang, P.M., 1975. Influence of citric acid on the crystallization of aluminium hydroxides. Clays Clay Min., 23 : 164-165.

Kwong Ng Kee, K.F. and Huang, P.M., 1978. Nature of hydrolytic precipitation products of aluminium as influenced by low-molecular weight complexing organic acids. Proc. Int. Clay Conf., Oxford, 27 : 527-536.

Lahodny-Šarc, O., Dragčević, Z. and Došen-Šver, D., 1978. The influence of the activity of water on the phase composition of aluminium hydroxides formed by reaction of amalgamated aluminium with water. Clays Clay Min., 26 : 153-159.

Lind, C.J. and Hem, J.D., 1975. Effect of organic solutes on chemical reactions of aluminium. Geolog. Survey Water-Supply Paper No.1827-G : Chemistry of Aluminium in Natural Water, Washington, D.C.,1-83.

Serna, C.J., White, J.L. and Hem, S.L., 1978. Structural survey of carbonate-containing antacids. J. Pharmac. Sc., 67 : 324-327.

Telzinskaja, N.N., Schwartz, E.M. and Nevint, A.F., 1971. Kompleksoobrazovanie mezdu solami glukonovoj i bornoj kislot. Zhurnal neorgan. himii, tom XVI, Vip. 4, 909-912.

Violante, A. and Jackson, M.L., 1978. Crystallization of norstrandite in citrate systems and in the presence of montmorillonite. Proc. Int. Clay Conf., Oxford, 27 : 517-525.

Violante, A. and Violante, P., 1980. Influence of pH, concentration and chelating power of organic anions on the synthesis of aluminium hydroxides and oxyhydroxides. Clays Clay Min., 28 : 425-434.

Volska, E., 1977. Conversion of an amorphous gel of aluminium hydroxide into norstrandite in the presence of aurintricarboxylic acid ammonium salt. Koll. Zhurnal, 39 : 539-542.

HYDROTHERMAL TRANSFORMATION OF SOME MINERALS INTO STEVENSITE

T. SAKAMOTO*, H. KOSHIMIZU**, S. SHINODA** and R. OTSUKA**
*Faculty of Science, Okayama University of Science, Okayama (Japan)
**Department of Mineral Industry, School of Science and Engineering, Waseda University, Shinjuku-ku, Tokyo (Japan)

ABSTRACT

In order to investigate the genesis of stevensite and allied minerals experimentally, naturally occurring minerals such as pectolite, wollastonite, xonotlite, and bustamite were treated hydrothermally with magnesium chloride solution and with basic magnesium carbonate for various Mg(in basic magnesium carbonate) to Si(in starting materials) molecular ratios ranging from 0.1 to 1.5. The products were examined by means of X-ray diffraction, thermal analysis, electron microscopy, and chemical analysis.

It was found that these minerals were altered to 'hydrated talc' and a small amount of serpentine by treatment with magnesium chloride solutions. On the other hand, by treatment with basic magnesium carbonate they were successively altered to stevensite, stevensite and serpentine, and serpentine with increasing Mg/Si ratios. In the latter case, the products were always associated with a large amount of calcite because of the presence of the carbonate ion. The amount of stevensite formed in these runs were rather small except in the runs for pectolite. However, when Na ion was added, a larger amount of stevensite was obtained. It is concluded that in addition to a low Mg/Si ratio, the removal of Ca ion from the solution and the presence of Na ion in it will play an important role in the formation of stevensite.

INTRODUCTION

Stevensite, a trioctahedral magnesium smectite, is a rather rare mineral with some cation deficiencies in the octahedral sheet. The following occurrences of stevensite have been reported:

(1) As authigenic minerals coexisting with carbonate and carbonaceous matter in the thin bed of the Green River Formation, Wyoming (Bradley and Fahey, 1962; Tettenhorst and Moore, 1978).

(2) As hypogene minerals fed from hydrothermal solution charged with magnesium, silica etc. (Sakamoto *et al.*, 1975; Yokomizo and Moriyama, 1976).

(3) As hydrothermal alteration products of pectolite (Faust and Murata, 1953; Faust *et al.*, 1959; Randall, 1959), of wollastonite (DeRudder and Beck, 1963; Otsu *et al.*, 1963; Imai *et al.*, 1966; Yokomizo and Moriyama, 1976), and of bustamite (Otsu *et al.*, 1963).

Among these, almost all known stevensites belong to the third group.

Roy and Roy (1955) synthesized stevensite in the system $MgO-SiO_2-H_2O$, and noted that at a water vapour pressure of 10000 psi (690 bar) 'magnesium montmorillonite' is stable below 250 °C, but is converted to talc at 275 °C. Mumpton and Roy (1956) reported similar results. Otsuka *et al.* (1972) studeid the stability relations of stevensite by hydrothermal treatment of natural stevensite, and showed that a 'dehydrated stevensite' and a 'hydrated talc' may exist as intermediate phases between stevensite and talc in their stability field. They subsequently treated several naturally occurring minerals hydrothermally in order to investigate the genesis of stevensite and allied minerals experimentally (Sakamoto *et al.*, 1975, 1981; Suzuki and Otsuka, 1976; Otsuka *et al.*, 1979). This paper summarizes previous investigations, adding new data on bustamite, and discusses the reaction mechanism of the alteration process.

EXPERIMENTAL

The starting materials used were the following: pectolite from Heguri, Chiba Pref,; wollastonite from the Akatani mine, Niigata Pref.; xonotlite from the Talnakh deposit, USSR; bustamite from Kanoiri, Tochigi Pref.. These materials were carefully hand picked and ground in an agate mortar. Powders in the 150 to 200 mesh range or below 100 were used in all experiments. Hydrothermal treatments were carried out by the following three methods:

Exp. 1: 60 mg of the starting materials were reacted in test-tube type reactors for 120 hours between 100 and 450 °C at a water vapour pressure of 250 kg/cm^2. In all runs, the starting materials were contained in small sealed silver tubes. Before welding the tubes about 0.8 cm^3 of magnesium chloride solution was added.

Exp. 2: 250 mg of the starting materials were reacted with 5 cm^3 of magnesium chloride solution in silver or teflon lined Morey type reactors for 3 to 240 hours at 250 °C.

Exp. 3: 100 mg of the starting materials were mixed with basic magnesium carbonate. The mixtures and 5 cm^3 of distilled water were reacted in teflon lined Morey type reactors for 360 hours at 250 °C.

In both Exp. 1 and Exp. 2, 0.5, 1.0, and 2.0 mol/dm^3 magnesium chloride solutions were used throughout the experiments. In Exp. 3, molecular ratios of Mg(in basic magnesium carbonate) to Si(in the starting materials) of the starting mixtures ranged from 0.1 to 1.5. In some runs, 15 mg of sodium hydrogen carbonate was added. All products were separated and washed repeatedly by centrifuging or by filtration using a micro filter (0.45 μm, Fuji Photo Film Co. Ltd. Type FM-45). The products in Exp. 3 were treated by 5 vol.% acetic acid for 3 hours at room temperature, in order to remove the large amount of calcite formed. The products were then examined by means of X-ray diffraction with Ni-filtered Cu radiation and in the case of bustamite with Mn-filtered Fe radiation. The relevant products were further examined by means of thermal analysis, infrared spectroscopy, transmission and scanning electron microscopy. In Exp. 2, the solutions separated from the products were chemically analyzed by atomic absorption spectroscopy. The formation of stevensite was confirmed by a X-ray peak shift from 15 to 17 Å after ethylene glycol saturation.

EXPERIMENTAL RESULTS

Approximately 260 runs were performed on pectolite, wollastonite, xonotlite, and bustamite. Representative runs and their products are summarized in Tables 1 and 2. Since the results for the first three minerals were reported previously (Sakamoto *et al.*, 1975, 1981; Suzuki and Otsuka, 1976; Otsuka *et al.*, 1979), only those for bustamite are described below.

Magnesium chloride solution treatment of bustamite

Bustamite was treated with 0.5 mol/dm^3 magnesium chloride solution (Exp. 1). The X-ray diffraction patterns of the products are shown in Fig. 1. Below 250 °C, bustamite remained unchanged. At 300 °C, the reflections of bustamite were weakened, and broad and diffuse reflections appeared at 10.1, 4.6, and 1.53 Å. These reflections seemed to correspond to those of clay minerals, and the 1.53 Å reflection suggested that the product had a trioctahedral structure. The 10.1 Å reflection was almost unchanged when treated with ethylene glycol. The patterns of the product were in good agreement with those of 'hydrated talc'*. With increasing temperature, the reflections of bustamite became weaker and disappeared above 400 °C. On the other hand, the reflections of the product became more distinct and sharper, and the 10.1 Å reflection shifted to 9.5 Å above 350 °C. The patterns above 350 °C were very similar to those of talc.

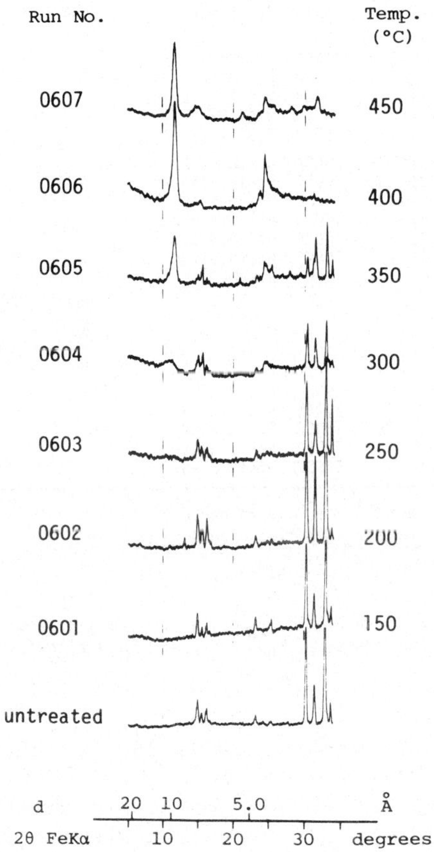

Fig. 1. X-ray diffraction patterns of bustamite, untreated and treated with 0.5 mol/dm^3 $MgCl_2$ solution at various temperature.

* In this paper, the name 'hydrated talc' is used to distinguish it from normal talc. It is a talc like mineral of low crystallinity, whose basal spacing is larger than 9.5 Å, and contains more water than normal talc (Imai *et al.*, 1973). We use this name in more wide sense than kerolite (Brindley *et al.*, 1977).

TABLE 1
Representative runs for magnesium chloride solution treatment

Conc. (mol/dm^3)	Method	Temp. (°C)	Press. (kg/cm^2)	Duration (hours)	Pectolite		Wollastonite		Xonotlite	
					Run No.	Products	Run No.	Products	Run No.	Products
0.5	Exp.1	100	250	120	0507	p.c.Ht+P+l.C	0499	W	-	-
"	"	150	"	"	0508	p.c.Ht+l.C	0500	p.c.Ht+W	-	-
"	"	200	"	"	0509	Ht(10.4)+l.C	0501	Ht(10.3)+Sr+W	-	-
"	"	250	"	"	0510	Ht(10.2)+l.C	0502	Ht(10.3)+Sr+l.W	-	-
"	"	300	"	"	0511	Ht(9.9)+l.C	0503	Ht(10.0)+Sr+l.W	-	-
"	"	350	"	"	0512	Ht(9.8)	0528	Ht(9.9)+Sr+l.W	-	-
"	"	400	"	"	0513	Ht(9.6)+Sr	0529	Ht(9.8)+Sr+l.W	-	-
"	"	450	"	"	-	-	-	-	-	-
1.0	Exp.1	100	250	120	0248	p.c.Ht+P	0213	W	6024	p.c.Ht+l.X
"	"	150	"	"	0249	p.c.Ht	0214	p.c.Ht+W	6025	p.c.Ht+l.Sr
"	"	200	"	"	0250	Ht(10.6)+l.C	0215	Ht(10.5)+Sr+l.W	6026	Ht(10.9)+Sr
"	"	250	"	"	0251	Ht(10.3)	0216	Ht(10.5)+Sr	6027	Ht(10.5)+Sr
"	"	300	"	"	0252	Ht(10.0)	0217	Ht(10.1)+Sr	6028	Ht(10.0)+Sr
"	"	350	"	"	0253	Ht(9.6)+Sr	0218	Ht(9.8)+Sr	6029	Ht(9.8)+Sr
"	"	400	"	"	0254	T(9.4)+Sr	0219	T(9.5)+Sr	6030	Ht(9.6)+Sr
"	"	450	"	"	-	-	-	-	6031	T(9.5)+Sr
2.0	Exp.1	100	250	120	0220	p.c.Ht+l.P	0259	p.c.Ht+W	-	-
"	"	150	"	"	0221	p.c.Ht	0260	p.c.Ht+W	-	-
"	"	200	"	"	0222	Ht(10.9)	0208	Ht(11.0)+Sr+W	-	-
"	"	350	"	"	0225	Ht(9.7)+l.Sr	0211	Ht(9.8)+Sr	-	-
"	"	400	"	"	0226	T(9.5)+Sr	0212	T(9.5)+Sr	-	-
0.5	Exp.2	250	s.v.p.	3	10070	Ht(10.8)+l.P	10069	p.c.Ht+W	-	-
"	"	"	"	6	10044	Ht(10.8)+l.P	10037	Ht(11.0)+Sr+W	-	-
"	"	"	"	12	10045	Ht(10.8)	10038	Ht(10.8)+Sr+l.W	-	-
"	"	"	"	24	10046	Ht(10.8)	10039	Ht(10.8)+Sr	-	-
"	"	"	"	72	10047	Ht(10.8)	10040	Ht(10.8)+Sr	-	-
"	"	"	"	120	10048	Ht(10.8)	10041	Ht(10.8)+Sr	-	-
"	"	"	"	240	10050	Ht(10.8)	10067	Ht(10.8)+Sr	-	-
1.0	Exp.2	250	s.v.p.	1	-	-	-	-	6075	p.c.Ht+X
"	"	"	"	3	-	-	-	-	6074	p.c.Ht+Sr
"	"	"	"	6	-	-	-	-	6072	Ht(11.3)+Sr
"	"	"	"	240	-	-	-	-	6081	Ht(10.6)+Sr
2.0	Exp.2	250	s.v.p.	3	10072	p.c.Ht+l.P	10071	p.c.Ht+W	-	-
"	"	"	"	6	10058	p.c.Ht	10051	p.c.Ht+l.W	-	-
"	"	"	"	12	10059	p.c.Ht	10052	p.c.Ht+l.W	-	-
"	"	"	"	24	10065	Ht(10.8)	10053	Ht(11.0)+Sr	-	-
"	"	"	"	240	10064	Ht(10.8)	10057	Ht(10.8)+Sr	-	-

P: pectolite; $(Ca_{3.96}Mg_{0.10})_{4.06}(Na_{1.95}K_{0.01})_{1.96}H_{2.82}(Si_{5.96}Al_{0.05}Fe_{0.01})_{5.88}O_{18.00}$

W: wollastonite; $(Ca_{3.84}Mg_{0.09}Fe^{2+}_{0.05}Mn_{0.02}Na_{0.01})_{6.01}(Si_{5.96}Al_{0.02}Fe^{3+}_{0.03})_{6.01}O_{18.00}$

X: xonotlite; $(Ca_{5.77}Mg_{0.07}Na_{0.07}K_{0.02})_{5.93}H_{2.38}(Si_{5.84}Al_{0.13}Fe_{0.04})_{6.01}O_{19.00}$

B: bustamite; $(Ca_{1.17}Mg_{0.06}Fe^{2+}_{0.94}Mn_{3.98})_{6.15}Si_{5.92}O_{18.00}$

Ht: 'hydrated talc', T: talc, Sr: serpentine, C: calcite, l.: a little, minor, p.c.: poorly crystallized, ( ): basal spacing of the product, s.v.p.: saturated vapour pressure, Exp.1, Exp.2,: see the text.

In addition, a weak reflection at 7.2 Å appeared above 350 °C indicating a small amount of serpentine. Similar results were obtained when the starting material was reacted with 1.0 mol/dm^3 magnesium chloride solution, but serpentine was more abundant. In Exp. 2, bustamite remained unchanged for up to about 6 hours. Between 12 and 24 hours 'hydrated talc' appeared. With longer duration, the reflections of bustamite became weaker and the formation of 'hydrated talc' and a little serpentine became more distinct.

TABLE 1
(continued)

Bustamite	
Run No.	Products
0601	B
0602	B
0603	B
0604	1.Ht(10.1)+B
0605	T(9.5)+1.Sr+B
0606	T(9.5)+1.Sr+1.B
0607	T(9.5)+1.Sr+1.B
-	-
0608	B
0609	B
0610	1.Ht(9.8)+B
0611	Ht(9.8)+1.Sr+1.B
0612	Ht(9.6)+1.Sr+1.B
0613	T(9.5)+1.Sr
0614	T(9.45)+Sr
-	-
-	-
-	-
-	-
-	-
10251	B
10252	B
10253	p.c.Ht+B
10254	p.c.Ht+B
10255	Ht(9.8)+1.Sr+B
10256	Ht(9.8)+1.Sr+1.B
10258	Ht(9.8)+1.Sr+1.B
-	-
-	-
-	-
-	-
-	-
-	-
-	-
-	-

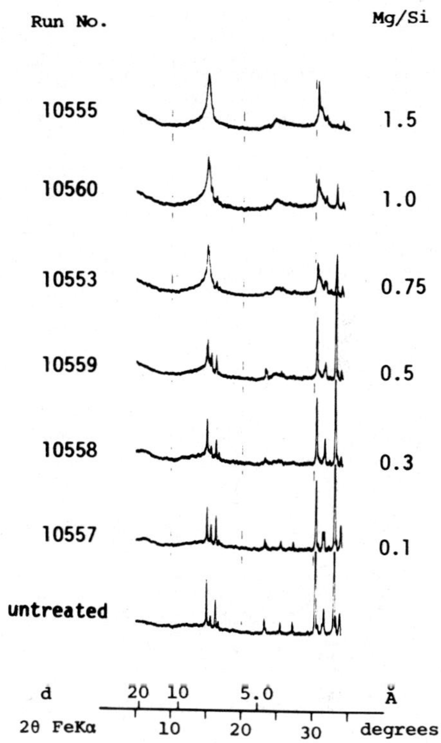

Fig. 2. X-ray diffraction patterns of bustamite, untreated and treated with basic magnesium carbonate at various Mg/Si ratios.

Basic magnesium carbonate treatment of bustamite

The X-ray diffraction patterns of the products are shown in Fig. 2. The product at a Mg/Si ratio of 0.1 showed a very weak and broad reflection at 13-18 Å. With ethylene glycol this reflection became more distinct and sharper, and shifted to about 17 Å, suggesting the formation of a small amount of stevensite. At a Mg/Si ratio of 0.3, in addition to the reflections of stevensite, a weak reflection at 7.4 Å appeared, which became more distinct with an increasing Mg/Si ratio. In the case of a Mg/Si ratio ≧ 0.5, no reflections of stevensite were observed in the products. The reaction product obtained at a Mg/Si

TABLE 2
Representative runs for basic magnesium carbonate treatment

Mg/Si ratio	$NaHCO_3$ addition	Pectolite		Wollastonite		Bustamite	
		Run No.	Products	Run No.	Products	Run No.	Products
0.1	-	PS-069	p.c.St+P+C	10354	p.c.St+W+C	10557	p.c.St+B+l.Bi
0.3	-	PS-073	St+P+C	10349	p.c.St+l.Sr+W+C	10558	p.c.St+l.Sr+B+l.Ku+Bi+C*
0.5	-	PS-076	St+l.Sr+P+C	10353	p.c.St+l.Sr+W+C	10559	Sr+B+l.Ku+Bi+C*
0.75	-	PS-095	St+l.Sr+P+C	10347	p.c.St+Sr+W+C	10553	Sr+B+l.Ku+Bi+C*
1.0	-	PS-096	St+Sr+l.P+C	10356	Sr+C+Mg	10560	Sr+l.B+l.Ku+Bi+C*
1.5	-	PS-097	Sr+C+Mg	10350	Sr+C+Mg	10555	Sr+l.B+Ku+l.Bi+C*+Mg
0.1	o	-	-	10361	St+W+C	10569	St+B+l.Bi+C*
0.3	o	-	-	10362	St+W+C	10570	St+B+l.Ku+l.Bi+C*
0.5	o	-	-	10372	St+l.Sr+W+C	10571	St+l.Sr+B+l.Ku+Bi+C*
0.75	o	-	-	10366	St+Sr+W+C	10564	Sr+B+l.Ku+Bi+C*
1.0	o	-	-	10367	St+Sr+W+C	10565	Sr+l.B+Ku+C*
1.5	o	-	-	10368	Sr+C+Mg	10566	Sr+Ku+C*+Mg

Experimental conditions: Exp. 3, 250 °C, saturated vapour pressure, 360 hours.

P: pectolite, W: wollastonite, B: bustamite, St: stevensite, Sr: serpentine, C: calcite, Mg: magnesite, Ku: kutnahorite, Bi:bixbyite (Mn_2O_3), C*: manganoan calcite, l.: a little, minor, p.c.: poorly crystallized.

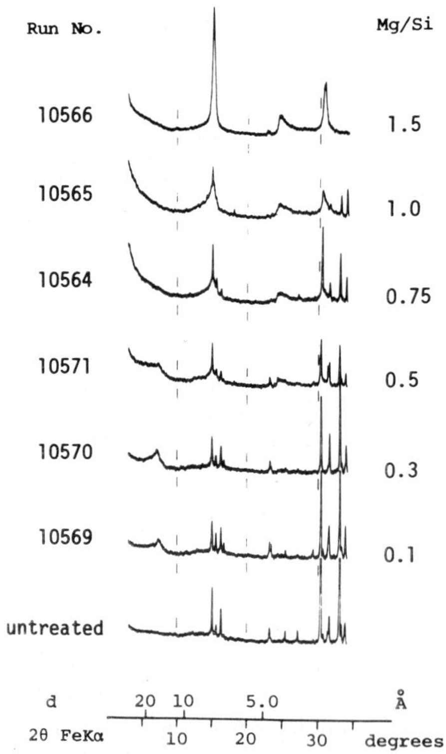

Fig. 3. X-ray diffraction patterns of bustamite, untreated and treated with basic magnesium carbonate and small amount of $NaHCO_3$ at various Mg/Si ratios.

Fig. 4. Representative SEM photograph of the product. Stevensite is formed on the surface of bustamite as curled flakes. Well developed crystals are of Mn-calcite.
(Run No. 10570, Exp. 3, Mg/Si=0. with a small amount of $NaHCO_3$, 250 °C, 360 hours)

ratio of 1.5 included serpentine and remaining bustamite. In addition to the magnesium clay minerals, carbonates such as manganoan calcite, kutnahorite, and magnesite (only in the case of a Mg/Si ratio of 1.5), as well as bixbyite were also observed in the products of almost all runs. When a small amount of sodium hydrogen carbonate was added, the formation of stevensite was promoted, as shown in Fig. 3. Fig. 4 shows the typical SEM photograph of stevensite formed in this run.

DISCUSSION AND CONCLUSIONS

We concluded that pectolite, wollastonite, xonotlite, and bustamite are altered to stevensite, 'hydrated talc', talc, and serpentine by hydrothermal treatment with magnesium bearing solutions. The formation of trioctahedral layer silicates from the starting materials can be considered as the removal of Ca ion and the subsequent addition of Mg ion. For example, the reactions for the formation of talc and serpentine from bustamite may be written as follows;

$$\underset{\textit{bustamite}}{2CaFeMn_4(SiO_3)_6}+9Mg^{2+}+6H^+ \rightarrow \underset{\textit{talc}}{3Mg_3Si_4O_{10}(OH)_2}+2Ca^{2+}+2Fe^{2+}+8Mn^{2+}$$

$$\underset{\textit{bustamite}}{2CaFeMn_4(SiO_3)_6}+18Mg^{2+}+18H_2O \rightarrow \underset{\textit{serpentine}}{6Mg_3Si_2O_5(OH)_4}+2Ca^{2+}+2Fe^{2+}+8Mn^{2+}+12H^+.$$

These equations show that the quantity of Mg ion required for the formation of the 1:1 type layer structure is twice as much as that for the formation of the 2:1 type. Hence, it is inferred that the 2:1 type is formed from the solution with low Mg/Si ratios, and the 1:1 type from that with high Mg/Si ratios. In treatment with magnesium chloride solutions, Si is introduced into the Mg containing solution by dissolution of the starting materials, so that the Mg/Si ratio of the solution is high in the initial stage and the 1:1 type will be formed, but when the reaction progresses this ratio decreases and the 2:1 type is formed. This may explain the presence of a small amount of serpentine in the products. The duration of the initial stage with a high Mg/Si ratio must depend on the dissolution rate of the starting material and the relative amounts of serpentine in the products are indeed closely related to this rate as shown in Table 1. In the case of the treatment with magnesium carbonate, Mg and Si are supplied to the solutions by dissolution of the carbonate and of the starting material, so that the Mg/Si ratio in the solution should in principle depend on the respective dissolution rates of these two materials and on their mixing ratios. However, the Mg/Si ratios are probably maintained at a level nearly equal to those of the starting mixtures. Therfore, the 2:1 type layer structure, both the 2:1 and 1:1, and the 1:1 type will be formed in order of increasing Mg/Si ratios in the starting mixtures.

The mechanism of the formation of the magnesium clay minerals from these starting materials may be interpreted as follows:
In the initial stage, by dissolution of the starting materials, Ca, Na, and Mn ions are brought into the solutions. Subsequently, remaining Si-O tetrahedra near the surface of the starting materials react with dissolved Mg ions and form a $MgO-SiO_2$ gel. This initially formed amorphous material shows an exothermic peak around 820 °C on the DTA curve.

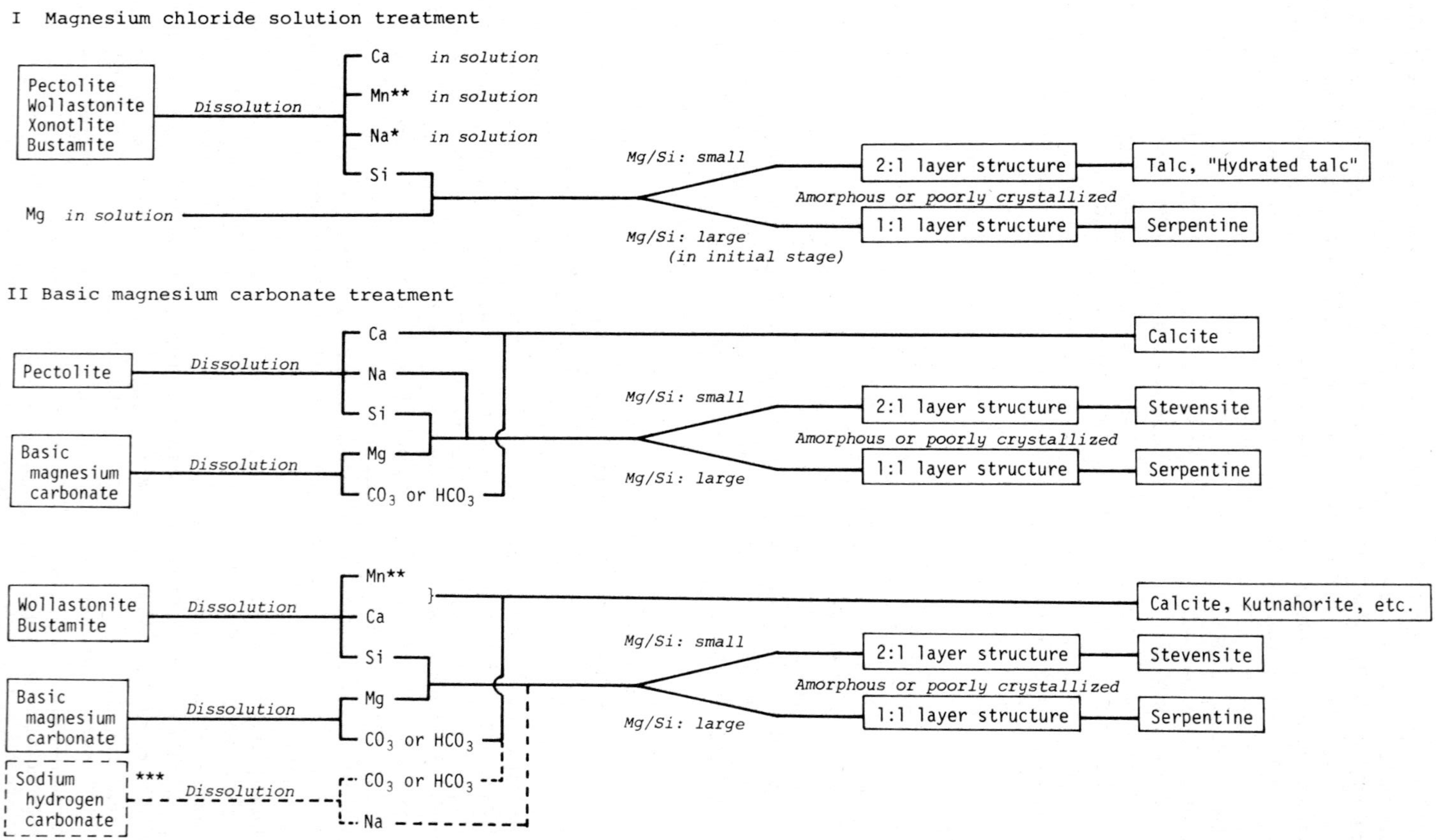

Fig. 5. Schematic diagrams showing the mechanism of the alteration process.

*: pectolite, **: bustamite, ***: in the case of the addition of a small amount of $NaHCO_3$ to the starting mixtures.

Quite possibly this material already has a poorly crystallized layer structure. With increasing temperature or longer duration of the experiment, the 2:1 type material will crystallize as stevensite, or 'hydrated talc' and talc, and the 1:1 material as serpentine. 'Hydrated talc' and talc are obtained by magnesium chloride solution treatment, and stevensite by magnesium carbonate treatment. In the latter treatment, the reaction proceeds under alkakine conditions and dissolved Ca ion is precipitated as calcite. With magnesium chloride solution treatment, the reaction proceeds under acidic conditions and Ca ion remains in solution. The amounts of stevensite formed in these runs were rather small, except in the runs for pectolite. In this connection, it is of interest that stevensite is formed only from the Na bearing gels and 'hydrated talc' from the Ca bearing gels, as reported previously (Otsuka *et al.*, 1972).

The reaction mechanism of the alteration processes are shown schematically in Fig. 5. It may be concluded that in addition to low Mg/Si ratios and alkaline conditions, the presence of Na ion promotes the formation of stevensite, whereas acidic conditions and the presence of Ca ion favour formation of 'hydrated talc'. These conclusions are consistent with the observation that almost all natural stevensites are closely associated with calcite and aragonite (*e.g.* Sakamoto *et al.*, 1975). Although in the stevensite produced in the present experiments the dominant exchangeable cation is Na and in natural stevensite, Ca (Faust and Murata, 1953; Randall, 1959; Otsu *et al.*, 1963; Imai *et al.*, 1970; Sakamoto *et al.*, 1975; Yokomizo and Moriyama, 1976), there is a strong possibility that the exchangeable cation of stevensite is initially Na and subsequently replaced by Ca through the action of Ca bearing ground water or hydrothermal solutions through the geologic age. The present experimental study supports Glenn's suggestion regarding the origin of stevensite; *i.e.* magnesium was probably first extracted by rain water from the basalt rock as bicarbonate, and that the resulting solution reacted with pectolite (Glenn, 1916).

ACKNOWLEDGEMENT

The authors wish to express their sincere thanks to Prof. Naoya Imai of Waseda University for many helpful suggestions during the course of this work. Thanks are also due to Mr. J. Yasui and Mr. M. Miyashita of Waseda University and Mr. H. Tatematsu of Railway Technical Research Institute, JNR for taking transmission and scanning electron micrographs.

REFERENCES

Bradley, W.H. and Fahey, J.J., 1962. Occurrence of stevensite in the Green River Formation of Wyoming. Am. Miner., 47: 996-998.

Brindley, G.W., Bish, D.L. and Wan, H.M., 1977. The nature of kerolite, its relation to talc and stevensite. Miner. Mag., 41: 443-452.

DeRudder, R.D. and Beck, C.W., 1963. Stevensite and talc — Hydrothermal alteration products of wollastonite. Clays Clay Min., 10: 188 -199.

Faust, G.T. and Murata, K.J., 1953. Stevensite, redefined as a member of the montmorillonite group. Am. Miner., 38: 973-987.

Faust, G.T., Hathaway, J.C. and Millot, G., 1959. A restudy of stevensite and allied minerals. Am. Miner., 44: 342-370.
Glenn, M.L., 1916. A new occurrence of stevensite, a magnesium-bearing alteration product of pectolite. Am. Miner., 1: 44-46.
Imai, N., 1966. Some aspects of clay mineralogy on the alteration products of hydrothermal-metasomatic deposits of "multiple-mineralization type" in the inner zone of northeastern Japan: a review (in Japanese). Nendo Kagaku (Jour. Clay Sci. Soc. Japan), 5: 49-70.
Imai, N., Otsuka, R., Nakamura, T. and Tsunashima, A., 1970. Stevensite from the Akatani mine, Niigata Prefecture, Northeastern Japan. Clay Sci., 4: 11-29.
Imai, N., Otsuka, R., Nakamura, T., Tsunashima, A. and Sakamoto, T., 1973. "Hydrated talc" — an alteration product of wollastonite by reaction with magnesium bearing hydrothermal solution. Clay Sci., 4: 175-189.
Mumpton, F.A. and Roy, R., 1956. The influence of ionic substitution on the hydrothermal stability of montmorillonite. Clays Clay Min., 4: 337-339.
Otsu, H., Shimazaki, Y. and Omachi, K., 1963. Stevensite from the Ohori mine (in Japanese). Bull. Geol. Sur. Japan, 14: 591-599.
Otsuka, R., Imai, N. and Sakamoto, T., 1972. Stability of stevensite under hydrothermal conditions. Mem. Sch. Sci. Eng. Waseda Univ., 36: 37-56.
Otsuka, R., Sakamoto, T., Suzuki, S., Shinoda, S. and Koshimizu, H., 1979. Hydrothermal treatment of pectolite (in Japanese). Kobutsugaku Zasshi (Jour. Miner. Soc. Japan), 14: S170-186.
Randall, B.A.O., 1959. Stevensite from the Whin Sill in the region of the North Tyne. Miner. Mag., 32: 218-225.
Roy, D.M. and Roy, R., 1955. Synthesis and stability of minerals in the system $MgO-Al_2O_3-SiO_2-H_2O$. Am. Miner., 40: 147-178.
Sakamoto, T., Suzuki, S., Otsuka, R. and Imai, N., 1975. Hydrothermal treatment of wollastonite and pectolite with $MgCl_2$ solution (in Japanese). Nendo Kagaku (Jour. Clay Sci. Soc. Japan), 15: 9-22.
Sakamoto, T., Otsuka, R. and Imai, N., 1975. Stevensite from the Kamaishi mine, Iwate Prefecture, Japan. Jour. Japan. Assoc. Miner. Petr. Econ. Geol., 70: 1-11.
Sakamoto, T., Otsuka, R. and Koshimizu, H., 1981. Hydrothermal treatment of wollastonite (in Japanese). Kobutsugaku Zasshi (Jour. Miner. Soc. Japan), 15: S55-65.
Suzuki, S. and Otsuka, R., 1976. Hydrothermal treatment of xonotlite with $MgCl_2$ solution (in Japanese). Nendo Kagaku (Jour. Clay Sci. Soc. Japan), 16: 98-111.
Tettenhorst, R. and Moore Jr, G.E., 1978 Stevensite oolites from the Green River Formation of Central Utah. Jour. Sed. Petr., 48: 587-594
Yokomizo, H. and Moriyama, Z., 1976. Stevensite from the Kiura mine, Oita Prefecture (in Japanese). Jour. Japan. Assoc. Miner. Petr. Econ. Geol., 71: 221-228.

THE ORIGIN OF CLAY MINERALS IN THE OCEANIC CRUST REVEALED BY NATURAL AND EXPERIMENTAL DATA

VICTOR B., KURNOSOV, IGOR V., KHOLODKEVICH, LUDMILA P., KOKORINA, *NICOLAI V., KOTOV, AND OLEG V., CHUDAEV
(FAR EAST GEOLOGICAL INSTITUTE, VLADIVOSTOK, 690022, USSR;
*LENINGRAD STATE UNIVERSITY, LENINGRAD, USSR)

ABSTRACT

Clay minerals are found to be widespread both in oceanic sediments and in basaltic basement, where they originate due to hydrothermal circulation in the oceanic crust. This remarkable phenomenon causes the release of chemical elements from basalts to the ocean, promoting the formation of hydrothermal clay sediments. Data from deep-sea drilling operations on clay mineral types and composition are related to their origin, on the basis of experimental modeling.

NATURAL OBSERVATIONS

Clay minerals from basalts resulting from deep-sea drilling in the Gulf of California, Emperor Chain, Nauru Basin, the Bering Sea, the Middle America Trench, and Costa Rica Rift were studied.

The principal results are the following:

Trioctahedral ferromagnesian smectite, originating in reducing alkali conditions, is the most widespread clay mineral in basalts, i.e. in groundmass, veins, and vesicles (Kurnosov et al., 1981). Occurrences of smectites in veins in association with minerals, formed at different temperatures, such as talc, chlorite, actinolite, magnetite, okenite, calcite, zeolites, apophyllite, pyrite, hematite, cristobalite, and quartz, allow the conclusion that smectites were formed at various temperatures and in various solutions, from sea water to juvenile solutions. The highest temperature of the formation of smectites is about 500 °C, as determined by the presence of associated talc and actinolite (Mottl et al., 1978).

Celadonite, dioctahedral smectite, and iron hydroxides were found in the upper part of the basaltic basement. The presence of these minerals are on indication of oxidizing conditions (Bass, 1976; Robinson et al., 1977). According to isotopic data the temperature of formation does not exceed 20 °C (Seyfried et al., 1978).

In basalts from the Middle America Trench, metamorphosed up to the greenschist facies, clay minerals consist of chlorite and corrensite.

Data obtained on clay minerals from oceanic basalts allow drawing a general diagram of their formation (Fig.1).

During slow cooling of basalts the formation of clay minerals began at 500°C, it was probably very rapid at 400 to 200°C and slowed down considerably at a final stage of cooling.

Halmyrolysis was not important in formation.

During warming up the formation of clay minerals proceeded in reverse order. Minerals of the chlorite group were formed due to the replacement of sea water by juvenile solutions at high temperatures. In alkaline basalts smectites were replaced by mica.

Dioctahedral micas and smectites were formed under oxidizing conditions, due to the penetration of fresh sea water into a previously formed basaltic complex.

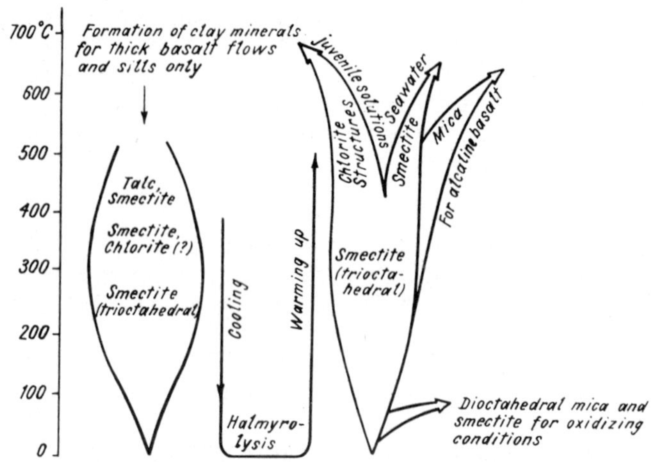

Fig. 1. Diagram showing the formation of clay minerals in oceanic environment from basalts based on natural data

The morphology of clay minerals derived from basalts is is peculiar. Glass and olivine were replaced by sponge-like smectite (Plate I-1,2). Rosett-like textures are common in veins (Plate I-3). Complex forms of seladonite occured in vesicles: lacy (Plate I-4,5), leaf-like (Plate I-6), tubes (Plate II-1), flower-like (Plate II-2). Hair-like corrensites also occured in vesicles (Plate II-3,4).

It seems impossible to test the supposed of origin of clay minerals under natural conditions. Therefore, model experiments were carried out.

EXPERIMENTAL MODELING

Experiments were carried out in the temperature interval of 100 to 700°C, 1 kbar pressure, with pure water, sea water, mineralized chloride, bicarbonaceous, sulphate and acidic waters (Kotov et al., 1978; Kholodkevich et al.,

1981; Kokorina et al.,1981),and sea water with the admixture of juvenile components: fluorine,sulphur,and carbon dioxide.

A schematic diagram of the synthesis of clay minerals shows the results of our experiments (Fig. 2). Water - magmatic rock interaction caused the formation of clay minerals in a wide temperature range. The composition of clay minerals and the rate of their formation depend on the composition of the primary rock. Tholeiitic basalts mainly yield trioctahedral smectite. Alkaline basalts at high temperatures yield trioctahedral mica, replacing smectites.

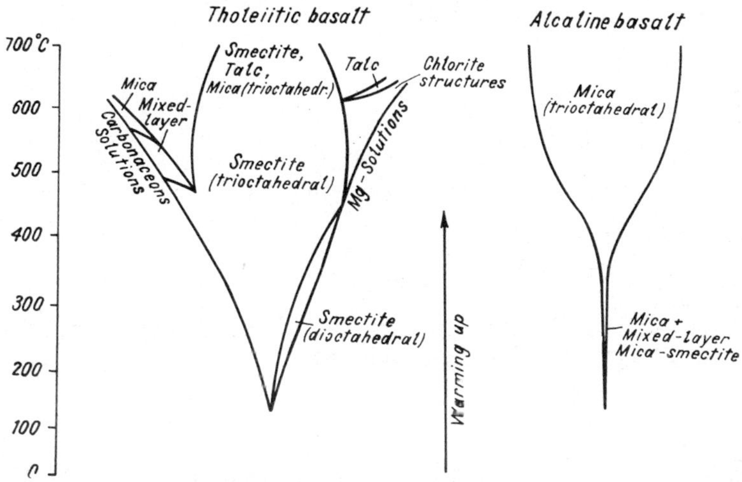

Fig. 2. Experimental data on the formation of clay minerals resulting from solution - basalt interaction.

The interaction of mineralized solutions and tholeiitic basalt (Kokorina et al.,1981) as well as the interaction of water and tholeiitic basalt caused primarily the formation of trioctahedral smectites. Chloride solutions were most effective for the smectite synthesis. An increased magnesium content in these solution promoted the formation of corrensite and talc at temperatures of about 500°C to 600°C. Chlorite structures appeared at lower magnesium contents in water,though their synthesis took place at higher temperatures.Bicarbonaceous solutions are effective for the formation of mica.

The alteration of tholeiitic basalt by sea water,enriched in fluorine,sulpher,and carbon dioxide,led also to the formation of trioctahedral smectite mainly.

Experiments with acid solutions did not yield clay minerals (Kokorina et al.,1981).

Variations of chemical composition within one type of

basalts for example of tholeiites,were also reflected in the chemical composition of smectites (Fig. 3),formed by water - basalt interaction.

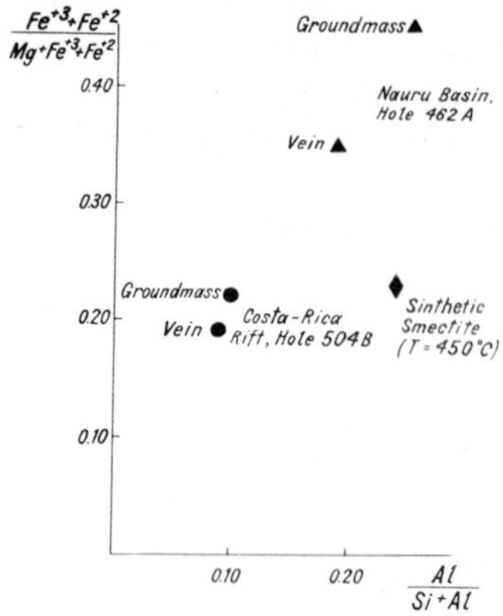

Fig. 3. Relation between iron and alumina content in natural and synthetic smectites.

The diagram "Relation between iron and alumina content in natural and synthetic smectites" shows,that more iron smectites were formed on the iron basalts of the Nauru Basin. Less iron smectites were formed on magnesium basalts of the Costa Rica Rift. Synthetic smectite,formed on iron basalt,corroborates this dependence.

Olivine (forsterite) was unstable in the temperature range from 200 to 600°C the interaction with water. Clay minerals deficient in aluminum were the newly formed minerals. Serpentine was the first mineral that replaced olivine. Smectites and talc appeared in the temperature range from 400 to 500°C. At 600°C part of the smectites transformed to chlorite. During the alteration of olivine,hair-like smectites were formed from the solutions in vesicles (Plate II-5).

Thus,the results of our experiments as well as of experiments by Mottl,and Bischoff (Mottl et al.,1978; Bischoff and Seyfried,1978) support that sea water is circulating in the upper part of the basaltic basement,where smectites are most widespread. In addition it was found that besides sea water mineralized solutions may circulate here,also forming smectites. In the very hot layers of the basaltic basement sea water did not promote the formation of chlorite structures. A supply of magnesium was necessary for their formation.

Our experiments aimed at revealing the release of chemical elements from basalts during formation of clay minerals. The results of the experiments are presented as histograms (Fig. 4). The histograms show the ratio of chemical elements, carried from the rock to the solution, to their content in a primary sample of tholeiitic basalt. Manganese, cobalt, titanium, and vanadium were strongly released. Iron and magnesium were primarily released at high temperatures. Earlier Mottle (1978), and Bischoff and Seyfried (1978) also showed experimentaly the strong release of calcium and silica from basalts interacting with sea water.

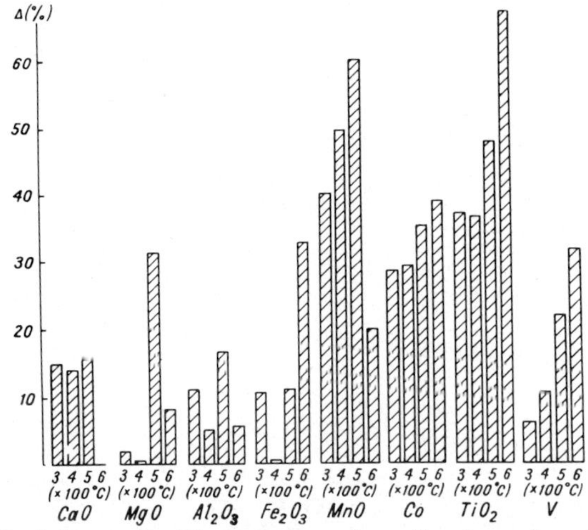

Fig. 4. Ratio of the amount of chemical elements released from rock to solution and their amount in primary tholeiitic basalt expressed as △ % (experimental data).

Chemical elements, extracted from basalts, may be released from the basaltic basement and used for the formation of hydrothermal clay sediments.

HYDEOTHERMAL CLAY SEDIMENTS

The influence of hydrothermal action on sediment formation is most pronounced in the area of the Galapagos spreading centre. Sediments in the Galapagos hydrothermal mounds with a very high heat flow consist mainly of nontronite and todorokite, precipitated from solutions. Juvenile components were also found in hydrothermal solutions. Their good correlation with rubidium suggest an endogenous origin of some elements, potassium first of all (Migdisov et al., 1981).

Nontronite is nearly lacking in aluminum and titanium. Potassium content is high, it reaches more than 3%. The morphology of nontronites varies. Hair-like structures were observed (Plate II-6).

Fig.5 shows the composition of nontronite minerals of different origin. The diagram shows that nontronites are very low in alumina according to both our data and data of other authors.

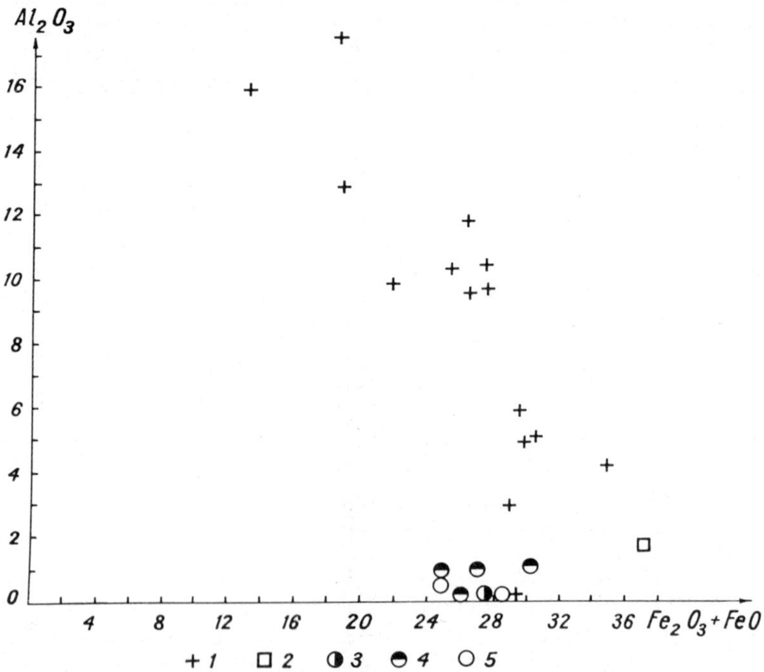

Fig. 5. Relation between Al_2O_3 and (Fe_2O_3 + FeO) in nontronites of different origin:
1 - nontronites (Isphording, 1975); 2 - nontronites from the Red Sea (Bischoff, 1974); 3 - Hole 506C, Leg 70 DSDP; 4 - Hole 509B, Leg 70 DSDP; 5 - Hole 507D, Leg 70 DSDP.

Various hydrothermal clay minerals were found in fault zones of the East Indian Ridge. Nearly monomineralic saponite, vermiculite, chlorite, palygorskite, and sepiolite were observed here (Skornyakova et al., 1978, 1979).

Hydrothermal clay minerals occur in basal sediments, studied through deep-sea drilling in the Gulf of California, Nauru Basin, Meiji guyot, and the Middle America Trench.

The main hydrothermal clay minerals are iron dioctahedral smectites. Hydromica - montmorillonite high in aluminum occurs in the Gulf of California.

Thus,the natural and experimental data show,that the formation of clay minerals caused by hydrothermal circulation,is a widespread phenomenon in the ocean. It occurs in a considerable part of the basaltic layer and is observed in the sediment cover. The formation of clay minerals is due to juvenile solutions in the deepest parts of the oceanic crust and the centres of eruptions.

REFERENCES

Bass,M.N.,1976. Secondary minerals in oceanic basalt,with reference to Leg 34,Deep Sea Drilling Project. In: R.S. Yeats et al.(Editors), Init. Repts. DSDP,34: Washington US Govt. Printing Office),34: 393-432.

Bischoff, J.l., 1974. Sediments of hydrothermal brines of the Red Sea. In: E.T. Degens and D.A. Ross (Editors), Hot brines and recent heavy metal deposits in the Red Sea (in Russian). Moscow, pp. 157-193.

Bischoff, J.L., Seifried, W.E., 1978. Hydrothermal chemistry of sea water from 25 to 350 C. Amer. J. Sci., vol. 278, 6: 838-860.

Isphording, W., 1975. Primary nontronite from Venezuelan Guayna. Amer. Miner., 60: 840-848.

Kholodkevich, I.V., Kotov, N.V., Kurnosov, V.B., 1981. Experimental study of secondary changes of volcanic glassy rocks in pure and model sea water at higher P - T parameters. In: A.G. Kossovskay (Editor), Mineral transformations of rocks of the oceanic substratum (epigenesis and initial metamorphism), Moscow, Nauka: pp. 67-71.

Kokorina, L.P., Kotov, N.V., Kurnosov, V.B., 1981. Hydrothermal transformations of oceanic basalts by highly mineralized model chloride,carbonate,sulphate and acid waters at higher P_{H_2O}-T parameters. In: A.G. Kossovskay (Editor), Mineral transformations of rocks of the oceanic substratum (epigenesis and initial metamorphism), Moscow, Nauka: pp. 57-66.

Kotov, N.V., Kurnosov, V.B.,Kholodkevich, I.V., 1978. Modeling of natural processes of volcanic rock transformation in pure and model sea water at higher P-T parameters. Lithol. and Miner. Res., 4: 78-89.

Kurnosov, V.B., Murdmaa,I.O., Skornyakova, N.S., 1981. Secondary changes of volcanic glasses in the ocean. In: A.G. Kossovskay (Editor), Mineral transformations of rocks of the oceanic substratum (epigenesis and initial metamorphism), Moscow, Nauka: pp. 50-56.

Migdisov, A.A., Belyi, V.M., Barskaya, N.B., Smirnova, O.N., 1981. On the geochemistry of Galapagos rift zone sediments in connection with the problem of formation of hydrothermal domes (Leg 70). Geology of the bottom of oceans based on data of deep-sea drilling. Abstracts of reports. Moscow, Nauka: pp. 45-46.

Mottl, M.J., Holland, H.D., 1978. Chemical exchange during hydrothermal alteration of basalt by sea water - I. Experimental results for major and minor components of sea water. Geochim. et Cosmochim. Acta, 42: 1103-1115.

Robinson, P.T., Flower, M.F.J., Schminke, H.V., Ohnmacht, W., 1977. Low temperature alteration of oceanic basalts, DSDP Leg 37. Init. Repts. DSDP, 37: Washington (US Govt. Printing Office), 37: 775-794.

Seyfried, W.E., Shanks, W.C., and Bischoff, J.L., 1976. Alteration and vein formation in Site 321 basalts. Init. Repts. DSDP, 34: Washington (US Govt. Printing Office), 34: 385-392.

Seyfried, W.E., Shanks, W.C., and Dibble, W.E., 1978. Clay mineral formation in DSDP Leg 34 basalt. Earth. and Planet. Sci. Lett., 41: 265-276.

Scornyakova, N.S., Kurnosov, V.B., Svalnov, V.N., Turanskaya, N.V., Tseitlin, N.Yu., Shevchenco, A.Ya., 1978. Sepiolites and palygorskites in the Indian Ocean. Lithol and Mineral. Res., 6: 29-45.

Skornyakova, N.S., Kurnosov, V.B., Shevchenko, A.Ya., 1979. Saponite and vermiculite in sedimentary rocks of the East Indian Ridge. Reports of Acad. Sci. USSR, vol. 244, 1: 185-188.

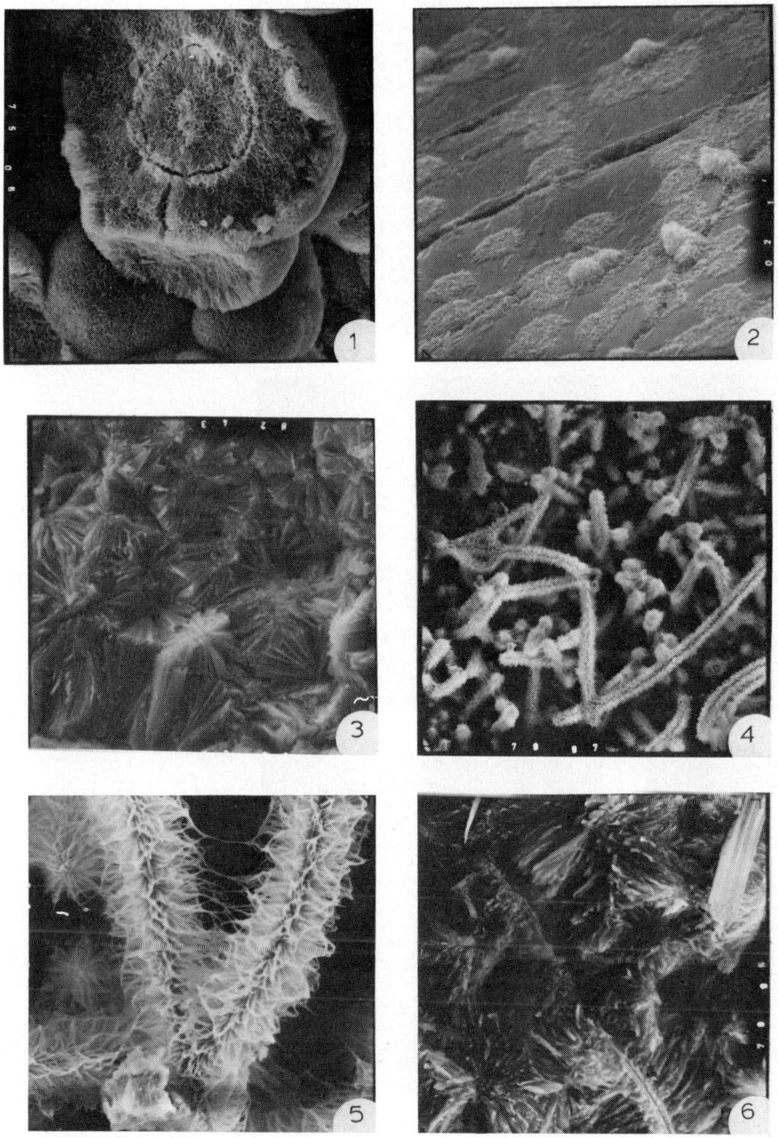

Plate I. Electron micrographs of clay minerals from basalts.

Figures 1,2. Sponge - like smectites: 1 - smectite replacing basalt glass, x 300; 2 - smectite from basalt groundmass,Nauru Basin, Leg 61, x 2000.
Figure 3. Rosette - like texture of vein smectite,Costa Rica Rift, Legs 69,70 DSDP, x 300.
Figures 4,5,6. Celadonite from Nauru Basin, Leg 61 DSDP: 4,5 - lacy celadonite (4 - x 300; 5 - x 1000); 6 - leaf - like celadonite, x 1000.

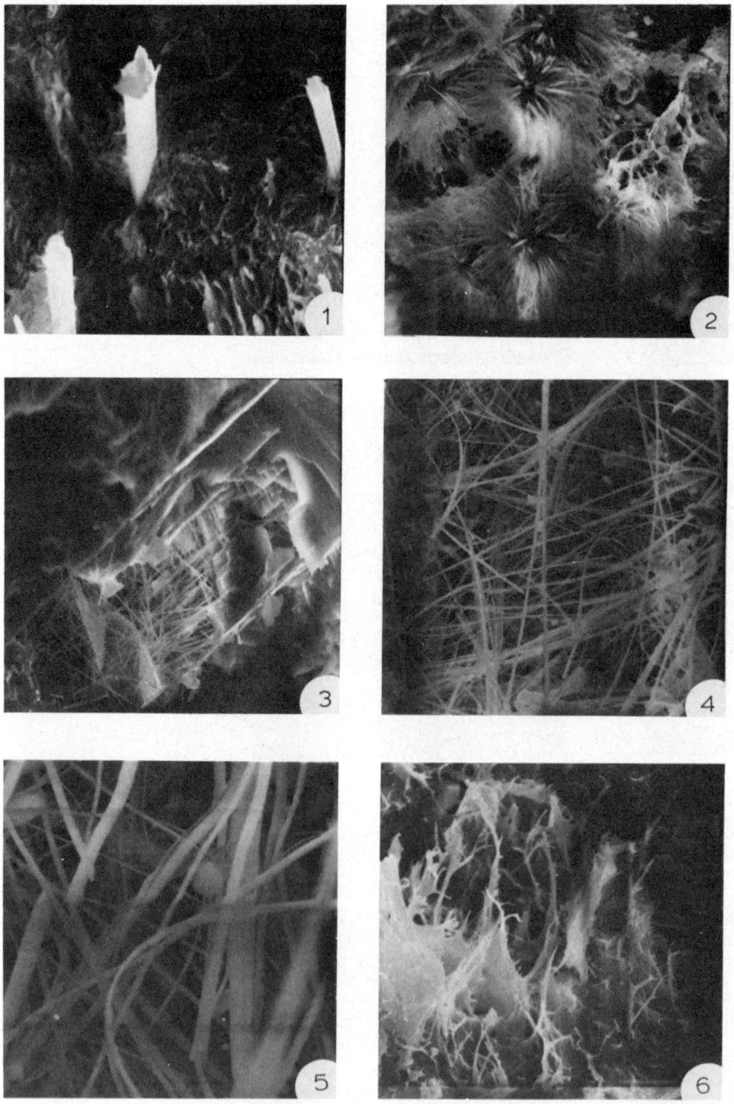

Plate II. Electron micrographs of clay minerals.

Figures ,2. Celadonite from Nauru Basin, Leg 61 DSDP:
1 - tubes, x 1500; 2 - flower - like celadonites, x 1500.
Figures 3,4. Hair - like corrensite, Middle America Trench, Leg 67 DSDP: 3 - x 1000; 4 - x 3000.
Figure 5. Synthetic smectite from olivine, x 10000.
Figure 6. Nontronite from the Galapagos Hydrothermal Mounds, Leg 70, x 1000.

1980 MT. ST.HELENS ERUPTION: CLAY MINERALS AND EARLY-STAGE ALTERATION

D.R.PEVEAR[1], D.P.DETHIER[2] AND D.FRANK[2]
1. Department of Geology, Western Washington University, Bellingham
2. U.S.Geological Survey, Seattle, Washington

ABSTRACT

Clay minerals from the 1980 Mt. St.Helens eruption and their early stage alteration were studied.

The debris avalanche deposit is derived from the old, altered rocks of the upper cone and contains saponite and zeolites formed by low temperature alteration, under mildly alkaline to neutral conditions, within the cone prior to the 1980 eruptions.

The air-fall tephra of May 18 and subsequent eruptions, which extended for hundreds of km east of the volcano, contains a magmatic component of fresh glass, plagioclase, hypersthene, hornblende and magnetite, and a lithic component of these same minerals and saponite, chlorite/smectite, chlorite, mica, tridymite, cristobalite and quartz. The lithics were stripped from the cone during violent eruption of tephra.

Warm surface waters in contact with the new deposits for three weeks showed higher than normal concentrations of most ions (especially Na^+, Ca^{2+}, SO_4^{2-} and Cl^-) but there was little evidence for extensive alteration.

Five months after emplacement, pumice from the pyroclastic flows, which were still quite hot 1 m below the surface (100-260°C) was being dissolved. New minerals formed near the surface include alunite, gypsum, halotrichite, hematite and several others. The pyroclastic flows are being altered by small-scale acid-sulfate systems where ascending S and H_2S are oxidized at the surface to form sulfuric acid.

INTRODUCTION-GEOLOGIC SETTING

The 1980 eruptions of Mt.St.Helens, the youngest volcano in the

Cascade Range of Washington State, produced a variety of deposits including a huge debris avalanche deposit consisting of partially altered rocks from the old cone, pyroclastic flows of new glassy pumice, and widespread air-fall tephra deposits.

Mt.St.Helens is a stratovolcano composed of andesite flows, dacite domes and pyroclastics (Verhoogen, 1937). Debris and pyroclastic flows extend many kilometers from the cone and tephras blanket extensive areas mainly east and northeast of the volcano (Mullineaux and Crandell 1962; Mullineaux et al., 1975). The oldest deposits were erupted about 38,000 years ago (Mullineaux et al., 1975) and the last period of activity was in the 1850's (Hoblitt et al., 1980). Since that time the volcano has been dormant with weak fumarolic activity along the margins of the summit dome (Friedman and Frank, 1977). The 1980 activity began on March 20 with seismic activity; mild ultravolcanian steam eruptions commenced on March 27. These phreatic eruptions produced tephra containing lithic components from the old, partially altered rocks within the cone. During April and early May these eruptions continued and the north flank of the volcano bulged outward at a rate of 1.5 m per day (Lipman et al., 1980), evidently due to intrusion of new magma. Large areas of red and yellow colored altered rocks were visible on the bulging north flank; these were produced by hydrothermal alteration during a previous eruptive episode.

The magmatic stage began on May 18 when a strong earthquake caused catastrophic avalanching of the bulging north flank to form a 2.5 km^3 debris avalanche deposit (Glicken et al., 1980) largely of old,altered rock. This uncapped the geothermal system within the cone resulting in a powerful, laterally-directed blast and eruption of new magmatic materials consisting of pumiceous pyroclastic flows (Melson et al., 1980) and a Plinian eruptive column which spread tephra for hundreds of kilometers to the east (Sarna-Wojcickiand Waitt, 1980). Several additional tephra eruptions and dome-building episodes occurred during the later part of 1980.

SAMPLES AND METHODS

Samples were collected from the debris avalanche deposit, the pyroclastic flows and the air-fall tephra. Tephras were obtained from the pre-May 18 phreatic stage, and from several of the magmatic eruptions; most of these were sampled within a day of falling. The magmatic-stage tephras were collected over an area extending several hundred kilometers east of the volcano. Pyroclastic flows were sampled on June 8 and in October. More detailed data on sample locations are given in Dethier et al. (1981).

Samples were disaggregated in distilled water by ultrasonic treatment, and the <2 μm (clay) fraction was separated using methods given in Jackson (1974). Oriented mounts for X-ray diffraction were prepared by suction of the clay suspension through porous ceramic plates. Bulk samples were ground to < 50 μm and packed in aluminum boxes. Samples were analysed with a G.E. XRD-5 diffractometer using Ni-filtered Cu radiation in the original condition, after ethylene glycol solvation, and after heating to 300 and to 550 °C.

RESULTS AND DISCUSSION

Composition of erupted materials.

The debris avalanche deposit is a mixture of old, partially altered rocks of the former north flank of the cone. The bulk mineralogy is dominated by plagioclase, tridymite, cristobalite, quartz, hornblende, and hypersthene; the clay fraction contains these same minerals but also large amounts of trioctahedral smectite. XRD analysis shows an 060 peak at 1.54 Å, and absence of a peak near 5 Å upon heating, and energy dispersive X-ray spectroscopy shows large amounts of Mg and Fe and only a minor amount of Al, indicating that this smectite is a Fe-saponite. Minor amounts of chlorite and mixed-layer chlorite/smectite are present as evidenced by peaks at 14 Å and between 14 and 10 Å for samples heated at 550 °C.

The clay minerals in the debris avalanche must have formed by hydrothermal alteration within the old cone prior to the 1980 activity. Similar minerals have been reported from geothermal wells in active volcanic areas such as Hawaii (Fan, 1979) and Iceland (Kristmannsdottir 1979) where measured temperatures suggest that smectite forms below about 200°C and mixed-layer chlorite/smectite and chlorite between 200 and 350°C. Large clasts of altered rock from the avalanche deposit show vugs filled with euhedral crystals of smectite and zeolites which have precipitated from interstitial fluids. Cristobalite and quartz are common in these rocks and probably formed by devitrification in lavas and dome rocks at high temperature (Keith and Muffler, 1978), but some may be hydrothermal.

Saponite and zeolites suggest alteration by neutral to mildly alkaline waters at relatively low temperature within the cone. Products of acid-sulfate alteration (Schoen et al., 1974) such as alunite, kaolinite and opal were found in minor amounts in only two of the old thermal areas and were totally absent from the debris avalanche and other new deposits just after their emplacement. Other Cascade volcanos, such as Mt.Baker (Frank, in press) exhibit extensive alteration to kaolinite, alunite, pyrophyllite, opal and other minerals by descending sulfuric acid produced by oxidation of sulfur species at the surface. Absence of these minerals at St.Helens indicated that prior to 1980 the hydrothermal system was well-plugged and did not allow much fluid discharge at the surface.

The pyroclastic flows consist of gravel, cobble and boulder-size rounded clasts of dacite pumice with considerable amounts of finer material. The pumice is a fresh vesicular glass with phenocrysts of plagioclase, hypersthene, hornblende and magnetite. When the coating of adhering tephra is broken off, the pumice was found to contain no clay minerals or silica polymorphs.

The phreatic-stage (pre-May 18) tephra contains no fresh glass and has a mineralogical composition identical to that of the debris avalanche. This indicates that this tephra was derived from old rock altered by a low temperature geothermal system within the upper cone.

The magmatic-stage (May 18 and after) tephras, like the pyroclastic flows, consist mainly of fresh glass with considerable amounts of plagioclase, hornblende, hypersthene and magnetite. But unlike the pyroclastic flows, they also contain cristobalite, tridymite, quartz and in the <2 μm fraction saponite, chlorite, chlorite/smectite, and mica. The silica polymorphs and clay minerals are lithic components stripped from the vent area during violent ejection of the tephra. Their presence suggests that a significant component of the tephra is lithic, including some of the plagioclase, hypersthene and hornblende. The ubiquitous saponite shows that temperatures in the vent area during eruption must have been below that for the upper stability limit of this mineral at atmospheric pressure. The presence of smectite and other clays as initial components of tephra complicates the interpretation of these minerals as products of pedogenesis in tephritic soils, a problem discussed by Borchardt (1977).

Early-stage alteration

Three weeks after the May 18 eruption the new deposits and warm ponds and surface waters on these deposits were sampled to determine whether any significant alteration was occurring. Compared with normal surface waters, the waters were enriched in most ions, especially Na^+, Ca^{2+}, SO_4^{2-} and Cl (Dethier et al., 1981), and incrusting precipitates of gypsum, thenardite, glauberite and other minerals were found on many of the deposits. The waters, with pH of 6-7.5 were calculated to be in equilibrium with kaolinite and in some areas with smectite, but no kaolinite or halloysite was detected and smectite, being widespread as an initial component, could not be detected if neoformed. Absence of kaolin minerals is probably due to the abundance of alkalis (Eberl and Hower, 1975) and to the short time involved. The surfaces of glass and mineral grains in contact with the warm waters were examined by scanning electron microscopy and appeared fresh and unetched. Adsorbed ions are probably the source of the dissolved constituents in the waters, but some dissolution of glass and minerals may be occurring.

Five months after emplacement, the pyroclastic flows were sampled by digging 1 m deep pits. Temperatures at this depth were in some places 100-260°C. The upper 25-50 cm of the deposits were stained pink to red by iron oxides and were commonly cemented by precipitated minerals and amorphous materials. Below this depth the deposits were hot, dry, unoxidized and grey in color. The surfaces of feldspar grains (examined by SEM) from the 260°C zone in the bottom of a hole were fresh and unpitted, indicating little alteration was taking place, but glassy pumice appeared rounded, dehydration-cracked and somewhat altered. Crystals of unidentified minerals resembling zeolites were found in some of the vesicles. The partially cemented red-colored zone in the upper parts of the holes contained alunite (as crusts and cementing material), gypsum and anhydrite. In some samples SEM images show the alunite to consist of 1 μm diameter plates with hexagonal outline. Hematite was identified in some of the deposits, but much of the iron oxide is probably amorphous. Treatment with boiling Na_2CO_3 solution (Jackson, 1974) greatly improved the quality of the XRD patterns from these samples, suggesting the presence of amorphous silica-alumina cement. Energy dispersive X-ray spectroscopy also showed grain coatings consisting primarily of Si and Al. No kaolin minerals were identified in these samples, and smectite was still present in samples containing alunite, but generally in much smaller amounts than in unaltered deposits.

Rootless fumaroles on the surface of the pyroclastic flows were surrounded by zoned incrustations of brightly colored minerals which include sulfur, gypsum, anhydrite, glauberite, thenardite, melanterite, halotrichite, halite, iron hydroxides and hematite.

The pyroclastic flows are being altered by small scale acid-sulfate systems in which ascending superheated steam contains unoxidized sulfur and H_2S as evidenced by lack of sulfate minerals at depth in the sample pits and presence of native sulfur at some rootless fumaroles. The sulfur must have been adsorbed originally onto the grain surfaces of the fresh deposits. Sulfur species are oxidized at the surface to form H_2SO_4 which reacts with the original minerals and especially the glass to form sulfates and amorphous cementing materials in the upper 30 cm of the deposits. As the deposits cool, reactions involving meteoric water will dominate these systems, and many of the early-formed minerals will dissolve.

REFERENCES

Borchardt, G.A.,1977. Montmorillonite and other smectite minerals. In: Minerals in soil environments. Soil Science Society of America, p.293-325

Dethier,D.P., Frank, D. and Pevear, D.R., 1971. Chemistry of thermal waters and mineralogy of the new deposits at Mount St.Helens - A preliminary Report. U.S.Geol.Survey Open-file report 81-80, 24 p.

Eberl, D. and Hower, J.,1975. Kaolinite synthesis - The role of the Si/Al and alkali/H^+ ratios in hydrothermal systems. Clays,Clay Min. 23: 301-309

Fan, P., 1979. Clays and clay minerals of hydrothermal origin in Hawaii. Proc.Int.Clay Conf. Oxford, 369-374

Frank, D., 1981. Origin, distribution and rapid removal of hydrothermally formed clay at Mount Baker, Washington. U.S.Geol.Survey Professional paper. (in press)

Friedman, J.D. and Frank, D., 1977. Heat discharge from Mount St. Helens, Washington. U.S.Geol.Survey Open-file report 77-541, 29 p.

Glicken, H., Janda, R. and Voight, B., 1980. Catastrophic landslide/debris avalanche of May 18, 1980 Mount St.Helens volcano.EOS,61:1135

Hoblitt, R.P., Crandell, D.W. and Mullineaux, D.R., 1980. Mount St. Helens eruptive behavior during the past 1,500 years. Geology,8:555

Jackson, M.L., 1974. Soil chemical analysis-Advanced course. Pub. by the author, Dept. of Soil Sciences, Univ. of Wisconsin, 895p

Keith, T.E.C., Casadevall, T.J. and Johnston, D.A., 1981. Fumaroles at Mount St.Helens: Occurrence and mineralogy and chemistry of incrustations. In: Lipman, P.W. and Mullineaux, D.R., Ed., The 1980 eruptions of Mt.St.Helens, Washington. U.S.Geol.Survey Professional paper 1250 (in press)

Keith, T.E.C. and Muffler, L.J.P., 1978. Minerals produced during cooling and hydrothermal alteration of ash flow tuff from Yellowstone drill hole Y-5. J.Volcanology and Geothermal Res.,3:373-402

Kristmannsdottir, H., 1979. Alteration of basaltic rocks by hydrothermal activity at 100-300°C. Proc.Int.Clay Conf.,Oxford, 369-374

Lipman, P.W., Moore, J.G. and Swanson, D.A., 1980. Bulging of the north flank of Mount St.Helens volcano before the May 18 eruption. Geodetic data: EOS 61, 1135

Melson, W.C., Hopson, C.A. and Kienle, O.F., 1980. Petrology of tephra from the 1980 eruption of Mount St.Helens. Geological Society of America, Abstracts with Programs, 12: no.7, 482

Mullineaux, D.R. and Crandell, D.R., 1962. Recent lahars from Mount St.Helens, Washington Geol.Soc.Am. Bulletin, 73: 855-870

Mullineaux, D.R., Hyde, J.H. and Rubin, M.,1975. Widespread late glacial and postglacial tephra deposits from Mount St.Helens volcano, Washington. U.S.G.S. Journal of Research, 3: 329-335

Sarna-Wojcicki, A.M. and Waitt, R.B.Jr., 1980. Areal distribution, thickness and composition,of volcanic ash erupted from Mount St. Helens on May 18, 1980. Geol.Soc.Am.,Abs.with Programs, 12: 515

Schoen, R., White, D.E. and Hemley, J.J., 1974. Argillization by descending acid at Steamboat Springs, Nevada. Clays,Clay Min.22:1-22

Verhoogen, J., 1937. Mount St.Helens, a recent Cascade volcano. Calif. University, Dept Geol.Sci.Bulletin, 24: 263-302

ALTERATION REACTIONS OF HALLOYSITE UNDER HYDROTHERMAL CONDITIONS WITH ACIDIC SOLUTIONS

Hideo MINATO,[1] Hirotatsu KUSAKABE,[2] and Atsuyuki INOUE[3]

1 Department of Earth Sciences, Hyogo University of Teacher Education, Yashiro-cho, Kato-gun, Hyogo (Japan)
2 Institute of Earth Science and Astronomy, College of General Education, University of Tokyo, Komaba, Meguro-ku, Tokyo (Japan)
3 Geological Institute, College of Arts and Sciences, Chiba University, Nishichiba, Chiba (Japan)

ABSTRACT

Halloysite, metahalloysite and kaolinite often occur in clay deposits of hydrothermal origin since field evidence of association of kaolinite with halloysite suggests some phase relation between the two minerals, the hydrothermal formation of kaolinite from halloysite was investigated using Morey-type reaction vessels. Halloysite (hydrated type) from Tawara granite quarry, Gifu Pref., Japan was treated with various acid solution at different temperatures and pressures. The products were tested by means of X-ray diffraction and electron microscopy. Kaolinite of relatively high crystallinity associated with alunite was formed at 400°C in a 0.1 M H_2SO_4 solution in 5 days. Halloysite was altered to an aggregate of pyrophyllite, kaolinite and quartz when treated at 400°C with a 0.1 M HCl solution in 3 days. Transformation of halloysite to metahalloysite was observed at low temperatures and low concentrations of acids. The influence of Ca and Na ions on these reactions was tested by addition of Na_2SO_4 or $CaSO_4$ to H_2SO_4 solutions. The Ca ion favors the preservation of halloysite (metahalloysite) and the presence of Na ion favors the formation of kaolinite, but at high concentrations of Na ion alunite is formed instead of kaolinite.

INTRODUCTION

Kaolinites in clay deposits of hydrothermal origin in Japan (Minato, 1976, 1977a) and of halloysite in such deposits in Japan (Minato et al., 1969; Minato, 1975), in Mexico (Keller et al., 1971), and in Korea (Minato, 1977a; Sang, 1981) have been reported. Syntheses of kaolinite have been carried out by many investigators (e.g., Siffert et al., 1972; Iglesa et al., 1972).

Although association of kaolinite and halloysite in mineral deposits is observed, the phase relations between the two minerals seldom have been studied. Therefore, we carried out laboratory experiments on hydrothermal formation of kaolinite from halloysite.

FIELD EVIDENCE OF ASSOCIATION OF KAOLINITE AND HALLOYSITE

In the sulfur ore deposit, Zaō Mine, Yamagata Prefecture, northern Japan, kaolinite occurs in the central part of the deposit, while halloysite is found in the marginal zone of the deposit (Mukaiyama, 1954). In the San José de Ranchos Mine, Mexico, hydrothermal kaolin deposits were formed by alteration of rhyolite which intruded the Cretaceous limestone (Keller et al., 1969; Minato, 1977b). The zonal distribution of minerals from the limestone contact zone (A) to the central part of the kaolin ore deposit (E) are as follows: (A) grossularite and montmorillonite as alteration mineral of grossularite, (B-C) metahalloysite, (C-D) kaolinite, and (D-E) kaolinite and alunite.

These occurrences of kaolinite and halloysite, suggest that the following physico-chemical conditions are important:

(1) The acidity of the aqueous solution, which may affect alteration of the minerals, was higher in the central part of the ore bodies than in the marginal zones.

(2) The thermal gradient decreased with the distance from the ore body to the marginal zone.

(3) The variations in the concentration of Ca^{2+} might also affect the formation of the two different clay minerals.

EXPERIMENTAL

Starting material: Halloysite from a granite quarry at Tawara, Naegi district, Gifu Prefecture, central Japan, was used as starting material. The material occurs in the weathered part of the surface of granite as veins of 1-2 cm width with white porcelain-like appearance (Plate I, Photograph 1). The material, associated with small amount of quartz, was separated by hand picking and examined by the following methods: X-ray powder diffraction, electron microscopy, electron diffraction (Plate II, Photographs 5 and 6), and thermal analysis (Fig. 2). Chemical analysis indicates that the material is a fully hydrated halloysite (Table 1).

Reacting solutions: The following acid solutions were prepared:

(1) 0.1 M HCl solution.

(2) 0.02 to 0.1 M H_2SO_4 solusions

(3) Mixtures of the H_2SO_4 solutions with 0.01 - 0.1 M Na_2SO_4 solutions.

(4) Mixtures of the H_2SO_4 solutions with 0.01 M to 0.1 M $CaSO_4$ solutions.

Apparaturs and procedure: Details of the experimental procedure are given in a paper by Kusakabe et al. (1981). Experiments in a closed system were carried out in a Morey-type pressure vessel enclosing a gold cup. Pulverized material, about 400 mg, was placed in the gold cup and 10 ml of a reacting solution was added. Experimental conditions of pH, temperature, duration, and solution compositions are shown in Table 2. The pressures of these runs were in the range of 50 to 650 bar (x 10^5 Pa). After the runs, the vessel was quenched. Immediately after the vessel was opened, the pH was measured at 20°C.

After products were washed and dried at room temperature, they were examined by X-ray diffractometer with Cu K_α radiation. Products were

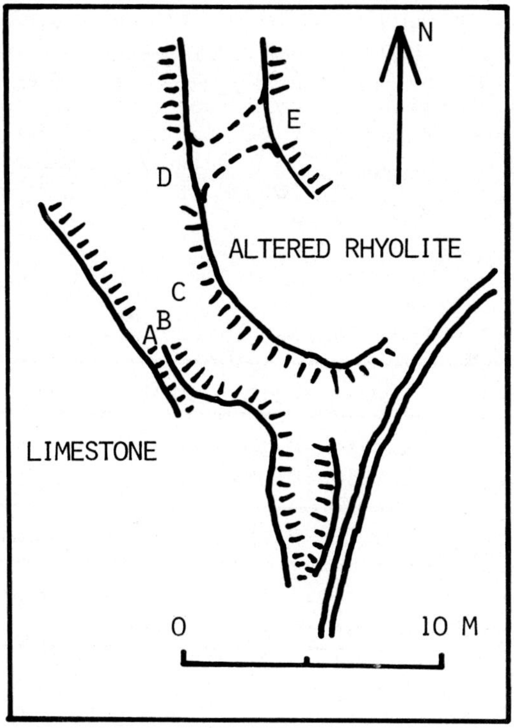

Fig. 1. Sample map of the San José de Rachos Mine.

A; Limestone contact, montmorillonite zone: B-C; Halloysite zone: C-D; Kaolinite zone: D-E; Kaolinite and alunite zone.

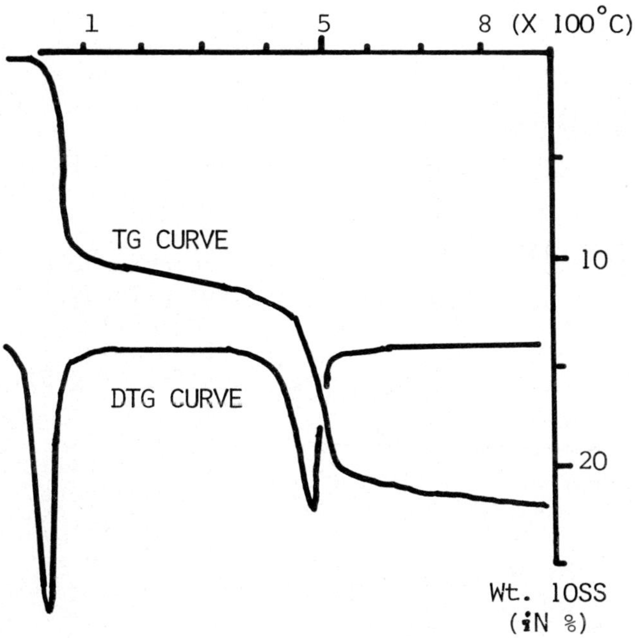

Fig. 2. Thermal behaviors of halloysite from Tawara.

also examined both by electron microscopic observation and by electron diffraction. Equilibration was not achieved in Run No. 4, as judged from the study of "Roseki" deposit in Japan and experimental results by Hemmi et al. (1975). According to X-ray powder diffraction, kaolinite was formed in Run No. 4 (with HCl solution) and Run Nos. 11, 20, 21, and 25 (with H_2SO_4 in the presence of Na^+ in the system). Metahalloysite was formed in Run Nos. 1, 6 and 7 with HCl solution and in Run Nos. 10, 12, 13, 14, 15, 18, 19, 20, 26 and 28 with H_2SO_4 solution in the presence of Ca^{2+} in the system.

TABLE 1
Chemical composition of halloysite from Tawara, Gifu Prefecture, Japan.

	SiO_2	TiO_2	Al_2O_3	Fe_2O_3	MnO	MgO
Wt. %	41.27	0.10	34.97	0.20	0.02	0.13
Mol. Prop.	0.6869	0.0013	0.3430	0.0012	-	0.0032
	CaO	**Na_2O**	**K_2O**	**H_2O^+**	**H_2O^-**	**Total**
Wt. %	0.73	0.21	0.39	11.15	10.04	99.29
Mol. Prop.	0.0130	0.0034	0.0041	0.6177	0.5562	

(Analyst: Minato, 1981)

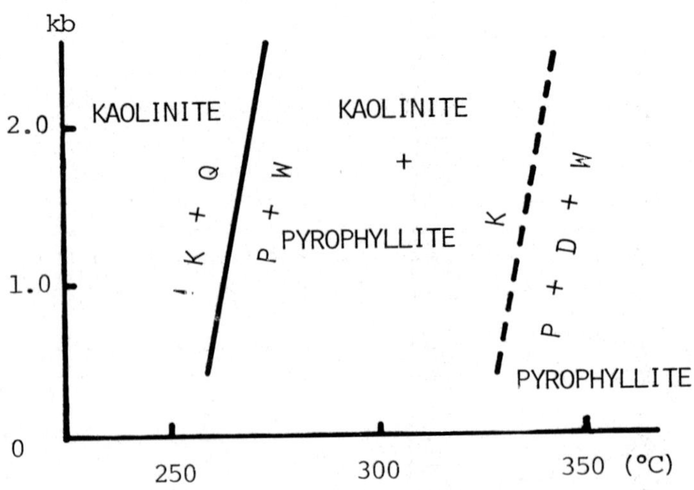

Fig. 3. Stability relationship of pyrophyllite and kaolinite by Hemmi et al. (1975).

Q: Quartz,
P: Pyrophyllite,
K: Kaolinite,
D: Diaspore,
W: Water.

TABLE 2
Experimental data.

Run No.	HCl (M)			pH	Temp (°C)	Days	Solid phase obtained Kao	M-H	Pyr	Mon	Alu	Gyp	Pl	Qu
1	0.1			1.07	300	3	?	S	–	–	–	–	–	–
4	"			5.06	400	3	S	?	S	–	?	–	–	M
6	"			4.80	300	3	–	S	–	–	–	–	–	–
7	"			2.25	300	9	?	S	–	–	–	–	–	M
8	"			3.9	300	13	?	S	–	–	–	–	–	W
	H_2SO_4 (M)	$CaSO_4$ (M)	Na_2SO_4 (M)											
10	0.1			4.3	250	5	?	S	–	–	S	–	–	–
11	"			1.4	300	5	S	W	–	–	S	–	–	–
12	0.02			2.2	250	10	?	S	–	–	–	–	–	–
13	"			3.8	300	10	?	S	–	–	–	–	–	–
14	"	0.1		1.86	240	24	?	W	–	M-W	S	VW	–	–
15	"	"		2.08	300	24	?	M	–	–	–	M-W	–	–
16	"		0.1	1.66	255	24	–	–	–	–	VS	–	–	–
17	"		"	1.73	275	24	–	–	–	–	VS	–	–	–
18	0.01	0.01		(4.0)	240	7	?	S	–	–	W	–	–	VVW
19	"	"		(4.5)	300	7	?	S	–	–	–	–	–	VVW
20	"		0.01	2.31	250	7	M	S	–	–	–	–	–	VVW
21	"		"	2.03	290	7	M	S	–	–	–	–	–	VVW
22	0.1	0.1		3.72	250	8	–	M-W	–	–	M-W	–	M	M
24	"	"		3.90	300	8	–	–	–	–	M	–	M	W
23	"		0.1	3.86	250	8	–	–	–	–	M-W	–	M	W
25	"		"	2.18	300	8	M	M	–	–	M	–	–	W
27	0.05		0.01	2.87	250	12	?	?	–	–	–	–	–	–
26	"	0.01		3.20	200	12	–	M	–	–	M	–	–	W
28	"	"		1.75	300	12	?	S	–	?	W	–	–	W
29	"		0.01	2.96	230	12	?	?	–	–	S	–	–	W
30	0.1			1.34	250	7	?	S	–	–	M	–	–	M

pH: pH of final solution. Abbreviations used for the designation of the phase in this table are defined as follows: Kao; kaolinite, M-H; metahalloysite, Pyr; pyrophyllite, Mon; montmorillonite; Alu; alunite, Gyp; gypsum, Pl; plagioclase; Qu; Quartz. X-ray powder diffraction pattern in intensity: VS > S > M > M-W > W > VW > VVW > ?.

SUMMARY AND CONCLUSION

The experimental data show the following effects of physico-chemical conditions on the formation of halloysite and kaolinite.

(1) The formation of kaolinite is favored at higher concentrations of H_2SO_4 below 300°C in the presence of Na^+ , but a high concentration of Na^+ in the system promotes the formation of alunite (natroalunite) instead of kaolinite.

(2) At higher temperatures both halloysite and kaolinite decompose to form pyrophyllite.

(3) In the present experiments, although metahalloysite presists in the system produced in the presence of Ca^{2+}, it is not clear how the formation of metahalloysite is affected by Ca^{2+}. Therefore further investigation will be necessary.

REFERENCES

Hemmi, K. and Matsuda, H., 1975. Stabilization relationships between kaolinite and pyrophillite in "Roseki" deposits in Japan. Contribution to Clay Mineralogy in honour of Professor Toshio Sudo, 153-161.

Iglesa, A. la F. and Martin Vivaldi, J.L., 1972. A contribution to the synthesis of kaolinite. Proc. Int. Clay Conf. Madrid, 1972, 1: 173-185.

Keller, W.P. and Hanson, R.F., 1969. Hydrothermal alteration of volcanic pipe in limestone in Mexico. Clays Clay Min., 17: 19-12.

Keller, W.D., Harison, R.F., Huang, W.H. and Cervantes, A., 1971. Sequented activation of rhyolitic volcanic rock to endellite and a precursor phase of it at spring in Michoacan, Mexico. Clays Clay Min., 19: 121-127.

Kusakabe, H., Minato, H., Utada, M. and Yamanaka, T., 1981. Phase relations of clinoptiolite, mordenite, analcime and albite with increasing pH, sodium ion concentration and temperature. Sci. Pap. Coll. Gen. Educ., Univ. Tokyo. 31: 39-59.

Minato, H. and Utada, M., 1969. Mode of occurrence and mineralogy of halloysite from Iki, Japan. Proc. Clay Conf., Tokyo, 1969, 1: 393-402.

Minato, H., 1975. Mineralogy and mode of occurrence of spherical halloysite from Japan. Contribution to Clay Mineralogy in honour of Professor Toshio Sudo, 73-81.

Minato, H., 1976. Mineralogy and genesis of hydrothermal kaolin in Japan. Proc. of the 7th Int. Symp. on Genesis of Kaolin, 17-31.

Minato, H., 1977a. Kaolin deposits of Japan and surrounding East Asia. Proc. of the 8th Int. Kaolin Symp. and Meeting of Alunite, K-14: 1-13.

Minato, H., 1977b. On the problem of genesis of kaolinite and halloysite by hydrothermal water. Jour. Miner. Soc. Jap. Special Issue No. 13: 197-206.

Minato, H. and Aoki, M., 1978. Rate of transformation of halloysite under hydrothermal condition. Proc. Int. Clay Conf., Oxford, 619-627.

Mukaiyama, H., 1954. On geology and ore deposits of Zaō sulfur mine, Yamagata Pref., Japan. Jour. Mining Geologist Jap., 3: 195-204.

Sang, K.N., 1981. Mode of occurrence and mineralogy of halloysite from Sang Chong area, Southern Korea. Doctoral Paper for the University of Tokyo.

Siffert, B. and Wey, R., 1972. Contribution à la connaissance de la synthése des kaolins. Proc. Int. Clay Conf., Madrid, 1972, 1: 159-172.

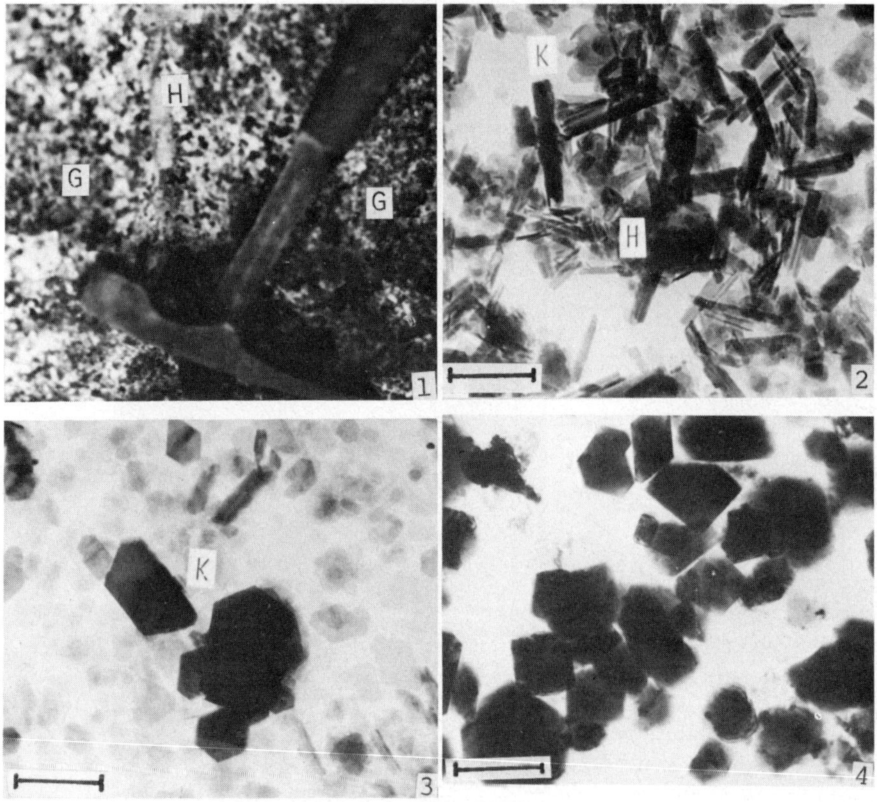

Plate I.

1. Halloysite (H) occurring as veinlets in weathered granite (G) at a granite quarry, Tawara, Naegi district, Gifu Pref., central Japan.

2. Electron micrograph of the product of Run No. 7. Fine hexagonal crystals observed in tubular halloysite. Grain (H) will be an opened tube of halloysite and electron diffraction shows a powder pattern.

3. Electron micrograph of the product of Run No. 11. Mainly composed of well crystallized kaolinite with rare tubular forms of halloysite.

4. Electron micrograph of the product of Run No. 17. Rhombohedral alunite crystals are the principal components.

(Each electron micrograph was scaled 1 $m\mu$ by length of bar.)

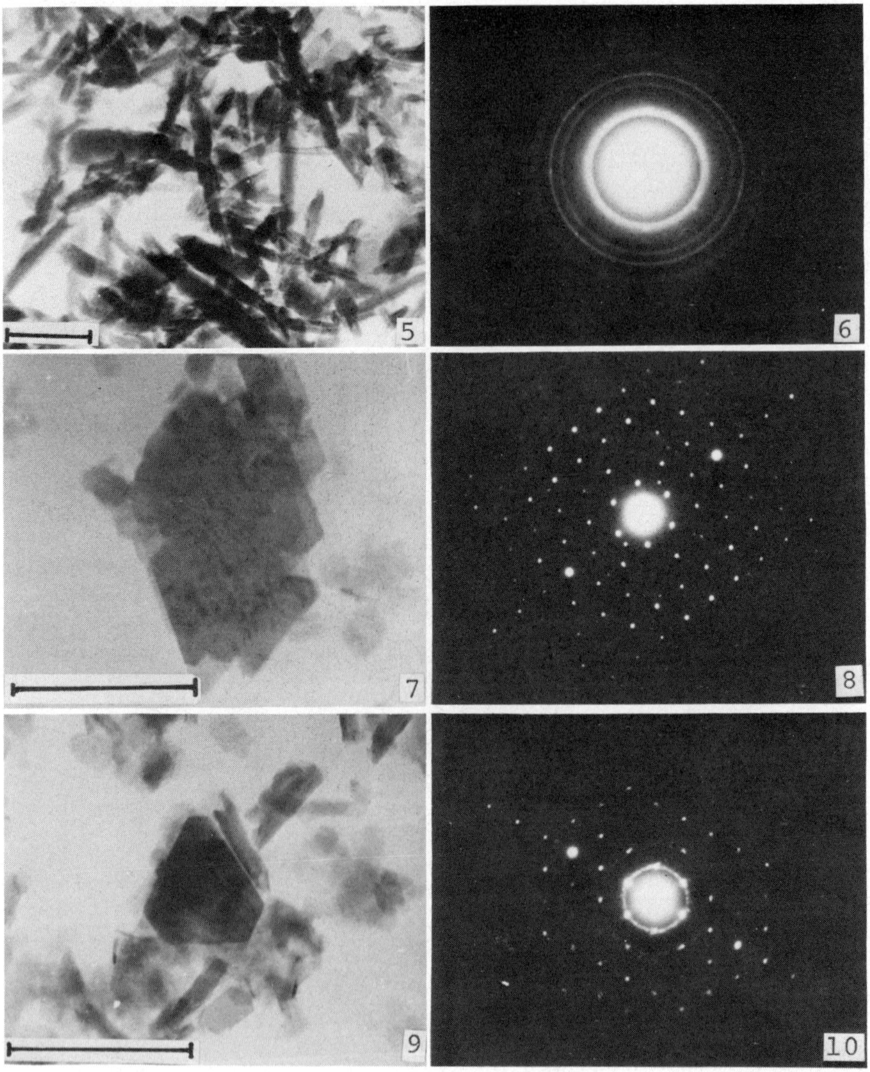

Plate II.

5. Electron micrograph of the starting material. Halloysite from Tawara.

6. Electron diffraction pattern of the halloysite.

7 and 9. Electron micrographs of kaolinite (Run No. 11).

8 and 10. Electron diffraction patterns of kaolinite shown in photographs of 7 and 9, respectively.

(Each electron micrograph was scaled 1 mμ by length of bar.)

ANALYTICAL ELECTRON MICROSCOPIC STUDIES OF HALLOYSITE FORMATION PROCESSES--MORPHOLOGY AND COMPOSITION OF HALLOYSITE--

KAZUE TAZAKI Institute of Sedimentary and Petroleum Geology, Geological Survey of Canada, 3303-33rd Street, N.W., Calgary, Alberta Canada. On leave from Institute for Thermal Spring Research, Okayama University, Misasa, Tottori-ken, 682-02 Japan.

ABSTRACT

The genetic relationships between morphology and chemical composition on a microdimensional scale of monomineralic halloysite were studied using a transmission electron microscope (TEM) equipped with a non-dispersive X-ray spectrometer.

Fourteen different halloysite samples were obtained by separation from weathered granite, weathered tuff breccia and volcanic ash soil. All have a basal 10 Å X-ray reflection and TEM observation shows mixtures of particles with tubular, spheroidal and film-like morphology.

The iron content of these particles varies. Crinkled films of samples containing more than 10.0% ΣFeO transform into spheroidal particles through reduction of iron. Tubular halloysite is mostly iron-free. This implies that the formation of the tubular form requires either the removal of iron from the crinkly halloyite film or the presence of material low in iron. The compositional difference between the long and short tube halloysites is very small compared with that between the spheroidal and tubular forms.

Successive transformations of halloysite morphology become apparent under electron microscopic examination. Sheet aggregates of allophane-like particles merge and become a crinkly film. The film then assumes a spheroidal shape and finally spheroidal particles form.

Electron microscopic microanalysis of natural halloysite shows that halloysite polymorphs are commonly and intimately intermixed on a microdimensional scale.

INTRODUCTION

By means of electron microscopic microanalysis, the character of a clay mineral specimen can be specifically defined in terms of crystal morphology, chemical composition, and other mineralogical properties (Hayashi et al., 1970; Tazaki, 1980). In the present work we applied electron microscopic microanalysis to natural halloysites in particular to study the relation between morphology and chemical composition of halloysite films, tubes and particles.

MATERIALS

The materials from which the fourteen halloysite samples studied were separated are given in the caption of Table 1. Samples No. 1, 10 and 11 were separated from weathered granite, samples No. 2, 12 and 13 from weathered tuff breccia. Other samples came from pumice grains and ashes separated from volcanic ash soils. The separated halloysites are easily identified by means of X-ray powder diffraction and electron microscopic observation. A 10 Å basal reflection is characteristic of halloysite and is easily distinguishable by expansion of the 10 Å - 11 Å spacing following ethylene glycol treatment. Also present are small amounts of quartz, cristobalite, plagioclase and gibbsite as demonstrated by X-ray powder diffraction peaks (see Table 1).

ANALYTICAL ELECTRON MICROSCOPY

The instrument used for this experiment is an analytical transmission electron microscope, the JEM-100C fitted with ORTEC's energy dispersive spectrometer and a scanning device. The maximum accelerating voltage of the JEM-100C is 100kV, and the energy resolution of the X-ray detector of ORTEC's energy dispersive spectrometer used for this experiment is 160 eV at the Mn Kα line. A JEM-100C electron microscope with decontamination devices was used under the following conditions: 40kV accelerating voltage

TABLE 1

Coexisting micro-morphologies of halloysite

	Samle No.	1	2	3	4	5	6	7	8	9	10	11	12	13	14
Micro-morphology	primitive f.													◎	
	allophane p.	○		○		○	●		●	●					◎
	crinkly film	●				◎	●	●	●	●	◎	◎	◎		●
	spheroidal p.	●	●	◎	●	○	●	●	●	●	●	◎	◎	○	
	short tube	●	◎	◎		●							◎	●	
	long tube			●		●					◎	◎	●	●	
associated non-clay min.		Qz	Cr	Qz, Pl, Gb				Cr, Pl		Cr, Pl	Pl				

○ rare; ◎ common; ● abundant; Qz = quartz; Cr = cristobalite; Pl = plagioclase; Gb = gibbsite.

1) Gyarome clay from Seto; 2) halloysite concretion in volcanic ash from Miyagi Pref.; 3) Volcanic ash of Daisen volcano from Okachi (01-31); 4) weathered obsidian from Kumi (Kumi 4); 5) altered agglomerate from Oita Pref.; 6) volcanic ash of Daisen volcano from Okachi (02-4); 7) volcanic ash of Daisen Volcano from Kamiodachi; 8) volcanic ash of Daisen volcano from Okachi (02-12); 9) volcanic ash of Daisen volcano from Okachi (02-6); 10) weathered granite from Misasa, Tottori Pref.; 11) weathered granite from Otani, Tottori Pref.; 12) halloysite concretion in volcanic ash from Hashizu, Tottori Pref.; 13) acid white clay from Ni-igata Pref.; 14) volcanic ash of Tarumae volcano, Hokkaido.

and 100 seconds counting time. A drop of the clay suspension was dried on a collodion film reinforced with carbon and observed at an electron optical magnification of X50,000.

The homogeneous long-tube halloysite from Biela Hora, Czechoslovakia, was used as a working standard for electron microanalysis. The method of quantitative analysis was essentially similar to that used by Hayashi et al. (1978).

RESULTS

All samples show a basal 10 Å X-ray reflection. TEM observations show the presence of a mixture of particles with tubular, spheroidal and film-like morphology.

Table 2 summarizes the mean values of microanalysis for halloysite particles with various morphologies. Films of primitive form, Figure 6A, contain 37.2 to 38.0% Al_2O_3, and 3.5 to 8.0% ΣFeO. The crinkly films from several localities show various ΣFeO concentration ranges from trace amounts to 15.0% (Figs. 5 and 6). Spheroidal particles have relatively low ΣFeO contents ranging from trace amount to 8.3% (Figs. 5A and 6A). Tubular halloysites are characterized by very low ΣFeO contents except for the No. 3 samples from Okachi volcanic ash (Fig. 7).

In Figure 1, SiO_2/Al_2O_3 ratios for the No. 5 sample from Oita are plotted. Close relationships between chemical composition and morphology can be seen from this diagram. Long tubes have higher Al_2O_3 contents and lower SiO_2/Al_2O_3 ratios than the short ones. The latter have SiO_2/Al_2O_3 ratios from 1.2 to 1.3. Spheroidal particles, (Fig. 1), have SiO_2/Al_2O_3 ratios over 1.3. Figure 2 shows the plot in terms of SiO_2-Al_2O_3-ΣFeO ratios for the coexisting primitive form of halloysite and the crinkly film type of halloysite from the No. 7 sample of Kamiodachi. Films of both the primitive and crinkly forms show similar chemical compositions.

In Figure 3A, the SiO_2-Al_2O_3-ΣFeO ratios of coexisting crinkly film and spheroidal particles are confined to a small area. A further inference may be drawn from Figure 3-A and 3-B in that the spheroidal particles have a lower SiO_2 content than the crinkly films, whereas the ΣFeO contents of both polymorphs of halloysite are similar. The crinkly film from Okachi (02-6) is slightly more enriched in iron oxides than the coexisting spheroidal particles but the SiO_2/Al_2O_3 ratios of these forms are the same (Fig. 3-C).

By contrast, the crinkly film from Okachi (02-12) shows a slightly lower SiO_2 content in comparison with that of the coexisting spheroidal particles (Fig. 4). The SiO_2, Al_2O_3 and ΣFeO contents of the sheets of allophane-like particles are shown for comparison with those of the crinkly film type of halloysite in Figure 4.

DISCUSSION

Morphological transformations of halloysite, observed by the electron microscope, occur as an orderly succession of changes. Sheet aggregates of allophane transform into a crinkly film which eventually grows into

TABLE 2

Average of electron microanalysis data of halloysites (% of sum of SiO_2 + Al_2O_3 + ΣFeO)

Film of primitive form

Sample No.	7	13
SiO_2	54.9	58.5
Al_2O_3	37.2	38.0
ΣFeO	8.0	3.5

Sheet of allophane-like particles

Sample No.	8
SiO_2	59.7
Al_2O_3	28.7
ΣFeO	11.6

Crinkly film

Sample No.	1	6	7	3	9	12	14
SiO_2	56.0	60.8	55.0	59.0	55.1	54.2	51.6
Al_2O_3	39.7	32.6	36.9	29.5	32.4	45.8	33.4
ΣFeO	4.3	6.6	8.1	11.5	12.5	tr	15.0

Spheroidal particles

Sample No.	1	2	3	4	5	6	8	9	12
SiO_2	54.0	53.1	54.9	54.4	59.5	55.9	64.3	57.2	58.6
Al_2O_3	42.0	44.1	39.8	40.2	40.5	37.2	27.6	34.5	41.4
ΣFeO	4.0	2.9	5.3	5.4	tr	6.8	8.1	8.3	tr

Short tube

Sample No.	5
SiO_2	58.1
Al_2O_3	41.9
ΣFeO	tr

Long tube

Sample No.	3	5	10	11	12	13
SiO_2	51.6	53.9	63.6	69.0	57.9	58.1
Al_2O_3	39.7	46.1	36.4	31.0	42.1	38.4
ΣFeO	8.7	tr	tr	tr	tr	3.5

tr = trace

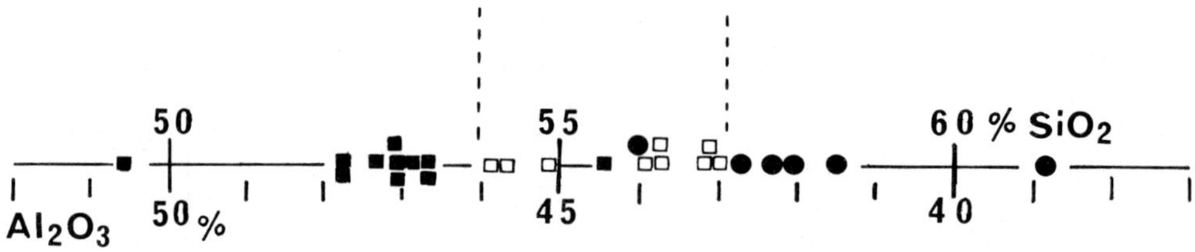

Fig. 1. Variation of SiO_2 and Al_2O_3 content of two types of tubular and one type of spheroidal halloysite derived from altered agglomerate rock from Oita W-75 (sample No. 5). Solid square = long tube. Open square = short tube. Solid circle = spheroidal particle.

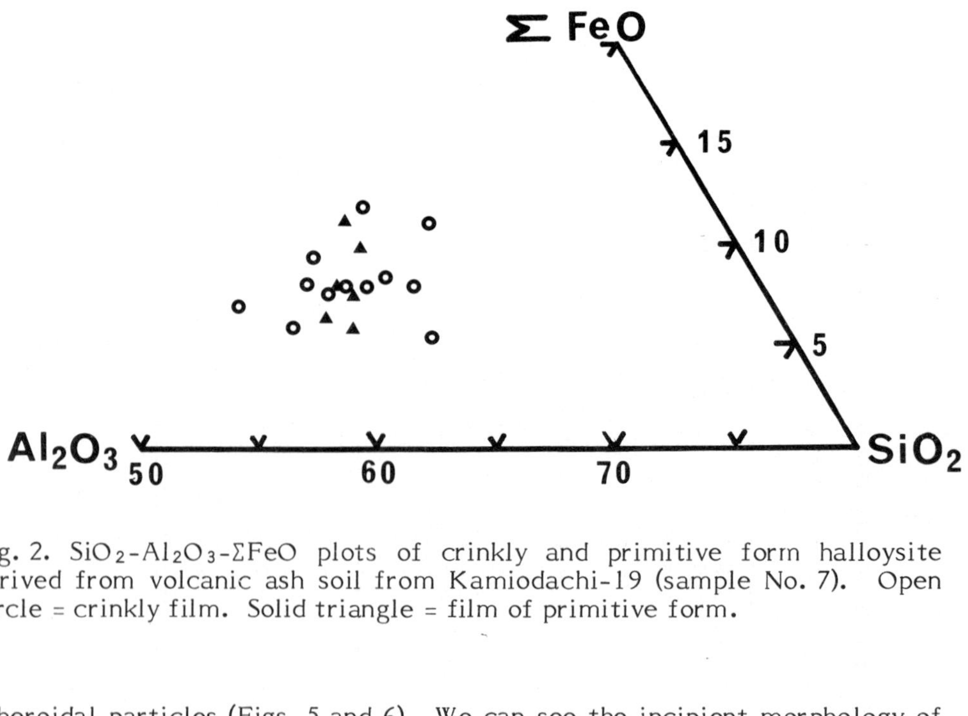

Fig. 2. SiO_2-Al_2O_3-ΣFeO plots of crinkly and primitive form halloysite derived from volcanic ash soil from Kamiodachi-19 (sample No. 7). Open circle = crinkly film. Solid triangle = film of primitive form.

spheroidal particles (Figs. 5 and 6). We can see the incipient morphology of crinkly film at the lower left in Figure 5A and the incipient stage of spheroidal particle development in Figure 6B.

The SiO_2-Al_2O_3-ΣFeO plots of coexisting crinkly film and spheroidal particles show a slight difference in chemical composition, but the role of chemical composition in morphological development is still ambiguous.

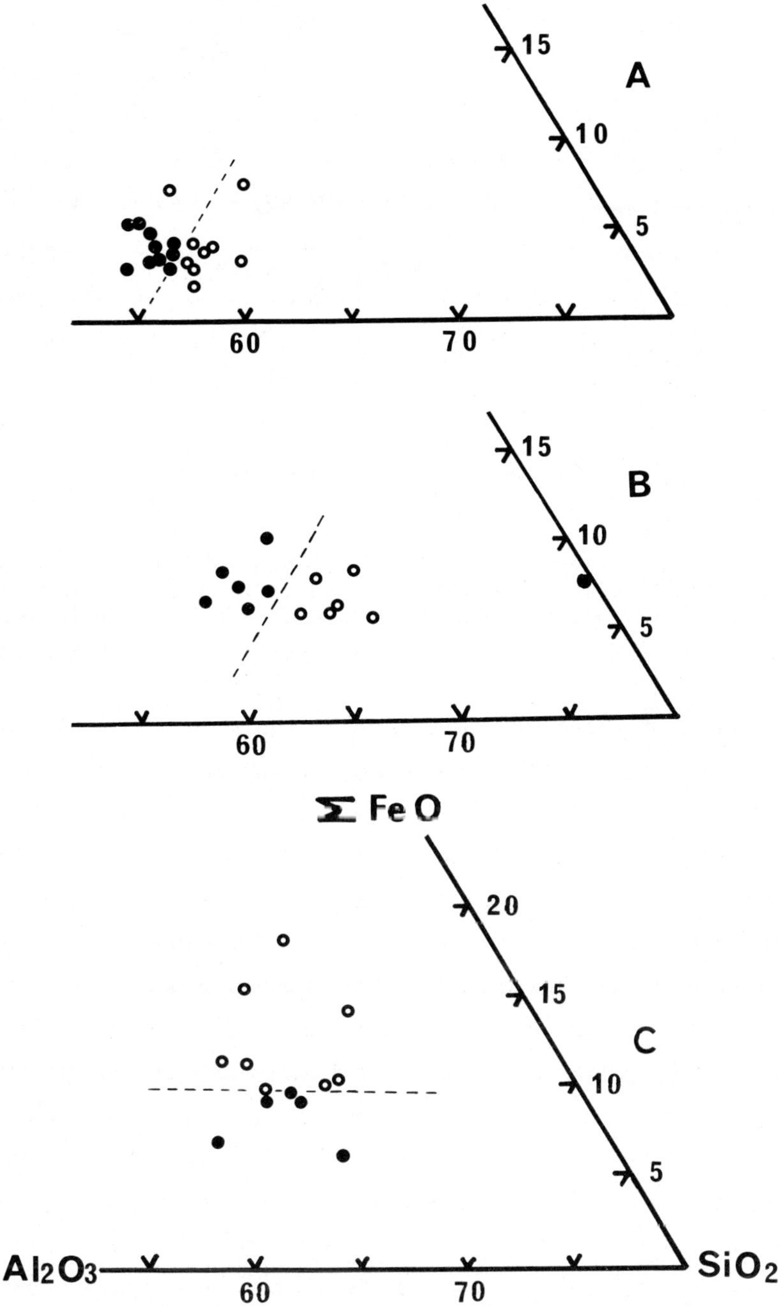

Fig. 3. SiO_2-Al_2O_3-ΣFeO plots of crinkly and spheroidal form halloysite derived from: A, clay deposits of Seto (sample No. 1); B, volcanic ash soil from Okachi O2-4 (sample No. 6); C, volcanic ash soil from Okachi O2-6 (sample No. 9). Open circle = crinkly film. Solid circle = spheroidal particle.

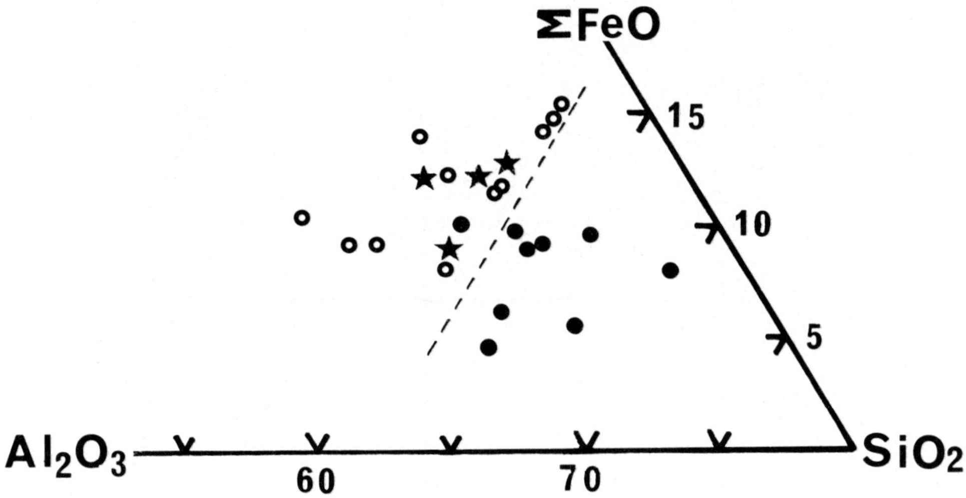

Fig. 4. SiO_2-Al_2O_3-ΣFeO plots of crinkly, spheroidal halloysite and sheet aggregate of allophane-like particles from the volcanic ash soil from Okachi O2-12 (sample No. 8). Open circle = crinkly film. Solid circle = spheroidal particle. Asterisk = sheet of allophane-like particles.

Halloysites of different morphology in the same specimen also show some differences in average chemical composition (Table 2).

Tentative genetic relationships between morphology and chemical composition of halloysite are shown in Figure 8. Crinkly films containing more than 10.0% ΣFeO transform into spheroidal particles with an accompanying loss of iron (samples 8 and 9, Table 2) and this decrease in iron content may be related to this morphologic change, although crinkly films with less than 7.3% ΣFeO (samples no. 1 and 6, Table 2) do not show a comparable decline in iron content during this morphologic transformation. The tubular halloysites are nearly free of iron. Thus the formation of tubular halloysite requires either that the raw material has a low iron content or that the iron is removed from the crinkly film halloysite in a manner similar to the loss of iron that accompanied the crinkly film-to-spheroidal particle transformation. The compositional difference observed between the long and short tube halloysites is very small compared with the differences between the spheroidal and tubular forms. Variation in morphology of tubular halloysite might be due to differences in rates of crystal growth.

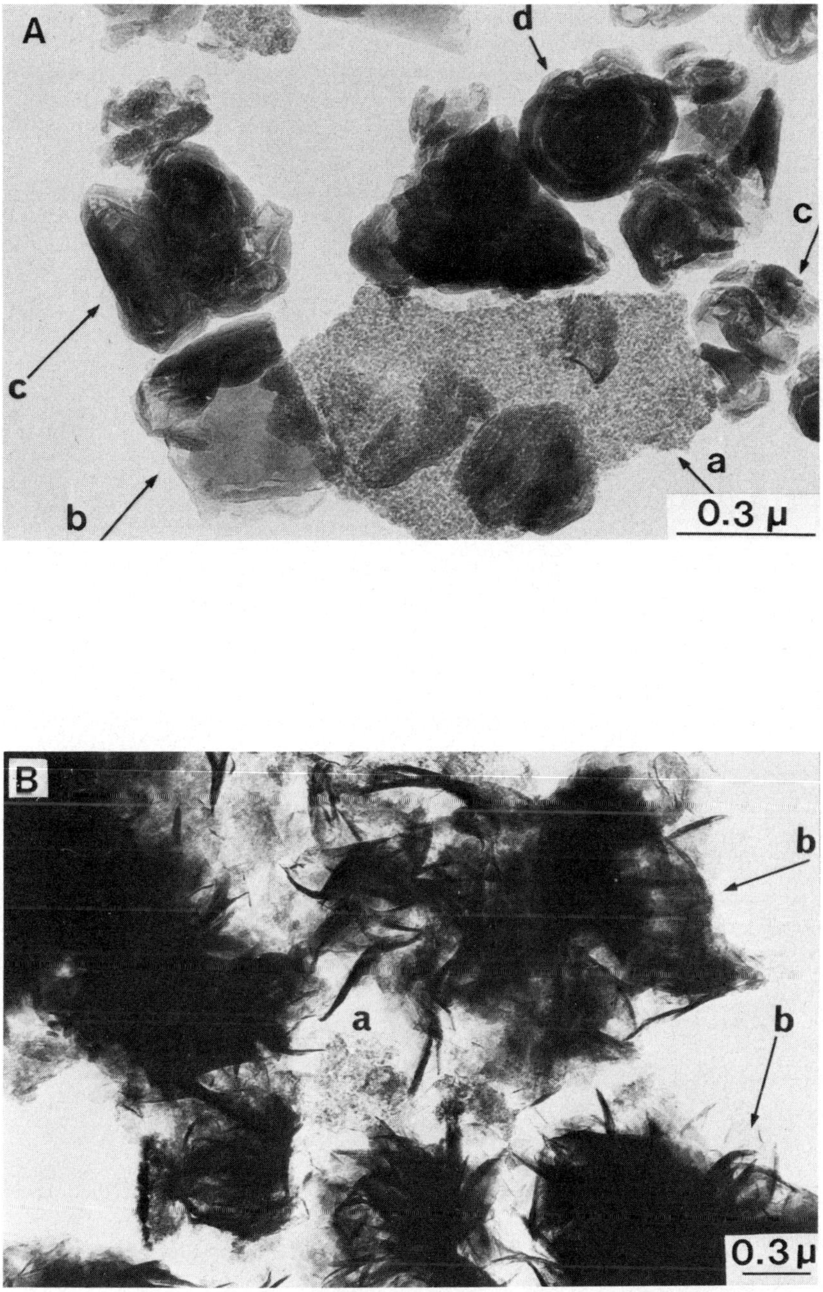

Fig. 5. Electron micrographs of halloysite from Okachi O2-4 (sample No. 6) (A) and Tarumae, Hokkaido (sample No. 14) (B). A-a: sheet of allophane-like particles; A-b: crinkly film; A-c: incipient stage of spheroidal particles; A-d: spheroidal particle; B-a: sheet of allophane-like particles; B-b: crinkly film.

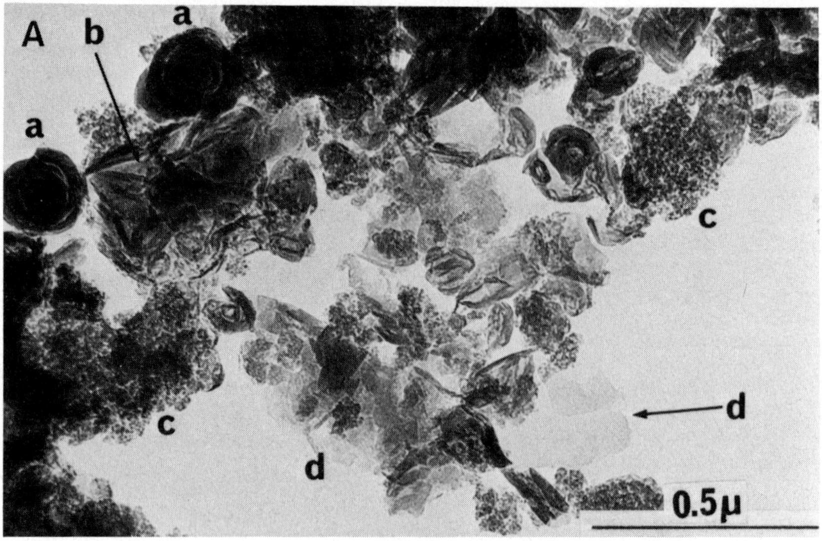

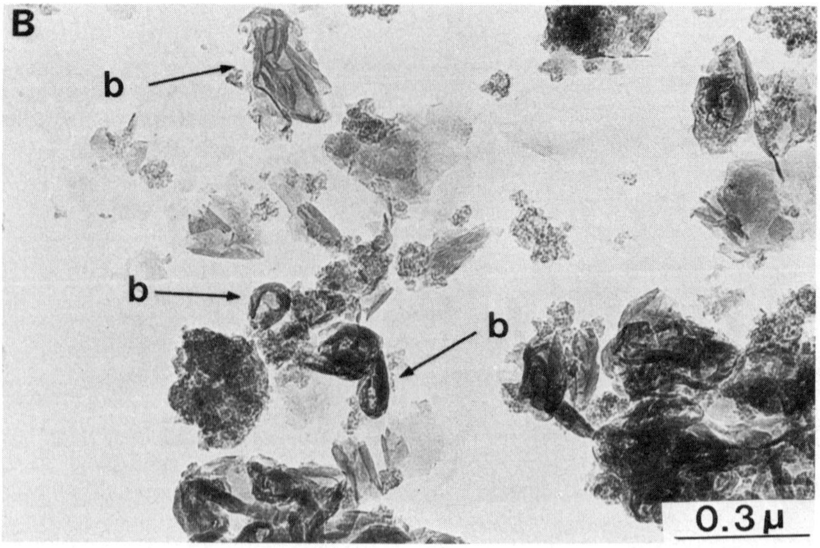

Fig. 6. Electron micrographs of halloysite from Okachi O2-12 (sample No. 8). A-a: spheroidal particles; A-b: crinkly film; A-c: sheet of allophane-like particles; A-d: film of primitive form; B: Arrow indicates crinkly film that is incipient stage of spheroidal particle.

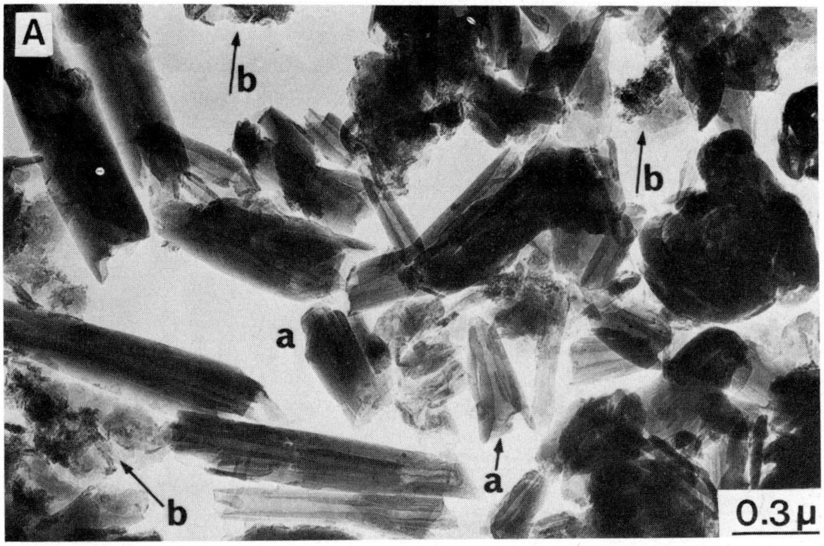

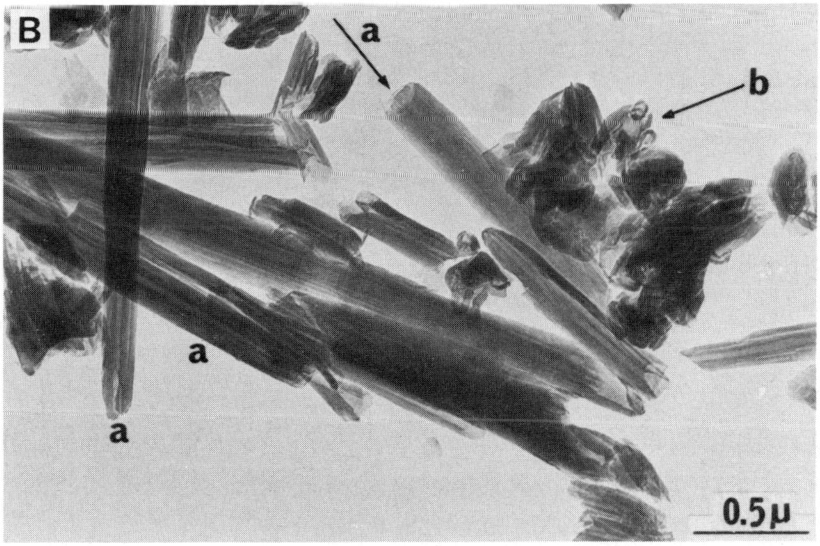

Fig. 7. Electron micrographs of halloysite from Okachi O1-31 (sample No. 3) (A) and from Oita W-75 (sample No. 5) (B). A-a: short tube; A-b: sheet of allophane-like particles; B-a: long tube; B-b: crinkly film that is incipient stage of spheroidal particle.

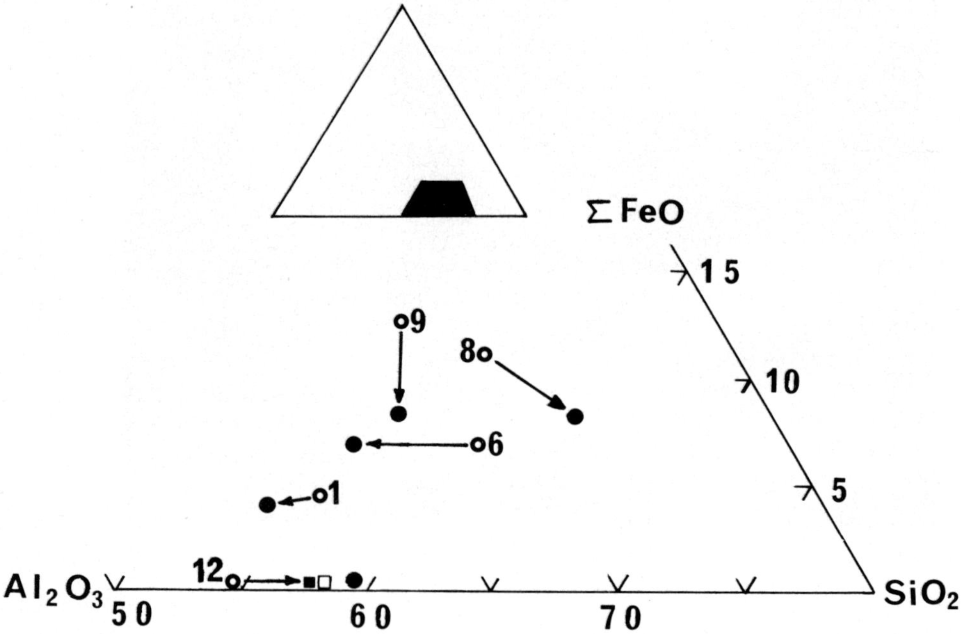

Fig. 8. Tentative genetical relationship between morphology and chemical compositions of halloysites. Arrows join data points for the same specimen. Open circle = crinkly film. Solid circle = spheroidal particle. Solid square = long tube. Open square = short tube.

ACKNOWLEDGMENTS

The writer thanks Dr. Koichi Tazaki for his valuable suggestions and instructions in the course of the study. Thanks are also due to Dr. A.E. Foscolos for his advice and critical reading of the manuscript and to Professor Hideo Minato for the use of a standard sample from Biela Haro, Czechoslovakia.

REFERENCES

1 Hayashi, H., Aita, S. and Suzuki, M., 1978. Semiquantitative chemical analysis of asbestos fibers and clay minerals with an analytical electron microscope. Clays Clay Min., 26: 181-188.

2 Tazaki, K., 1980. Micromorphology and chemical composition of kaolin minerals -- Observation with an analytical electron microscope--. Clay Sci. Soc., Japan, 20: 83-90 (in Japanese, with English abstract).

WEATHERING SEQUENCES. "CLIMATIC"PLANATIONS. LEVELED SURFACES AND PALEOSURFACES.

Georges MILLOT
Institut de Géologie, Centre de Sédimentologie et Géochimie de la Surface (CNRS), Strasbourg, France.

ABSTRACT
Weathering sequences of Western Africa yield genetic models of leveled surfaces or paleosurfaces, developed on hard rocks of old crystalline or metamorphic basements. "Climatic" planation results from the succession and interaction of three processes : 1- basal weathering with regularization of bed rocks; 2- pedogenesis which generates superficial leachy and sandy horizons; 3- superficial planation by rain wash giving way to pediments and pediplains.

INTRODUCTION

This paper deals with the problem of the extended planations of the old continental basements which are found nowadays on the center of old continents under tropical, subtropical and desert climates. These planations by weathering developed in the five continents and especially in South America, Africa, Arabia, Middle-East, India and Australia. Some evidences of ancient tropical weathering mantles are frequently observed in these landscapes of leveled continental basements. The problem may already be predicted not to be actual but historical, i.e. caused by a succession of climates. Moreover, such gigantic planations are found at the bottom of numerous Primary, Secondary and Tertiary marine transgressions. Finally, our Proterozoic and Paleozoic ranges were frequently submitted to such generalized planations which constituted the old peneplains dominating our landscapes under the form of dissected plateaus.

This problem of geomorphology and paleogeomorphology is thus a vexing problem of historical geology. The common factors of fluviatile erosion cannot generate such planations on hard rocks over some hundred and thousands kilometers. Moreover, no trace of these presumed rivers is found.

The purpose of the present chairman's introduction is to suggest a working hypothesis, developed after fifteen years of collaboration between pedologists, geochemists, geologists and geomorphologists in the vast areas of savanna, sahel and desert, from Sudan and Chad to Senegal, Mauritania and Morocco.

In this short paper, I shall only mention the recent studies performed by our team in Western Africa. These studies refer to the international literature about weathering sequences. Moreover, thirty-two "historic" papers were collected in Adams (1975).

The three following mechanisms are thought to operate in combination and to interact : 1- geochemical weathering; 2- pedogenetic transformation; 3- superficial erosion.

The following four landscapes will be discussed : 1- the calcrete landscape; 2- the landscape with smectitic mantle; 3- the landscape with kaolinitic mantle; 4- the iron crust landscape lacking equilibrium.

VOCABULARY : ABLATION GLACIS, PEDIMENT, PEDIPLAIN, RAIN WASH

- The elementary form of planation is the "glacis", which is a smooth or slightly concave landform with very low longitudinal slope (between 1 and 4-5 %). No lateral slope is observed.

- "Accumulation glacis" and "ablation glacis" are distinguished. Only the ablation glacis will be considered. For many authors, there are two types of ablation glacis : the ablation glacis s.s. cut in soft rocks, such as clays, marls, sands, etc..., and the ablation glacis shaped in hard rocks or "pediments".

- Only the pediments or ablation glacis on hard rocks, without rocky upslope and without sedimentary cover will be described here.

- A group of joined together pediments constitutes a landform called "pediplain", commonly found on old continental basements such as the African or Australian ones. These pediplains which were generated by the removal of weathering mantles are called "etchplains". An "inselberg", a relic, is a relief with steep sides rising over a peneplain.

- Many authors deal with the shaping of pediments (Adams, 1975). All investigators of geomorphology point out that planations under the form of glacis or pediments coincide with the climatic zones characterized b alternating seasons, i.e. semi-humid, subtropical, mediterranean and su arid ones. These climatic zones have more and more scattered vegetation The shaping of pediments is attributed to rain wash.

- For soft rocks, rain wash takes place in these climates in the case of strong and short rainfall. The superficial rills do not come together but because of obstacles such as stones or vegetal tufts they disperse to an infinity of rills which may form a sheet. This rain wash carries particles and fine-grained sands and scatters the latter, leading to a general planation of slope.

- For hard rocks, the potential of rain wash is too weak. Rain wash will tend to emphasize the differences in rocks, whether they are softe or harder, in contrast to planation.

- In the following sections, we discuss weathering and pedogenesis which transform nature, structure and erosion resistance of rocks.

CALCRETE LANDFORMS

We consider mediterranean and tropical landscapes presenting beautiful glacis with meteoric calcretes which may invade soils, weathering products and even parent rocks. Such glacis, cut in hard rocks, which are thus calcrete pediments, were especially studied in Mauritania, Morocco and Spain (Ruellan, 1970, Nahon, 1976, Millot et al., 1977, Ruellan et al., 1977, Nahon et al., 1977). The present-day mean annual rainfall is between 300 and 800 mm.

- Calcrete is shown to develop by epigenesis (replacement with constant volume) of soils, weathered products or parent rocks. This can be demonstrated geometrically on an outcrop, sample or mineral scale.

- In subarid and arid areas, calcitization may reach bed rock directly, even if the latter is composed of granites, metamorphic rocks, pelites or sandstones very poor in calcium. Limestone may be supplied to the profile by wind in the form of particles trapped by the scattered vegetation and introduced in the profile by dissolution.

- Epigenesis of parent rock with conservation of petrographic features exists, but with transformation of the nature of the rock (Millot et al., 1977). This epigenesis, which is ruled by the underground water level, attenuates the bed rock irregularities and is deeper up-slope than down-slope.

- Near the top of calcrete, biological and pedological processes occur more frequently and the original structures become disorganized. Finally, at the top of the profile, calcite is unstable and dissolves during wet seasons. Calcrete is thus degraded from the top, where a lixiviated, leachy, sandy and loose horizon develops, which may contain more or less numerous relicts such as minerals, rock or calcrete fragments. Transformation of structure and loss of volume are observed.

- Then, superficial erosion by rain wash takes place. The loose horizon is mobile, it is transported over the slope and the erosion results in planation of the calcrete pediment.

- Three mechanisms thus operate in succession as follows : 1- weathering which ensures epigenesis of parent rock into calcite and gives way to a subterranean "hypodermic" regularization of bed rock; 2- reorganization by pedobioturbation accompanied by dissolution of calcite and followed by development of a loose horizon; 3- planation of the slope into pediment by rain wash.

SMECTITIC MANTLE LANDSCAPES

- These landscapes develop also on basements composed of crystalline, metamorphic, volcanic or pelitic non calcareous rocks, submitted to still contrasting but slightly more humid climatic conditions (mean annual rain fall between 500 and 800 mm). The landscapes are characterized by smectite-rich weathering mantles and soils (Paquet, 1969, Bocquier, 1971, Boulet, 1974, Boulet et al., 1977).

- In Africa, these are the landscapes of savanna and sahel which constitute a double weathering belt around the earth (Paquet, 1969). Vegetation becomes more and more scattered and discontinuous from savanna to sahel. Slopes are leveled into pediments.

- In equilibrium, in savanna areas, when this smectite-rich weathering mantle is living, isovolumic weathering develops on the bottom at the expense of bed rock. This weathering which is governed by the water level is stronger up-slope than down-slope and attenuates the irregularities in the rock. Near the surface, pedo- and bioturbation take place and then, skeleton quartz grains are separated from clay matrix by alternation of humectations and desiccations. Because of volume losses, a leachy, sandy and loose horizon develops which is vulnerable to erosion by rain wash.

- When a climatic change occurs, for instance when the dry season becomes more marked, this smectitic mantle is no longer in equilibrium, as may be observed when going from savanna to sahel, where ancient weathering mantles formed under more humid climates are submitted to subarid ones. The lower part of the profile is no longer supplied with water. Weathering stops except in special periods. In contrast, the development of the loose horizon continues and superficial erosion also. The weathering mantle becomes thinner, interfluves sink, pediments join together and the stage of the pediplains of sahel is reached.

- In desert areas, this process continues as well as superficial erosion. Here and there, bedrock outcrops in a form which we called "whale backs". The roots of the ancient weathering mantle are present between two outcrops. Planation is generalized : there are inappreciable dislevelments. So, the stage of the Saharian "reg" (of Touareg people) or "rag" (of the Moorish people) is reached.

- Thus, in the course of time, three mechanisms operate in succession 1- clay weathering, with hypodermic and concealed regularization of bed rock; 2- pedogenesis,i.e. the development of a superficial loose horizon; 3- generalized planation by rain wash, reaching the stage of the pediplain.

LANDSCAPES WITH KAOLINITIC MANTLE LACKING EQUILIBRIUM

- In tropical zones with contrasting seasons, the landscape frequently inherits a kaolinitic mantle from past periods. Under these conditions, the weathering mantle is characterized by a lack of pedobioclimatic equilibrium (Boulet, 1974). This kaolinitic mantle (ferrallites, latosols, ferrisols, etc...) which was formed in a humid climate and forest vegetation, is then submitted to the inclemency of dry seasons in a savanna area.

- The weathering mantle changes. Briefly, this transformation is the following (Chauvel, 1976, Chauvel et al., 1977). In a ferrallitic soil, structure is created by a porous assemblage of argillo-ferric micronodules (about 100 μm). The skeleton quartz grains are trapped therein. When submitted to alternation of strong desiccations and humectations, this structure breaks down and iron is mobilized. Disjunction of skeleton grains from the matrix takes place, followed by leaching of clay . Skeleton grains are released and the material becomes sandy. A sandy organization succeeds an argillo-ferric structure, with all obvious consequences.

- For a long time, it has been very difficult to learn to understand the secret of these processes. They were progressively illuminated by lateral and vertical models.

- Lateral models were presented by Bocquier (1971) and Boulet (1974) from examples observed on the periphery of inselbergs in Chad and Upper Volta, in an intermediary climatic zone called sahelo-sudanian (mean annual rainfall between 800 and 900 mm, with 7 to 8 months of dry season On the periphery of inselbergs which are oversupplied with water by rain wash, proofs of lateral transformation of a ferrallitic mantle , such as pedological transformations, particle migrations, structural changes, decrease of volume and change of the pediment slope (3.7 % before-and 1 % after transformation) can be observed.

- Vertical models were studied in Senegal and Cameroon by Chauvel

(1976) and Humbel (1976) respectively. Observations could be made in places were forest was cut, or northward from the humid zone into the savanna area. The transformation of the argillo-ferric structure into a sandy organization generates volume losses of about 30 %, depression and dissection of plateaus, and development of lateral glacis (Chauvel et al., 1977).

- The disjunction of skeleton grains and matrix accompanied by leaching of the clay matrix and appearance of loose horizons is a major geodynamic phenomenon, which may be generalized. In the case of more and more scattered vegetation, superficial erosion by rain wash progresses and develops pediments which are initially short and then longer and longer, when going toward dry climates. The effects of cropping are similar (Roose et al., 1981). The transportation of superficial leachy horizons over pediments supplies fluviatile system and eolian erosion with sand and gives way to dunes system and planation.
- Thus, four mechanisms operate in succession : 1- weathering into ferrallitic soils; 2- pedological transformation into "tropical ferruginous soils"; 3- development of leachy, sandy, loose and mobile horizons; 4- planation.

IRON CRUST LANDSCAPES

- Numerous kaolinitic mantles show an iron crust. When submitted to climates prevailing in savanna or sahel areas, they are no more in equilibrium and generate landforms with slightly dipping slopes or nomad rocks. These iron crusts are not stiff.
- The recent studies performed in Upper-Volta and Senegal (Leprun, 1972, 1977, 1979, Nahon, 1976, Nahon et al., 1977) demonstrates that, when iron crusts are submitted to alternation of humid and dry seasons, their evolution continues. They are affected by lack of equilibrium and by degradation.
- Initially, iron accumulation does not attack original rock structures and volumes, but develops by filling of voids and by epigenesis of primary minerals. Then, pedological transformations induce restructuration, nodulation and internal reworking. Finally, degradation increases. By iron dissolution, clay hydrolysis and leaching, the residual skeleton quartz grains are released, giving way to basal, internal and superficial leachy horizons. Superficial erosion by rain wash attacks the superficial horizon. Iron crusts are transformed into glacis. Planation is posterior and discordant upon iron crusts.
- Thus, by combination of weathering and iron accumulation with constant volume, pedological restructurations and development of leachy horizons with volume losses, an iron crust landscape is leveled. All the stages of this planation can be followed from the long glacis to the desert "black regs", ultimate relicts of a more humid past.

INTERPRETATION

- Our problem was to explain how the massifs of hard rocks such as granites, metamorphic rocks, pelites, sandstones and even limestones were submitted to planation, so that they gave rise to pediments and then to very extended pediplains.

- Superficial erosion by rain wash is efficient on soft rocks such as clays, marls, sands and loose sedimentary rocks, as confirmed by all authors. These authors emphasize the relation of the planed land forms to climates with marked dry seasons.
- But this superficial erosion by rain wash is not active on hard rocks. Rain wash would rather emphasize the differences in erosion resistance of rocks.
- All the studies briefly summarized here provide a solution to the problem. Weathering and pedogenesis are very active intermediary factors between the resistant basements and the superficial erosion of loo sened material.
- Weathering does not affect structures and conserves the volume of original material, but transforms its nature. Weathering does not change the external land form. But when the parent rock is homogeneous, wea thering acts deeper up-slope than down-slope, reducing the bed rock ir regularities and regularizing basement in a hypodermic, concealed and preparatory way.
- The specific mechanisms of pedogenesis are stronger in the uppermos part of the weathering mantle. Dissolution or hydrolysis and lixiviation are more active near the surface. The nature of weathered product is less transformed than their proportions and especially their structures are, with sometimes very important volume losses.
- At the top which is subjected to the alternation of humid and more and more severe dry seasons, these transformations result in the development of a residual, loose and mobile horizon covered with scattered vegetation.
- This loose horizon is submitted to planation by rain wash, typical of climates with marked dry seasons. This horizon is transported over the slope and supplies fluviatile and eolian systems with relict sandy material.
- At the end of evolution, dislevelments decrease, interfluves sink, the thickness of the weathering mantle decreases, leaving here and there remnants. The stage is reached of pediplain spread with inselbergs.

GENERALIZED DISCORDANCE OF "CLIMATIC"PLANATION (see Fig. 1)

- By definition, the <u>front of weathering</u> and the <u>pedological system</u> are discordant on geological formations. Weathering and pedogenesis commonly generate clay accumulations, but ferruginous, aluminous, siliceous and calcareous accumulations may relay or replace the last ones. The <u>first geometric discordance</u> takes place at the interaction between the earth crust and the weathering mantle under climatic conditions with <u>marked humid seasons</u>.
- In areas with alternating humid and dry seasons, the <u>leaching front</u> which separates skeleton sand grains from clay matrices is <u>discordant</u> on alterites and soils. As mentioned above, this leaching front cuts all these products. All the weathering mantles, should they be argillaceous, calcareous, ferruginous and even aluminous (bauxites) and siliceous (silcretes) are affected. This <u>second geometric discordance</u> develops through alternating "humidity-drought" action on the weathering mantle under climatic conditions with <u>marked dry seasons</u>.

- These two geometric discordances are ruled by water levels promoting the horizontality due to both hypodermic regularization of bed rock by weathering and superficial planation by lateral erosion by rain wash.

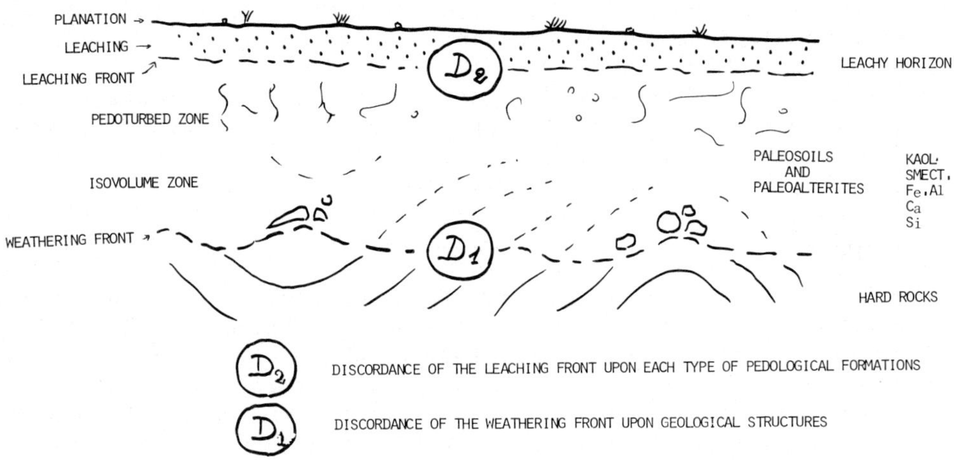

Fig. 1. Schematic representation of the "double discordance"

- Under regularly alternating tropical climates, the three following mechanisms act in succession : geochemical weathering, pedological transformation and superficial erosion. Humid periods or seasons favour an increasing depth of weathering. Drier periods or seasons favour erosion and lateral removal. At equilibrium, the two fronts progress together. So, planations form, develop and give way to pediments.
- Under more and more arid climatic conditions, the weathering front stops. The deepening of the leaching front continues, so that, little by little, the leaching front joins the weathering front. The two planes of discordance thus finally make only one which is the plane of the gigantic desert pediplain, like in the Sahara. This generalized planation which has no sedimentary cover shows only alterite and soil relicts.
- This generalized planation with variable intensity may apply :
to numerous actual continental basements which have been submitted to alternating climates with marked dry season for a long time;
to numerous perched paleosurfaces or ancient peneplains whose appearance is reconstructed in our landscapes from plateaus to plateaus;
to numerous leveled bottoms of geological transgressions.

Weathering sequences thus permit to explain numerous present and past land forms of the earth's surface.

REFERENCES

Adams, G.F., 1975. Planation Surfaces. Peneplains, Pediplains, and Etch plains. Benchmark Papers in Geology, 22, Dowden, Hutchinson & Ross Inc., Stroudburg, Pennsylvania, 476 pp.

Bocquier, G., 1971. Genèse et évolution de deux toposéquences de sols tropicaux du Tchad. Interprétation biogéodynamique. Thèse Sci. Strasbourg et Mém. O.R.S.T.O.M., 62, 1973, 350 pp.

Bocquier, G., Rognon, P., Paquet, H., and Millot, G., 1977. Interprétation pédologique des dépressions annulaires entourant certains inselbergs. Sci. Géol. Bull., 30, 4 : 245-253.

Boulet, R., 1974. Toposéquences de sols tropicaux en Haute-Volta : équilibres dynamiques et bioclimats. Thèse Sci. Strasbourg et Mém. O.R.S.T.O.M., 85, 1978, 272 pp.

Boulet, R., Bocquier, G., and Millot, G., 1977. Déséquilibre pédobioclimatique dans les couvertures pédologiques de l'Afrique tropicale de l'Ouest et son rôle dans l'aplanissement du relief. Sci. Géol. Bull., 30, 4 : 235-243.

Chauvel, A., 1976. Recherches sur la transformation des sols ferrallitiques dans la zone tropicale à saisons contrastées. Thèse Sci. Strasbourg et Trav. et Doc. O.R.S.T.O.M., 62, 1977, 532 pp.

Chauvel, A., Bocquier, G., and Pédro, G., 1977. Les mécanismes de la disjonction des constituants des couvertures ferrallitiques et l'origine de la zonalité des couvertures sableuses dans les régions intertropicales de l'Afrique de l'Ouest. Sci. Géol. Bull., 30, 4:255-263

Humbel, F.X., 1976. L'espace poral des sols ferrallitiques du Cameroun caractéristiques et comportements en relation avec les régimes hydriques et les bioclimats. Thèse Sci. Paris VI, 301 pp.

Leprun, J.C., 1972. Cuirasses ferrugineuses autochtones et modelé des bas reliefs des pays cristallins de Haute-Volta orientale. C.R.Acad Sci.Paris, 275, D : 1207-1210.

Leprun, J.C., 1977. La dégradation des cuirasses ferrugineuses. Etude et importance du phénomène pédologique en Afrique de l'Ouest. Sci. Géol. Bull., 30, 4 : 265-273.

Leprun, J.C., 1979. Les cuirasses ferrugineuses des pays cristallins de l'Afrique occidentale sèche. Genèse. Transformation. Dégradation Thèse Sci. Strasbourg, 222 pp.

Millot, G., 1977. Géochimie de la Surface et Formes du Relief. Présentation. Sci. Géol. Bull., 30, 4 : 229-233.

Millot, G., Nahon, D., Paquet, H., Ruellan, A., and Tardy, Y., 1977. L'épigénie calcaire des roches silicatées dans les encroûtements carbonatés en pays subaride, Antiatlas, Maroc. Sci. Géol. Bull., 30, 4 : 129-152.

Millot, G., 1980. Les grands aplanissements des socles continentaux dans les pays subtropicaux, tropicaux et désertiques. Mém. h. sér. Soc. géol. Fr., 10 : 295-305.

Nahon, D., 1976. Cuirasses ferrugineuses et encroûtements calcaires au Sénégal occidental et en Mauritanie. Systèmes évolutifs : géochimie structures, relais et coexistence. Thèse Sci. Marseille III et Sci. Géol. Mém., 44, 232 pp.

Nahon, D., and Millot, G., 1977. Enfoncement géochimique des cuirasses ferrugineuses par épigénie du manteau d'altération des roches mères gréseuses. Influence sur le paysage. Sci. Géol. Bull., 30, 4 : 275-282.

Nahon, D., Millot, G., Paquet, H., Ruellan, A., and Tardy, Y., 1977. Digestion et effacement des cuirasses ferrugineuses par les encroûtements calcaires en pays aride, Sahara et Mauritanie. Sci. Géol. Bull., 30, 4 : 289-296.

Paquet, H., 1969. Evolution des minéraux argileux dans les altérations et les sols des climats méditerranéens et tropicaux à saisons contrastées. Thèse Sci. Strasbourg et Mém. Serv. Carte géol. Als. Lorr., 30, 1970, 210 pp.

Roose, E.J., Fauck, R., Lelong, F., and Pédro, G., 1981. Modifications fondamentales de la dynamique actuelle de sols ferrallitiques et ferrugineux d'Afrique occidentale sous l'influence de la mise en culture. C.R.Acad.Sci.Paris, 292-II : 1457-1460.

Ruellan, A., 1970. Les sols à profil calcaire différencié des plaines de la Basse Moulouya (Maroc oriental). Thèse Sci. Strasbourg et Mém. O.R.S.T.O.M., 54, 302 pp.

Ruellan, A., Nahon, D., Paquet, H., and Millot, G., 1977. Rôle des encroûtements et épigénies calcaires dans le façonnement du modelé en pays aride. Sci. Géol. Bull., 30, 4 : 283-288.

VARIATIONS IN THE COMPOSITION OF PHYLLOSILICATES MONOPARTICLES IN A WEATHERING PROFILE OF ULTRABASIC ROCKS

Hélène PAQUET*, Joëlle DUPLAY* and Daniel NAHON**

* Centre de Sédimentologie et de Géochimie de la Surface, Institut de Géologie, Strasbourg, France ** Laboratoire de Pétrologie de la Surface, Université de Poitiers, France

ABSTRACT

The ultrabasic massif of Biankouma (Western Ivory-Coast) was weathered yielding nickeliferous laterites. A weathering profile, developed on olivine pyroxenites, was studied by means of X-ray diffraction and electron microscopy. Single particles of weathering phyllosilicates were also studied by elemental microanalysis. Talcs and smectites represent two population groups with variable compositions.

INTRODUCTION

About thirty profiles of ultrabasic rocks were studied by Nahon (1977) and Nahon et al. (1981) in the Western Ivory-Coast in the context of a Ni indices research by the "Société pour le Développement minier de Côte-d'Ivoire" (SODEMI).

To complete the information on the size fraction smaller than 2μ and especially to specify more precisely the nature and chemical composition of the newly formed phyllosilicates, elemental microanalyses were carried out on material from one profile (R_{17}). This profile developed on an olivine pyroxenite (orthopyroxenes : between 70 and 90 %) in a climate characterized by a mean annual rainfall of 1600 mm.

This profile can be subdivided into four main layers (horizons) which are from bottom to top : the parent rock (I), the transitional layer (II) the ocherous ferruginous layer (III) and the reddish-purple ferruginous layer (IV). The principal structural and petrographic features of these layers were described elsewhere (Nahon, 1977).

CLAY MINERALOGY OF THE LESS-THAN-2-MICROMETRE FRACTION

Results

The results obtained by X-ray diffraction are given on the right part of Fig. 1.

In the transitional layer (II), the clay fraction consists of smectites, generally predominant over talc and serpentine and chlorite, the latter occurring only in traces. Amorphous material is observed, accompanied by goethite in the upper part of the transitional layer (II+III).

In the ferruginous layers (III) and (IV), kaolinite prevails largely over talc and traces of smectite which disappear very rapidly from the

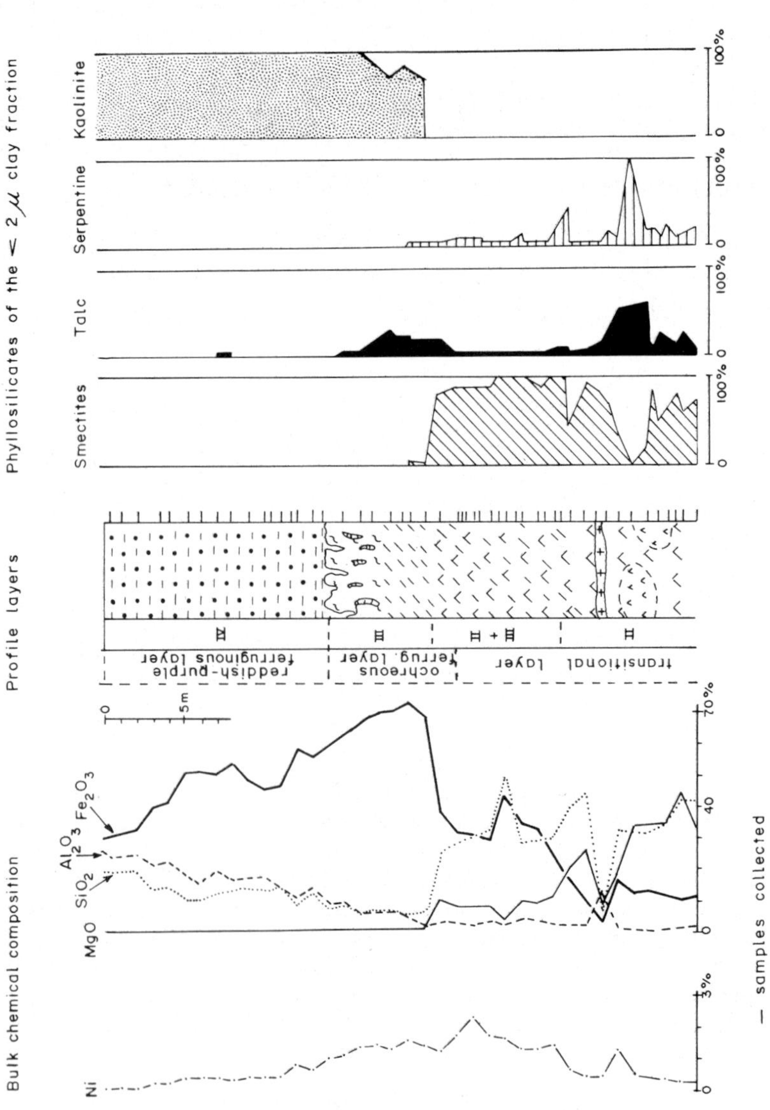

Fig.1. Data on profile (R_{17})

top of the transitional layer (II). In the ocherous ferruginous layer (III), the phyllosilicates are accompanied by goethite and in the reddish-purple ferruginous layer (IV) by goethite and gibbsite.

Phyllosilicates of the transitional layer (II)

The phyllosilicates of the transitional layer (II) are serpentine, talc and smectites.

- Serpentine, which was identified as a retrodiagenetic mineral is of the lizardite type. Precise identification based on X-ray data only is difficult (Brown, 1961). This variety of nickeliferous serpentine is recognized by the platy morphology of its particles and its characteristic electron diffraction pattern of points (Whittaker and Zussman, 1971)

- Although the profile represents a nickeliferous laterite, the observed talc is a normal form of talc, with a basal spacing near 9,35 Å. Its plates are more or less elongated. Its crystallinity is good but there are indications of the presence of some swelling layers in some of the levels in the profile. X-ray study shows that this talc is not a mineral of the β-kerolite - pimelite series as described by Maksimovic (1966) and Brindley and Maksimovic (1974), or a willemsite such as pointed out by De Waal (1970), or a talc-stevensite mixed layer of the type characterized by Trescases (1973), or a talc-saponite mixed layer of the type observed by Veniale and Van der Marel (1968) in similar profiles in Transvaal, Yugoslavia, New-Caledonia and Northern Italy respectively.

- The smectites occur as thin particles with folded edges. Their true nature can be specified only by chemical analysis. The nature of the smectites of nickeliferous laterites has been studied before : Trescases (1973) found it very difficult to specify whether weathering smectites were nontronites or strongly Fe-substituted saponites. Kühnel et al. (1978) presented a "smectite zone" including Mg-Fe smectites in their descrption of weathering profiles developed on Indonesian peridotites. And Melfi et al. (1979-1980) described the transformation of pyroxenes to montmorillonitic clay minerals of nontronite type in weathering horizons of Brasilian ultramafic rocks.

In summary, the phyllosilicates observed in the transitional layer (II) consist of talc and smectites, generated by the weathering of the minerals of parent rock, talc resulting from weathered orthopyroxenes (Trescases, 1973), and these two families of weathered minerals are always mixed with retrodiagenetic serpentines and chlorites. Therefore, their composition must be determined by elemental microanalysis of single particles.

ELEMENTAL MICROANALYSIS OF SINGLE PARTICLES

Method

The method of elemental microanalysis was described by Badaut and Weber (1977) later refined by Duplay (1982). A Philips EM 300 transmission electron microscope and a Si (Li) energy-dispersive spectrometer coupled to a Cameca scanning electron microscope are used.

In a given suspension, the particles are first carefully selected and located. They must be isolated, not be superposed and should be clean. Since our transmission electron microscope was not fitted to a X-ray spectrometer, as in the method described by Namae (1975) and Hayashi et al. (1978), the same grid was used for the study by means of stereoscan. For each element of the particle there are typical peaks in the

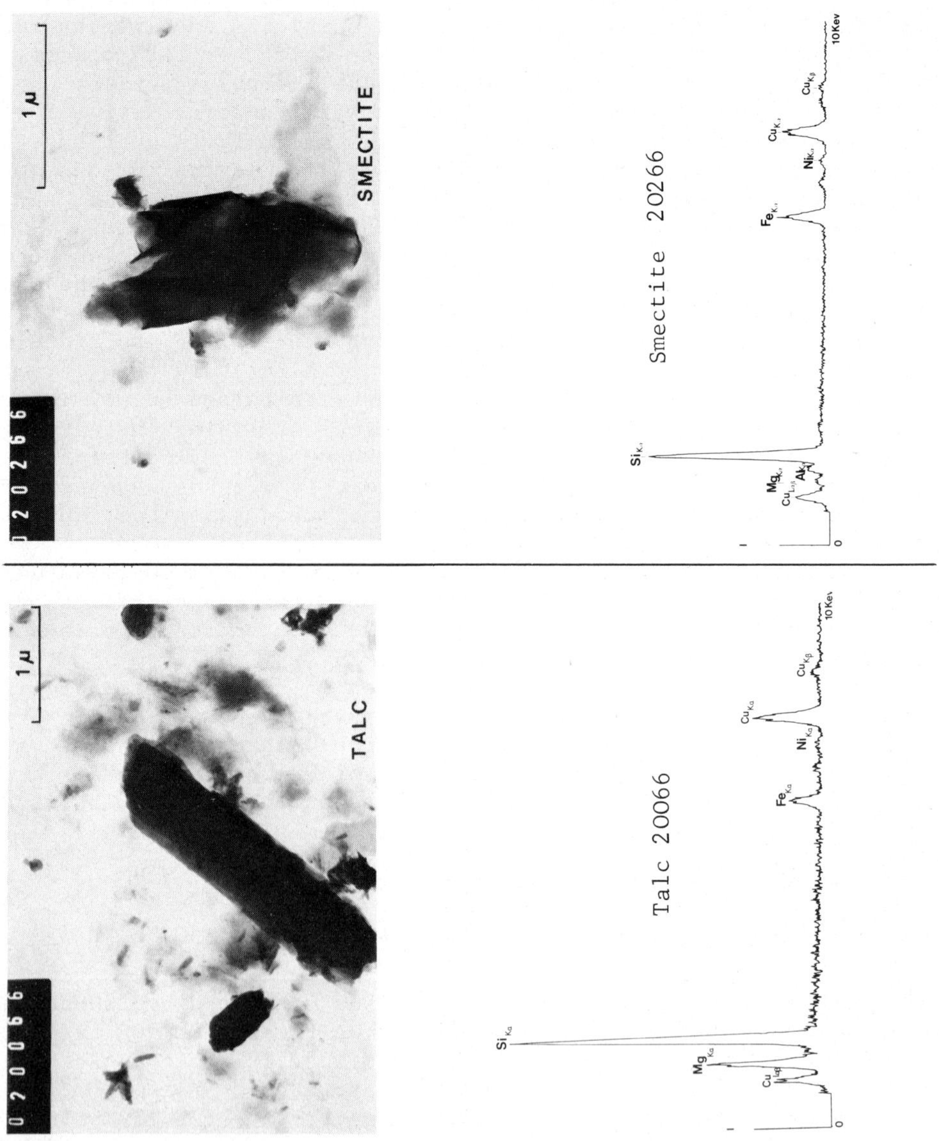

Fig. 2. Transmission electron micrographs of talc and smectite particles and corresponding X-ray emission spectra. Energy of electron beam:20 kV, Counting time:100 sec., Sample current:0,5nA. Grid is made of copper and the supporting film of carbon.

X-ray emission spectra, as shown on Fig. 2.

Contents of each element are calculated from peak areas applying the Castaing relation (Tixier, 1978), with reference to a standard. This standard should be similar to the analyzed mineral, which permits to cancel disturbing effects. This gives :

$$K_A = \frac{(I_A)\ \text{sample}}{(I_A)\ \text{standard}} = \frac{(C_A)\ \text{sample}}{(C_A)\ \text{standard}} \cdot \frac{(\rho_x)\ \text{sample}}{(\rho_x)\ \text{standard}}$$

where K_A is the relative intensity of an element A
I_A is the measured intensity
C is mass concentration
ρ_x is mass thickness

$$\text{and}\ \frac{k_A}{k_B} = \left(\frac{C_A}{C_B}\right)\text{sample} \Big/ \left(\frac{C_A}{C_B}\right)\text{standard}$$

where $C_A, C_B \ldots C_N$ are the unknown quantities of elements A, B...N
$C_A + C_B \ldots + C_N$ being equal to 100.

This method permits the analysis of particles with areas between 1 and 2 μm^2 and with thickness not exceeding 1000 Å.

According to studies on single particles of pure standard monominerals there is good agreement between this technique and standard chemical analysis. Therefore, we presented a tentative distribution of elements in tetrahedral and octahedral sheets of individual phyllosilicates from the chemical data for single particles.

Results

The distribution of the elements in tetrahedral and octahedral sheets are presented in Table 1 for talc particles and in Table 2 for smectite particles.

For 18 among 22 analyzed talc particles, tetrahedral Si is between 3.8 and 4; in the octahedral sheet, Fe is above all substituted for Mg, Fe values being between 0.12 and 0.67. Ni is not measurable but it is observable in the X-ray emission spectra (Fig. 2), demonstrating its low content in these phyllosilicates (Table 1).

For smectite particles, tetrahedral Si is comprised between 3.22 and 3.97; in the octahedral sheet, Fe and Mg values are between 1.00 and 2.00 and between 0.09 and 1.14 respectively. Thus the compositions range between those for nontronites, ferriferous beidellites and Fe saponites in two cases. Ni contents are too low to be measured as in the case for talc, but this element can be observed in X-ray emission spectra (Fig. 2). Ni may be accompanied by Cr.

The plot of Fe contents versus Mg contents in the octahedral sheet of all particles measured (Fig. 3) shows that two different population groups of phyllosilicates exist under the denomination of talcs and smectites.

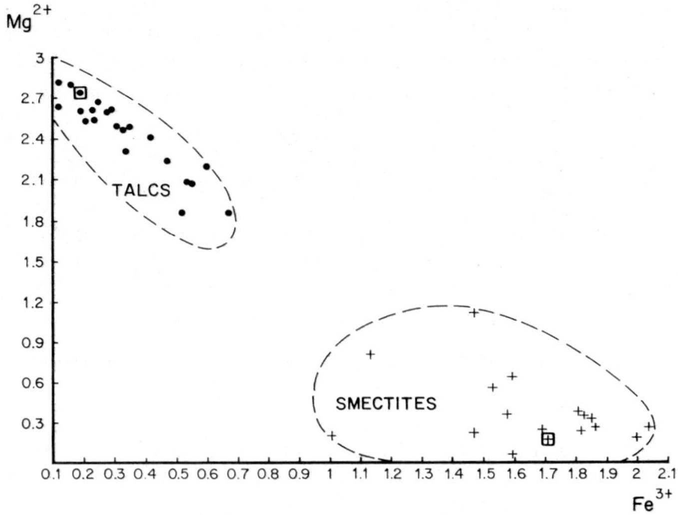

Fig. 3. Mg contents versus Fe contents in the octahedral sheets of talc and smectite particles (Profile R_{17}, transtional layer (II)

⊡ talc particle of Fig. 2 ⊞ smectite particle of Fig. 2

Proust and Velde (1978) already described the genesis of Fe beidellite from hornblende crystals and of Wyoming beidellite from plagioclases in an amphibolite of the Massif Central, France. When studying the weathering profiles in the Ivory-Coast, Colin et al. (1980) pointed out that smectites generated by pyroxenes were different from the smectites formed from peridots. And Le Brusq and Ildefonse (1981) indicated that the smectites developing from olivine phenocrystals were distributed around three poles, magnesian, ferriferous and aluminous.

We have showed here that talcs and smectites originating from weathered orthopyroxenes and peridots within a population group have a range of variable compositions.

CONCLUSIONS

The study of single particles of weathered phyllosilicates newly formed in a weathering profile of ultrabasic rocks leads to the following conclusions :

1. Talcs and smectites do not represent two well-defined clay mineral types, but two mineral population groups with variable chemical compositions.

2. Talcs show Al-substitutions in the tetrahedra and Fe-substitution

TABLE 1

Results of elemental analysis of talcs particles

Tentative distribution of elements in tetrahedral sheet - octahedral sheet

Talc 20066	Tetrahedral sheet			Octahedral sheet			
Nr particle	Si	Al tetra.	Fe^{3+} tetra.	Al octa.	Fe^{3+} octa.	Mg	Ti
9	3,83	0,10	0,07	–	0,25	2,69	0,01
10	3,73	0,13	0,14	–	0,42	2,43	0,03
11	3,82	0,13	0,05	–	0,29	2,63	0,01
14	3,96	0,04	–	0,06	0,19	2,62	–
15	3,90	0,10	–	0,05	0,31	2,51	–
16	4,0	–	–	0,15	0,52	1,88	0,05
19	3,91	0,05	0,04	–	0,28	2,62	–
25	3,74	0,25	0,01	–	0,67	1,87	0,[illegible]1
26	3,92	0,08	–	0,10	0,55	2,09	–
27	3,88	0,12	–	–	0,35	2,51	0,01
28	3,88	0,12	–	0,08	0,54	2,10	–
30	3,97	0,03	–	0,08	0,21	2,55	0,01
36	3,91	0,09	–	0,02	0,33	2,49	–
39	3,77	0,22	0,01	–	0,60	2,21	–
41	3,93	0,07	–	0,13	0,34	2,33	–
43	3,96	0,04	–	0,07	0,12	2,66	0,03
45	3,90	0,09	0,01	–	0,19	2,76	–
48	3,86	0,14	–	0,04	0,23	2,63	–
49	3,66	0,34	–	0,12	0,47	2,26	–
50	3,88	0,09	0,03	–	0,12	2,84	0,01
54	3,80	0,12	0,08	–	0,24	2,56	0,07
57	3,86	0,10	0,04	–	0,16	2,82	–

TABLE 2

Results of elemental analysis of smectites particles

Tentative distribution of elements in tetrahedral sheet - octahedral sheet

Smectite 20266	Tetrahedral sheet			Octahedral sheet			
Nr particle	Si	Al tetra.	Fe^{3+} tetra.	Al octa.	Fe^{3+} octa.	Mg	Ti
3	3,44	0,56	–	0,02	2,00	0,21	0,01
1	3,78	0,22	–	0,14	1,53	0,58	–
4	3,78	0,22	–	0,89	1,00	0,23	0,01
17	3,97	0,03	–	0,32	1,59	0,09	–
18	3,85	0,15	–	0,19	1,71	0,20	–
20	3,80	0,20	–	0,17	1,69	0,27	0,02
22	3,59	0,41	–	0,06	1,86	0,29	0,01
23	3,48	0,52	–	0,12	1,59	0,66	–
24	3,57	0,40	0,03	–	1,85	0,35	0,03
33	3,65	0,35	–	0,08	1,82	0,27	0,03
34	3,22	0,47	0,31	–	1,47	1,14	0,02
37	3,56	0,44	–	0,06	1,83	0,37	0,01
38	3,26	0,48	0,26	–	2,04	0,28	–
46	3,90	0,10	–	0,31	1,13	0,83	–
47	3,54	0,56	–	0,04	1,81	0,40	0,01
52	3,51	0,49	–	0,52	1,47	0,25	–
55	3,73	0,27	–	0,21	1,58	0,38	0,03

in the octahedra. These substitutions are likely to contribute to the swelling tendency of interlayers, but the mixed layers observed by Veniale and Van der Marel (1968) are still far away.

3. The chemical compositions of smectites are most frequently close to the ferriferous pole of nontronites, but the others range from those for Fe beidellites to Fe saponites.

4. These phyllosilicates may be slightly nickeliferous and chromiferous. In fact, Ni and Cr contents are lower than the measurement limits of the elemental microanalysis of single particles, but give shoulders on the X-ray emission spectra. It is thus confirmed that Ni and Cr are rather related to iron hydroxides in the nickeliferous laterites of the Ivory-Coast (Colin et al., 1980, Nahon et al., 1981).

5. This study demonstrates that since the beginning of weathering, as soon as secondary phyllosilicates form, population groups of different particles are observed. Each particle has its own chemical composition characterizing its formational micro-environment, either intra- or extracrystalline. Each micro-environment depends on the parent minerals, the nature of percolating solutions, the diameter of pores, the flow rate in the fissure walls, cracks, cleavages and micropores, etc.. (Tardy et al., 1973, Clauer and Millot, 1978).

ACKNOWLEDGMENTS

The authors aregreatly indebted to S. Chaudhuri and N. Clauer for language corrections and critical reading of the manuscript.

REFERENCES

Badaut, D. and Weber, F., 1977. Etude par des techniques de microscopie électronique (microscopie électronique à transmission et spectromètre équipant un microscope à balayage) de deux échantillons de serpentinites de Tiébaghi. Rapport interne, Institut de Géologie, Strasbourg, 12 pages.

Brown, G., 1961. The X-ray identification and crystal structures of clay minerals. Mineralogical Society (Clay minerals group), London, 544 pages.

Brindley, G. and Maksimovic, Z., 1974. The nature and nomenclature of hydrous nickel-containing silicates. Clay miner., 10 : 271-277.

Clauer, N. and Millot, G., 1978. Genèse des minéraux argileux et géochimie isotopique du strontium. Docum. Labo. Géol. Fac. Sci. Lyon, H.S. 4 : 129-142.

Colin, F., Parron, C., Bocquier, G. and Nahon, D., 1980. Nickel and chromium concentrations by chemical weathering of pyroxenes and olivines. U.N.E.S.C.O., Int. Symp. Metallogeny of mafic and ultramafic complexes, Athens, Oct. 9-11 1980, In Press.

De Waal, S.A., 1970. Nickel minerals from Barberton, South Africa : III. Willemsite, a nickel-rich talc. Am. Miner., 55 : 31-42.

Duplay, J., 1982. Thèse de 3ème cycle, Poitiers, In prepration.

Hayashi, H., Aita, S. and Suzuki, M., 1978. Semi-quantitative chemical analyses of asbestos fibers and clay minerals. Clays Clay Min., 26 : 181-188.

Kühnel, R.A., Roorda, H.J. and Steesma, J.J.S., 1978. Distribution and partitioning of elements in nickeliferous laterites. Bull. B.R.G.M., Section II, 3 : 191-206.

Le Brusq, J.Y. and Ildefonse, P., 1981. Altération météorique de cristaux d'olivine dans des basaltes du Cameroun, In preparation.

Maksimovic, Z., 1966. β kerolite - pimelite series from Golès Mountain, Yugoslavia. Proc. Int. Clay Conf. Jerusalem, I : 97-105.

Melfi, A.J., Trescases, J.J. and Barros de Oliveira, S.M., 1979-1980. Les latérites nickélifères du Brésil. Cah. O.R.S.T.O.M., sér. Géol., XI : 15-42.

Namae, T., 1975. A method of quantitative analysis for thin specimens by energy dispersive spectrometer fitted to transmission electron microscope. Electron Microsc., 24 : 1-6.

Nahon, D., 1977. L'altération des roches ultrabasiques de la région de Biankouma (Côte-d'Ivoire). I. Le massif de Sipilou nord. Rapport interne, Fac. Sci. Abidjan, 22 pages.

Nahon, D., 1979. Cuirasses siliceuses ou silcretes nickélifères dans les profils d'altération de roches ultrabasiques de Côte-d'Ivoire. Rôle des épigénies. Sci. Géol. Bull., Strasbourg, 32 : 189-197.

Nahon, D., Paquet, H. and Delvigne, J., 1981. Ultramafic rocks and nickel concentration in Western Ivory-Coast. Econ. Geol., In press.

Proust, D. and Velde, B., 1978. Beidellite crystallization from plagioclase and amphibole precursors. Clay miner., 13 : 199-209.

Tardy, Y., Bocquier,G., Paquet, H. and Millot, G., 1973. Formation of clay from granite and its distribution in relation to climate and topography. Geoderma, 10 : 271-284.

Tixier, R., 1978. Microanalyse sur échantillons minces. In : Microanalyse et microscope à balayage. Les Editions de Physique, Orsay, Chapitre 5 : 433-448.

Trescases, J.J., 1973. L'évolution géochimique supergène des roches ultrabasiques en zone tropicale. Formation des gisements nickélifères de Nouvelle-Calédonie. Thèse Sci.Strasbourg et Mém. O.R.S.T.O.M., 78, 1975, 259 pages.

Veniale, F. and Van der Marel, H.W., 1968. A regular talc-saponite mixed layer mineral from Ferriere, Nure Valley (Piacenza Province, Italy). Contr. Mineral. and Petrol., 17 : 237-254.

Whittaker, E.J. and Zussman, J., 1971. The serpentine minerals. In : J.A. Gard, editor, The electron-optical investigations of clays, Mineralogical Society (Clay minerals group), Chapter 5 : 159-191.

BASALTIC AND RHYOLITIC ROCKS AS PARENT MATERIALS OF HALLOYSITE IN ARGENTINE DEPOSITS

A. M. IÑIGUEZ RODRIGUEZ

Centro de Investigaciones Geológicas
Universidad Nacional de La Plata - CONICET
1900 La Plata, Argentina

ABSTRACT

In northeastern Argentina there is a vast area of deeply weathered tholeiitic basalts of Cretaceous age. Studies based mainly on X-ray diffraction, electron microscope and chemical analyses have shown that zeolites and montmorillonite are formed at depth while halloysite and kaolinite associated with gibbsite and iron oxides occur near the surface. Halloysite is abundant at the contact between weathered basalt and soil (oxisol) and decreases towards the surface where kaolinite predominates.

On the other hand, rhyolitic vulcanites and ignimbrites of the Jurassic, Bahia Laura Group, outcropping in Patagonia have undergone weathering and hydrothermal alteration forming halloysite and kaolinite with abundant silicification and transformation of biotite into hydrobiotite. These processes have produced important clay deposits in the provinces of Chubut and Santa Cruz.

In the case of the basalts, halloysite is derived by weathering action on labradorite, while in the case of the rhyolites it is produced by hydrothermal action on sanidine. Operations to beneficiate the Chubut and Santa Cruz kaolins are necessary for applications other than ceramics.

INTRODUCTION

Two major areas of basaltic and rhyolitic rocks provided the parent materials for the formation of halloysite (7Å). One of these areas is located in northeastern Argentina, on the border of Brazil and Paraguay (Fig. 1), where cretaceous tholeiitic basalts outcrop, which have been deeply weathered, to form clayish red soils (oxisoils). The relief is slightly undulating, cut by the Paraná and Uruguay Rivers. The annual rainfall averages 1.755 mm and temperatures are between 39.6°C in summer and below 0°C in winter.

The basalts are typically flow-banded. Bands consist of vesicular and massive basalt flows, varying in color from dark gray to reddish gray and grayish green. Intergranular and hyaloophitic textures are common in vesicular basalts, while intergranular and subophitic textures are present in the massive varieties.

Optical microscopy (Teruggi 1955; Iñiguez 1978), indicates that plagioclases are the dominant constituents, as randomly oriented, well developed tabular crystals, with slightly altered edges. The composi-

Fig. 1. Map showing the location of the study areas.

tion corresponds to labradorite (An_{50} - An_{65}). The second mineral in importance is augite, with prismatic or granular habit, and associated with minor amounts of pigeonite and diopside.

Opaque minerals are abundant, occurring as large crystals or small granular aggregates of magnetite and hematite. Only a trace amount of volcanic glass remains as interstitially, deeply altered clay products. Accessory olivine, apatite and zircon have been observed.

Surface weathering has deeply altered these basalts (Riggi et al. 1964), leading to the formation of clayish red soils (oxisoils), which are spread over an extensive area.

A cross section of deeply weathered basalt to the soil zone, developed in situ, shows a distinctive mineralogical zonation (Fig. 2) Halloysite (7Å) is the dominant clay mineral in the limit between the deeply weathered basalt and the soil zone.

The second area previously mentioned, is of greater importance and comprises volcanic outcrops of the Serie Tobifera (Bahía Laura Group),

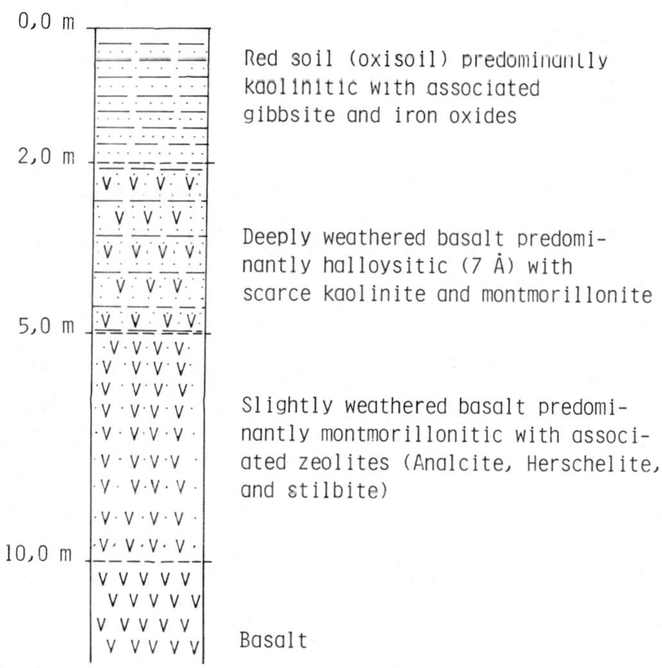

Fig. 2. Diagrammatic section showing grading alteration.

in the Patagonia Argentina (Fig. 1).

Climate is arid and cold, the annual rainfall averages 223 mm and the mean annual temperature is 9°C. Topographically, the area is a peneplain, with several rivers running in a west-east direction such as the Chubut and Santa Cruz Rivers, and others.

These volcanic rocks, Jurassic in age, (Feruglio 1949), are for the most part rhyolites and ignimbrites, which by weathering and hydrothermal alteration formed secondary kaolin deposits (Hayase, K y Maiza 1969-1970) and (Romero, A. et al. 1975), which are mainly halloysitic. Petrographically, porphyritic rhyolitic rocks contain well developed quartz, sanidine and biotite crystals. Euhedral quartz crystals are dominant, showing corrosion effects, embayments and secondary growths of abundant euhedral phenocrystals of typically twinned sanidine, also showing corrosion effects and embayments. Sanidine usually alters to halloysite (7Å), and kaolinite. No plagioclases were detected. Biotite alteration to hydrobiotite was frequently observed.

The ground mass has a felsitic to microgranular texture indicating a significant silicification process.

Mineralogical Analyses of the Clay Fractions

The clay fractions were analyzed by X-ray powder diffraction, scan electron microscopy, differential thermal analyses and chemical analysis

Samples from the basalts were collected starting with the fresh slightly weathered basalt, bearing smectite, through a deeply weathered intermediate zone with halloysite (7Å), up to the kaolinitic red soil zone (oxisoils), (Fig. 2). Fig. SEM 3A shows a plagioclase crystal (labradorite) completely altered to halloysite (7Å). Fig. 3B is a view of part of Fig. SEM 3A, enlarged to higher magnification, and shows tubes which form a randomly oriented matted mass. X-ray patterns and differential thermal analyses confirm the presence of halloysite (7Å). X-ray powder diffraction and scan electron microscopy of clay deposits from Chubut and Santa Cruz Provinces, in the Patagonia region, show the presence of halloysite (7Å) and kaolinite. In Fig. SEM 4A and SEM 4B (suspension in water), mixtures of the two minerals can be seen.

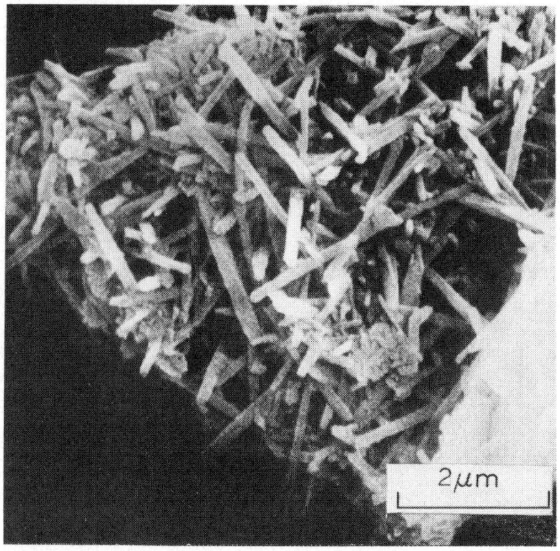

Fig. 3A. Matted fibers of halloysite tubes randomly oriented growing up from plagioclase crystals in basalts.

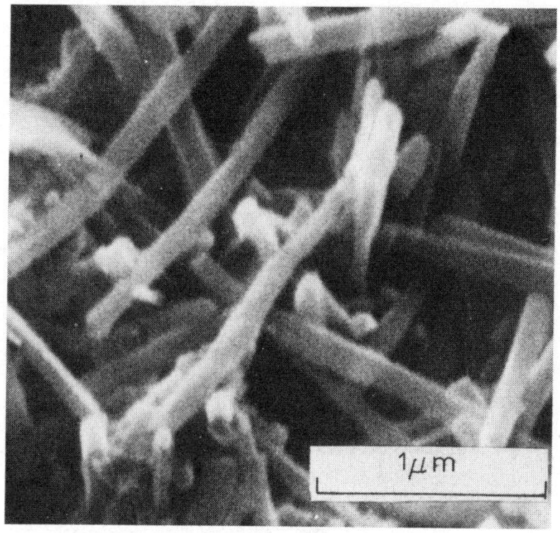

Fig. 3B. Views of part of Fig. 3A with higher magnification.

Fig. 4A. Subhedral kaolinite flakes and halloysite elongates in rhiolites.

Fig. 4B. Subhedral hexagonal flakes of kaolinite prevailing ones halloysite (7Å) tubes.

Allophane was not observed. Crystals of sanidine, probably superficially altered to halloysite (7Å) are commonly observed (Fig. SEM 5A) and bipyramidal volcanic quartz crystals (10 μm long), are important components (Fig. SEM 5B).

Chemical Analyses

Fig. 6 is a diagrammatic presentation of the results of chemical analyses of the fresh basalts and rhyolites on the left, and of the oxisoils and kaolinized rhyolites on the right. The compositions are in general quite similar: silica decreases and Al_2O_3 increases for both. In basalts, FeO decreases and Fe_2O_3 increases; TiO_2 increases for both and alkaline earths and alkali content, all decrease with a notable decrease of the CaO and MgO contents for the basalts and of K_2O for the rhyolites. MnO decreases slightly in the basalts and remains constant in the rhyolites. A similar behavior has been reported by Murray (1978) for the alteration of a granite to kaolin.

Genesis of Halloysite

Weathering of basaltic rocks, as a result of movement of rain water

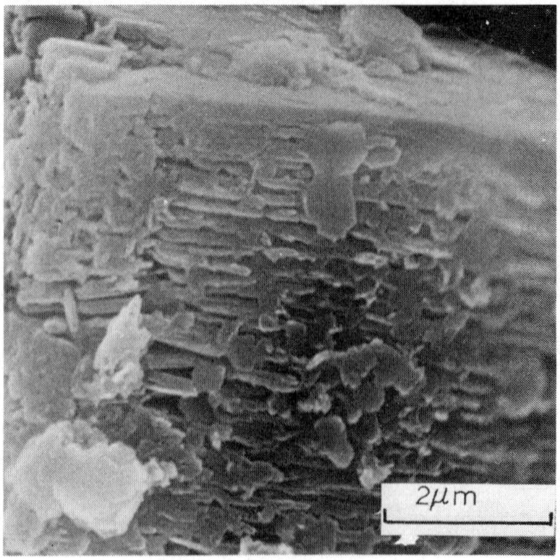

Fig. 5A. Altered sanidine crystal leading to the formation of halloysite (7Å).

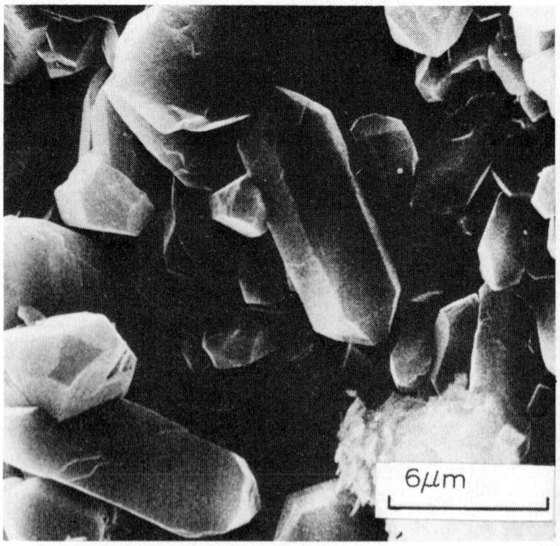

Fig. 5B. Fine grained bypyramidal volcanic quartz crystals.

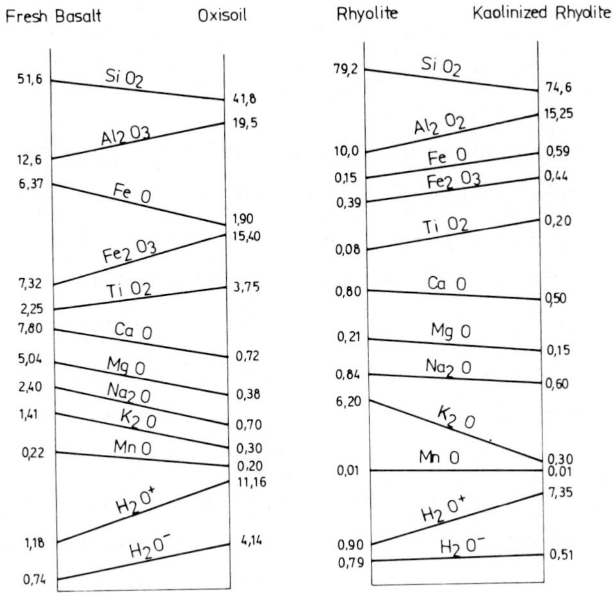

Fig. 6. Diagrams showing chemical analysis of fresh rocks (left) and kaolinized rocks (right).

through the rocks, was evidently responsible for the alteration of feldspars (labradorite) and consequently for the formation of halloysite (7Å). This process occurs less with increasing depth. Chemical analyses of fresh and weathered basalts (soils), show various changes of composition as a result of alteration processes. From SEM evidence, halloysite seems to have transformed into kaolinite towards the top zone. Formation of gibbsite, as well as concentration of iron oxides, are also significant.

According to field and petrograpical observations, the formation of halloysite in rhyolitic rocks is due to hydrothermal alteration of sanidine. Parallel bands of well defined alteration zones show that biotite has changed to hydrobiotite. Silica veins are abundant and feldspars (sanidine) are completely altered. Chemical analyses of fresh and weathered rhyolites confirm the occurrence of this process, which was responsible for the formation of residual and sedimentary deposits in the Patagonia Argentina.

Economic Implications

Red soils formed as weathering products of basaltic rocks are not high quality raw materials, they are used locally in the red brick industry. Patagonian hydrothermally altered rhyolites are important deposits comprised of halloysitic and kaolinitic admixtures with good whiteness. The acceptable reserves are especially in Chubut River Valley, Chubut Province, and in San Julián, Santa Cruz Province. These deposits contain not more than 30-40% clay size fraction (<4 μm) with varying Al_2O_3 contents, between 15 and 20%. Mineralogically they consist essentially of halloysite (7Å) and kaolinite with associated quartz and feldspars. As Fig. SEM 5B shows, quartz has a very small particle size. These features limit their industrial applications to white ware products. Processing operations of purification could make these clays suitable for use as fillers; but not for paper coatings because of the high content of halloysite and fine particle quartz.

REFERENCES

Feruglio, E., 1949. Descripción geológica de la Patagonia. Dirección de YPF. Bs. As. Argentina.

Hayase, K. y Maiza, P., 1969. Génesis del Yacimiento de caolin de la mina Villegas. Prov. de Chubut, Républica Argentina, Rev. Assoc. Geol. Arg., XXIV, 1, 65-71, Buenos Aires.

Hayase, K. y Maiza, P., 1970. Génesis del yacimiento de caolin de la mina Eguivocada, Provincia de Chubut, República Argentina, Rev. AMPS, 1, 1-2, 33-47, Buenos Aires.

Iñiguez, A. M., 1978. Los minerales secundarios y su influencia en el comportamiento geotécnico de los basaltos del Rio Uruguay. Area de Garruchos-Garabi (Provincia de Corrientes), República Argentina. Obra del Centenario del Museo de La Plata, Tomo IV, 137-151. La Plata.

Murray, H. H., 1978. Alteration of a granite to kaolin. Mineralogy and geochemistry. Schriftenr. Geol. Wiss. Berlin, 197-208.

Riggi, J. C. y Feliu de Riggi, N. A., 1964. Meteorización de basaltos en Misiones. Rev. Asoc. Geol. Arg. XIX, 1, 57, 70, Buenos Aires.

Romero, A., Dominguez, E. y Whewell, R., 1975. Génesis de los yacimientos de Caolin del Rio Chubut inferior. II Congr. Iberoamericano de Geologia Económica. V, 423, Buenos Aires.

Teruggi, M. E., 1955. Los basaltos tholeiiticos de Misiones. Notas Museo de La Plata. XVIII, Geol. N° 70, 259-278. La Plata.

ON THE ROLE OF SURFACE COMPLEXATION IN WEATHERING REACTIONS. DISSOLUTION KINETICS OF HYDROUS ALUMINA IN THE PRESENCE OF ORGANIC LIGANDS.

Vera ZUTIC* and Werner STUMM
Federal Institute for Water Resources and Water Pollution Control, Swiss Federal Institute of Technology, Zurich, Switzerland.

ABSTRACT

Complex-forming organic ligands affect the extent and the rate of dissolution of minerals, in particular aluminum oxides and aluminosilicates, but the understanding of kinetics of dissolution is still incomplete. We attempted to distinguish between equilibrium and kinetic effects of organic ligands in the dissolution process of hydrous alumina using the aluminum/aluminum oxide rotating disc electrode technique. In the pH range 3 - 6, dissolution of hydrous alumina is a surface-controlled process. The presence of organic ligands does not increase the rate of dissolution. If the dissolution rate is enhanced by addition of 5×10^{-5}M fluoride ions, the organic ligands block the surface release process by displacing fluoride ions from the surface.

The results are compared with simulated weathering experiments and interpreted in terms of a surface complexation model. The release of the surface complex is the rate-limiting step.

INTRODUCTION

Complex-forming organic solutes affect the extent of dissolution and the rate of weathering of minerals, in particular aluminum oxides and aluminosilicates.

Water-soluble anions of organic acids of low molecular weight are important complexing agents (Smith and Martel, 1976), and they are present in top soils in concentrations as high as 10^{-5} - 10^{-4}M, oxalate being the most abundant (Graustein, 1977). The complexing by oxalate and other ligands tends to raise the Al concentration to levels that allow significant mass transfer in soils.

---

*) On leave from "Rudjer Boskovic" Institute, Zagreb, Yugoslavia

However, the understanding of the kinetics of mineral dissolution in the presence of organic ligands is still very incomplete because of the complexity of the reaction steps that lead to dissolution and also because of a lack of simulated weathering experiments under controlled and constant conditions.

In general, complexing agents increase the equilibrium solubility of hydrous oxide phases due to formation of complex species in solution (Lind and Hem, 1975; Kragten, 1980). Stumm and coworkers (Stumm et al., 1980; Kummert and Stumm, 1980; Sigg and Stumm, 1981) have demonstrated that there is a tendency of ligands to form surface complexes at the oxide/water interface analogous to the formation of metal complexes in the solution phase. The specifically adsorbed ligands may be viewed as "inner sphere" complexes. The surface equilibrium constants can be used to quantify the extent of adsorption and the resulting net charge of the oxide surface as a function of pH and solute activity.

Surface complexation plays an important role in the kinetics of nucleation and dissolution since it changes the surface charge and hence the approach of protons, and it modifies active surface sites chemically.

We propose, using a corrosion technique (Zutic and Stumm, 1981), to distinguish between equilibrium and kinetic effects of complexation in the dissolution process of hydrous alumina as a convenient model which also has geochemical relevance. There is a certain analogy in the behaviour of an aluminum anode and a primary mineral in the aquatic environment:

- both are thermodynamically unstable and undergo a series of kinetically controlled elementary reaction steps;
- both dissolve incongruently; different layers are involved with different transport characteristics (solid state diffusion, semiconductor properties, aqueous diffusion);
- the reactions are primarily controlled by surface properties and interfaces;
- the concept of formation of a protective surface layer during weathering of aluminosilicates (Wollast, 1967; Paces, 1973) is strongly reminiscent of passivity phenomena in corrosion.

METHODOLOGY

Using the rotating disc methodology (Levich, 1962) with an oxide film covered aluminum electrode (Wiedmer and Grauer, 1974; Zutic and Stumm, 1981), it is possible to simulate an ideal open system at environmentally relevant concentrations of solution constituents (dissolved aluminum ions and organic ligands). By recording the dissolution current as a function of the transport parameter (rotation rate), it is possible to distinguish among transport-controlled, surface-controlled and mixed dissolution processes.

The oxide film formed by the prolongated conditioning of the aluminum disc in aqueous solution of pH = 4.0, with 0.1 M $NaClO_4$ as supporting electrolyte, was characterized by ISCA and ESM analysis. Thickness of the film was about 40 Å, and the surface properties correspond to those of amorphous hydrous alumina.

RESULTS AND DISCUSSION

In the pH range 3 - 7, the anodic dissolution of hydrous alumina is a typical surface-controlled process. The experiments with the rotating disc electrode revealed no increase in the dissolution rate in the presence of 10^{-6} - 10^{-3} M highly complexing anions such as oxalate ions (Fig. 1). The ESCA date indicate no formation of a protective oxalate layer over the initial 40 Å thick oxide layer.

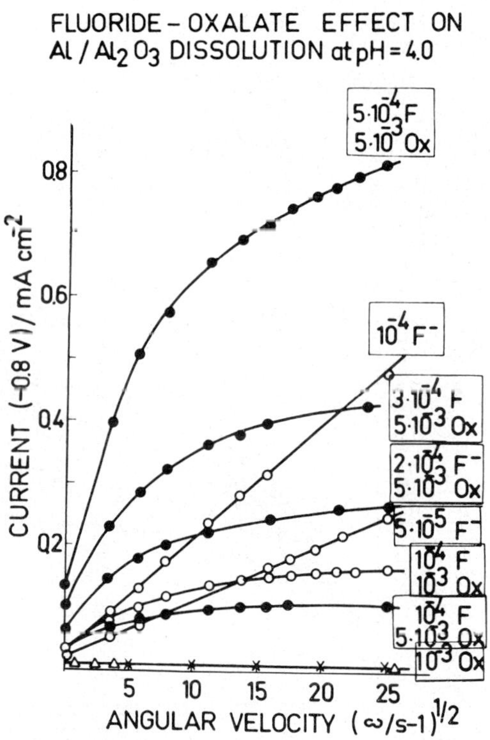

Fig. 1.

Effect of oxalate and fluoride ions on anodic dissolution of hydrous alumina. The dissolution current is plotted against the square root of the rotation rate. The linear dependence is characteristic of a transport-controlled process (5×10^{-5}M and 10^{-4}M F^-) and independence of the transport parameter (no complexing ligands [x] and 10^{-3}M oxalate [Δ]) is typical of a surface-controlled process. All other compositions (indicated on the corresponding curves) showed a mixed transport-surface controlled dissolution.

The rate of the surface reaction is increased only in the presence of fluoride ions at fluoride concentrations of 10^{-5} - 10^{-3} M, the dissolution process is controlled by the rate of transport of fluoride ions (Fig. 1). A steady state is established between the veolcity of the surface reaction and the mass transfer through the boundary layer, allowing the determination of dissolution stoichiometry at very high undersaturations. The rate constant of the dissolution process in the presence of fluoride is $k_{F,o} \geqslant 3x10^2 cm\ s^{-1}$. Addition of organic ligands in the concentration range 10^{-5} - 10^{-2} M (Figs. 1 - 4) decreases the dissolution rate due to the displacement (ligand exchange) of fluoride from the surface. Concentration step experiments indicate that the surface complex formation and ligand exchange reactions are reversible and diffusion-controlled. The extent of displacement depends on the concentration of free ligand and its affinity for Al atoms in the oxide.

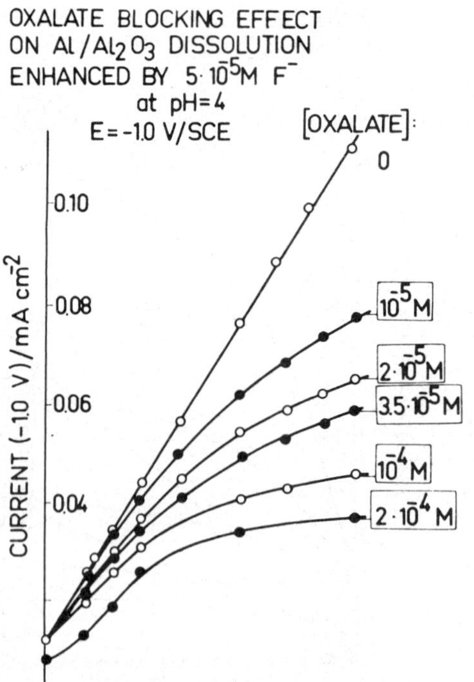

Fig. 2.

Effect of oxalate on the dissolution current in $5x10^{-5}M\ F^{-}$ + 0.1 M $NaClO_4$ solution at pH = 0.4, 25 ^{o}C

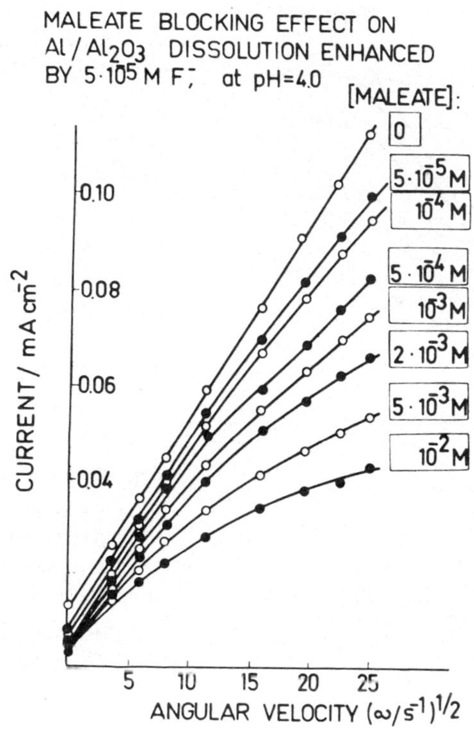

Fig. 3. Effect of maleate on the dissolution current in $5x10^{-5}$M F^- + 0.1 M $NaClO_4$ solution, pH = 4.0, 25 °C.

The displacement can be quantitatively interpreted in terms of competitive adsorption. The blocking effect (Fig. 5) obeys the Langmuir adsorption isotherm; and for example for oxalate, at pH = 4.0 the adsorption equilibrium constant is calculated to be $K_L = 2x10^4\ mol^{-1}$.

At pH = 4, the affinities of the ligands for the surface increase in the sequence: formiate ≲ carbonate < sulphate < salicylate < fumarate < maleate < malonate << ocalate ≲ citrate.

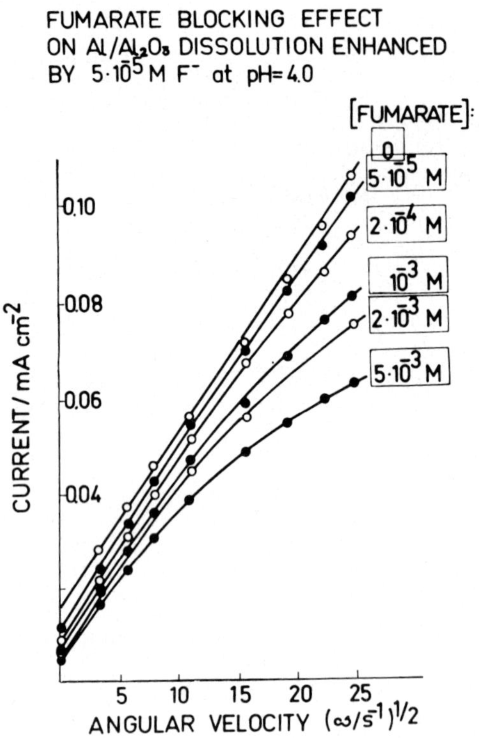

← Fig. 4.

Effect of fumarate on the dissolution current in $5x10^{-5}$ M F^- + 0.1 M $NaClO_4$ solution, pH = 4.0, 25°C.

↓ Fig. 5.

Effect of organic acids (total concentration is plotted) on the dissolution current in $5x10^{-5}$M F^- + 0.1 M $NaClO_4$ solutions at pH = 4.0, 25°C, angular velocity = 600 s^{-1}.

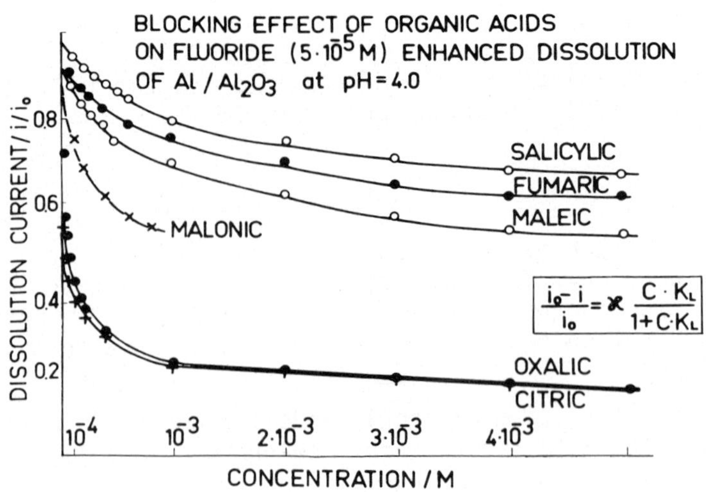

It is well-known (Berner, 1977; Petrovic, 1976; Graustein, Stumm and Morgan, 1981) that ligands which form complexes with Al(III) (or other mineral constituents) in solution tend to increase the solubility and the extent of dissolution. Correspondingly, the extent of weathering of soil minerals is usually enhanced by such ligands because (1) the higher equilibrium levels of Al(III) (or Fe(III)) allow significant mass transfer in the soil; and (2) dissolution processes that are otherwise incongruent (e.g. of feldspars) may become more congruent. We need to distinguish between the extent of reaction and the instantaneous (incipient) dissolution rate. The latter is also affected by surface complexes formed by the same ligands that may control the rate of the dissolution process at the oxide surface. The presence of F^- obviously aids the detachment of the surface complex, while oxalate and other organic ligands have little effect on the instantaneous dissolution rate.

CONCLUSION

The effect of organic ligands and fluoride on the anodic dissolution of hydrous alumina (summarized in Table 1) could be interpreted in terms of the surface complexation model. Detachment of the surface complex represents the rate-determining step.

Our results are in a qualitative agreement with adsorption measurements on gibbsite and imogolite (Parfitt et al., 1977) and simulated weathering experiments with γ-Al_2O_3 dispersions (Furrer and Stumm, 1981) and with aluminosilicates (Robert et al., 1980). They also agree with fluoride-oxalate displacement data in suspensions of montmorillonite, kaolinite and hydrous oxide (Samson, 1952).

These results clearly show common general features of the hydrous oxide/solution interface and of the predominant role of chemical reactions at interfaces in the dissolution process.

ACKNOWLEDGEMENT

This research has been supported by the Swiss National Science Foundation. Valuable advice given by R. Grauer and G. Furrer is acknowledged.

TABLE 1. DISSOLUTION RATE CONTROLLING MECHANISMS FOR HYDROUS ALUMINA AT $3 \leq pH < 7$, 25°C

Ligand	Surface equilibria	Rate-controlling step	Rate expression
-	$\equiv AlOH_2^+ \rightleftarrows \equiv AlOH + H^+$	surface reaction	$\frac{d[Al_{aq}]}{dt} = k_o[H^+]^n$
organic acid (H_2X)	$\equiv AlOH + H_2X \rightleftarrows \equiv AlHX + H_2O$ $\equiv AlOH + HX^- \rightleftarrows \equiv AlX^- + H_2O$ $2\equiv AlOH + H_2X \rightleftarrows \equiv Al_2X + 2H_2O$	surface reaction	$\frac{d[Al_{aq}]}{dt} = k_{o,}\ (1-\Theta)[H^+]^n + k_{X,o}\Theta[H^+]^m$ $k_o \geq k_{X,o}$
	$\equiv AlOH + HF \rightleftarrows \equiv AlF + H_2O$ $\equiv AlOH + F^- \rightleftarrows \equiv AlF + OH^-$	transport-controlled ($k_{F,o} \geq 3x10^{-2} cm\ s^{-1}$)	
fluoride + organic acid	$\equiv AlHX + F^- \rightleftarrows \equiv AlF + HX^-$ etc.	mixed surface + transport	for $\Theta_X + \Theta_F \approx 1$: $\frac{d[Al_{aq}]}{dt} = k_F(1-\Theta_X)[F^-][H^+]^p + k_X\Theta_X[H^+]^m$

REFERENCES

Berner, R.A., 1977. Rate Control of Mineral Dissolution under Earth Surface Conditions. Am. J. of Science, 278: 1235-1252.

Furrer, G. and Stumm W., 1981. Unpublished results.

Graustein, W.C., 1977. Calcium Oxalate; Occurence in Soils and Effect on Nutrient and Geochemical Cycles. Science, 198: 1252-1254.

Kragten, J., 1980. Atlas of Metal-Ligand Equilibria in Aqueous Solutions. John Wiley, Chichester GB, 781 pp.

Kummert, R. and Stumm, W., 1980. The Surface Complexation of Organic Acids on Hydrous γ-Al_2O_3. J. Colloid. Interf. Sci., 75: 373-385.

Levich, V.G., 1962. Physiochemical Dynamics. Prentice Hall, Englewood Cliffs., N.J., 700 pp.

Lind, C.J. and Hem, J.D., 1975. Effect of Organic Solutes on Chemical Reactions of Aluminum. Geol. Survey Watersupply Paper no. 1827 g.

Paces, T., 1973. Steady-state Kinetics and Equilibrium between Groundwater and Granitic Rock. Geochim. Cosmochim. Acta,37: 403-409.

Parfitt, R.L., Fraser, A.R., Russel, J.D. and Farmer, V.C., 1977. Adsorption on Hydrous Oxides; Oxalate, Benzoate and Phosphate on Gibbsite. J. Soil. Sci.,28: 40-47.

Petrovic, R., 1976. Rate Control in Feldspar Dissolution. II. Protective Effect of Precipitates. Geochim. Cosmochim. Acta,40: 1509-1521.

Robert, M., Berrier, J. and Veneau, G., 1980. Solubilisation comparée des silicates, carbonates et hydroxides en fonction des conditions du milieu. Bull. Minéral.,103: 324-329.

Samson, H.R., 1952. Fluoride Adsorption by Clay Minerals and Hydrated Alumina. Clay Bull.,1: 266-270.

Sigg, L.and Stumm, W., 1981. The Interactions of Anions and Weak Acids with the Hydrous Geothite Surface. Colloids and Surfaces, 2: 101-117.

Smith, R.M. and Martell, A.E., 1976. Critical Stability Constants; Other Organic Ligands (vol. 3). Plenum Press, New York, 495 pp.

Stumm, W., Kummert, R. and Sigg, L., 1980. A Ligand Exchange Model for the Adsorption of Inorganic and Organic Ligands at Hydrous Oxide Interfaces. Croat. Chem. Acta, 53: 291-312.

Stumm, W. and Morgan, J.J., 1981. Aquatic Chemistry, 2nd ed.(p.532-34) Wiley Interscience, New York, 780 pp.

Wiedmer, E. and Grauer R., 1974. Das Elektrodenverhalten von Aluminium in Fluoridlösungen. Aluminium,50: 265-270.

Wollast, R., 1976. Kinetics of Alteration of K-Feldspar in Buffered Sollutions at Low Temperature. Geochim. Cosmochim. Acta,31: 635-648.

Zutic,V. and Stumm W., 1981. Effect of Organic Ligands and Fluoride on Anodic Dissolution of Aluminum Oxide. Submitted to J. Electroanal. Chem.

FORMATION OF NICKEL SILICATE ORES BY WEATHERING OF ULTRAMAFIC ROCKS

W. SCHELLMANN

Federal Institute for Geosciences and Natural Resources
Hannover, Fed. Rep. of Germany

ABSTRACT

Analyzed nickel silicate ores from various deposits contain a few percent nickel which is evenly distributed within the ore. The samples consists mostly of primary serpentine which is strongly depleted in magnesium by incipient rock weathering and slightly enriched in nickel derived from overlying nickel limonite. The chemical and mineralogical differences between the nickel-enriched serpentinites and unmineralized parent rocks are discussed.

INTRODUCTION

Lateritic nickel ores are increasing in importance world wide as a source of nickel. About 75 % of terrestrial nickel resources were formed by tropical weathering of ultramafic rocks; important deposits occur in New Caledonia, Indonesia, the Philippines, Australia, Brazil, Cuba, Burundi and other tropical countries. Two types of ore can be distinguished: nickel limonite ore (laterite) at the surface, underlain by nickel silicate ore (saprolite), which in turn is underlain by unweathered ultramafic rock.

Nickel is residually concentrated with Fe in the nickel limonite zone; it is bound in the goethite lattice, which can incorporate up to 1.5 % Ni (Schellmann, 1978). Excess nickel is leached and carried downwards and is immobilized in the nickel silicate zone, which normally contains 1.5 - 2.5 % Ni. In most deposits only a small part ($<$ 1o %) of the nickel is incorporated in high-grade nickel phyllosilicates (garnierite ore) precipitated in pockets and open fractures in the silicate ore zone. The bulk of the nickel is evenly distributed at low concentration in this zone which consists of slightly weathered ultramafic rock, usually serpentinite. This altered rock is economically the most important tropical nickel ore.

While it is generally accepted that the nickel in the Ni-serpentinites is leached from the overlying nickel limonite, its distribution and fixation in the Ni-serpentinites is still unclear. It is suggested that the nickel is fixed in extremely small inclusions of neoformed garnierite minerals or in amorphous components in the Ni-serpentinites.

In the present study, representative samples of nickel silicate ore from various tropical ore deposits were subjected to detailed analysis and compared with that of the unmineralized ultramafic parent rock. Although in some deposits nickel-bearing smectites are present in considerable amounts, the main constituent of nickel silicate ores is usually serpentine. The ore samples analyzed were therefore slightly weathered Ni-serpentinites with serpentine as their main constituent.

COMPOSITION AND PROPERTIES OF NI-SERPENTINITES AND THEIR UNMINERALIZED PARENT ROCK

The analyzed ultramafic rocks and ores (Table 1) can all be defined as serpentinites. X-Ray diffraction analysis (XRD) shows that serpentines (lizardite and chrysotile) is the main constituent of each sample. Besides serpentine, the samples contain relics of olivine and pyroxene (mostly orthorhombic) in varying amounts; in addition, some nickel serpentinites show small amounts of secondary quartz. No differences in the (oo1) spacing of various serpentines could be detected and the half peak widths all lie in the same range (mean value for nickel serpentinites o.26 ^{o} 2 θ, for serpentinites o.23 ^{o} 2 θ). The (oo1) integral intensities show, on average, a tendency to lower values for the nickel serpentinites, but this difference is not significant.

Although the Ni-serpentinites cannot be distinguished from their unmineralized equivalents by XRD, there are marked differences in their physical appearance.

While the normal serpentinites are hard and predominantly greyish in color, the nickel serpentinites are less hard, and have lighter colors varying between grey, brown and greenish. Their bulk density (Table 1) is normally lower due to fine pores.

The unmineralized and the nickeliferous serpentinites exhibit distinct differences in chemical composition (Table 1) only for some elements. While the bed rock normally contains between o.19 and o.35 % nickel, the nickel values of the ore samples increase to between 1.22 and 2.52 %. This increase is accompanied by an appreciable loss of magnesium. The Mg content of the nickel serpentinites varies between 16.8 and 34.8 % MgO, whereas the parent rock samples, on average, contain 38 % MgO. No difference could be found in the silica contents, so that unmineralized serpentinites can be distinguished on the basis of their MgO/SiO_2 ratio which is near unity for normal serpentinites but <1 in the ores (Fig. 1). The major differences between the two groups can thus be explained by

No.	Origin	Ni	MgO	SiO_2	Fe_2O_3	Al_2O_3	IL	pH	D
		Ni-Serpentinites							
7	Morro do Niquel, Brazil	1.39	29.o	48.8	9.7	o.21	1o.3	8.2	1.7o g/cm^3
8	Niquelandia, Brazil	1.48	25.6	39.1	18.7	1.85	11.2	7.3	1.67
17	New Caledonia	2.11	3o.3	44.3	12.5	o.61	8.4	7.4	2.12
2o	INCO	1.69	29.7	36.1	15.6	1.41	13.6	7.6	2.31
23	Barro Alto, Brazil	2.26	3o.3	36.7	15.o	1.42	12.6	7.3	2.25
25	Nonoc, Philippines	1.81	34.8	38.6	1o.2	o.67	12.8	8.2	2.53
473	Palawan, Philippines	2.52	16.8	36.2	24.o	1.o9	16.3	7.o	1.5o
1187	Barro Alto, Brazil	1.5o	31.2	37.o	12.6	1.85	13.9	7.7	2.45
1442	Mayagüez, Puerto Rico	1.22	32.8	37.1	12.3	1.o8	13.9	8.1	1.74
	Average	1.78	28.9	39.3	14.5	1.13	12.6	7.6	2.o3
		Serpentinites							
397	New Caledonia	o.28	4o.5	39.5	8.4	o.58	1o.o	8.7	2.63
474	Palawan, Philippines	o.35	37.o	4o.9	7.2	o,37	14.3	7.8	2.23
58o	New Caledonia	o.28	42.8	4o,1	8.3	o.46	7.5	8.5	2.82
1189	Barro Alto, Brazil	o.35	39.6	4o.8	8.1	1.25	8.6	9.1	2.88
1143	Mayagüez, Puerto Rico	o.22	35.1	37.7	8.1	2.54	13.7	8.8	2.51
1623	Conakry, Guinee	o.19	32.3	34.7	17.o	o.33	12.7	9.4	2.53
	Average	o.28	38.o	39.o	9.5	o.92	11.1	8.7	2.6o

Table 1: Chemical composition, loss of ignition (IL), pH and bulk density (D) of Ni-serpentinites and serpentinites

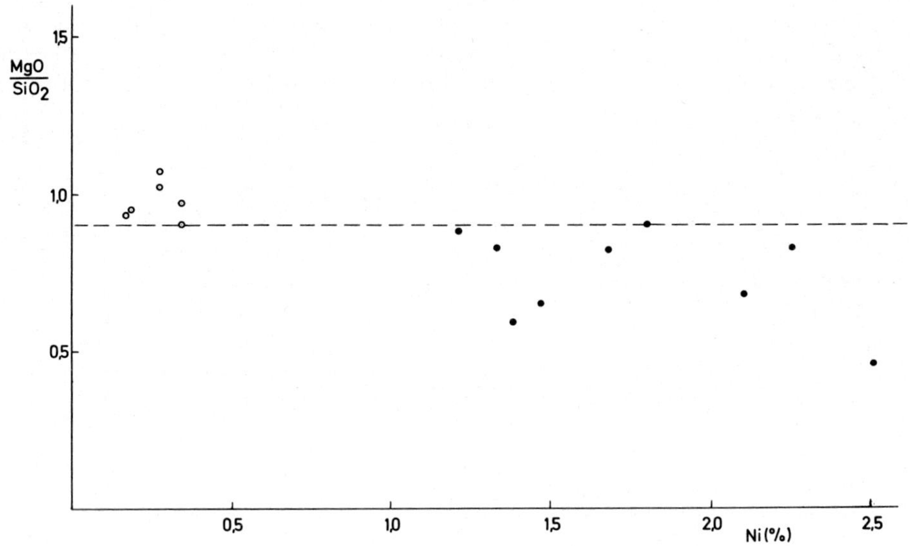

Fig. 1. MgO/SiO_2 weight ratio versus Ni for serpentinites (o) and Ni-serpentinites (.)

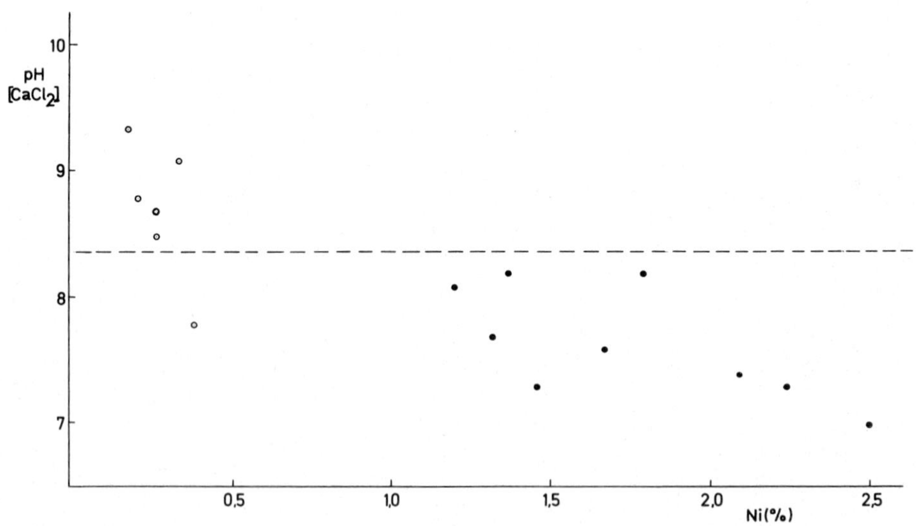

Fig. 2. pH (in o.o1 M $CaCl_2$) versus Ni for serpentinites (o) and Ni-serpentinites (.)

weathering of the nickel serpentinites. Besides certain physical effects, weathering caused depletion of magnesium and a slight enrichment in iron. The influence of weathering is also reflected by the pH values of the samples (1o g of sample in 25 cm^3 o.o1 M $CaCl_2$) which are lower in the nickel serpentinites (Table 1, Fig. 2) probably because hydrolysis had already occurred during weathering.

Weathering of the serpentinite might cause a decrease in the thermal stability of their main constituent serpentine. The thermal properties of the samples were compared using differential thermal analysis and thermogravimetric analysis. DTA indicated only a small, not very characteristic decrease in the dehydroxylation temperature of serpentine from about 69o ^{o}C in the serpentinites to about 65o ^{o}C in the nickel serpentinites. This tendency was confirmed by the thermogravimetric analysis. The main weight loss began at about 5oo ^{o}C; however a small, but continuous loss was recorded over the total range between 2o and 1ooo ^{o}C. The nickel serpentinites show a more pronounced weight loss at lower temperatures. The relative weight loss in the 2o - 5oo ^{o}C interval was therefore calculated as a percentage of the total weight loss between 2o and 1ooo ^{o}C. While the relative weight loss $\Delta G_{2o - 5oo\ ^{o}C}$ of the serpentinites (with one exception) does not exceed 25 %, this percentage increases in the nickel ores (Fig. 3). A clear inverse correlation was found between $\Delta G_{2o - 5oo\ ^{o}C}$ and the MgO/SiO_2 ratio (Fig. 4). Decrease in the MgO/SiO_2 ratio reflects decreasing stability of serpentine, which in turn corresponds to an increased weight loss in the lower temperature range.

NICKEL DISTRIBUTION

The nickel distribution was studied by microprobe analysis of polished thin sections. This was correlated with other elements present, associated minerals and textural features.

Several kinds of microprobe analysis were carried out: X-Ray micrographs of the main elements, line scans and quantitative energy dispersive analysis at selected points in the thin sections. In addition, we used "microprobe correlation analysis", a very informative method for fine grained material (Schellmann, 1977). By this method, two elements are analyzed at several thousand points along a line on the thin section. The two elements are continuously analyzed by two spectrometers while the section is moved at a minimum rate of 5o μm/min along a line 1 - 2 cm long. The analyses are displayed and automatically recorded on an x-y recorder for which the printing sequence can be adjusted to suit requirements. This enables continuous correlation to be made between the two elements.

The correlation of Ni and Mg in samples of parent rock and asso-

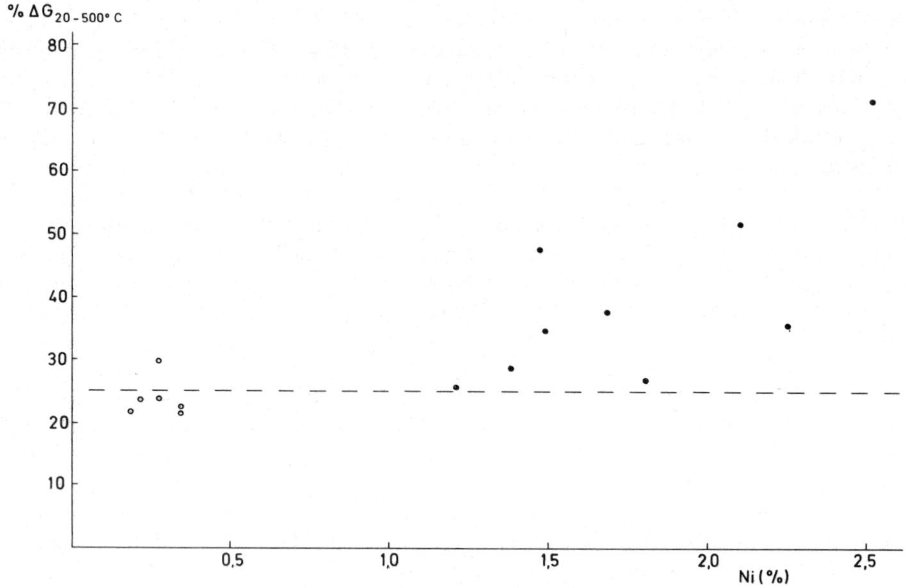

Fig. 3. Relative weight loss between 2o and 5oo ^{o}C versus Ni for serpentinites (o) and Ni-serpentinites (.)

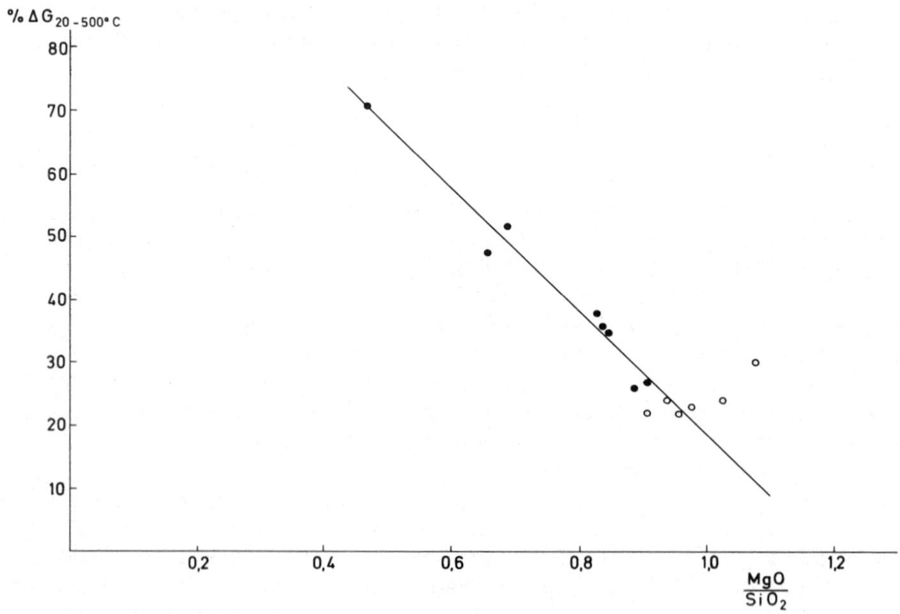

Fig. 4. Relative weight loss between 2o and 5oo ^{o}C versus the MgO/SiO_2 weight ratio in serpentinites (o) and Ni-serpentinites (.)

ciated nickel ore from Barro Alto, Brazil (Nos. 1189 and 1187, respectively) is shown in Fig. 5. The correlation pattern of the parent rock shows three clusters of points which reflect serpentine, olivine and pyroxene, the three main constituents. The differences in their Ni and Mg contents can be clearly seen. The nickel ore also contains olivine and pyroxene but no low-nickel serpentine; this is replaced by a spread of serpentine compositions with relatively high nickel content showing a steady decrease in magnesium with increasing nickel.

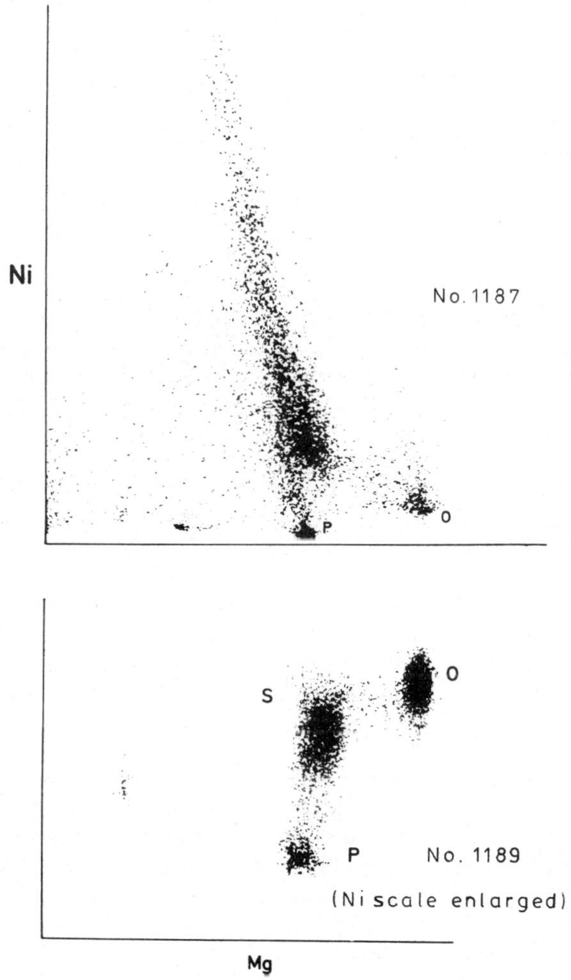

Fig. 5: Microprobe correlation analysis plots of Ni versus Mg for serpentinite (No. 1189) and Ni-serpentinite (No. 1187), both from Barro Alto, Brazil

O = olivine P = pyroxene S = serpentine

The Ni-Mg plot also demonstrates that nickel is not concentrated in few particles strongly enriched in nickel but distributed more or less evenly over the parts of the sample which consists of serpentine. Under the microscope, no significant difference between rock and ore can be detected; both exhibit the typical serpentinite mesh texture and numerous relics of olivine and enstatite.

The same nickel serpentinite sample (No. 1187) was used for microprobe correlation analysis of Ni/Si and Ni/Fe (Fig. 6), and shows no major change in silica and iron with increasing nickel concentration.

The correlation graphs of Ni/Mg, Ni/Si and Ni/Fe for a nickel serpentinite sample (No. 25) from the Philippines (Fig. 7) exhibit the same relationship observed in the nickel ore from Barro Alto. Microprobe correlation analysis of the other nickel serpentinite samples gave similar results: the nickel is contained in the serpentine, which always shows a more or less pronounced inverse correlation between Ni and Mg.

The inverse correlation between Ni and Mg, however, cannot be explained as a simple replacement of Mg by Ni, because the loss of Mg ist considerably greater than the gain in Ni. This was confirmed by quantitative microprobe analysis of all the nickel serpentinite samples.

Serpentine normally contains 14 % H_2O; the other elements of the analyzed serpentines add up to about 86 %. However, this is not the case with microprobe analyses of Ni-serpentines; with increasing Ni, the sum of NiO, MgO, FeO, Al_2O_3 and SiO_2 drops continuously to values as low as 5o %, although several exceptions were also observed. This behaviour can be explained by a very fine porosity which is higher in the nickel-enriched serpentines more strongly affected by weathering. A pronounced microporosity was confirmed by SEM.

The oxide percentages determined by microprobe were recalculated to the theoretical sum of 86 %. These converted values also clearly show the inverse correlation between Ni and Mg. This is presented in detail in Fig. 8 for the sample No. 1187 from Barro Alto and in the form of regression lines in Fig. 9 for all other samples. The atomic ratio ($-\Delta Mg/\Delta Ni$) of magnesium loss to nickel gain is tabulated for the various Ni-ore deposits in Fig. 9. It varies between 2.5 and 11.5. This indicates strong depletion of the cations in the octahedral sheets and is reflected by the decrease of the mole ratio (NiO + FeO + MgO)/SiO_2 from 1.5 in normal serpentine to about 1.o in nickel-rich serpentines.

In serpentinites with pronounced mesh texture, nickel is commonly more enriched in the serpentine of the mesh centers than in the

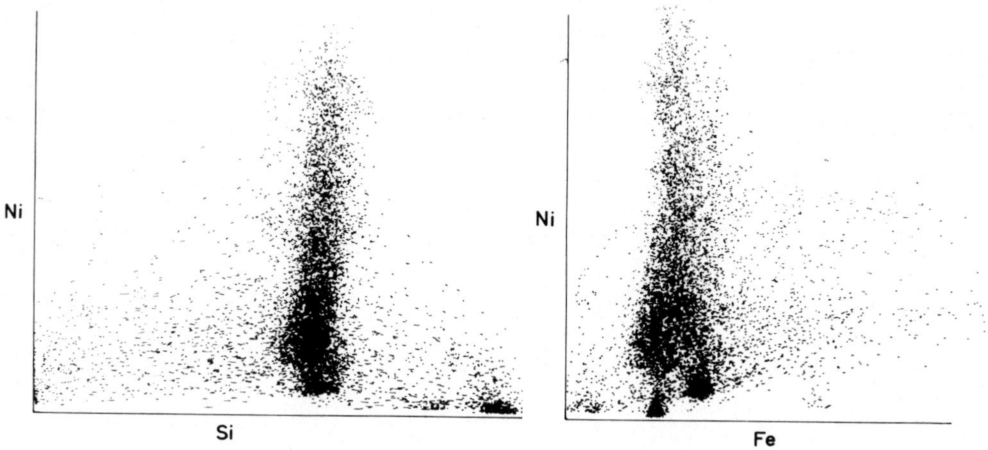

Fig. 6. Microprobe correlation analysis plot of Ni versus Si, and Ni versus Fe for Ni-serpentinite (No. 1187), Barro Alto, Brazil

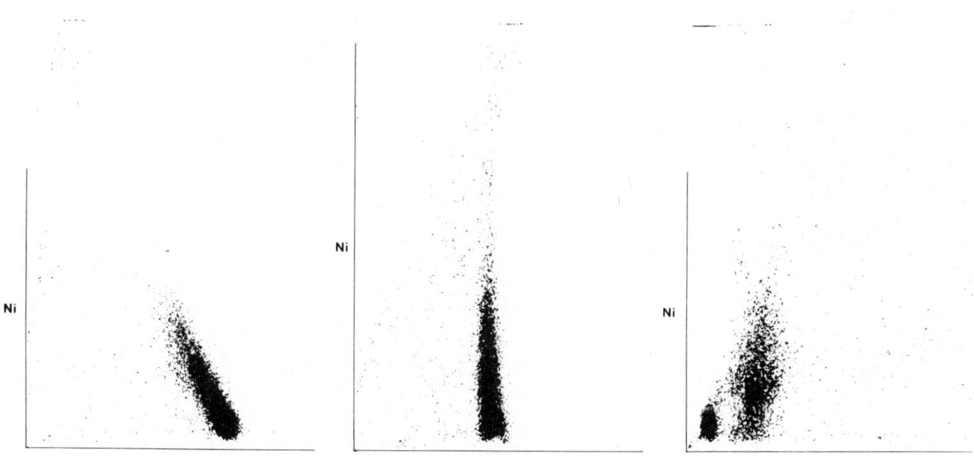

Fig. 7. Microprobe correlation analysis plots of Ni versus Mg, Ni versus Si, and Ni versus Fe for Ni-serpentinite (No. 25) from Nonoc, Philippines

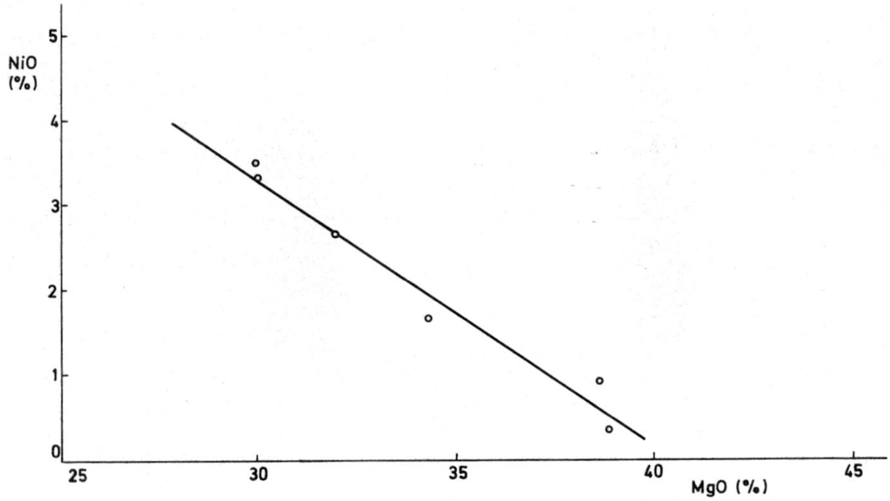

Fig. 8. Relationship between NiO and MgO in serpentine for Ni-serpentinite (No. 1187) from Barro Alto, Brazil

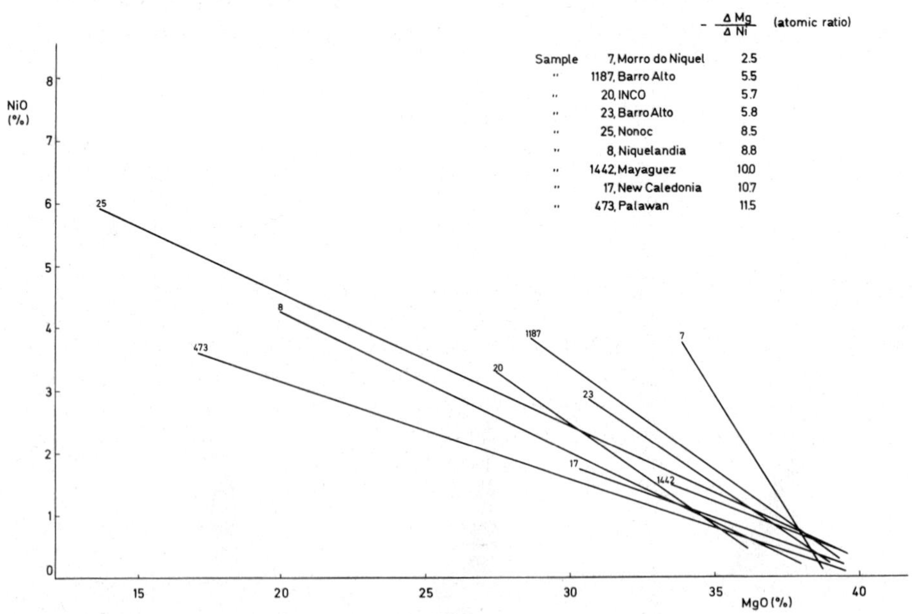

Fig. 9. Regression lines showing the relationship between the NiO and MgO in serpentine for various Ni-serpentinites. Numbers denote specimen numbers; for localities see Table 1

better crystallized serpentine forming the rims. Bastites are also enriched in nickel and show an irregular distribution of nickel within any one large pseudomorph of serpentine. They also yielded an inverse correlation between Ni and Mg.

In a few ore samples Ni-bearing smectite was detected and analyzed by microprobe. The smectite inclusions have similar Ni-conents to the serpentine and were formed by weathering of olivine and pyroxene relicts.

DISCUSSION

The question arises as to whether the Ni-bearing constituents, being highly depleted in magnesium but showing a continuous transition to normal primary serpentine, can still be called serpentine. As has already been stressed, no differences in texture between the Ni-bearing and normal serpentine were detectable under the microscope, nor could they be distinguished by their X-Ray diffraction powder patterns. On the other hand, the particles with the highest Ni and lowest Mg contents are mostly located at the centers of the serpentine meshes and are commonly more or less isotropic. We therefore suggest that under the influence of continuous weathering and addition of nickel, primary serpentine is first converted to Mg-depleted, Ni-bearing serpentine which becomes more and more unstable until it is finally decomposed to "nickel-rich amorphous constituents".

No definite conclusions can be drawn about the physical nature of the Mg-depleted serpentine. Magnesium, which is only partially substituted by Ni, may be replaced by H^+, but total compensation of the charge only appears possible if the OH^- in the octahedral sheets is partially converted to H_2O.

It is certain that nickel serpentine cannot be formed without addition of Ni in solution from the limonite zone. However, fixation of Ni^{2+} from descending solutions takes place only if the rock has already been affected by weathering, causing depletion of the Mg in the serpentines. Between the nickel limonite and the parent rock, a broad zone of slightly weathered serpentinite must exist to fix the dissolved nickel. This demands moderate weathering conditions; if the parent rock weathers too quickly, it would be directly transformed into limonite.

Continuous, but slow penetration of the rock by water is taken to be the most important factor favouring deep but moderate weathering.

Nickel smectite can only play a major role in formation of nickel silicate ores if the parent rock is relatively weakly serpentinized.

REFERENCES

Schellmann, W., 1977. Korrelationsnalysen mit der Mikrosonde am Beispiel einer pazifischen Manganknolle. Geol. Ib., D 23: 1o5 - 117.

Schellmann, W., 1978. Behaviour of nickel cobalt and chromium in ferruginous lateritic nickel ores. Bull. BRGM, Section II, 3: 275 - 282.

SMECTITES IN ACIDIC SOILS

LOWELL A. DOUGLAS, Dept. of Soils and Crops, Cook College, New Jersey Agricultural Experiment Station, Rutgers University, New Brunswick, New Jersey 08903, U.S.A.

SUMMARY

Smectites were often found in three highly acidic soils and sediments of New Jersey. These smectites have tetrahedral substitution, and some of them meet the criteria for beidellite. The formation of these smectites is highly dependent on both the type of parent material and on a soil environment that discourages the formation of aluminum polymers, i.e., a low pH or the presence of organic Al-complexing compounds.

INTRODUCTION

A viewpoint held by many clay mineralogists, and which is only partially supported by the literature, is that smectites are not formed in strongly acidic environments and that they are so unstable under these conditions that they are not found in strongly acidic environments. In this paper, two soils and one sediment in highly acidic environments, their distribution, and the smectites found in these materials are discussed.

Major summaries on stability or formation of smectites emphasize basic environments. Millot (1970, p. 305) stated: "In weathering, montmorillonite is rapidly destroyed as soon as lessivage produces sufficient desaturation of cations and silica." This statement implies that montmorillonite is rapidly destroyed in soils where leaching has been intensive enough to produce low base saturations and accompanying low pH's. Millot (1970, p. 324 and 325) limits the formation of montmorillonite to basic environments. Buol et al. (1980, p. 88), when discussing conditions of formation and the persistence of smectites states, "Montmorillonite is unstable under conditions of high hydronium concentration and rapid leaching. However, it often occurs in dense clay layers in which the rate of leaching is slow, and thus it persists under conditions of high weathering intensity when inherited from the parent material." Buol et al. imply that smectites are formed only in basic environments and are very unstable in sandy soils at low pH's. Several investigators (Barshad, 1964; Grim, 1968, p. 490) have recognized that acidic conditions favor the formation of kaolinite and basic conditions favor the formation of smectites. It is widely recognized that the stability of smectites decreases with decreasing pH (Douglas and Fiessinger, 1971; Grim, 1968, p. 435-439; Kittrick, 1971).

Several investigators (Douglas, 1965; Douglas and Trela, 1979; Gjems, 1967; Nash, 1979; Plaster and Sherwood, 1971) have, however, reported the occurrence of smectites in highly acidic (pH $<$ 4.5) soils and indicate that these smectites were not inherited from the parent materials. Vicente et al. (1977) describe the formation of smectites in the laboratory at pH's of 3.24 to 3.90. These investigations contradict the generalizations quoted above, but many authors consider that the reports of smectites in highly acidic soils or sediments represent rare or isolated occurrences. In this paper it will be shown that smectites are not rare in an acidic environment.

MATERIALS AND METHODS

The following clays that are representative of three different environments (Figure 1, Table 1) were selected for detailed evaluation:

Table 1.

Sample Characteristics

Parent Material	Sample Location	Soil Classification	Soil Series	pH (water)
Diabase	B2t horizon	Ultic Hapludalf	Neshaminy	4.2
Glauconitic Sediments	Black Marl	--	--	3.6
Cohansey Sand	A1 Horizon	Spodic Quartzipsamment	Lakewood	3.9

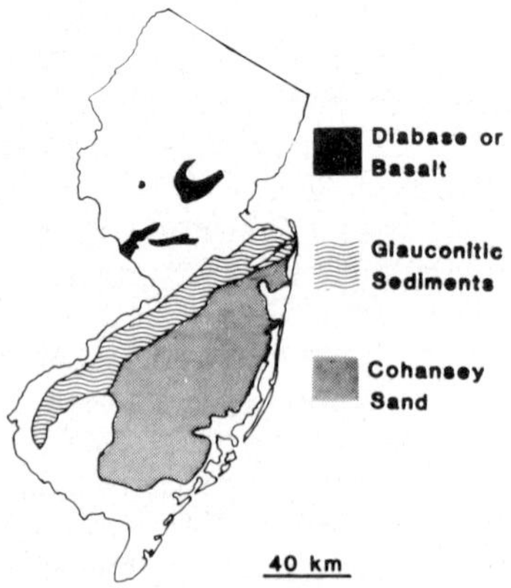

Figure 1. Soil parent materials upon which smectite bearing soils form in New Jersey.

(1) Smectites occurring throughout the profiles of Ultic Hapludalfs which were formed on both the basalts and diabases of Central New Jersey. (2) Smectites which are common in the A and B horizons and upper portions of C horizons of soils with moderate to large amounts of greensand (glauconite) on the inner coastal plain of New Jersey. Within this area sedimentary facies sometimes contain considerable amounts of organic matter and pyrite, and are locally called "black marl." These black marls typically have pH values between 1.5 and 4, which is considerably below the pH of the soils developed on typical greensand containing

sediments. Smectites derived from these black marls will be discussed. (3) Smectites, in soils developed on the Cohansey sand, the major geologic formation of New Jersey's outer coastal plain (silt plus clay < 5%). Spodic Quartzipsamments (Podzols) have formed on these sands. These soils have a pH range of 3.7 to 4.4. Smectites are common in A1 and A2 horizons, but not in the lower horizons of these soils. Ross and Mortland (1966) found beidellite in a similar soil in Michigan.

Samples were air dried and passed through a 2 mm. sieve. Organic matter (black marl and Quartzipsamment samples) was destroyed with sodium hypochlorite (Clorox), and excess soluble salts were removed by filtering and washing with distilled water. The samples were dispersed with sodium hexametaphosphate and the clay fraction (<2 μm) was separated by centrifugation and dried by freeze-drying. Easily reducible iron was removed with citrate-bicarbonate-dithionite (Mehra and Jackson, 1960).

Smectite species were identified by treatment with ethylene glycol or glycerol (Harward and Brindley, 1965) and with molten lithium nitrate at 270° C (Greene-Kelly, 1952). To study the effects of these treatments on expansion, oriented samples for X-ray diffraction were prepared by transferring clays from membrane filters to glass slides (Drever, 1973). Ethylene glycol was adsorbed from the vapor phase (Kunze, 1955) and glycerol from a 1:4 (glycerol:water) solution.

RESULTS AND DISCUSSION

In Fig. 2a, b and c, X-ray diffractograms are shown for the treated clays from the B2 horizon from an Ultic Hapludalf developed from diabase (Fig. 2a), from a highly acidic glauconitic (black marl) sediment (Fig. 2b), and from the A1 horizon of a Spodic Quartzipsamment (Fig. 2c). The patterns of the ethylene glycol-treated clays are typical of those for many samples from the areas shown in Figure 1. In addition to smectite, these clays contain vermiculite (usually dioctahedral with aluminum-interlayers), kaolinite, and mica.

Harward and Brindley (1965) prepared synthetic beidellites and montmorillonites and showed that the expansion characteristics of the beidellites were intermediate between vermiculite and montmorillonite. The expansion of the smectites from the acidic sediment and from the A1 horizon of the Spodic Quartzipsamment with glycerol is similar to that of Harward and Brindley's montmorillonites while the smectite in the B2 of the Hapludalf admits only one layer of glycerol in the interlayer space, similar to the behavior of Harward and Brindley's beidellites.

Greene-Kelly (1952) showed that montmorillonite could be distinguished from beidellite or nontronite by heating a lithium clay. When a smectite has tetrahedral substitution, lithium does not enter vacant octahedral positions. Upon glycerol treatment, a lithium montmorillonite shows an (001) diffraction line at 9.5A while beidellite or nontronite will expand to 17.9A. All of the smectites in our samples expanded with glycerol after treatment with molten lithium nitrate, indicating that all have tetrahedral substitution.

The smectites in the soils studied were formed in place. Although it cannot be shown directly that the smectite in the black marl was formed in place, other data show that the smectites in highly acidic

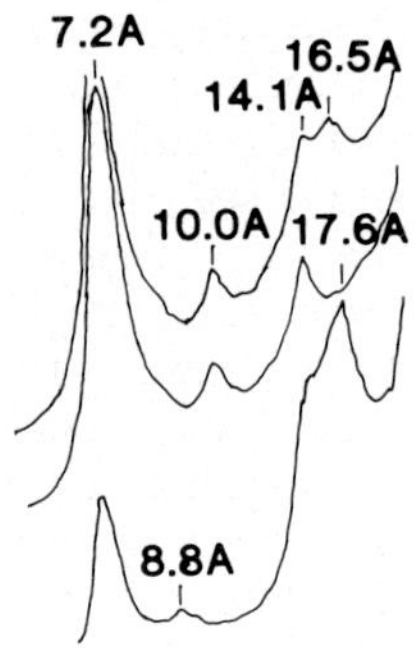

Fig. 2a

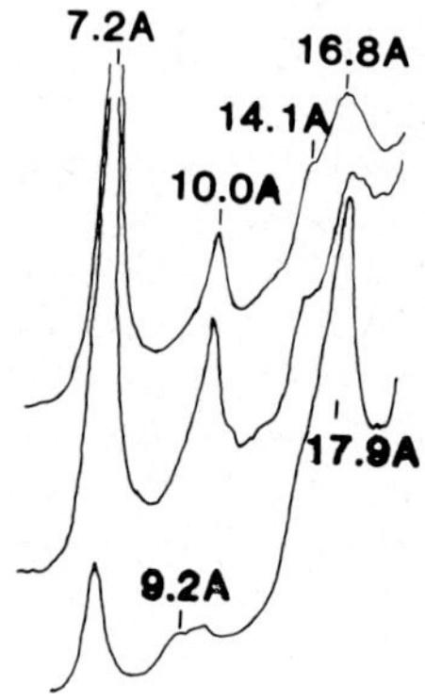

Fig. 2b

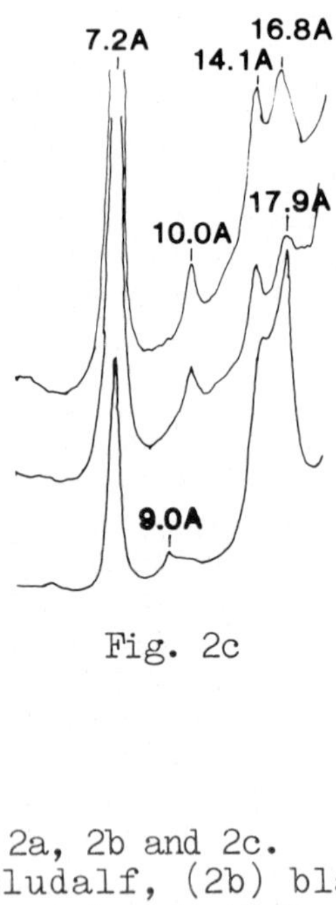

Fig. 2c

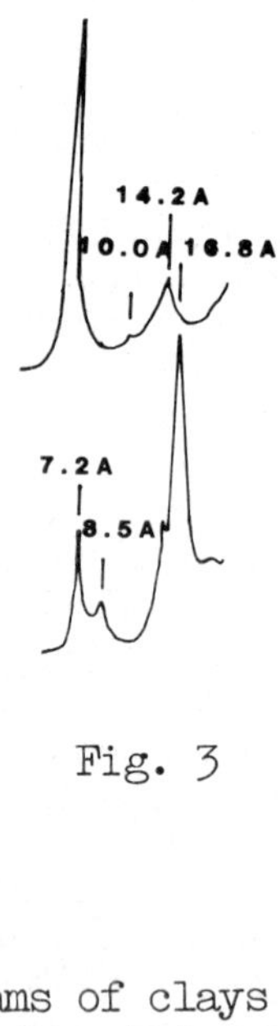

Fig. 3

Figure 2a, 2b and 2c. X-ray diffractograms of clays from the (2a) Hapludalf, (2b) black marl and (2c) Quartzipsamment. Top, Mg saturated and ethylene glycol solvated; middle, Mg saturated and glycerol solvated; bottom, molten $LiNO_3$ and glycerol solvated

Figure 3. X-ray diffractograms of clay from the B2 horizon of soil developed on: top, granite gneiss, and bottom, biotite gneiss. Clays were Mg saturated and ethylene glycol solvated.

soils formed on glauconitic sediments are formed in place. This strongly suggests that the smectites in the black marls are not inherited.

The environments in which these smectites were formed are highly acidic, pH below 4.3. Hsu (1977) pointed out that at these low pH's one would not expect polymeric aluminum to form. Considering the clays in soils on the Cohansey sands, aluminum-interlayered dioctahedral mica is a major component of the B2 and C horizons of the Spodic Quartzipsamments, and of most other soils of New Jersey. Smectites are found in the A horizons of these Podzols only when mica is present in the parent material. Gjems (1967) studied many Podzols (Spodosols) and found a similar clay mineral distribution, smectites in A horizons and aluminum-interlayered vermiculites in B horizons. He found that the amount of smectite increased as pH decreased and as organic matter in the A horizons increased. This implies that smectite is formed in these very sandy soils at low pH's when aluminum is complexed by organic matter so that the aluminum is not available in large amounts for formation of polymers. In New Jersey smectites do not constitute an important portion of the clay fraction of soils with low pH's formed on acidic (siliceous) parent materials unless there has been considerable translocation of soluble organic matter within the soil.

The environment of smectite formation in both the Ultic Hapludalf and the black marl is somewhat different from that of Spodic Quartzipsamments. The pH values of the Hapludalf and black marl are in the range that would suppress polymer formation. On these parent materials the low pH's should result in a comparatively high concentration of mafic elements, in the soil solution, for formation of smectites. The importance of parent material is illustrated in Figure 3. These clays are from the B2 of Paleudults from Lousia County, Virginia. Parent materials are upturned gneisses. The two soils are only 23 meters apart. The clay fraction of the soil formed on granite gneiss is dominated by kaolinite while the clay of the soil formed on biotite gneiss is dominated by smectite. While high soil organic matter levels, as well as low pH's, are essential for smectite formation in Quartzipsamments, only low pH's and a leaching environment are needed for smectite formation in soils on mafic parent materials.

The areas of New Jersey where one would expect to find smectites in highly acidic soils or sediments are shown in Figure 1. Beidellite (Nash, 1979; Ross and Mortland, 1966) has been reported in similar soils at other locations. The occurrence of smectites with trioctahedral substitution in acidic soils is a rather common phenomenon.

ACKNOWLEDGMENT

Paper of the Journal Series, New Jersey Agricultural Experiment Station, Cook College, Rutgers University, New Brunswick, New Jersey. This work was performed as a part of NJAES Project No. 15107, supported by the New Jersey Agricultural Experiment Station.

REFERENCES

Barshad, I., 1964. Chemistry of Soil Development. In: Bear, F., Ed., Chemistry of the Soil. Reinhold, New York. p. 1-70.

Buol, S.W., Hole, F.D. and McCracken, R.J., 1980. Soil Genesis and Classification, Second Edition. The Iowa State University Press, Ames. 404 pp.

Douglas, L.A., 1965. Clay mineralogy of a Sassafras soil in New Jersey. Soil Sci. Soc. Amer. Proc. 29: 163-167.

Douglas, L.A. and Fiessinger, F., 1971. Degradation of clay minerals by H_2O_2 treatments to oxidize organic matter. Clays Clay Min. 19: 67-68.

Douglas, L.A. and Trela, J.J., 1979. Mineralogy of Pine Barrens soils. In: Forman, R.T.T., Ed., Pine Barrens - Ecosystem and Landscape. Academic Press, Inc., New York. p. 95-109.

Drever, J.I., 1973. The preparation of oriented clay mineral specimens for X-ray diffraction analysis by a filter-membrane peel technique. Amer. Mina. 58: 553-554.

Gjems, O., 1967. Studies on clay minerals and clay-mineral formation in soil profiles in Scandinavia. Medd. Nor. Skogforsoksven. 81: 303-415.

Greene-Kelly, R., 1952. A test for montmorillonite. Nature 170: 1130.

Grim, R.E., 1968. Clay Mineralogy, Second Edition. McGraw-Hill Book Company, New York. 596 pp.

Harward, M.E. and Brindley, G.W., 1965. Swelling properties of synthetic smectites in relation to lattice substitutions. Clays Clay Min. 13: 209-222.

Hsu, Pa Ho, 1977. Aluminum hydroxides and oxyhydroxides. In: Dixon, J.B. and Weed, S.B., Eds., Minerals in the Soil Environments. Soil Sci. Soc. Amer., Madison. p. 99-143.

Kittrick, J.A., 1971. Montmorillonite equilibria and the weathering environment. Soil Sci. Soc. Amer. Proc. 35: 815-820.

Kunze, G.W., 1955. Anomalies in the ethylene glycol solvation techniques used in X-ray diffraction. Clays Clay Min. 3: 88-93.

Mehra, O.P. and Jackson, M.L., 1960. Iron oxide removal by a dithionite-citrate system buffered with sodium bicarbonate. Clays Clay Min. 7: 317-327.

Millot, G., 1970. Geology of Clays. Springer-Verlag, New York. 429 pp.

Nash, V.E., 1979. Mineralogy of soils developed on Pliocene-Pleistocene terraces of the Tombigbee River in Mississippi. Soil Sci. Soc. Amer. Proc. 43: 616-623.

Plaster, R.W. and Sherwood, W.C., 1971. Bedrock weathering and residual soil formation in Central Virginia. Geol. Soc. Amer. Bull. 82: 2813-2326.

Ross, G.J. and Mortland, M.M., 1966. A soil beidellite. Soil Sci. Soc. Amer. Proc. 30: 337-343.

Vicente, M.A., Razzaghe, M. and Robert, M., 1977. Formation of aluminum hydroxy vermiculite (intergrade) and smectite from mica under acidic conditions. Clay Miner. 12: 101-111.

THE ROLE OF CLAY MINERALS IN ENVIRONMENTAL SCIENCES

M. L. JACKSON and C. H. LIM
Soil Science, College of Agricultural and Life Sciences, University of Wis., Madison, USA

INTRODUCTION

The purpose of this paper is to focus on the vital role clays play in the environmental sciences, encompassing the chain: soil clays, plants, animals, and human beings and their health. Increasing intensity of agricultural management has impoverished some essential trace elements in soil, while increased industrialization has raised some toxic trace elements, organic toxins, and radioactive elements to acute levels in some environments. Clays provide ion adsorption and exchange to roots and thus are the basis of agriculture and the food and fibre supplies (Mortvedt et al., 1972).

CLAYS IN AGRICULTURAL ENVIRONMENTS

Clays in soils adsorb both major and trace elements which are taken up by plants through desorption from the clays. As crop yields have been pushed to higher levels under intensified agriculture (Table 1) by applications of major elements and by irrigation (Evans, 1980), a greater removal of trace elements from soils occurs (Walker and Peck, 1980). Where needed for plant growth, ground limestone and essential major and trace elements are applied to soil. Liming lowers the toxicity of aluminum in acid soils to plants and human

TABLE 1

As intensive soil management has increased crop yields, a lower supply of trace elements not applied as fertilizer, such as Se from clay, results. Crops take up Se, but do not require it.

	Highest yields[a] (bushels/acre)	
	Corn	Soybeans
Theoretical	580	165
Farmer	352	104
Researcher	312	99
U.S.A., average	109	32

[a]1980. Foundation for Agronomic Research, 2801 Buford Highway, NE Atlanta, GA 30329, U.S.A.

beings (Liss, 1980; Jackson and Huang, 1981). Excess nitrates appear in groundwater under feedlots (Keller and Smith, 1967). Element accumulator plants take up large amounts of elements such as Al and F (tea) from acid soils or Se from neutral soils (lupines), to nutritionally toxic levels (National Academy of Sciences, 1974).

With increasing industrialization and urbanization, environmental contamination of soils, by disposition of wastes thereon has become of increasing concern in agriculture. Lead, Hg, and Cd are carried in wastes (Lagerwerff, 1972). The possible toxicity of bioorganic and inorganic trace elements in wastes applied to soils must be considered (McCalla et al., 1977; Allaway, 1977). Soils, particularly the clays in soils, to some extent adsorb excess contaminants and interrupt their flow into the food chain, provided proper limitations of total applications, pH adjustments by liming, and other precautions (such as attention to element ratios) are taken. Constant monitoring is required, lest elements become deficient, toxic, or attain unfavorable ratios to one another in the plant-animal-food-human nutrition-health linkage (Mortvedt et al., 1972; Underwood, 1979).

Besides C, H, and O from air and water, N, P, K, Ca, Mg, S, and B are commonly added to soils for plant use. Most, but not all, of the trace elements needed by animals are also needed by plants. The 16 trace elements Fe, Cl, I, Cu, Mn, Zn, Co, Mo, Se, Cr, Ni, Si, Sn, F, V, and As are accepted as essential to animals (Underwood, 1977, 1979) and perhaps others such as Cd likewise may be essential (Mertz, 1981). The element, Si (second most abundant element in the earth's crust, but considered a trace element in animal nutrition), functions in stiffening stems (plants) and cross-linking in elastin (animals). One trace element may serve in one, several, or dozens of different metalloenzymes or tissue constituents (Hoekstra et al., 1974; Schrauzer, 1979).

<u>Depletion of essential elements from soils</u>

Intensive cropping lowers the content of soil clay-held elements that are taken up by plants. Also, the clay fractions of soils is selectively eroded, leaving coarser particles (Massey and Jackson, 1952), which are less active nutrient suppliers. Although lime and fertilizers are applied as needed for elements essential to crop plants, feed crops for livestock may not supply enough of certain trace elements essential to animals, such as Co, Zn or Se. For example, direct supplementation of Se in livestock eliminated the sheep disease (myopathy) of the white muscle (WMD) in Oregon, U.S.A. (Muth et al., 1958; Rosenfeld and Beath, 1964) and is being used for this purpose in New Zealand (Hupkens van der Elst and Watkinson, 1976). Several elements needed by animals are being supplied as supplements in animal feed (Allaway, 1975). Serious soil-based trace element deficiencies also appear in human beings (Hoekstra et al., 1974; Underwood, 1979; Schrauzer, 1979; Schamberger, 1980; Mertz, 1981).

<u>Sources of selenium for animals</u>

Direct application of compounds of the element Se to soils is a fairly inefficient way to supply livestock because of high cost and the strong retention of $SeO_3^=$ and $SeO_4^=$ by clay (Cary and Allaway, 1969); the $Fe(OH)_3$-$HSeO_3$ form (colloidal iron-fixed selenite) applied to soil is more available to alfalfa than soil-adsorbed water-soluble selenite. Plant-available selenium in volcanically-derived soils increases as the soil clay content increases during hundreds of years of weathering (Watkinson, 1981). Selenium accompanies sulfur in volcanic rocks and slowly becomes available to plants as the rocks weather. Clays contain a wide variety of trace elements, but Se is deficient for livestock in two-thirds of the U.S.A. (Kubota et al., 1967). Copper sulfide ore emplacements contain Se and these ores are the principal source of Se supplements (Oldfield et al., 1974, p. 57).

GEOCHEMICAL ABUNDANCE OF ELEMENTS IN THE ENVIRONMENT

A natural geochemical variation occurs in the supply of elements in various areas of the earth, with long-term cycling between land and sea (Fig. 1a). Diversification of food sources and supplementation are the main methods available to overcome element imbalances and deficiencies in the food supply of a particular soil area (Fig. 1b). For example, iodine has been supplied for decades to overcome

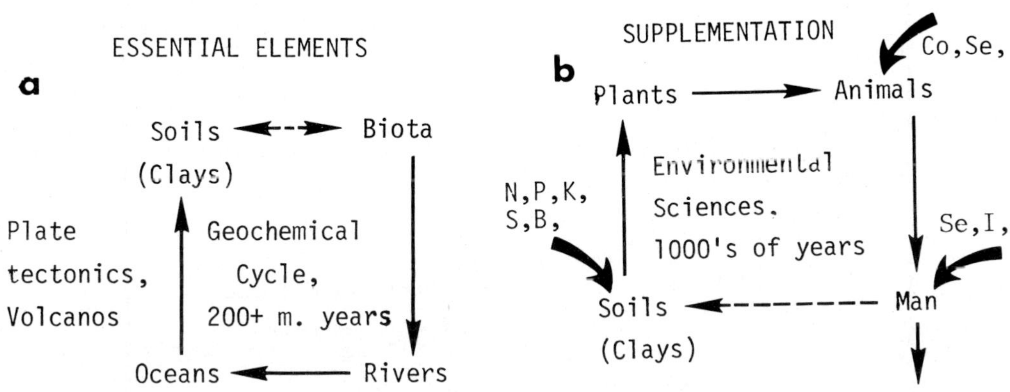

Fig. 1. Essential elements for human beings: (a) geochemical cycle of crustal elements, largely closed on a 200-million year scale; (b) supplementation needed on the environmental sciences scale of a few thousands of years or less during intensifying agriculture.

the thyroid disease known as "goiter" which frequently occurred in some areas (Fleischer et al., 1974, p. 27). This disease was virtually eliminated in developed countries by widespread inclusion of a form of iodine in common salt, but some 200 million people are still affected by goiter in vast areas where supplementation is not applied (Shamberger, 1980). In the past several decades, deficiency for animals and human beings of selenium adsorbed on soil clays has been recognized, as documented above.

Environmental anticarcenogenicity

Schrauzer (1979, p. 228) believes that "susceptibility to cancer development is enhanced by Se deficiency", analogous to goiter occurrence with iodine insufficiency. Selenium was found to decrease carcinogenicity of *p*-dimethylaminoazobenzene (Clayton and Baumann, 1949). An inverse correlation ($r = -0.8$; $p < 0.0001$) was found between human breast cancer and Se intake in 27 countries (Fig. 2). Similar inverse correlations were found between leukemia and cancer of the intestine, rectum, ovary, prostate, lung, skin, and bladder with Se intake or Se levels in blood (Schrauzer, 1979) within

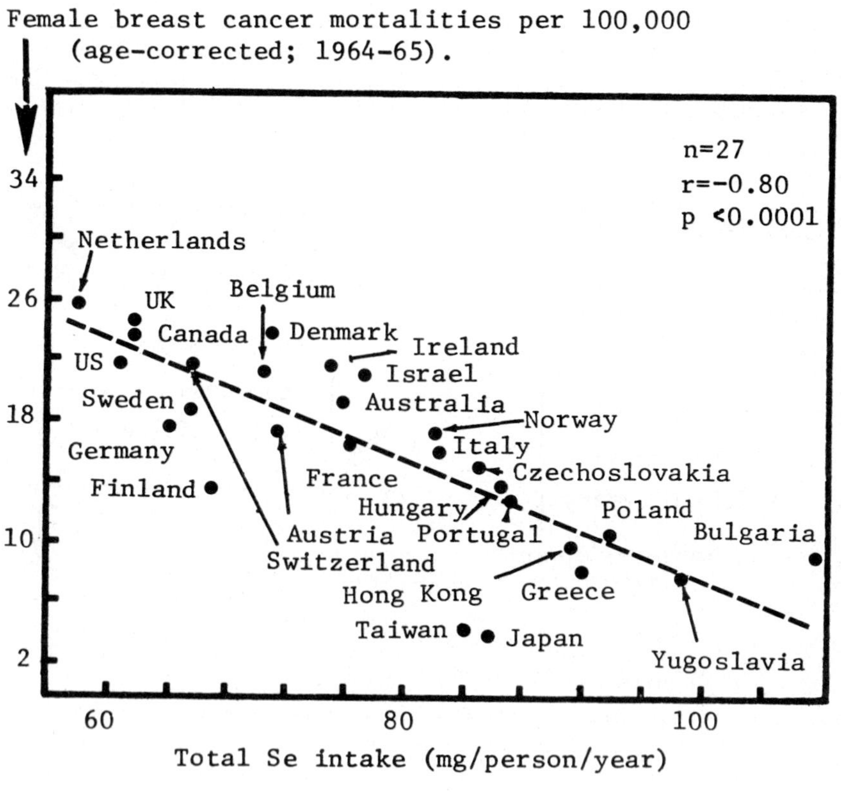

Fig. 2. Female breast cancer mortalities per 100,000 (age-corrected, 1964-1965). Most other cancers follow the same trend. After Schrauzer et al. (1977).

countries and among countries. More than a 10-fold variation in cancer-death rate is indicated (Fig. 2). The possible actions of Se as an anticarcinogen, particularly at Se levels in blood of 0.3 or 0.4 μg/cm^3 or slightly above this level have been summarized by Schrauzer as "cell environmental protection". Such protective actions are, for example, antagonism of carcinogenic actions of toxic elements such as Cd, aid in the metabolism of organic carcinogens such as aflatoxin B_1, retardation of cell division so that the action of DNA repairase is completed where needed, enhancement of interferon production, and stimulation of phagocytosis. The statistics are remarkable when the 20-fold variations of Se content in a given food and in blood levels are considered (Underwood, 1979, p. 236). A positive Chi square association was observed between blood bank Se and forage crop Se content, a measure of available Se levels in soil clays (Table 2).

TABLE 2

Selenium in human blood, and feed or food.

Area	Blood Se (μg/cm^3)	Blood Se rank	Forage or food Se rank
Keshan, Manchuria	0.01[a]	Damaging[a]	Deficient
New Zealand	0.07[b]	Low	Low
Illinois	0.12[c]	+[c]	+[c]
Ohio	0.16[c]	+	+
California	0.16[c]	+	++
Wyoming	0.23[c]	+++	+++
Japan	0.28[e]	++++	High[d]
Villa Bruzual, Venezuela	0.7-1.5	Toxic[f]	--

[a]Heart disease beginning with children reported by <u>Academia Sinica</u> (Cheng et al., 1980; Frost, 1981).
[b]Watkinson (1974); range, 0.02-0.14 (Underwood, 1979, p. 236).
[c]Shamberger et al. (1979), after Kubota et al. (1967).
[d]From ocean fish foods (Schrauzer et al., 1977).
[e]Schrauzer et al. (1977).
[f]Toxicity limited to fingernails (Jaffé et al., 1972; Schrauzer, 1979, p. 227).

Environment and disease

Children with severe Se deficiency (Keshan disease) were identified in two soil areas in China with endemic human heart disease (Cheng et al., 1980) and Se levels in blood of 0.01 μg/cm^3 or less were reported for the children in these areas (Frost, 1981).

Favorable response to supplementing Se in the form of selenite in large populations reduced the disease to near zero in a 5-year experiment according to Chen Xioashu (Shamberger et al., 1979; Shamberger, 1980). The age-specific heart death rate (separately for hypertensive, coronary infarct, and cardiovascular-renal types) per 100,000 persons age 55-64 has a significant inverse relationship ($r = -0.8$; $p<0.001$) to the Se content of the blood in blood banks in 19 states in U.S.A. (Shamberger et al., 1979). Earlier, hypertensive persons were shown to have a 50-fold higher Cd content in urine than persons free of hypertension (Perry, 1969). A high Se content of blood may protect against toxicity effects of Cd (Chen et al., 1975).

A sufficiency of selenoglutithione peroxidase and sufficient vitamin E are required for optimal metabolism of polyunsaturated fats in human diets (Hoekstra, 1974). Selenoglutithione peroxidase activity increased rapidly with increase of Se in blood of New Zealanders in the range from 0.02 to 0.14 $\mu g/cm^3$ of Se (Underwood, 1979). Supplementation of Se (Se-yeast type) raised the average Se level in human blood from 0.17 to 0.23 $\mu g/cm^3$ (Schrauzer and McGinnes, 1979) without affecting the level of selenoglutithione peroxidase activity (Schrauzer and White, 1978).

In Jamaica, persons suffering from the protein deficiency, kwashiorkor, responded to selenoaminoacid administration (Schwartz, 1961). Cystic fibrosis in young children has been reported to be a result of Se deficiency (Wallach and Garmaise, 1979).

At the present time Se is likely to be deficient in foods from countries with exhaustively cropped land (China) or with a highly productive agriculture (the U.S.A., New Zealand, and other developed countries). Elimination of a large fraction of cancer, heart deaths, and perhaps other diseases of human beings could result from intake of adequate amounts of Se in food and supplements (Schrauzer, 1979; Shamberger, 1980). Selenium intake in the food in the U.S.A. ranges from 100 to about 160 μg/day for persons with about 0.15 $\mu g/cm^3$ Se in their blood. Yeast Se supplementation for human beings (provisionally proposed at 50 to 200 μg/day of Se by the National Academy of Sciences) is available commercially in the U.S.A., usually combined with vitamin E (Pechter, 1981). A total intake of Se of somewhat over 300 μg/day (Fig. 2 extrapolates to zero cancer at 330 μg/day of Se) results in a Se content in blood of 0.3 to 0.4 $\mu g/cm^3$, which has been suggested as protective and commensurate with good health (Schrauzer, 1979). The Japanese Se intake is almost adequate (Table 2) through their high ocean-fish diet, and their cancer rate is less than 0.1 of that in most industrial countries (Fig. 2).

The immunocompetence role of Se is not unique. For example, experiments on the effects of Cu-deficiency in mice showed (Prohaska and Lukasewyca, 1981) that Cu, like certain other trace elements, was necessary ($p < 0.01$) for immunocompetence (production of antibodies to foreign protein, i.e. rejection of sheep red cells). Association of Cu deficiency with increased susceptibility to bacterial infections and many physiological abnormalities had been recognized earlier

(Underwood, 1979). In arid lands such as Iran, with highly calcareous soils, Zn deficiency is observed in human beings (Mertz, 1981).

Balance between the elements

The balance between major and trace elements is of key importance in agriculture (Olsen, 1972) and in human beings (Levander and Cheng, 1980). Although trace amounts of Ni and perhaps of Cd are biologically essential, these elements may be carcinogenic at higher levels unless balanced by other trace elements (Schrauzer, 1979; Mertz, 1981). These elements, as well as the toxic elements Hg (Ganther et al., 1972) and Pb, are counteracted by Se and they are also excreted in the urine by exposed persons, particularly as the PbEDTA chelate (Tsuchiya, 1971). In contrast, Cd accumulates in body parts such as bone of the young ("Itai, Itai"; "ouch, ouch" disease in Japan) and in lung, liver and kidney leading to a health-damaging Cd-proteinuria (protein that can precipitate in urine).

A reciprocal relationship known as nutrient antagonism is found between Cd and Se (Helmke et al., 1979). The total Se content of a Wisconsin soil was 0.3 μg/g; applications of 15 to 60 metric tons/hectare of sludge (Se content, 0.5 μg/g and Cd content 102 μg/g) resulted in a progressive increase of the Cd content of earthworm tissue from 7 μg/g (control) to 118 μg/g. The Se content of the earthworm decreased from 0.2 to 0.1 μg/g. Conversely, a higher Se intake caused a redistribution of Cd in tissue probably providing some protection against the toxic effects of Cd (Chen et al., 1975).

Cadmium and lead contents of old compared to modern glacial ice showed several-fold increases from 1200 AD to 1965-1972, particularly in the industrialized areas of Poland and Norway and was lower in ice of the Himalayas of Nepal (Jaworowski et al., 1977). Car tires and Pb from gasoline are indicated as main sources. Lead intake by humans has reached subacute to marginally acute toxic levels of 30 to 150 μg/day in the U.S.A. (Settle and Patterson, 1980). Increasing Se intake will provide an important protection against toxic metals mobilized by industrialization (Frost, 1981).

The balance between available Mg in soils and other elements is important. Low Mg and low K/(Ca+Mg) ratios in vegetation often result in fatal grass tetany in cattle, especially in cool regions and in cool seasons (Kubota et al., 1980). In Michigan a low Mg to Ca ratio (calcitic limestone) causes a greater incidence of not only grass tetany but also of fluorosis in human beings (Walbott et al., 1978). In contrast, treatment of acid soil with pulverized dolomitic limestone is common in Wisconsin. Available Mg in soils (adjusted with $MgSO_4$ and $MgCO_3$) in association with Al colloids in acid soils (Ultisols and Oxisols) of Georgia and of South Africa became unavailable when they were limed with CaO (Grove et al., 1981). An inverse relationship was found between Mg content in soils and human cancer deaths (mostly epithelioma) in various regions of France (Delbet et al., 1951). Magnesium deficiency has been recognized to be related to human cardiovascular, skeletal, and renal abnormalities including osteoporosis and tissue calcification (Seelig, 1980). In senile dementia of the Alzheimer type, Al accumulates in neural fibrillary tangles of the brain as a co-factor with unidentified biological factors (Liss, 1980). A high fluoride content of drinking water yields high statistics ($p < 0.0001$) against senile dementia compared to control areas with low-fluorine drinking water (Still and Kelley, 1980). Certain fatal central nervous system diseases, are endemic in areas where Al in food from acid soils and other sources, with a low Mg content are combined, as summarized by Liss (1980) and by Jackson and Huang (1981).

Clay, the environment and longevity

The elements Pb, Hg, Ni, Cd, Mg, Ca, Sr, and Ba are adsorbed by clays and by hydrous oxides in soils (Kinniburgh et al., 1977) when released to the environment as a result of man's industrial activities. In the specific industries, these elements enter human beings directly. They also move up the food chain via soil and waters, for example, Cd in sewage sludge applied to soils is accumulated particularly in leafy parts of vegetables such as lettuce (Giordano and Mays, 1977).

A reflection on where life developed over time is helpful in understanding why adequate Se, I, and Mg (with low Al) are so essential in nutrient balance promoting longevity (Table 3). Life

TABLE 3

Evolutionary periods leading up to the present intensive agriculture.

Time[a]	Event	Episode
100+ y.	NPK fertilization (Se DEFICIENCY)	late Holocene
12,000 y.	Civilization, agriculture	late Pleistocene
15[b] m.y.	Early human forms	Miocene plus
250 m.y.	Animals to land, 12% O_2	Permian
420 m.y.	Plants to land	Silurian
3800[c] m.y.	Photosynthesis banded Fe_2O_3	Precambrian

[a]y = years; m.y. = million years.
[b]Lovejoy (1981).
[c]Cloud (1980).

developed nine-tenths of the way in the sea, where Se, I and Mg were abundant and played an important role in the development of characteristics, including immunocompetence and neural health. Banded iron formations in Greenland, 3,800 million years old, contain evidence of photosynthesis by one-celled organisms (Breuer, 1978, p. 52). During the Silurian period, 420 million years ago, photosynthesizing plants were established on land and by 350 million years ago the buried carbon age began so that free O_2 could accumulate.

The warm blooded animals appeared on land during the Permian, 250 million years ago, when the O_2 level was about two-thirds that of the present. By the time civilization arrived, life had passed more than 99.99% of its existence with little more than minor adverse effects on soil Se, I, and Mg supplies or of increases of Al contents in acid soils. Nor had much accumulation occurred of environmentally damaging metal elements and organic substances (Fig. 1a), during the 15 million years since the origin of mankind (Lovejoy, 1981). Simultaneously, DNA repairase support for longevity was ever improving, according to George Sacher of Argonne National Laboratory (Lewin, 1981).

Longevity statistics (Fig. 3) vary geographically (Sauer, 1980; National Research Council, 1981) with the corresponding soil clay environments. Moreover, the highest death rates by stroke (Sauer,

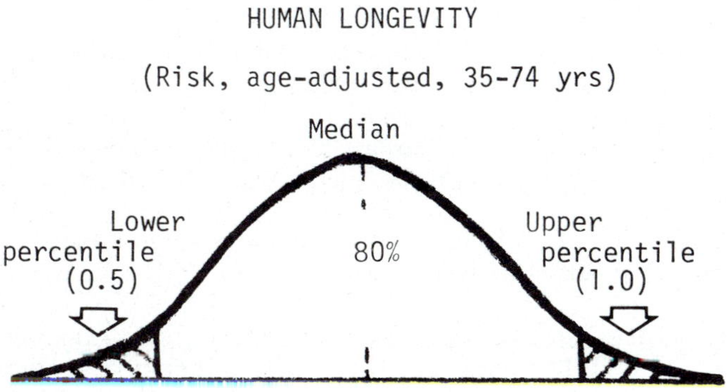

Fig. 3. Human longevity [1-(risk of death)] in U.S.A. varies relatively from 0.5 for the lowest percentile with mainly highly weathered soils in the Southeast to 1.0 in the upper percentile in more nearly neutral soils in the central High Plains (higher clay-associated available Se from weathered pyroclastics). The area under the curve represents longevity for white males (age-adjusted, 35-74 year age group).

1980, p. 40) are concentrated in southeastern U.S.A. where high Al, low-Si clays predominate and the soils are acid. Studies on the elimination from the food chain of excess Al by liming acid soil clays; the control of environmental imbalances of Mg/Ca, P, and F; and the adequate supplies of Se, I, Cu and Zn and the other trace elements and their ratios in soils, waters, and foods should be fostered by further interdisciplinary environmental research.

ACKNOWLEDGEMENTS

This research was supported in part by the College of Agricultural and Life Sciences of the University of Wisconsin, under project 1123; and in part by the National Science Foundation, EAR-76-19783-Jackson; through an International Consortium for Interinstitutional Cooperation in the Advancement of Learning (ICICAL). Appreciation is extended to W. D. Keller, and J. G. Bockheim for their valuable suggestions and to Mrs. C. M. Jackson for her care in editing and typing the manuscript.

REFERENCES

Allaway, W. H., 1975. The Effects of Soils and Fertilizers on Human and Animal Nutrition. Agric. Inform. Bull. 378. U.S.D.A., Washington D.C.

Allaway, W. H., 1977. Food chain aspects of the use of organic residues. Soils for Management of Organic Waste Waters. Eds., L. F. Elliott and F. J. Stevenson, pp. 281-298. Am. Soc. Agronomy, Madison, WI 53711.

Breuer, G., 1978. Air in Danger. Cambridge University Press, Cambridge.

Cary, E. E., and Allaway, W. H., 1969. The stability of different forms of selenium applied to low-selenium soils. Soil Sci. Soc. Am. Proc., 33:571-574.

Chen, R. W., Whanger, P. D. and Weswig, P. H., 1975. Selenium-induced redistribution of cadmium bonding to tissue proteins: A possible mechanism of protection against cadmium toxicity. Bioinorg. Chem., 4:125-133.

Cheng, B., Ju, S., Yue, S., He, R. and Sheng, S., 1980. Studies on the selenium in soils of northeastern China. Acta Pedologica Sinica, 17:56-61.

Clayton, C. C. and Baumann, C. A., 1949. Diet and azo-dye tumors. Effect of diet during a period when the dye is not fed. Cancer Res., 9:575-582.

Cloud, P., 1980. Beyond plate tectonics. Am. Scientist, 68:381-387.

Delbet, P., Depeyre, G. P. and Heinemann, H., 1951. Appauvrissement du sol en magnesium et ses conséquences. Bull. Acad. Natl. Med. (Paris), 135:169-173.

Evans, L. T., 1980. The natural history of crop yield. Am. Scientist, 68:388-397.

Fleischer, M., Forbes, R. M., Harriss, R. C., Krock, L. and Kubota, J., 1974. Iodine. Geochemistry of the Environment: I. The Relation of Selected Trace Elements to Health and Disease. Eds., H. L. Cannon and H. C. Hopps, pp. 26-28. National Academy of Sciences. Washington D.C.

Frost, D. V., 1981. Selenium in biology and medicine. Feedstuffs, 53(28):23-24.

Ganther, H. E., Goudie, C., Sunde, M. L., Kopecky, M. J., Wagner, P., Oh, S. H. and Hoekstra, W. G., 1972. Selenium: relation to decreased toxicity of methylmercury added to diets containing tuna. Science, 175:1122-1124.

Giordano, P. M. and Mays, D. A., 1977. Yields and heavy metal content of several vegetable species grown in soil amended with sewage sludge. Biological Implications of Metals in the Environment. Eds., H. Drucker and R. E. Wildung, pp. 417-425. U.S.E.R.D.A., Natl. Technical Information Service, Springfield, VA, 22161.

Grove, J. H., Sumner, M. E. and Syers, J. K. 1981. Effect of lime on exchangeable magnesium in variable surface charge soils. Soil Sci. Soc. Am. J., 45:487-500.

Helmke, P. A., Robarge, W. P., Korotev, R. L. and Schomberg, P. J., 1979. Effects of soil-applied sewage sludge on concentrations of elements in earthworms. J. Environ. Qual., 8:321-327.

Hoekstra, W. G., 1974. Biochemical role of selenium. Trace Element Metabolism in Animals-2. Eds., W. G. Hoekstra, J. W. Suttie, H. E. Ganther and W. Mertz, pp. 61-77. University Park Press, Baltimore, MD.

Hoekstra, W. G., Suttie, J. W., Ganther, H. E. and Mertz, W. (Editors), 1974. Trace Element Metabolism in Animals-2. University Park Press, Baltimore, MD.

Hupkens van der Elst, F. C. C. and Watkinson, J. H., 1977. Effect of topdressing pasture with selenium prills on selenium concentration in blood of stock. N.Z. J. Exp. Agri., 5:79-83.

Jackson, M. L. and Huang, P. M., 1981. Aluminum of acid soils in the food chain and senility. 21st Hanford Life Sciences Symposium. Biological Availability of Trace Metals. Oct. 4-8, Richland, WA.

Jaffé, W. G., M. Ruphael, D., Mondragon, M. C. and Cuevas, M. A., 1972. Estudio clinico i bioquimico en ninos escolares de una zona selenifera, Arch. Latinam. de Nutr., 22:595-611.

Jaworowski, Z., Bilkiewicz, J., Bysiek, M., Grzybowska, D., Kownacka, L. and Wlodek, S., 1977. The influence on man and the environment of natural radionuclides and heavy metals from industrial operations. Biological Implications of Metals in the Environment. Eds., H. Drucker and R. E. Wildung, pp. 628-647. U.S.E.R.D.A., Natl. Technical Information Service, Springfield, VA 22161.

Keller, W. D. and Smith, G. E., 1967. Ground-water contamination by dissolved nitrate. Geol. Soc. Am. Spec. Paper, 90:47-59.

Kinniburgh, D. G., Sridhar, K. and Jackson, M. L., 1977. Specific adsorption of zinc and cadmium by iron and aluminum hydrous oxides. Biological Implications of Metals in the Environment. Eds., H. Drucker and R. E. Wildung, pp. 417-425. U.S.E.R.D.A., Natl. Technical Information Service, Springfield, VA 22161.

Kubota, J., Allaway, W. H., Carter, D. L., Cary, E. E. and Lazar, V. A., 1967. Selenium in crops in the United States in relation to selenium-responsive diseases of animals. J. Agr. Food Chem., 15:448-453.

Kubota, J., Oberly, G. H. and Naphan, E. A., 1980. Magnesium in grasses of three selected regions in the United States and its relation to grass tetany. Agronomy J., 72:907-914.

Lagerwerff, J. V., 1972. Lead, mercury, and cadmium as environmental contaminants. Micronutrients in Agriculture. Eds., J. J. Mortvedt, P. M. Giordano and W. L. Lindsay, pp. 593-636. Soil Sci. Soc. Am., Madison, WI 53711.

Levander, O. A. and Cheng, L. (Editors), 1980. Micronutrient interactions: vitamins, minerals, and hazardous elements. Ann. N.Y. Acad. Sci., vol. 355.

Lewin, R., 1981. Is longevity a positive selection? Science, 211:373.

Liss, L. (Editor), 1980. Aluminum Neurotoxicity. Pathotox Pub., Forest Park South, IL 60466.

Lovejoy, C. O., 1981. The origin of man. Science, 211:341-350.

Massey, H. F. and Jackson, M. L., 1952. Selective erosion of soil fertility constituents. Soil Sci. Soc. Am. Proc., 16:353-356.
McCalla, T. M., Peterson, J. R. and Lue-Hing, C., 1977. Properties of agricultural and municipal wastes. Soils for Management of Organic Waste Waters. Eds., L. F. Elliott and F. J. Stevenson, pp. 11-43. Am. Soc. Agronomy, Madison, WI 53711.
Mertz, W., 1981. The essential trace elements. Science, 213:1332-1338.
Mortvedt, J. J., Giordano, P. M. and Lindsay, W. L. 1972. Micronutrients in Agriculture. Soil Sci. Soc. Am., Madison, WI 53711.
Muth, O. H., Oldfield, J. E., Remmert, J. F. and Schubert, J. R., 1958. Effects of selenium and vitamin E on white muscle disease. Science, 128:1090.
National Academy of Sciences, 1974. Geochemistry and the Environment: I. The Relation of Selected Trace Elements to Health and Disease. Washington D.C.
National Research Council, 1981. Aging and the Geochemical Environment. Washington D.C.
Oldfield, J. E., Allaway, W. H., Laitinen, H. A., Lakin, H. W. and Muth, O. H., 1974. Selenium. Geochemistry of the Environment: I. The Relation of Selected Trace Elements to Health and Disease. Eds., H. L. Cannon and H. C. Hopps and others, pp. 57-63. National Academy of Sciences, Washington D.C.
Olsen, S. R., 1972. Micronutrient interactions. Micronutrients in Agriculture. Eds., J. J. Mortvedt, P. M. Giordano and W. L. Lindsay, pp. 243-264. Soil Sci. Soc. Am., Madison, WI 53711.
Pechter, K., 1981. Selenium, vitamin E and cancer. Prevention, April, pp. 72-76.
Perry, H. M., 1969. Hypertension and trace metals, with particular emphasis on cadmium. Trace Substances in Environmental Health-2. Ed., D. D. Hemphill, pp. 101-122. Univ. of Missouri, Columbia.
Prohaska, J. R. and Lukasewycz, O. A., 1981. Copper deficiency suppresses the immune response in mice. Science, 213:559-561.
Rosenfeld, I. and Beath, O. A., 1964. Selenium Geobotany, Biochemistry, Toxicity, and Nutrition. Academic press, New York.
Sauer, H. I., 1980. Geographical Patterns in the Risk of Dying and Associated Factors Ages 35-74 Years. Series 3, No. 18. U.S. Dept. of Health and Human Services, Hyattsville, MD.
Schrauzer, G. N., 1979. Trace elements of carcinogenesis. Adv. Nutritional Res. Vol. 2. Ed., H. H. Draper, pp. 219-244. Plenum press, New York.
Schrauzer, G. N. and McGinness, J. E., 1979. Observations on human selenium supplementation. Trace Substances in Environmental Health-13. Ed., D. D. Hemphill, pp. 64-67. University of Missouri, Columbia.
Schrauzer, G. N. and White, D. A., 1978. Selenium in human nutrition: dietary intakes and effects of supplementation. Bioinorg. Chem., 8:303-318.
Schrauzer, G. N., White, D. A. and Schneider, C. J., 1977. Cancer mortality correlation studies-III: statistical associations with dietary selenium intakes. Bioinorg. Chem., 7:23-34.
Schwartz, K., 1961. Development and status of experimental work on Factor 3-selenium. Federation Proc., 29, Part I:666-673.

Seelig, M. S., 1980. Magnesium Deficiency in the Pathogenesis of Disease: Early Roots of Cardiovascular, Skeletal, and Renal Abnormalities. Plenum, New York.

Settle, D. M. and Patterson, C. C., 1980. Lead in Albacare: Guide to lead pollution in Americas. Science, 207:1167-1176.

Shamberger, R. J., 1980. Trace metals in health and disease. Nutrition Elements and Clinical Biochemistry. Eds., M. A. Brewster and H. K. Naito, pp. 241-275. Plenum Press, New York.

Shamberger, R. J., Willis, C. E. and McCormack, L. J., 1979. Selenium and heart disease. III-blood selenium and heart mortality in 19 states. Substances in Environmental Health-13. Ed., D. D. Hemphill, pp. 59-63. University of Missouri, Columbia.

Still, C. N. and Kelley, P., 1980. On the incidence of primary degenerative dementia vs. water fluoride content in South Carolina. Aluminum Neurotoxicity. Ed., L. Liss, pp. 125-131. Pathotox Pub., Park Forest South, IL 60466.

Tsuchiya, K., 1971. Some industrial problems in absorption and excretion of metal ions (lead, cadmium and mercury). Intestinal Absorption of Metal Ions, Trace Elements and Radionuclides. Eds., S. C. Skoryna and D. Waldron-Edward, pp. 399-408. Pergamon Press, Oxford.

Underwood, E. J., 1977. Trace Elements in Human and Animal Nutrition. Fourth Edition. Academic Press, New York.

Underwood, E. J., 1979. New findings with trace elements including effects of processing on supplies and availability. Intl. Review Biochem. Vol. 27 Biochemistry of Nutrition-1. Eds., A. Neuberger and T. H. Jukes, pp. 207-243. University Park Press, Baltimore, MD.

Waldbott, G. L., Burgstahler, A. W. and McKinney, H. L., 1978. Fluoridation: The Great Dilema. Coronado Press, Lawrence, KS.

Walker, W. M. and Peck, T. R., 1980. Nutrient content of corn plants increases faster than grain yields. Better Crops with Plant Foods, 64 (Fall):3-5.

Wallach, J. D. and Garmaise, B. 1979. Cystic fibrosis: a perinatal manifestation of selenium deficiency. Trace Substances in Environmental Health-13. Ed., D. D. Hemphill, pp. 469-476. University of Missouri, Columbia.

Watkinson, J. H., 1974. The selenium status of New Zealanders. N.Z. Med. J., 80:202-205.

Watkinson, J. H., 1981. Movement of selenium, sulphur and phosphorus during the pedogenesis of volcanic soils in New Zealand. Soils with Variable Charge. International Society of Soil Science, Commissions IV, V, VI and VII. Programme and Abstracts. 1981:163-164.

THE ROLE OF THE CLAY MINERAL MATRIX IN PETROLEUM GENERATION DURING BURIAL DIAGENESIS

W.D. JOHNS
DEPT. OF GEOLOGY, UNIVERSITY OF MISSOURI
COLUMBIA, MO. 65211 U.S.A.

INTRODUCTION

During the last two decades, largely through the worldwide growth of research in organic geochemistry, a great deal has been learned about the chemical aspects of petroleum formation and accumulation. The results of this research have been summarized recently in two excellent monographs by Tissot and Welte (1978) and by Hunt (1979). While both emphasize the geochemical aspects, the latter does so in a more geological context. Together these monographs summarize what must now be regarded as the concensus view of petroleum geochemistry.

At the same time clay mineralogists, particularly those studying the diagenetic processes involved in the generation of claystones and shales, have suggested a number of ways in which the clay mineral matrix is involved in petroleum generation and migration. These include:

1. Selective preservation of polar lipid components, including fatty acids, of the complex organic residues buried at the time of deposition of marine muds (Johns, 1979). The lipids now are believed to be the ultimate precursors of the kerogen material from which petroleum is generated diagenetically (Silverman, 1967).
2. The flushing of hydrocarbons from the source rock by a water-drive mechanism during late-stage dehydration accompanying smectite to illite transformations (Powers, 1967; Burst, 1969).
3. The use of clay mineral indicators in evaluating oil generating potential of shales (Foscolos et al., 1976; Foscolos and Powell, 1979).
4. The compositional modification of hydrocarbon mixtures, resulting from chromatographic partitioning by the clay-kerogen complex during migration from source to reservoir rocks (Young and McIver, 1977).
5. A fifth direct involvement of clays is as natural catalysts, promoting the alteration of dispersed organic matter into liquid and gaseous hydrocarbons at low temperatures (Johns, 1979).

This paper will confine itself to this catalytic

role of clays.

KEROGEN MATURATION

The immediate precursor of petroleum hydrocarbons is kerogen, the major organic component of sediments and sedimentary rocks. Kerogen is a complex high molecular weight, insoluble organic polymer formed largely from plant lipids early in the diagenetic history of the sediment.

Figure 1, modified after Tissot and Welte (1978) shows schematically a kerogen particle within a silty clay matrix. It is seen to consist of aromatic-,

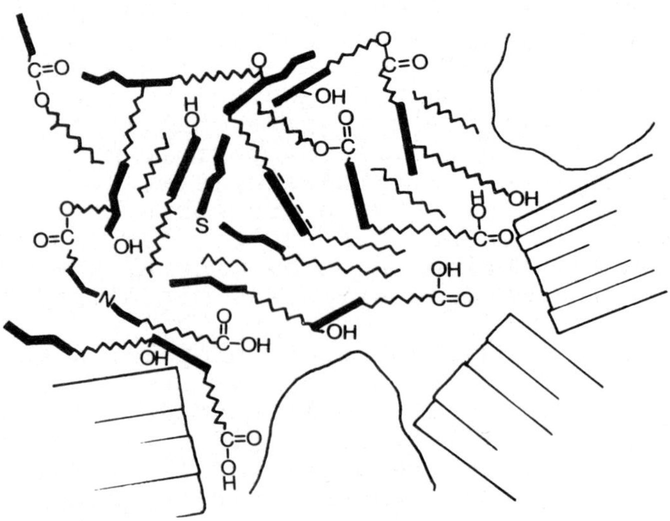

Fig. 1. Structure of kerogen particle in silty-clay matrix. Bars represent polyaromatic nuclei; bent bars, polycyclic alkanes; zig-zag lines, alkyl and isoalkyl chains. Modified after Tissot and Welte (1978).

alkanic-, and heterocyclic-polycyclic nuclei linked together through ester and ether linkages by alkane chains of moderate to long length. Additional alkane and isoalkane chains terminating in carboxyl and OH groups are attached to the polycyclic nuclei. Additional free hydrocarbons and fatty acids are included within the polymeric complex. Polar functional groups are attracted to the clay mineral surfaces, as indicated in the diagram.

As the kerogen-bearing sediment is gradually buried to greater and greater depths, the kerogen progressively is decarboxylated and dehydroxylated and alkane chains and fragments of polycyclic nuclei gradually yield a complex mixture of liquid hydrocarbons. At the same time the kerogen residue becomes more aromatic or graphite-like.

The elemental chemical changes during this maturation process are shown in figure 2.

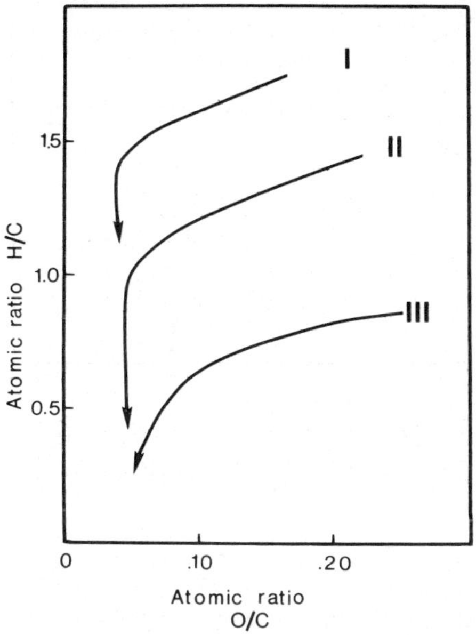

Fig. 2. Diagenetic paths of kerogen. Types I, II, and III kerogens represent decreasing lipid and increasing terrestrial plant components.

If one follows this process by analyzing the soluble organic content in a sediment profile, very often one observes a distribution of bitumenous hydrocarbons as a function of depth as shown in figure 3: Especially in Tertiary and late Mesozoic basins a maximum occurs in the bitumen distribution-depth curve. With increasing depth and temperature there is a loss of bitumenous hydrocarbons as they "crack" and form methane gas. There is a petroleum or liquid hydrocarbon "window" defined approximately by the temperature limits of 60°-110°C (Pusey, 1973).

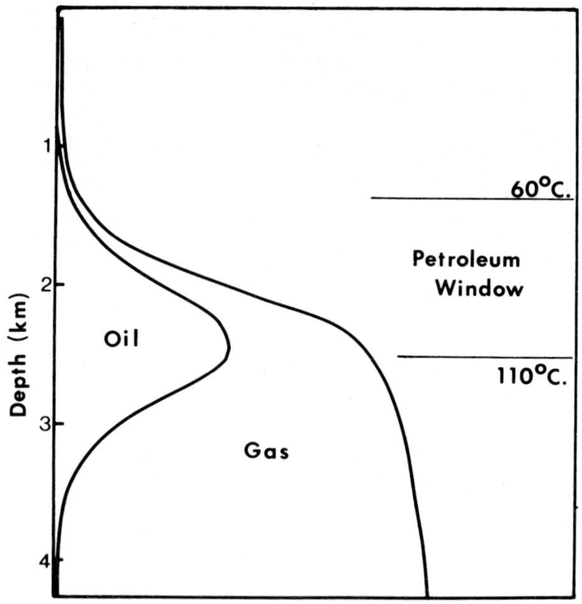

Fig. 3. Liquid and gaseous bitumenous hydrocarbon formation with depth during diagenesis.

For decarboxylation and C-C bond cleavage to take place in young sedimentary rocks at these low temperatures a catalyst is required to lower sufficiently the activation energies of these decomposition reactions (Hunt, 1979).

CLAY-DECARBOXYLATION REACTION KINETICS

The most obvious natural catalysts are the clay minerals of the matrix in which the kerogen is dispersed. Since smectite clays are common in young (Tertiary) sediments, attention has been concentrated on these clays and laboratory studies of smectite-fatty acid decarboxylation have been carried out in an attempt to simulate the natural process.

Figure 4 summarizes the results of a kinetic study of the decarboxylation of behenic acid (C_{22}), utilizing the following clays: KGa-1 kaolinite, SWy-1 and SWa-1 smectites and NG-1 nontronite (using the designations of the Source Clay Minerals Repository of the Clay Minerals Society). The Arrhenius plots of figure 4

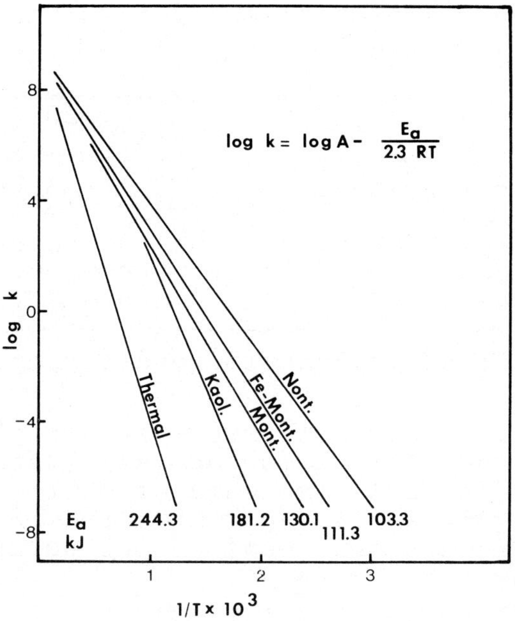

Fig. 4. Arrhenius plots of thermal and clay-catalyzed decarboxylation of behenic acid (C_{22} fatty acid).

show graphically the marked decrease in activation energy of decarboxylation for smectite-catalyzed reactions when compared to the E_a value of 244.3 kJ/mol (58.4 kcal/mol) for thermal reaction alone. These data also show that the catalytic activity is enhanced with increasing Fe^{+++} content of the smectite catalyst.

Table 1 shows even more dramatically the effect of these catalysts by calculating the time required for 90% decarboxylation at 60°C. The greater effectiveness of the catalyst as Fe^{+++} content increases is readily apparent.

Almon and Johns (1977) have shown that the catalytic sites are located at smectite crystallite edges where Al and Fe^{+++} ions can act as electron acceptors. Their interpretation of the reaction mechanism suggests that carboxyl groups in proximity to exposed acceptor ions lose a pi electron, causing bond rearrangement and loss of CO_2. The resulting free radical can undergo further rearrangement reactions including further bridging, but also cleavage of C-C bonds with the formation of free alkanes of lesser chain length.

TABLE 1

Decarboxylation reaction times (for 90% conversion of behenic acid at 60°C)

Thermal	2.9×10^{20} years
Montmorillonite (SWy-1*)	413 years
Fe-Montmorillonite (SWa-1*)	10 years
Nontronite (NG-1*)	.03 years

* Source Clay Minerals Repository designations

These studies show that smectites are effective catalysts for decarboxylation and simultaneous carbon bond cleavage. In fact, the data of table 1 indicate that the catalytic activity as measured in these laboratory studies surpasses that of the natural shale-kerogen systems.

NATURAL DIAGENESIS

We can turn to the rocks themselves to see why this may be so. By studying the kerogen composition and distribution in selected sedimentary basins, Tissot (1969) and Connan (1974) could calculate apparent kinetic constants for bitumen formation and kerogen decarboxylation in rocks of known age and temperature regime. Using their data, Arrhenius plots are presented in figure 5. Here we see activation energies (41.8-57.7 kJ/mol; 10.0-13.8 kcal/mol), even lower than the experimentally determined values discussed earlier. More significant are the much lower values for the Arrhenius constants ($A=10^{-7}-10^{2}$ sec^{-1}), in contrast to the high experimental values ($A=10^{9}-10^{10}$ sec^{-1}). Recalling that A is a constant representing the frequency with which molecules can collide in proper orientation to enable reaction, we can see that the natural system differs significantly from that represented by experiment.

Kerogen is a solid, relatively immobile polymer. Decarboxylation can occur only when carboxyl groups are brought into contact with clay surfaces. Thus mechanical movement, e.g. during continuous sediment compaction, is necessary to activate carboxyl groups. Thus the low kinetic frequency factor may be related to the solid character of kerogen and catalytic reaction

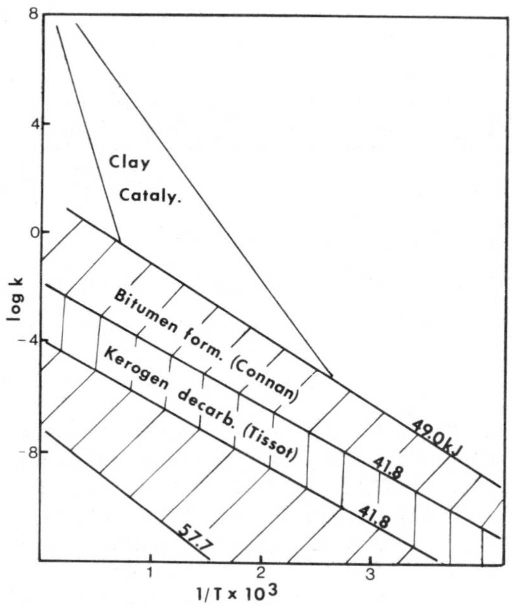

Fig. 5. Arrhenius plots from kinetic data of Tissot (1969) and Connan (1974).

continues only as compaction takes place. Reed and Oertel (1978) have presented evidence that compaction of shales does promote organic diagenesis by creating contact with fresh catalytic sites.

This picture of the involvement of natural catalysis in the petroleum-generating process is further complicated when it is recalled that the initial smectite catalyst is itself undergoing change as it transforms gradually through smectite-illite mixed phases to illite. Involved in this transformation are: 1. dehydration, 2. increased interlayer and, 3. increased tetrahedral charge.

Mortland and Raman (1965) showed that all these factors result in increasing the dissociation (therefore, acidity) of residual interlayer water in smectites. Figure 6, after Fripiat et al. (1974) shows schematically the nature of this interlayer reaction, where M is the exchangeable cation of charge z; x represents the water coordinated to the cation, and nz the interlayer charge. The changes accompanying illitization would increase the value of n and, therefore, the acidity of residual

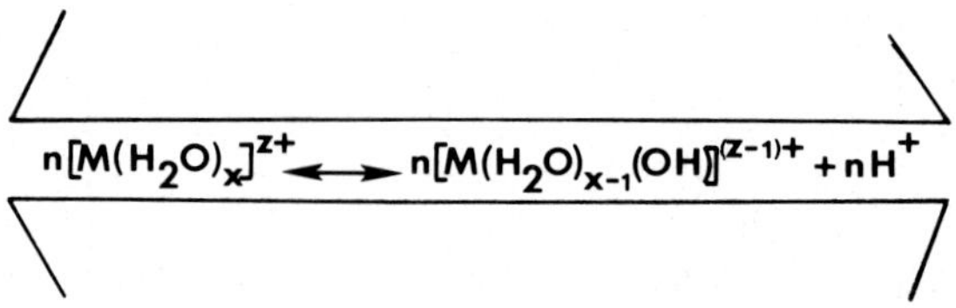

Fig. 6. Development of interlayer water acidity (after Fripiat et al., 1974)

smectite. As a result of diagenesis the smectite can assume proton donating capability, become naturally activated and enter into carbonium ion reactions involved in bitumen hydrocarbon cracking. Evidence of these processes should be seen in natural sediments.

Figure 8 shows the results of combined organic geochemical and clay mineralogical studies of a profile of a clayey silt section sampled over a depth of 2,000-16,000 ft (600-5000 m) from the U.S. Gulf Coast Tertiary. The bitumen and alkane distributions exhibit

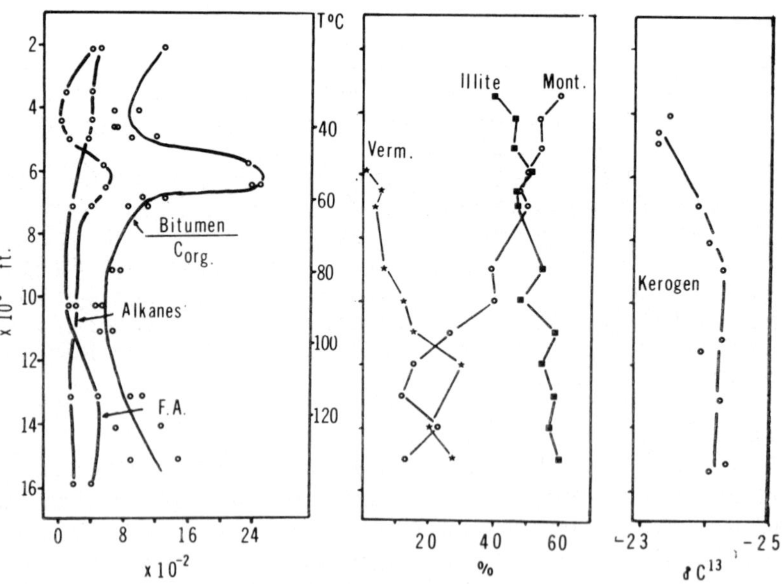

Fig. 7. Organic and clay mineral distribution in a U.S. Gulf Coast Tertiary section.

a maximum at about 6,000 ft. Below this depth the bitumen content decreases sharply, presumably as bitumen hydrocarbons are cracked to methane.

This is the point also at which a "vermiculite" or high charged expandable phase appears in the clay mineral matrix. Could this phase be the acid catalyst that is responsible for the bitumen cracking and conversion to gaseous hydrocarbons? Further study is necessary to confirm this interpretation and to unravel additional details of these very complex relationships. Such work is still in progress in the clay mineralogy laboratories of the University of Missouri.

ACKNOWLEDGEMENTS

Support for portions of this work by an American Chemical Society-Petroleum Research Fund Grant (10052-AC2) is gratefully acknowledged.

REFERENCES

Almon, W.R. and Johns, W.D., 1977. Petroleum forming reactions: The mechanism and rate of clay catalyzed fatty acid decarboxylation. In: R. Campos and T. Goni (Editors), Proc. Int. Meet. Org. Geochem., 7th, 157-171.

Burst, J.F., 1969. Diagenesis of Gulf Coast clayey sediments and its possible relation to petroleum migration. Am. Assoc. Petrol. Geol. Bull., 53, 73-93.

Connan, J., 1974. Time-temperature relation in oil genesis. Am. Assoc. Petrol. Geol. Bull., 58, 2516-2521.

Focolos, A.E., Powell, T.C. and Gunter, P.R., 1976. The use of clay minerals and inorganic and organic indicators for evaluating the degree of diagenesis and oil generating potential of shales. Geochim. Cosmochim. Acta, 40, 953-966.

Focolos, A.E. and Powell, T.G., 1979. Mineralogical and geochemical transformation of clay during burial diagenesis (catagenesis): Relation to oil generation, Proc. Int. Clay Conf., Oxford, 1978.

Fripiat, J.J. and Cruz-Camplido, M.I., 1974. Clays as catalysts for natural processes, Ann. Rev. Earth Planetary Sci., 2, 239-256.

Hunt, J.M., 1979. Petroleum Geochemistry and Geology. W.H. Freeman and Co, San Francisco, 617 pp.

Johns, W.D., 1979. Clay mineral catalysis and petroleum generation. Ann. Rev. Earth Planetary Sci., 7: 1983-1998.

Mortland, M.M. and Raman, K.V., 1968. Surface acidity of smectites in relation to hydration, exchangeable

cation and structure. Clays Clay Miner., 16: 393-398.
Powers, M.C., 1967. Fluid release mechanisms in compacting marine mud rocks and their importance in oil exploration. Am. Assoc. Petrol. Geol. Bull., 51: 1240-1254.
Pusey, W.C., 1973. Paleotemperatures in the Gulf Coast using ESR-kerogen method. Gulf Coast Assoc. Geol. Trans., 23: 195-202.
Reed, W.E. and Oertel, G., 1978. Is compaction a factor in organic diagenesis?, Geol. Soc. Amer. Bull., 89: 658-662.
Silverman, S.R., 1967. Carbon isotope evidence for the role of lipids in petroleum. J. Am. Oil Chem. Soc., 44: 691-695.
Tissot, B., 1960. Premières données sur les mécanismes et la cinétique de la formation du pétrole dans les sediments: Simulation d'un schema reactionnel sur ordinateur. Inst. Francais Pétrole Rev. 24: 470-501.
Tissot, B. and Welte, D.H., 1978. Petroleum Formation and Occurrence. Springer/Verlag, Berlin-Heidelberg-N.Y., 538 pp.
Young, A. and McIver, R.D., 1977. Distribution of hydrocarbons between oils and associated fine-grained sedimentary rocks-Physical chemistry applied to petroleum geology, Am. Assoc. Petrol. Geol. Bull., 61: 1407-1436.

CLAY DIAGENESIS AND ORGANIC MATURATION IN NORTHERN NORTH SEA SEDIMENTS

M.J. PEARSON, D. WATKINS and J.S. SMALL

Department of Geology, The University, Marischal College, Aberdeen, AB9 1AS, Scotland.

ABSTRACT

Mineral distribution and composition of the interstratified clay minerals have been studied in sediments of Middle Jurassic to Oligocene age from exploration wells in the Inner Moray Firth and Viking Graben areas of the Northern North Sea Basin. Organic maturation parameters including vitrinite reflectance, miospore colour, total soluble extract and carbon preference index have been determined on samples mainly from the Middle and Upper Jurassic of the same wells.

Interim results indicate that changes in the composition and structure of interstratified illite-smectite clays and resultant expulsion of interlayer water occur with increasing temperature during burial and can be correlated with temperature-related changes in the organic component of the sediments. Significant illitisation of detrital smectite begins at a vitrinite reflectance level of about 0.4% Ro and effectively ceases after the structure becomes ordered at about 0.6% Ro.

Jurassic sediments from Inner Moray Firth wells contain immature organic matter. Interstratified clays are rich in smectite but show evidence of illitisation in the deepest samples. A decrease in carbon preference index of n-alkanes down the section indicates incipient organic maturation. Sediments of the same age from Viking Graben wells are mature being at about the peak stage of oil generation and contain ordered interstratified illite-smectite from which water loss (clay dehydration) is nearing completion.

Clay dehydration begins in the Palaeocene or Eocene in Viking Graben wells and occurs at progressively greater depths in a southerly direction. The depth of clay ordering and comparable levels of organic maturity also deepen to the south. This trend is more probably related to late tectonic movements than to differences in geothermal gradient.

INTRODUCTION

Following the work of Burst (1969) and Perry and Hower (1972) on burial diagenesis of clay minerals in sediments of the Gulf Coast of America several excellent studies have been carried out in other sedimentary basins. Notable among these have been the work of Heling (1978) on the Rhine Graben of Germany and of Foscolos et al. (1976) and Powell et al. (1978) on basins in N.E. British Columbia and the N.W. Territories of Canada. In each of these studies the disappearance with depth of discrete smectite and the appearance of neo-formed discrete phases such as kaolinite and chlorite were reported. In the majority of cases the loss of smectite has also been accompanied by the appearance of interlayered minerals, most commonly illite-smectites in which illitic layers appear to form by replacement of Si by Al in the smectite tetrahedral sheet and fixation of K^+ at interlayer sites with expulsion of water. Boles and Franks (1979) constructed an approximate temperature scale for inorganic diagenetic changes occurring in interbedded permeable and impermeable sediments in which they proposed that illitisation of smectite clays occurred principally over the range of 50-100°C. Foscolos and Powell (1979) also studied the temperature dependence of this reaction by reference to standard organic parameters used in the estimation of maturity of petroleum source beds. They found that illitisation was active over the depth range of petroleum generation but that the main pulses of water loss occurred above and below this zone. The importance of this expelled water to the facilitation of oil migration is therefore somewhat ambiguous.

We have undertaken a study of both clay diagenesis and organic maturation in the northern North Sea basin with the aims (i) of establishing whether a pattern of clay diagenetic changes comparable to those in previously studied basins exists; (ii) whether there is a consistent relationship between time-temperature (reflected in the organic maturity) and stage of clay reaction; (iii) whether the compositions of aqueous pore fluids and/or detrital mineral components are limiting factors in the illitisation reaction and whether any estimate of their effects can be made, and (iv) whether the water released by dehydration during illitisation is of importance in oil migration. This paper presents interim results of the study and tentative answers to some of these questions.

MATERIALS AND METHODS

Samples were provided from exploration wells by Total Oil Marine Ltd in the form of ditch cuttings. Obvious contaminating material was removed by hand picking and where mixed lithologies occurred shale fragments were selected. Fragments were briefly washed ultrasonically (to remove drilling mud coatings) and examined under the microscope before disaggregation in distilled water by prolonged ultrasonic agitation. For routine clay investigation a <0.7μm fraction was prepared by centrifugation, this being a compromise to ensure sufficient material could be collected. After Mg saturation oriented XRD specimens were prepared by suction of the dispersed fine fraction onto plastic micropore film and subsequently transferred to glass by pressing the film onto a slide. Powder specimens were prepared from another portion of the fragments for determination of the bulk mineralogical composition. Diffractograms of the Mg-saturated fine fractions were obtained for all samples at 50% RH and after glycolation. Where fine-grained calcite was abundant a separate sample was treated with HCl prior to Mg saturation to enhance clay reflections and a similar treatment was applied where chlorite was present to aid identification of kaolinite. A number of representative samples was heated to 550°C whilst others were K-saturated and glycolated to investigate the possible presence of vermiculitic interlayers.

Kerogen concentrates, palynology slides and resin mounted blocks for vitrinite reflectance measurements were prepared by standard methods. An alkane component was separated from the dichloromethane-soluble extract and analysed by gas chromatography using methods generally similar to those described in Powell et al. (1978).

Results reported in this paper relate to eight exploration wells of which A-A', B, C, D and H are located in the Viking Graben within quadrants 3, 9 and 16 and progress alphabetically in a southerly direction. Wells M, N and O are from quadrant 12 in the Inner Moray Firth basin and may be identified as 12/21-1, 12/22-1 and 12/24-1 respectively. The two areas are strongly contrasting both in terms of facies and burial depth for sediments of equivalent ages. Only Jurassic sediments which occur at depths as shallow as 700m have been examined in the Moray Firth wells, whilst complete burial sequences of up to 4700m from Quaternary to Middle Jurassic have been studied in wells A-A', C, D and H.

INTERPRETATION OF ILLITE-SMECTITE IN DIFFRACTOGRAMS

In the vast majority of samples (<0.7μm fractions, Mg-saturated) significant amounts of non-discrete clay minerals have been detected (Fig.1). Following Reynolds and Hower (1970) these have been interpreted as follows: (i) randomly interstatified illite-smectite (I-S) where the first basal peak fully expands to 17 Å on glycolation and contracts to 10 Å on heating to 550°C as does smectite but is somewhat broadened with a higher low angle background than for discrete smectite (Fig.1a); (ii) as ISIS - ordered illite-smectite (rectorite or allevardite) where a single intense basal peak at 11-12 Å splits into two peaks, one at 10 Å and one at 13-14 Å on glycolation. The latter collapses to 10 Å on heating (Fig.1b).

Using the criteria of Reynolds and Hower (1970) it is also possible in many cases to estimate the proportions of expandable (smectite) layers present in both random and ordered minerals. They demonstrated that the 5.6 Å smectite-glycol 003 reflection migrates with progressive illitisation towards the 5.0 Å illite 002 position. The d-spacing of this reflection can be related directly to the % smectite layers in the mixed-layer mineral although in practice difficulties arise through interference by the 5.0 Å reflection of discrete mica which is usually present. In the case of random structures the visual appearance of the 17 Å reflection (which does not migrate) as either sharp or broad and diffuse is used as an additional subjective criterion. In the case of ordered structures the 13-14 Å reflection which also migrates towards 12 Å is used as an additional criterion. Reynolds and Hower (1970) indicate that minerals with between 100% and 40% smectite layers are generally random, whilst those between 40% and 15% smectite layers are ISIS ordered. All the North Sea material examined appears to fit this pattern and the transition to pure illite was not observed at the depths studied.

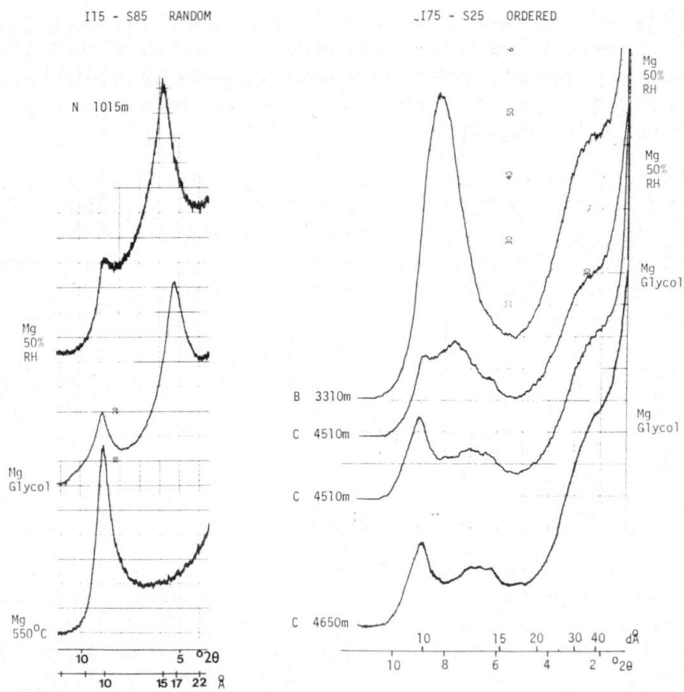

Fig.1

Diffractograms of <0.7μm fractions of Jurassic sediments illustrating random and ordered interstratified illite-smectites. Discrete illite is present except in B3310.

Examples of partial diffractograms, all of Upper Jurassic age samples, are shown in Fig.1. Compositions estimated by the above criteria are approximately as indicated. The three deeply-buried samples included in the diffractegrams all show a low intensity reflection around 30-35 Å which if interpreted as a superlattice reflection, appears to be somewhat displaced from the expected value of 27 Å. The reason for this displacement is not fully understood but Foscolos and Powell (1979) cited limited re-expandability after K-saturation as evidence for an additional vermiculitic interlayer component. The present ordered mineral shows identical behaviour on K-saturation but the random mineral fully re-expands on glycolation. The authors intend to study this phenomenon further before giving a reasoned interpretation.

The sample from well N (Inner Moray Firth) at 1015m shows a random I-S pattern which is typical of many shallowly-buried samples from all wells irrespective of age. The interlayered mineral here is close to a pure smectite and such minerals have been identified by some previous authors (Heling, 1978 for example) simply as smectite. Thus care must be exercised in comparing the work of different authors regarding depths of smectite disappearance.

MINERALOGY, LITHOLOGY AND STRATIGRAPHY OF WELL A-A'

Stratigraphy, lithology and mineralogical variations are similar for all Viking Graben wells investigated though thicknesses vary. The Tertiary sequence comprises mainly shale, though important sands occur in the Upper Danian and Oligocene. The top Palaeocene tuff band marks a widespread volcanic event traceable throughout the Viking Graben. Increased calcite abundance in the Cretaceous reflects the dominantly marly lithology terminated by chalk at the top and limestone at the base, which lies unconformably on pyritic black Kimmeridgian shales. Sideritic Callovian-Oxfordian siltstones overlie Middle Jurassic shales which occur up to the total depth of drilling.

Plagioclase is found throughout the sequence in amounts between 5 and 10%. K-feldspar occurs in similar amounts in the Tertiary but is absent below 2500m. The pattern observed in wells C and H is broadly similar though K-feldspar is scarce even in shallow samples. The deepest noted occurrences are at 2400m and 2800m for wells C and H respectively. In each case the depth noted falls within the zone of smectite layer loss (Fig.3). These observations are therefore consistent with a model in which K and Al for illitisation of smectite are derived at least partly from destruction of K-feldspar.

Clay mineral abundances in the $<0.7\mu m$ fraction (estimated from X-ray heights) also show broadly similar variation through the Viking Graben wells. Random I-S is dominant throughout the Tertiary reaching a peak of over 80% in the Eocene sediments and Palaeocene tuff. Volcanic activity in the North Sea is known to have continued until the lower Miocene and altered ash may well have been the major contributor of smectite to these sediments. Fluctuations may indicate nothing more than varying levels of volcanic activity swamping the detrital input to a greater or lesser extent. A decrease in random I-S abundance observed through the Cretaceous may be more apparent than real since serious difficulties arise in estimating the abundance of a mineral whose characterisitc peak decreases in intensity with progressive illitisation. A further relative decrease in I-S abundance after ordering may be partly due to a lower reflection efficiency of the 12-13 Å ordered peak. Significant kaolinite is present in virtually all samples whilst chlorite is generally absent or undetected though significant chlorite does occur towards the base of well H. Discrete illite is generally present though it is less abundant or absent in Eocene - Oligocene samples.

COMPOSITIONAL CHANGES IN INTERSTRATIFIED ILLITE-SMECTITE

Estimates of % smectite layers in I-S and presence or absence of ordering are plotted against depth and stratigraphy for well A-A' in Fig.2. The scatter of values in the Tertiary and Cretaceous may be attributed partly to the large error ($\pm$ 10%) in estimating the composition but also partly to variations in source material. Weathering of volcanic ash and formation of soil on the contemporary land surface would have produced K-poor smectite relative to that which altered in-situ in contact with sea water. Large differences at shallow burial

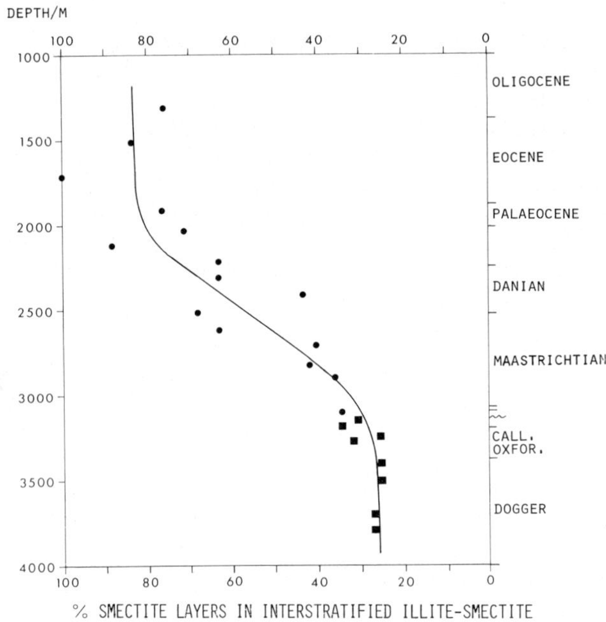

Fig.2

Compositional and structural changes in interstratified illite-smectite through the sedimentary sequence in well A-A'. Solid circles indicate random inter-layering; solid squares indicate IS-ordered inter-layering.

depths are thus not difficult to account for. The convergence of points towards the illitisation curve with increasing burial tends to support this view. Whilst the upper point of inflexion on the curve (corresponding to the onset of active illitisation or diagenetic reaction) is somewhat poorly defined the lower inflexion point (cessation or hiatus in illitisation) is in most cases more sharply defined because of the convergence of data points. The change from random to ordered structure can easily be recognised from diffractograms and is a reliable reference point on the curves lying close to or just above the lower inflexion point. In well A-A' this change occurs at the Cretaceous/Jurassic unconformity and the discontinuity is emphasized by the presumed gap in the geological record here. Below 3100m no further random I-S is seen.

The curve in Fig.2 is interpreted as indicating that the reaction:

$$\text{SMECTITE} + \text{K} + \text{Al} \longrightarrow \text{ILLITE} + \text{Si} + \text{H}_2\text{O}$$

occurs over the interval 2000-3100m and that through the resulting clay dehydration water is released into newly created pore space and may be subsequently expelled by compaction. Chemical evidence for the validity of this reaction is provided by the data in Table I in which chemical analyses of <0.7μm fractions corrected for kaolinite, pyrite and apatite for two shallowly buried Upper Jurassic samples from the Inner Moray Firth (M1543m and O1858m) containing random I-S are compared with two deeply buried samples of similar age from the Viking Graben (B3500m and A3512m) containing ordered I-S. These data indicate that a pronounced depletion of Si and an enrichment of K, Al has occurred during burial accompanied by a significant if modest water los

TABLE 1

Chemical analyses of <0.7μm fractions corrected for kaolinite, pyrite and apatite and normalised to 100%

well and depth (m)	SiO_2	Al_2O_3	Fe_2O_3	FeO	MnO	MgO	CaO	Na_2O	K_2O	H_2O^+
M 1543	55.1	16.2	3.0	1.9	0.4	3.2	5.4	0.5	2.8	11.7
O 1858	59.3	14.5	2.9	1.1	0.4	2.8	6.8	1.1	3.2	7.9
B 3500	49.8	23.4	0.9	5.9	0.5	2.8	3.5	0.4	7.4	5.4
A 3512	42.0	24.2	5.3	10.1	0.6	3.0	1.0	0.9	5.8	7.1

The reaction is assumed to occur progressively over the interval indicated although no chemical data are yet available within the dehydration interval to confirm this view. From the X-ray evidence little further change in composition occurs after ordering within the depths studied and the limit of 80% illite layers noted by Hower et al. (1976) and others seems to be applicable here also

Fig.3 is a composite plot of smectite reaction curves and stratigraphy against depth for the four Viking Graben wells A-A', C, D and H arranged in geographical order. In each case reaction of smectite begins within or just above the Palaeocene which occurs at progressively greater present burial depths in the range 2000-2500m increasing to the south. The more sharply defined lower inflexion point is located over a stratigraphic range from Upper Cretaceous to Upper Jurassic, at depths varying between 3200 and 3800m. The zone of active illitisation is therefore somewhat variable in thickness but within the limits of 1100-1400m.

There is no doubt from these data that the change in composition and structure of the interstratified mineral with depth is real and independent of stratigraphy or facies. It is a diagenetic temperature-related reaction though the precise importance of temperature relative to time or aqueous fluid composition has yet to be determined. Present day geothermal gradients determined from measured bottom hole temperatures (Small, 1980) do not appear to differ greatly between the four wells and lie within the range 2.9 to 3.2 ^{o}C/100m. We consider that the differences in apparent diagenetic level (see also Fig.4) may be explained partly by tectonic movements during the Tertiary. Subsidence by about 500m of H relative to D has occurred during post-Miocene times which is about the same as the depth difference in lower inflexion point between the two wells. Such an interpretation is undoubtedly an oversimplification but it does illustrate a possible approach to an estimation of equilibration times when more data become available. The apparent reversal at 2700m on the reaction curve for well H is not yet fully understood but the anomalous values relate to coarser-grained

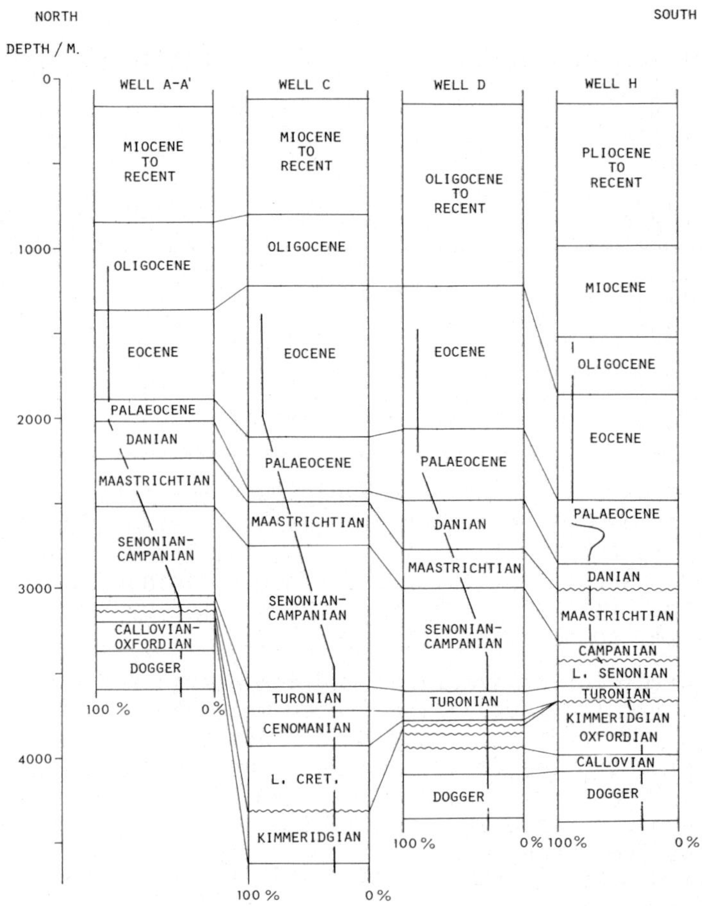

Fig.3

Variation of % smectite layers in interstratified illite-smectite with depth and stratigraphy for four wells from the Viking Graben.

siltstones lying between tuffaceous shales of the top Palaeocene and marls of the Danian. One possible interpretation is that the more porous and permeable sediments contain authigenic I-S forming in equilibrium with fluids migrating from depth.

RELATION TO ORGANIC MATURATION PARAMETERS AND PETROLEUM GENERATION

In Fig.4 vitrinite reflectance (mean random, oil immersion), spore colour (seven point scale of Batten, 1980 with oil generation threshold at 3/4), dichloromethane-soluble extract (as parts per mil relative to total organic carbon) and CPI (carbon preference index for n-alkanes over the range nC_{23}-nC_{33}) have been plotted against the smectite reaction curve for a limited range of Jurassic sediments in wells A and H. These limited data suggest that the I-S ordering point and the lower inflexion point on the I-S reaction curve occur at similar levels of organic maturity in the two wells. Spore colour indices of 3/4 or 4 and solvent extract ratios in excess of 50 per mil for oil-prone kerogens at this depth indicate a maturity level within the zone of intense oil generation. A predominance of catagenic alkanes is

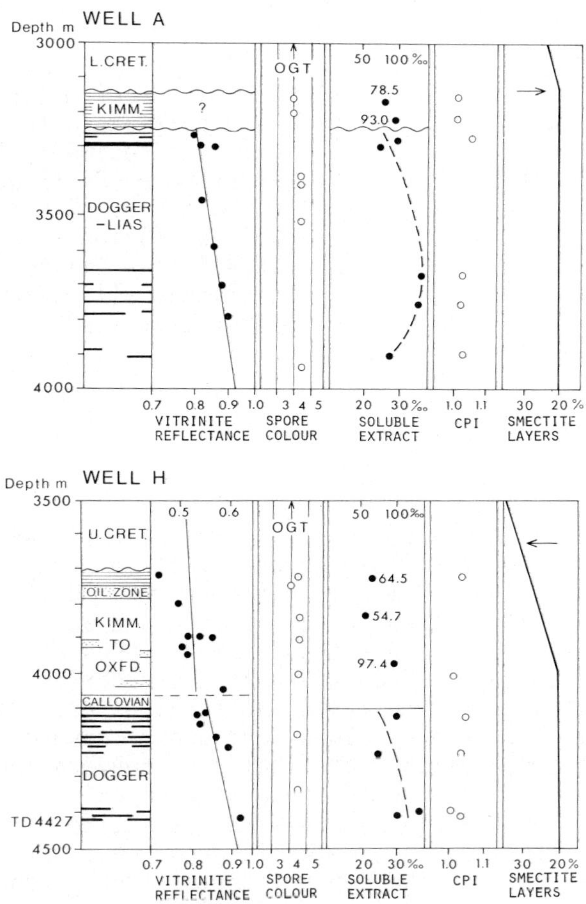

Fig.4 Comparison of maturation parameters for Jurassic sediments from Viking Graben wells A and H. O.G.T. = Oil Generation Theshold. Arrow indicates clay ordering.

confirmed by CPI values close to unity. Vitrinite reflectance measurements for well A are as yet confined to Middle Jurassic coals and significant thermal breaks may occur at either of the unconformities above or below the Kimmeridgian. The Middle Jurassic is at a similar maturity level in both wells although burial depths are about 500m greater in well H. A large thermal break between Middle and Upper Jurassic in well H is tentatively attributed to contemporary volcanic activity in the Forties area. The Upper gradient is drawn as a best fit line between Tertiary and Kimmeridgian-Oxfordian reflectances and suggests a value of 0.55% Ro at the lower inflexion point or 0.52% Ro at the I-S ordering point on the Cret./Jur. unconformity. Current work on other North Sea wells suggests a reflectance value of around 0.55% Ro as an average for the first appearance of ordering and 0.60 to 0.65% Ro for the lower inflexion point. There is thus no doubt that the smectite to illite reaction and by inference clay dehydration are active during the early mature stage of oil generation (onset about 0.5% Ro, peak about 0.8% Ro, Heroux et al., 1979).

Fig.5 shows comparable data for Jurassic sediments from the Inner Moray Firth wells where organic maturity levels are much lower. The few clay data available suggest that active illitisation commences around 0.4% Ro.Continuing organic geochemical studies of younger sediments from other wells (including H) tend to confirm this observation.

DISCUSSION AND CONCLUSIONS

Smectite which is ubiquitously present in shallowly-buried Viking Graben sediments has been demonstrated to undergo progressive illitisation commencing around 2000-2500m and proceeding via randomly-

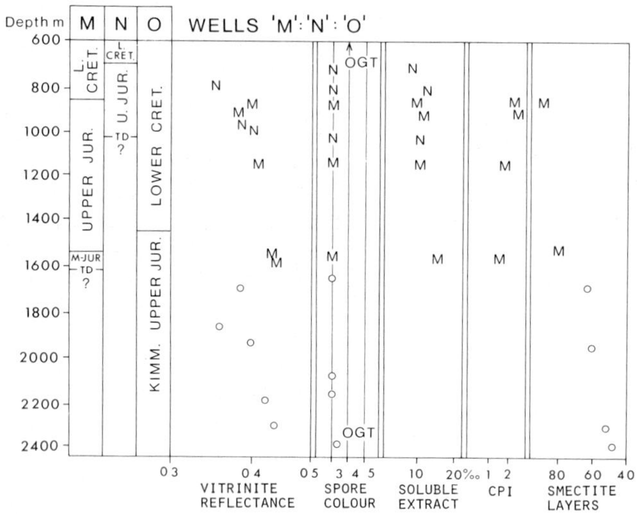

Fig.5

Comparison of maturation parameters for Jurassic sediments from Inner Moray Firth wells M, N, and O. O.G.T. = Oil Generation Threshold. Illite-smectites all randomly interstratified.

interlayered minerals to around 3200-3800m where by the criteria of Reynolds and Hower (1970) a change to ISIS-ordered structure occurs. Significant further reaction beyond I80-S20 is not observed. Using an average present day geothermal gradient of 3.0°C/100m and an assumed sea water temperature of 5°C, present day temperatures at the onset of smectite reaction are around 65-80°C and at the ordering point 100-120°C. Corresponding vitrinite reflectance values are around 0.4% Ro and 0.6% Ro respectively. These values, and other measured organic parameters (Fig.4) clearly demonstrate that clay dehydration is taking place in northern North Sea sediments containing organic matter that is either immature or at most early mature with respect to oil generation. Nevertheless we consider that water is produced by clay reaction within the uppermost part of the depth interval over which petroleum generation occurs and may therefore be available to play a role in hydrocarbon migration. It is significant that the Kimmeridge clay which is widely regarded as the main source rock for North Sea Oil is close to or below the clay ordering point. Water released during illitisation would therefore have been available at the time of hydrocarbon expulsion from the source rock.

Comparison with the work of other authors is hindered by the somewhat different methods of clay interpretation. Foscolos and Powell (1979) envisage clay dehydration occurring in two distinct steps with their first dehydration corresponding to the ordering point in the present study. Hower et al. (1976) obtained similar smectite reaction curves in Gulf Coast wells to those presented here and estimate a temperature of 95°C for the ordering point at a depth around 3700m but no organic maturity data were given for comparison. Srodon (1979) compared vitrinite reflectance values with stages of clay diagenesis in the carboniferous coal basin of Silesia. Lowest reflectance values found were around 0.8% Ro at which level a change from ISIS to ISII-ordered illite-smectite was observed. The latter type of ordering has not been identified in Viking Graben sediments at the depth studied, though comparable maturity levels are reached.

ACKNOWLEDGEMENTS

The authors are indebted to Total Oil Marine for provision of material and for financial support of one of us (JSS). Financial support by the N.E.R.C. is also gratefully acknowledged.

REFERENCES

Batten D.J., 1980. Use of transmitted light microscopy of sedimentary organic matter for evaluation of hydrocarbon source potential. Proc. 4th Internat.Palynol.Conf., Lucknow (1976-77), 2, 589-594.

Boles, J.R. and Franks, G.S., 1979. Clay diagenesis in Wilcox sandstones of Southwest Texas: implications of smectite diagenesis on sandstone cementation. Journ.Sedim.Petrol., 49: 55-70.

Burst, J.R., 1969. Diagenesis of Gulf Coast clayey sediments and its possible relation to petroleum migration. Bull.Amer.Assoc.Pet.Geol., 53: 73-93.

Foscolos, A.E., Powell, T.G. and Gunther, T.R., 1976. The use of clay minerals and inorganic and organic geochemical indicators for evaluating the degree of diagenesis and oil generating potential of shales. Geochim.Cosmochim.Acta., 40: 953-966.

Foscolos, A.E. and Powell, T.G., 1979. Mineralogical and geochemical transformation of clays during burial diagenesis (catagenesis): relation to oil generation. Proc.Int.Clay Conf., Oxford, 261-270.

Heling, D., 1978. Diagenesis of illite in argillaceous sediments of the Rhinegraben. Clay Miner., 13: 211-219.

Heroux, Y., Chagnon, A. and Bertrand, R., 1979. Compilation and correlation of major thermal maturation indicators. Bull.Amer.Assoc. Pet.Geol., 63: 2128-2144.

Hower, J., Eslinger, E.V., Hower, M.E. and Perry, E.A., 1976. Mechanism of burial metamorphism of argillaceous sediment: 1. Mineralogical and chemical evidence. Bull.Geol.Soc.Amer., 87: 725-737.

Perry, E.A. and Hower,J., 1972. Late stage dehydration in deeply buried pelitic sediments. Bull.Amer.Assoc.Pet.Geol., 56:2013-2021.

Powell, T.G., Foscolos, A.E., Gunther, P.R. and Snowdon, L.R., 1978. Diagenesis of organic matter and fine clay minerals: a comparative study. Geochim.Cosmochim.Acta., 42: 1121-1197.

Reynolds, R.C. and Hower, J., 1970. The nature of interlayering in mixed-layer illite-montmorillonites. Clays Clay Miner., 18: 25-36.

Small, J.S., 1980. Clay mineral diagenesis in Jurassic and Cretaceous sediments from the Moray Firth and Viking Graben, Northern North Sea. M.Sc. Thesis, University of Aberdeen.

Srodon, J., 1979. Correlation between coal and clay diagenesis in the Carboniferous of the Upper Silesia coal basin. Proc.Int.Clay Conf., Oxford, 251-260.

THE CLAY MINERALOGY OF JURASSIC SHALES FROM BRORA, N.E. SCOTLAND

ANDREW HURST

Geologisk Laboratorium, Statoil, Forus, Stavanger, Norway.

ABSTRACT

Jurassic shales are hydrocarbon source rocks in the North Sea area. The shales are kaolinitic and contain large proportions of ordered illite/smectites derived from the reworking of older Devonian sediments. Kaolinite is thought not to be derived solely from weathering reactions, but primarily from a kaolinite-rich source rock by erosion. The ordered illite/smectite is typical of clays which have been buried at temperatures of ~ 100°C. The maximum burial temperature of the Brora section is 50-60°C, therefore these illite/smectites are metastable. The presence of some more expanadable illite/smectites allows estimation of the Jurassic thermal maximal, this estimate correlates with an organic maturation index. The combination of kaolinite and high temperature polymorph illite/smectite as detrital clays implies that most of the shales would be stable up to burial temperatures of at least 100°C, or ~ 40°C higher than those attained during burial. Therefore, the shales have a low potential to produce cation-rich fluids as is commonly the case in the thermal alteration of clays. Similarly, the expulsion of large volumes of interlayer water, frequently proposed to aid primary migration of hydrocarbons, is unlikely.

INTRODUCTION

At Brora, (Fig.1) an inlier of Mesozoic sediments, mainly of Jurassic age, represents onshore correlatives of hydrocarbon reservoir sandstones and source rocks in the North Sea.

The stratigraphy of the section is well known (Neves and Selley, 1975). At the base of the sequence in the L.Jurassic there is a transition from continental to marine sedimentation. During the M. Jurassic a period of regressive sedimentation occurred. A major marine transgression took place in the M.Jurassic (L.Callovian) and marine conditions persisted at least until the end of Jurassic time. In general, this sedimentary sequence is similar to that in the Inner Moray Firth (Hurst, 1980) and elsewhere in the Northern North Sea (Deegan and Scull, 1977). The clay mineralogy of the shales is used to examine the provenance and diagenetic history of the succession.

CLAY MINERALOGY

Semi-quantitative analysis of the clay mineralogy by X-ray diffraction was done following a method similar to that of Weir et al. (1975). X-ray analyses were made using Cu Kα radiation through a Ni filter. Kaolinite is common in most shale intervals (Fig. 2), generally accounting for >30% of the < 2 μm fraction.

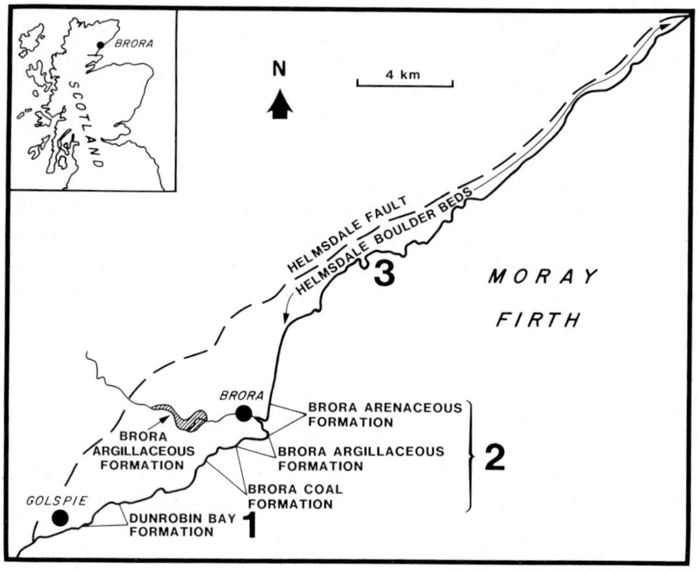

Fig. 1. Locality map of the Brora area.

Smectite was only detected in shales at the base of the section where it is common. Mixed-layer illite/smectites are common, but illite is also present as a discrete 10 Å phase. Chlorite is rarely detected. The general trends shown in Fig. 2 mask many small-scale variations present within the sequence. These small-scale variations are not discussed here.

Two characteristic suites of illite/smectites are present:
1) highly expandable, randomly-ordered phases (Fig. 4);
2) less expandable ordered phases (Fig. 5).

The pattern of the fully-expandable smectite typical of the basal Jurassic is shown in Fig. 3. The randomly-ordered illite/smectites have varied expandability, frequently less than shown by the pattern in Fig. 4, but in general with between 40% and 65% smectite interlayers as estimated by comparison with the data of Reynolds and Hower (1970). Air-dried untreated samples of this expandable illite/smectite are typified by a 12-13 Å basal spacing, or a 14.5-15 Å basal spacing when Mg^{2+}-saturated. The less expandable illite/smectites (Fig. 5) generally form broad, frequently indistinct peaks, between 12 Å and 13.5 Å when glycolated, implying interlayer compositions of between 20% and 30% smectite. The ordered phases are more common than the randomly-ordered phases.

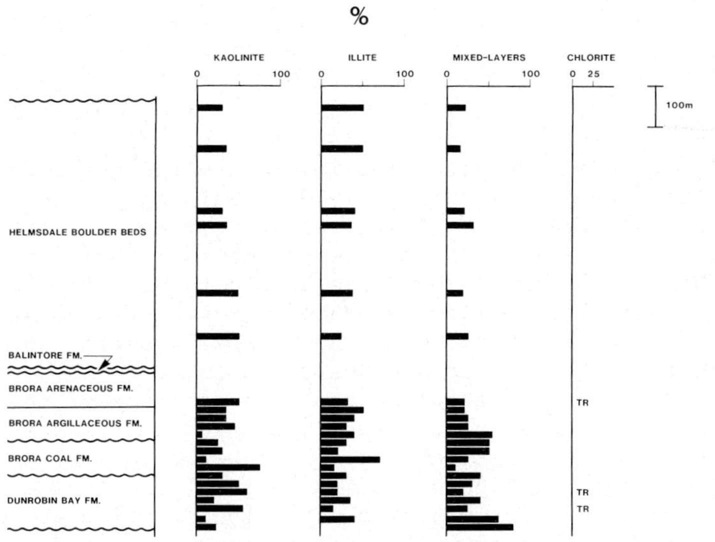

Fig. 2. Semi-quantitative clay mineralogy of shales from Brora. The figure includes representative data from more than 300 analyses (Hurst, 1980). TR = trace.

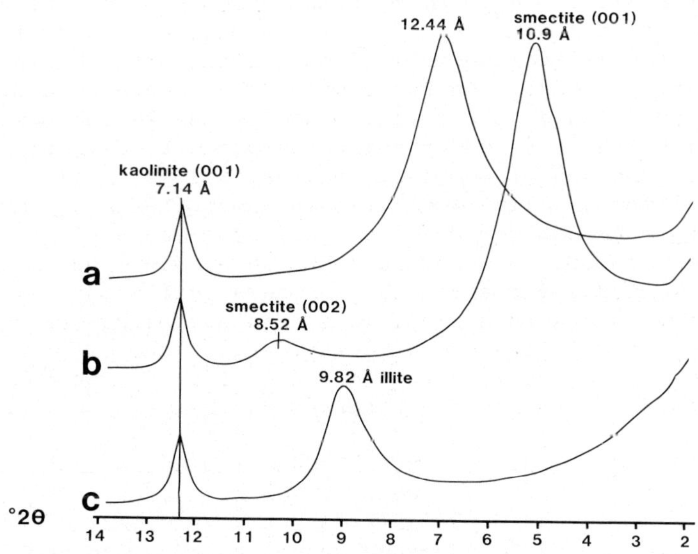

Fig. 3. Fully expandable smectite from the base of the Jurassic section at Brora.
a=air-dried untreated, b=glycolated, c= heated to 375°C for 1 hr.

ORIGIN OF DETRITAL CLAYS

When compared with other Mesozoic shales (Foscolos, et al. 1976; Frey, 1978) these Brora shales are relatively kaolinite-rich. Kaolinite distribution in modern oceans tends to be concentrated near to major landmasses, particularly adjacent to areas where leaching is the dominant weathering process (Griffin, et al. 1968). More specifically, abundant detrital kaolinite is often associated with major estuaries and deltas (Gibbs, 1977). Palaeogeographic reconstructions of the Jurassic landmass in NE Scotland (Anderton, et al. 1979) show that Brora was proximal to a landmass throughout the Jurassic. No palaeoclimatic evidence exists to confirm the presence of lateritic weathering during the Jurassic at Brora, although the climate was more humid than during the Triassic, sufficiently so to support forest growth (Harris and Rest, 1966). From the available data, it is probable that the high kaolinite content of these shales is mainly caused by erosion of a kaolinite-rich source rock and only partly caused by the prevailing probably sub-tropical climatic conditions. The ordered illite/smectites (Fig. 5) are typical of illite/smectites which have undergone diagenetic modification at temperatures in excess of 100°C during burial (Hoffman and Hower 1979). Devonian sediments to the NE of Brora contain similar ordered illite/smectites (Wilson 1971) and so constitute probable source rocks for the illite/smectites found in the Jurassic shales. Conclusive evidence of the origin of the ordered illite/smectite clays is found in Kimmeridgian Helmsdale Boulder Beds (Fig. 1). The boulder beds contain large boulders of Devonian sandstone which form thick sandstone units within black shales. The clay mineral composition of the Devonian boulders is the same as that of the sandy matrix between them. Furthermore, the illite/smectite compositions in the boulders, in the sandy matrix, and in the surrounding shales are identical. The variable composition of these illite/smectites (20-30% smectite interlayers) is interpreted as being the product of minor alteration during weathering and of compositional variations in the source rocks.

Randomly-ordered illite/smectites and smectites are common products of weathering. Randomly-ordered, highly expandable illite/smectites (Fig. 4) are recorded in Jurassic rocks in W. Scotland (Amirri-Garroussi, 1977), and their abundance in these shales may reflect periods during the Jurassic when a specific source was being weathered. Alternatively, the illite/smectite may reflect the existence of specific chemical conditions in the weathering environment.

On the basis of present-day outcrops and by using palaeogeographic reconstructions, the probable Jurassic source-area for these shales may be identified. Sedimentological data indicates that the detrital input came from the north and west of the exposed section, an area containing Triassic and Devonian sediments, Precambrian metamorphic and igneous rocks, and Caledonian igneous rocks. Ordered illite/smectites were derived from the weathering and erosion of Devonian sediments (Wilson, 1971). Basic igneous and metamorphic rocks are possible sources for the randomly-ordered illite/smectites. No kaolinite-rich source rock can be found, but typical subtropical

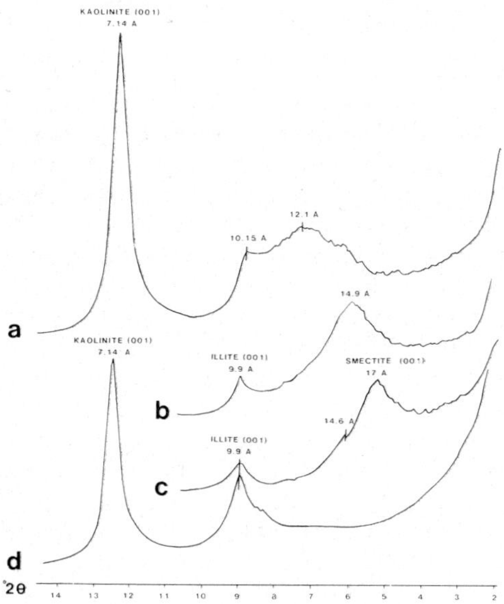

Fig. 4.
< 2 μm fraction of a shale (Lady's Walk Member) from Brora typified by the presence of highly expandable illite/smectite. a=air-dried untreated sample, b=Mg^{2+} saturated, c=glycolated, d= heated to 375°C for 1 hr.

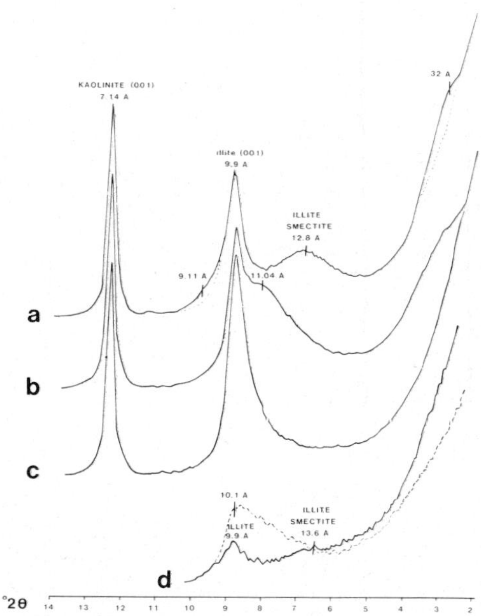

Fig. 5.
< 2 μm fractions from shales typified by the presence of low expandability, ordered illite/smectites. a=glycolated sample, b=air-dried untreated, c=heated to 375°C for 1 hr. a,b and c are from a Kimmeridgian shale, d is a poorly defined illite/smectite typical of shales at Brora. Solid line = glycolated, broken line - air-dried untreated.

weathering of igneous terrain produces kaolinite along with smectite (Drever, 1971). The question of the possible source of high kaolinite content of the shales remains unanswered.

POSSIBLE DIAGENETIC ALTERATION OF CLAYS

Illite/smectite compositions may be useful in determining the extent of thermal metamorphism of shales (Hoffman and Hower, 1979). A plot of illite/smectite compositions against the stratigraphic section (Fig. 6) shows a large variability throughout the section. The maximum burial temperature for the section is estimated from the illite/smectites which contain the highest percentages of smectite interlayers (A,B,C and D, Fig. 6). Using this data, the thermal maxima attained in the section is unlikely to have been greater than 65°C. From independant organic maturation data (D.G.Batten, pers.comm.) the maximum thermal maturity of the shales is estimated between 50 and 60°C. Many of the points for illite/smectites are above the 95°C isotherm (Fig. 6); these are the reworked clays from Devonian sediments. The reworked illite/smectites have not re-attained thermal equilibrium, which was reached during their previous burial history. Effectively, the ordered illite/smectites have been metastable interstratified species since their deposition in the Jurassic.

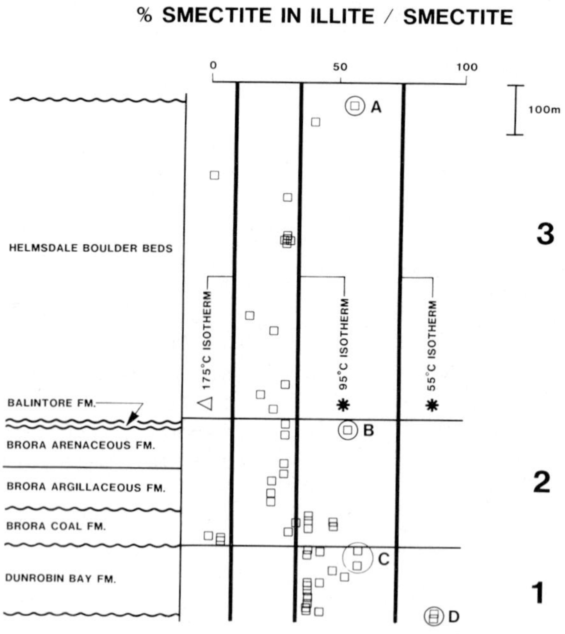

Fig. 6. Estimation of thermal maxima for the Brora section, presented as a vertical log, using the % smectite content of illite/smectite as a geothermal index. Areas 1,2 and 3 are geographical areas of outcrop (Fig. 1). A,B,C and D are the values used to estimate thermal maxima. * after Perry and Hower (1970). Δ after Hoffman and Hower (1979).

DISCUSSION

The mixture of illite/smectites of high and low expandability in these shales produces a superficially confusing picture of the thermal maturation of the section. High temperature polymorph illite/smectites, interpreted to be reworked from Devonian sediments, are unlikely to have undergone any major structural changes since their redeposition. Therefore, thermal reactions are unlikely to have caused the expulsion of interlayer cations and bound water during the burial of these illite/smectites.

Absence of any trends of thermal maturation in the stratigraphic section (Fig. 6) is interpreted to mean that the estimated thermal maxima were not reached when the rocks formed a vertical column of sediment, but rather when they were buried at similar depths after disruption by faulting. No comparable section exists to confirm this hypothesis.

Diagenetic reactions in shales are often considered to be important sources for authigenic cements in sandstones (Curtis, 1978) and for expelling water which aids the primary migration of hydrocarbons (Cordell, 1972). The high content of kaolinite plus ordered illite/smectite of these shales implies a low source potential for the diagenetic (thermal) evolution of structurally bound cations and water. In general, the clay mineral composition of the Brora shales will be stable up to temperatures of ~ 100°C, or 40°C higher than those attained during burial.

CONCLUSIONS

1) Brora shales are rich in detrital kaolinite, the presence of which cannot be directly correlated with palaeoclimatic data.

2) Ordered illite/smectites occur as thermally metastable mixed-layer clays which have been reworked from Devonian sediments.

3) The estimate of the thermal grade using illite/smectite compositions is confused by an overprinting effect caused by the presence of the ordered (high temperature) illite/smectites.

4) During the burial diagenesis of the shales since the Jurassic they have been poor sources for expulsion of cation-rich fluids and structural water. Therefore, these shales are unlikely to have a high potential for authigenic cement production or for primary flushing of hydrocarbons.

REFERENCES

Amirri-Garroussi, K., 1977. Origin of montmorillonite in the early Jurassic shales of NW Scotland. Geol. Mag. 114, 281-290.

Anderton, R., Bridges, P.H., Leeder, M.R. and Sellwood, B.W., 1979. A dynamic stratigraphy of the British Isles, A study in crustal evolution. George Allen and Unwin, London. 301 pp.

Cordell, R., 1972. Depths of oil migration and primary migration: a review and critique. Amer. Ass. Pet. Geol. Bull. 56, 2029-2067.

Curtis, C.D., 1978. Possible links between sandstone diagenesis and depth-related geochemical reactions occuring in enclosing mudstones. J. geol. Soc. London 135, 107-118.

Deegan, C.E. and Scull, B.J., 1977. A standard lithostratigraphic nomenclature for the Central and Northern North Sea. I.G.S. Report 77/25; Bull. Norw. Petrol. Direct., No. 1.

Drever, J.I., 1971. Chemical weathering in a subtropical igneous terrain, Rio Ameca, Mexico. J. Sedim. Petrol. 41, 951-961.

Foscolos, A.E., Powell, T.G. and Gunther, P.R., 1976. The use of clay minerals and inorganic and organic geochemical indicators for evaluating the degree of diagenesis and oil generating potential of shales. Geochim. cosmochim. Acta 40, 953-966.

Frey, M., 1978. Progressive low-grade metamorphism of a black shale formation, Central Swiss Alps, with special reference to pyrophyllite and margarite bearing assemblages. J. Petrol. 19, 93-135.

Gibbs, R.J., 1977. Clay mineral segregation in the marine environment. J. sedim. Petrol. 47, 237-243.

Griffin, J.J., Windom, H. and Goldberg, E.D., 1968. The distribution of clay minerals in the world ocean. Deep Sea Res. 15, 433-459.

Harris, T.M. and Rest, J.A., 1966. The flora of the Brora coal. Geol. Mag. 103, 101-109.

Hoffman, J. and Hower, J., 1979. Clay mineral assemblages as low grade metamorphic geothermometers: application to the thrust disturbed belt of Montana, U.S.A. In: P.A. Scholle and P.R. Schluger(Editors), Aspects of Diagenesis, SEPM Spec. Publ. No. 26, p. 55-80.

Hurst, A.R., 1980. The diagenesis of Jurassic rocks of the Moray Firth, NE Scotland. Univ. Reading, Ph. D. thesis.

Neves, R. and Selley, R.C., 1975. Review of the Jurassic rocks of NE Scotland. Jurassic Northern N. Sea Symp. Stavanger, Sept. 1979. JNNSS/5, p. 1-29.

Perry, E. and Hower, J., 1979. Burial diagenesis in Gulf Coast pelitic sediments. Clays Clay Min. 18, 165-177.

Reynolds, R.C. and Hower, J., 1970. The nature of interlayering in mixed-layer illite/montmorillonites. Clays Clay Min. 18, 25-36.

Weir, A.H., Ormerod, E.C. and ElMansey, I.M.I., 1975. Clay mineralogy of sediments of the Western Nile Delta. Clay miner. 10, 369-368.

Wilson, M.J., 1971. Clay mineralogy of the Old Red Sandstone (Devonian) of Scotland. J. sedim. Petrol. 431, 995-1007.

MINERALOGICAL AND TECHNOLOGICAL CHARACTERISTICS OF CERAMIC CLAYS

HEINRICH KROMER

Staatl. Forschungsinstitut für Angewandte Mineralogie bei der Technischen Universität München, 8400 Regensburg, West-Germany

INTRODUCTION

Clays have been among the first natural raw materials used for the cultural development of mankind and they have become an important factor in the history of civilization. Clay sculptures have been known for more than 10.000 years. Certain epochs of ancient history are dated according to specific ceramic technological achievements. Up to now clay has been an important natural raw material and has a large variety of uses. As industrial minerals, clays are a complex group which consists of several mineral commodities, each having a somewhat different mineralogy, geological occurrence, technology and industrial use. The reason for the great importance of clays is that they have outstanding properties, e.g. high plasticity when mixed with water, high dry strength and the ability to form hard and chemically resistent bodies upon heating.

The author has been involved in some research projects on German clays, their origin, mineralogical composition and technical properties for the ceramic industry (Kromer 1968-1978, 1978, 1980, 1980a, Kromer et al. 1980). This report is based on some data and results obtained in the projects mentioned above. The conclusions are generally valid for such materials.

GEOLOGICAL ORIGIN OF CERAMIC CLAYS

Clays as secondary sedimentary materials are found in many parts of the world. They occur in various geological formations of the Mesozoic and Neozoic era. Most of the best clays are of the Tertiary age. They are found e.g. in Germany, France, Great Britain, CSSR, USA, etc.. Clayey sediments from the Quaternary appear mainly as loess loams or loams. As cheap raw materials they are mainly used in ordinary ceramic.

NOMENCLATURE

The term clay is used differently by earth scientists and technicians. The scientists consider clay to be a sedimentary rock with a very high percentage of finest particles $< 2\ \mu m$. In ceramic application clays in general are selected exclusively on the basis of their technical properties. Special products require certain clays with special properties; accordingly, clays are classified e.g. as brick clays, stone ware clays, white ware clays or refractory clays.

RELATION BETWEEN MATERIAL COMPOSITION AND TECHNICAL PROPERTIES

The technical properties of clays are principally determined by their mineralogical composition and particle size distribution. The most important feature is the type of phyllosilicate minerals in a clay.

Phyllosilicates affect the plasticity and binding abilities because of their fine particle sizes and surface properties. Kaolinite controls the refractoriness; illite, mica, montmorillonitic minerals and chlorites promote sintering. Dry strength depends partly on particle sizes, but the presence of montmorillonite minerals is still more important. Illite, feldspar and fine grained carbonates influence the vitrification. Coarse grained carbonates lead to bursting ("lime splits"). Feldspar and iron minerals are still more effective for vitrification even if present only in small quantities. Iron minerals, titanium and lime control the development of the fired colour. Quartz opens the structure of a body and thus it facilitates drying processes. On the other hand it can cause difficulties during the firing process because of its phase transformations which effect volume changes.

The reactions which occur during firing are partly reactions in the solid state and partly interactions with a melting phase. The former type of reactions takes place at low temperatures, since a melt develops only at about 900°C. The predominant crystal phase is mullite, besides a glass phase when fired at sufficiently high temperatures. But the formation of cristobalite, hematite or cordierite can also be observed and they have an important influence on the properties of a body. The formation of cristobalite competes with the dissolution of quartz in the melting phase. The presence of residual quartz or of newly developed cristobalite causes some problems in cooling after firing because of phase transformations which bring about volume changes.

CLAYS FOR ORDINARY CERAMICS

The main products of ordinary ceramics are bricks, roof tiles, flower-pots, wall and floor tiles etc..The basic raw materials used in Germany are mostly loams or loess loams. These materials are of a silty nature with low amounts of particles $< 2\ \mu m$. In addition to large amounts of quartz ($\pm$ 50%) and some feldspars, the following phyllosilicates are present: illites (+sericite), mixed-layer minerals, rarely kaolinite, varying amounts of montmorillonite and some chlorite which determine plasticity and sintering behavior. The latter depends also on the content of iron minerals such as limonite, goethite, siderite. They determine in conjunction with some carbonates (calcite, dolomite) the red firing colour of the bodies. The firing temperature is of the order of about 1000° C.

Coarse loams mostly cause some difficulties in workability; therefore the body is plasticized by addition of finer grained clays. Most of them are red firing and they are often also used exclusively for the production of ceramics of this type.

Characteristic data of these materials are listed in Table 1. The clays are characterized by high contents of iron minerals (5-15%), a high percentage (20-63%) of illite (+sericite), various amounts of mixed-layer minerals and smectites; the amount of kaolinite is almost always very small. The quartz contents are commonly low; quite often some feldspar and carbonates are present. Chemical analysis shows large amounts of Fe_2O_3 and K_2O. The particle fraction $< 2\ \mu m$ in these raw materials is normally very large.

The dry strength increases with a higher fine particle content, but increases even more with a higher percentage of smectites and mixed-layer minerals. Even the silty loams show a high dry strength due to the high montmorillonite content.

The mineral composition of the red firing clays is favourable for good sintering behavior, as demonstrated by low water absorption, already observed at 1050° C.

Only Tertiary clays containing higher amounts of kaolinite show high porosity in the fired bodies and a greater refractoriness in comparison with other clays.

Red firing clays with high contents of illite, smectite and iron minerals are the typical materials used for the production of bloating clays.

WHITE WARE CLAYS - STONE WARE CLAYS

The required properties of these materials are listed in Table 2. Data for typical German clays used as raw materials for these ceramic products are shown in Table 3 and 4. White ware clays are characterized by lower amounts of fine particles $< 2\ \mu m$ (20-44%), higher contents of quartz and kaolinite, various amounts of illite, and a low iron content as required for the white colour of the products.

The casting properties of the white ware clays are good. Glazing asks for high porosity, which is met easily in these bodies. After heating processes white ware clays with high amounts of coarse grained quartz and a small content of fluxes develop only little cristobalite, but much mullite. The quartz minerals remain in the bodies; only very small amounts of quartz are dissolved.

Stone ware clays must have developed dense hard bodies at about 1200° C. A high percentage of particles $< 2\ \mu m$ and a high content of illite, besides kaolinite and finer grained quartz are characteristic for these raw materials. This material composition is favourable for the high density (low porosity) of the fired bodies as well as for good refractoriness. When firing these clay bodies, the formation of mullite and cristobalite starts at temperatures of about 1050° C; a maximum amount of cristobalite and mullite is obtained at 1200° C; at the same time fine quartz particles are dissolved.

REFRACTORY CLAYS

Refractory clays are used for the production of chamotte or as binding clays in refractory industries. Minor quantities are applied for the production of technical porcelains.

The main requirement of such raw materials is refractoriness. Typical clays of that type are listed in Table 5. They are characterized by very high contents of kaolinite, which determine the refractoriness. Besides there are present illite and mixed-layer minerals and often some small amounts of smectites. The contents of quartz, feldspar, and iron minerals should be small.

In some cases quartz can improve refractoriness, especially when the clay contains a bit more illite. In this case the coarser grained quartz stabilizes the structure of the body and the softening temperature is unusually high. The Seger cone data for the refractory clays range between 30-36. Due to the high content of kaolinite, the Al_2O_3 contents is also high and this is characteristic for these materials. Even somewhat higher amounts of Fe_2O_3 and K_2O (Table 5) not essentially impair the refractoriness.

If there are more fluxes in bodies fired at about 1200°C the formation of mullite increases beside the formation of some cristobalite and a melting phases. Refractory clays with not many fluxes exhibit crystallization of cristobalite besides other phases at these firing temperatures (Kromer & Schüller, 1974). In both cases quartz is partly dissolved.

A high percentage of particles $< 2\ \mu m$ is also characteristic for the refractory clays, which improves plasticity and binding properties. Illite, mixed-layer minerals and occasionally small amounts of pure montmorillonite together with finer dispersed organic matter promote good dry strength of the materials. Dependent on mineralogical composition and particle size distribution the porosity of the bodies fired at about 1200° C ranges between 0 and 25%.

KAOLIN

Kaolin is a very important technical product and has a large variety of uses. It is produced in huge industrial washing plants from disintegrated granitic rocks or kaolin sands. Properties and applications of kaolins are only briefly discussed in this report (Kromer et al. 1973).

In the ceramic industry kaolins are mainly used for the production of high grade porcelains and white wares. Mineralogically these kaolins, apart from their major component kaolinite, sometimes contain halloysite, and further minor amounts of illite, mixed-layer minerals illite montmorillonite, quartz and feldspar. Metal oxides of iron or titanium would impair the white colour and the translucency of high grade porcelains, but they are mostly eliminated in these materials during the washing and cleaning (bleaching) processes. The required properties of kaolins are: good dry strength, plasticity, easy deflocculation and an excellent white firing colour. Dry strength and plasticity are improved by small amounts of mixed-layer minerals such as illite-montmorillonite, or pure montmorillonite, and in some cases by halloysite; but higher amounts of montmorillonite cause problems with deflocculation. A certain amount of halloysite in the materials improves the translucency of the bodies.

REFERENCES

Kromer, H., 1968-1978. Rohstoffmerkblätter der Deutschen keramischen Gesellschaft; Ber. Dt. Keram. Ges.

Kromer, H., 1978. Tertiär-Tone in NE-Bayern, 1. Petrographie-Mineralogie-Geochemie; Fortschr. Miner. 56, S. 1-104

Kromer, H., 1980. Tertiary Clays in Northeastern Bavaria (Oberpfalz); Geol. Jh. D. 39, S. 7-23

Kromer, H., 1980a. Tertiary Clays in the Westerwald Area; Geol. Jh. D 39, S. 69-84

Kromer, H. & Schüller, K.H., 1980. Farbigbrennende Tone als keramische Rohstoffe; Keram. Zt. 32 Jhg. Nr. 10, S. 574-578.

Kromer, H., 1977. Blähton, Handbuch der Keramik; Keram. Zt., Gruppe II M, 1-14

Kromer, H. & Schüller, K.H., 1974. Sintervorgänge bei Kaolinen und Tonen; N. Jb. Miner. Abh. 122, S. 145-185

Kromer, H. & Schüller, K.H., 1973. Eigenschaften von Kaolinen für die Keramik und als Füllstoff; Ber. Dt. Keram. Ges. 50, Nr. 1, S. 15-19, S. 39-41

TABLE 1

RED-FIRING CLAYS

	Mineral Comp. %							Clay %	Dry Strength	W_A %	
	I	K	X_{ML}	Q	Y_{fe}	F	C	< 2 µm	N/cm^2	1050^oC	SK
Tertiary/Bavaria	20	50	4	15	8	3	tr.	66	120	15.5	28
Rhät-Lias/Bavaria	40	28	5	15	6	3	3	65	680	0.2	14
Tertiary/Westerw.	30	35	-	20	15	-	-	64	340	0.3	17
Tertiary/Elsass	20	35	10	20	11	4	-	70	253	5.4	15
Keuper/Bavaria	60	-	10	20	6	4	tr.	72	540	2.6	4a
Tertiary/Bavaria	35	3	15/25	15	6	1	-	43	1050	0.1	4a
Loam/Bavaria	20	-	20	50	5	3	2	35	470	8.7	8
	Chemical Comp. %				SiO_2	48 - 57		Fe_2O_3	6 - 14		
	(dried 110° C)				Al_2O_3	11 - 25		K_2O	2,4 - 6,4		

Symbols:
K = kandite, I = mica, illite minerals, X_{ML}+ M = mixed-layer + smectite minerals,
Q = quartz, Y_{fe} = iron minerals, F = feldspar, C = carbonate minerals
W_A = water absorption of the fired clay
SK = pyrometric cone (seger cone)

TABLE 2

WHITE WARE CLAYS	PROPERTIES	STONE WARE CLAYS
white-creme	colour of firing	mostly ligth-grey-beige red, too
porous porosity apparent at 1200° C: 10-20% easy to glaze	structure of body	dense porosity apparent at 1200° C: 0-2% problems in glazing salt-loam-glazes
small, high accuracy	shrinkage	high variety in accuracy
extremely good allows extra ordinary shapes in sanitary ware, dishes, etc.	casting properties	only exceptionally possible
many low firing temperature two firing processes 1st high, 2nd low (glaze)	glaze properties	few one firing temperature
no	acid proof	yes
no	frostproof	yes
little	mechanical stress	very high, various techni- cal applications

TABLE 3

WHITE WARE CLAYS

Clay •	Mineral Comp. %						Porosity %		Clay %	Chemical Comp. %		
	Q	K	I	X_{ML}	F	Y_{fe}	1200°C	SK	< 2 µm	SiO_2	Al_2O_3	Fe_2O_3
Giesston 502	50	23	22	2	2	1	14.6	29	36.6	76.6	19.0	0.84
CHR 20	71	10	13	2	6	1	16.3	26	27.8	85.4	11.7	0.31
Kreuzton	55	7	30	-	8	1	19.0	23	21.8	74.9	18.1	0.55
Ton I 20	51	23	21	3	2	1	19.8	29	48.1	80.1	16.8	0.58
GTM 8012	65	22	8	-	3	2	15.4	26	41.9	81.9	14.2	0.94
Ton 303	50	30	13	4	-	1	9.9	30	35.5	76.3	18.9	0.81
Ton 1403	57	22	17	-	3	1	10.4	31	44.4	78.3	17.1	0.45

• Westerwald, BRD

TABLE 4

STONE WARE CLAYS

	Mineral Comp. %						Porosity %		Clay %	Chemical Comp. %			
Clay˙	K	I	X_{ML}	Q	F	Y_{fe}	1200°C	SK	<2 µm	SiO_2	Al_2O_3	Fe_2O_3	K_2O
P 203	39	30	2	25	2	2	0	31/32	74.6	64.2	28.2	1.6	3.2
KMG I	51	32	tr.	16	tr.	1	0	32	83.9	60.8	32.2	1.3	3.4
Ton 2001	46	25	3	22	tr.	3	0	32	76.1	62.5	29.8	2.1	1.9
Ton 1351	38	20	-	36	5	1	0	30/31	61.2	69.5	24.4	0.9	2.5
P 161	40	32	4	20	3	1	0.9	31	78.5	61.6	31.3	1.2	3.5
Steinmühle˙˙	32	36	3	26	-	2	0.3	32	50.7	63.2	29.0	1.0	3.9
Ponholz ˙˙	35	25	2	32	3	3	1.4	32	59.2	64.8	29.3	2.4	1.5

˙ Westerwald, BRD

˙˙ Oberpfalz, BRD

TABLE 5

REFRACTORY CLAYS

Clay *	Mineral Comp. %						SK	Clay %	Chemical Comp. %		
	K	I	X_{ML}	Q	F	Y_{fe}		< 2 µm	Al_2O_3	Fe_2O_3	K_2O
Ton VI*	85	9	3	3	-	1	35/36	96.8	42.8	1.8	0.7
Ton XIII	90	4	4	-	-	2	35/36	97.8	43.2	2.3	0.5
HFF Ton A	83	7	4	4	1	1	34/35	84.0	41.8	2.0	1.0
Uni 1	74	12	4	7	3	1	32	67.2	38.1	2.2	1.7
Uni 2	66	14	3	12	4	1	33	59.2	37.0	1.9	1.7
HB 40/42**	70	13	8	6	-	3	33/34	96.4	38.3	1.8	1.11
P 202	51	27	-	17	3	2	33	80.2	33.9	2.9	3.2
Blauton HFF	66	20	8	5	-	1	33	98.7	38.9	1.6	2.4
TK 8054	44	20	5	26	3	2	30	77.8	27.7	1.4	2.7

*Oberpfalz, ** Westerwald/Rhein, BRD.

GEOLOGICAL AND PHYSICO-CHEMICAL CHARACTERIZATION OF BELGIAN NON-REFRACTORY CLAY DEPOSITS AND ITS IMPLICATIONS FOR INDUSTRIAL USE

J. DECLEER[1], R. OTTENBURGS[1], N. VANDENBERGHE[2] & W. VIAENE[1]

[1] Fysico-chemische geologie, K.U.Leuven, Belgium
[2] Geologische Dienst, Brussel, Belgium

ABSTRACT

The geographical distribution and geological characteristics of non-refractory Belgian clays and loams are described. Chemical analysis, mineralogy, grain size analysis, total specific surface and plasticity of fifteen clays and loams are given.

The chemical and physical properties can be classified on the basis of their correlations into four groups. The main variables in each of the four groups are : silica and alumina content; iron oxide and adsorbed water content; Mg, Ca and Na; and sulphur and organic matter.

INTRODUCTION

A geological and physico-chemical characterization of different genetic types of Belgian post-palaeozoic clays and loams, exploited by the structural clay product industry, has been carried out by means of a total chemical analysis, a mineralogical analysis, a determination of the grain size, of specific surface and plasticity.

Three topics will be discussed : first, the structure of the analytical data; secondly, a comparison with some other European clays; and thirdly, an evaluation of the suitability of the Belgian deposits.

OCCURRENCE OF NON-REFRACTORY CLAYS IN BELGIUM

Structural clay products are made from geological strata different both in age and in sedimentary environment. They are listed in Table 1 (°) and the geographical distribution is represented in Figure 1. Most data are taken from Calembert et al. (1947).

1- Holocene Meuse alluvial plain (Limburg-Meuse area)

The deposits are situated within the recent alluvial plain, consisting of a series of abandoned and filled river channels; between these channels sandy to clayey silts occur (less than 2 m thick). Only in a few cases these silts are sufficiently clayey to interest the local brick industry. More often the infill sediment of the abandoned channels is clayey enough for brick production. The channels, however, are in general only 1 m thick and 100 m wide. Some of them are visible as slight depressions in the landscape.

TABLE 1 - Non-refractory clays in Belgium

Geological strata	Exploitation area	Sedimentary facies
1 -Holocene Meuse alluvial plain (1)	Limburg-Meuse area	fill-up of abandoned channels
2 -Holocene Scheldt alluvial plain	Oudenaarde	alluvial sediments
3 -Upper & Middle Weichsel loam and loess (Brabant & Hesbaye member) (4)	Middle Belgium (geomorphologically)	windblown
4 -Dunkerque clay (Holocene)	coastal plain	tidal clay deposit
5 -Campine formation (Lower Pleistocene) (2)	Antwerp-Campine	coastal environment (fluvial-tidal flat)
6a-Rupelian Boom clay b-Lower Rupelian clay c-Altered Boom clay (1)	Mechelen-Boom- Sint-Niklaas S-Limburg Waasland-Hasselt	marine shelf deposit
7 -Lower Oligocene Henis clay	Tongeren	lagoon or coastal lake
8 -Bartonian Asse clay (1)	E of Brugge NW of Brussels	marine shelf clay
9 -Lower Panisel formation (2) a.Lower member b.Basal member	 S-Flanders Kortrijk	 marine shelf clay shallow or lagoonal
10-Ypres clay (2)	S-Flanders N-Hainaut,S-Brabant	marine shelf deposit
11-Upper Landen clay	NE-Hainaut SE-Hesbaye	fluvio-lagoonal
12-Hergenrath clay (Santonien) (2)	Welkenraedt	lagoonal
Weald clay (Cretaceous)	Hautrage-Baudour	deltaic lake deposit
13-Shales of Ethe (Jura)	Arlon-Virton	marine deposit
Keuper	Houdemont	
14-Palaeozoic shale alteration from : a.Carboniferous shale b.Upper Devonian shale c.Lower Devonian shale	 Charleroi,Liège Fagne & Famenne Bastogne	alteration
Upper Oligocene & Lower Miocene preserved in karstic depressions	Andenne,Condroz (Entre Sambre et Meuse)	continental
Decalcification clay from Palaeozoic limestone	Flönnes (Welkenraedt)	

(°) Strata, still of significance today, are underligned.
( ) Number of analysed samples.

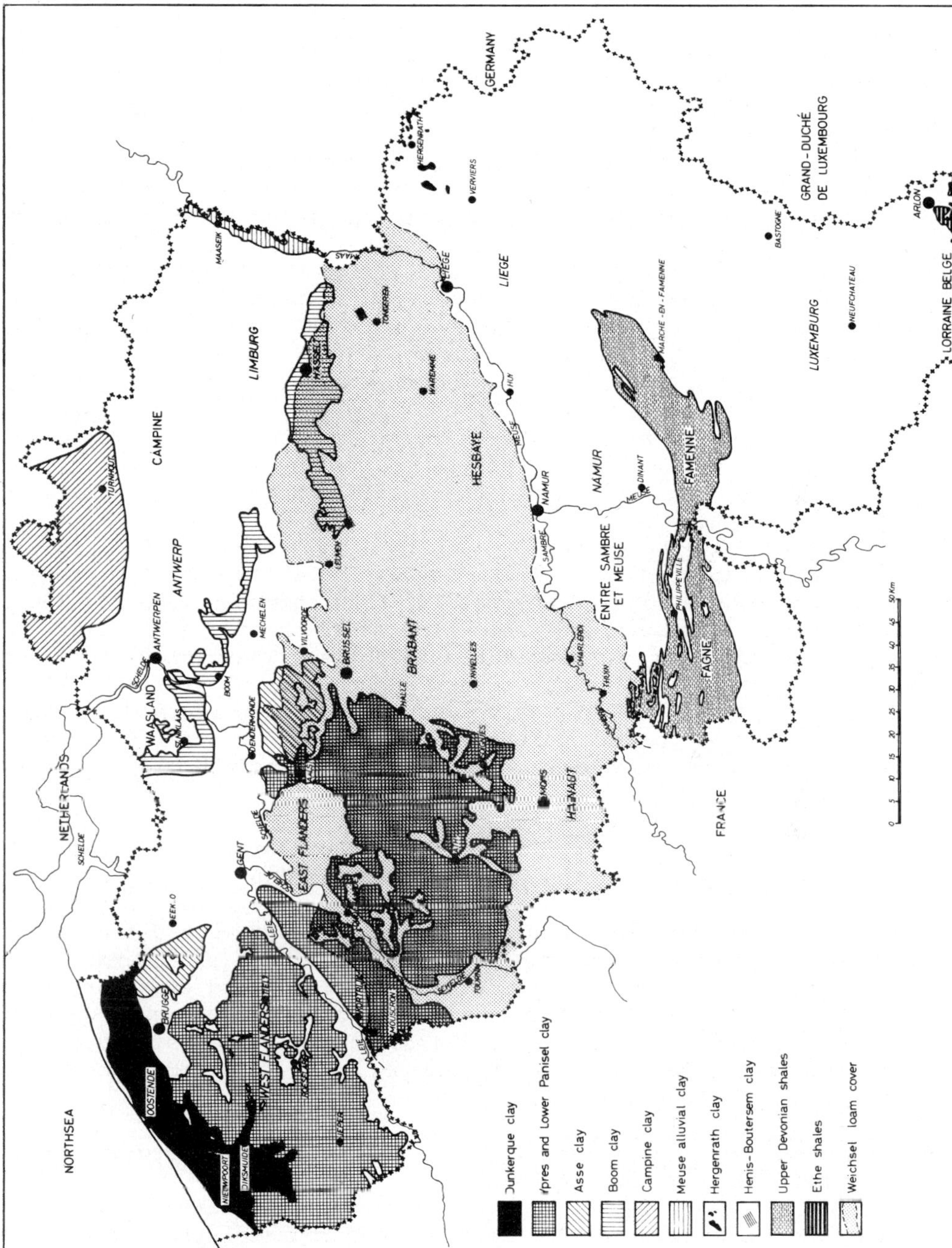

Fig.1 OCCURRENCE OF CLAY RAW MATERIALS FOR STRUCTURAL CLAY PRODUCT INDUSTRY IN BELGIUM.

2- Holocene Scheldt alluvial plain (N of Oudenaarde)

In parts of the actual alluvial plain of the Scheldt and Leie rivers loamy deposits were formed due to periodical flooding of plain. In the Scheldt alluvial plain thicknesses of 1 to 1.5 m are recorded.

3- Upper and Middle Weichsel loam and loess belt

Mainly during the Upper and Middle Weichsel a layer of wind-blown loess was deposited essentially over Middle Belgium and spread over the land-surface as a blanket. In general, the loess layer is a few meters thick, but in the river valleys it can reach more than 10 m. Meteoric water and ground water gradually decalcified the loess. The term loam or 'Lehm' refers to this decalcified loess. Loess and loam are for the largest part made up of silt-sized quartz grains with only minor sand and clay admixtures. The carbonate content will not only move the red fired-brick colour towards yellow colours but it lowers considerably the firing temperature (from ± 1200°C to ± 1050°C). Loam shows a greater plasticity than loess. The valley loams show greater lithological variability since they have been transported by water action. In the loess deposits, the carbonates are finally dispersed but small concretions do frequently occur as well.

4- Dunkerque clay (Holocene)

A wadden depositional environment was installed over the present coastal plain by a major transgressive phase around 2000 BP. The tidal flat clays are soft and shell-rich while the marsh clays are more oxidized and contain more small plant debris. These clays occur together with sandy flat, small creeks and periodical storm deposits. The colour of the fired bricks varies between pale red and yellow according to the carbonate content; carbonates are generally present in finely dispersed state. The maximum thickness of the clay deposit is about 3 m.

5- Lower Pleistocene Campine clay

The Campine clays are sedimented in the transition area between land and sea. In some areas they are described as fluvial clay deposits, in other areas they belong to a tidal wadden system. Different clay types are present ranging from sandy to very fine clays with different colours and varying organic matter content; besides, fine sands and peat occur frequently. Some clays contain carbonates and even carbonate concretions. The different clay and sand layers occur as wedges, as rather continuous layers or in gullies. The clay deposits are rarely exploited over more than 10 m and mostly in the S of their occurrence area where the clays are wedging out.

6- Rupelian Boom clay

The Boom clay (6a) is a marine shelf deposit of very wide-spread occurrence. Silty clay and heavy clay layers of only some tens of cm thickness alternate regularly. Besides these vertical grain size variations, also organic rich layers and carbonate rich layers of some tens of cm thickness occur; septaria are developed in the carbonate rich layers. It has been demonstrated that the sequence of these different layers is constant over the whole present exploitation area. The clay thickness in this area varies between 10 and 50 m. A major division in the clay is between an upper blackish grey clay, which is silty and rich in plant debris, and a lower grey clay which is made up of heavy clays which, towards the base, are silty and contain carbonates. Pyrite frequently occurs as concretions (mm up to 10 cm).

Especially in the Waasland area, it has been observed that the upper part of the clay (6c) can be altered down to 2 m deep. The altered clay is brown to yellow, oxidized and more plastic. This clay is especially sought for pottery production.

7- Lower Oligocene Henis clay

This is a green clay of about 5 m thick, containing some disseminated pyrite. It is a lagoon or coastal lake deposit and occurs as a restricted but continuous layer in the Tongeren area. An equivalent layer is present to the E of Louvain. Sand lenses and peaty horizons can occur. The clay is extremely fine grained and very plastic.

8- Bartonian Asse clay

The Asse clay is a relatively homogeneous very plastic grey clay, deposited as a marine shelf clay. The thickness of the clay in the former exploitation area rarely exceeded 10 m.

9- Lower Panisel Formation

The lower part (9a) contains a clay member with many thin sandy streaks or patches. It is rich in glauconite; pyrite is present as well. In general quartzitic sandstone layers are present. The deposit can reach thicknesses between 10 m and 20 m.

The basal layer (9b) is a very plastic homogeneous clay occurring as a continuous layer in S-Flanders and only a few meters (max. 7) thick. It is used particularly for roofing tile production.

10-Ypres clay

The Ypres clay is a marine shelf clay, reaching more than 150 m thickness in W-Belgium. In Brabant and Hainaut the Ypres clay is 10 to 30 m thick, partly because of erosion and partly because of thinner development. The clay is subdivided in a lower very heavy plastic clay and an upper much more sandy part. Both parts can be traced over the main area of occurrence, the lower part being approximately 100 m thick. On top of this sequence a very heavy clay ($\pm$ 10 m) occurs, used for roofing tile manufacturing.

11-Upper Landen clay

The Upper Landen clays are clay lenses of limited vertical (a few m) and horizontal extension between mainly sandy deposits in a fluvio-lagoonal complex.

12-Hergenrath clay

The Hergenrath clay occurs in patches in E-Belgium where it is never thicker than 10 m: in fact, the clay is made up of several thinner clay bands with different lithological properties. The facies is reported as lagoonal. Sand layers occur within the clay. Pyrite is present. It was noted that the basal part of the Hergenrath clay was more refractory than the upper parts. This basal part is possibly related to the alteration clays of the palaeozoic shales.

13-Shales of Ethe (Jura)

The grey Ethe shales represent a marine deposit of 20 to 25 m thickness. Especially the somewhat altered clay near the surface was used.

14-Alteration clays from Palaeozoic shales

A variety of Palaeozoic altered shales were exploited for brick production. The main shale types were Upper Carboniferous shales and Frasnian-Famennian shales. In general, only a few metres were exploited. Maybe it is appropriate here to mention the frequent use of spoil heap fragments as additives in different brick factories.

MINERALOGICAL, CHEMICAL AND PHYSICAL CHARACTERISTICS

In Table 2, some important mineralogical, chemical and physical data of representative samples of the different studied clays and loams, are listed.

DISCUSSION AND IMPLICATIONS

1- Structure of the analytical data

In a first attempt Pearson-correlation coefficients were calculated between 47 variables in order to find probable mutual relationships. In Table 3, only the most significant are given. The data of the clay mineralogy could not be used in the statistical analysis because of their semi-quantitative nature.

TABLE 3 - Pearson-correlation coefficients

	SiO_2	Al_2O_3	K_2O	Fe_2O_3	S	OrgC	$H_2O^{\pm}$	H_2O^+	Sp. Su.	Pla.	<2μm
SiO_2	1	-.76	-.72	-.61	-.60	-.65	-.78	-.49	-.55	-.52	-.70
Al_2O_3		1	.71	.37	.46	.59	.91	.58	.57	.60	.92
K_2O			1	.74	.56	.56	.84	.64	.71	.64	.73
Fe_2O_3				1	.61	.61	.63	.71	.81	.71	.53
S					1	.95	.59	.37	.61	.52	.50
OrgC						1	.69	.41	.65	.61	.64
$H_2O^{\pm}$							1	.64	.70	.63	.87
H_2O^+								1	.78	.79	.70
Sp.Su.									1	.92	.76
Pla.										1	.80
<2μm											1

(N = 15 / Correlations significant above .59 (99% confidence level), above .64 (99.5% confidence level) and above .73 (99.9% confidence level)

All the variables are inversely correlated with SiO_2. These are in fact, the components of the clay minerals, or are associated with the clay sediment. Al, K and H_2O are the main constituents of the clay minerals. The clay minerals are the finest particles and are indeed the bulk composition of the minus-2μm-fraction. Organic matter is a

TABLE 2 - Characteristics of the clays and loans

([1])	Clay Minerals Ill:Sm:Kao([2])	Theor. Cl.Min.	SiO_2	Al_2O_3	K_2O	Fe_2O_3	H_2O^{+}–	H_2O^{+}	S	F	OrgC	Na_2O	CaO	MgO	Sp. Su. ([3])	Pla. ([4])	<2μm	<20μm
1.	5 : 5 : <1	59.1	69.7	11.1	2.1	5.1	7.9	1.4	0.0	298	0.4	0.9	0.8	0.9	110	28.2	32.5	63.3
3.	4 : 5 : 1	34.3	71.0	7.6	1.9	3.5	3.4	0.9	0.0	350	0.1	1.1	5.5	1.4	60	23.1	18.6	37.1
3.	6 : 4 : <1	32.9	73.8	6.5	1.7	3.3	4.2	1.1	0.0	300	0.1	0.9	4.5	1.3	55	23.0	12.8	22.9
3.	6 : 4 : <1	36.4	66.0	7.4	1.8	3.6	4.2	0.9	0.0	298	0.1	1.2	7.9	1.8	67	24.6	14.5	35.1
3.	7 : 2 : 1	39.7	78.7	7.4	2.1	3.8	4.7	1.2	0.0	241	0.0	0.8	0.4	1.0	77	24.1	19.4	34.0
5.	2 : 7 : 1	61.4	66.5	14.2	2.9	4.4	8.8	1.7	0.1	600	0.3	0.4	0.2	0.9	108	33.2	43.2	59.2
5.	3 : 6 : 1	76.8	58.8	15.4	2.9	6.2	10.3	2.4	1.9	740	1.5	0.4	0.4	1.2	201	43.0	50.0	85.9
6c.	4 : 5 : 1	43.7	78.0	8.9	2.2	3.2	5.4	1.6	0.0	448	0.1	0.3	0.3	0.9	101	28.2	32.5	44.7
8.	2 : 8 : <1	75.0	54.2	16.5	3.1	6.4	9.9	2.7	1.1	539	1.1	0.8	2.4	2.0	187	47.4	68.0	96.2
9.	1 : 9 :<<1	52.7	69.4	10.9	2.3	5.8	6.6	2.6	0.9	450	1.0	0.3	0.8	1.3	160	52.5	49.9	70.5
9.	1 : 9 :<<1	59.2	71.6	10.7	2.2	5.2	6.4	2.8	0.0	463	0.2	0.4	0.5	1.2	219	52.5	48.0	60.1
10.	1 : 9 :<<1	53.0	63.9	13.9	2.8	6.7	8.9	5.1	0.4	447	0.5	0.6	0.9	1.4	185	46.7	50.2	78.6
10.	1 : 9 :<<1	44.0	77.4	7.2	2.2	3.9	4.6	1.5	0.3	212	0.4	0.5	1.0	1.1	115	36.0	16.7	27.9
12.	2 : 7 : 1	57.8	63.1	19.0	2.4	2.9	8.8	2.5	0.0	620	0.3	0.3	0.5	0.6	118	41.8	61.7	96.1
12.	4 : 3 : 3	42.5	74.5	12.0	1.6	2.1	6.0	1.3	0.4	365	0.7	0.2	0.3	0.6	78	28.0	38.2	71.2
		%	%	%	%	%	%	%	%	ppm	%	%	%	%	m^2/g	%	%	%

([1]) Numbers correspond with numbers of table 1.

([2]) Peak areas are calculated. The results are given in parts per ten (Vandenberghe, 1978).

([3]) Specific surface is related to the equilibrium moisture content (van der Velden, 1978).

([4]) Plasticity is determined by the Pfefferkorn impact test (van der Velden, 1979).

(Current methods were used for the other analyses.)

preferential site for pyrite formation, hence the mutual positive correlation. The plasticity and the specific surface are directly correlated with the grain size and the H_2O-content.

The theoretical clay mineral content (Keeling, 1962) shows interesting correlations with SiO_2 (-.75), Al_2O_3 (.81), Fe_2O_3 (.69), S (.67), K_2O (.84), OrgC (.75), <2μm (.83), F (.78), specific surface (.80), and $H_2O^{\pm}$ (.93). This determination has proved useful as a quick guide to clay types and as a routine control of variations in the clay.

There is a mutual relationship between MgO, Na_2O and CaO, but no correlations exist with other variables.

F, which is an important pollutant in the structural clay product industry, has also been analysed and shows important positive correlations with the clay content, i.e. with Al_2O_3, <2μm, OrgC, $H_2O^{\pm}$ and K_2O, but an inverse correlation with SiO_2.

In a second stage a R-mode factor analysis (Davis, 1973) was carried out on the data of 16 variables, in order to group these variables with mutual correlations. After calculating the principal components, those with an eigenvalue larger than .99 were transformed into factors by applying a varimax rotation. The results of the factor analysis are given in Table 4. The four factors explain more than 90 % of the total variance.

TABLE 4 - Grouping of the variables as a result of the factor analysis

Factor 1 :	SiO_2, Al_2O_3, K_2O, $H_2O^{\pm}$, F, <2μm and theoretical clay mineral content (32.1% variance)
Factor 2 :	Fe_2O_3, $H_2O^{\pm}$, plasticity and specific surface (25.3% variance)
Factor 3 :	MgO, Na_2O and CaO (17.6% variance)
Factor 4 :	S and organic C (15.8% variance)
The communalities of the 16 variables are all above .80.	

The first factor reflects the mutual correlations between the seven variables which are directly connected with the clay mineral components.

The second factor is dominated by the plasticity and the specific surface which are narrowly correlated with the adsorbed water. Fe_2O_3 is positively correlated with these three; hence they are grouped together. As can be seen from Table 3, the group is less strongly correlated with the clay content but shows high mutual correlations.

Factor 3 groups Mg, Na and Ca which are indeed mutually correlated, but not with other variables. Finally S and organic C form the fourth factor. In the given set of data this two parameters vary more or less indepently.

This statistical treatment allows us to group the different variables and to select a limited set of parameters to characterize the clay for its evaluation or to control the raw material in a production process. For a same type of clay deposit, the correlations may become enhanced, so that the selection of parameters to be analysed can be simplified.

2- Comparison of clays and loams

Mineralogically most of the Belgian clays and loams are rich in smectite and poor in kaolinite. The Holocene loams and the alluvial clays are relatively rich in illite. The estuarine clays show an intermediate illite content, while the marine clays are low in illite. This facts explain the high specific surface and consistency values of the heavy Belgian clays. The important contribution of smectite increases the flux content of the clays (up to 15%). On the other hand, their low K_2O, MgO, Fe_2O_3 and $CaCO_3$ content compensates the negative influence on the firing temperature. The eolian loams owe their high flux content (13-16%) to a relatively high carbonate percentage.

Table 2 shows in a general way that the illite content becomes lower with increasing age and that the smectite shows a reversed trend. However, these trends are not reflected in one of the chemical parameters.

Because of their high Fe_2O_3/Al_2O_3 and Fe_2O_3/CaO ratios most of the clays and loams give red coloured products on firing. The sulphur content of the majority of the studied materials does not exceed 0.1%; two clays scoring above 1% S show relatively high organic carbon content. The values for F vary between 200 and 740 ppm. The eolian loams and the alluvial clays show the lowest F-values, the estuarine clays the highest.

In comparison with 23 Dutch clays and loams (van Amerongen & van der Velden, 1970), the Belgian ones in general are poorer in Fe_2O_3, organic C and $CaCO_3$. Compared with 61 German clays and loams (Schmidt, 1973) the Belgian materials show a lower MgO, organic C and $CaCO_3$ content.

3- Suitability of Belgian clays and loams

Figure 2 shows the grain size distribution of Belgian clays and loams. They are spread over all the zones defined by Winkler (1955).

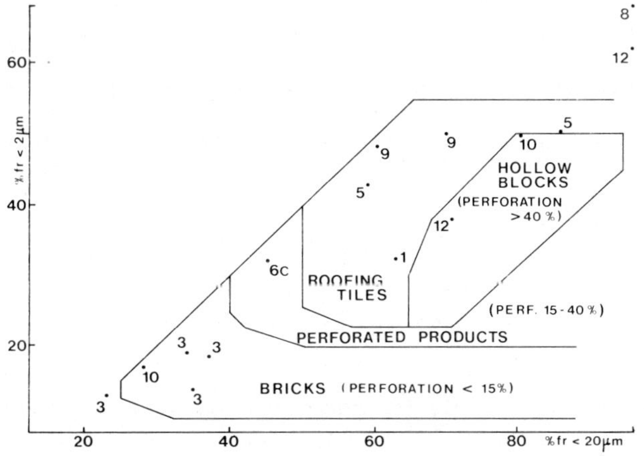

(Numbers correspond with numbers in Table 1.)

Fig. 2 - Grain size diagram according to Winkler (1955)

According to their grain size half of them are suitable for the production of roofing tiles. A few heavy clays are even too fine and have to be mixed with lighter clay, schist, porphyry or sand. The eolian loams are used to make bricks (perforation <15%). In spite of the high smectite content of most Belgian clays, good roofing tiles, hollow products and/or bricks were made in the past and are still made today. However, these high smectite contents necessitate much water to make the clay plastic enough for shaping. The drying process lasts longer and requires more care because of the high degree of shrinkage.

If a yellow coloured product is wanted, CaO must be added. The minimum amount of CaO (wt%) needed to yield a yellow colour is mentioned in Table 5.

TABLE 5 - Minimum amount of CaO to be added for yellow colour

	%CaO		%CaO
Eolian loams (3)	1.5	Estuarine clays (5)	8.0-12.0
(3)	2.0	(12)	0.0
(3)	7.0	Marine clays (6c)	6.0
(3)	0.0	(8)	11.0
Alluvial clay(1)	9.5	(9)	10.0-11.0
		(10)	7.0-13.0
Numbers between ( ) correspond with numbers in Table 1.			

From the point of view of energy conservation, it is interesting to be able to make a good approximation of the percentage of water a clay or loam needs in modelling by extrusion, in shaping of roofing tiles or for hand moulding. The data in Table 6 are derived from the Pfefferkorn consistency curves. For the extrusion method a Pfefferkorn residual depth of 28.5 mm was chosen, for roofing tiles 22.5 mm and for hand moulded bricks 5 mm.

TABLE 6 - Range of modelling waterpercentage

Genetic type of material	Extrusion	Roofing tiles	Hand moulding
Eolian loams (3)	17 - 20	not suited	26 - 28
Alluvial clay (1)	19 - 25	20 - 26	27 - 35
Estuarine clays (5,12)	21 - 40	24 - 41	31 - 44
Marine tertiary clays (6,8,9,10)	32 - 66	33 - 70	43 - 96
Numbers between ( ) correspond with numbers in Table 1.			

ACKNOWLEDGMENTS

The technical assistance by C. Moldenaers and D. Coetermans is acknowledged. The first author likes to thank the I.W.O.N.L. for financial support.

REFERENCES

Calembert, L. (avec la collaboration de M. Gulinck, G. Manil, R. Marlière et R. Tavernier), 1947. Les roches argileuses. Centenaire de l'A.I.Lg. Congrès '47, Section Géologie, pp. 245-314.

Davis, J.C., 1973. Statistics and Data Analysis in Geology. J.Wiley & Sons, Inc., 550 pp.

Keeling, P.S., 1962. Clay Min. Bull., 5(28): 155.

Schmidt, H., 1973. Ziegel Industrie, 6: 212-216.

van Amerongen, H. & van der Velden, J.H., 1970. Samenstelling en eigenschappen van 31 kleisoorten. Rapport CTI, TNO-Apeldoorn, Deel I & II, 129 pp.

Vandenberghe, N., 1978. Sedimentology of the Boom clay (Rupelian) in Belgium. Verhandelingen Kon. Acad. WLSK België, Jrg. XL, 147, 137 pp.

van der Velden, J.H., 1978. Evenwichtswatergehalte en specifiek oppervlak van kleien. Rapport CTI, TNO-Apeldoorn, 16 pp.

van der Velden, J.H., 1979. Ziegel Industrie, 9: 532-541.

Winkler, H.G.F., 1955. Ziegel Industrie, 8: 307-311.

THE CERAMIC PROPERTIES OF KAOLIN FROM EMPRESA DE MINERAÇÃO HORII, VARINHAS, MOGI DAS CRUZES, SÃO PAULO, BRASIL

I.R.WILSON [1] AND PERSIO DE SOUZA SANTOS [2]

(1) English China Clays, St.Austell, Cornwall, England

(2) Departamento de Engenharia Química, Escola Politécnica, University of São Paulo, and Instituto de Pesquisas Tecnológicas de São Paulo S/A, São Paulo, SP, Brasil

ABSTRACT

Data are presented on kaolins produced commercially from a mine near Sao Paulo, Brasil, by Empresa de Mineração Horii Ltda. Mineralogically, these kaolins consist primarily of kaolinite with some mica (muscovite) and quartz, but the composition varies with the level of refining. Morphologically, the kaolins are dominantly lamellar(kaolinite) with only a small percentage of tubes (halloysite). Chemically, the kaolins contain variable amounts of iron which are controlled at the mine to a level of 0.7 % Fe_2O_3 by washing at the mine face, and small amounts of titania, calcium and magnesium. The alkali content varies from 0.8 to 1.2 %. The coarse grained material shows good casting concentrations and casting rates. Deflocculant demands are low. Due to the absence of expanding minerals the modulus of rupture is low and plasticity is poor. Fired brightness at 1180°C and 1280°C, along with all other pysical properties makes the clay suitable for use in sanitaryware. The material is also suitable for use in frits, glass fibre and paints. Selective mining and further beneficiation could yield a kaolin suitable for earthenware and tableware.

INTRODUCTION

For the past eleven years, kaolins for the Brasilian ceramic and related industries are being produced by Empresa de Mineração Horii Ltda. The current annual production is about 60,000 tons. Pertinent Brasilian literature is listed under "References". This paper presents a brief summary of the geology of the area in the vicinity of the mine, the nature of the kaolin and its physical, mineralogical, morphological and chemical properties.

LOCATION, GEOLOGY, AND MINING

The mine is situated 50 km east of São Paulo, near the towns of Mogi das Cruzes and Suzano (Fig. 1).

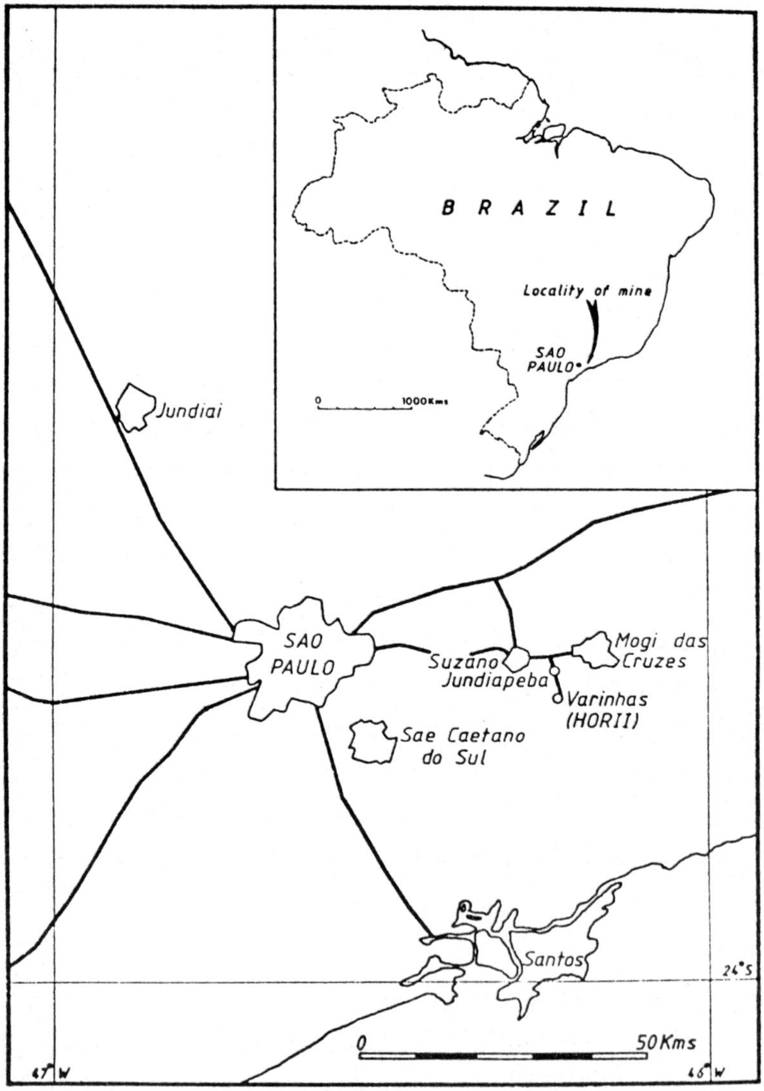

Fig.1 Locality of Empresa de Mineração Horii, Varinhas, São Paulo, Brasil

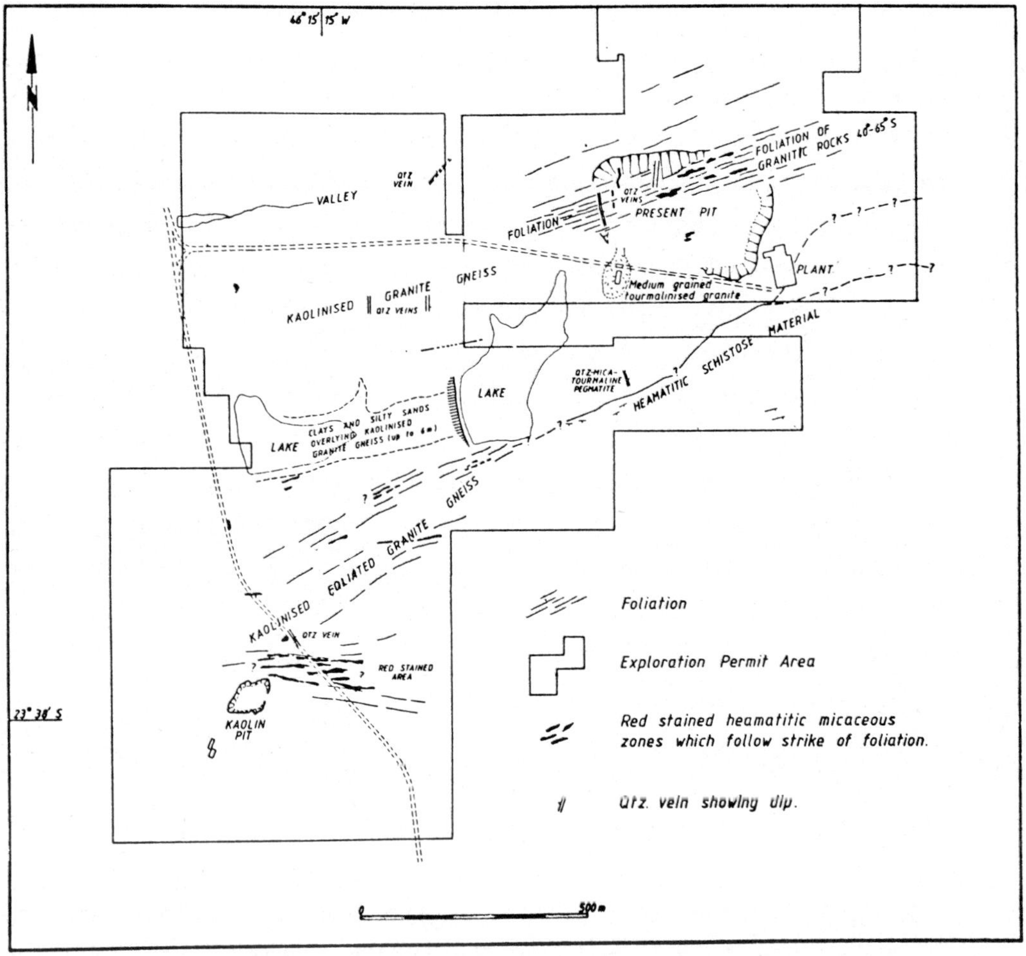

Fig.2 Simplified geology of the area in the region of the Horii kaolin mine, Varinhas, São Paulo, Brasil

Geologically the area consists of rocks of Precambrian age (Inferior) of the São Roque group. In the area of the mine the rocks have a distinct foliation striking 40-65° with a dip to the south. This foliation follows the regional trend of the Sao Roque group of rocks. The rocks are foliated granitic-type gneisses, migmatites and mica schists. In one area the granite is unfoliated, medium-grained and contains tourmaline - it is possibly an intrusive granite, but since no contacts are visible this assumption is problematical. Generally, with a north-south trend, vertical quartz-pegmatites have been emplaced, displacing the foliated gneisses. Mineralogically, the pegmatites consist of quartz, feldspar, and mica(muscovite books), and weathered tourmaline crystals which crumble easily. The kaolinization, as observed in a drilling program has affected all the granitic gneisses. The depth of kaolinization is not known exactly since drilling was limited to 40 m. Later activity has resulted in filling of fracture plains, which follow the regional trend, with a reddish hematite micaceous material. These red zones must be avoided during mining because of the high iron content. (Fig. 2)

Mining is carried out in an open-cast pit by first washing the face by hosing with water. The kaolin suspension is pumped to sand boxes where the sand settles. The overflow containing the kaolin is then pumped to a processing shed where it is passed through 8 hydrocyclones and a 180 screen to remove coarse particles. The slurry is then pumped via a large storage tank to the plant where it is filter pressed and dried. The filter cake is sold as "Bruto", and the dried clay is sold as "Beneficiado".

METHODS

Mineralogical composition and morphology were studied by X-ray diffraction and by various electronmicroscopic techniques, i.e. EMMA, TEM, and SEM.

The commercial ceramic properties were assessed following the test recommendations in Bulletin 6 of English China Clays, St.Austell, 1971, "Routine testing of ceramic materials". Measured are particle size, deflocculation behavior, casting concentration and rate, modulus of rupture (transverse strength), fired brightness, firing shrinkage, and chemical composition.

RESULTS

Mineralogical composition Table 1 shows the results of X-ray diffraction of some representative samples designated H1, H6, and I11, refined to different levels of particle size, and of Horii "Bruto". The clays are predominantly kaolinite, with mica, some quartz and feldspar and occasional gibbsite. The mica content drops when the clay is refined to the below 5 micrometre level, the quartz content

Table 1. Mineralogical composition of Horii kaolins at different levels of refining

Refining level	-150 mesh			-300 mesh			-30 micrometre				-15 micrometre				-5 microm.		
Sample	H1	H6	I11	H1	H6	I11	H1	H6	I11	BR.	H1	H6	I11	BR.	H1	H6	
Kaolinite	84	88	87	88	91	88	89	93	91	91	88	93	91	92	89	100	)
Mica	11	8	11	9	6	10	10	6	8	8	11	6	8	7	7	tr	)
Quartz	3	4	1	2	2	2	1	1	1	1	1	1	1	1	?tr	–	) % wt
Feldspar	2	–	–	1	–	–	?tr	–	–	–	–	–	–	–	–	–	)
Gibbsite	–	–	1	–	–	tr	–	–	–	–	–	–	?tr	–	4	–	)

Table 2. Ceramic and other property data for Horii kaolins A,B,C and "Bruto", and English China Clays' Traviscoe(T) and Remblend(R)

Refining level	-150 mesh			-300 mesh			-30 microm.			-15 microm.			-5microm.				
Sample	A	B	C	A	B	C	A	B	C	A	B	C	A	C	BR.	T	R
% +10 micrometre	63	46.8	66	51	46.3	63	44	35.0	56	17	32.6	52	–	–	47	18	18
% – 2 micrometre	14	25.0	9	17	25.0	10	21	28.2	13	25	30.4	14	78	71	20	44	39
Modulus of rupture (kgf/cm^2 at 80% RH) (o)	2.5	5.6	–	2.8	5.6	–	4.1	7.6	2.8	5.4	7.8	2.8	6.7	6.8	1.2	7.5	5.0
Plasticity	poor	poor	none	poor	poor	none	poor	poor	poor	poor	poor	poor	poor	poor	poor	fair	fair
Cast.conc.,% solids	62.1	63.3	59.6	61.6	63.0	60.4	61.8	63.3	61.0	62.4	64.4	61.0	62.1	62.9	61.3	62	64–67
Cast.rate,mm^2/min,2CC	14.5	4.92	45.3	15.2	4.53	42.2	–	4.40	25.4	8.90	4.04	19.8	2.94	4.3	17.5	2.0	2.0
Min.amount P84	0.50	0.45	0.60	0.30	0.45	0.60	0.20	0.35	0.55	0.25	0.85	0.32	0.90	0.82	0.65	0.60	0.55
Fired prop.-1180°C																	
% Brightness	93	82	80	94	83	82	95	83	80	95	84	81	98	79	81	84	86
% Absorption	24	23	24	23	23	24	24	22	25	23	22	25	27	24	25	15	16
% Contraction	3	4	3	4	4	4	4	4	4	4	5	4	8	9	5	9	8
Fired prop.-1280°C																	
% Brightness	94	88	85	95	87	86	96	87	87	97	86	87	99	89	86	84	87
% Absorption	21	19	23	21	19	22	21	19	22	21	19	22	19	18	26	6	9
% Contraction	4	6	4	5	6	5	5	6	5	6	6	5	11	12	5	13	11
Chem.analysis % wt XRF																	
SiO_2	48	46	47	47	46	46	47	46	46	47	46	46	46	44	46	48	48
Al_2O_3	39	39	38	39	39	39	39	39	39	39	39	39	40	40	39	36.5	37
Fe_2O_3	.42	.71	.77	.41	.71	.72	.38	.71	.73	.37	.71	.75	.35	1.1	.82	.75	0.8–1.0
TiO_2	.03	.05	.07	.02	.05	.06	.02	.05	.06	.02	.05	.06	.02	.07	.08	.03	.05
CaO	.04	.08	.04	.05	.08	.05	.04	.07	.04	.04	.08	.04	.08	.29	.03	.10	.06
MgO	.00	.05	.05	.01	.05	.04	.00	.04	.03	.00	.04	.03	.00	.01	.00	.30	.30
K_2O	1.1	1.4	1.5	.97	1.4	1.3	.83	1.3	1.2	.83	1.2	1.1	.22	.32	1.1	2.2	1.9–2.2
Na_2O (wet anal.)	.02	.00	.00	.03	.00	.02	.07	.03	.05	.09	.03	.06	.04	.03	.00	.10	.10
L.O.I.	12.4	12.7	12.2	12.8	12.6	12.7	13.0	12.8	13.0	13.1	12.8	13.0	14.1	14.6	12.8	12.0	11.9

(o) 1 kgf = 9.80665 N

decreases with decreasing particle size, and feldspar is only present in small amounts in the -150 and -300 mesh material.

Morphology. Electron microscopy of refined clay fractions show predominantly lamellar particles (kaolinite) with only a few tubular particles (halloysite), Fig. 3. The scanning electron microscope shows that the kaolinite forms lamellar stacks which in the coarser fractions are up to 50 micrometre long measured at right angles to the basal plane, Fig. 4.

Ceramic testing. Table 2 shows test results for three typical samples covering the range of clays mined in the area, and refined to different levels of particle size. Included is a typical analysis of Horii's "Bruto" production and of two English China Clay products, Traviscoe and Remblend, for comparative purposes. The three samples are:

A: White matrix with kaolinite, quartz, mica and tourmaline;

B: Slightly iron-stained biscuit-cream colored matrix with kaolinite, quartz, mica and tourmaline;

C: Iron stained matrix with kaolinite, quartz, mica and tourmaline, and possibly some gibbsite.

The results on these samples may be summarized as follows:

Sample A is a coarse grained ceramic kaolin with low strength, good casting properties and excellent fired properties due to the low iron content. Properties vary little in the coarser fractions, but at the refining level below 15 micrometre the strength increases slightly, and the casting rate at this level (8.90) is good. The clay refined to the below 30 micrometre level is similar to the Horii "Bruto" clay.

Sample B is a coarse grained kaolin with low strength, good casting properties and good fired brightness. The strength is higher than that of the samples A and C because it is less coarse. The casting concentration is higher than that of the other samples while the casting rate is lower but still good. The iron content is remarkably constant at all levels of refining.

Sample C is the coarsest of the three samples, and even at the below 15 micrometre refining level it contains 52 % particles larger than 10 micrometre and only 14 % particles smaller than 2 micrometre.

This coarse grained kaolin has poor plasticity and low strength, and a very high casting rate. At the under 5 micrometre level the strength increases while the casting rate drops. The fired brightness is not very high because of high iron contents-up to 0.75% Fe_2O_3. The increase of the iron content to 1.1% Fe_2O_3 in the under 5 micrometre fraction indicates that the iron species are fine grained. The clay refined to the below 15 micrometre level closely match the properties of the "Bruto" product.

Potential uses of the clay in ceramics. It may be concluded from the above results that at the below 30 micrometre refining level the clay is coarse, has poor strength and plasticity, a good casting concentration and casting rate. Fired brightness is variable depending on the iron content. At this refining level the clay will be useful as a component in sanitaryware bodies when mixed with finer grained more plastic clays. At the below 15 or 10 micrometre level the clay will be suitable for use in a tableware body as the strength increases while casting and fired properties are also good.

The clays are also considered suitable for other applications, such as for glass fibres, frits and paints.

ACKNOWLEDGEMENTS

We wish to thank Fumio Horii for allowing access to the mine to carry out this work and for permission to publish this paper. IRW acknowledges the work of various colleagues in the assessment of the ceramic properties of the clay. We also thank Dr.Helena Souza Santos for preparing the TEM micrographs. Aid by a grant from CAPES (Proc. 2343/81) is recognized.

REFERENCES (BRASILIAN LITERATURE)

Souza Santos,P and Souza Santos,H.(1969). Cerâmica(Brasil),15: 92-135

COPAMI, 1970. Visit to Mineracao Horii, Mogi das Cruzes. 15th Brasilian Geological Congress, Sao Paulo, 83-87

Souza Santos, P. and Pimentel, C.A. (1971). Cerâmica , 17: 258-295

Argentière, R. (1972) Mineração Metalurgia, 36: 14-17

Kiyohara, P.K., Souza Santos, P. and Souza Santos, H. (1974), Rev. Bras. Tecnol. Sup. 1: 711-720

Araújo, R., Jye, C.T. and Souza Santos, P.(1976). Proc. 9th Bras. Conf. Pulp Paper, 17-22

Correa, W.L.P. and Souza Santos, P.(1978). Cerâmica 24: 169-175

Trindade, E.D., Amarante, A.Jr. and Souza Santos, P. (1981). Cerâmica 27: 161-166
Amarante, A.Jr., Neves, L.M. and Pierro, J. (1981). Cerâmica 27:219-224

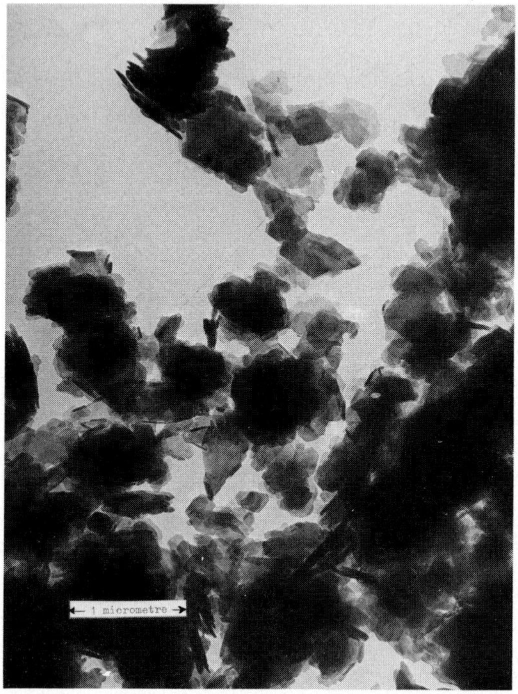

Figure 3. Horii kaolin. Borehole 37 (7470)

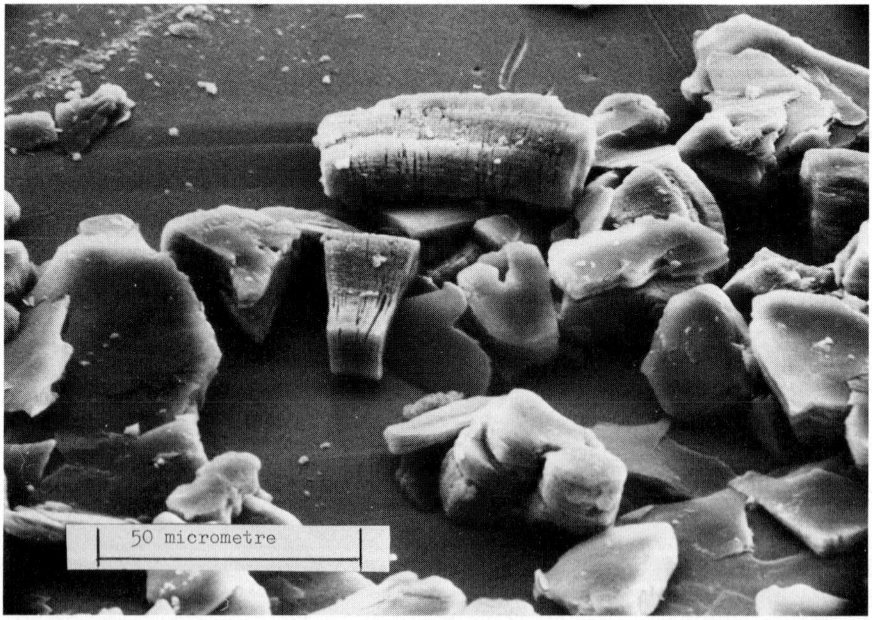

Figure 4. Horii kaolin, coarse fraction.
Borehole 37

KAOLIN DEPOSITS OF CHINA

ZHENG ZHI, LU DAREN, FENG MOLIN, FENG BAOHUA, AND JIN TAIQUAN

INTRODUCTION

China has a long history of exploitation and utilization of kaolin. The deposits are excellent in quality and highly varied in genetic type. Current knowledge allows the kaolin deposits of China to be preliminarily divided into four classes:

1. Hydrothermal-alteration type
2. Weathering-residual type
3. Weathering-leaching type
4. Sedimentary type

Deposits of Class 1 - Hydrothermal type are formed by the action of meso and epithermal solutions, related to magmatic intrusion and volcanic eruptions, on the parent rocks. In China there are many kinds of parent rocks on which hydrothermal kaolin deposits are formed including granitic intrusive rocks, intermediate and acid volcanic rocks, pyroclastic rocks, kaolinitic and hydromicaceous claystones, and even basic dike rocks. The shape of these hydrothermal kaolin deposits depends on the parent rocks with some being layered, lenticular, or vein shaped. They are associated as outer zones of alteration of endogenic metallic mineral deposits or with such non-metallic mineral deposits as alunite and pyrophyllite. The size of the deposits range from very small to exceedingly large with some of the large deposits having reserves of over 20 million tons. The large and good quality hydrothermal deposits result from two necessary prerequisites: (1) frequent magmatic activity of intermediate acid and alkaline extrusive types and (2) a well developed structural fault zone.

Deposits of Class 2 - Weathering residual type are the result of weathering of granites and hypabyssal dikes of rhyophorphyry and granophyre. The factors controlling the formation of kaolin deposits of this type include structural, parent rocks, and climatic conditions. The composition and grain size of the parent rocks is very important.

Deposits of Class 3 - Weathering leaching type are formed by the action of ground water on pyrite bearing rocks resulting in acidic solutions which react with underlying limestones. The solutions infiltrate along fissures, joints, and fractures forming kaolinitic rock alteration zones. The presence of a pyrite bearing kaolinite clay rock is necessary for the formation of this type of deposit. Changes in the oxidizing and reducing conditions determine the presence of the deposit.

Deposits of Class 4 - Sedimentary type are sedimentary rocks that contain over 90 percent kaolin minerals and are called kaolins. There are two basic types related to the conditions of their formation. The first type is the kaolins associated with coal measures usually forming the floor of the coal or occurring as bands within the coal and those associated with bauxites. These kaolins were formed by the transportation of terrigenous sediments by water and were deposited in the collodial form in a sedimentary basin. The second type is kaolin deposited near weathered crystalline rocks, usually large weathered grainte areas. This type of kaolin was formed by in situ differentiation and deposition after the weathered granite was eroded by water.

Table 1 is a summary of the main kaolin deposits in China showing the geographic location, size, genetic type, and the general geology. In some deposits there are multiple geological processes associated with kaolin formation. In some cases the hydrothermal process is dominant with weathering superimposed and later there has been some erosion, transportation, and deposition. In other cases weathering processes dominate and others are the result of sedimentary processes.

TABLE 1 Main kaolin deposits of China

Locality							General Geology
Province	County	Size	I	II	III	IV	
Beijing		•		0			Parent rock is dikes
Hebei	Shahe	θ				0	
Shanxi	Yangquan	•		0			
Lianoning	Faku	0	0				
Tilin	Antu	•				0	Quaternary sediments
	Yongji	θ				0	Quaternary riverchannel sediments
	Hunchun	0		0			
Heilongjiang	Linkou	0				0	Tertiary sediments
	Tangyuan	•				0	Tertiary sediments
Shandong	Wendeng	•		0			Parent rocks are granite and pegmatite
Jingsu	Suzhou	X	0		0		
Anhui	Suxian	0	0				Altered quartz porphyry dikes
	Qimeng	•		0			Parent rock is quartz syenite
	Fanchang	•		0			Parent rock is felsite
Xinejiang	Yongjia	0		0			Parent rock is rhyolite porphyry
	Wenzhou	0		0			Parent rock is rhyolite porphyry
	Hangzhou	•		0			Parent rocks are tuffy feloitic porphyry and tuff
	Changxing	•		0			Parent rock is tuff
Jiangxi	Jingdezhen	0		0			Parent rock is granite
	Linchuan	0		0			Parent rock is granite
	Jiujiang	0		0			Parent rock is granite
	Shangyao	•		0			
Eujian	Minqing	0		0			Quartz porphyry
	Liancheng	0		0			Quartz porphyry
	Dehua	•	0				
Henan	Bo'ai	•				0	
	Gongxian	0				0	
Hubei	Huangshi	•		0			Tuff
Hunan	Hengyahg	θ		0			Schist and gneiss
	Changsha	θ		0			Granite
	Qianyang	•		0			Granite
	Liling	X		0			Granophyre and quartz porphyry
Guangxi	Yongning	•		0			Granite
	Binyang	•		0			Granite
	Pingle	•		0			Granophyre and quartz porphyry
Guangdong	Qingyuan	θ				0	Quaternary

Locality							General Geology
Province	County	Size	I	II	III	IV	
	Dianbai	θ		0			Quartz porphyry
	Yangshan	•				0	Deposit associated with Permian asbolite deposit
Shanxi	Tongchuan	0				0	Permian
Ningxia	Pingluo	0					Upper Carbonaceous fire clay
Xinjiang	Yining	0					Middle-lower Jurassic lacustrine sediments
	Wulumuqi	0					Middle-lower Jurassic lacustrine sediments
	Hami	0					Middle-lower Jurassic lacustrine sediments
	Nilka	0					Middle-lower Jurassic lacustrine sediments
Sichuah	Xuyong	•			0		Base of Upper Permian
	Gulin	•			0		Base of Upper Permian
	Dechang	•		0			Syenite
	Hanyuan	•		0			Ancient erosion surface of the Sinian
Guizhou	Renhui	•			0		Base of Upper Permian
	Tongzi	•			0		Base of Upper Permian
	Qianxi	•			0		Base of Upper Permian
	Xishui	•			0		Base of Upper Permian
	Zunyi	•			0		Base of Upper Permian
	Wengan	•			0		Base of Upper Permian
	Zhijin	•			0		Base of Upper Permian
	Guiding	•			0		Base of Upper Permian
	Xingyi	•				0	Upper Triassic sediments
Yunnan	Anning	0		0			Muscovite granite and felsite
	Jianshui	•				0	Tertiary lacustrine sediments
	Gejiu	•		0			Two-mica monzonitic granite, aplite and pegmatite
	Long-lin	•		0			Pegmatite
Xiang (Tibet)	Lhasa	X		0			

Explanation:
- • – small deposits
- 0 – deposits of intermediate size
- θ – large deposits
- X – exceedingly large deposits
- I – hydrothermal – alteration type
- II – weathering-residual type
- III – weathering-leaching type
- IV – sedimentary type

SUZHOU DEPOSIT - A HYDROTHERMAL ALTERATION TYPE

One of the largest and highest quality kaolin deposits in China is located near Suzhou in Jingsu province. The geology of the Suzhou kaolin mining area is very complex. There are Carboniferous, Permian, and Triassic carbonate and clastic rocks in the area along with igneous rocks including granites, volcanic quartz latites, adamellite intrusives, and multistage acid and alkaline dikes, which filled large high angle faults trending NNE and NWW. Jointing is very prominent and as a result of the multiple intrusion of granite rocks and the Jurassic volcanism, intermediate and low temperature hydrothermal alteration is very widespread. Kaolin deposits associated with small alunite and siderite occurrences and some copper, lead, and zinc mineralization are the products of this alteration. Some high temperature pneumato-hydrothermal alterations as skarns are only locally developed. The skarn zone is at the contact of limestones with intermediate acid and acid hypabyssal porphyry and dike rocks and consists of andradite, salite, and less commonly small amounts of epidote, wollastonite and vesuvianite. The limestone is recrystallized and silicified into a marble associated with isolated lenses of siderite. The kaolin zone is positioned above the skarns corresponding to the shapes of the parent rocks which may be bed-like, lenticular, vein-shaped, etc. The kaolin quality and quantity is variable from one area to another in the Suzhou deposit. In addition to the various rock types that are present, there are a number of structural elements including a thrust fault, a series of imbricate structure sets, brecciation caused by intense volcanic activity in the late Jurassic, and a series of high angle faults associated with the Yanshanian movement in the late Mesozoic.

Intermediate and low temperature hydrothermal alteration is widespread throughout the area. In the kaolin zone of alteration, the minerals consist chiefly of kaolinite and halloysite associated with small quantities of alunite and pyrite. The thickness of the kaolin zone generally ranges from 5 to 30 metres with some local thicknesses as much as 60 to 80 metres. In one locale the thickness of the kaolin zone is more than 130 metres. The top and bottom of this kaolin zone of alteration are gradational upward into a siliceous sericite zone and downward into the country rocks with a higher montmorillonite content when the country rock is marble. There are several mines in the Suzhou district.

The major portion of the Suzhou deposit is hydrothermal but there are localized secondary kaolin deposits that were eroded and deposited in sinkholes in the limestone karst. Also some of the hydrothermal kaolins have been altered by weathering.

The hydrothermal kaolin at Suzhou is divided into four natural types:

1. Kaolin with relict phenocrysts - This kaolin is generally white with some brownish-yellow staining by ferruginous compounds and alunite. The phenocrysts are relicts of feldspar which are kaolinized and include small amounts of secondary quartz and alunite. Some primary quartz phenocrysts have not been altered.

2. Compact kaolin - This kaolin is white, compact, and massive and normally contains small amounts of secondary quartz and alunite. No relict textures are visible.
3. Banded kaolin - This kaolin is comprised of alternating white and gray or grayish-black bands. There are two examples of this type of kaolin; one is where small quantities of microcrystalline pyrite 0.01 to 0.06 mm in grain size occurs as bands in the white kaolin, and the other is where organic matter constitutes the dark bands. This latter type occurs in the Permian age clay rocks. Under the microscope relict plant textures are observed in this organic rich band.
4. Sandy compact kaolin - This kaolin is white, often iron stained, and sandy. There is a large amount of primary or secondary quartz which is relatively coarse.

The secondary kaolin deposits contain three natural ore types that can be recognized:

1. Compact kaolin - This kaolin is fine textured, white, compact, and smooth feeling and appears free from impurities.
2. Brecciated banded kaolin - The ribbons or bands have been broken caused by collapse from solution cavities in the underlying limestone or by small faults or slumps in the sedimentary kaolin.
3. Clastic-sedimentary kaolin - This kaolin occurs as clasts that have been transported by water and deposited along the walls of solution cavities or in fissures and caves. Clear sedimentary rhythms can often be observed.

Table 2 gives a summary of the mineral composition of both genetic types of kaolin occurring at Suzhou.

TABLE 2 Mineral composition

Genetic type	Dominant minerals	Accessory minerals	Trace minerals
Hydrothermal kaolin	Kaolinite Halloysite	Halloysite, alunite, pyrite, sericite, quartz, chalcedony	Galena, rutile, sphalerite, sphene
Secondary kaolin	Halloysite	Kaolinite, alunite, limonite, chalcedony, montmorillonite, gibbsite, quartz	

The kaolinite formed as a result of hydrothermal alternation is well crystallized with a grain size of 0.5 - 2μ. Halloysite (7Å) is the dominant type but there is some halloysite (10Å) also present. Transmission electron micrographs of the Suzhou kaolin are shown on Figure 1 which shows (a) well crystallized kaolinite, (b) tubular halloysite, and (c) a mixture of kaolinite and halloysite.

Fig. 1 a - Crystallized kaolinite

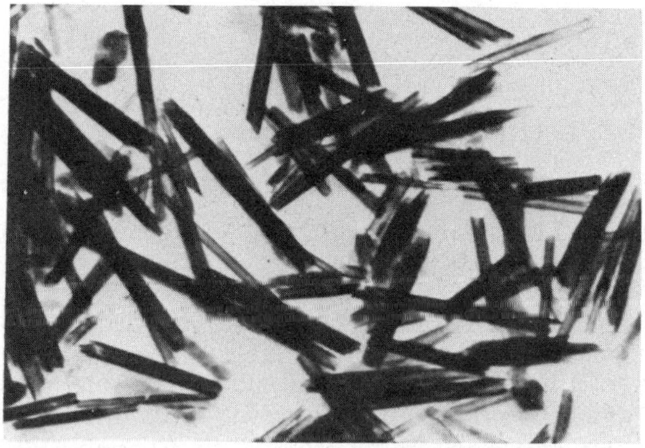

Fig. 1 b - Tubular halloysite

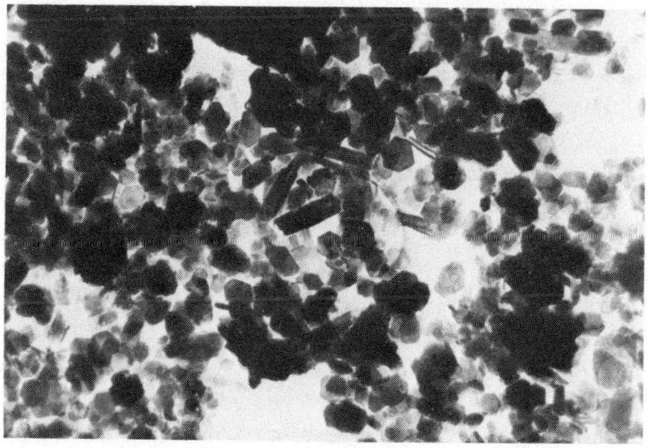

Fig. 1 c - Kaolinite and halloysite

The chemical composition of four types of kaolin from the Suzhou deposits are shown in Table 3.

TABLE 3 Chemical composition

Oxide	Kaolin ore type			
	1	2	3	4
SiO_2	57.26	47.77	46.59	49.20
Al_2O_3	29.04	36.00	36.81	33.51
Fe_2O_3	1.09	0.79	1.17	1.64
TiO_2	0.42	0.33	0.10	0.13
MgO	0.14	0.13	0.29	0.42
CaO	0.31	0.27	0.35	0.55
Na_2O	0.02	0.08	0.04	0.06
K_2O	0.21	0.21	0.11	0.23
SO_3	1.48	0.88	0.24	0.35
LOI	10.93	13.76	14.18	13.20
TOTAL	100.90	100.22	99.88	99.29

1 - Kaolin with relict phenocrysts; 2 - Compact kaolin; 3 - Brecciated banded kaolin; 4 - Clastic-sedimentary kaolin.

The Suzhou kaolin is beneficiated using a wet process scheme. The kaolin is mixed with water and blunged in a large agitator. The sand is separated from the kaolin using hydrocylones. The kaolin is then pumped to settling vats where the coarse and fine particle sizes are separated by sedimentation. The kaolin products are then dewatered using filter presses and dried in tunnel dryers. The products are used for electrical porcelain, filler for paper, rubber, and paint, special ceramics including refractories, in white cement, and other special uses. The whiteness ranges from 65 to 82 percent. The kaolin is very plastic and has a melting temperature of about 1770°C. The vitrification or sintering temperature ranges between 900°C and 1500°C depending upon the purity. The kaolin disperses readily and burns very white. The black and white banded kaolins also burn very white because the organic matter burns out during firing. The Suzhou kaolin deposit is very large with reserves indicated of over 20 million tons.

KAOLIN IN GAOLING VILLAGE - A DEPOSIT OF WEATHERING RESIDUAL TYPE

Deposits of this type can be classified into two subgroups: (a) residual deposits formed from weathering of granites, and (b) residual deposits formed from weathering of hypabyssal dikes of rhyoporphyrys and granophyres. The first is exemplified by the weathering-residual kaolin deposit in Gaoling Village, Jingdezhen, Jiangxi province.

This kaolin deposit was formed by the alteration of medium grained granites and pegmatite dikes. The depth of weathering varies from place to place but is generally 30 to 50 metres. The ore body is 400 metres long and 5 to 20 metres wide. The factors controlling the kaolin formation include structural, parent rock composition, and climatic conditions. The best kaolin is generally located in the upper part of the medium-grained granite. The upper weathered zone consists of poorly crystallized kaolinite (Fig. 2a), the middle zone a mixture of poorly crystallized kaolinite and halloysite (Fig. 2b), and in the lower zone the kaolin content decreases and the quartz and feldspar content increases accordingly.

Fig. 2a - Poorly crystallized kaolinite

Fig. 2b - Poorly crystallized kaolinite and halloysite

The mineral composition of the raw and washed kaolin is shown in Table 4.

TABLE 4 Mineral composition of Gaoling Village kaolin

Ore	Kaolins	Quartz	Feldspar	Micas	Others
Raw	56	17	3	23	1
Washed	73	11	3	12	1

The chemical compositions are shown in Table 5.

TABLE 5 Chemical composition of Gaoling Village kaolins

Ore	SiO_2	Al_2O_3	Fe_2O_3	TiO_2	CaO	MgO	K_2O	Na_2O	LOI	TOTAL
Raw	49.65	33.82	1.13	0.05	0.33	0.23	2.70	1.03	10.84	99.78
Washed	47.77	37.18	0.74	-	0.45	0.38	2.10	0.53	11.31	100.46

The kaolin in Gaoling Village is used primarily for the production of fine china and the ceramic properties are very good. The kaolin is mixed with feldspar to make the fine china body. The reserves of the Gaoling deposit are rather small because of the long history of mining, but large quantities of similar quality are available at Dazhou and Xingsi which are near Gaoling Village.

The second type of residual kaolin deposit formed by the weathering of hypabyssal dikes is exemplified by a deposit in Wenzhou, Zhejiang province. The main parent rocks of this deposit are acid dikes of rhyolite porphyry. The dikes cut through volcanic and sedimentary rocks. The dikes are weathered to depths of 10 to 30 metres. The kaolin ore can be divided into two types depending on the degree of weathering. One is a deeply weathered kaolin called "potter's clay" which is white or cream colored and the other is "china clay" which is partially altered grayish to milky-white rhyolite porphyry. The two ores differ in mineral varieties and quantities. Feldspars in "potter's clay" have been completely kaolinized, so the ore has a high content of kaolinite and a small amount of corroded quartz. "China clay" contains quartz and feldspar which is mostly replaced by kaolinite and sericite. The chemical composition is shown in Table 6.

TABLE 6 Chemical composition

Ore	SiO_2	Al_2O_3	Fe_2O_3	K_2O	Na_2O	LOI	TOTAL
China clay	70.96	19.67	1.08	2.10	0.09	6.27	100.17
Potter's clay	67.37	21.80	1.15	0.99	0.06	8.36	99.73

The Wenzhou kaolin is used mainly in the manufacture of porcelains. The fired color is very white and the other ceramic properties very good.

The Xuyong kaolin - a deposit of the weathering-leaching type. Deposits of this type are formed by the action of ground water between pyrite bearing kaolin clay rock and underlying limestone. The kaolin deposit of Xuyong in Sichuan is an example. The limestone is Lower Permian in age and the gray to dark-gray pyrite bearing clay rock is Upper Permian in age. The pyrite-bearing kaolinite clay rock is 2 to 11 metres thick and is comprised of kaolinite, pyrite, hydromica, organic matter, and small amounts of marcasite and nacrite. The limestone is very thick ranging between 130 to 342 metres. The ore is compact, massive, banded, and pseudo-brecciated with a pelitic texture. The mineral content of the ore is dominantly halloysite (Fig. 3), with subordinate kaolinite along with trace amounts of calcite, hydrogoethite, hematite, gibbsite, and alunite.

Fig. 3 - Xuyong kaolin

The chemical composition is shown in Table 7.

TABLE 7 Range of chemical composition

	SiO_2	Al_2O_3	Fe_2O_3	TiO_2	MnO_2	MgO
Xuyong Kaolin	42.54-46.44	37.07-39.46	0.01-2.04	0-0.55	0-0.37	0.0.41
	CaO	Na_2O	K_2O	LOI	SO_3	^-C
	0-0.49	0-0.13	0-0.27	11.38-17.44	0.08-0.51	0.04-0.14

The grade 1 kaolin is used in the manufacture of high grade electical porcelain and optical glass crucibles. Grade 2 and 3 kaolins are used in the ceramics and enamel industries. The substandard and black kaolins are used in the manufacture of refractories.

KAOLIN DEPOSITS OF THE SEDIMENTARY TYPE

Sedimentary kaolins are widespread in China. They are roughly classified into two basic types. The first is associated with the coal measures and the kaolin usually constitutes the floor of a coal seam and is dark gray in color because the layer usually contains carbonaceous matter. Kaolinite can account for more than 90 percent of the minerals present and impurities include quartz, mica, pyrite, coal fragments and chlorite. These underclays are the principle raw material for ceramics in the northern part of China.

The second type of sedimentary kaolin is deposited near weathered crystalline rocks often in areas where vast tracts of granite are exposed. The sedimentary kaolin in Qingyuan county, Guangdong province is an example of this type. The area is surrounded on all sides by Yanshanian granites and the sedimentary basin is filled with quaternary sediments represented from the top down by the following sequence.

1. Yellow clay layer - 1.5 metres thick.
2. Mud and sand layer - 1 to 5 metres thick.
3. Kaolin layer - the upper part consists of alternating purplish-red, brownish-yellow, and grayish-white clay ranging from 1 to 6 metres thick. Sometimes the clay layers are separated by a quartz sand layer up to 2 to 4 metres thick.
4. Grayish-black clay layer - It is sort of a sapropel comprised of large amounts of plant remains occasionally containing sand and white clays generally 2 to 6 metres thick.
5. Quartz sand and gravel layer - 0 to 5 metres thick.

The kaolin ore body occurs in an intermontane basin and is several thousand metres long, over 1,000 metres wide, and up to 10 metres thick. Prior to Quaternary time, the granite was subjected to intense weathering and the feldspars were altered to kaolinite and halloysite. This weathered mantle of kaolinite, halloysite, and quartz were scoured off by rivers and streams resulting in deposition in the basin of alternating layers of kaolin and quartz sand. The kaolin ore occurs as white and grayish-white masses. Kaolinite is the dominant mineral along with accessory amounts of hydromica and halloysite. The quartz content is about 1 percent and there is trace amounts of sericite. The chemical composition is shown in Table 8.

TABLE 8 Chemical composition of two Qingyuan kaolins

	SiO_2	Al_2O_3	Fe_2O_3	TiO_2	CaO	MgO	Na_2O	K_2O	SO_3	LOI
1.	N.d.	31.30	2.30	Tr	0.46	0.46	Tr	Tr	0.31	9.75
2.	N.d.	35.64	1.96	Tr	0.86	0.67	Tr	Tr	0.24	9.81

This kaolin has good ceramic properties and is used mainly for that purpose.

SUMMARY

Four examples of various types of kaolin deposits are described based on their origin. These are as follows:

1. Hydrothermal alteration - Example: Suzhou
2. Residual weathering - Example: Gaoling kaolin deposit
3. Weathering - Example: Xuyong kaolin deposit
4. Sedimentary - Example: Qingyuan kaolin deposit

The largest deposit being mined is the Suzhou kaolin which is used as a high quality filler clay and for ceramic uses. The Suzhou kaolin is beneficiated in a relatively modern kaolin plant. The other kaolins described are used primarily for ceramics of various types. The Gaoling deposit and others in that vicinity are used to make fine china at Jingdezhen which is the center of the whiteware industry in China.

* This article was rewritten by Haydn H. Murray who visited the Suzhou and Gaoling deposits in 1980.

ENGINEERING PROBLEMS ASSOCIATED WITH SEDIMENTATION, WEATHERING AND DIAGENESIS OF SOILS AND ROCKS

J. E. Gillott, Department of Civil Engineering, The University of Calgary, Calgary, Alberta, Canada T2N 1N4

INTRODUCTION

In the early part of this century engineers modified the theories of Coulomb and Rankine to take account of cohesion and recognized that in soils and sediments cohesion was due mainly to the presence of clay. Failure on a curved rather than a plane surface was proposed and increasing use of mechanics and hydraulics was made in design. In determining engineering behaviour the role of large-scale features such as sedimentary bedding, soil profile, joints, fissures, slickensides, etc., is clearly recognized but it has become equally clear that a fundamental understanding of behaviour of soils and rocks depends upon knowledge of genesis, composition and microstructure.

The object of the present paper is to show that knowledge of microstructure and composition of soils and rocks helps to make engineering behaviour more intelligible, particularly when coupled with an understanding of genesis.

GEOLOGICAL BACKGROUND

Soils and rocks formed by sedimentation, eluvial processes - weathering and hydrothermal action - and diagenesis all may contain clay minerals. A proper understanding of fabric requires consideration of the conditions of soil and rock genesis.

In Canada much of the surficial cover developed within about the last 10 or 15 thousand years during and after the retreat of the Wisconsin continental glacier. Many of these deposits are often referred to as soft soils. The properties of these materials are determined by their origin and depend on the conditions at the time of formation. Three factors are of major importance. Firstly, the crust was depressed by the weight of ice during the Pleistocene. When the ice melted the crustal depression together with the eustatic rise of sea-level led to extensive marine transgression. Secondly, during retreat of the ice many present drainage channels remained blocked by ice dams, and thirdly, large volumes of fresh-water were released from melting ice. Consequently extensive fresh-water lakes formed, the extent and positions of which shifted as the ice retreated northwards and as drainage channels opened. As a result much of the hard basal till and bedrock is covered by soft sediments which were deposited in the seas or lakes which existed for the few thousand years before the present. Quigley (1980) has recently summarized the main features and distribution of these lakes and seas (Figure 1).

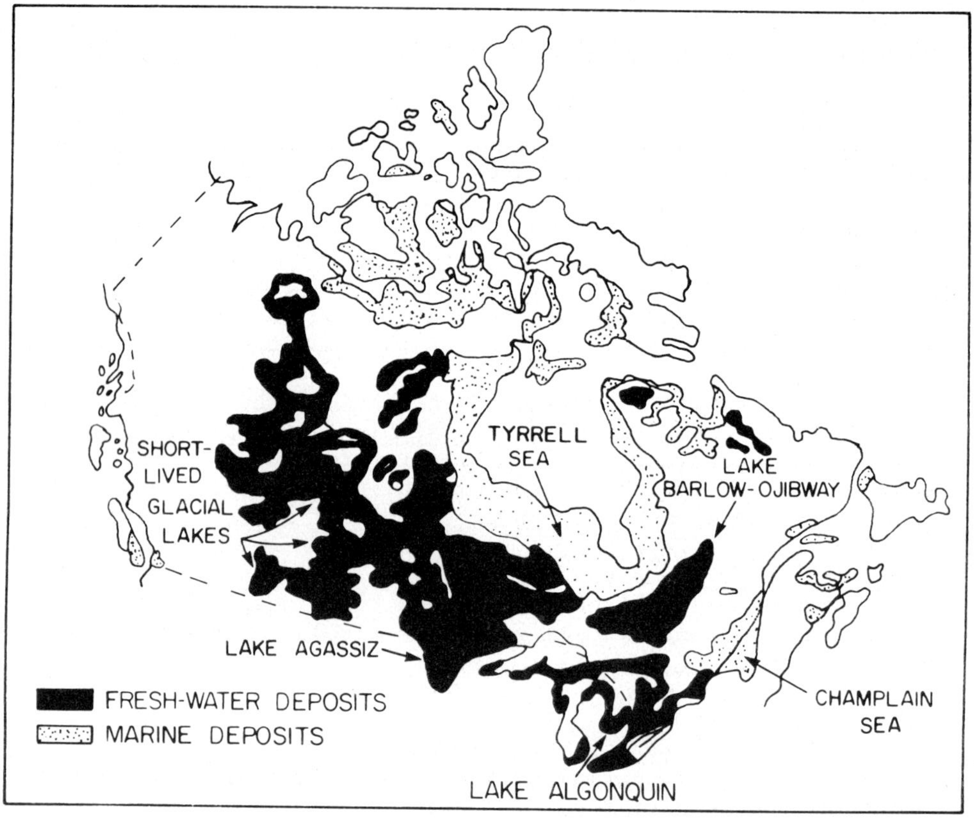

Fig. 1. Glacial and post glacial sediments in Canada (Adapted from Quigley, Can. Geotech. J. 1980).

Elsewhere extensive wind-blown deposits are found. Loess, (Figure 2A), a wind-blown silt containing 10 to 20% clay and derived from glacial floodplains, is an important soil in the Midwest of North America, western Europe, the Soviet Union and China. Earthquake induced flows of loess, said to be in the form of a silt-in-air dispersion, have caused very heavy loss of life in Asia.

In many areas of the world, such as large parts of Africa, superficial deposits are residual since they were formed by weathering of the bedrock with little lateral movement. Their composition and fabric reflect that of the parent material, weathering regime, topography, age, and other factors. Soviet scientists (Sergeyev, et al., 1980) recognized that soils formed by hydrothermal alteration (Figure 2B) have a distinct microstructure.

PROBLEMS OF SLOPE INSTABILITY

Slope stability failure is probably the most serious problem with which the engineer has to deal since slides often occur abruptly with catastrophic effects. Slow downhill movements due to creep, though less dramatic, can nonetheless entail costly maintenance. Natural or artificial processes which result in an increased shear stress or a decrease of shear strength of the formation are involved in all slope stability failures. Typical causes of increased shear stress are steepening of slopes by excavation as in quarries, pits, canals, loading at the head of slopes, and undermining at the foot of the slope by excavation or erosion. Causes of a decrease in shear strength of the formation are an increase of pore water pressure, moisture uptake by swelling clay or other materials, dissolution of intergranular cement, change of pore water composition in clays, frost action and excavation, particularly when strata dip towards the free face.

Slope movements are classified on the basis of types of movement and classes of materials involved (rock or engineering soil). Six groups of movements are distinguished, termed falls, topples, slides, spreads, flows and complex movements. Slides are subdivided into rotational slides, where failure occurs on a concave surface of rupture, and translational slides where the failure surface is more or less planar; these slides are controlled by surfaces of weakness. Spreads and flows are partly overlapping groups since flow of liquified mud belongs to both.

Flows also include slow downhill movements termed solifluction and creep (Varnes, 1978). These effects are attributed to gravity combined with frost-action, heating and cooling, burrowing and wedging by animals and plants and plastic deformation of clays. Creep normally precedes failure. The rate and magnitude of these time dependent movements is influenced by many factors including the resistance offered by the microstructure, the mineralogical composition, the pore pressure, water content and permeability, and the stress history of the soil. Saturated sensitive clays are particularly susceptible to creep and to stress relaxation phenomena.

An example of this sort of problem occurred at Oakville, Ontario where a bridge had been constructed in the 1920's across the mouth of a 100 ft. wide river. Significant deformation had resulted from intermittent movements of the 50 ft. high river banks. A new bridge was built in the early 1960's, while the banks were flattened to about 3 (horizontal) to 1 (vertical) and the upper portion of the existing fill replaced by light weight slag fill to further stabilize the slopes. The fill materials are, however, underlain by up to 15 ft. of organic silt which overlies a thin glacial till deposit and shale bedrock. Despite the reduction in the shear stress level, continued creep movements occurred in the organic silt and caused the bridge abutments to close by as much as 1/4 to 1/2 inch per annum (Crooks, 1981). In this area sediments were deposited in the southern part of Lake Algonquin where glacial erosion of Palaeozoic bedrock supplied carbonates, illite, chlorite and some smectite. The clay grade material was found by particle size analysis to comprise about 6 percent of the silt. Analysis by x-ray diffraction showed that the fine fraction was composed of illite, chlorite, quartz and feldspar. Electron micrographs showed that the silt was composed mainly of primary minerals, with some clay and microfossils (Figure 2C, D). Clay minerals were sometimes present as agglomerations which were probably formed as fecal pellets by zooplankton (Figure 2E). Such organisms aid agglomeration of clay minerals and contribute to the high organic content of the sediment. Microfossils tend to trap organic constituents in the void space of the test and in the intergranular pores (Figure 2F). In some places diatomite is oilbearing with more than 60% of the pore volume being oil saturated (Earnest, 1981).

Creep rate versus time curves are commonly subdivided into primary, secondary and tertiary branches. The secondary stage or "steady state" may be followed by strain softening with an increase in creep rate and abrupt failure. Strength loss in soils has been attributed to rupture of cemented junctions, changes in pore water pressure and effective stress and increases in water content. An example involving strain softening of sensitive soil followed by failure occurred at Atikokan, Ontario, to the west of Thunder Bay. At the site, about 10 ft. of loose sands were overlying soft, firm sensitive clay. Construction involved excavation to a depth of about 25 ft. in these materials. The sides of the excavation were supported by light sheet piles with only nominal shoring and bracing. As a result, there was significant deformation of the sides of the excavation and associated strain softening of the retained soil mass. When the excavation was completed, a full-scale failure occurred. A second excavation was made using a much stiffer sheet pile section and a much stiffer bracing system. Thus only small deformations were permitted within the retained soil mass and strain softening did not occur. This second excavation was successfully completed with less than 1 in. maximum deflection of the walls (Crooks, 1981). The clay fraction made up about 70% of the soil and contained chlorite, illite, small amounts of montmorillonite, quartz and feldspar. Some of the electron micrographs showed a relatively open texture (Figure 3A) but in others the clay minerals had a preferred orientation with a wave-like and arching pattern which may have resulted from shear-movements rather than being of sedimentary origin (Figure 3B).

The apparent abrupt movement and the often disastrous consequences of slides make them of major concern to the geotechnical engineer. Fabric and composition of the soil can sometimes help in understanding the nature of the problem and in devizing suitable remedial measures.

In the Ashcroft area of British Columbia particulary at Hat Creek mineral deposits including coal occur. The coals are found in sediments of Miocene or earlier Tertiary age consisting of conglomerates, sandstones, siltstones and claystones; andesites, basalts, breccias, tuffs and ash also occur (Duffell and McTaggart, 1951).

Open cast methods of coal recovery have been used and geotechnical interest has been concerned with slope stability, and slope susceptibility to erosion among other matters. Detailed information exists concerning the engineering behaviour of the materials and their mineralogy and petrography. Some of the volcanic deposits have been converted almost completely to bentonite while in other beds primary silicates such as quartz and feldspar were found; precipitation of siderite appears to have inhibited alteration of the original silicates to swelling clay minerals (B.C. Hydro and Power Authority, 1978). The electron micrographs show that the phyllosilicates are oriented and have the characteristic "curly" appearance of smectites (Figure 3C, D, E). The shale contains silt-sized (or larger) inclusions, possibly of tuffaceous origin. The inclusions are altered to varying degrees and sometimes developed an approximation to "onion-skin" texture (Figure 3F).

In Canada the most serious and also the most interesting problems of slope instability involve the sensitive post-glacial sediments in the Ottawa-St. Lawrence lowland. That area includes the largest cities and greatest urban development and slope failures have resulted in loss of life and costly destruction of property. The region is underlain by gently dipping Palaeozoic limestones, shales and orthoquartzites with some igneous instrustives in the Monteregian Hills south and east of Montreal. The geotechnical problems however involve the soft soils which formed at the end of the Pleistocene. These reach thicknesses of 30 to 60 m in the valleys of the Ottawa and St. Lawrence rivers. The principal sources of the detritus were the crystalline rocks of the Canadian Shield, the sedimentary and metamorphic rocks of the Appalachians and the local Palaeozoic sediments. A significant proportion of the material is rock flour - primary minerals reduced by mechanical abrasion to the clay grade (< 2 μm). Quartz, feldspar, amphibole and mica are common minerals of this sort; illite, chlorite and sometimes smectite are the commonest clay minerals. Carbonates are relatively common and non-crystalline minerals are now considered significant (Gillott, 1971; McKyes et al., 1947; Yong et al., 1979). Coarser more permeable layers interbedded with the clays raise the water table and may generate artesian pressures (Fransham and Gadd, 1977). Slides have been correlated in many areas with periods of heavy rainfall linked to high pore pressures in the sediments. Rutter (1980) has stated that early slides were broad, shallow and extensively retrogressive implying a high degree of liquefaction, whereas more recent slides tend to be narrow and steep with relatively low liquefaction. One proposal which may account for this observation is that amorphous cementitious coatings are precipitated at interparticle junctions. Amorphous iron may be produced by in situ chemical weathering resulting from changes in enviroment with time (Bentley et al., 1980). Soils of this sort commonly have a liquidity index greater than one because the natural moisture content is higher than the liquid limit. This may result from changes in the chemical composition of pore solutions favouring dispersion together with precipitation of cementitious compounds at interparticle contacts. Such strengthening of junctions would enhance the mechanical difficulty of fabric rearrangement to a closer packed configuration with corresponding reduction in void ratio and water content by consolidation or syneresis. In addition to high moisture content sensitive soils have other features in common such as similar index properties, higher than normal content of primary minerals in the clay size range and an open fabric (Figure 4); agglomeration of clay minerals are sometimes present (Figure 4C).

The bent flake of silt-size mica shown in Figure 4D is rather unusual. Mica is normally elastic and presumably residual stress in the underlying soil is holding it in this configuration. It appears that chemical deposits and fine clay are adhering to the surface. Precipitates are thought to strengthen mineral contacts by acting as cementitious agents and it seems probable that their presence on the underlying surface of the mica accounts for the strong bond to the soil and the retention of the distorted shape against the restoring elastic forces in the mica crystal. Sensitive soils occur on the west coast and detailed studies of the deformation behaviour of Haney clay were reported a few years ago (Campanella and Vaid, 1972); electron micrographs of this soil are shown in Figure 4E, F.

Various physical and chemical theories have been proposed to account for the behaviour of sensitive soils and the nature of the problem has been extensively reviewed (Penner and Burn, 1978; Smalley, 1979, 1980). It seems certain that because of the many factors which may increase sensitivity either by increasing undisturbed strength or decreasing remoulded strength, there are a number of different sequences of events which can set the stage for an abrupt failure.

SETTLEMENT AND HEAVE

Differential settlement and heave cause engineering problems which for the affected structures may be as severe as those due to slope failure. While the effects of microfabric are emphasized in this paper it is worth reiterating that macrofabric is often extremely important. Understanding conditions of soil formation can also alert the engineer to possible problems originating from causes at either structural level. For example, varves deposited close to a sediment source ("proximal") are more likely to contain irregular lenses of coarse sand leading to higher permeability than are "distal" varves deposited farther out (Kenney and Chan, 1973). Fissures similarly reduce effective drainage path length as does anisotropy resulting from clay mineral orientation. The increase in permeability can increase consolidation (as well as affecting strength and other properties).

Microstructure is thought to be particulary important when soft sensitive soils are involved and problems of this sort have been encountered in construction on the island of Trinidad in the Carribean. Trinidad is linked geographically and geologically with Venezuela to the west and indeed it has been claimed that the island was joined to the South American mainland until as recently as 10,000 years B.P. Submarine slides and chaotic deposits termed "wildflysch" have been described and comparisons made to sedimentary conditions in the Swiss Molasse basin and in Italy. Petroleum, natural gas and asphalt from Pitch Lake are the most important products (Mattson, 1977; Barr and Saunders, 1970). Major industrial plants are being constructed on the west coast of Trinidad about 20 miles south of the main city of Port of Spain. In that region there are low-lying terraces, alluvial plains and swamps. The geological structure is that of a synclinal basin filled with marine or brackish water sediments of Teritary age which were folded and uplifted during Pliocene times. These older sediments were overlain by soft lightly overconsolidated interbedded clays, sands and silts which are 40 ft. thick at the site.

The major heavy structures associated with the industrial plants are supported on piles driven into the lower more competent materials. However, the majority of the large storage tanks associated with these facilities are supported on granular pads and are stage constructed; alternatively the tank sites are preloaded prior to erection of the tanks. Both of these operations involve the prediction of the rate of dissipation of excess porewater pressures. Conventional soil mechanics analyses provide a reasonably accurate assessment of the time of dissipation in the silts and sands. However, the clay deposits are relatively sensitive and the behaviour of the clay structure has a significant influence on the stress-strain-time behaviour of the soil to the extent that conventional methods of analysis do not satisfactorily model the field behaviour of these materials (Crooks, 1981).

The samples examined from the recent deposits consist of soft clayey silts containing about 26 percent clay grade material composed largely of montmorillonite with chlorite, illite, quartz and feldspar. The fabric studies show angular silt composed of primary minerals, mica and small clay agglomerations or fecal pellets in a clay matrix with open fabric (Figure 5A, B, C). Some of the mica crystals show exfoliation (Figure 5D) of the type described as "layer-weathering" (Tarzi and Protz, 1978). Potash is said to be released more readily from trioctahedral micas than from dioctahedral micas, in part as a result of different orientations of the hydroxyl groups. In trioctahedral structures hydroxyl is oriented near normal to (001) but other orientations are possible in dioctahedral minerals (Giese, 1979). The course of the weathering reaction is thought to be affected by other factors such as particle size and environmental conditions including potassium concentration, the nature of other ions in pore solution, pH and redox potential. The concentration of potassium influences the diffusion gradient and the tendency of the mineral to release this element. Under oxidizing conditions biotite releases potassium less readily, possibly because of re-orientation of the hydroxyl groups away from the interlayer potassium with a corresponding increase in bond strength. The presence of authigenic pyrite in framboids (Figure 5E) indicates that weathering occurred under reducing conditions so exfolation should be favoured.

Heave susceptible soils often contain swelling clay minerals. In Canada the responsible clay minerals are generally detrital and have been inherited from older rocks such as the Bearpaw shale of Cretaceous age (Figure 5F) in western Canada. Elsewhere swelling clay minerals have formed by in situ weathering of parent rocks as for example in South Africa. Special methods of construction and design must be adopted (Van der Merwe et al., 1980; Williams and Donaldson, 1980; Pidgeon, 1980). Other causes of heave include frost action (Ueda and Penner, 1978; Highway Research Board, 1969) and biogeochemical weathering of black shale (Quigley and Vogan, 1970; Gillott et al., 1974).

CONCLUSIONS

Knowledge of the fabric of soils and rocks plays a large part in understanding the mechanical behaviour of these materials. Information about the fine constituents is only now becoming routinely available because of the increased use of the electron microscope. This knowledge is used to best advantage by the engineer when it is combined with information about the composition and geological history of the deposit. The examples in this paper show that qualitative studies can be of use in understanding behaviour of slope instability, heave, settlement and flow. Engineers are accustomed to work with quantitative information and the future is likely to see an increased trend to satisfy this demand.

ACKNOWLEDGEMENTS

Many engineers and geologists have supplied samples and information and I wish particularly to thank the following for their generous help: Andy Baracos, Dennis Becker, Richard Beddoes, Jack Clarke, Jack Crooks, Jim Graham, Victor Osipov, Don Shields, Nigel Skermer, V. N. Sokolov, Fernando Veniale, B. C. Hydro, Curtis Engineering and Testing, R. M. Hardy and Associates, Golder and Associates, Robertson Research. The paper would not have been possible but for the conscientious technical help of Mrs. Luci Jermy to whom I extend my sincerest thanks. I wish also to acknowledge the financial support of the National Science and Engineering Research Council of Canada.

REFERENCES

Barr, K. W. and Saunders, J. B., 1970. An outline of the geology of Trinidad. IN: J. F. Tomblin (Editor), Field guides to the geology of Trinidad. Caribbean Guidebook - 1970. Trans. Fourth Caribbean Geol. Conf., Trinidad, 1965. Am. Geol. Inst., Natl. Sci. Foundation.

Bentley, S. P., Clark, N. J. and Smalley, I. J., 1980. Mineralogy of a Norwegian postglacial clay and some geotechnical implications. Can. Mineral., 18, 535-547.

British Columbia Hydro and Power Authority, 1978. Hat Creek geotechnical study.

Campanella, R. G. and Vaid, Y. P., 1974. Triaxial and plane strain creep rupture of an undisturbed clay. Can. Geotech. Journ. 11, 1, 1-10.

Crooks, J. H. A., 1981. Personal communication.
Duffell, S. and McTaggart, K. C., 1951. Ashcroft map-area, British Columbia. Geol. Surv. Canada, Dept. Mines Tech. Surveys Memoir 262.
Earnest, L. J., 1981. In California diatomite may yield petroleum. Geotimes 26, 3, 17-19.
Fransham, P. B. and Gadd, N. R., 1977. Geological and geomorphological controls of landslides in Ottawa valley, Ontario. Can. Geotech. Jour. 14, 531-539.
Giese, Jr., R. F., 1979. Hydroxyl orientations in 2:1 phyllosilicates. Clays Clay Min. 27, 3, 213-223.
Gillott, J. E., 1971. Mineralogy of Leda clay. Can. Mineral. 10, 5, 797-811.
Gillott, J. E., Penner, E. and Eden, W. J., 1974. Microstructure of Billings shale and biochemical alteration products, Ottawa, Ontario. Can. Geotech. Jour., 11, 4, 482-489.
Highway Res. Board, 1969. Effects of temperature and heat on engineering behaviour of soils. Spec. Rept. 103, Washington, D.C., 300 pp.
Kenney, T. C. and Chan, H. T., 1973. Field investigation of permeability ratio of New Liskeard varved soils. Can. Geotech. Jour. 10, 473-488.
Mattson, Peter H. (Editor), 1977. West Indies island arcs. Benchmark papers in geology, 33, Dowden, Hutchinson and Ross Inc., Stroudsburg, Penn., 226-262.
McKyes, E., Sethi, A. J. and Yong, R. N., 1974. Amorphous coatings on particles of sensitive clay soils. Clays Clay Min. 22, 427-433.
Penner, E. and Burn, K. N., 1978. Review of engineering behaviour of marine clays in Eastern Canada. Can. Geotech. Jour. 15, 2, 269-282.
Pidgeon, J. T., 1980. A comparison of existing methods for the design of stiffened raft foundations on expansive soils. Proc. Seventh Regional Conf. for Africa Soil Mech. Foundation Eng. Accra. A. A. Balkema, Rotterdam, 277-290.
Quigley, R. M. and Vogan, R. M., 1970. Black shale heaving Ottawa, Canada. Can. Geotech. Jour. 7, 106-112.
Quigley, R. M., 1980. Geology, mineralogy and geochemistry of Canadian soft soils: a geotechnical perspective. Can. Geotech. Jour. 17, 2, 261-285.
Rutter, N. W., 1980. Geology in urban areas of Canada and its relation to landslides. Specialty Conf. on Slope Stability Problems in Urban Areas. Can. Geotech. Soc., 1-8.
Sergeyev, Y. M., Grabowska-Olsqewska, B., Osipov, V. I., Sokolov, V. N. and Kolomenski, Y. N., 1980. The classification of microstructures of clay soils. Jour. Microscopy 120, 3, 237-260.
Smalley, I. J. (Editor), 1979. Sensitive soils and quick clays. Eng. Geol. Spec. issue 14, 2/3, 217 pp.
Smalley, I. J., Craig, Ross, W. and Whitton, J. S., 1980. Clays from New Zealand support the inactive particle theory of soil sensitivity. Nature 299, 576.

Tarzi, J. G. and Protz, R., 1978. Characterization of morphological features of soil micas using scanning electron microscopy. Clays Clay Min. 26, 5, 352-360.

Ueda, T. and Penner, E., 1978. Mechanical analogy of a constant heave rate. Proc. Intl. Symp. Frost Action in Soils. University of Lulea, Sweden, 1, 57-67.

Van der Merwe, D. H., Hugo, F. and Steyn, A. P. The pretreatment of clay soils for road construction. Proc. 4'th Int. Conf. on Expansive Soils Vol. 1, 361-382.

Varnes, D. J., 1978. Slope movement types and processes. In.: Schuster, R. L. and Krizek, R. J. (Editors), Landslides analysis and control. Transp. Res. Board., Natl. Res. Council, Natl. Acad. Sci., Washington, D.C., 11-33.

Williams, A. A. B. and Donaldson, G. W., 1980. Developments relating to building on expansive soils in South Africa: 1973-1980. 4'th Intl. Conf. on Expansive Soils, Denver, U.S.A.

Yong, R. N., Sethi, A. J. and LaRochelle, P., 1979. Significance of amorphous material relative to sensitivity in some Champlain clays. Can. Geotech. Jour. 16, 511-520.

Figure 2. Scanning Electron Micrographs of Soils from the U.S.S.R. and Oakville, Ontario.

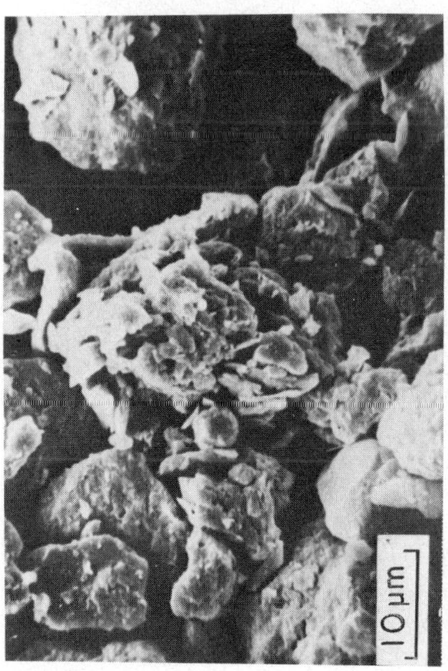

2A. Silty clay microaggregate in loess, Minusinsk, U.S.S.R. Photograph courtesy Profs.V.I.Osipov and V.N.Sokolov.

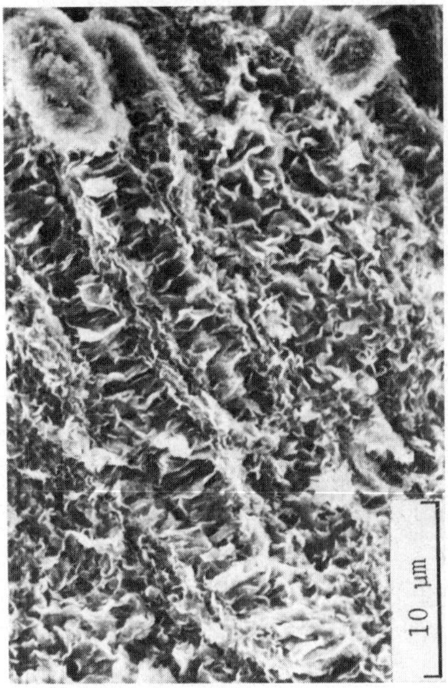

2B. Montmorillonite - hydrothermal, Oskania, Georgia, U.S.S.R. Sample courtesy Profs. V.I.Osipov and V.N.Sokolov

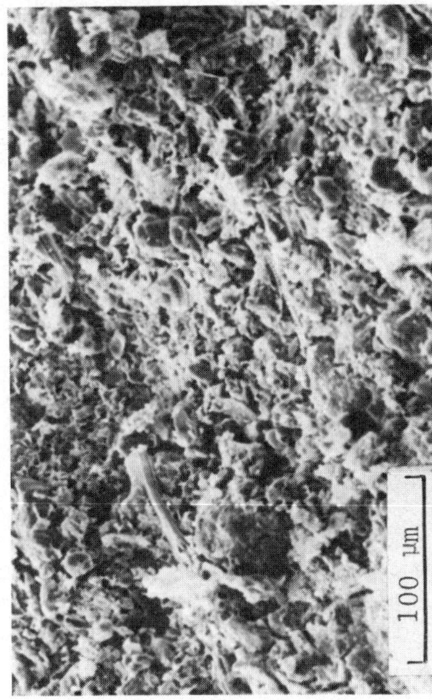

2C. Oakville, Ontario, Silt and clay

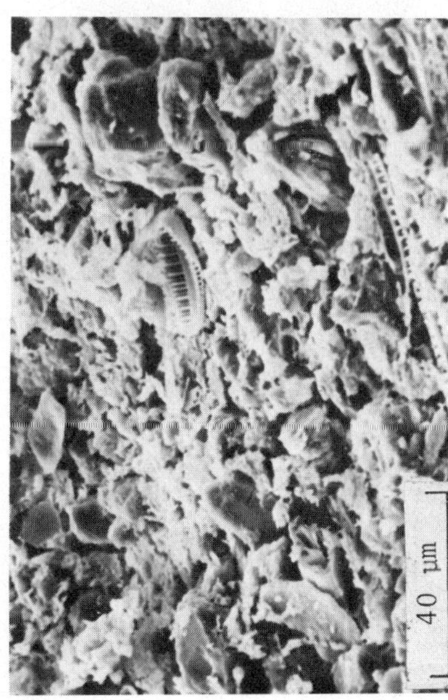

2D. Oakville, Ontario. Primary minerals, clay, and microfossils.

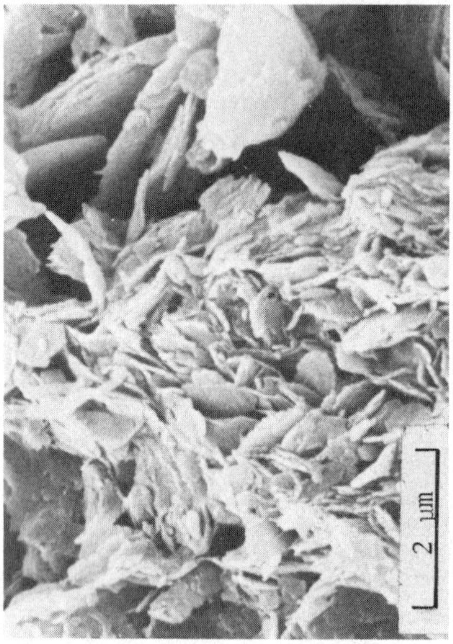

2E. Oakville, Ontario. Fecal pellet or agglomeration

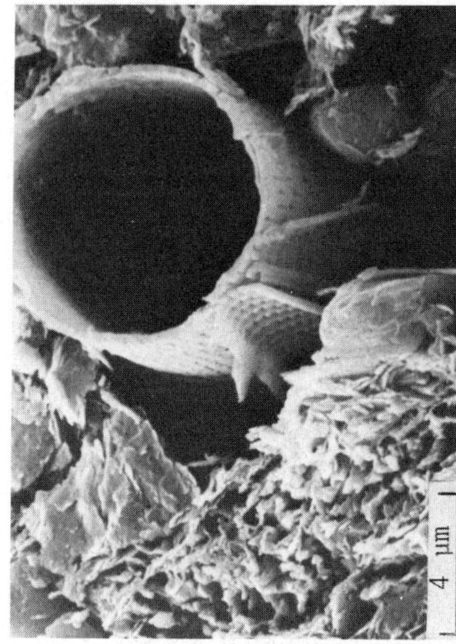

2F. Oakville, Ontario. Microfossil.

Figure 3. Scanning Electron Micrographs of Soil from Atikokan, Ontario, and Hat Creek, British Columbia.

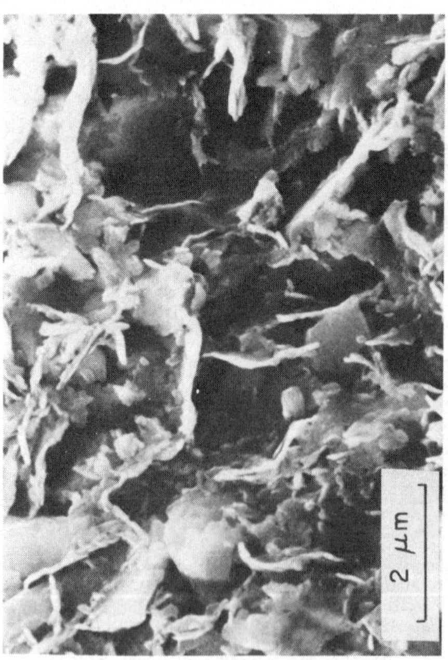

3A. Atikokan, Ontario. Open fabric

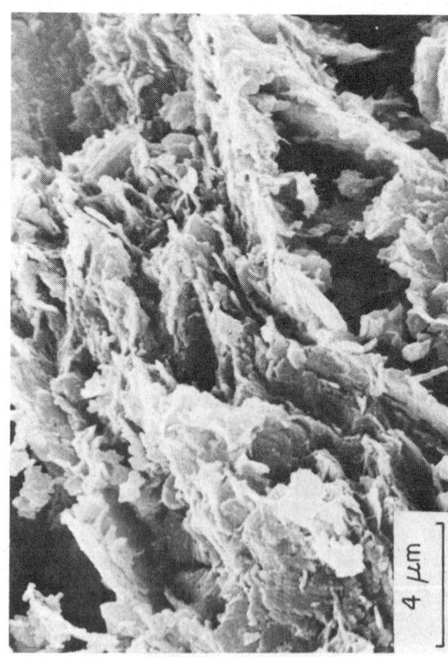

3B. Atikokan, Ontario. Arching patterns

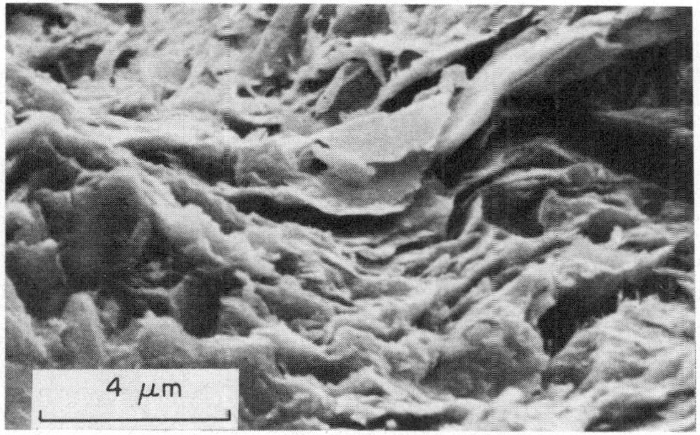

3C. Hat Creek. Preferred orientation of phyllosilicates.

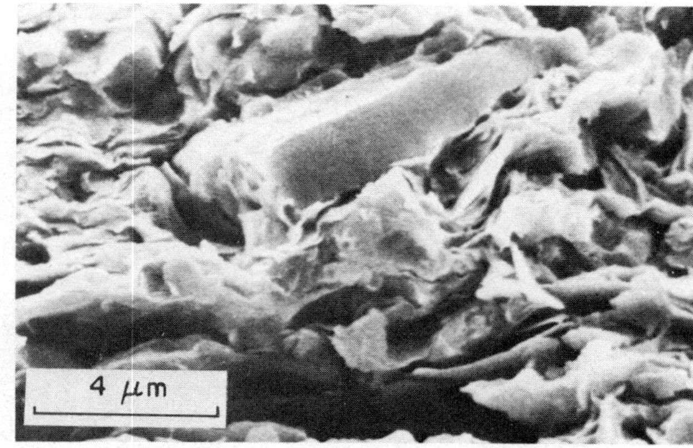

3D. Hat Creek. Preferred orientation

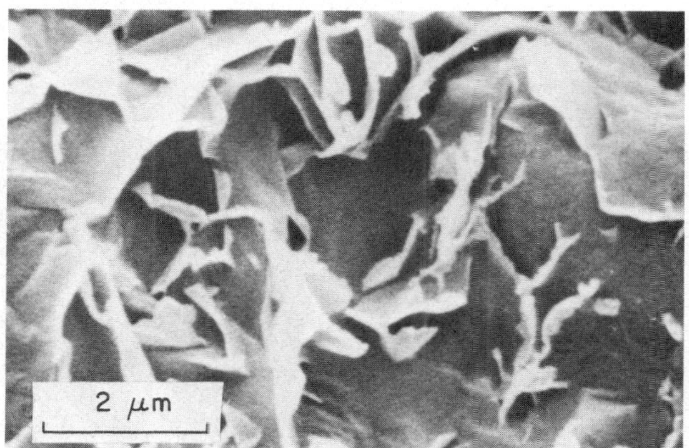

3E. Hat Creek. Clay faces - detail

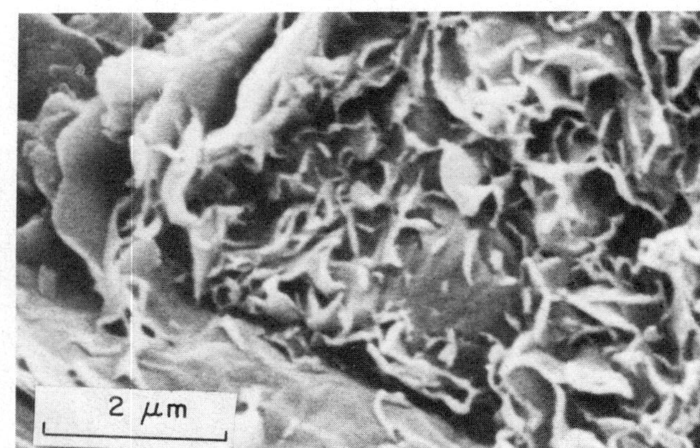

3F. Hat Creek. Alteration texture

Figure 4. Scanning Electron Micrographs of Sensitive Soils from Different Parts of Canada

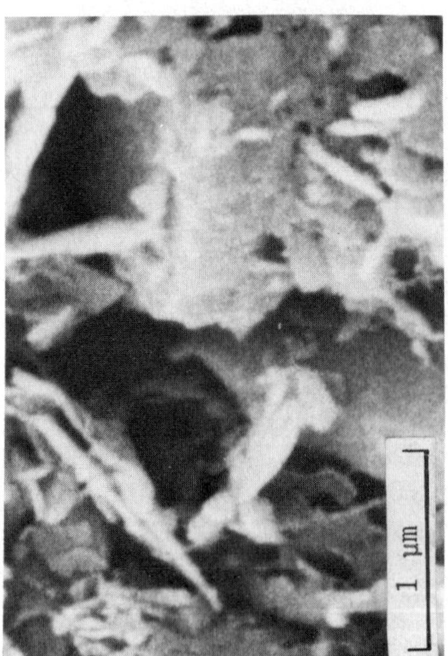

4A. Kars Bridge, Ontario. Primary minerals and clay - open fabric

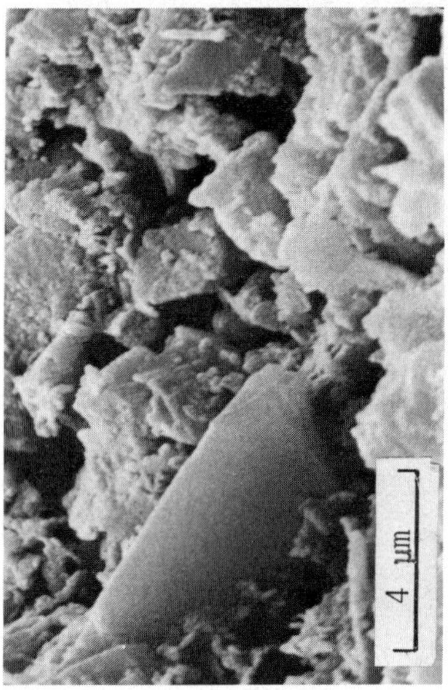

4B. North Bay, Ontario. Primary minerals and clay - open fabric

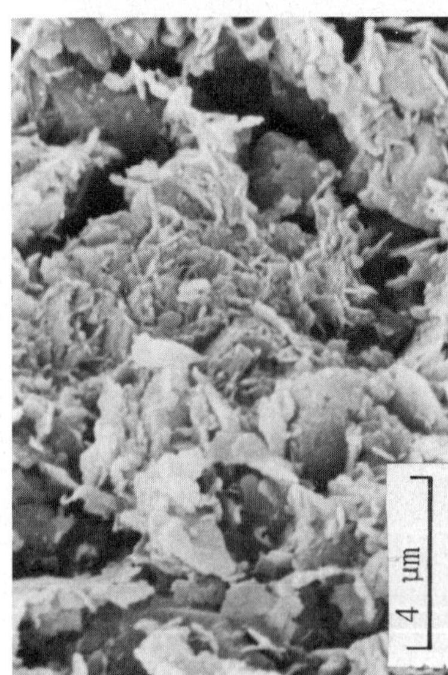

4C. Thunder Bay, Ontario. Silt-sized clay agglomeration.

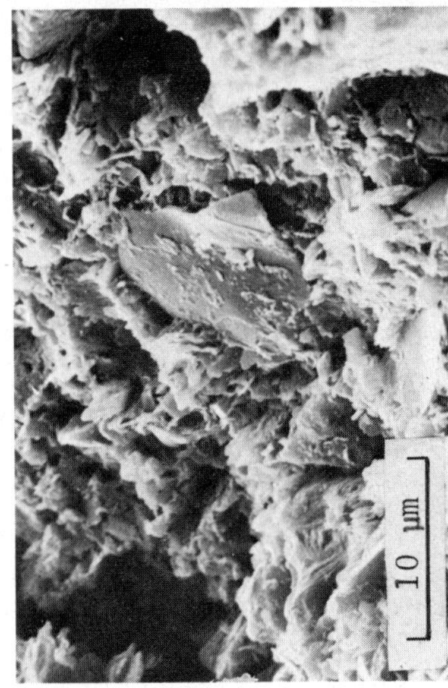

4D. Thunder Bay, Ontario. Deposits on surface of bent mica.

4E. Haney clay, B.C. Fresh fractures in primary minerals

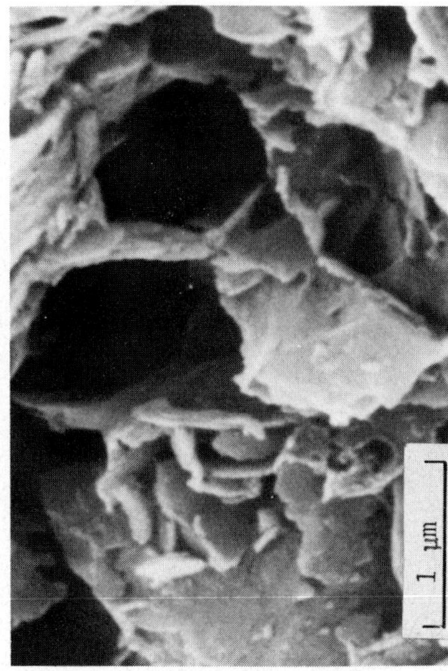

4F. Haney clay, B.C. Open fabric

Figure 5. Scanning Electron Micrographs of Soft Soil from Trinidad and Bearpaw Shale, Canada

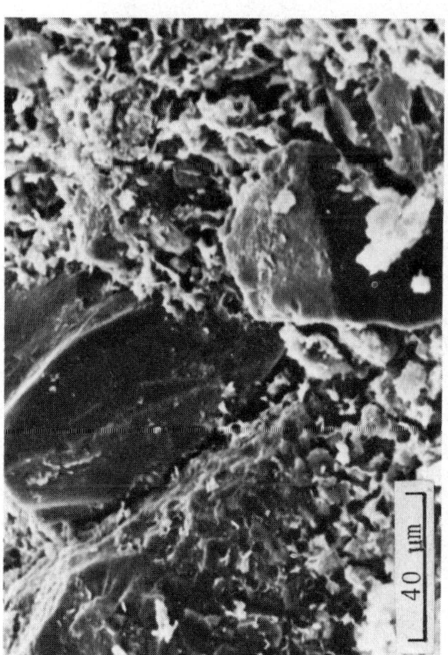

5A. Trinidad. Angular silt in clay matrix

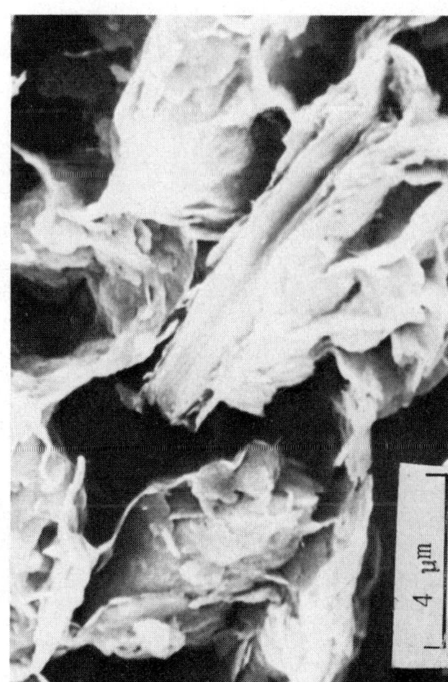

5B. Trinidad. Fecal pellet and mica

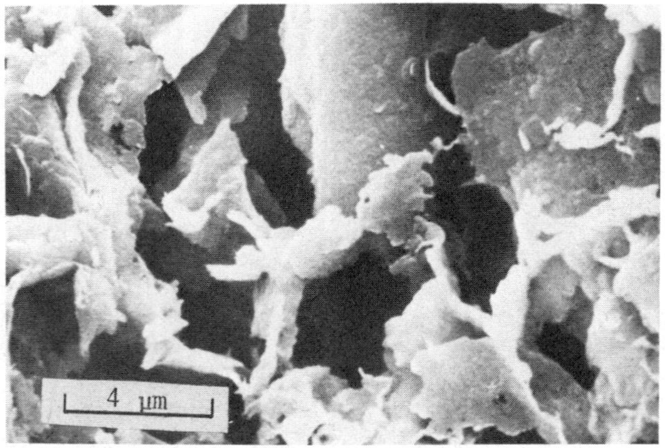

5C. Trinidad. Clay and silt-open fabric

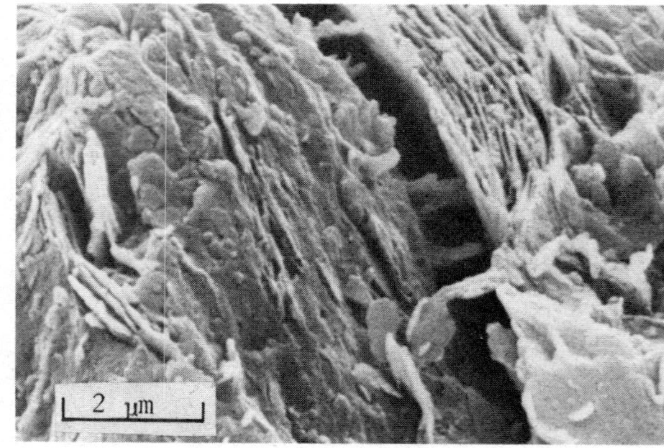

5D. Trinidad. Phyllosilicate showing

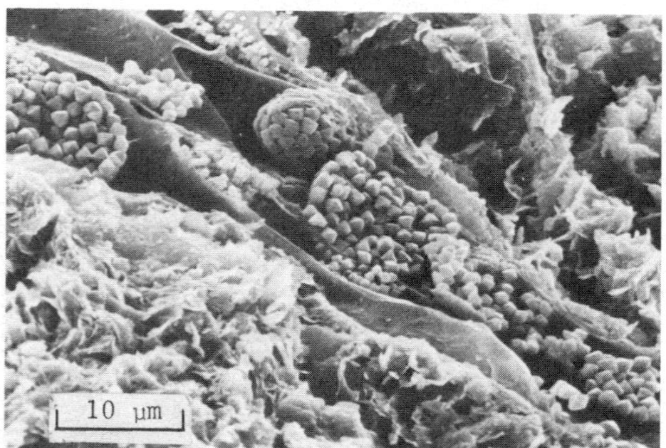

5E. Trinidad. Pyrite framboids

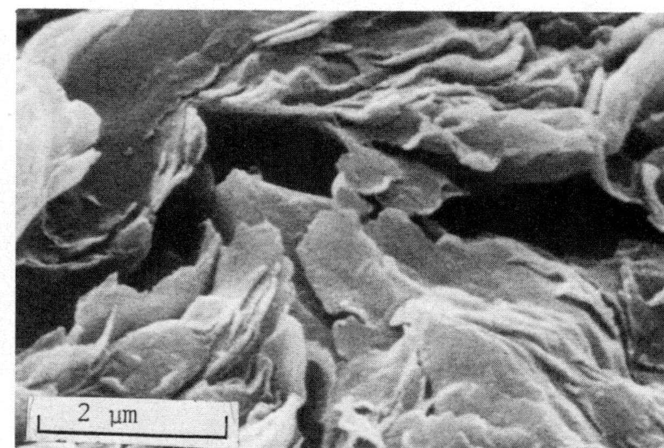

5F. Alberta. Bearpaw shale-oriented phyllosilicates

COLOUR IN SOILS AND SEDIMENTS - A REVIEW

R.M. TAYLOR
CSIRO Division of Soils Glen Osmond, South Australia

ABSTRACT

The colour of soils and sediments has for a long time been one of the readily observable morphological characteristics that have been used for description and classification. In both sediments and soils the dominant spectral colour is produced by the combination of the colours of the individual constituents, their relative concentrations, their particle size ranges and their degree of dispersion or homogeneity. In igneous rocks also colours are due to the same factors but are not darkened by the presence of organic matter as are soils and sediments. Some mineral phases impart almost diagnostic colours to soils and sediments, and marked changes in colour can arise from the isomorphous substitution of small amounts of some elements in these mineral phases. Sometimes this can be an important factor in geochemical exploration. Some of the work that has been done in trying to quantify observed or measured colours with composition is also reviewed. Generally the colours of both soils and sediments are influenced by physical and chemical factors that have applied in the environment. In summary it can be stated that:

(1) The colour of a soil or sediment depends upon the quantity and type of pigmented mineral phases present, or the masking effect of dark coloured components also in the system.

(2) The colour of a sediment may be different from that of the parent soil from which it was derived, due to changed environmental factors, and

(3) Some normally white or neutral coloured minerals can develop quite marked colours due to the incorporation of small quantities of transition metals.

COLOUR IN SOILS AND SEDIMENTS - REVIEW

R.M. TAYLOR
CSIRO Division of Soils, Glen Osmond, South Australia.

In the Journal of Chemical Education, Mary Orna (1978) states that colour science is one of the few disciplines that cuts across the boundaries of art, biology, physics, psychology, chemistry, geology and mineralogy, and that colour provides the perfect link between an easily observed and described property and an underlying theory.

The colour of sedimentary rocks is stressed most in relation to shales, where pigmentation results from a mixture of constituents, darker shades often being due to organic matter. Red shales, which generally contain hematite, need not necessarily contain more Fe than those which are black, grey or green, where the Fe is possibly present in minerals with less pigmentation power, or present in the ferrous rather than the ferric state.

Although it appears to be generally accepted that red beds owe their colour to the presence of hematite, there are many, and possibly equally correct, theories as to what were the dominant factors in their formation. Friend (1966) and Danilov (1968) noted that in variegated sediments the red hematitic beds were confined to the silt or fine grained material, while the green or non-red beds were generally coarser, sandy sediments. Friend considered that the coarser sandstone material was deposited in channel beds, where, being permanently wet, the iron could be leached out in the soluble Fe(II) form, whereas in the finer sediments in the overbank area, being subject to wet and dry conditions, solution and reprecipitation, or direct alteration of brown oxidic material to hematite could occur. Friend (1966) suggested that the original alluvial deposits of the Devonian red beds in the Catskill mountains of America were brown and that post-depositional alteration to hematite occurred, and he quotes Fischer (1963) as commenting that there is a notable lack of ancient brown sediments. Possibly the initial brown material was ferrihydrite which, under certain conditions, ages to hematite. In this regard McBride (1974) also considers that reddening of sediments is a post-depositional effect.

However Friend realised that the factors primarily responsible for red bed formation in some areas need not apply in other cases. Dmitrievskii (1977) submits that red beds in the Caspian Sea Basin result directly from weathering of red parent material. Danilov (1968) reached a similar conclusion for red sediments in the Donetz Basin, but remarked that the formation of the green colours in associated beds was post-depositional due to the epigenic alteration of the original red-brown material. Here the Fe(III) exceeded the Fe(II) content in the red beds but not in the green sediments. The alternation of green and red beds can be interpretated as reflecting climatic changes producing alternative oxidative and reducing environments. Under the reducing conditions the mobilised Fe(II) is incorporated and partially retained in this divalent state in the associated clay minerals. Subsequently, in an oxidative environment, the red-brown ferruginous phase is aged or transformed into more crystalline forms of fe oxide. McBride proposes that the colour variation in some Mexican sediments arises from the depositional environment. In an examination of deltaic, fluvial and shelf sediments the colours were restricted to the deltaic deposits. The red and purple colours were attributed to variations of size of particles and the greens to the presence of illite and chlorite and the absence of hematite, organic matter and sulphides.

Post-depositional reddening is considered to be responsible for desert sand dune coloration. Their redness has been noted to increase with distance of transport and increasing age and there has been much speculation as to the factors responsible (Norris, 1969). Walker (1979) offered an explanation for the increasing redness of the eolian deposits in western Libya. The initial weathering of ferromagnesian minerals in an oxidising and alkaline environment gives rise to a surface coating on sand grains of what Walker refers to as chemically-precipitated, amorphous, yellow ferric hydrate or limonite. It would appear from our present knowledge that this coating is most probably ferrihydrite, which, when naturally formed, has a yellow to yellow-brown rather than red colour.

During subsequent eolian transport, these surface coatings are more effectively abraded from the larger grains, and persist to a greater extent on the indented surfaces on the smaller particles which are transported over much longer distances. If the surface phase is ferrihydrite it is meta-stable and can transform via the solution phase to goethite or it can coalesce in suspension to form hematite. Hematite formation is favoured by greater concentrations of the ferrihydrite suspensions or by the presence of compounds which accelerate coalescence of the particles (Schwertmann and Fischer, 1966). Even the aridity of desert regions could be expected to provide variations in the diurnal climatic environment which would favour this transformation of ferrihydrite to hematite. The high daytime temperatures could cause slow dehydration of the ferrihydrite to hematite, and during the low nocturnal temperatures small amounts of dew (the importance of which has often been stressed as an agent of mineral alteration (Margolis and Krinsley, 1971)) condense and further promote the coalescence of ferrihydrite particles. Moreover, unweathered, transported ferro-magnesian minerals could give rise to the formation of further hematite in situ.

In most sediments and the various soils of different climatic zones, colours are often due to the free iron oxide content, the most common pigmenting phase in soils. Consequently, especially in earlier systems of soil classification based on morphology, soils were often classified and grouped by the colours imparted by these oxides, or by the changes or absence of these colours in certain parts of the profile.

One of the ways in which colours have been factually described, especially in soils, is the use of the Munsell notation, which allows soil scientists to give reproducable colour codes that are immediately recognised by others. This notation is based on three-number symbols, the first expressing hue, the second the value and the third the chroma.

The code number for hue refers to the dominant spectral colour and arises, as has been stated, by the combination of colours and other properties of the individual minerals and other components present. The value, which is expressed as a graduation from 1-10, refers to the apparent lightness as compared to an absolute white. The apparent lightness can be influenced by the degree of moisture (Shields *et al.* 1968) which imparts a darker shade. Leger *et al.* (1979) have attributed this effect of moisture to either an interaction with soil components, especially humic material, causing multiple reflections from water films thus favouring selective absorption, or to a greater dispersion of the organic material over the surface of other colloids. The number for chroma is a measure of the purity of the hue and would be influenced by the relative amounts of the coloured minerals present and their relative pigmenting powers.

Various degrees of success have been obtained in attempts to correlate Munsell notations with different soil components, for example the hue with mineral composition and value with organic matter content (see Krishna Murti and Satyanarayana, 1971). McKeague *et al.* (1971) who examined a large number of soil samples claimed that no general relationship existed between chroma and dithionite extractable Fe, or between the Munsell value and organic matter content. Therefore, although a mineralogical analysis may not indicate what the resultant colour or Munsell notation should be, the colour can often be indicative of the kind of minerals present.

Winters (1930), recognising that a precise method for quantifying soil colour was needed, tried a technique based on spectral reflectivity, and this method has been used with success by other workers (Shields *et al.* 1966; Karmanov, 1968). The wavelength and intensity of reflected colours could be significantly correlated with the constituents of some brown soils from Dagestan. In this work Karmanov and Rozhkov (1972) concluded that the intensity and spectral distribution of reflected radiation depended on three basic groups of compounds: (1) humic substances which reflect light weakly and evenly throughout the spectral range, (2) light coloured compounds, carbonates and alumino-silicates etc., which reflect light strongly and evenly throughout the spectrum, and (3) iron compounds, different forms of which reflect light at different intensities and quite unevenly throughout the spectral range. Leger *et al.* (1979) presented a new interpretation of the spectral reflectivity of soils, and related their measurements to organic matter, iron oxide and water content. They suggested that colour variations were the result of changes throughout the total range of reflected wavelengths, rather than changes in specific wavelength ranges. The nature of the organic matter is also important since only the humic fraction darkens the soil. Organic matter and iron oxides (the majority of which was oxalate-soluble (Schwertmann, 1964)) were found to be more effective in darkening sandy soils than clays. The opposing pigmenting effects of Fe oxides and organic matter were also investigated by Stace (1956). He studied the chemical differences between the red to brown terra rossa soils and the associated black to brown-black rendzinas. Both soils are formed on carbonate and the colour difference was found to be due to the larger amounts of organic matter in the latter group. A division in the gradation between these commonly associated groups could be made on the value of the ratio of percent free iron oxide in the < 2 μm fraction and percent nitrogen, where the latter is taken as a measure of the organic content. Stace found that this ratio was generally < 10 for the rendzinas and > 10 for the terra rossa soils.

It is generally accepted that the yellow and red colours of soils are derived from the presence of goethite and hematite respectively, the two most common soil iron oxides. For example, a colour redder than 7.5YR indicates the presence of hematite whereas goethite induces a colour between 2.5Y and 7.5YR, (Schwertmann, personal communication, 1980).

Other iron oxides with distinctive colours can also occur in reasonable amounts in soils that are subject to certain pedogenic conditions. Lepidocrocite, which can vary in colour between orange and yellow, often occurs as mottles in association with goethite in soils subject to hydromorphic conditions. Natural ferrihydrite, as opposed to rapidly precipitated ferric hydroxide, is also yellow-brown and may persist for long periods in soils, especially if stabilised by adsorbed impurities which inhibit further crystallisation or transformation. Ferrihydrite and lepidocrocite generally range between 5YR and 7.5YR although the value for ferrihydrite is usually lower (3-5) than that of lepidocrocite (5-6), (Schwertmann, personal communication, 1980).

The common soil iron oxide minerals may originate from the same source of mobilised Fe, and the relative proportions formed during pedogenesis will depend on the environmental soil forming factors. Quite small variations in the micro-environment can be responsible for a change in the direction of mineralisation. Hematite, however, has a greater pigmenting power than goethite, and soils which appear red can actually have dominant amounts of goethite rather than hematite. Segalen (1969), Schwertmann and Lentze (1966) and Soileau and McKracken (1967) in their respective studies virtually concluded that colour is not an infallible indication of the dominant iron oxide phase present. Torrent *et al.* (1980) demonstrated this effect in a series of river terrace soils from Spain. Even in soils where goethite was up to six times more abundant than hematite, a high correlation was obtained between the calculated percentage of hematite in the fine earth fraction and the colour or redness rating, given by the product of the chroma/value ratio of the Munsell rating and a numerical value for the hue.

Soils are generally black due to the presence of humic materials associated with various clay minerals and other non-pigmenting components. Black soils often occur in poor drainage situations associated with the presence of smectite. In Australia, Norrish and Pickering (1981) have made a general observation that black earths at latitudes north of 31-32° South generally contained smectite, whereas at more southerly latitudes illite was the dominant clay mineral.

Black colours on ped faces and along cracks in soils and sediments often may be due to thin coatings of Mn oxides. Generally these oxides do not contribute to the overall soil colour because the average concentration of Mn in soils is only around 0.1%, much less than that of the oxides of Fe. The Mn oxides generally contain large amounts of foreign cations, which, however, do not alter their colour. These minerals in soils are generally poorly crystalline lithiophorite, members of the hollandite family or related to the birnessite group of minerals.

Grey horizons, generally owe their colour to the lack of pigmenting minerals and small amounts of organic matter. Often, when wet, these horizons can appear green and can resemble, to a slight extent, the green colours observed under the reducing conditions of a gleyed soil.

The green-blue and related colours often encountered in gleyed soils were suggested to come from a form of "green rust", Fe(II) Fe(III) hydroxy-chlorides and -sulphates described by Bernal *et al.* (1959) after the earlier work of Feitknecht and Keller (1950). Ponnamperuma *et al.* (1967) suggested that the theoretical compound hydromagnetite, $Fe_3(OH)_8$, could be responsible for these dark green phases.

It has been established, however, that these green-blue colours are associated with Fe(II) since Naohara and Usami (1976) found that the blue-green sites around plant residues in waterlogged soils were centres of high Fe^{2+} concentrations. The most obvious feature of these green-blue soil phases is the rapidity with which they become yellow or yellow-brown on exposure to air. Recently Taylor and McKenzie (1980) and Taylor (1980) showed that small amounts of Al(III) and Fe(III) hydroxy species in solution could cause the precipitation of Fe^{2+} present under anaerobic conditions around pH 7, resulting in compounds such as Fe(II)Al(III) or Fe(II)-Fe(III) hydroxy-carbonates. These compounds are members of the pyroaurite group and are isostructural with hydrotalcite. They exhibit the same blue-grey to green-blue colours found in gleys and also oxidise just as readily on exposure to air to give the more common yellow-brown Fe oxides found in soils. Because these meta-stable compounds can be synthesised so readily from components and under conditions that could be expected in soils, it was postulated that these green-blue phases in gleyed soil horizons are composed of the above double hydroxy compounds (basic-carbonates).

Green colours in layer lattice silicate minerals are also quite common and are usually due to the presence of Fe in the octahedral sheet. Glauconite is an illite type mineral which is thought to form under marine diagenesis conditions through the slow deposition of soluble materials in shallow water. It generally has a uniform Mg content and a restricted range for the Fe(III)/Fe(II) ratio (Grim, 1968). However, there are green glauconitic illites that are not of marine origin. Porrenga (1968) found that many of the green illites from USA and Europe are of Oligocene age and are of a non-marine origin, presumed to have formed in lacustrine or lagoonal environments often in association with carbonates. He proposed that they formed from the alteration of detrital smectite or illite, but Jung *et al.* (1954) assumed that they formed through chemical precipitation. A similar genesis is proposed for the green Muloorina illite, also of Oligocene age, from Australia (Norrish and Pickering, 1981). Green illites generally contain more Fe(III) than Fe(II), both of which are in the octahedral sheet (Porrenga, 1968). Manganese can also influence the colour of illites, and Tien (1969) attributed the purple colour of some illitic clays to trace amounts of this element.

The colours of smectites also depend on composition, especially of the content of transition metals. Nontronite nominally contains Fe(III) replacing a major part of the octahedral Al but very small changes in the Fe(III)/Fe(II) ratio can cause the colour to change from yellow to green. This observation is in accord with the very many observations that an association of Fe in both valence states gives rise to green colours. Linares *et al.* (1972) displayed samples of coloured Spanish bentonites at the Madrid conference of this Society. Mn was considered responsible for the red colour of a bentonite deposit, even though its concentration was only 0.7%. Mn was also found to be responsible for the pink mottles in some veins of smectite in Tasmania. The outer edges of these veins had thin Mn oxide coatings suggesting that the smectite was losing its Mn and becoming bleached, while the Mn on reaching the surface face was oxidised and precipitated.

Linares *et al.* (1972) reported the occurrence of green bentonites which had far less Fe than is usually associated with nontronites. The chemical composition of these bentonites appeared to differ from that of the associated uncoloured material: in the green layers the Fe(II)/Fe(III) ratio was much higher (average 1.16 vs. 0.063) and the Cr content was higher. Possibly, combinations of elements can also be responsible for colour variations. Maksimovic and White (1972) found that the blue colour in some Yugoslavian halloysites increased with increasing Cr content, where the Cr(III) replaced Al in the octahedral sheet. Maksimovic *et al.* (1978) also reported blue kaolinite in a zone of hydrothermally altered rocks in Yugoslavia. Again the colour was due to the presence of Cr but at much lower levels than reported for the halloysites (kaolinites 0.26 - 0.70% Cr_2O_3). Siderite often imparts a grey to bluish tinge on freshly exposed surfaces, but hydrolysis and oxidation soon cause darker brown colours to develop (Pettijohn, 1957). Although a pure white sediment or deposit could consist of any mineral or a mixture of a variety of minerals, the bleached zones of laterites in the Australian continent are considered as a diagnostic feature of kaolinite.

REFERENCES

Bernal, J.D., Dasgupta, D.R. and Mackay, A.L., 1959. The oxides and hydroxides of iron and their structural interrelationships. Clay Min. Bull., 4: 15-30.

Danilov, I.S., 1968. Nature and geochemical formation conditions of a variegated colouring of Triassic and Permian deposits of the western Donetz Basin. Litol Polez. Iskop., 5: 100-106.

Dmitrievskii, A.N., 1977. Origin of red and variegated colours in upper Permian and lower Triassic formations in the eastern part of the Caspian Sea Basin. Mosk. Inst. Neftekhim. Gazov. Prom - sti. im. I.M. Gubkina, 130: 115-120.

Feitknecht, W. and Keller, G., 1950. Über die dunkelgrünen hydroxyverbindungen des Eisens. Z. Anorg. Allg. Chem., 262: 61-68.

Fischer, A.G., 1963. Descriptive Paleoclimatology. Am. J. Sci., 261: 282-293.

Friend, P.F., 1966. Clay fractions and colours of some Devonian red beds in the Catskill Mountains, U.S.A. Q. J. Geol. Soc. Lond., 122: 273-292.

Grim, R.E., 1968. Clay Mineralogy. McGraw-Hill Book Co./New York.

Jung, J., Alexian, C., Blot, P., Oberlin, A. and Rey, M., 1954. Les illites due bassin oligocene de Salins (Cantal). Bull. Soc. Fr. Mineral. Cristallogr., 77: 1231-1249.

Karmanov, I.I., 1968. Use of spectrophotometric coefficients in the study of soil forming processes. Pochvovedeniye, 2: 13-28.

Karmanov, I.I. and Rozhkov, V.V., 1972. Experimental determination of quantitative relationships between the colour characteristics of soils and soil constituents. Pochvovedeniye, 12: 71-79.

Krishna Murti, G.S.R. and Satyanarayana, K.V.S., 1971. Influence of chemical characteristics in the development of soil colour. Geoderma, 5: 243-248.

Leger, R.G., Millette, G.J.F. and Chomchan, S., 1979. The effects of organic matter, iron oxides and moisture on the colour of two agricultural soils of Quebec. Can. J. Soil Sci, 59: 191-202.

Linares, J., Huertas, F., Lachica, M. and Reyes, E., 1972. Geochemistry of trace elements during the genesis of coloured bentonites. Proc. Int. Clay Conf. Madrid, pp. 351-360.

Maksimovic, Z. and White, J.L., 1972. Infrared study of chromium-bearing halloysites. Proc. Int. Clay Conf. Madrid, pp. 61-73.

Maksimovic, Z., White, J.L. and Logan, M., 1978. A study of chromium-bearing kaolinite from Teslić, Yugoslavia. Abstr. Int. Clay Conf. Oxford, p. 754.

Margolis, S.V. and Krinsley, D.H., 1971. Submicroscopic frosting on aeolian and subaqueous quartz sand grains. Geol. Soc. Amer. Bull., 82: 3395-3406.

McBride, E.F., 1974. Significance of colour in red, green purple, olive brown and gray beds of Difunta Group, Northeastern Mexico. J. Sed. Pet., 44: 760-773.

McKeague, J.A., Day, J.H. and Shields, J.A., 1971. Evaluating relationships among soil properties by computer analysis. Can. J. Soil Sci., 51: 105-111.

Naohara, T. and Usami, S., 1976. Some observations on the colour changes of soils. 1. Microbial processes in paddy soils. Nippon. Dojo-Hinjogaku Zasshi, 47: 367-372.

Norris, R.M., 1969. Dune reddening and time. J. Sed. Petr., 39: 7-11.

Norrish, K. and Pickering, J.G. Clay Mineralogy. In: "Soils: An Australian Viewpoint" in Press, CSIRO Melbourne.

Orna, M.V., 1978. The Chemical Origins of Colour. J. Chem. Education, 55: 478-484.

Pettijohn, F.J., 1957. Sedimentary Rocks, Harper and Brothers, New York, p. 145.

Ponnamperuma, F.N., Tianco, E.M. and Loy, T., 1967. Redox equilibria in flooded soils: 1. The iron hydroxide system. Soil Sci., 103: 374-382.

Porrenga, D.A., 1968. Non-marine glauconitic illite in the lower Oligocene of Aardebrug, Belgium. Clay Min., 7: 421-430.

Schwertmann, U., 1964. Differenzierung der Eisenoxide des Bodens durch photo-chemische Extraktion mit saurer Ammoniumoxalat-Lösung. Z. Pflanzenernähr Düng. Bodenkd., 105: 194-202.

Schwertmann, U. and Fischer, W.R., 1966. Zur Bildung von α-FeOOH und α-Fe_2O_3 aus amorphen Eisen (III)-hydroxid. III. Z. Anorg. Allg. Chem., 346: 137-142.

Schwertmann, U. and Lentz, W., 1966. Bodenfarbe und Eisenoxidform. Z. Pfanzenernähr. Düng Bodenkd., 115: 209-214.

Segalen, P., 1969. Contribution à la connaissance de la couleur des sols à sesquioxides de la zone intertropicale des sols jaunes et sols rouges. Cah. O.R.S.T.O.M. Pédologie, 7: 113-236.

Shields, J.A., St. Arnaud, R.J., Paul, E.A. and Clayton, J.S., 1966. Measurement of Soil Colour. Can. J. Soil Sci., 46: 83-90.

Shields, J.A., Paul, E.A., St.Arnaud, R.J. and Head, W.K., 1968. Spectrophotometric measurement of soil colour and its relationship to moisture and organic matter. Can. J. Soil Sci., 48: 271-280.

Soileau, J.M. and McCracken R.J., 1967. Free iron and coloration in certain well-drained coastal plain soils in relation to their properties and classification. Soil Sci. Soc. Am. Proc., 31: 248-255.

Stace, H.C.T., 1956. Chemical characteristics of terra rossas and rendzinas of South Australia. J. Soil Sci., 7: 280-293.

Taylor, R.M., 1980. Formation and properties of Fe(II)Fe(III) hydroxy-carbonate and its possible significance in soil formation. Clay Miner, 15: 369-382.

Taylor, R.M. and McKenzie, R.M., 1980. The influence of Al on iron oxides. VI. The formation of Fe(II) Al(III) hydroxy-chlorides, -sulphates and -carbonates as new members of the pyroaurite group and their possible significance in soils. Clays Clay Min., 28: 179-187.

Tien Pei-Lin, 1969. A purple coloured 1M mica clay from Silverton Colorado, U.S.A. Clays Clay Min., 17: 245-249.

Torrent, J., Schwertmann, U. and Schulze, D.G., 1980. Iron oxide mineralogy of some soils of two river terrace sequences in Spain. Geoderma, 23: 191-208.

Walker, T.R., 1979. Red colour in dune sand. In: D. McKee (Editor), Study of Global Sand Seas, U.S. Geological Survey Paper 1052, chapter D.

Winters, E. 1930. The measurement of soil colour. Amer. Soil Survey Assoc. Bull., 11: 34-37.

EFFECT OF LITHIUM TREATMENT ON THE BULK DENSITIES OF EXFOLIATED VERMICULITES

P. SOUZA SANTOS and A.R. RIEGO
Department of Chemical Engineering, Escola Politécnica, University of São Paulo and Instituto de Pesquisas Tecnológicas, São Paulo - São Paulo - Brasil

ABSTRACT

Walker and Milne have shown that vermiculite crystals, after treatment with aqueous solutions of lithium chloride, swell in water.

Two Brazilian vermiculites, of particle sizes used for exfoliation in industrial kilns, were treated with aqueous lithium chloride solutions (saturated solution for 6 hours at 25°C or 1M solution for 24 hours at 85°C). Swelling values from 9 to 14 cm^3/g were obtained when the swollen samples, after draining, were exfoliated in an electric furnace at temperatures between 550°C and 1050°C.

The measured bulk densities of the lithium-swollen exfoliated vermiculites are significantly smaller than those of non-treated exfoliated vermiculites, at the same exfoliation temperatures.

INTRODUCTION

Walker and Milne (1950) have shown that vermiculite crystals from Kenya, after treatment for four days with saturated aqueous solutions of lithium chloride, swell in distilled water; in other papers (Norrish, 1954; Norrish and Rausell-Colom, 1962, 1963; Walker, 1950, 1959, 1961, 1975; Walker and Garrett, 1961; Garrett and Walker, 1962; Myers, 1969), the conditions for swelling were discussed and it was shown that other substances (N-alkylammonium ions; hydrogen peroxide; weak acids) can also produce swelling.

The objective of this paper is to show that some Brazilian commercial vermiculites, that have been previously treated with aqueous lithium chloride solutions and swollen in water, can be thermally exfoliated to bulk densities values significantly smaller than those obtained from the non-treated vermiculite at the same temperature.

MATERIALS AND METHODS

Samples:- The samples used were crude commercial loosefill vermiculites from two mines in Brazil; both were supplied by Minebra Minérios Brasileiros S/A. All samples were received in closed 50 kg polyethylene bags; they were air dried at 90°C-100°C prior to grinding and sieving.

Sample I, called "type 23", is from the locality Brumado, State of Bahia and contains 90% wt (in crude form) of particles in the size range between ASTM sieve numbers 10 and 20; samples II, III and IV are from the locality of São Luiz do Monte Belo, State of Goiás; both samples II and III are of "type 23", from different parts of the mine; sample IV, "type 4", contains about 90% wt in the range between ASTM sieve numbers 20 and 80. The chemical analysis of samples I and II is shown in Table I.

TABLE I

Chemical Analysis of Vermiculite Samples

Sample	I. Brumado, State of Bahia* "Type 23"	II. São Luiz de Monte Belo* State of Goiás "Type 23"
Loss on Ignition (%):	5.90	7.30
SiO_2 (%)	37.1	36.4
Al_2O_3 (%)	19.2	16.2
Fe_2O_3 (%)	5.14	11.3
TiO_2 (%)	0.33	2.73
MgO (%)	23.0	18.5
CaO (%)	2.04	0.59
Na_2O (%)	1.35	0.10
K_2O (%)	5.47	6.78
Total	99.53	99.90
By powder XRD	Vermiculite+Mixed Layers	Essentially vermiculites

* Dried 105°C-110°C

X-ray diffraction analysis:- Sample I, from Brumado, contains a small amount of a mixed-layer mineral with a 24Å-25Å basal spacing and samples II, III and IV traces of a 10Å mica. The vermiculite reflexions were confirmed by tests described by Walker (1961).

Bulk density:- The bulk density of both crude and exfoliated of dry loosefill vermiculite was measured according to the British Standards method B.S. 1460 (1961) ("Determination of apparent density after compaction of precipitated calcium carbonate").

Lithium treatment:- Portions of about 1.0 gram of vermiculite were swollen in 10 or 50 cm^3 of 1M, 6M or saturated lithium chloride solutions for various periods at either room temperature (varying from 15°C to 25°C); at 25.0°C (in a thermostat) or 85°C ± 2°C (in a electric oven) with reflux condenser if desired.

Swelling in water:- After the desired time of contact with the lithium chloride solution, the vermiculite crystals were drained, washed three times with 100 cm^3 distilled water. The known weight of vermiculite was quantitatively transfered to a graduate cylinder by a jet of distilled water and allowed to swell. As in the Foster's swelling test for smectites (1953), a proportion of 1.0 gram of original crude loose fill vermiculite to 100 cm^3 of water was selected. After 24 hours, the swelling was constant and the swelling volume was read and expressed in cm^3/g. The swollen vermiculite crystals were drained in a Büchner funnel and quantitatively transfered to a stainless steel plate for exfoliation in a laboratory electric muffled kiln. All experiments were conduced at least in duplicate.

Exfoliation:- When the selected temperature of the electric kiln was stabilized, the door was opened and a slice of refractory brick pushed out; the stainless steel plate with the vermiculite sample was placed on the brick and introduced in the kiln; the pyrometer temperature was observed, until the original temperature was attained and after 4 minutes the stainless steel plate was taken out and placed in a dessicator to cool down. Industrial exfoliation is usually done a temperature between 1600°F and 2000°F, with a retention time of 4 to 8 seconds. If the sample in left too long in the hot zone, additional combined water is removed after achieving optimum exfoliation and a product with reduced toughness and pliability results (Schroeder, 1970). Results of the laboratory procedure can be directly coordinated: see Navajas and Souza Santos, 1976, 1977, 1981 , for the correlation equation with the results of exfoliation conducted on a pilot scale in a vertical baffled gas fired furnace.

After cooling, the bulk density is measured and expressed in g/cm^3. The results are presented, either in terms of swelling (in cm^3/g) or as bulk densities in (g/cm^3); the reported values are arithmetic means of at least duplicate experiments.

RESULTS AND DISCUSSION

Swelling in lithium chloride aqueous solutions:- The effect of several variables in the swelling experiments was investigated. Table II shows the effect of LiCl concentration, at constant temperature and time (6 hours at 25°C), on sample I, Brumado vermiculite, using 1 g of sample and 50 cm^3 of LiCl solution.

TABLE II

Effect of the concentration of lithium chloride solution on swelling in water of Brumado vermiculite "type 23" (6 hours at 25°C; 1 g of vermiculite in 50 cm^3 LiCl solution)

Concentration	Swelling (cm^3/g) in water
saturated (± 20M)	7.0
18M	8.0
16M	8.0
14M	7.0
12M	4.0
10M	2.0
5M	1.5

The Table shows that swelling decreases significantly with decreasing concentration of the LiCl solution. Table III shows above 14M LiCl the swelling amounts to 7-8 cm^3/g. The swelling is also dependent on the time of contact (for diffusion) between the LiCl solution and the vermiculite crystals. The same swelling values of 7.0-7.5 ml/g can be obtained with 6M solutions at a time of contact of 32 hours, as with

14M-20M solutions when contacted for 6 hours; both observation were made with sample I. The swelling values are also dependent on the particle size distribution of the loosefill crude vermiculite in accordance with the results of Walker and Milne (1950).

TABLE III

Effect of time of contact between 6M-lithium chloride solution and Brumado vermiculite (at 25°C) on swelling in water (1 g of vermiculite in 50 cm^3 LiCl solution)

Time of contact	Swelling (cm^3/g) in water
5 s	1.0
15 min	2.0
30 min	2.5
1 h	3.5
2 h	4.5
4 h	5.0
8 h	5.5
16 h	6.5
32 h	7.5
124 h	7.0
128 h	7.0
256 h	7.0

Table IV shows swelling data at 85°C using vermiculite-LiCl solution proportions of 1.0 g/10 cm^3 and 1.0 g/50 cm^3, times of contact of 6 and 24 hours and a 1M LiCl concentration. Evaporation was allowed with drying of the mixture in most of the experiments.

TABLE IV

Swelling in water of vermiculite samples after 6 and 24 hours contact with a 1M LiCl aqueous solution at 85°C, with free evaporation

	6 hours contact of 1 g sample		24 hours contact of 1 g sample	
	in 10 cm^3 LiCl	in 50 cm^3 LiCl	in 10 cm^3 LiCl	in 50 cm^3 LiCl
Sample	sol. cm^3/g	sol. cm^3/g	sol. cm^3/g	sol. cm^3/g
I	14.0 *	10.0	17.0 *	18.0 *
II	12.0 *	0.5	12.0 *	12.0 *
III	12.0 *	0.5	10.0 *	9.0 *
IV	12.0 *	0.5	9.0 *	9.0 *

* After drying of the vermiculite - LiCl solution mixture

The following comments can be made: a) with sample I (Brumado,Bahia) higher values of swelling (14 - 17 - 18 cm^3/g) were obtained when the vermiculite-LiCl solution mixture was dried; however, the particles became very fragile; b) without drying, no swelling was obtained with samples II,III and IV; c) using 1.0 g of vermiculite with 10 cm^3 of 1M LiCl aqueous solution seems satisfatory for obtaining the swelling; d) higher temperatures for the LiCl solution lead to higher swelling values but the particles became fragile, in accordance with the tendency of separation of the individual lamellae and the extreme fragility of the Kenya Li-vermiculite flakes observed by Walker and Milne (1950).

Table V shows results for swelling in a saturated LiCl solution without drying; 6 hours contact time, at room temperatures and at 85°C. Smaller swelling values were obtained than above but the particles were not so fragile.

TABLE V

Swelling in water of vermiculite after a 6 hour contact with a saturated LiCl solution without drying (1 g sample on 10 and 50 cm^3 LiCl solution)

	in 10 cm^3 LiCl (solution)		in 50 cm^3 LiCl (solution)	
Sample	Room temperature cm^3/g	at 85°C cm^3/g	Room temperature cm^3/g	at 85°C cm^3/g
I	8.0	8.0	7.0	10.0
II	8.0	9.0	8.0	8.0
III	8.0	10.0	8.0	10.0
IV	10.0	10.0	9.0	9.0

From these results, it was decided to conduct the exfoliation with samples treated as follows: proportion of 1.0 g of vermiculite to 10 cm^3 of saturated ($\pm$ 20M) aqueous lithium chloride solution; time of contact of 6 hours; temperature 25°C $\pm$ 1°C.

Exfoliation in the laboratory kiln:- Table VI shows the values of the bulk density, for several exfoliation temperatures, for sample I (Brumado, Bahia); significant differences are obtained when compared with the bulk densities of untreated vermiculite exfoliated at the same temperature. The bulk density obtained with the untreated sample at a given exfoliation temperature can be obtained at much lower temperatures with the lithium treated swollen sample.

TABLE VI

Bulk density of Brumado vermiculite exfoliated at several temperature ranges with and without lithium treatment and swelling in water

	Bulk density in g/cm^3	
Temperature	Non-Treated	Li-Treated
460°C	0.420	0.350
510°C	0.341	0.220
560°C	0.250	0.125
660°C	0.210	0.133
760°C	0.200	0.135
850°C	0.211	0.143
960°C	0.210	0.143
1000°C	0.250	0.150

Table VII shows, at a given exfoliation temperature, the bulk densities of the four samples.

TABLE VII

Bulk densities (g/cm^3) of vermiculites with and without lithium treatment, and exfoliated at 760°C

Sample	Crude	Non-Treated	Li-Treated
I-Bahia 23	0.88	0.200	0.133
II-Goiás 23a	1.13	0.180	0.111
III-Goiás 23b	1.12	0.280	0.118
IV-Goiás 4	1.09	0.400	0.111

Table VIII shows the bulk densities obtained for different exfoliation temperatures for the four samples.

TABLE VIII

Bulk densities (g/cm^3) of vermiculite samples after lithium treatment and exfoliated at several temperature

Temperatures	I	II	III	IV
560°C	0.125	0.105	0.143	0.133
600°C	0.133	0.111	0.133	0.125
760°C	0.133	0.110	0.125	0.118
850°C	0.143	0.110	0.125	0.118
960°C	0.143	0.118	0.125	0.125

CONCLUSIONS

The bulk density of the LiCl swollen exfoliated vermiculites are significantly smaller than that of the untreated vermiculite exfoliated at the same temperature.

ACKNOWLEDGEMENT

The first author acknowledge a grant from CAPES (Proc. 2343/81).

REFERENCES

Foster, M.D., 1953. Geochemical studies of clay minerals: (II) - Relation between ionic substitution and swelling in montmorillonite. Am. Miner. 38: 994-1006.

Garrett, W.G. and Walker, G.F., 1962. Swelling of some vermiculite organic complexes in water. Clays Clay Min. 9: 557-567.

Myers, J.B., 1960. Vermiculite. In: AIMMPE's Industrial Minerals and Rocks, 3rd Ed., N.Y. pp. 889-895.

Navajas, R. and Souza Santos, P., 1976. Preliminary tests of exfoliation of vermiculite from Bahia in a pilot vertical kiln. Rev. Bras. Tecnol. 7: 415-431.

Navajas, R. and Souza Santos, P., 1977. Comparative tests of the pyroexpansion in electric kilns of Brazilian vermiculites. Proc. 1st Latin. Amer. Conf. Mineralurgy 1: pp. 121-134.

Navajas, R. and Souza Santos, P., 1981. Studies on pyroexpansion of Brazilian vermiculites: a review. Cerâmica 27: 423-449.

Norrish, K., 1954. The swelling of montmorillonite. Disc. Faraday Soc. 18: 120-134.

Norrish, K. and Rausell-Colom, J.A., 1962. Effect of freezing on the swelling of clay minerals. Clay Min. Bull. 5: 9-16.

Norrish, K. and Rausell-Colom, J.A., 1963. Low angle X-ray diffraction studies of the swelling of montmorillonite and vermiculite. Clays Clay Min. 10: 123-131.

Schroeder, H.J., 1970. Vermiculite. In: Mineral Facts and Problems, U.S. Bureau of Mines. Bull. 650, Washington, pp. 1283-1291.

Walker, G.F. and Milne, A., 1950. Hydration of vermiculite saturated with various cations. Trans. Int. Congress Soil Sci., II: 62-67.

Walker, G.F., 1950. Vermiculite-organic complexes. Nature, 166: 695-696.

Walker, G.F., 1959. Diffusion of exchangeable cations in vermiculite. Nature, 184: 1392.

Walker, G.F., 1960. Macroscopic swelling of vermiculite crystals in water. Nature, 187: 312-313.

Walker, G.F. and Garrett, W.G., 1961. Complexes of vermiculite with amino-acids. Nature, 191: 1389.

Walker, G.F., 1961. Vermiculite minerals. In: Brown, G. (Editor) X-Ray Identification and Crystal Structure of Clay Minerals. Miner. Soc. London, pp. 297-324.

Walker, G.F., 1975. Vermiculite. In: Gieseking, J.E. (Editor) Soil Components II, Springer, Berlin, pp. 155-189.

ROLE OF CLAY MINERALS AS BACKFILL IN RADIOACTIVE WASTE DISPOSAL[1,2]

R. DAYAL and R. J. WILKE
Department of Nuclear Energy
Brookhaven National Laboratory
Upton, New York, U.S.A.

ABSTRACT

The use of clay minerals as backfill materials in high level radioactive waste isolation systems is presently a subject of great interest and research in the United States and abroad. This paper describes the potential role of clay minerals as backfill in a high level waste repository and presents a review of the research and development activities in this field. A technical evaluation of the existing data generated in the United States and abroad indicates that clay minerals, especially the smectites, can serve as an extremely effective barrier to radionuclide migration for periods of thousands of years. However, more data are required on the stability, the sorption characteristics, and the swelling behavior of clay minerals under anticipated repository conditions to assess their long-term performance in a waste repository.

INTRODUCTION

From the viewpoint of radioactive waste isolation, the low permeability and high sorption capacity of argillaceous materials would provide an effective barrier to radionuclide migration. As a result, argillaceous materials are being actively investigated in the United States and abroad for radioactive waste isolation with the following principal strategies in mind (O.E.C.D., 1980): (1) emplacement of the waste in deep sea sediments, (2) emplacement of the waste in a shale formation, and (3) use of clay minerals as backfill in a high level waste repository.

---

[1]Support for this work was provided by the U.S. Nuclear Regulatory Commission. Contract No. DE-AC02-76CH00016; BNL-NUREG-30204R.

Ocean disposal of low level wastes as an alternative to land disposal is a subject of great interest and research in Europe and Japan. In the United States, environmental surveys of abandoned federal waste disposal sites in the Atlantic and Pacific Oceans have been conducted (Dyer, 1976). At the 2800 m deep Atlantic site and the Pacific Farallon sub-sites the barrier characteristics of sediments with respect to the migration of radionuclides have been investigated (Dayal et al., 1979a,b). Dayal et al. (1979b) reported that both ^{137}Cs and ^{134}Cs were present in significant concentrations in sediment sampled close to a waste container at the 2800 m Atlantic site and that their presence, concentration levels, and distribution patterns indicate the waste container to be the source of the observed radionuclides. The highly localized distribution of the radionuclides around the container indicates that the clayey sediment at the study site serves as an effective barrier to radionuclide migration. Furthermore, the radionuclide retention properties of deep sea clays are being studied as part of on-going research in the United States concerning the sub-seabed emplacement of high level wastes in the pelagic red clays of the North Pacific (Talbert, 1980).

The most important studies on the feasibility of high level radioactive waste emplacement in clay formations are under way in Europe (O.E.C.D., 1980). Specifically, the Boom clay formation which underlies the Nuclear Center at Mol, Belgium and the marly clay layer at the Trisasia Nuclear Research Center in Italy are being studied, among other rock types, as potential repositories for high level waste disposal.

The purpose of this paper is to consider the third strategy involving the use of clay minerals as potential backfill materials surrounding a high level waste container emplaced in a cavity excavated in a geological formation (Figure 1).

FUNCTIONS AND PROPERTIES OF BACKFILL

Basically, the purposes of backfill are twofold: 1) to assist or enhance the ability of the waste package to isolate radioactive wastes and 2) to function independently as a barrier to radionuclide migration should failure of the waste package occur. It is envisioned that a backfill can fulfill the first purpose by controlling groundwater ingress and modifying the chemical composition of the near field groundwater so that the likelihood of waste container corrosion would be reduced. The backfill can act as an independent barrier to radionuclide migration by (i) sorption of radionuclides, (ii) reduction of near field groundwater flow thereby mitigating radionuclide transport away from the repository, and (iii) controlling groundwater pH and Eh so that multivalent radionuclides may be kept in their most favorable valence state for sorption.

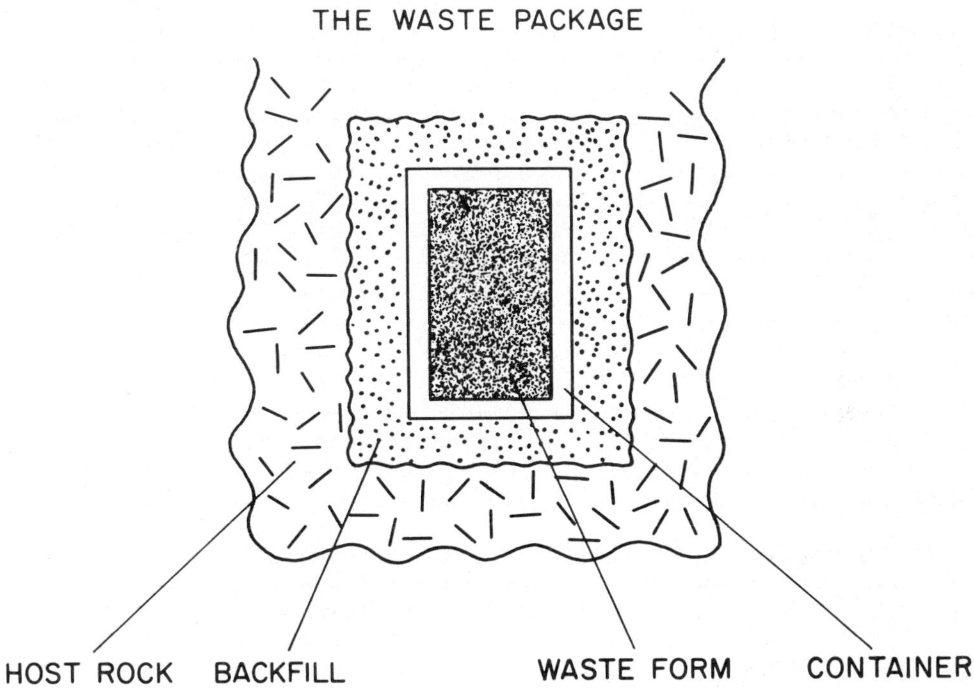

Fig. 1. A schematic diagram of the waste package, showing the relative position of discrete backfill in the system.

At the same time, as Wheelwright (1980) has pointed out, a backfill is required to withstand mechanical stress, high radiation, above ambient temperature, groundwater flow, and the influence of various groundwater constituents for up to 10^6 years and still be able to function adequately. In this context, backfill must be thermally stable and have adequate thermal conductivity during the thermal period of waste disposal. The backfill must be stable under hydrothermal conditions and be able to withstand alternating wet-dry cycles, which can be expected during early repository flooding, and not crack or fissure to prevent ready access of groundwater to the waste container.

CANDIDATE BACKFILL MATERIALS

Given the various functions of backfill, it is no surprise that many materials of divergent characteristics have been suggested for use in nuclear waste repositories. Some proposed backfill designs are given in Table 1. Basically, each design incorporates an expandable clay, usually bentonite, as a principal component for its high sorption capacity, plasticity, swelling character, and low permeability. Quartz or crushed host rock are included in some backfill systems for mechanical strength and thermal conductivity and depending upon the design, various accessory components are added for either sorption of particular radionuclides or control of the redox environment. In general, most backfills are designed to possess the previously discussed required properties within a single layer around the waste package. Since our paper deals specifically with clay minerals, we have not considered in our review the role and properties of non-clay materials such as those listed in Table 1. Further details are presented elsewhere (Dayal et al., 1981).

TABLE 1.

Some Proposed Backfill Designs

Swedish Plan[1]	U.S. Salt Repository[3]
Pure Compacted Bentonite*	30% Bentonite + 70% Charcoal
10-20% Bentonite + Quartz**	30% Bentonite + 20% Charcoal + 40% Mordenite + 10% Na-Titanate
Canadian Plan[2]	**U. S. Basalt Repository[4]**
10-20% Bentonite + Crushed Granite + Quartz Sand	Bentonite + Crushed Basalt + Zeolite + Quartz Sand + Apatite + Accessory Minerals

* Spent Fuel
**Reprocessed Waste

[1]Swedish Plan (Pusch, 1979).
[2]Atomic Energy of Canada Limited (1980).
[3]Nowak (1981).

SWELLING PROPERTIES AND PERMEABILITY

Research on these properties of backfill has focused primarily on bentonite clays. Pusch (1977, 1978, 1979), as part of the Swedish high level nuclear waste program, studied the swelling properties and permeability of highly compacted Na-bentonite clays as a function of bulk density (Figures 2 and 3). His findings indicated that a maximum swelling pressure of 10 MPa could be achieved (at a bulk density

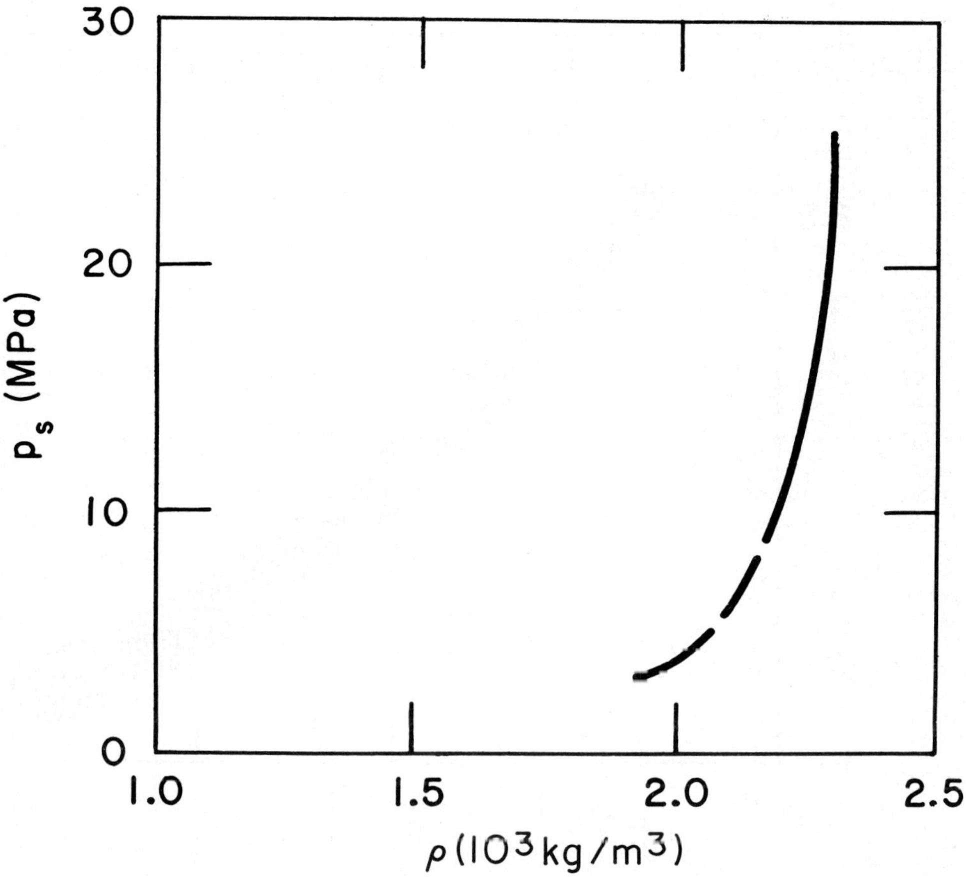

Fig. 2. Swelling pressure as a function of bulk density for sodium bentonite at 100% saturation (after Pusch, 1978; 1979).

of 2.1×10^3 kg/m^3) and that at compaction equilibrium, this would decrease to 5 MPa. Backfill with such properties would, according to Pusch, accommodate host rock stresses without fissuring and could also fill cracks in the repository wall to a certain extent. In addition, permeabilities of 10^{-13}–10^{-14} m/s were achieved for compacted bentonite which are so low that any radionuclide transport from the repository, presuming breach of containment, would occur through the backfill strictly by means of molecular diffusion.

While these properties are functions of bulk density, they are also dependent upon such factors as composition of the water phase,

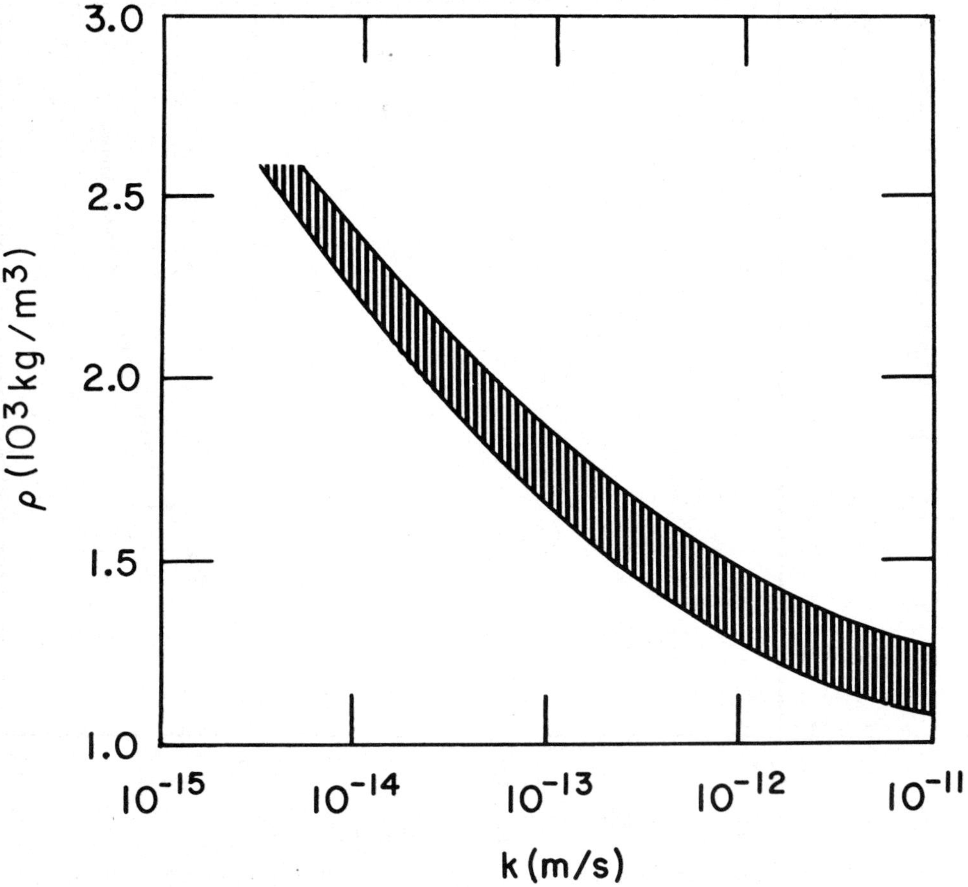

Fig. 3. Range of experimentally determined coefficients of permeability for sodium bentonite as a function of bulk density (after Pusch, 1978; 1979).

particle orientation, and the type and quantity of adsorbed cations. For example, the swelling volume of the calcium form of montmorillonite is less than that of the sodium form by a factor of 2.5 (see Figure 4, taken from Beall et al., 1981). This phenomenon is an important factor considering that different groundwater compositions are encountered in different rock types. Given the importance of swelling clays in all of the proposed backfill designs, these problems and the properties of the clays in general need to be more fully investigated.

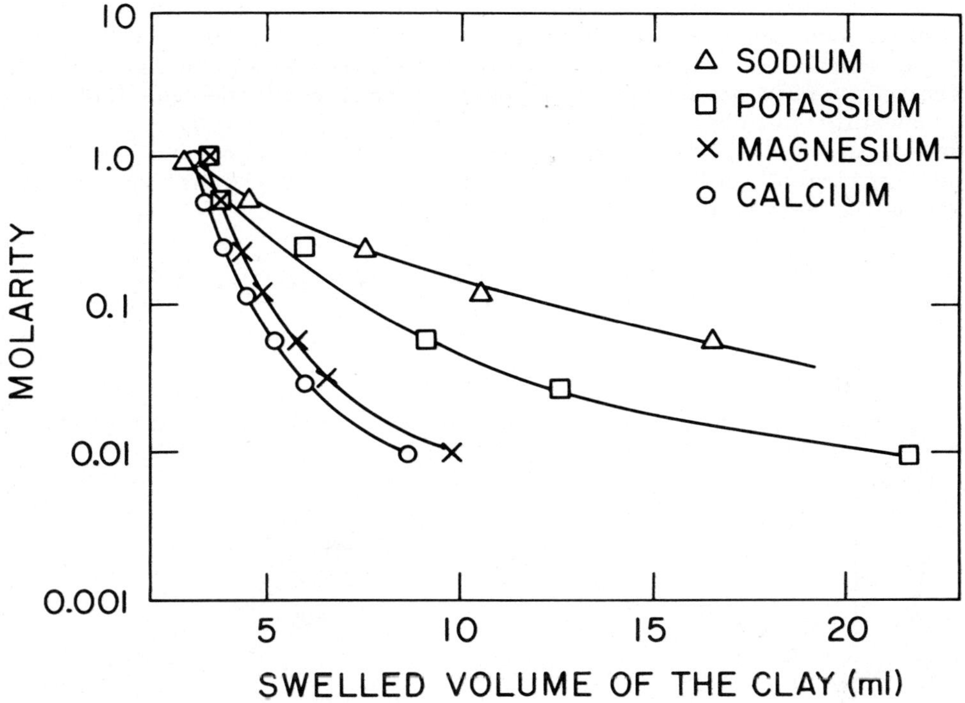

Fig. 4. The influence of various cations upon the swelling volume per gram of montomorillonite clay (after Beall et al., 1981).

THERMAL CONDUCTIVITY AND STABILITY

Unless pre-burial cooling is employed in the high level radioactive waste disposal scheme, backfill materials are likely to encounter reasonably high temperatures (up to 300°C). An important parameter in this respect is the thermal conductivity of the backfill barrier mass. The backfill must have adequate thermal conductivity so as to conduct heat from the waste container and maintain its own temperature at levels low enough to minimize any structural damage or alteration which could cause failure of one or more of its functions. Radhakrishna (1981) recently determined the thermal conductivities of various potential backfill mixtures. Other work currently underway concerns the effects of temperature and pressure on the permeability, strength, and swelling characteristics of the fill.

Preliminary results indicate that illite based backfill mixtures had the best overall thermal conductivity followed by kaolinite backfills and then bentonite based backfills, but Radhakrisha (1981) concluded that they would all be adequate. The thermal conductivity of all the mixtures tested decreased with increasing clay content and with higher temperature. It increased significantly with increasing backfill moisture content up to a "critical" value after which only a slight increased was observed (Figure 5). The figure shows that the thermal conductivity of an illite/quartz backfill falls off rapidly below a moisture content of about 1% in this case. During the thermal period of a high level waste repository, the backfill would presumably be subjected to dehydration. If the moisture content would fall below the 1% level, the backfill would become thermally less efficient during the time when the highest thermal conductivity would be desirable.

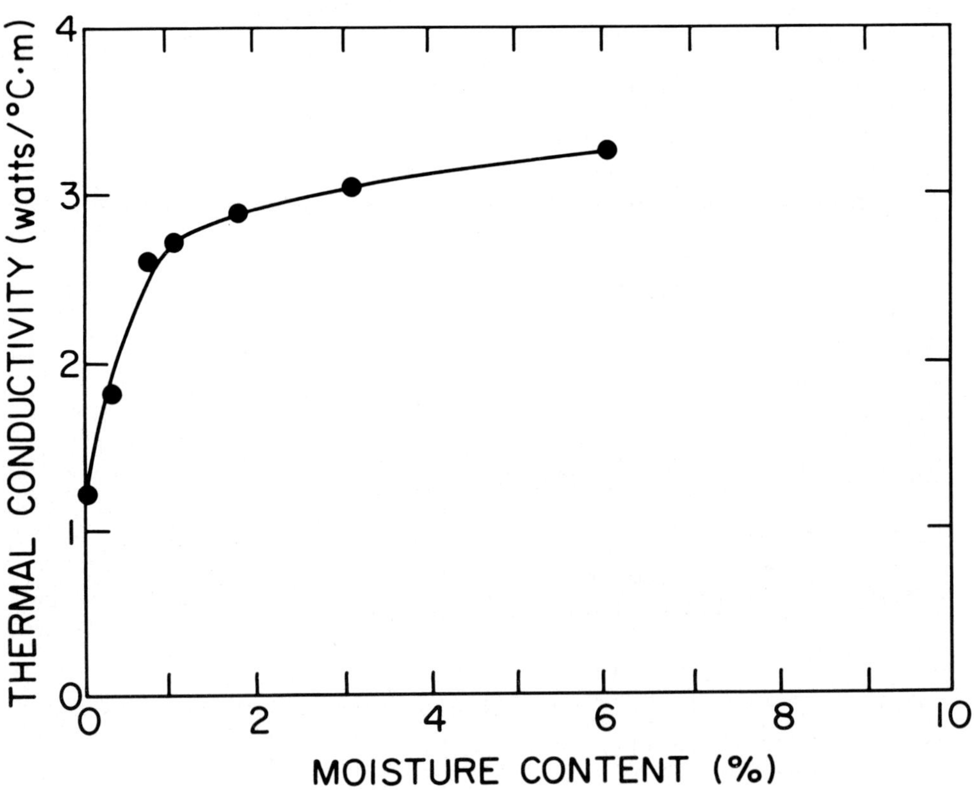

Fig. 5. Thermal conductivity of an 85% graded silica + 15% illite backfill as a function of moisture content (after Radhakrisha, 1981).

Komarneni and Roy (1981) studied the effect of hydrothermal treatment on the structure and the sorption capacity for Cs and Sr of several backfill materials at 200°C, and showed that only the vermiculites and phillipsite exhibited layer collapse with a definite decrease in exchange capacity. At 300°C, only two montmorillonites and mordenite did not exhibit any detectable structural changes, therefore from the point of view of thermal stability, the authors felt that these materials would be preferable as a backfill. It is indeed well known that inter-layer water in bentonite is lost reversibly up to 300°C and that structural water is not lost until temperatures of 450°C are reached (e.g., see Wood and Schultz, 1980). Therefore, the montmorillonite structure should be thermally stable up to 300°C and presumably retain its properties.

Based on observed diagenetic alteration of clays in the U.S. Gulf Coast sediments, Weaver (1979), however, reported that montmorillonite alteration to chlorites and illites can occur at temperatures as low as 40°C, but as Wood and Schulz (1980) point out, such alterations to illite, for instance, would require the presence of potassium which, in the case of basalt repositories, is only present at very low concentrations in the groundwater. The presence and concentration of groundwater constituents, therefore, is a very significant factor in clay mineral alteration. For instance, montmorillonite alteration to chlorite has been reported to occur in the presence of magnesium and iron. More experimentation simulating repository conditions is needed.

During the thermal period of high level waste disposal, the possibility exists that the backfill could be subjected to alternating wet and dry cycles which could result in cracks or fissures in the backfill, permitting groundwater access to the waste package. Little information is available on this topic. Sethi et al. (1981) conducted wet/dry cycling experiments on three montmorillonitic clays at 75°C, 100°, and 150° in basaltic and granitic groundwater. Preliminary result indicated that, while X-ray diffraction studies did not show significant structural alteration, the gel strength, liquid limit, and the flow index of the Na-montmorillonite clays decreased with an increasing number of wet and dry cycles. In contrast, in the Ca-Mg montmorillonite clays these properties increased with the number of wet-dry cycles. These results indicate that microcracking may occur in the Na-montmorillonites should these conditions exist. This has direct significance for the montmorillonite based backfills since, in general, it has been proposed to use the sodium form of the clay in these backfills. These results, while preliminary, along with the findings of Beall et al. (1981) reported earlier suggest that efforts should be directed towards tailoring the cation composition of the clays used in backfills to the specific groundwater in which they will be utilized.

RADIONUCLIDE SORPTION, DIFFUSIVITY, AND RETENTION

Research on the radionuclide sorption properties of potential backfill materials is important since this interaction represents the most effective means of radionuclide retardation aside from the control of groundwater transport by a highly impermeable clay barrier. The standard means used to evaluate the sorption of radionuclides by a solid phase is the determination of the equilibrium distribution coefficient (K_d value) of the system. This coefficient is a measure of the partitioning of a radionuclide between the solid and liquid phase of the system. Many experimental variables can influence the K_d of a system (see for example Ames and McGarrah, 1979; Dosch, 1980; Nowak, 1980a,b; Strikert, 1980; Winslow, 1981). These variables must be kept in mind both when experimentally determining these coefficients and when attempting to evaluate the performance of clay minerals as backfill. We shall not present all available backfill/radionuclide sorption data. The data selected in this section only illustrates significant aspects of backfill sorption research.

The distribution coefficients for secondary smectite clay in basaltic groundwater at 60°C (ambient temperature of potential repository host rock) are reported by Wood and Schultz (1980). Not surprisingly, in light of its high cation exchange capacity, the smectite possessed high sorptive properties for the radionuclide cations ^{90}Sr, ^{137}Cs, ^{226}Ra, and ^{241}Pu under these conditions. Significantly, however, the sorption of toxic radionuclides such as ^{75}Se, ^{99}Tc, and ^{125}I was very poor as was the sorption of the actinide ^{237}Np hence smectites are not effective for the full range of radionuclides. For further data see e.g., Allard et al. (1977) and on the sorption of anionic species of I, Tc, Se, Np, and Pu: Dayal et al. (1981).

Although a screening of various materials for their sorption properties is useful, an evaluation of the performance of clay minerals for backfill under dynamic conditions would be more appropriate. Neretnieks and Skagius (1967a,b) and Allard and Kipatsi (1977) determined the diffusivities of various radionuclides and other species in compacted bentonite and bentonite/quartz backfills (Table 2). This type of information is particulary meaningful in cases where molecular diffusion is expected to be the primary mechanism of radionuclide transport, a situation that has been suggested to exist in backfills based on expanding clays where low permeability is achieved upon clay hydration (Beall et al., 1981; Nowak, 1981).

The influence of tortuosity upon the migration of radionuclides through backfill can be seen by comparing the diffusivities for strontium and cesium in Table 2 which were corrected for sorption, with typical bulk solution diffusivities. The latter are of the order of 10^{-9} to 10^{-10} m^2/s, an order of magnitude greater than those for cesium and strontium in Table 2 demonstrating the value of low permeability materials in retarding radionuclide migration in

TABLE 2.

Experimentally Determined Diffusivities of Various Dissolved Species in Backfill

Species	Diffusivity (m^2/s)	Medium	Ground-water	T(°C)	Bulk Density (g/cm^3)	Source
Sr^{2+}	3.3×10^{-11} [a]	Compacted Bentonite	Granitic	50	2.1	Neretnieks and Skagius (1978b)
Cs^{+}	5.4×10^{-11} [a]	"	"	50	2.1	"
CH_4	3.9×10^{-11}	"	Distilled	50	2.1	Neretnieks and Skagius (1978a)
H_2	1.8×10^{-11}	"	"	50	2.1	"
^{110m}Ag	8.0×10^{-13}	Bentonite + Quartz	Brine	85	Not Given	Allard and Kipatski (1977)
^{110m}Ag	7.0×10^{-13}	"	"	25	"	"
^{110m}Ag	7.0×10^{-13}	"	Granitic	85	"	"
^{110m}Ag	6.0×10^{-13}	"	"	25	"	"

[a]Adjusted for sorption effects.

backfill. Similarly, the effect of sorption can be illustrated with the diffusivities for silver ion in bentonite/quartz backfill in the table, which were not adjusted for sorption and which are nearly three orders of magnitude lower than those for the bulk solution diffusivity. As shown later, diffusivity data can be utilized to calculate radionuclide breakthrough times through backfill.

Neretnieks (1977) calculated the retention times of various radionuclides in a one-meter thick bentonite backfill, assuming diffusion controlled transport. The results, given in Table 3, show that ^{90}Sr, ^{137}Cs, and ^{241}Am would decay to insignificant levels ($C/C_o = 10^{-9}$) within the barrier. With the exception of the anionic radionuclides ^{99}Tc and ^{125}I, all the radionuclides studied would be retained within the barrier for more than about 2000 years. Using radionuclide retardation factors, Neretnieks (1977) has also estimated retention times for one-meter thick barriers of 10% clay + 90% quartz and of clinoptilolite (Table 3). While these barriers are also incapable of retaining the toxic anions for any reasonable period of time, the other radionuclides, with the exception of ^{137}Cs and ^{237}Np, are retained within the barriers of more than 1000 years. These values are an important qualitative measure of the potential of backfills to aid in compliance with regulatory criteria.

Nowak (1980a,b, 1981) applied mass balance principles to modeling radionuclide migration through backfill barriers. From the solutions of mass balance differential equations, Nowak calculated the breakthrough times of various radionuclides through backfills of different composition. Knowing the barrier thickness, using empirically determined retardation factors, and assuming a bulk diffusion coefficient of 1×10^{-9} m^2/s, Nowak arrived at the retention times given in Table 4. These results essentially represent a confirmation of Neretnieks' findings but under more rigorous conditions and also demonstrate the effectiveness of tailoring backfills for particular environments, in this case salt brine. Nowak, in using this approach to evaluate backfill performance, defined that radionuclide breakthrough occurred when the radionuclide concentration outside the backfill reached 1% of the radionuclide concentration of the source. Such calculations could be performed for any desired definition of radionuclide breakthrough. Regulatory criteria may define breakthrough at a much lower concentration.

CONCLUSIONS

Clay minerals are being actively investigated as potential backfill materials in high level radioactive waste isolation systems. Most researchers have considered the use of montmorillonite clay as the primary component of a backfill barrier in a high level waste repository. The swelling nature of the clay would limit groundwater access to the waste container thereby delaying the onset of container corrosion. Reduced water flow near the waste package would also restrict radionuclide migration to diffusional transport. Some pre-

TABLE 3.

Radionuclide Retardation Factors and Retention Times for One Meter Thick Backfill Barriers[a]

		Compacted Bentonite		10% Bentonite + 90% Quartz	Clinoptilolite
Nuclide	Half-life (yr)	Retardation Factor	Retention[b] Time (yr)	Retention[c] Time (yr)	Retention[c] Time (yr)
^{90}Sr	2.77×10^{1}	600	1.3×10^{4}[d]	30	600 -1400
^{99}Tc	2.12×10^{5}	1	22	1	1
^{129}I	1.70×10^{7}	1	22	1	1
^{137}Cs	3.00×10^{1}	400	9.0×10^{3}	20 - 30	2200 -5200
^{226}Ra	1.60×10^{3}	800	1.8×10^{4}	40 - 50	600-1400[e]
^{229}Th	7.34×10^{3}	1000	2.0×10^{4}	50 - 300	Unknown
^{237}Np	2.14×10^{6}	200	4.4×10^{3}	10	"
^{239}Pu	2.44×10^{4}	1200	2.6×10^{4}	60 - 80	"
^{240}Pu	6.58×10^{3}	1200	2.6×10^{4}	60 - 80	"
^{241}Am	4.33×10^{2}	4000	9.0×10^{4}	200 - 4000	1000-3000[f]
^{243}Am	7.95×10^{3}	4000	9.0×10^{4}	200 - 4000	1000-30000

[a]Granitic groundwater.
[b]Flow rate $\approx 1.4 \times 10^{-7}$ cm/s.
[c]Flow rate $\approx 3 \times 10^{-6}$ cm/s.
[d]Underlined values indicate radionuclide decay to below 10^{-9} of the original activity.
[e]Based on analogy to Sr.
[f]Based on analogy to Eu.

TABLE 4.

Calculated Retention Times for Various Backfill Compositions Assuming Diffusion Controlled Transport and 1% Radionuclide Breakthrough[1]

Backfill Composition	Bed Thickness (m)	Retention Times (Years)					
		Pu	Am	Cs	Sr	I	TcO_4
Compacted Bentonite[a,c]	0.3	$1x10^4$	$1x10^4$	$1x10^3$	$1x10^3$	--	--
Compacted Bentonite[a,c]	1.0	$1x10^5$	$1x10^5$	$1x10^4$	$1x10^4$	--	--
70% Bentonite[b] 30% Charcoal	1.0	$5x10^4$	$2x10^4$	--	--	$1x10^3$	$1x10^3$
30% Bentonite[b] 40% Mordenite 10% Charcoal 10% Na-Titanate	1.0	$5x10^4$	$2x10^4$	$6x10^2$	$7x10^2$	$4x10^2$	$4x10^3$

[1]From Nowak (1980a,b, 1981).
[a]Groundwater not specified.
[b]Waste Isolation Pilot Plant A Brine.
[c]On basis of estimated retardation factors for actinides and fission products.

liminary research has been iniated with regards to the parameters which control the swelling nature of these clays. The cation form of the clay and the composition of the groundwater are important factors.

Assuming diffusion controlled transport and 1% radionuclide breakthrough, the retention times calculated for a one-meter thick bed of compacted bentonite for Pu, Am, Sc, and Sr are on the order of tens of thousands of years. Such data indicate that certain clay minerals as backfill can serve as an extremely effective barrier to radionuclide migration.

There are, however, several critical factors in the use of clay minerals in high level waste isolation systems. The stability and performance of clay minerals under anticipated hydrothermal conditions and radiation fields, and alternating wet/dry cycles must be established. The effect of particle orientation in compacted clay systems on such properties as thermal conductvity, groundwater flow, and swelling and collapse behavior should be known so that the long-term performance of clay minerals can be predicted under repository conditions. A thorough knowledge of varying groundwater composition on the sorption and swelling properties of clay minerals is essential if the clay minerals are required to function as backfill in a repository for long time periods.

ACKNOWLEDGEMENTS

We thank G. W. Beall for reviewing the manuscript. We are also grateful to Sharon Moore, Cathy Van Noy, and Eileen Pinkston for their patience in the typing of this report, and to Grace Searles for her skillful editing and preparation of the manuscript for publication.

REFERENCES

Allard, B. and Kipatski, H., 1977. Measurement of the Rate of Diffusion of Silver in a Clay-Sand Mixture. KBS Technical Report No. 53. Swedish Nuclear Safety Project, 18 pp.

Ames, L. L. and McGarrah, J. E., 1979. Geomedia-Radionuclide Distribution Coefficients. Fiscal Year 1979 Annual Report. Pacific Northwest Laboratories, PNL-3146.

Atomic Energy of Canada Limited, 1980. Irradiated Nuclear Fuel and Immobilized Waste Vaults: Backfilling Systems. Acres Consulting Services Ltd., TR-66.

Beall, G. W., Lamkin, A. G., and Kelly, K. L., 1981. Critical Factors Affecting the Use of Montmorillonite as a Nuclear Backfill Material. Presented at Workshop on Research and Development Needs Relating to Backfill. National Bureau of Standards, Gaithersburg, Maryland, April 13-14, 1981.

Dayal, R., Lee, B. S., Wilke, R. J., Swyler, K. J., Soo, P., Ahn, T. M., McIntyre, N. S., and Veakis, E., 1981. Nuclear Regulatory Commission, Nuclear Waste Management Technical Support in the Development of Nuclear Waste Form Criteria: Task 1: Waste Package Overview. Brookhaven National Laboratory, BNL-NUREG-51477, Washington, DC, 321 pp.

Dayal, R., Duedall, I. W., Fuhrmann, M., and Heaton, M. G., 1979a. Sediment and Water Column Properties at the Farallon Islands Radioactive Waste Dump Sites. Final Report to the Office of Radiations Programs, U.S. Environmental Protection Agency, Washington, DC, 88 pp.

Dayal, R., Okubo, A., Duedall, I. W., and Ramamoorthy, A., 1979b. Radionuclide Redistribution Mechanisms at the 2800-m Atlantic Nuclear Waste Disposal Site. Deep Sea Res. 26A: 1329-1345.

Dosch, R. G., 1980. Assessment of Potential Radionuclide Transport in Site-Specific Geologic Formations. Sandia National Laboratory, SAND 79-2468, 31 pp.

Dyer, R., 1976. Environmental Surveys of Two Deep Sea Radioactive Waste Disposal Sites Using Submersibles. In: Proceedings of an International Symposium on Management of Radioactive Waste From the Nuclear Fuel Cycle, International Atomic Energy Agency, Vienna, 2: 317-338.

Komarneni, S. and Roy, R., 1981. Hydrothermal Transformation in Candidate Overpack Materials and Their Effects on Cesium and Strontium Sorption. Nuclear Technology, 54: 118-122.

Neretnieks, I., 1977. Retardation of Escaping Radionuclides From a Final Repository. KBS Technical Report No. 30, Swedish Nuclear Safety Project, 55 pp.

Neretnieks, I. and Skagius, C., 1978a. Diffusivity Measurements of Methane and Hydrogen in Wet Compacted Clay. KBS Technical Report No. 86, Swedish Nuclear Safety Project, 31 pp.

Neretnieks, I. and Skagius, C., 1978b. Diffusivity Measurements in Wet Clay for Sr^{2+}, Cs^{+}, and La-Lignosulfonate, KBS Technical Report No. 87, Swedish Nuclear Safety Project, 14 pp.

Nowak, E. J., 1981. Backfill Barriers for Nuclear Waste Repositories in Salt. Sandia National Laboratory, SAND 81-0377A, 7 pp.

Nowak, E. J., 1980a. Radionuclide Sorption and Migration studies of Getters for Backfill Barriers. Sandia National Laboratory, SAND 79-1110, 54 pp.

Nowak, E. J., 1980b. The Backfill as an Engineered Barrier for Nuclear Waste Management. In: C. J. Northrup (Editor), Sientific Basis for Nuclear Waste Management, Vol. 2, Plenum Press, New York, 403-410.

Organization for Economic Co-operation and Development, 1980. Use of Argillaceous Materials for the Isolation of Radioactive Waste. Proceedings of the Nuclear Energy Agency Workshop, Paris, 275 pp.

Pusch, R., 1979. Highly Compacted Sodium Bentonite for Isolating Rock Deposited Radioactive Waste Products. Nuclear Technology, 45: 153-157.

Pusch, R., 1978. Properties of Bentonite Based Buffer Material. KBS Technical Report No. 2, Swedish Nuclear Safety Project.

Pusch, R. 1977. Required Physical and Mechanical Properties of Buffer Masses. KBS Technical Report No. 33, Swedish Nuclear Safety Project, 10 pp.

Radhakrishna, H. S., 1981. Evaluation of Thermal Properties of Buffer Materials for a Deep Underground Nuclear Waste Disposal Vault. Presented at Workshop on Research and Development Needs Relating to Backfill. National Bureau of Standards, Gaithersburg, Maryland, April 13-14, 1981.

Sethi, A. J., Bird, G. W., and Yong, R. N., 1981. Effect of Hydrothermal and Cyclic Wetting-Drying Treatment on Properties and Characteristics on Montmorillonitic Clays. Presented at Workshop on Research and Development Needs Relating to Backfill. National Bureau of Standards, Gaithersburg, Maryland, April 13-14, 1981.

Strikert, G. R., 1980. Sorption of redox Sensitive Elements - Critical Analysis. Pacific Northwest Laboratories, PNL-3432, 46 pp.

Talbert, D. M., 1980. Subseabed Radioactive Waste Disposal Feasibility Program: Ocean Engineering Challenges for the 80's. Sandia National Laboratory, SAND 80-0305, 26 pp.

Weaver, C. E., 1979. Geothermal Alteration of Clay Minerals and Shales: Diagenesis. Office of Nuclear Waste Isolation. Technical Report, ONWI-21, 176 pp.

Wheelwright, E. J., 1980. Development of Backfill Material. Nuclear Waste Management Quarterly Project Report. (A. M. Platt and A. J. Powell, Editors.) Pacific Northwest Laboratories, PNL-3000-6, 8 pp.

Winslow, C. D., 1981. The Sorption of Cesium and Strontium From Concentrated Brines by Backfill Barrier Materials. Sandia National Laboratory, SAND 80-2046, 57 pp.

Wood, M. I. and Schultz, W. W., 1980. Engineered Barrier Development for a Nuclear Waste Repository in Basalt. (M. J. Smith, Editor.) Rockwell International Corporation, RHO-BWI-ST-7.

REMOVAL OF CESIUM AND STRONTIUM BY NATURAL ZEOLITES AND THE IMPROVED ZEOLITES FROM AQUEOUS SOLUTION AND THEIR FIXATION IN THE ZEOLITES

YOSHIAKI GOTO, JUN MATSUZAWA, AND SHUNJI MATSUDA
Department of Chemistry, Gunma University, Kiryu, Gunma, 376, Japan

ABSTRACT

The removal of Cs and Sr by natural and improved zeolites from aqueous solution and the fixation of the elements in these zeolites were studied. The natural zeolite were improved by treatment with NaOH solution and with a mixed solution of NaOH and NaCl. With increasing time of treatment, clinoptilolite changes to zeolite P. Adsorption of Cs increased in the order: sodalite $<$ zeolite X $\simeq$ clinoptilolite $\simeq$ mordenite $<$ zeolite P and of Sr in the order: clinoptilolite $\simeq$ mordenite $\leq$ sodalite $<$ zeolite X $\simeq$ zeolite P. Effective fixation of these elements could be realized in the zeolites by heating above 1000°C to 1100°C. The fixation is directly related to structural changes of the zeolites.

INTRODUCTION

Natural zeolites, in particular clinoptilolite and mordenite occur abundantly in sedimentary rock of volcanic origin in Japan and are low-cost natural resources. Ames (1959, 1960) demonstrated that the removal of radioactive Cs and Sr by clinoptilolite is based on its ion-exchange property. Similar studies have been carried out by many investigators (Mathers and Watson, 1962, Adam et al., 1971, Kakasy et al., 1973, and Kato, 1974).

The most popular procedure for fixation of radioactive Cs and Sr is the incorporation of the elements into a glass. The fixation of Cs and Sr in ceramic material (McCarthy, 1977) or synthetic rock (Ringwood, 1978) has been studied. Kanno and Mimura (1977) reported that Cs and Sr in zeolite A, zeolite X, and zeolite Y become fixed by heating. They also reported fixation in natural clinoptilolite and mordenite (Mimura and Kanno, 1977).

Unpublished work of Goto showed that the natural zeolites treated with alkali solution sorbed heavy metals (Pb, Cu, Cd, Zn) 7-50 times more than the original zeolites. In this paper removal of Cs and Sr by natural and solution treated improved zeolites, and their fixation in the zeolites is reported.

EXPERIMENTAL

Materials

The natural zeolites used are: clinoptilolite rock from Futatsui-machi, Akita Pref., Japan, which consists of 69% of clinoptilolite,

TABLE 1

Preparation of improved zeolites and adsorption of Cs and Sr

Sample No.		Treating conditions[1] Solution	Sample/Solution (g/cm^3)	Time (h)	Yield[2] (%)	Zeolite species[3]	Amounts adsorbed (meq/100g) Cs	Sr
Clinoptilolite rock	C-0	---	---	--	--	Cp	76	32
	C-1			1	53	Cp	101	79
	C-2	3M NaOH	8.0/200	3	53	Cp	104	93
	C-3			8	53	Cp,P	120	123
	C-4			24	58	P	122	128
	C-5			3	59	Cp,S,X	92	119
	C-6		8.0/200	6	55	S,Cp,X	80	121
	C-7	3M NaOH		48	60	S,P	48	93
	C-8	3M NaCl		3	66	Cp,X	88	152
	C-9		15.0/100	8	58	X,P	84	159
	C-10			12	58	X,P	77	173
Mordenite rock	M-0	---	---	--	--	Md	72	24
	M-1			3	71	Md,P	81	116
	M-2	3M NaOH	8.0/200	8	65	P,Md,H	88	147
	M-3			24	58	P,H	106	195
	M-4			36	57	P,H	106	197
	M-5	3M NaOH		3	70	Md,X,S	69	121
	M-6	3M NaCl	8.0/200	6	60	S,X	46	110
	M-7			48	57	S,X	19	43

[1]Temperature: 100°C. [2]Insoluble products. [3]Cp: clinoptilolite, P: zeolite P, S: sodalite, X: zeolite X, Md: mordenite, H: herschelite.

15% of amorphous materials, 8% of feldspars, 5% of cristobalite and minor amounts of mica and quartz, and mordenite rock from Shirasawa, Miyagi Pref., Japan, which contains 64% of mordenite, 24% of amorphous materials, 8% of quartz, and minor amounts of feldspars and mica.

The natural zeolites were ground and the fraction below 200 mesh was used. Solution treatment was carried out as follows: the natural zeolite (8.0 g or 15.0 g) was treated with a solution (100 cm^3 or 200 cm^3) of 3M NaOH or a mixed solution of 3M NaOH and 3M NaCl for 1 to 48 h at 100°C. The products were washed with distilled water until the filtrate reached a pH of about 9, and dried at 50°C. Their mineral compositions were identified by X-ray diffraction. The conditions of preparation of the 17 different kinds of improved zeolites studied were shown in Table 1.

<u>Measurements of adsorption of Cs and Sr</u>

A 0.2 g of the natural or the improved zeolite was immersed in 40 cm^3 of CsCl or $SrCl_2$ solution at a controlled pH of 3 to 4, at room temperature. Adsorption reaches equilibrium after 2 to 3 h. After 5 h, the amount of Cs or Sr remaining in solution was analysed respectively by flame photometry and atomic absorption spectroscopy, from which the amounts adsorbed were calculated.

<u>Measurements of fixation of Cs and Sr</u>

To study fixation of Cs and Sr in the zeolites, the sample were heated at various temperatures of 200°C to 1300°C for 3 h. The degree of fixation was determined by leaching 0.2 g of the heated zeolite in 40 cm^3 of 0.1M HCl solution for 5 h. The leached amounts of Cs and Sr were determined with the same method as described for the measurements of adsorption. The degree of leaching was defined as the percent ratio of the amount removed by leaching to the amount adsorbed.

RESULTS AND DISCUSSION

<u>Adsorption of Cs and Sr</u>

The effects of treatment are shown in Table 1 and Figures 1-3. The cristobalite in the clinoptilolite rock was removed by the treatment with 3M NaOH solution. Clinoptilolite changes to zeolite P as the time of treatment increases. In the case of mordenite, zeolite P is produced as well as a minor amount of herschelite.

Adsorption of Cs and Sr versus time of treatment with 3M NaOH solution is plotted in Figure 1. The clinoptilolite rock and the mordenite rock show higher amount of adsorption for Cs than for Sr. Adsorption of Cs and Sr increases with increasing time of treatment. The adsorption levels off after about 24 h. Adsorption of Cs for the improved zeolites from clinoptilolite and mordenite rock is about 1.5 times that for the corresponding natural zeolites. Adsorption of Sr, for the treated clinoptilolite rock is about 6 times that for the untreated zeolite, and 8 times for mordenite. After treatment for 3 h, the zeolites do not contain zeolite P. At this stage they have a high adsorption capacity for Sr, i.e. 3 times the amounts of adsorption

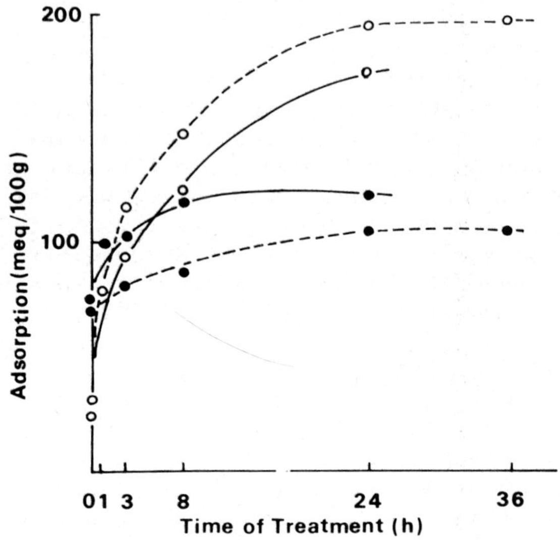

Fig. 1. Relationships of the amount of adsorption versus time of treatment with 3M NaOH solution: ——, clinoptilolite rock; ---, mordenite rock; •, Cs, and o, Sr.

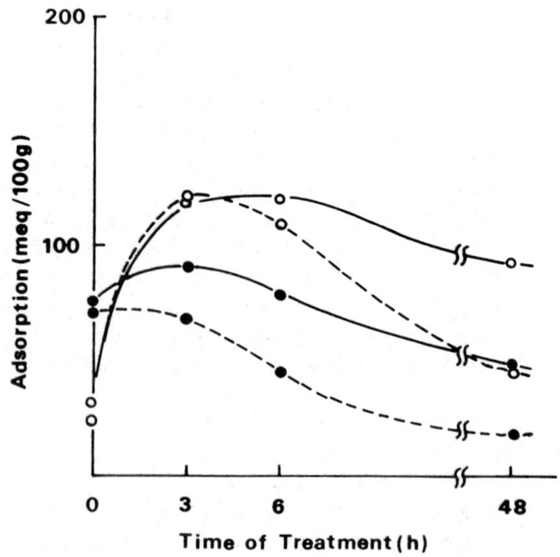

Fig. 2. Relationships of the amount of adsorption versus time of treatment with a mixed solution of 3M NaOH and 3M NaCl: Sample/Solution ratio = 8.0 g/ 200 cm^3; ——, clinoptilolite rock; ---, mordenite rock; •, Cs, and o, Sr.

on the untreated zeolites for clinoptilolite and 4.5 times for mordenite. This is probably due to removal of either the amorphous materials or cristobalite covering the crystal of clinoptilolite and mordenite present in the natural zeolites.

The natural zeolites change to sodalite and zeolite X in the mixed solution of 3M NaOH and 3M NaCl. The amount of sodalite is increased with increasing time of treatment. In the case of clinoptilolite, zeolite X changes to zeolite P after treatment for 48 h. The adsorption of Cs and Sr tends to increase with increasing time of treatment, followed by a decrease, and this behaviour corresponds fairly well with the increase of the amount of zeolite X and with the increase of the amount of sodalite thereafter (Figure 2).

Treating conditions of clinoptilolite were selected to increase the content of zeolite X. The results are shown in Table 1, samples C-8 to C-10. The adsorption curves for Cs and Sr on these samples are shown in Figure 3. No significant change is observed in the adsorption of Cs. The adsorption of Sr is remarkably increased initially, and then decreases gradually. After treatment for 12 h, adsorption of Sr becomes 6 times that of clinoptilolite rock.

As the factors which govern the adsorption behaviour, the following will be considered (i) cation exchange capacity, (ii) size of aperture formed by rings of tetrahedra in the zeolite structure, and (iii) cation-exchange selectivity.

Physical adsorption of cations on zeolites is known to be small and their high adsorption capacity is governed by cation exchange. The aperture size in the sodalite is reduced by the presence of Cl in the aperture. The decrease of the adsorption capacity under condition, at which much sodalite is formed may be explained by the fact that the reduced size of aperture in the sodalite structure is 2.2 Å (Zhdanov, et al., 1964, Shishakova and Dubinin, 1965) as compared with 3.38 Å and 2.26 Å for Cs^+ and Sr^+, respectively. A high selectivity for Cs of clinoptilolite, mordenite and zeolite P, and a high selectivity for Sr of zeolite P, zeolite X and herschelite have been reported (Frysinger, 1962, Ames, 1964, 1966, Kato, 1971, Barrer and Munday, 1971, Sherry, 1968, Barrer et al., 1969). The results shown in Figures 1-3 are in agreement with these observations reported in the literature.

In summary, for Cs, the ease of adsorption increased in the order sodalite $<$ zeolite X $\simeq$ clinoptilolite $\simeq$ mordenite $<$ zeolite P, and for Sr in the order clinoptilolite $\simeq$ mordenite $\leq$ sodalite $<$ zeolite X $\simeq$ zeolite P.

Fixation of Cs and Sr

The natural zeolites and the two improved zeolites, M-4 and C-10 were selected as typical examples for the study of fixation. Relationships of the degree of leaching versus heating temperatures for clinoptilolite and improved zeolite M-4 are shown in Figure 4. Higher degrees of leaching for Cs than for Sr are observed below around 1000°C. For leaching of Cs from clinoptilolite rock, a maximum is observed about 800°C. Beyond this temperature, leaching is greatly reduced and reaches about 0.4% at 1100°C. The degree

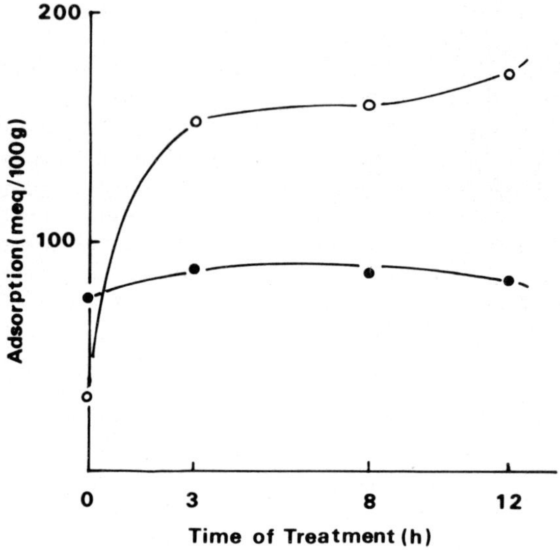

Fig. 3. Relationships of the amount of adsorption versus time of treatment with a mixed solution of 3M NaOH and 3M NaCl; Sample/Solution ratio = 15.0 g/100 cm^3; ——, clinoptilolite rock; •, Cs and o, Sr.

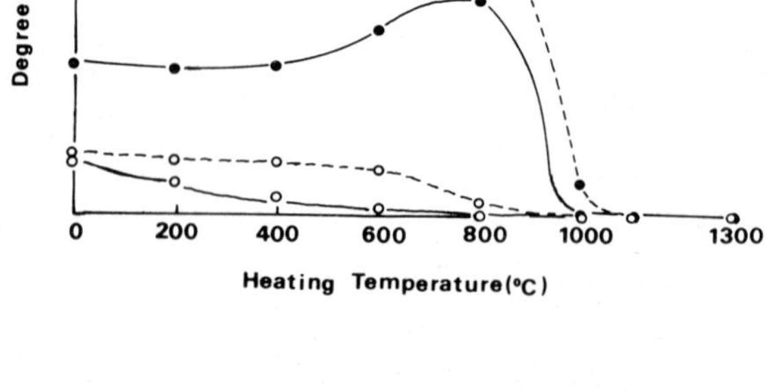

Fig. 4. Relationships of the degree of leaching of Cs and Sr from the clinoptilolite rock and the improved zeolite, M-4, versus heating temperature: ——, clinoptilolite rock; ---, improved zeolite; •, Cs, and o, Sr.

of leaching of Sr decreases gradually with increasing temperature, to 0.2% at 1000°C. Similar tendencies were observed for mordenite rock. The degree of leaching of Cs from the improved zeolite is steeply reduced beyond about 800°C which may be attributed to structural changes and approaches 0.5% at 1100°C. The degree of leaching of Sr becomes about 0.03% at 1000°C. The leaching behaviour observed for the improved zeolite, C-10 was quite similar. These experiments show that effective fixation of Cs and Sr can be realized also in the untreated zeolites by heating above 1000°C to 1100°C.

Loss of the weight of Cs and Sr due to sublimation during heating was examined. The results obtained are shown in Table 2. The losses for both elements are quite small i.e. on the average 4.5% of the total amount of adsorption for Cs and 4.0% for Sr.

TABLE 2

Cs_2O and SrO contents in unheated and heated zeolites

Samples	Heating conditions	Cs_2O (wt%)	SrO (wt%)
C-0	No heating	12.5	1.60
	1300°C	12.3	1.57
M-0	No heating	15.1	n.d.
	1300°C	14.4	n.d.
M-4	No heating	19.4	8.64
	1000°C		7.83
	1300°C	19.1	
C-10	No heating	17.7	10.76
	1000°C		10.66
	1300°C	16.1	

n.d. : not determined

Structural changes during heating were determined by X-ray diffraction. The results are schematically shown in Figure 5, from which it is evident that the type of structure associated with effective fixation is different for the natural and the improved zeolites. The natural zeolites change to amorphous materials capable of incorporating Cs and Sr. Similar results have been reported by Mimura and Kanno (1977). The effective fixation of Cs will be enhanced by the formation of pollucite via the amorphous state from the improved zeolite and Sr is successfully fixed in the finally produced Sr-feldspar through the intermediate phase of Sr-paracelsian. The temperature at which the structure of the natural zeolite begins to disrupt is consistent with the temperature at which the degree of leaching reaches a maximum. This suggests that the amorphous

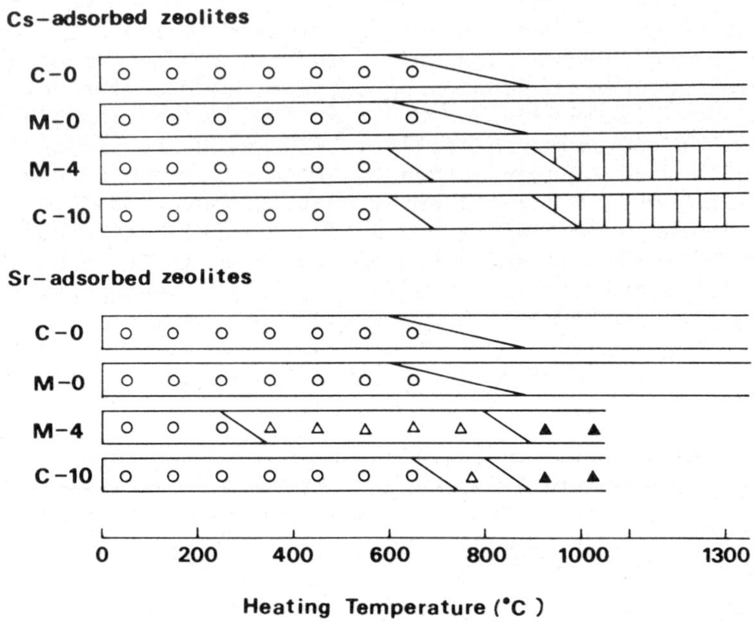

Fig. 5. Schematic representation of structure change of natural and improved zeolites during heating at various temperatures: C-0, clinoptilolite rock; M-0, mordenite rock; M-4, mordenite rock treated with 3M NaOH solution for 36 h; C-10, clinoptilolite rock treated with mixed solution of 3M NaOH and 3M NaCl for 12 h. The corresponding structure is illustrated as follows:
, zeolite; , pollucite; , amorphous material;
, Sr-paracelsian; and , Sr-feldspar.

materials formed during the structural change are less effective for the fixation of Cs than the finally produced amorphous materials.

The most popular procedure for fixation is the incorporation of radioactive elements in a borosilicate glass. Ringwood (1978) has stated that the glass may gradually crystallize to form a stable phase, resulting in increased leachability of the elements. The same may be true for the amorphous materials produced by heating of natural zeolites. On the other hand, for improved zeolites, Cs and Sr are bound with in the framework of the crystal structure of pollucite and Sr-feldspar. However, we cannot be sure whether or not the radioactive elements incorporated in the crystal structure would remain fixed during the radioactive decay period.

There would be an economic advantage of the use of available zeolites for adsorption and fixation of Cs and Sr rather than ceramic materials or synthetic rock.

ACKNOWLEDGEMENT

We wish to express our sincerest thanks to Dr. Koichi Aoyagi of the Central Technical Laboratory, Japan Petroleum Exploration Co. Ltd. for encouragement. This work was partially supported by Scientific Research Fund provided by the Education Ministry of Japan, Project No. 56540497.

REFERENCES

Adam, L., Kakasy, Gy. and Pallots, I., 1971. Removal of radioactive pollutants from water using mineral substances of natural domestic origin. Hung. Mining Res. Inst., 14: 209-212.

Ames, L.L., 1959. Zeolitic extraction of caesium from aqueous solutions. U. S. Atomic Energy Comm. HY-62607, 23pp.

Ames, L.L., 1960. The cation sieve properties of clinoptilolite. Am. Miner., 45: 689-700.

Ames, L.L., 1961. Strontium and cesium loading characteristics of Decalso, Linde 4A and clinoptilolite columns. U. S. Atomic Energy Comm. HW-70768, 13pp.

Barrer, R.M., Davis, J.A., and Ree, L.V.C., 1969. Comparison of the ion exchange properties of zeolite X and zeolite Y. J. Inorg. Nucl. Chem., 31: 2599-2609.

Barrer, R.M. and Munday, B.M., 1971. Cation exchange reaction of zeolite NaP. J. Chem. Soc. (A)., 2909-2914.

Frysinger, C.R., 1962. Cesium-sodium ion exchange on clinoptilolite. Nature, 28: 351-353.

Kakasy, I., Pallos, I. and Adam, L., 1973. Etude des roches clinoptilolitiques de la Hongrie au point devue de la fixation des dechets radioacifs liquedes. Publ. Hung. Min. Res. Inst., 16: 169-175.

Kanno, T. and Mimura, H., 1977. Processing of radioactive waste solution with zeolite, (II) Leachabilities of Cs and Sr from calcined zeolites. Nippon Genshiryoku Gakkaishi. 19: 113-118 (in Japanese).

Kato, K., 1974. Removal of Cs^{137} from aqueous solutions by zeolite. Hoken Butsuri, 9: 11-16.

Mathers, W.G. and Watson, L.C., 1962. A waste disposal experiment using mineral exchange on clinoptilolite. Atomic Energy of Canada, Ltd., AECL-1521.

McCarthy, G.J., 1977. High-level waste ceramics: Materials considerations, process simulation and product characteristics. Nuclear Technol., 32: 92-105.

Mimura, H. and Kanno, T. 1978. Processing of radioactive waste solution with zeolite, (III) Acid resistance of zeolites and leachabilities of Cs and Sr from calcined natural zeolites. Nippon Genshiryoku Gakkaishi. 19: 170-176 (in Japanese).

Ringwood, A.E., 1978. Safe disposal of high level nuclear reactor wastes: A new strategy. Australian National University Press, 64pp.

Sherry, H., 1968. Ion-exchange and properties of zeolites IV. Alkaline earth ion exchange in the synthetic zeolite Linde X and Y. J. Inorg. Nucl. Chem., 31: 2599-2609.

Shishakova, T.N. and Dubinin, M.M., 1965. The structure of synthetic zeolite Zh. Izv. Ada. Nauk. Ser. Khim., 7: 1303-1305.

Zhdanov, S.P., Buntar, N.N. and Egarova, E.N., 1964. Structure and adsorption properties of zeolite Zh. Dokl. Akad. Nauk SSR, 154: 419-422.

INVESTIGATION ON THE BOOM CLAY, A CANDIDATE HOST ROCK FOR FINAL DISPOSAL OF HIGH LEVEL SOLID WASTE*

A.A. BONNE, R.H. HEREMANS
Studiecentrum voor Kernenergie/Centre d'Etude de l'énergie Nucléaire
Mol, Belgium

ABSTRACT

In contrast with other underground disposal host rock types (salt, granite, basalt,...) argillaceous formations exhibit the particularity to function simultaneously as a geological and a geochemical barrier.

The investigations performed to determine the effectiveness of the Boom clay in these respects are discussed in the present paper.

The rock characteristics are determined on specimens sampled by coring through the Boom clay in the underground of the nuclear site of Mol.The mineralogical composition of the clay fraction is fairly constant over the whole thickness of the formation : illite (2-3/10), smectite (2/10), vermiculite (?) (3/10), illite-montmorillonite interstratified (1-2/10), chlorite + chlorite-vermiculite (?) interstratified (1/10).

From the average hydraulic conductivity of the clay samples and the local hydraulic gradient a water flow velocity of ∿ 7 metres per 10^5 years is estimated.

A probabilistic approach for assessing the failure of the geological barrier by events and processes originating from outside the Boom clay, revealed very low probability values for failure by these causes.

The effectiveness of the geochemical barrier of the Boom clay can best be understood by modelling radionuclide migration within that formation and by introducing in the model specific "in situ" values for the different parameters.

From this model it is calculated that for sorbed radionuclides the migration edge, where the concentration equals the maximum permissible concentration, reaches out at a maximum of ∿ 5 metres. On the other hand non sorbed long lived radionuclides, like ^{129}I, may leave the clay layer by diffusion. It is therefore advisable to consider a proper separation and/or conditioned form for these radioelements.

In order to obtain "in situ" repository data, an underground experimental laboratory is under construction at the Mol site.

*Work performed in the frame of a contract between the European Community of Atomic Energy and the CEN/SCK of Belgium.

INTRODUCTION

In 1974 the decision was taken in Belgium to start R&D-work on geological disposal of conditioned reprocessing wastes, arising from electronuclear power production. A total annual production of 5,400 MW(e) by nuclear plants (± 45 % of the electricity production) is scheduled for the mid-eighties. Taking into account this nuclear capacity and a possible slight increase in the future, a final repository should have a capacity of roughly 1,000 m^3 vitrified high level waste.

A national inventory of potential geological disposal formations indicated that only clays and shales could be considered in Belgium. Seven potential areas were identified, where such rocks are known to be present and to meet the selection criteria. One of these areas is situated in the North-eastern part of Belgium, where the Boom clay is present at sufficient depth and with satisfactory thickness. In this area several nuclear facilities are located. It was thus decided to perform the R&D-programme concerning the disposal possibilities on the Boom clay at the specific site of Mol.

The R&D-programme of SCK/CEN at Mol aims at obtaining the knowledge of the system behaviour and of appropriate engineering technologies in order to assess the safety and the feasability of the disposal system.

THE DISPOSAL SYSTEM AND BARRIER CONCEPT

Schematically a generic disposal system can be broken down into the following parts : waste form, repository, geosphere and biosphere (see Table 1). From the standpoint of safety assessment several of these parts or their components have the function of "barrier", to block the return of radionuclides to mankind. The barrier function of the waste package (waste form and container) is provided by its physical and chemical stability, which must ensure the occlusion of the radionuclides in the waste matrix. The backfilling and the geochemical barrier obstruct migration of radionuclides by retaining them. The geological barrier guarantees segragation of the wastes from the biosphere and screens them from the influences of disruptive processes or events (e.g. water flow in surrounding geomedium erosion, etc.). In our R&D-programme the performance of such a disposal system with clay as host formation is studied by investigating the different system components and their mutual interactions. With regard to other disposal host rock types (salt, granite, basalt, ...) argillaceous formations exhibit the particularity simultaneously to match a geological and a geochemical barrier function. This can very well be demonstrated by the results from our investigations on the Boom clay.

THE OCCURRENCE OF THE BOOM CLAY AT THE MOL SITE

A compilation of available geological data demonstrated that over a large area in the North-eastern part of Belgium the Boom clay is situated at a depth at more than 100 m and at least 100 m thick. The dip is gently north-east. At the Mol site the top of the Boom clay is situated at $\sim$ -160 m, its thickness is $\sim$ 110 metres. A repository could be build in the mid-plane of this formation at $\sim$ -220 metres (Manfroy et al, 1980).

Intensive geological sampling was performed by drilling. Geophysical bore-hole logging, especially resistivity logging with short electrode spacings could reveal slight lithological variations of decimetre scale (Neerdael at al, 1981). This technique also demonstrated the continuity of these small lithological variations over longer distances (more than 10 km). In the uppermost part of the Boom clay, the so called transition zone, resistance variations of $\sim$ 0.5 Ωm are observed corresponding to slight variations of the silt content. The lithological variations mainly concern silt, carbonate and organic matter content. These variations are reflected in the physical and chemical characteristics of the Boom clay. Table II gives an overview of the main characteristics of the Boom clay, as they were identified in the laboratory on samples from the Mol site. The data on chemical composition are indicatory values for about 50 samples taken over the whole section of the Boom clay (-160 to -265 m). Before analysis the samples were preheated at 110 °C ; the results are normalised to 100 %. The organic matter content was determined on 14 samples. About 50 % of the total organic carbone belongs to the "humic" fraction. The clay mineral composition is quite constant for the whole section of the formation.

The hydraulic conductivity has been determined by percolation tests and the average value is 3.6×10^{-3}m.yr^{-1}. For the Boom clay between -178.85 m and -260.29 m the overall hydraulic conductivity is calculated to be 1.39×10^{-3}m.yr^{-1}. The samples used were "undisturbed" samples.

The average Atterbergh limits (limits of liquidity and plasticity) were determined according to the DIN 18122 norms. For 50 samples (geotechnical core drilling) the averages are : liquid limit : $\sim$ 76 % ; plastic limit $\sim$ 27 % ; plasticity index : $\sim$ 49 %. According to the Casagrande classification these values indicate that the Boom clay belongs to the very plastic clay. The values obtained for the cation exchange capacity are variable, depending on the sample, sample treatment and the determination method. The highest values were obtained for samples of the lower-most part of the formation, exposed to air and for the Ag-thiourea method.

THE GEOLOGICAL BARRIER FUNCTION OF THE BOOM CLAY AT THE MOL SITE

The value of the geological barrier is based on its "imperviousness" to water flow and its stability in the long term. Water flow is generally accepted to be the most important potential agent for leaching of the waste. However, theoretically almost no geologic formation is totally impervious. The stability in the long term depends on the compatibility of the formations and the waste repository and its screening ability against natural disruptive processes or events and human actions.

For the assessment of the effectiveness of the Boom clay as a geological barrier all these factors must be considered.

From hydrological investigations it was learned that a downward gradient of 1-2 % presently exists in the area. Assuming the validity of Darcy's law for water flow in the Boom clay and combining a 2 % hydraulic gradient with the average hydraulic conductivity, a water drainage velocity of ∿ 7 metres per 10^5 years is estimated. It is however expected that "in situ" the water drainage velocity will be several times lower for the following reasons :

1) the overall hydraulic conductivity of the whole clay formation is 2-3 times lower than the average value ;
2) calculations of the hydraulic conductivity from reconsolidation tests, simulating "in situ" conditions (20 $kg.cm^{-1}$ loading) indicate a lower conductivity than the figures obtained by percolation tests on samples loaded normally at ∿ 2.9 $kg.cm^{-1}$ (Figures on Table II) ;
3) the actual hydraulic gradient is probably lower than 2 % ;
4) it is questionable whether Darcy's law is applicable to clay containing an important amount of swelling clays. Other authors show water drainage velocities in clays, lower than those calculated by Darcy's law (Thirriot et al., 1979, Srivastava, 1975).

This drainage velocity is thus insignificant because the time needed by the water to seep completely through the clay (∿ 50 m half thickness) is several times longer than the required containment period.

Another assessment to be done concerning the effectiveness of geological isolation is the behaviour of the host rock under "in situ" repository conditions. What will be the impact of the heat released by the high level solid waste, of γ-irradiation, of corrosion, etc.? Several of these aspects are studied now for a disposal concept in the Boom clay (Heremans et al., 1981).

Heat impact

A heat transfer model was developed in order to calculate heat dissipation through the host rock and the surrounding geological medium. Thermal conductivity and specific heat of the Boom clay was determined first on laboratory scale (see Table II). This conductivity decreases very rapidly to a value of 0.3-0.5 $W.m^{-1}K^{-1}$ at 100 °C and as a function of decreasing water content. Dehydration of confined clay under an hydrostatic pressure of 20 bar is however theoretically impossible even at temperatures up to 200 °C. This was qualitatively confirmed during a one year laboratory test performed on a clay lump 30 cm diameter, 50 cm high at 6 bar and 160 °C.

In order to compare small scale conductivity results to actual conditions, a near-surface experiment with a full size simulated HLW canister (30 cm in diameter and 150 cm length) was performed in the area where the Boom clay is outcropping (south of Antwerp). Heat output is obtained by electrical wiring. The heater was burried at a maximum depth of 650 cm and the space between clay and cylinder filled up with a sand having nearly the same thermal conductivity as the clay. In total 36 thermoprobes were installed at various distances from the heat source in order to measure the temperature distribution around it. The experiment ran for one year at a 1,000 W(e) output and for another 6 months at 1,500 W(e). Due to ground surface weathering phenomena and quarrying works, water infiltrations near to the heat source could occur. It was possible to take into account these water movements in the mathematical model. The average value found for "in situ" heat conductivity is 1.69 $W.m^{-1} K^{-1}$ and for heat diffusivity 18.8 $m^2.yr^{-1}$.

Temperature distribution computations were performed for several assumptions and conditions : cooling time, thermal conductivity of the clay... In Figure 1 the expected temperature increase around a stack of 12 HLSW containers is represented. From this figure it can be learned that the rock volume with a significant temperature increase of 100 °C is very limited (60 cm distance from the stack).

As far as water convection streams induced by a thermal gradient are concerned and taking into account the local known conditions, a preliminary mathematical assessment has shown that such a phenomenon might be neglected. So the heat transfer model developed at CEN/SCK considers that the thermal flow is only transmitted by conduction.

Gamma-irradiation

In order to identify the effects of γ-irradiation on natural clay, orientative tests were performed. The samples were sealed in quartz tubes in the absence of air and placed in the centre of spent BR2 fuel assemblies. For an integrated dose of 10^8 rad, releases of hydrogen and carbon dioxide were detected together with small amounts of methane in some cases. The following reactions are supposed to take place :

$$2H_2O \rightarrow 2H_2 + O_2$$

$$\text{organic } 2(CH) + 2O_2 \rightarrow H_2 + 2CO_2.$$

Additional laboratory investigations are going on in order to confirm and quantify the various phenomena and also to evaluate the effects which could result from these reactions : for instance the physico-chemical behaviour of H_2 and CO_2 in natural clay and the consequences for material corrosion and sorption capacity. For the actual repository concept, calculations indicated that the affected zone in the vicinity of the buried waste could be 1.5 m thick ; beyond this distance the dose rate is dropped to the natural background dose rate of 5.5×10^{-3} mrad h^{-1}.

Corrosion

In our corrosion programme the present objective is to select a metallic material which could resist corrosion by clay for several centuries (700 to 1000 years). This material, if not suitable at the same time as container material for vitrified waste, can then be used as overpack for the classical stainless steel container.

We only summarize here some trends observed thus far in our experiments :

- a first comparison between laboratory and "in situ" tests indicated good agreement ;
- Ti and Ti alloys, Hastelly C, Inconel 625 and UHB 904L gave the best resistance to local and preferential corrosion in a clay medium ;

- heat pretreatment of the samples generally increase their corrosion rate ;
- the presence of sulphur and some soluble sulphur components is one of the mean causes of the corrosion detected.

Oxidation

It is conceivable that during the operational phase, oxygen abundantly available by ventilation of the mined cavities, will have contant with the clay. Observations on clay samples exposed to air revealed important physico-chemical changes : a pH-drop from 10 to 5-3 in extreme cases and a increase of solute content in the interstitial solution. Estimation of the oxygen penetration depth is thus important. Assuming a gallery diameter of 3.5 m, a free exchange of oxygen at the interface air-clay (absence of lining), a temperature of 60 °C at the interface, an operation time of 50 years and diffusion constant of 4.8×10^{-6} cm^{-2} s^{-1} it was estimated that at a distance of about 4 metres the oxygen in the interstitial solution water is $\sim$ 1/1000 of the oxygen concentration in the clay at the interface with the air in the gallery.

Long term stability with respect to the impact of disruptive natural processes or events and human actions is also considered for the Boom clay at the Mol site.

The assessment of this stability must be done within the existing overall geological context and geographical position. The overburden and underlying formations being aquiferous sands, a direct release to the upper or lower aquifers was considered as a failure of the geologica isolation barrier.

A first study aimed at constructing possible future evolution scenarios (Vandenberghe et al., 1980). Reviewing the quaternary and tertiary geological history it was concluded that the most important processes were related to epeirogenetic movements and climatic changes implying tidal gully erosion, river and glacial erosion. The study demonstrated that within about 2×10^5 years these processes are not likely to disrupt the repository.

In the frame of risk assessment studies a particular methodology, developed by the Joint Research Centre (Girardi et al., 1978 ; Bertozzi et al., 1978), was applied to the specific case of the Boom clay at the Mol site. This study (D'Alessandro et al., 1981) did'nt aim directly at demonstrating the safety of a repository but rather at validating the Fault Tree Analysis methodology.

The possible events which were considered and which could be treated in the fault tree analysis are the following : 1) presence of ground water above ; 2) presence of pressurized ground water below ; 3) migration with retention (outside the clay) ; 4) migration without retention (outside the clay) ; 5) 3rd order fault ; 6) 2nd order fault ; 7) 1st order fault ; 8) extrusive magmatic activity ; 9) explosive magmatic activity ; 10) diapirism ; 11) glacial overburden (causing loss of plasticity) ; 12) permafrost ; 13) 1st order glacial action ; 14) 2nd order glacial action ; 15) subsidence ; 16) 1st order stream erosion ; 17) 2nd order stream erosion ; 18) 1st order denudation; 19) 2nd order denudation ; 20) pumping of contaminated ground water ; 21) drilling exposing water to aquifers ; 22) 2st order human activities causing waste rise ; 23) 2nd order human activities causing waste rise ; 24) 1st order meteorite (causing release to atmosphere) ; 25) 2nd order meteorite (causing release to ground water).

The most important release receptor appeared to be the groundwater and land surface.

The result of application of the Fault Tree Analysis method to the Boom clay at the Mol site is that the phenomenon governing the probability of geological barrier failure is a faulting phenomenon with a displacement greater than 5-10 m. However in the relatively shorter time span of $2x10^3$ yrs, human action have the same weight.

The overall probability of geological barrier failure is estimated by attributing to each intervening event a probability range. This probability range accounts for the uncertainty in the probability estimation.

The overall probability (mean value) of barrier failure obtained for the exercise on the Boom clay is very low : $\sim 4x10^{-8}$ yr^{-1}.

An analysis of the consequences of such failures is in progress (collaboration between SCK/CEN Mol and JRC-Ispra).

THE GEOCHEMICAL BARRIER FUNCTION OF THE BOOM CLAY

Depending upon the nature of the different waste package components, the waste form will come eventually in contact with the clay or at least with interstitial clay fluids and corrosion and leaching of the waste matrix will occur. The migration distance within the clay of the thus released radionuclides can best be assessed by modelling taking into account "in situ" conditions. Starting from the classical mass transfer equation

$$R \frac{\delta c}{\delta t} = D \frac{\delta^2 c}{\delta x^2} - V \frac{\delta c}{\delta x} \qquad \text{(x= distance ; c = concentration ; t = time)}$$

and with proper boundary conditions a model and computational programme was developed for the three-dimensional migration of radionuclides within clay (Put et al., 1977). Also an intensive research is done and is still in progress for estimating the "in situ" value of the parameters to be introduced in the model. The drainage velocity within the clay has already been discussed before. The retardation factor (R) and the diffusion constant (D) for the Boom clay are two parameters which were investigated by different techniques.

The retardation factor can be calculated from the distribution coefficient (K_d) by the following relation $R = 1 + rK_d$, in which r is the ratio of the bulk density and the effective porosity of the clay and K_d is the distribution coefficient. For estimating the "in situ" K_d, numerous laboratory tests (batch and column) have been performed. For instance the K_d-dependence on the concentration of the radionuclides has been investigated (Fig. 2). This figure shows decreasing K_d-values with increasing nuclide concentrations as a general trend. Another aspect studied is the dependence of the K_d on the solution/clay ratio. Results of such experiments are represented in Figure 3 which shows that under "in situ" conditions (low solution/ clay ratios) the K_d-values may be much different from the values determined by laboratory batch experiments with relatively high solution/clay ratios (50 or more). Extrapolation to low ratios points to K_d-values of 1-10 for Sr, 10-100 for Cs and 10^4-10^5 for Eu.

D(diffusion constant) and D/R (apparent diffusion coefficient) can also be estimated from diffusion experiments. Diffusion tests of several types were carried out. The techniques used are static. A filterpaper or clay slab is loaded with radiotracer, sandwiched between two clay plugs and placed in a thermostat (at 25 °C). For ensuring good contact between source and clay plugs, a pressure of 1.5 $kg.cm^{-2}$ was applied on the clay plugs. After several weeks the samples are unloaded and the two clay plugs sliced (1 mm thick). From the activity of the element in each slice a concentration profile can be drawn and the apparent diffusion (D/R) calculated. For the different elements studied the results of such experiments are given on Table III.

The diffusion constant (D) for the Boom clay, $4x10^{-6}$ cm^2 s^{-1}, was obtained by the same technique, using a non sorbed element (I). The figures obtained for D were confirmed by a quite similar diffusion technique, but using a ^3H- containing supernatant solution as source ($D_{^3H}$: $4.1x10^{-6}$ cm^2 s^{-1}).

K_d-values for Cs, Sr and Eu, calculated from diffusion experiments agree with extrapolated data from K_d-batch tests with decreasing solution clay ratios (see above).

Diffusion cells have been developed allowing diffusion experiments on reconsolidated clay plugs, simulating in situ conditions (∿ 2 MPa reconsolidation pressure). Preliminary results obtained with tritiated water show no significant influence of reconsolidation.

We consider now more in detail results of the radionuclide migration model and more specifically the results of a sensitivity analysis (Bonne et al., 1980) for the different intervening parameters.

The distance from the source, beyond which the concentration in the interstitial clay solution is lower than the C_{mpc} (maximum permissible concentration for drinking water) is considered as a yardstick for assessing the migration distance. The C_{mpc} used is ten times lower than the ICRP drinking water norms for occupationally exposed individuals.

All parameters intervening in the migration model may be grasped in the following variables : D/R (apparant diffusion coefficient), C_o/C_{mpc} (ratio of source concentration to C_{mpc}) and v (drainage velocity).

The sensitivity of the model was studied by calculating the migration distance changes as a function of parameter variations with respect to base case conditions. The base case conditions considered are presented in Table IV. The results of the computations of the maximum migration distance as a function of parameter variations are given in

Table V.

In drawing our main conclusions from the figures in this table we have to make a distinction between extremely long-lived, non-sorbed radionuclides (e.g. ^{129}I) and sorbed long-lived radionuclides.

(i) For non-sorbed, extremely long-lived radionuclides the maximum migration distance is very sensitive to the C_o/C_{mpc}-ratio. In the base case conditions, we considered 100 % of the I in the vitrified high level waste. In the present waste treatment options,a separation of ^{129}I is foreseen and alternative waste packages are studied.

(ii) For sorbed long-lived radionuclides the maximum migration distance is very sensitive to the parameters determining the apparant difussion (D/R). Research concerning migration in clay should thus emphasize the determination and identification of parameters and phenomena controlling the D/R-ratio. The maximum migration distance of these isotopes is not sensitive to drainage velocity.

The computations show the slight influence of the concentration parameters on the migration distance which implies that from a pure migration point of view a given repository could be loaded more densely without increased risk for the environment.

With a distance of 25 metres between the stacks, and taking into account the maximum migration edge for the reference case, it may be assumed that, from a migration point of view each stack is an independent source, not affected by the presence of another stack. The maximum migration distance of the sorbed not extremely long-lived radionuclides is very limited and of the same order of magnitude as the influence zone of impact from wastes (heat, γ-irradiation,...).

Also corrosion and/or leaching products of the waste package and the structural material will be released and compete with the radionuclides. With the scope of such "in situ" repository conditions the sensitive parameters D and R must be investigated very carefully. Preliminary K_d-measurement of Sr and Cs have been performed for "cocktail" solutions containing different concentration levels of corrosion products. They demonstrate a decrease of K_d for increasing corrosion product concentrations.

To estimate the influence of temperature and γ-irradiation on sorption behaviour, samples of the Boom clay (-220 and -240 m) were submitted to pretreatments and K_d-measurements (at pH = 3) were performed. The influence of temperature was studied by heating the samples at 300 °C and 500 °C. This pretreatment very weakly affected the K_d-values of Cs. For Sr and Eu, at lower concentration levels, there is a decrease of the K_d-values. At 500 °C the K_d-values for Sr abd Eu are respectively 5 and 10 times lower than at 300 °C.

To test the influence of γ-irradiation, samples were submitted to doses of 3.10^9 to 3.10^{10} rad, prior to K_d-batch measurements. Also some samples were irradiated and heated at 110 °C, 300 °C and 500 °C before K_d-determinations. For samples which were irradiated only, results were the same as for single heat pretreatment. It is believed that the destruction of organic constituants is responsible for the observed K_d-decreases. Irradiation of samples pretreated at 500 °C does not influence K_d-values as compared with samples pretreated at 500 °C only.

These tests for identifying the sorption characteristics of the Boom clay under "in situ" repository conditions are now complemented by supporting fundamental research, e.g. on characterisation of interstitial solution water, studies on the selectivity of the Boom clay (Baeyens et al., 1981) and specification of released elements. These supporting studies will also contribute to the phenomenological understanding and to the development of adequate models (reversibility of sorption, colloïd formation,a.o.).

In situ conditions will be studied at real depth in a last phase of our R&D-programme. The acces shaft to an underground laboratory is under construction now. The in situ laboratory will be located at -220 m into the Boom clay.

CONCLUSIONS

A clay formation (e.g. the Boom clay), as host for radioactive waste disposal combines two important barrier functions : geological segregation and geochemical retention. The barrier effectiveness in regard to processes originating from the repository and the host rock can be assessed deterministically. Reliable assessments can only be done on the basis of "in situ" repository data, which can be obtained by "in situ" experimentation and appropriate laboratory research.

The geological barrier effectiveness in regard to events and processes originating from outside the repository and the host formation can be assessed by defining failure scenarios probabilistically and on a site specific basis.

REFERENCES

Baeyens, B., Maes, A. and Cremers, A., 1981. Geological storage of radioactive waste : I. Cesium "fixation" in Boom clay. Abstracts, Int. Clay Conf., Bologna/Pavia, 36.

Bertozzi, G., D'Alessandro,M., Girardi, F. and Vanossi, M., 1978. Safety Assessment of Radioactive Waste Disposal into Geologic Formations ; A Preliminary Application of Fault Tree Analysis to Salt Deposits, EUR-report 5901 EN.

Bonne, A., Put, M., Heremans, R. and Baetslé, L.H., 1980. Migration of radionuclides in clay : a sensitivity analysis. Int. J. Energy Systems, 1 : 147-152.

D'Alessandro, M. and Bonne, A., 1981. Radioactive Waste disposal into a plastic clay formation (a site specific exercise of assessment of geological containment),Harwood Academic Publishers, London, 150pp

Girardi, F., Bertozzi, G. and D'Alessandro, M., 1978. Long-term risk assessment of radioactive waste disposal in geological formations, EUR-report 5902 EN, 42 pp.

Heremans, R., Bonne, A. and Manfroy, P., 1981. Experimentation on and evaluation of near-field phenomena in clay. The Belgian approach. In : Near-field Phenomena in Geologic Repositories for Radioactive Waste; Nuclear Energy Agency, OECD, Paris, pp. 183-203.

Manfroy, P., Heremans, R., Put, M., Vanhaelewyn, R. and Mayence, M., 1980. Conception d'une installation pour l'enfouissement dans l'argile de déchets radioactifs conditionnés. In : Underground Disposal of Radioactive Wastes, Vol. II, Int. Atom. Energy Agency, Vienne, pp. 59-74.
Neerdael, B., Vandenberghe, N., Bonne, A., Fierens, E., Laga, P. and Manfroy, P., 1981. Utilisation des diagraphies dans la détermination des caractéristiques lithologiques dans l'argile de Boom et leur intérêt pour les corrélations dans l'argile. In : Siting of Radioactive Waste Repositories in Geological Formations, Nuclear Energy Agency, OECD, Paris, pp. 133-149.
Put, M. a,d Heremans, R., 1977. Modélisation mathématique de la migration de radionuclides dans une formation argileuse homogène. In : Risk Analysis and Geologic Modelling in Relation to the Disposal of Radioactive Wastes into Geological Formations, Nuclear Energy Agency, OECD and Comm. Europ. Comm., Ispra, pp. 116-129.
Srivastava, R.C., 1975. Water Flow through Mixed Clays. J. of Hydrol., 26 : 295-301.
Thirriot, C. and Habib, J., 1979. Problèmes hydrogéologiques posés par les argiles et les couches argileuses. In : 95ème Congrès national des Sociétés savantes, T.I., Reims, pp. 331-346.
Vandenberghe, N., Bonne, A. and Heremans, R., 1980. Scénarios d'évolution géologique lente, appliqués au site argileux de Mol (Belgique). In : Radionuclide Release Scenarios for Geologic Repositories, Nuclear Energy Agency, OECD, Paris, pp. 169-180

TABLE I. DISPOSAL SYSTEM - BARRIER SYSTEM

PARTS	COMPONENTS	BARRIERS
WASTE PACKAGE	WASTE FORM	WASTE FORM
	CONTAINER	CONTAINER
	OVERPACK	
	(MIGRATION RETARDANT)	
	EVENT.	BACKFILLING
REPOSITORY	BACKFILLING	
	ENGINEERING STRUCTURES	
	SHAFT SEAL	
	EMPLACEMENT GEOLOGY	
		GEOLOGICAL ISOLATION
GEOSPHERE	HOST ROCK	
	SURROUNDING GEOMEDIUM	GEOCHEMICAL BARRIER
(BIOSPHERE)	SOIL	
	LAND SURFACE	
	SURFACE WATER	ENVIRONMENT
	ATMOSPHERE	

TABLE II: SOME CHARACTERISTICS OF THE BOOM CLAY (MOL SITE)

Chemical composition of dry material (%)	∼ 64 SiO_2, ∼ 14 Al_2O_3, ∼ 5.9 Fe_2O_3, ∼ 2.2 K_2O, ∼ 1.4 Na_2O, ∼ 0.6 CaO, ∼ 0.5 TiO_2, ∼ 0.7 MgO weight loss at 1000 °C: ∼ 10
Natural water content (weight %)	∼ 26
Mineralogical composition of the fraction < 2 µm (parts per ten)	illite (2-3), smectite (2), vermiculite-like (3), illite-montmorillonite interstratified (1-2), chlorite + chlorite-vermiculite-like interstratified (1)
Organic matter (%, 14 samples)	2.3 - 5.5
Granulometric composition (%)	d < 2 µm : 49 2 µm < d < 60 µm : 47 60 µm < d < 200 µm : 3.5 d > 200 µm : 0.5
Bulk density	∼ 1.93
Dry density	∼ 1.53
Hydraulic conductivity ($cm\ s^{-1}$)	between 1.4 10^{-8} and 4.7 10^{-10}
Porosity (%)	between 43.6 and 44
Saturation degree (%)	bewteen 88.4 and 100
Plastic limit (%), average of 50 samples	∼27
Liquid limit (%), ibidem	∼76
Index of plasticity (%), ibidem	∼49

TABLE II: CONTINUED

Thermal conductivity ($W\ m^{-1}K^{-1}$)	: 0.9 - 1.3 (from 20 to 90°C) 0.3 - 0.5 at 100°C and 0.6 - 0.8 at 500 °C
Natural radioactivity ($\mu Ci\ 100\ g^{-1}$ dry sample)	: $^{40}K \sim 2.10^{-3}$ $^{226}Ra \sim 2.1\ 10^{-4}$ $^{232}Th \sim 1.2\ 10^{-4}$
Cation exchange capacity (meq. per 100 g dry clay)	: 20 - 40, depending on sample and technique used

TABLE III. DIFFUSION COEFFICIENTS IN BOOM CLAY ($cm^2.s^{-1}$)

	SOURCE : FILTER PAPER	SOURCE : CLAY
Cs	$1.1\text{-}3 \times 10^{-8}$	9.6×10^{-9}
Sr	-	$2.9 - 5.3 \times 10^{-7}$
Eu	2.6×10^{-10}	-
Pu	2.3×10^{-9}	-
I	-	4×10^{-6}

TABLE IV. REFERENCE CASE CONDITIONS

ISOTOPE	HALF-LIFE years	INT.CONC.(C_o) Ci/m^3	C_{mpc} Ci/m^3	DIF.CONST.(D) $m^2/year$	RET.FACT.(R)
^{154}Eu	16	6.5×10^4	2×10^{-5}	0.0126	2×10^4
^{90}Sr	28	7.5×10^5	4×10^{-7}	0.0126	15
^{137}Cs	30	1×10^6	2×10^{-5}	0.0126	300
^{240}Pu	6600	26	5×10^{-6}	0.0126	1600
^{239}Pu	24000	16	5×10^{-6}	0.0126	1600
^{129}I	1.6×10^7	0.38	2×10^{-5}	0.0126	1
$dh/ds = 0.02$ m H_2O		$k = 1.39 \times 10^{-3}$ m/yr		$v = 2.78 \times 10^{-5}$ m/yr	

TABLE V : SENSITIVITY OF MAXIMUM MIGRATION DISTANCE $C = C_{mpc}$ TO VARIATIONS OF C_o/C_{mpc}, D/R AND V

Isotope		C_o/C_{MPC}	X_{max} (meters)		D/R (m^2/y)	X_{max} (meters)		Water velocity (m/y)	X_{max} (meters)
Eu^{154}	1	3.25×10^7	0.065	1	5.04×10^{-7}	0.065	1	1.39×10^{-5}	0.075
	2	3.25×10^9	0.075	2	6.30×10^{-7}	0.075	2	2.78×10^{-5}	0.075
	3	3.25×10^{11}	0.095	3	8.40×10^{-7}	0.085	3	5.56×10^{-5}	0.075
Sensitivity		$S_1 = 3.26 \times 10^{-3}$ m			$S_2 = 3.92 \times 10^{-2}$ m			$S_3 < 3.4 \times 10^{-7}$ m	
Sr^{90}	1	1.875×10^{10}	3.38	1	6.3×10^{-4}	3.65	1	1.39×10^{-5}	4.17
	2	1.875×10^{12}	4.17	2	8.4×10^{-4}	4.17	2	2.78×10^{-5}	4.17
	3	1.875×10^{14}	4.97	3	1.26×10^{-3}	5.02	3	5.56×10^{-5}	4.17
Sensitivity		$S_1 = 0.17$ m			$S_2 = 2.00$ m			$S_3 < 8 \times 10^{-4}$ m	
Cs^{137}	1	5×10^8	0.73	1	3.15×10^{-5}	0.80	1	1.39×10^{-5}	0.91
	2	5×10^{10}	0.91	2	4.20×10^{-5}	0.91	2	2.78×10^{-5}	0.91
	3	5×10^{12}	1.10	3	6.30×10^{-5}	1.11	3	5.56×10^{-5}	0.91
Sensitivity		$S_1 = 0.04$ m			$S_2 = 0.45$ m			$S_3 < 4 \times 10^{-5}$ m	
Pu^{240}	1	5.2×10^4	1.72	1	6.3×10^{-6}	2.54	1	1.39×10^{-5}	2.79
	2	5.2×10^6	2.79	2	7.875×10^{-6}	2.79	2	2.78×10^{-5}	2.79
	3	5.2×10^8	3.92	3	1.26×10^{-5}	3.40	3	5.56×10^{-5}	2.79
Sensitivity		$S_1 = 0.24$ m			$S_2 = 1.24$ m			$S_3 < 8 \times 10^{-4}$ m	
Pu^{239}	1	3.2×10^4	2.56	1	6.3×10^{-6}	4.15	1	1.39×10^{-5}	4.54
	2	3.2×10^6	4.54	2	7.875×10^{-6}	4.54	2	2.78×10^{-5}	4.54
	3	3.2×10^8	6.66	3	1.26×10^{-5}	5.48	3	5.56×10^{-5}	4.54
Sensitivity		$S_1 = 0.45$ m			$S_2 = 1.92$ m			$S_3 < 2.4 \times 10^{-3}$ m	
I^{129}	1	1.9×10^4	19.68	1	1.26×10^{-1}	85.62	1	1.39×10^{-5}	84.25
	2	1.9×10^6	85.47	2	1.26×10^{-2}	85.47	2	2.78×10^{-5}	85.47
	3	1.9×10^8	429.12	3	6.3×10^{-3}	85.25	3	5.56×10^{-5}	87.90
Sensitivity		$S_1 = 28.6$ m			$S_2 = 0.11$ m			$S_3 = 2.61$ m	

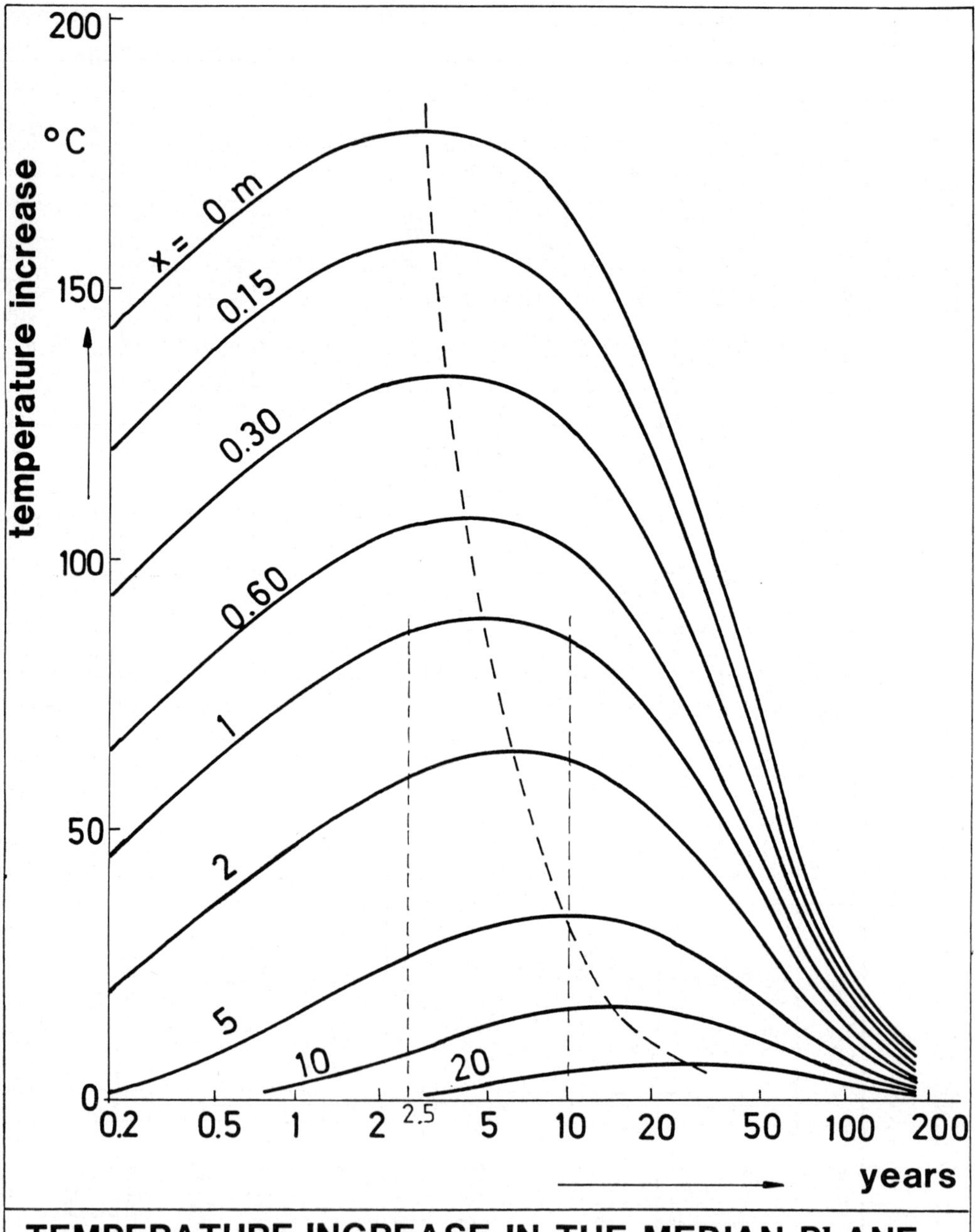

TEMPERATURE INCREASE IN THE MEDIAN PLANE PERPENDICULAR TO THE STACK AXIS, AS FUNCTION OF TIME AND DISTANCE FROM THE STACK AXIS cool. time fifty yrs

110 / 81

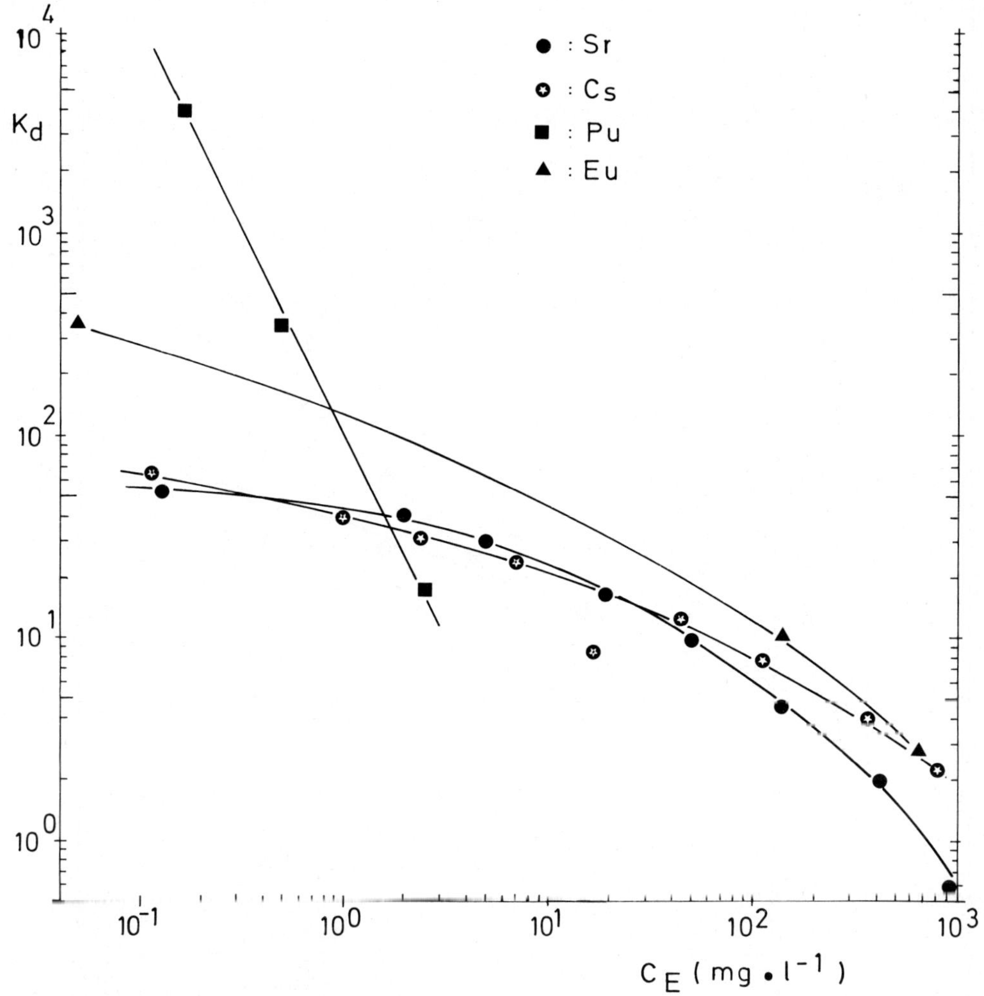

K_d OF BOOM CLAY VS. ELEMENT CONCENTRATION

(samples: Pu:220m / Cs, Sr, Eu:220.65m (depth)
pH: Cs, Sr:3 / Eu:6–8.3 / Pu±7
Solution type : anversian water)

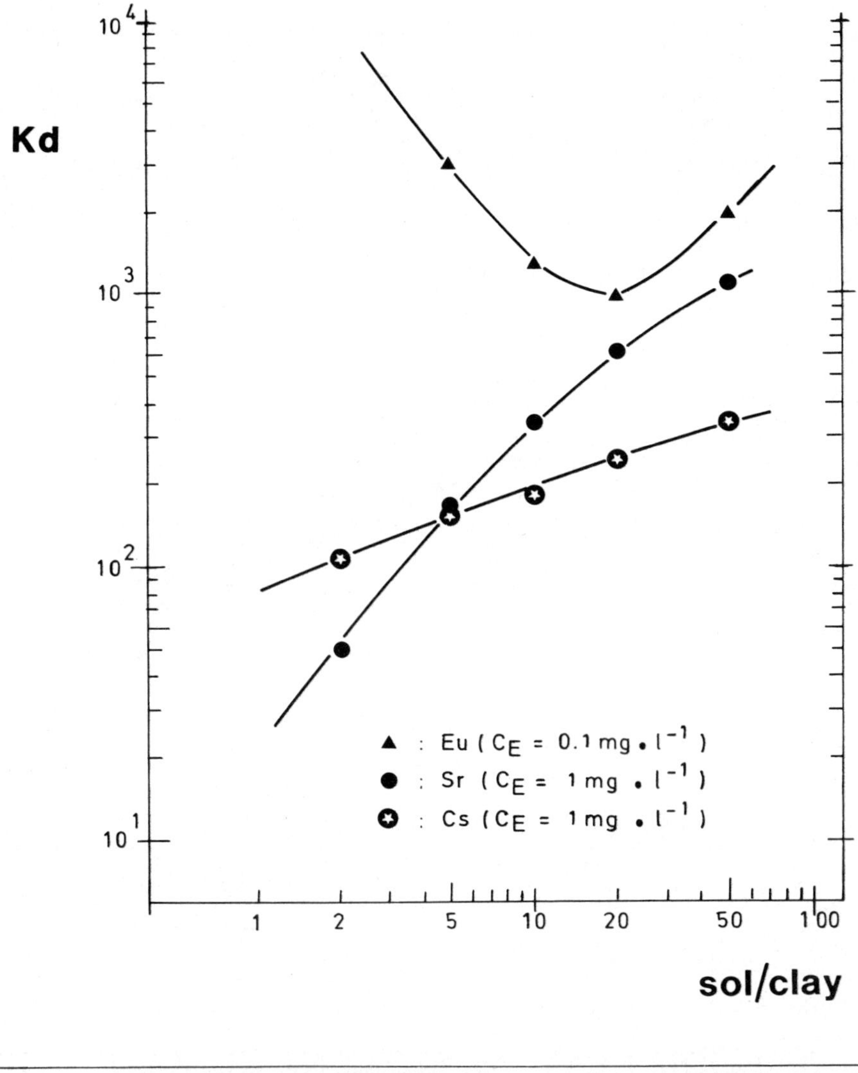

Kd OF BOOM CLAY AS A FUNCTION OF SOLUTION-CLAY RATIO

samples: Cs:−210,6 m/Sr:−210 m/Eu:−230,6 m

AUTHOR INDEX

SUBJECT INDEX

This subject index is compiled from key words drawn from titles and long abstracts which were submitted prior to the Conference.

The numbers are those of the first pages of the papers concerned.